ALLE ZEIT WACH
1842

Erik Flügel

Microfacies Analysis of Limestones

Translated by K. Christenson

With 53 Plates, 78 Figures, and 58 Tables

Springer-Verlag
Berlin Heidelberg New York 1982

Professor Dr. ERIK FLÜGEL
Institut für Paläontologie, Universität Erlangen-Nürnberg
Loewenichstraße 28, 8520 Erlangen/FRG

Translator: KAREN CHRISTENSON
Glaserstr. 14, 8590 Kraftshof/FRG

Translation of a revised and expanded version of
E. Flügel, Mikrofazielle Untersuchungsmethoden von Kalken

ISBN-13:978-3-642-68425-8 e-ISBN-13:978-3-642-68423-4
DOI: 10.1007/978-3-642-68423-4

Library of Congress Cataloging in Publication Data. Flügel, Erik. Microfacies analysis of limestones. An enlarged and updated translation of the 1978 ed. of: Mikrofazielle Untersuchungsmethoden von Kalken. Includes bibliographies and indices. 1. Facies (Geology) 2. Rocks, Carbonate. I. Title. QE651.F5813 552'.5 81-23318 AACR2

Softcover reprint of the hardcover 1st edition 1982

2132/3130-543210

Preface

Reviewers of the German edition of this book found that the text deals with facts and descriptions of limestones rather than with global speculations on facies models or large-scale sedimentation patterns. The book is neither a picture-book nor a recipe-book for facies interpretation of carbonates, but an attempt to summarize the present "state of the art" of a rather small but increasingly more important part of geology.

The book is written for advanced undergraduate and post-graduate students as well as for research workers and exploration geologists who need rapid and intensive training in modern methods of microfacies analysis. The book should facilitate decisions about which methods to use in one's own investigations, and where to look for comparative studies.

Microfacies interpretation of carbonate rocks can not rely solely on the investigation of sedimentological and paleontological thin-section data, but must also consider geological and palecological criteria. It is beyond the scope of this book to describe all of these aspects. However, broader applications are indicated in the chapters dealing with the relationships between geochemical as well as physical data and the depositional and diagenetic fabric of limestones.

The original German edition has been expanded and brought up to date. Changes have been made in the chapters on carbonate diagenesis, facies-controlled porosity, deeper-marine environments, and fossils in thin-sections. Case histories, taken from different depositional environments and based mainly on the author's own facies studies in Paleozoic and Mesozoic rocks, may illustrate the potential of microfacies analysis.

The modifications in the English edition reflect the growing interest in these fields, and also the need to put more stress on the practical applications of microfacies studies in applied geosciences and integrated facies research.

I wish to thank the technical staff of the Institut für Paläontologie der Universität Erlangen for their continuous help. Special thanks are extended to H. Hagn, München, R. Schröder, Frankfurt, and J. Remane, Neuchâtel who contributed thin-section photographs, and to all colleagues who offered suggestions for the improvement of the book. I appreciate the help of all persons and institutions who gave permission to use their figures and tables.

I am especially grateful to Karen Christenson for the translation and for her constantly growing interest in microfacies problems.

Erlangen, February, 1982 ERIK FLÜGEL

Cover: Lower Permian limestone breccia, Southern Alps: Microfacies types of lithoclasts indicate erosion of different parts of a marine shallow-water carbonate platform. The bioclastic grainstone (left) with dasycladacean algae (*Connexia carniapulchra, Mizzia* sp., *Gyroporella* sp.) and fusulinid foraminifera (*Pseudofusulina* cf. *P. paragregaria*) was derived from protected, high-energy environments within the shelf-lagoon; primary high interparticle porosity was reduced by submarine cementation. The bioclastic packstone (right) yields poorly sorted skeletal grains (echinoderms, foraminifera) which have undergone selective dolomitization prior to the deposition of the lithoclasts. Fusulinids are represented by *Pseudoschwagerina* sp. (center) and *Pseudofusulina* cf. *P. paragregaria,* indicating a Lower Artinskian age (Zone with *Pseudofusulina lutugini*). Both lithoclasts were deposited together with quartz pebbles (center and upper right) within a lacustrine environment characterized by the homogeneous dark grey micritic matrix with dewatering structures. During burial diagenesis the breccia was affected by strong pressure-solution. – The sample comes from the Tarvis Breccia of the Sexten Dolomites. The lithoclasts of this Permian breccia were eroded from the Tresdorf Limestone of the Trogkofel Formation (Artinskian). Width of photograph 2 cm.

Preface to the German Edition

„The geological study of facies has been carried out scientifically in the strictest sense of the word since the 1840s, yet one finds even today that the methods used reveal a noticeable lack of clarity and order and that great gaps are present in the results of earlier work in this field . . . One finds side by side the most varied descriptions of the way in which certain sedimentary layers are formed and new views are expressed each day without those already in existence being refuted. . . . It is high time that the principles behind facies research were clearly laid down and that the extensive literature dealing with such questions were revised."

These introductory words from *Geologische Fazieskunde* by Laszlo Strauss (1928) are still valid today. They also hold true for the microfacies analysis of carbonate rocks in thin-sections. Here, too, there are methodological difficulties, for the range of possible interpretations is subject to continual change, owing to the rapid advances in the sciences of paleoecology and carbonate sedimentology, whilst a comprehensive overview of the extensive literature from the various fields of the geosciences does not exist.

The following work is not intended as a synthesis, but rather as a presentation of the current methods of microfacies research as well as a pointer to promising new developments. This applies to the exact description of microfacies types and the facies models based upon them as well as to the reciprocal relationships between biogenic and abiogenic factors in carbonate sedimentation. Important possibilities for development exist with regard to research into the connections between physical parameters of carbonate rocks and facies types, on the one hand, and on the other, between geochemical criteria and sedimentary as well as diagenetic characteristics found in microfacies. The paleontological description and ecological or stratigraphic evaluation of many "thin-section fossils" is just now beginning (e.g., reef associations), while in some cases new, more detailed studies are needed (e.g., dasycladaceans). Compounded with this is the fact that certain individual time periods, such as the Tertiary, have so far hardly been examined from a microfacies point of view.

The structure of this book owes much to important questions that have arisen in the inter-university courses on microfacies held in Erlangen, Germany, over the last few years as well as to the author's own results from microfacies studies of the Permian and the Mesozoic in the Alps and the Malm (Upper Jurassic) found in Franconia, Germany. The literature which has been evaluated covers a broad range of stratigraphic problems, drawn from all over the world, and the same holds for the thin-section samples selected for the Plates.

The intent behind the book was not to produce a "thin-section picture book"; neither does it claim to provide quick and easy recipes for facies interpretation – even if the tables and the condensed presentation of the material may sometimes give this impression. The object of this book is to make possible independent study of carbonate rocks with regard to paleontological and sedimentological aspects, by clarifying definitions and compiling information (in the form of annotated bibliographies and reference tables). For this reason it seemed necessary to include a synopsis of Recent carbonates and carbonate diagenesis, despite the fact that Bathurst (1971) and Milliman (1974) have already published excellent surveys of these topics.

The "Key to Identification" and the exercises should facilitate the direct transfer of what has been read in theory to the practical application of microfacies techniques. This should not be limited to mere description and classification, but rather the attempt should be made to correlate critically one's own observations with standard Recent and Ancient values. It is only in this way and in close association with the various disciplines of the geosciences that microfacies analysis can contribute to facies studies.

I would like to thank the Deutsche Forschungsgemeinschaft for the support it has given my work, the staff at the Springer Verlag for its concern for my wishes, and all the colleagues and institutions who kindly gave me permission to use their illustrations and tables.

I am also very grateful to my co-workers at the Institut für Paläontologie der Universität Erlangen for their help in preparing this book, especially Frau U. Scholl, Frau Chr. Sporn and Herr F. Grimmer. The critical questions asked by students participating in the Erlangen microfacies courses played no small part in my decision to write the book.

Erlangen, January, 1978 ERIK FLÜGEL

Contents

1 Introduction to Facies Analysis

"Every facies of a deposition shows well-defined petrographic, geognostic and paleontological properties which can be clearly differentiated from the properties of other facies in the same geological period." The necessity for interdisciplinary studies is clearly expressed in this definition given by Amanz Gressly in 1838. In facies analysis paleontological, sedimentological, geological, and geochemical data provide the basic information about the sedimentary environment, the lithogenesis, and the biotopes of organisms preserved as fossils.

1.1 The Microfacies Concept

A. Name: The term "microfacies" was suggested by Brown (1943: 325): "In thin section the rock is seen to be composed of ... microfacies." In other words, microfacies refers to the criteria appearing in thin-sections under the microscope.

The various objections raised to the term microfacies have not gained a foothold: Calkins (1943 – a "microfacies" must be counterbalanced with a "megafacies"); Campbell (1944 – instead of "microfacies," simply "under the microscope"); Alling (1945 – "microfacies" is difficult to define, therefore preferably "microlithology" as a result of quantitative microscopic studies).

Apparently without any knowledge of Brown's suggestion, Cuvillier (1952, 1958, 1961) re-introduced the name microfacies to characterize paleontological and petrographic criteria in thin-sections. The *International Sedimentary Petrographical Series,* which first appeared, upon the request of Cuvillier, at the Third International World Petroleum Conference in Paris in 1951, contributed substantially to the rapid adoption of the term microfacies. The distinction between micro-lithofacies and micro-biofacies called for by Fairbridge (1954) has gained little support.

B. Definition: Microfacies is the total of all the paleontological and sedimentological criteria which can be classified in thin-sections, peels, and polished slabs.

C. Discussion: This definition assumes the following:

a. Study of thin-sections, peels, and polished slabs at magnifications of up to approximately ×200.

b. Equal consideration of paleontological and sedimentary criteria.

c. Classification of data, taking into account the qualitative and quantitative criteria (e.g., organic associations, limestone classification, and modal composition).

d. Testing of microfacies classifications by comparison with geological field data, paleoecological interpretations and, perhaps, geochemical criteria.

e. Application of small-scale observations to larger dimensions (outcrop: meter to kilometer range). The methodological relationship between thin-sections and outcrop areas comes from the use of limestone classifications based on textural and structural criteria (Dunham, 1962; see 6.2.2) and from the statistical evaluation of data (see 7.3).

The definition presented here signifies a methodological refinement of Gressly's classical facies concept (see Teichert, 1958; Franke, 1963; Lützner et al., 1974). Because microfacies takes into consideration biological as well as lithological criteria, the arguments presented by W. Schäfer (1962: 546) opposing a "micro-biofacies" (third order biofacies) do not seem to be applicable.

D. Development of the concept: Even the earliest thin-section studies of carbonates aimed for genetic interpretations (Sorby, 1879; Bornemann, 1886) as well as for stratigraphic evaluation and ecological interpretations of fossils (Gümbel, 1873). Probably the oldest microfacies studies originated from K. Peters at the University of Graz, Austria, where in 1863 he published a paper entitled *Über Foraminiferen im Dachsteinkalk* (Foraminifera in Dachstein limestone); here thin-sections were evaluated to clarify paleoecological and paleogeographical questions. Hovelacque and Kilian (1900) published the first illustrated volume of thin-section photographs. The practical application of limestone structures in thin-sections was demonstrated by Udden and Waite (1927) for oil exploration in the Pennsylvanian of Texas.

Microscopic studies of carbonates were given substantial impetus by Bruno Sander (1936 – description of the depositional fabric; English translation 1951) and by Julius Pia (1933), who presented one of the first comprehensive general surveys of the Recent carbonates.

As illustrated in Fig. 1 by the Alpine Triassic, microfacies studies coupled with sedimentological and paleontological objectives did not begin in earnest until the 1960s. The rapid advances made in microfacies since then are a result of: (1) the exploration for oil in carbonate rocks (approximately 50% of the world's crude oil production); (2) creation of useful limestone classifications (Folk, 1959; Ham, 1962); and (3) paleontological problems (e.g., paleoecology of fossil reefs, biostratigraphy of "monotonous" limestone series by means of fossils in thin-sections). At the moment an increasing number of papers is available in which the multitude of microfacies criteria are compiled and interpreted in the form of "facies models" (J. L. Wilson, 1975, see Sects. 8 and 10). In addition, there are studies on the submicroscopic criteria of carbonate rocks (ultra-facies), on the possibilities offered by multivariate facies analyses, and on the mutual relationships between microfacies criteria and technological properties (see Sect. 9.2.3).

E. Divergent definitions: Gübler et al. (1967: 55) defined a "primary microfacies" on the basis of criteria observable in a sedimentation unit (=a rock lamina whose thickness does not exceed the largest diameter of the largest particle). According to this, the term is restricted to criteria which are determined in thin-sections parallel

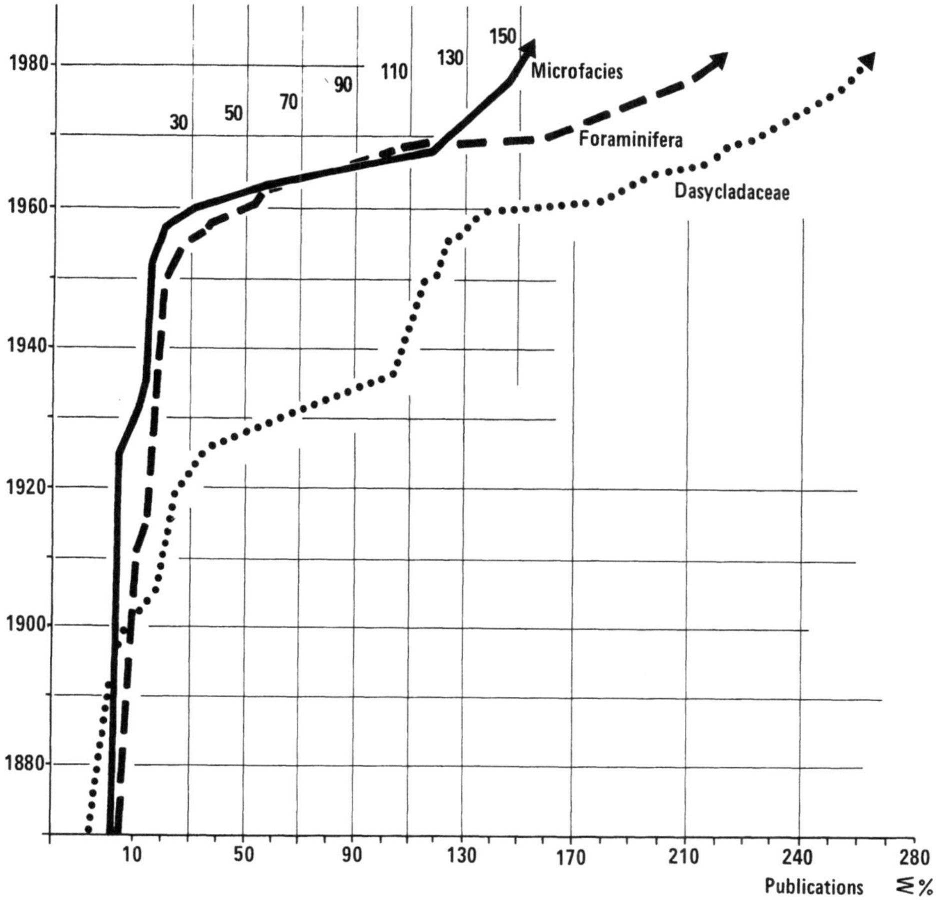

Fig. 1. The interest in microfacies analyses of Triassic sediments in the Calcareous Alps, expressed by the annual number of publications. Calcareous algae were studied relatively early because of their value for biostratigraphy. The number of papers on foraminifera studied in thin-sections and on microfacies characteristics do not begin to grow appreciably until about 1960, as a result of new micropaleontological and sedimentological problems

to bedding planes. "Microsequences" result from a vertical succession of primary microfacies types. These correspond to the microfacies types determined in vertical sections (perpendicular to the bedding plane) in the usual sense of the definition. Cloud and Barnes (1957) use the term microfacies to characterize minor or less important facies developments within larger facies environments. We cannot accept the use of the term microfacies to characterize certain mineral parageneses (e.g., "stress-controlled microfacies", Shaw and Walton, 1963) or to describe textures of sedimentary boulders ("microfacies analyses of till," Ostry and Deane, 1963).

1.2 Methods of Facies Analysis

Basically we can distinguish between studies carried out in the field and those carried out in the laboratory (Fig. 2).

1.2.1 Fieldwork

Prerequisites for microfacies analysis are geological field studies and profiles, with special consideration of facies criteria (lithology, rock colors, bedding and lamination, sedimentary structures and textures, fossil content, stratigraphic relationships and geometric shapes of rocks). Introductions to the methods of field studies are to be found in Krumbein and Sloss (1963), Geyer (1973), Falke (1975) and in Vossmerbäumer (1976).

The description of recognizable criteria for carbonate rocks both in hand specimens and outcrops can be made easier by using the following *check list:*

1. Mineralogical composition (calcite, dolomite) and rock name (limestone, calcareous dolomite, dolomite, dedolomite, limestone-marl cycles, etc.). Methods: HCl, staining with alizarin red S, etc. (see 1.2.3.3 and Bouma, 1969).

2. Rock colors (fresh fracture and weathering colors). Methods: standard color chart (e.g., ROCK-COLOR-CHART, Geol. Soc. Amer. 1951 or MUNSELL SOIL COLOR CHART; see Folk, 1969).

3. Fracture patterns (conchoidal, subconchoidal, uneven, splintery, fracturing through particles, cleavage along particle boundaries).

4. Main constituents at ×10 magnification (fine-grained carbonate matrix, sparite, particles, pores; see Leighton and Pendexter, 1962, Sect. 6.2.3):

- Matrix (grain size and color: homogeneous or non-homogeneous);
- Sparite (equant or non-equant grain sizes; cement fabric, see Sect. 3.3);
- Main particle types (ecological and systematic grouping of the skeletal grains; peloids, aggregate grains, oncoids, ooids, lithoclasts, terrigenous minerals, e.g. detrital quartz, see Sect. 4.1.3.8;
- Particle sizes (siltite up to 0.063 mm, arenite to 2 mm, rudite >2 mm; possible differentiation of fine, medium, and coarse arenite <0.25, <0.50 and >0.50 mm). Methods: Müller hand lens, G. Müller (1967);
- Particle shape (angular, subangular, rounded, etc.);
- Particle frequency and matrix/particle ratio;
- Particle orientation (not oriented, s-fabric, imbricate structures, graded bedding, etc.);
- Particle packing (open or closed fabric, see Sects. 6.2.2 and 4.2.3.1);
- Pores and open-space structures (see 4.2.3.1 and 9.2.1).

5. Bedding and sedimentary textures:

- Bedding types (horizontal bedding, cross-bedding, flaser and lenticular bedding, deformation bedding, biogenic bedding);
- Features of the bedding surface (even, uneven; marks – ripple marks, scour marks, pits and imprints; shrinkage cracks and injection cracks; traces and trails; films);

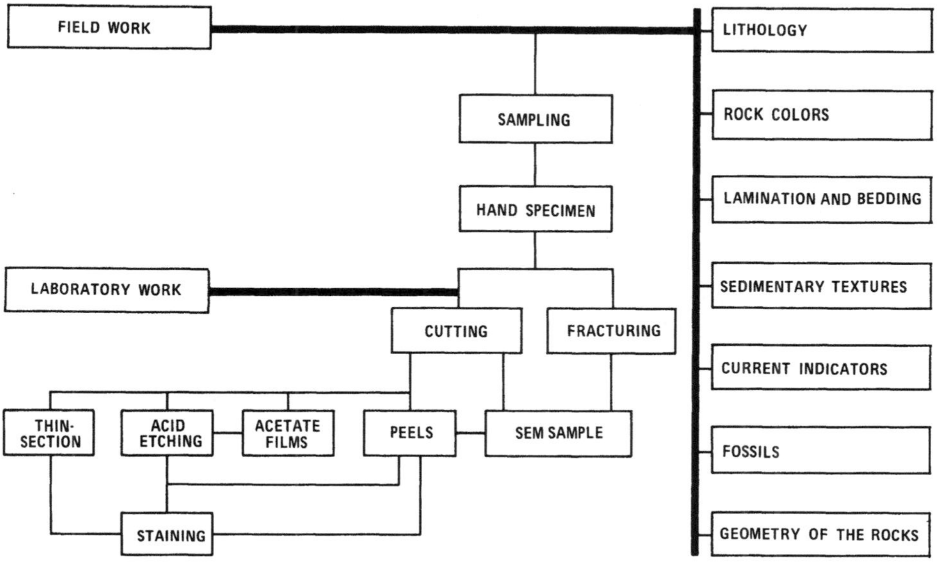

Fig. 2. Flow chart for microfacies analyses

- Features of the underside (even, uneven; load casts, current marks, traces);
- Features within the bedding (small scale bedding, e.g., lamination, see 4.2.3.3 and Lützner et al., 1974; bioturbate fabric and trace fossils; parting lineation; sedimentary dikes, etc.);
- Bed thickness (standardized thicknesses useful, e.g., Ingram, 1954; Kelley, 1956).

6. Dolomitization
7. SiO_2 (silicification, chert, etc.);
8. Concretions
9. Stratigraphic relationships and geometric shapes of rocks (see Krynine, 1948; Füchtbauer, 1974).
10. Fossil content (species, preservation, abundance, ecological criteria; microfossils and trace fossils).

For graphic logs refer to Bouma and Nota (1961) and Lützner et al. (1974); general surveys of the sedimentary structures are given in Potter and Pettijohn (1963), Pettijohn and Potter (1964), Ricci Lucchi (1970) and Reineck and Singh (1973). In Sect. 1.3.3 literature is compiled on the possibilities of interpreting sedimentary structures in carbonate rocks. Particular mention should be made of the work done by Imbrie and Buchanan (1965) and Schwarz (1970 and 1975).

1.2.2 Sampling

The number and kinds of samples are dependent on the geological settings (e.g., outcrop conditions; sampling units indicated by types and thickness of bedding), on the approach to the problem (statements about the depositional environment and biostratigraphy), and on the amount of time and energy to be invested in processing the samples.

Introductions to the methods of statistical sampling can be found in Cochran (1963), Krumbein and Graybill (1965), Griffith (1967) and Marsal (1967).

In facies analyses, sampling units should be selected for which approximately similar or analogous sedimentation conditions can be assumed. This would be, given an ideal case, the "sedimentation phase" (Apfel, 1938) or the "sedimentation unit" (Otto, 1938) – that is, a sediment layer originating under essentially constant physical conditions. However, grain size analyses of the finest sand layers (Walger, 1961) have shown the practical difficulties involved in recognizing such a unit. Analogous textures and structures do not necessarily derive from similar physical factors (Jopling, 1964). If a single large thin-section encompassing the entire thickness of a carbonate bed is investigated closely, one sees, as in clastic rocks, millimeter- to centimeter sized layers (primary microfacies, see Gubler et al., 1967) which have originated under different genetic and diagenetic conditions (example in Tietz, 1976). Sedimentation units of carbonate rocks are, thus, not segregated by bedding planes in all cases.

1.2.2.1 Selection and Number of Samples

Microfacies (MF) samples may be selected under two aspects: (1) Search sampling – used for limestone beds conspicuous because of their texture or fossil content, and (2) Statistical sampling – random samples are collected to estimate the statistical probability of facies categories.

In well-defined studies the rock populations are found to be grouped hierarchically in terms of their positions and physical limits: quarry – limestone bed with a "sequence of beds" (=evenly spaced samples) – rock sample (hand specimen) – sample piece (slice vertical to the bedding plane, showing the variability) – thin-section.

As the methods of statistical sampling strongly influence the statistical evaluation of homogeneity and the recognition of the dominant factors producing variability, a "sampling plan" should be devised in terms of the objectives of the study. This plan can be adapted to one of the four statistical sampling models listed below (Krumbein, 1965; Marsal, 1967). First, however, the area to be studied should be divided into equal intervals (e.g. square grid upon the bedding planes; the sampling interval corresponds to the thinnest sedimentary unit in the vertical section); the randomness (Cadigan, 1962) of the sampling points can be ensured by using random numbers. Sampling occurs then as follows:

a. Simple random sampling: Samples are taken from the sampling unit at unevenly spaced intervals (determined by random numbers). The variability of the samples is checked by variance tests.

b. Systematic sampling: Sampling takes place at evenly spaced intervals. If the characteristics studied occur periodically, there is the danger of biased samples. A test of significance is carried out using the Chi-square test or cumulative frequency distribution.

c. Stratified sampling: At least two samples are taken from each sampling unit or, for bedded carbonates, from each bed. The number of samples per bed should be proportional to the bed thickness.

d. Cluster sampling: After determining the sample points with the aid of random numbers, two adjacent samples are taken at each point. The sampling interval remains constant (see Fig. 54).

Often a *pilot study* is necessary to select the appropriate sampling model (see E. Flügel and H. Hötzl, 1976). Hence, for a series of random samples the sampling error, which is dependent upon the variability in the materials studied and upon the number of random samples, is determined by using the Student's t-distribution (Vogeltanz, 1970). Stratified sampling is advantageous for the study of vertical profiles, systematic sampling for facies units distributed evenly over a horizontal area (McCammon, 1975). If the microfacies data are to be statistically grouped (see Sect. 7.3.1), an early check of sample homogeneity by sequential analyses (Eltgen, 1969) is recommended.

For microfacies analyses most authors use a large *number of samples:*

For 45% of 100 non-selectively chosen examples the calculated interval between the samples (= profile thickness/number of samples) is 0.10 to 0.5 m; for 14% the

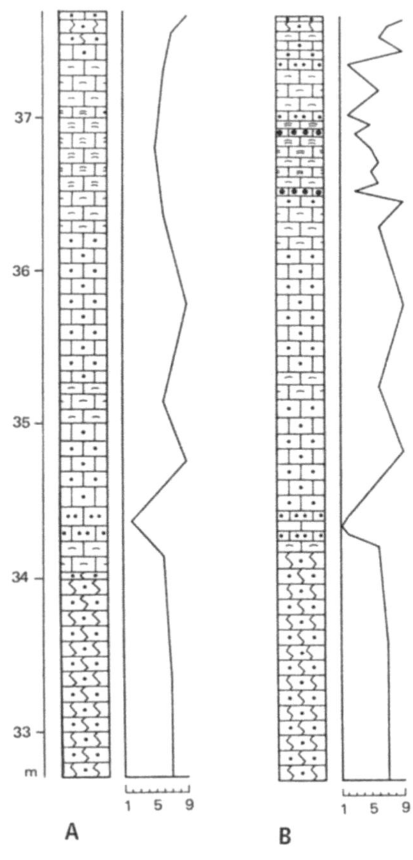

Fig. 3 A, B. Various sample densities (Pennsylvanian/Permian, Nevada; after Nowak and Carozzi, 1973). Example **A:** The 5 m thick profile was studied by using thin-section samples taken at intervals of 20 cm. Example **B:** The profile was sampled as thoroughly as possible, so that a continuous series of samples was available. Although this provides more information and leads to a better separation of genetically related microfacies types (indicated by numbers along the abscissa), the trend in the chronological sequence of the microfacies types is clearly recognizable even for sampling intervals of 20 cm

interval is either under 0.10 m or between 0.50 and 1.00 m. A sample interval larger than 1 m (seldom more than 5 m) is used in only 27% of the examples. The figures show that, in general, bed-to-bed sampling has been conducted. Shelf carbonates are generally analyzed using at least 100, usually, however, more than 250 and frequently more than 1000 thin-sections. The calculated sample interval for these carbonates usually lies between 0.10 and 0.30 m. The average number of thin-sections used for the study of reef carbonates lies between 100 and 350; the calculated sample interval between < 0.10 and 0.20 m. The number of thin-sections used during the analysis of deep water carbonates is generally quite low (50–150), because of the often relatively monotonous microfacies spectra; the calculated sample interval lies between 1.00 and 2.50 m. The influence sampling intervals can have on the results of microfacies analyses is evident in Fig. 3.

1.2.2.2 Size and Orientation of Samples

The MF sample must provide enough material to allow for the production of thin-sections or peels of different sizes and, under certain circumstances, for geochemical analyses (non-carbonate insoluble residues, trace elements, see Sect. 9.1). The sizes 5×4 cm, 5×5 cm, and larger sections of up to 20×15 cm have proved useful as section formats.

The samples should be marked before they are taken (top, perhaps dip and strike) in order to permit a later check on the stratigraphic relationships with geopetal fabrics (Sect. 4.2.3.9) in sections oriented vertically to the bedding plane. The strike can be important for the interpretation of particle orientation as an indicator of current.

When taking larger samples from thick bedded or massive sedimentary rocks, an angle grinder may be useful (see Kasig, 1974).

1.2.3 Laboratory Work

Good introductions to the important preparation methods in MF analyses are found in Bouma (1969), Kummel and Raup (1965) and Müller (1964).

1.2.3.1 Thin-Sections

If polished and ground sections do not provide sufficient information (for production of sections, see Bouma 1969: 133), the preparation of thin-sections is necessary: the polished cut surface of the sample is mounted with the aid of polyester resins or plastics on a glass plate. The sample is then cut again parallel to the glass and finally polished with a machine or by hand to the desired thickness. Technical instructions are found in Völkel (1967). For petrographic samples the thickness of the section is < 30 microns. For MF analyses section thicknesses between 30 and 70 microns are usually preferred, because such "thick sections" often allow the limestone textures and structures to be identified more easily.

For the analysis of very fine grained micritic limestones, ultra-sections (0.5 to 5.0 microns) are necessary; see Honjo and Fischer (1965) and Lindholm and Dean (1973). The thickness of these sections must allow for observations at magnifications of between × 150 and × 1000.

Covering the section is not always recommended, because staining (determination of carbonate minerals) or etching (indication of distribution of non-carbonate components, see Hedberg, 1963; Sainsbury, 1965) is prevented. Moussa (1976) suggests plastic spray as a cover for the section.

1.2.3.2 Peels

The imprint of the etched microreliefs on acetate sheet peels is considered to be the fastest and simplest method for studying cut, polished, and etched surfaces of carbonates (dry-peel technique, Sternberg and Belding, 1942; McCrone, 1963; Swarbrick, 1964; Germann, 1965); the replica can also be made on a very thin plastic sheet which is poured on in liquid form (acetate solution or nitrocellulose film, see Gwinner, 1963; Kampa, 1967). The procedure is as follows: cut the sample, grind with a coarseness of 100, polish with a coarseness of 600 to 800, dry, etch for a few seconds, spray with acetone, press on the acetate film (recommended thickness 0.01 mm), pull off the peel. The peel must then be pressed or mounted between glass slides. For very fine grained textures, the etched cut surface can be flooded with ethylene chloride and pressed on to a small 3 mm thick plexiglass plate (Frank, 1965).

Peels can be studied and observed just like thin-sections using a binocular microscope or a microfilm reader; the peel can also be placed in an enlarger and its image projected on sheet film to produce a negative, which is then contact printed.

1.2.3.3 Staining and Acid Etching

Carbonates and other minerals can be distinguished by using a combination of peels (Brissell, 1957; Katz and Friedman, 1965; Gegman, 1965; Davies and Till, 1968) and cut surfaces or sections and a preferential staining technique. Methods are found in Friedman (1959), Warne (1962), Ayan (1965), Dickson (1966) and Wolf et al. (1967) as well as in Evamy (1963, identification of dedolomitization). Interference factors are the effect of different grain sizes and grain orientations, and the color shades of the stains, which are often only minimally distinguishable. Moreover, staining methods in general allow only a qualitative differentiation of the carbonate minerals. Semi-quantitative determinations can be made by combining stain intensity and chemical reaction (spot test – Feigl, 1954; identification of dolomite with alizarin red S, after Schnitzer, 1958).

1.2.3.4 Scanning Electron Microscopy

Grain fabric in the micron range can only be resolved with the aid of the scanning electron microscope. SEM analyses are extremely important for classifying "ultrafacies" (e.g., origin of micrites, role of blue-green algae in carbonate formation, interpretation of cortoids). In order to ensure the agreement between microfacies and ultrafacies, acetate peels which have been coated with Au, C, or Cu can be studied. Good to excellent results can be obtained from previously untreated fracture surfaces of limestones (Plates 7 and 8) and from polished limestone surfaces etched with Titriplex-III (distinguishes cement types and microstructures in skeletal grains). The size of the SEM samples lies around 15×15 mm.

1.3 References

1.3.1 References: Microfacies Concept

Alling, H. L. (1945): Use of microlithologies as illustrated by some New York sedimentary rocks. Bull. Geol. Soc. Amer. **56,** 737–757, 6 Figs., New York

Bornemann, I. G. (1886): Beiträge zur Geologie des Muschelkalkes, insbesondere der Schichtenfolge und Gesteine des Unteren Muschelkalkes in Thüringen, Jb. preuß. geol. Landesanst. **1885,** 267–321, 8 Pls., Berlin

Brown, J. S. (1943): Suggested use of thè word microfacies. Econ. Geol. **38,** 325, New Haven

Calkins, F. C. (1943): The word "Microfacies". Econ. Geol. **38,** 608–609, New Haven

Campbell, I. (1944): The word "Microfacies". Econ. Geol. **39,** 70–71, New Haven

Chrobok, S., Kautzsch, E., Meier, R. (1968): Fazies und Milieu – Grundprobleme ihrer Individualisierung und Typisierung. XXIII Internat. Geol. Congr. **8,** 137–146, Prague

Cloud, P. E. Jr., Barnes, V. E. (1957): The Ellenberger Group of Central Texas. Univ. Texas Publ. **4621,** 473 pp.

Cuvillier, J. (1952): La notion de "microfacies" et ses applications. VIII Congr. Naz. Mètano e Petrolio, sect. I, 1–7

Cuvillier, J. (1958): Micropaléontologie moderne. Rev. Micropaléont. **1**/1, 5–8, Paris

Cuvillier, J. (1961): Etude et utilisation rationelle de microfacies. Rev. Micropaléont. **4**/1, 3–6, Paris

Dunham, R. J. (1962): see 6.4

Fairbridge, R. W. (1954): Stratigraphic correlation by microfacies. Amer. J. Sci. **252**/11, 683–694, 1 Fig., New Haven

Folk, R. L. (1959): see 6.4

Franke, D. (1963): Zu Fragen geologischer Terminologie und Klassifikation (II). Der Begriff der Fazies (3. Teil). Z. angew. Geol. **1963**/3, 153–157, Berlin

Gressly, A. (1838): Observations géologiques sur le Jura Soleurois. N. Denkschr. allgem. schweiz. Ges. Naturwiss. 2, Neuchâtel

Gubler, Y., Bertrand, P., Mattavelli, L., Rizzini, A., Passega, R. (1967): Petrology and petrography of carbonate rocks. In: Chilingar, G. V., Bissel, H. J., Fairbridge, R. W. (eds.): Carbonate Rocks. Dev. Sediment. **9** A, 51–86, 13 Figs., 1 Tab., Amsterdam-London-New York: Elsevier

Gümbel, C. W. (1873): Mikroskopische Untersuchungen alpiner Triaskalke und Dolomite. Verh. geol. Reichsanstalt 7 (8), 141–144, Vienna

Ham, W. E. (ed., 1962): see 1.4.1

Hovelacque, M., Kilian, C. C. W. (1900): Album de microphotographies de roches sédimentaires, d'après les échantillons recueillis et choisis par W. Kilian. 14 pp. 69 Pls., Paris (Gauthier-Villars)

Lützner et al. (1974): see 1.3.2.1

Ostry, R. C., Deane, R. E. (1963): Microfacies analyses of till. Bull. Geol. Soc. Amer. **74,** 165–168, New York

Peters, K. (1863): Über Foraminiferen im Dachsteinkalk. Jb. geol. Reichsanst. Wien **13,** 293–298, Vienna
Pia, J. (1933): see 2.4.2
Rey, M. (1954): Sur l'emploi des photo-logs de microfacies dans les séries indurées sédimentaires traversées par les sondages. Rev. inst. Franç. Petrol. et Ann. Combustibles liquides **9**/2, Paris
Sander, B. (1936): see 1.4.1
Schäfer, W. (1962): see 1.4.1
Shaw, Ch. E. Jr., Walton, M. (1963): Stress-controlled microfacies in a metamorphosed conglomerate. Geol. Soc. Amer. Abstr. 1962, 239, Houston
Sorby, H. C. (1879): The structure and origin of limestones. Geol. Soc. London Proc. **35,** 56–95, 11 Figs., 2 Tabs., London
Teichert, C. (1958): Concept of facies. Bull. Amer. Ass. Petrol. Geol. **42**/11, 2718–2744, 1 Fig., Tulsa
Udden, J. A., Waite, V. V. (1927): Some microscopic characteristics of the Bend and Ellenburger limestones. Texas Univ. Bull. **2703,** 8 pp, 9 Pls., Austin
Wilson, J. L. (1975): see 1.4.1

1.3.2 References: Methods of Facies Analysis

In this and the following lists of references more articles are cited than are referred to in the text. This additional reading list should help those who want to become better acquainted with the special topics discussed.

1.3.2.1 References: Fieldwork

Banks, J. E. (1950): Particle-types well logging. Bull. Amer. Ass. Petrol. Geol. **34,** 1729–1739, Tulsa
Bouma, A. H. (1962): Sedimentology of some flysch deposits. A graphic approach to facies pattern. 168 pp., 31 Figs., 3 supps., 17 Tab., 8 Pls., Amsterdam–New York: Elsevier
Bouma, A. H. (1969): see 1.3.2.3
Bouma, A. H., Nota, D. J. G. (1961): Detailed graphic logs of sedimentary formations. Intern. Geol. Congress, Report 21st Sess., Norden **23,** 52–74, 6 Pls., Copenhagen
Dunham, R. J. (1962): see 6.4
Falke, H. (1975): Anlegung und Ausdeutung einer geologischen Karte. 210 pp., 120 Figs., Berlin: de Gruyter
Folk, R. L. (1969): Toward greater precision in rock-color terminology. Bull. Geol. Soc. Amer. **80,** 725–728, New York
Forgotson, J. Jr. (1960): Review and classification of quantitative mapping techniques. Bull. Amer. Ass. Petrol. Geol. **44**/1, 83–100, 7 Figs., Tulsa
Füchtbauer, H. (1974): see 1.4.1
Geyer, O. F. (1973): see 1.4.1
Ginsburg, R. N. (1975): see 1.4.1
Ingram, R. L. (1954): Terminology for the thickness of stratification and parting units in sedimentary rocks. Bull. Geol. Soc. Amer. **65,** 937–938, New York
Kelley, V. C. (1956): Thickness of strata. J. Sed. Petrol. **26,** 289–300, Tulsa
Krumbein, W. C., Sloss, L. L. (1963): see 1.4.1
Krynine, P. P. (1948): The megascopic study and field classification of sedimentary rocks. J. Geol. **56,** 130–165, Chicago
Leighton, M. M., Pendexter, C. (1962): see 6.4
Lützner, H. (1968): Tabellarische Dokumentation von Sedimentgesteinen auf feldgeologischer Grundlage. Z. Angew. Geol. **13,** 11/12, 635–642, 3 Figs., Berlin

Lützner, H., Falk, F., Ellenberg, J., Grumbt, E. (1974): Tabellarische Dokumentation klastischer Sedimente. Akad. Wiss. DDR, Forschungsbereich Geo- und Kosmowiss., Zentralinst. Phys. Erde **20,** 153 pp., 7 encls., 36 Figs., Potsdam
Müller, G. (1967): A pocket microscope for grain size measurement in the field. J. Sed. Petrol. **37**/2, 703–704, 2 Figs., Tulsa
Müller-Jungbluth, W. V., Toschek, P. H. (1969): see 6.4
Pettijohn, F. J., Potter, P. E. (1964): see 1.4.1
Potter, P. E., Pettijohn, F. J.: (1963): see 1.4.1
Reineck, H. E. (1963): Parameter von Schichtung und Bioturbation. Geol. Rdsch. **56,** 420–438, Stuttgart
Reineck, H. E., Singh, I. B. (1973): see 1.4.1
Ricchi Lucchi, F. (1970): see 1.4.1
Rizzini, A., Mattavelli, I. (1964): Logs pétrographiques et cartes de lithofacies pratiques des roches carbonatées. Rev. Inst. franç. Pétrole **19**/5, 720–728
Shrock, R. R. (1948): see 1.4.1
Vossmerbäumer, H. (1976): see 1.4.1
Wilson, J. L. (1975): see 1.4.1

1.3.2.2 References: Sampling

Apfel, E. T. (1938): Phase sampling of sediments. J. Sed. Petrol. **8,** 67–78, Tulsa
Cadigan, R. A. (1962): A method for determining the randomness of regionally distributed quantitative geologic data. J. Sed. Petrol. **32**/4, 813–818, 1 Fig., Tulsa
Cochran, W. G. (1963): Sampling technique. 2nd ed., 333 pp., New York: Wiley & Sons
Dennison, J. M. (1962): Graphical aids for determining reliability of sample means and an adequate sample size. J. Sed. Petrol. **32**/4, 734–750, 5 Figs., Tulsa
De Vries, N. (1970): On the accuracy of bed-material sampling. J. Hydraul. Res., **8,** 523–534
Eltgen, H. (1969): see 7.3.2
Flügel, E., Hötzl, H. (1976): Palökologische und statistische Untersuchungen in mitteldevonischen Schelf-Kalken (Schwelmer Kalk, Givet, Rheinisches Schiefergebirge). Bayer. Akad. Wiss. Math.-naturwiss. Kl. Abh., N. F. **156,** 70 pp., 29 Figs., 17 Tabs., München
Griffith, J. C. (1967): Scientific Methods in Analysis of Sediments. 508 pp., New York: McGraw Hill Int. Ser. Earth and Planetary Sci.
Gubler, Y., et al. (1967): see 1.3.1
Jopling, A. V. (1966): Some applications of theory and experiment to the study of bedding genesis. Sedimentology **7,** 71–102, 3 Pls., 5 Figs., Amsterdam
Kasig, W. (1974): Eine Methode der Probeentnahme aus massigen Sedimentgesteinen. N. Jb. Geol. Paläont. Mh. **1974**/2, 65–71, 3 Figs., Stuttgart
Krumbein, W. C. (1965): Sampling in paleontology. In: Kummel, B., Raup, D. (eds.): Handbook of Paleontological Techniques, 137–150, 1 Fig., San Francisco–London: Freeman
Krumbein, W. C., Graybill, F. A. (1965): An Introduction to Statistical Models in Geology. 475 pp., New York: McGraw Hill
McCammon, R. B. (1975): see 7.3.2
Marsal, D. (1967): see 7.3.2
Nowak, F. J., Carozzi, A. V. (1973): see 7.3.2
Otto, G. H. (1938): The sedimentation unit and its use in field sampling . J. Geology **46,** 569–582, Chicago
Tietz, G. F. (1976): Petrographische und geochemische Untersuchungen an einer kondensierten Kalkbank des unteren Silurs der Karnischen Alpen (Österreich). Verh. Geol. Bundesanst. **1976**/2, 207–242, 7 Figs., 1 Tab., Wien
Walger, E. (1961): Die Korngrößenverteilung von Einzellagen sandiger Sedimente und ihre genetische Bedeutung. Geol. Rundsch. **51**/2, 494–507, 8 Figs., Stuttgart

1.3.2.3 References: Laboratory Work

Ayan, T. (1965): Chemical staining methods used in the identification of carbonate minerals. Bull. Min. Res. Explor. Inst. Turkey **65,** 133–147, Ankara

Bissell, H. (1957): Combined preferential staining and cellulose peel technique. J. Sed. Petrol. **27,** 417–420, Tulsa

Bouma, A. H. (1969): Methods for the Study of Sedimentary Rocks. 458 pp., New York–London–Sidney–Toronto: Wiley

Davies, P. J., Till, R. (1968): Stained dry cellulose peels of ancient and recent impregnated carbonate sediments. J. Sed. Petrol. **38**/1, 234–237, 1 Tab., Tulsa

Dickson, J. A. D. (1966): Carbonate identification and genesis as revealed by staining. J. Sed. Petrol. **36**/2, 491–505, 10 Figs., Tulsa

Evamy, B. D. (1963): The application of a chemical staining technique to a study of dedolomitization. Sed. **2,** 164–170, Amsterdam

Feigl, F. (1960): Tüpfelanalysen. 1. Anorganischer Teil. 4. deutsche Auflage, 594 pp., Frankfurt/M.

Fenninger, A. (1968): Das Kalzitgefüge der sparitischen Kalke des Plassen (Tithonium, nördliche Kalkalpen, Oberösterreich). Sed. **10**/4, 273–291, 9 Figs., 3 Tab., Amsterdam

Frank, R. M. (1965): An improved carbonate peel technique for high-powered studies. J. Sed. Petrol. **35**/2, 1 Fig., Tulsa

Friedman, G. M. (1959): Identification of carbonate minerals by staining methods. J. Sed. Petrol. **29**/2, 87–97, 2 Figs., 4 Tab., Tulsa

Gensmer, R. P., Weiss, M. P. (1980): Accuracy of calcite/dolomite ratios by X-ray diffraction and comparison with results from staining techniques. J. Sed. Petrol. **50**/2, 626–629, 2 Figs., Tulsa

Germann, K. (1965): Die Technik des Folienabzuges und ihre Ergänzung durch Anfärbemethoden. N. Jb. Geol. Paläont. Abh. **121,** 293–306, Stuttgart

Gwinner, M. P. (1963): Lackfilme als Hilfsmittel bei der Herstellung von Gesteinsschnittbildern, Photographische Korrespondenz **99,** 188–190, 4 Figs., Darmstadt–Wien–

Hedberg, R. M. (1963): Dip determination in carbonate cores. J. Sed. Petrol. **33,** 680–693, Tulsa

Honjo, S., Fischer, A. G. (1965): Thin section and peels for high-magnification study and phase-contrast microscopy. In: Kummel, B., Raup, D. (eds.): Handbook of Palaeontological Techniques, 241–247, 1 Fig., San Francisco–London: Freeman

Kampa, E. (1967): Ein neues Nitrocellulosefilm-Verfahren in der Paläontologie. Der Präparator **13**/3, 205–206, Bonn

Katz, A., Friedman, G. M. (1965): The preparation of stained acetate peels for the study of carbonate rocks. J. Sed. Petrol. **35,** 248–249, Tulsa

Kummel, B., Raup, D. (eds.) (1965): Handbook of Paleontological Techniques. 852 pp., San Francisco–London: Freeman

Lees, A. (1958): Etching techniques for use on thin sections of limestones. J. Sed. Petrol. **28**/2, 100–202, 3 Figs., Tulsa

Lindholm, R. C., Dean, D. A. (1973): Ultra-thin sections in carbonate petrology: a valuable tool. J. Sed. Petrol. **43**/1, 295–297, 1 Fig., Tulsa

McCrone, A. W. (1963): Quick preparation of peel-prints for sedimentary petrography. J. Sed, Petrol. **33**/1, 228–230, Tulsa

Moussa, M. T. (1976): Plastic spray thin-section cover. J. Sed. Petrol. **46**/1, 252–253, Tulsa

Müller, G. (1964): Methoden der Sedimentuntersuchung. 303 pp., 91 Figs., 29 Tab., 3 Suppl., Stuttgart: Schweizerbart

Müller, G., Raith, M. (1976): Methoden der Dünnschliffmikroskopie. 2. Auflage. Clausthaler Tektonische Hefte **14,** 151 p., 38 Figs., 6 Tables, Clausthal-Zellerfeld

Sainsbury, C. L. (1965): Improved method for studying carbonate rocks in thin sections. J. Sed. Petrol. **35**/2, 491–494, 3 Figs., Tulsa

Schneidermann, N., Sandberg, P. A. (1971): Calcite-Aragonite differentiation by selective staining and scanning electron microscopy. Transact. Gulf Coast Ass. Geol. Soc. **21,** 4 pp., 6 Figs., New Orleans

Schnitzer, W. A. (1958): Eine anwendbare Methode quantitativer Dolomitbestimmung in Weißjura-Kalken mittels Tüpfelreaktion. Geol. Bl. NO-Bayern **8,** 53–112, Erlangen
Schnitzer, W. A. (1967): Bromphenolblau zur Unterscheidung von Kalkstein und Dolomit. Zement-Kalk-Gips **1967**/1, 31–32
Sternberg, R. M., Belding, H. (1942): The dry-peel technique. J. Paleont. **16,** 135–136, Tulsa
Swarbrick, E. E. (1964): A peel technique for the study of sedimentary structures. Sedimentology **3**/1, 75–78, 1 Pl., Amsterdam
Voigt, E. (1949): Die Anwendung der Lackfilmmethode. Mitt. geol. Staatsinst. **19,** 111–129, Hamburg
Völkel, H. (1967): Allgemeines über die Anfertigung von Dünnschliffen und Anschliffen. Der Präparator **13,** 155–169, 7 Figs., Bonn
Warne, J. (1962): A quick field or laboratory staining scheme for the differentiation of the major carbonate minerals. J. Sed. Petrol. **32**/1, 29–38, 2 Pls., 3 Figs., Tulsa
Warth, M. (1969): Selektive Anfärbung von Gesteinsoberflächen. Der Präparator **15**/1, 11–17, Bochum
Wolf, K. H., Easton, A. J., Warne, S. (1967): Techniques of examining and analyzing carbonate skeletons, minerals and rocks. Dev. Sedimentology **9** B, Carbonates 253–341, 21 Figs., Amsterdam: Elsevier

1.3.3 References: Sedimentary Structures in Carbonate Rocks

Bacelle Scudeler, L. (1968): Strutture torbitiche e frane sottomarine nel "S. Cassiano" di Carbonin (Dolomiti nord-orientali). Accad. Naz. Lincei, Rend. Cl. Sci. Fis. Mat. Naturali, Fasc. 6, Ser. 8, **44,** 821–826, Pl. 1, 1 Fig., Roma
Ball, M. M. (1967): Carbonate sand bodies of Florida and Bahamas, J. Sed. Petrol. **37,** 556–591, Tulsa
Bouma, A. H. (1972): Fossils contourites in lower Niesenflysch, Switzerland. J. Sed. Petrol. **42**/4, 917–921, 3 Figs., Tulsa
Broadhurst, F. M. (1968): Large scale ripples in Silurian limestones. Lethaia **1,** 28–38, 6 Figs., Oslo
Campbell, Ch. V. (1967): Lamina, laminaset, bed and bedset. Sedimentology **8,** 7–26, 4 Figs., Amsterdam
Carozzi, A. V. (1960): Microscope arched flow structures and spiral structures in sedimentary rocks. Bull. Inst. Nat. Genevois **60,** 4–23, Geneva
Davies, D. K. (1968): Carbonate turbidites, Gulf of Mexico. J. Sed. Petrol. **38,** 1100–1109, 8 Figs., Tulsa
Dott, R. H. Jr. (1958): Cyclic patterns in mechanically deposited Pennsylvanian limestones of northeastern Nevada. J. Sed. Petrol. **28**/1, 3–14, 3 Figs., 1 Pl., Tulsa
Engel, W. (1970): Die Nummuliten-Breccien im Flyschbecken von Ajdovščina in Slowenien als Beispiel karbonatischer Turbidite. Verh. geol. Bundesanst. **1970**/4, 570–582, 5 Figs., Wien
Grunau, H. (1959): see 5.5
Gwinner, M. P. (1962): Subaquatische Gleitungen und resedimentäre Breccien im Weißen Jura der schwäbischen Alb (Württemberg). Z. dtsch. geol. Ges. **113,** 2/3, 571–590, 6 Figs., Pls. 7–10, Stuttgart
Hamblin, W. K. (1969): Marine paleocurrent directions in limestones of the Kansas City Group (Upper Pennsylvanian) in Eastern Kansas. State Geol. Surv. Kansas Bull **194**/2, 24 pp., 19 Figs., Lawrence
Harbaugh, J. W. (1959): Small scale cross-lamination in limestones. J. Sed. Petrol. **29**/1, 30–37, 6 Figs., Tulsa
Imbrie, J., Buchanan, H. (1965): Sedimentary structure in modern carbonate sands of the Bahamas. Soc. Econ. Paleont. Min. Spec, Publ. **12,** 149–172, Tulsa
Kuenen, P. H., ten Haaf, E. (1956): Graded bedding in limestones. Proc. Nederl. Akad. Wetensch., Ser. B. **59,** 314–317, Amsterdam

Lindström, M. (1963): Sedimentary folds and the development of limestone in early Ordovician Sea. Sedimentology **2**/3, 243–292, 15 Figs., Amsterdam
Lumsden, D. N. (1971): Facies and bed thickness distributions of limestones. J. Sed. Petrol **41**/2, 593–598, 4 Figs., Tulsa
MacKenzie, F. T. (1964): Bermuda Pleistocene eolianites and paleowinds. Sedimentology **3,** 52–64, 6 Figs., Amsterdam
Marcinowski, R. (1970): Turbidites in the Upper Oxfordian Limestones at Jaskrow in the Polish Jura Chain. Bull. Acad. Polonaise Scien. Ser. sci. geol. **18**/4, 219–225, Pls. 1–6, 2 Figs., Warsaw
Meischner, K. D. (1964): Allodapische Kalke, Turbidite in riffnahen Sedimentations-Becken. Dev. Sediment. **3,** 156–191, 3 Pls., 5 Figs., Amsterdam
Radwánski, A., Roniewicz, P. (1962): Submarine slumping in the Famennian of the Holy Cross Mts. (Central Poland). Acta Geol. Polonica **12**/2, 295–305, Pls. 12–15, Warsaw
Remane, J. (1970): Die Entstehung der resedimentären Breccien im Obertithon der subalpinen Ketten Frankreichs. Eclogae geol. Helvetiae **63**/3, 685–740, 20 Figs., 6 Pls., Basel
Rigby, K. J. (1958): Mass movements in Permian rocks of Trans-Pecos Texas. J. Sed. Petrol. **28**/3, 298–315, 14 Figs., Tulsa
Roehl, P. O. (1967): Stony mountains (Ordovician) and interlake (Silurian) facies analogs of recent low energy marine and subaerial carbonates, Bahamas. Amer. Ass. Petrol. Geol. Bull. **51**/10, 1979–2032, 48 Figs., Tulsa
Schuler, G. (1967): Zur Kalk-Mergel-Sedimentation stromatometrisch vergleichbarer Profile im Malm Alpha in der mittleren Frankenalb. Geol. Bl. NO-Bayern **17**/3, 127–136, 1 Tab., 4 Figs., Erlangen
Schwarz, H.-U. (1970): Zur Sedimentologie und Fazies des unteren Muschelkalkes in Südwestdeutschland und angrenzenden Gebieten. Diss. Univ. Tübingen, 267 pp., Tübingen
Schwarz, H.-U. (1975): Sedimentary structures and facies analysis of shallow marine carbonates (Lower Muschelkalk, Middle Triassic, Southwestern Germany). Contrib. Sed. **3,** 100 pp., 11 Pls., 35 Figs., 1 Tab., Stuttgart: Schweizerbart
Sedimentation Seminar (1966): Cross-bedding in the Salem limestone of Central Indiana. Sedimentology **6,** 95–114, 12 Figs., Amsterdam
Seibold, E. (1952): Chemische Untersuchungen zur Bankung im unteren Malm Schwabens. N. Jb. Geol. Paläont. Abh. **95,** 337–370, Stuttgart
Sorauf, J. E. (1965): Flow rolls of Upper Devonian rocks of south-central New York State. J. Sed. Petrol. **35,** 553–563, Tulsa
Swinchatt, J. P. (1967): Formation of large scale cross-bedding in a carbonate unit. Sedimentology **8,** 93–120, 18 Figs., 1 Tab., Amsterdam
Szulczewski, M. (1968): Slump structures and turbidites in Upper Devonian limestones of the Holy Cross Mts. Acta Geol. Polonica **18**/2, 303–324, Pls. 1–6, 8 Figs., Warsaw
Vortisch, W. (1964): Die Entwicklung unserer Anschauungen über nichttektonische Verformung der Sedimente. Geol. Rdsch. **53**/2, 650–679, Stuttgart
Wendt, J. (1971): Genese und Fauna submariner sedimentärer Spaltenfüllungen im mediterranen Jura. Palaeontographica, A, **136,** 121–192, Pls. 15–18, 20 Figs., 7 Tab., Stuttgart

1.4 Supplementary Reading

1.4.1 Annotated Bibliography – Methods and Interpretation of Facies Analysis

The abbreviations in parentheses preceding the standard works listed below refer to the general topics dealt with. The key to the abbreviations is as follows: FI – facies interpretation, C – Modern and/or Ancient carbonates, MSA – methods of sediment analysis, E – ecology and paleoecology, RS – Recent sedimentation, S – sediment description. Monographs on microfacies have not been mentioned here (see Sect. 5.5).

(S) Allen, J. R. L. (1977): Physical processes of sedimentation. 248 p., London: Allen and Unwin
Important discussion of the physical background of sedimentation; the origin of sedimentary structures; and the influence of winds, river flows, waves, tides, oceanic circulation, turbidity currents, and glaciers on sedimentation.

(FI) Anstey, R. L., Chase, T. L. (1974): Environments through time. A laboratory manual in the interpretation of Ancient sediments and organisms. 136 pp., Minneapolis: Burgess
Material for a well-organized introductory course; sedimentology, paleontology and paleoecology, stratigraphy, geotectonics. New edition: 1979.

(FI) Asquith, G. B. (1979): Subsurface carbonate depositional models: A Concise Review. 121 p. Tulsa: Petrol. Publ. Company
Written for exploration geologists, the book deals with the criteria for five carbonate depositional models (Regressive carbonate shoreline; carbonate tidal flat; oolite shoal; reef sand-bank) using case histories as examples. Special attention is given to the relationships between carbonate lithologies and physical logs.

(C) Bathurst, R. G. C. (1971): Carbonate sediments and their diagenesis. Dev. Sediment. **12,** 620 pp., 359 Figs., Amsterdam: Elsevier
Excellent survey of Recent carbonate sedimentation and diagenesis. Chapters: Petrography of carbonate grains, skeletal structures – ooids, pisolites, peloids, and other micritic fabrics; Recent carbonate environments – Great Bahama Bank, Florida, Gulf of Batabano, Persian Gulf, British Honduras; Recent carbonate algal stromatolites; Chemical considerations; Growth of ooids, pisolites and grapestones; Diagenesis in the subaerial freshwater environment; Diagenesis on the sea-floor; Cementation, pressure-solution; Neomorphic processes in diagenesis; Recent dolomites.

(S) Bentor, Y. K. (ed.) (1980): Phosphorites. Soc. Econ. Paleont. Min., Spec. Publ., **29,** 247 pp., Tulsa
A collection of very important papers dealing with the genesis, geochemistry and diagenesis of, usually fossil, phosphorites.

(S) Berner, R. A. (1971): Principles of chemical sedimentology. 256 pp., New York: McGraw Hill

(S) Blatt, H., Middleton, G., Murray, R. (1972): Origin of sedimentary rocks. 653 pp., New Jersey: Englewood Cliffs
Very stimulating description of sedimentation processes and methods of analysis. Annotated bibliographies.

(MSA) Bouma, A. H. (1969): Methods for the study of sedimentary structures. 458 pp., New York: Interscience
Introduction to the methods of sediment analysis. Peels, impregnation, X-ray photographs, staining methods, etc., sampling. Many practical tips.

(A) Bouma, A. H., Brouwer, A. (eds.) (1964): Turbidites. Dev. Sed. **3,** 264 pp., Amsterdam–London–New York: Elsevier
Important collection of articles about Recent and Ancient turbidites. Important for carbonates: allodapic limestones – turbidites in near-reef areas

(S) Carozzi, A. V. (1960): Microscopic sedimentary petrography. 485 pp., 88 Figs., New York–London: Wiley
Introduction to thin-section analysis of sedimentary rocks. Chapters: Clastic rocks; Biochemical rocks (carbonate rocks, siliceous rocks, ferruginous rocks, phosphorites); Chemical rocks (evaporites).

(C) Cayeux, L. (1970): Sedimentary rocks of France. Carbonate rocks (Limestone and dolomites). Translated and updated by A. V. Carozzi, 506 p., 26 Plates, Darien: Hafner
A classic description of carbonate rocks, still very interesting, especially the chapter dealing with freshwater limestones.

(C) Chilingar, G. V., Bissell, Fairbridge, R. W. (eds.) (1967): Carbonate rocks. Dev. Sedimentology, **9** A (origin, occurrence and classification). 471 pp., 131 Figs., 20 Tab.; **9** B (Physical and Chemical Aspects), 413 pp., 80 Figs., 70 Tab., Amsterdam: Elsevier
Interesting but now somewhat outdated monographs. Chapters: Volume 9 A – Modern carbonate sediments; Petrology and petrography of carbonate rocks; Classification; Origin and occurrence of limestones; Origin and occurrence of dolomites; Carbonate oil reservoir rocks; Carbonate rocks and paleoclimatology. Volume 9 B – Elemental composition of carbonate skeletons, minerals and sediments; Physical chemistry of formation of carbonates; Chemistry of dolomite formation; Stable isotope distribution in carbonates;

Influence of pressure and temperature on limestones; Origin of petroleum in carbonate rocks; Techniques of examining and analyzing skeletons, minerals and carbonate rocks; Properties and uses of the carbonates.

(FI) Conybeare, C. E. B. (1979): Lithostratigraphic analysis of sedimentary basins. 555 p. New York: Academic Press
The main focus of this book is on the concepts of depositional environments, paleogeographic settings, structural framework, and geometry of Recent and Ancient sedimentary basins. Special attention is paid to the evolution and development of various types of basins, to the analysis of outcrops, drill cores, and cuttings, and the construction of maps and sections. The application of basin analysis to economic geology is described by examples from the exploration for oil and gas, coal, mineral deposits, and for aquifers and storage reservoirs.

(C) Cook, H. E., Enos, P. (eds.) (1977): Deep-water carbonate environments. Soc. Econ. Paleont. Min. Spec. Publ. **25,** 336 p., Tulsa
A very important book containing 15 stimulating papers on Paleozoic and Mesozoic deeper-water carbonates.

(E) Crimes, T. P., Harper, J. D. (eds.) (1970): Trace Fossiles. Symposium. Geol. J. Spec, Issue **3,** 547 pp., 114 Figs., 88 Pls., London

(RS) Davies, R. A. Jr., Ethington, R. L. (eds.) (1976): Beach and nearshore sedimentation. Soc. Econ. Paleont. Min. Spec. Publ. **24,** 187 p., Tulsa
A collection of papers dealing with weather patterns and coastal processes, a continental shelf wave climate model, longshore current velocities and sand transport rates, wave modeling, sand fountains in the surf zone, vortices in surf zone prediction, interactions between swash-ground and water-beach profile, conceptual models of wave-formed sedimentary structures, facies relationships on a barred coast, interaction of biological and geological processes in the beach and near-shore area. All the examples taken from clastic environments.

(S) Dean, W. E., Schreiber, B. Ch. (1978): Marine evaporites. Soc. Econ. Paleont. Min. Short Course, No. 4, 188 p., Oklahoma City
Well-written introduction to the classification, distribution, chemistry and depositional environments of evaporites.

(S) Degens, E. T. (1968): Geochemie der Sedimente. 282 pp., 75 Figs., 20 Tab., Stuttgart: Enke
General survey geared to geologists: Weathering residues of igneous and metamorphic minerals; Minerals of low temperature and aqueous formation; Mobile phases; Organic geochemistry.

(RS) Dietrich, G. (ed.) (1970): Erforschung des Meeres. 318 pp., Frankfurt a.M.: Umschau
Worthwhile introduction to the basic oceanographic data of ocean water, the ocean floor, marine life, ocean technology.

(S) Dott, R. H. Jr., Shaver, R. H. (eds.) (1974): Modern and ancient geosynclinal sedimentation. Soc. Econ. Min. Paleont., Spec. Publ. **19,** 380 pp., Tulsa
Very stimulating articles about Recent and Ancient deep-sea sediments, submarine vulcanism, and the problems of palinspastic reconstructions.

(S) Doyle, L. J., Pilkey, O. H. (eds.) (1979): Geology of continental slopes. Soc. Econ. Paleont. Min. Spec. Publ., **27,** 374 p., Tulsa
Very important contributions on Recent and Ancient continental slope deposits, especially important for the study of sediment gravity flows. Carbonate environments are represented by the description of deep carbonate margin structures in the Northern Bahamas, and of the platform – basin transition in the Tongue of the Ocean area, Bahamas.

(S) Duff, P., McL. D., Hallam, A., Walton, E. K. (1967): Cyclic sedimentation. Dev. Sed. **10,** 280 pp., Amsterdam–London–New York: Elsevier
Excellent description of cyclic sedimentation in various depositional environments.

(FI) Dunham, R. J. (1972): Guide for study and discussion for individual reinterpretation of the sedimentation and diagenesis of the Permian Capitan geologic reef and associated rocks, New Mexico and Texas. Permian Basin Section, Soc. Econ. Paleont. Min., Publ. **72–14,** 235 pp.

Excellent and very stimulating, unconventional description of sedimentological and paleontological criteria of the Permian reef complex. Guide to a critical interpretation particularly of the modes of diagenesis.

(C) Elf-Aquitaine (1975): Essai de caractérisation sédimentologique des dépôts carbonatés. (An attempt at sedimentological characterisation of carbonate deposits). 1. Eléments d'analyse (Analytic elements). 172 pp., 61 Pls., Boussens-Pau: Elf-Aquitaine
This very well illustrated book, which was originally the working basis for French petroleum geologists, provides a quick introduction into the distinguishing features of carbonate rocks in thin-sections, the possibilities for classification as well as a few characteristics of fabrics.

(C) Elf-Aquitaine (1977): Essai de Caracterisation sédimentologique des dépots carbonatées, **2.** Eléments d'interprétation. 231 p., 76 Figs., 17 Plates, Boussens-Pau: Elf-Aquitaine
Very important discussion of the depositional models of carbonate rocks in relation to the possibilities for reservoir rocks. Explanation of porosity classifications. Interesting case studies from the Mesozoic and Paleozoic.

(S) Engelhardt, W. v. (1973): Die Bildung von Sedimenten und Sedimentgesteinen. In: Sediment-Petrologie, **3,** 378 pp., 133 Figs., 55 Tab., Stuttgart: Schweizerbart
General survey of the principles of sediment and rock forming processes. Chapters: Provenance of the sedimentary material; Subaerial weathering; Transport and deposition of clastic components; Formation of chemical sediments; Diagenesis.

(S) Fairbridge, R. W., Bourgeous, J. (eds.) (1978): The encyclopedia of sedimentology. Encyclopedia of Earth Sciences Series, vol. VI, 901 p., Stroudsburg: Dowden, Hutchinson and Ross
A very useful comprehensive discussion of the main aspects of sedimentology and sedimentation.

(S) Folk, R. L. (1961): Petrology of sedimentary rocks. Univ. Texas, Geol. 370 K, 383 L, 383 M, 154 pp., Austin: Hemphill's
Excellent introduction to the methods of grain-size analysis, mineralogical composition of sedimentary rocks, petrology of sandstones, mudrocks, and carbonates.

(E) Frey, R. W. (ed.) (1975): The study of trace fossils. 625 pp., 308 Figs., Berlin: Springer

(C) Friedmann, G. M. (ed.) (1969): Depositional environments in carbonate rocks. Soc. Econ. Paleont. Min. Spec, Publ. **14,** 208 pp., Tulsa
Important articles about Ancient deep-water and shallow-water carbonates as well as the identification of surfaces of subaerial exposure, with the aid of vadose cement types.

(S) Friedman, G. M., Sanders, J. E. (eds.) (1978): Principles of sedimentology. 792 p. New York: John Wiley and Sons
Comprehensive description of sedimentary particles, sedimentary processes, classifications of sediments, modern sedimentary environments and their products in the rock record, as well as the stratigraphic relationships of sedimentary strata. The book contains a lot of important information, not only about the description of sediments but also about the physical processes of sedimentation.

(E) Friedrich, H. (1965): Meeresbiologie. 436 pp., 6 Pls., 206 Figs., Berlin: Borntraeger
Good description, well worth reading: The history of biological oceanography, morphology of the oceans, ecological factors, pelagic plants and animals, benthic plants and animals; Distribution of marine organisms, near-coastal inhabitants of the ocean; Economic relations to oceanographic biology; General biological picture of the oceans.

(S) Füchtbauer, H. (1974): Sediments and sedimentary rocks 1. – In: Engelhardt, W. v., Füchtbauer, H., Müller, G.: Sedimentary Petrology, part II, 464 pp., 199 Figs., 39 Tab., Stuttgart: Schweizerbart
Modern description of sedimentary rocks, stemming from Füchtbauer, H., Müller, G. 1970. Chapters: The exogene cycle; Principles of nomenclature; Sandstones; Conglomerates and breccias; Pyroclastic rocks; Carbonate rocks; Cyclic sedimentation.

(S) Füchtbauer, H., Müller, G. (1970): Sedimente und Sedimentgesteine. In: Sediment-Petrologie **2,** 726 pp., 326 Figs., 66 Tab., Stuttgart: Schweizerbart
Important survey of the origin of sediments. Chapters: The exogene cycle; Principles of nomenclature; Sandstones; Conglomerates and breccias; Siltstones and mudrocks;

Carbonate rocks; Special sediments (salt rocks, siliceous sediments, sedimentary phosphate rocks, sediments rich in Fe and Mn, ores in sedimentary rock, residual rocks, pyroclastic rocks, peat-lignite); Sedimentary cycles. 3rd edition has 57 pages of addenda.

(E) Funnel, B. R., Riedel, W. R. (eds.) (1971): The micropaleontology of the oceans. 838 pp., Cambridge

(S) Geyer, O. F. (1973): Grundzüge der Stratigraphie und Fazieskunde, Band 1: Paläontologische Grundlagen I. Das geologische Profil, Stratigraphie und Geochronologie. 279 pp., 166 Figs., 7 Tab., Stuttgart: Schweizerbart
Introduction to the methods and problems of "applied paleontology" (stratigraphy and facies analyses).

(FI) Ginsburg, R. N. (ed.) (1975): Tidal deposits. A casebook of recent examples and fossil counterparts. 428 pp., Berlin–Heidelberg–New York: Springer
Important survey of Ancient and Recent tidal sedimentation in silicate and carbonate environments.

(MSA) Griffiths, J. C. (1967): Scientific method in analysis of sediments. 508 pp., New York: McGraw Hill
Very stimulating description of the statistical treatment of petrographic data. Chapters of special interest for microfacies analysis: Methods of sampling; Measurement of grain size, grain shape, grain fabric, and grain packing; Modal analysis. The method of frequency distribution as well as the possible statistical description of variable correlation are dealt with in detail.

(FI) Hallam, A. (ed.) (1967): Depth indicators in marine sedimentary environments. Marine Geol. **5,** 5/6, 327–567, Amsterdam
Important survey of how the depth of depositional environments can be reconstructed with the aid of indices such as distribution of organisms, sedimentological or geochemical criteria.

(C) Ham, W. E. (ed.) (1962): Classification of carbonate rocks. A symposium. Mem. Amer. Ass. Petrol. Geol. **1,** 279 pp., Tulsa
Absolutely essential for the description of carbonate rocks; contains the most important methods of classification (Leighton and Pendexter, Folk, Plumley et al., Dumham, and others).

(RS) Hay, W. W. (ed.) (1974): Studies in paleo-oceanography. Soc. Econ. Paleont. Min., Spec. Publ. **20,** 218 pp., Tulsa
Very important articles about carbonates and carbonate solution in deeper oceans; Chemistry of ocean water; Paleogeographic evolution of the oceans.

(E) Hedgpeth, J. W. (ed.) (1957): Treatise on marine ecology and paleoecology. Vol. 1. Ecology. Mem. Geol. Soc. Amer. **67**/1, 1296 pp., New York
Short list of the methods of analysis applicable in the reconstruction of Ancient biotopes and Ancient depositional environments with the aid of sedimentological, lithological, and geochemical criteria.

(RS) Hill, M. N. (ed.) (as of 1962): The sea. **1,** 864 pp., **2,** (1963), 554 pp., **3,** (1963), 963 pp., **4,** (1970), 802 pp., published by A. E. Maxwell
Standard work about all aspects of oceanography and marine sedimentation.

(C) Hsü, K. J., Jenkyns, H. C. (eds.) (1974): Pelagic sediments: on land and under sea. Spec. Publ. Intern. Ass. Sedimentologists **1,** 447 pp., many Figs., Oxford: Blackwell
Results of an international symposium held in Zurich, Switzerland, in 1973 about Recent and Ancient pelagic sediments. Very important studies of the genesis of chalk, silicification, and chert formation as well as carbonate solution in the deep sea. Theoretical models of the origin of nodular limestones as well as of the connection between "plate-stratigraphy" and various carbonate solution levels throughout geologic time.

(C) Humbert, L. (1977): Eléments de pétrologie dynamique des systèmes calcaires. 1. Typologie. Description. Diagenese, Applications. 2. Atlas photographique. Paris: Editions Technique
Ways of describing limestones in hand specimens, thin-sections, and SEM preparations. Richly illustrated survey of limestone particles and diagenetic types.

(C) James, N. P., Ginsburg, R. N. (1979): The Seaward Margin of Belize Barrier and Atoll Reefs. Spec. Publ. Int. Ass. Sed., **3,** 191 p., Oxford

Detailed description of morphology, sedimentology, diagenesis, organism distribution and Late Quaternary history of the Belize barrier reef east of Honduras. Of special interest is the description of the reef front and the fore-reef slope.

(S) Klein, George deVries (1977): Clastic tidal facies. 149 p., 92 Figs. Champaign, Illinois: Continuing Education Publication Company
Tidal sedimentary structure process-response models; Tidal paleocurrent models; Textural depositional models; Facies models; Occurrence of energy minerals (Petroleum, Uranium) in clastic tidal facies; Clastic tidal facies and geological history.

(S) Krumbein, W. C., Sloss, L. L. (1963): Stratigraphy and sedimentation. 2. ed., 660 pp., San Francisco–London: Freeman
Stimulating description of the methods of sedimentary rock analysis. Field work, facies mapping, didactically good description of sedimentation principles.

(E) Krumbein, W. E. (ed.) (1978): Environmental biogeochemistry and geomicrobiology. Vol. 1. The aquatic environment. 394 p. Ann Arbor: Ann Arbor Sci.
This important volume is based on papers given at the Third International Symposium on Environmental Biogeochemistry in Wolfenbüttel, Germany. The contributions deal with: Microbiology, chemistry, and global cycles in the aquatic environment; Element cycles, budgets and transfer rates in lakes and rivers and their alteration by man; Shallow photic waters and sediments with special reference to stromatolitic environments; Diagenesis.

(RS) Kukal, Z. (1970): Geology of recent sediments. 490 pp., Prague: Publ. House Czechosl. Acad. Sci.
Good listing of Recent sedimentation in terrestrial, lacustrine, and marine environments. Many tables with examples.

(E) Ladd, H. S. (ed.) (1957): Treatise on marine ecology and paleoecology. Vol. II. Paleoecology. Mem. Geol. Soc. Amer. **67**/2, 1077 pp., New York
Major work on paleoecological findings and methods, with annotated bibliography of significant groups of Ancient marine organisms.

(FI) Laporte, L. F. (1968): Ancient environments. 117 pp., Englewood Cliffs: Prentice-Hall
Stimulating introduction to the interpretation of facies criteria.

(FI) Laporte, L. F. (ed.) (1974): Reefs in time and space. Selected examples from the Recent and Ancient. Soc. Econ. Paleont. Min. Spec. Publ. **18**, 256 pp., Tulsa
Important articles for analyzing fossil reefs, in particular the shape of the reef structure. In addition: General survey of "carbonate buildups" in the history of the earth; Devonian carbonate complexes in Central Europe; Triassic reefs in the Dolomites of Southern Tyrol; Reef diagenesis.

(FI) Laporte, L. F. (1978): Ancient environments. 2nd Edition. 176 p. Englewood Cliffs: Prentice Hall

(S) Larsen, G., Chilingar, G. V. (eds.) (1967): Diagenesis in sediments. Dev. Sed. **8**, 551 pp., Amsterdam–London–New York: Elsevier
Important compilation of aspects of diagenesis in sandstones, argillaceous sediments, carbonates, organic substrates, coals, sedimentary mineral deposits, etc.

(C) Leblanc, R. J., Breeding, J. G. (eds.) (1957): Regional aspects of carbonate deposition. Soc. Econ. Paleont. Min., Spec, Publ. **5**, 178 pp., Tulsa
Survey of Recent carbonate sedimentation; Sedimentation in the Bahama Banks; Early diagenesis and lithification of shallow-water carbonates in Southern Florida; Dolomite problems; Mississippian limestones.

(RS) Lisitzin, A. P. (1972): Sedimentation in the world ocean. Soc. Econ. Paleont. Min. Spec. Publ. **17**, 218 p., Tulsa
Important survey of the major types of sediments in the oceans, the conditions of ocean sedimentation, quantitative distribution of suspended sedimentary material and of sedimentary material affected by ocean bottom currents; size composition of suspensions and bottom sediments; terrigenous and biogenous sedimentation; distribution of eolian material; volcanogenic and authigenic sedimentation; Cosmic material in sediments.

(C) Matter, A., Tucker, M. E. (eds.) (1978): Modern and Ancient lake sediments. Spec. Publ., **2**, Int. Ass. Sed., 290 pp., Oxford

The most important book on Recent and Ancient lacustrine sedimentation, containing 14 articles, some of which deal with lacustrine carbonates (Comparison of the Permian Saar-Nahe Basin and the Recent Lake Constance, Germany – A. Schäfer and Stapf; Holocene Carbonate sedimentation in Lake Balaton, Hungary – G. Müller and F. Wagner; Carbonates of the Eocene Green River Formation, Wyoming – Smoot).

(RS) Menzies, R. J., George, R. Y., Rowe, G. T. (1973): Abyssal environment and ecology of the world oceans. 488 pp., New York: Wiley
One of the few summaries of the sedimentation and distribution of organisms in the deep sea.

(FI) Merriam, D. F. (ed.) (1964): Symposium on cyclic sedimentation. State Geol. Surv. Univ. Kansas **169**/1, 1–380, **169**/2, 381–636, Lawrence
Fundamental articles about cyclic sedimentation in carbonate and silicate environments. Of particular importance are the articles about the Upper Paleozoic cyclothems in North America and the Lofer cyclothems in the Upper Triassic of the Alps.

(C) Middleton, G. V. (ed.) (1965): Primary sedimentary structures and their hydrodynamic interpretation. Soc. Econ. Paleont. Min., Spec, Publ. **12,** 265 pp., Tulsa
Contains, among other things, extremely important articles about Recent (Bahamas) and Ancient (Jurassic "Korallenoolith," England) sedimentary structures in shallow-water carbonates.

(C) Milliman, J. D. (1974): Marine carbonates, recent sedimentary carbonates, Part 1. 375 pp., 39 Pls., 94 Figs., Berlin–Heidelberg–New York: Springer
Excellent survey of Recent carbonate sediments and sedimentation as well as the diagenesis of carbonates. Chapters: Mineralogy and chemical composition of carbonates in the ocean; Methods of analysis; Carbonate components (here a very good description of the skeletal mineralogy and microstructures of Recent marine organisms); Shallow-water carbonates in coral reef environments; Sublittoral carbonate sedimentation in shelf environments; Deep-sea carbonates; Carbonate diagenesis.

(MSA) Milner, H. B. (1962): Sedimentary petrography, 1. Methods in sedimentary petrography. 643 pp., 2. Principles and Applications. 715 pp., New York: McMillan

(MSA) Müller, G. (1964): Methoden der Sedimentuntersuchung. In: Sediment-Petrologie, 1. 303 pp., 91 Figs., 29 Tab., 3 Supps. Stuttgart: Schweizerbart
The first description in German of the methods of analysis of sediments. Chapters: Geophysical methods; Sediment analysis and sampling in the field; Sediment analysis in the laboratory (sample preparation, granulometry, fabric analyses, obtaining pure mineral fractions, determining the mineral content in sediments, determining the organic material content in sediments, pore spaces and pore content.

(C) Müller, G., Friedman, G. M. (eds.) (1968): Recent developments in carbonate sedimentology in Central Europe. 255 pp., Berlin–Heidelberg–New York: Springer
Results of a symposium held in Heidelberg in 1967. Processes of carbonate formation and diagenesis; Microstructure and microporosity; Geochemistry of carbonates and carbonate rocks; Regional carbonate petrology; Applied carbonate petrology.

(C) Müller-Jungbluth, W. U., Toschek, P. H. (1969): Karbonatsedimentologische Arbeitsgrundlagen (Begriffe, Erläuterungen, Hinweise). 2. ed., Veröff. Univ. Innsbruck 8, Alpenkundliche Studien **4,** 32 pp., 3 Pls., Innsbruck
Good, brief description of the important criteria found in carbonate rocks, which are useful for facies interpretation in the field and the laboratory. Chapters: Matrix; Particles; Fabric; Open-space structures; Recording sedimentogenesis; Diagenesis; Documentation. The methods were tested out on the analysis of Triassic carbonates from the Northern Alps.

(S) Pettijohn, F. J. (1957): Sedimentary rocks. 2nd ed., 718 pp., 173 Figs., 40 Pls., New York: Harper

(S) Pettijohn, F. J., Potter, P. N. (1964): Atlas and glossary of primary sedimentary structures. 370 pp., 117 Pls., Berlin–Göttingen–Heidelberg: Springer
Excellent photographs serve to describe the most important primary sedimentary structures. The pictures in the atlas are subdivided into four groups: Bedding – external form; Bedding – internal organization and structure; Bedding plane marking and irregularities; Bedding deformed by penecontemporaneous processes. A German-English/French-English glossary of primary sedimentary structures is of great value.

(S) Pettijohn, F. J., Potter, P. N., Siever, R. (1972): Sand and sandstone. 618 pp., 258 Figs., Berlin–Heidelberg–New York: Springer
Fundamental description of sand sedimentation and sandstone formation. Chapters: mineralogical and chemical composition; Petrographic classification of sandstones; Petrography of sand and sandstones; Vulcanoclastics; Production and provenance of sand; Transport, deposition, and deformation of sand; Diagenesis; Sand bodies and environment; Sandstones, sedimentary basins and continental evolution; Petrographic analysis of sandstones (a good clear introduction to the description of thin-sections).

(C) Pia, J. (1933): Die rezenten Kalksteine. Min. Petrograph. Mitt., N. F., Ergänzungsband, 420 pp., 22 Figs., 1 Tab., 4 Pls., Leipzig
Extensive and still very worthwhile description of the genesis of "abiogenous", "physiological" (origin controlled by plants), and "organic" limestones. Of particular interest is the treatment of terrestrial and freshwater carbonates.

(S) Potter, P. E., Maynard, J. B., Pryor, W. A. (1980): Sedimentology of shale. Study guide and reference source. 303 p., 154 Figs., New York–Heidelberg–Berlin: Springer
First comprehensive review of the origin of shales, sedimentary structures, mineralogy and geochemistry, and organisms in mud environments. The book consists of three parts: Overview; Question Set (for field and laboratory studies); and Annotated and Illustrated Bibliography.

(S) Potter, P. N., Pettijohn, F. J. (1963): Paleocurrents and basin analysis. 269 pp., 30 Pls., 130 Figs., Berlin–Göttingen–Heidelberg: Springer
Very important description of the sedimentary criteria significant for the interpretation of sedimentary basins. Chapters: History of paleocurrent investigations; Fabrics and geophysical properties; Cross-bedding and ripple marks; Linear structures; Deformational structures; Shape of sedimentary bodies; Current systems; Basin analysis and sedimentary models; Methods of study.

(C) Pray, L. C., Murray, R. C. (eds.) (1965): Dolomitization and limestone diagenesis – A Symposium. Soc. Econ. Paleont. Min., Spec. Publ. **13,** 180 pp., Tulsa
Important articles about the calcite-aragonite problem; Recrystallization and diagenesis of limestones; Recent dolomitization (Bonaire, Persian Gulf, Bahamas); Interrelationships between facies, diagenesis and properties of reservoir rocks (Upper Jurassic, NW Germany).

(C) Purser, B. H. (ed.) (1973): The Persian Gulf. Holocene carbonate sedimentation and diagenesis in a shallow epicontinental sea. 471 pp., Berlin–Heidelberg–New York: Springer
Standard work about Recent and Subrecent carbonate sedimentation in the Persian Gulf.

(C) Purser, B. G. (1980): Sédimentation et diagenèse des carbonates néritiques récents. Tome 1. 366 p., 192 Figs., Paris: Soc. Editions Technip
This important book deals with the sedimentation and diagenesis of Recent neritic platform carbonates. Special attention is given to the diagenetic phenonema. Chapters: Particles, matrix and sedimentary textures; Structures and sedimentary sequences of the littoral environment; Submarine diagenesis of carbonate sediments; Diagenesis of carbonate sediments in the intertidal zone; Diagenesis of carbonate sediments in the continental environment; Conclusions and possibilities for the interpretation of Ancient carbonates.

(FI) Reading, H. E. (ed.) (1978): Sedimentary environment and facies. 576 p., 382 Figs., Oxford (Blackwell Scientific Publications)
A collection of very important and stimulating papers on the total spectrum of Recent and Ancient environments, written by competent authors. The book includes chapters on: Facies (Reading), Alluvial Sediments (Collinson), Lakes (Collinson), Deserts (Collinson), Deltas (T. Elliott), Clastic Shorelines (T. Elliott), Arid Shorelines and Evaporites (R. Till), Shallow Siliciclastic Seas (H. D. Johnson), Shallow-water Carbonate Environments (Sellwood), Pelagic Environments (H. C. Jenkyns), Deep Clastic Seas (Rupke), Glacial Environments (Edwards), Sedimentation and Tectonics (Mitchell and Reading).

(RS) Reineck, H. E. (1970): Das Watt. Ablagerungs- und Lebensraum. Senckenberg-Buch **50,** 142 pp., 75 Figs., Frankfurt/M.: W. Kramer

Very good description of the tidal flats in Northwest Germany. Takes into consideration sediments and sedimentary fabric, biotopes, ecofactors, flora and fauna, relationships between tidal flat organisms and sediment, and tidal flats as a depositional environment.

(S) Reineck, H. E. (1973): Bibliographie geologischer Arbeiten über rezente und fossile Kalk- und Silikatwatten. Courier Forschungsinst. Senckenberg **6,** 57 pp., Frankfurt/M.
Important bibliographical compilation. Chapters: Definitions; Recent tidal flats; Ancient tidal flats (categorized according to geologic age). Special articles about hydrography, sediment transport, petrography, mineralogy, geochemistry, sedimentary fabric and marks, biology and actuopaleontology.

(RS) Reineck, H. E., Singh, I. B. (1973): Depositional sedimentary environments. 439 pp., 579 Figs., Berlin–Heidelberg–New York: Springer
Very good description of Recent sedimentary structures and environments. Chapters: Part I – Primary structures and textures: Depositional environments; Physical parameters; Current and wave ripples; Surface markings and imprints; Current marks; Bedding; Sediment grain parameters; Biological parameters. Part II – Recent sedimentary environments; Glacial environment; Desert environment; Lake environment; Fluvial environment; Deltaic environment; The coast; The shelf; Examples of beach-shelf profiles; Coastal lagoons; Tidal flats; Continental margin; Slope and ocean basin.

(S) Ricchi Lucchi, F. (1970): Sedimentografia. Atlante fotografico delle strutture primarie dei sedimenti. 288 pp., 170 Pls., 6 Figs., Bologna: Zanichelli
Good pictures of the most frequent sedimentary structures. Italian/English and English/Italian glossary.

(FI) Rigby, J. K., Hamblin, W. K. (eds.) (1972): Recognition of Ancient sedimentary environments. Soc. Econ. Paleont. Min., Spec. Publ. **16,** 340 pp., Tulsa
Important articles about criteria used for the identification and interpretation of facies environments. The following areas are dealt with: Aeolic sediments; Alluvial depositions; Fluvial sediments; Lacustrine sediments; Tidal sediments; Coastal sedimentation in evaporite carbonate environments; Marine shallow-water environments; Sedimentation on the continental slope.

(S) Ruchin, L. B. (1958): Grundzüge der Lithologie. 806 pp., 304 Figs., Berlin: Akademie-Verlag
Systematic description of sedimentary rocks and methods of facies analysis. Chapters: Sedimentary rocks; Conditions for the formation of sedimentary rocks; Facies types and methods of facies analysis; Sedimentary formations; Recent sediments.

(C) Sander, B. (1936): Beiträge zur Kenntnis der Anlagerungsgefüge (Rhythmische Kalke und Dolomite aus der Trias). Tschermaks Min. Petrograph. Mitt. **48,** 27–139, Leipzig
Classic work about mechanical and chemical depositional textures in thin-sections.

(E) Schäfer, W. (1972): Ecology and paleoecology of marine environments, 568 pp., Edinburgh–Chicago: Oliver and Boyd, Chicago Press
For paleoecological work a fundamental survey of biostratonomy, fossilization, and facies fabrics of Recent organisms.

(C) Scholle, P. A. (1978): A color illustrated guide to carbonate rocks constituents, textures, cements, and porosities. Mem. Amer. Ass. Petrol. Geol. **27,** 241 p., many Figures, Tulsa
A very interesting introduction to thin-section analysis of carbonate rocks. Chapters: Skeletal Grains, Other Carbonate Grains, Other Minerals, Carbonate Cements, Carbonate Textures, Porosity Classification.

(S) Scholle, P. A., Schluger, P. R. (eds.) (1979): Aspects of diagenesis. Soc. Econ. Paleont. Min. Spec. Publ., **26,** 443 pp., Tulsa.
Deals especially with the diagenesis of clastic rocks with respect to the relationships between diagenetic processes and reservoir properties of rocks. Another part of the book is devoted to the determination of diagenetic paleotemperatures.

(E) Schumacher, H. (1976): Korallenriffe. Ihre Verbreitung, Tierwelt und Ökologie. 275 pp., 205 Fotos, 58 Figs., München–Bern–Wien: BLV Verlagsgesellschaft
Very worthwhile description of the reef environments on the earth, reef types and shapes, reef areas, corals, coral destroyers, reef formation, theories of the origin of atolls, origin of a coral island, ecology of biotopes in coral reefs, ecological niches in coral colonies, ecology and behavior patterns. Excellent color pictures!

(M) Schwarzacher, W. (1975): Sedimentation models and quantitative stratigraphy, Dev. Sed. **19,** 382 pp.
A very stimulating discussion of statistical models which can be used to describe sedimentation processes. Chapters: Sedimentation and environment; Sediment and time; Deterministic models; Stochastic models; Markov chains; Renewall processes and semi-Markov processes; Stationary stochastic processes and time-series analysis; Spectral analysis of geological data; Analysis of stratigraphical trends and the smoothing of data; Sedimentary cycles; Sedimentary cycles and time; Stratigraphical correlation; Problems of three-dimensional stratigraphical analysis.

(FI) Schwarzbach, M. (1974): Das Klima der Vorzeit. Eine Einführung in die Paläoklimatologie. 3rd. ed., 380 pp., 191 Figs., 41 Tab., Stuttgart: Enke
Survey of general and historical paleoclimatology. Important description of climatic indicators for facies interpretation.

(RS) Seibold, E. (1974): Der Meeresboden. Ergebnisse und Probleme der Meeresgeologie. Hochschultext, 183 pp., Berlin–Heidelberg–New York: Springer
Good survey of marine sedimentation, marine organisms, and the origin of oceans.

(FI) Selley, R. C. (1970): Ancient sedimentary environments. A Brief Survey. 327 pp., Ithaca: Harper and Row
Brief description of Ancient facies environments; worthwhile reading. The author starts in each case with a definition of the environment being dealt with and relates examples from the history of the earth; this is then followed by a discussion of the common characteristics and remarks about the economic significance of the sedimentation types. Described are: River deposits; Aeolic sediments; Lake deposits; Deltas; Linear clastic shorelines; Mixed clastic/carbonate shorelines; Carbonate shorelines and shelf deposits; Reefs; Flysch and turbidites; Pelagic deposits.

(S) Selley, R. C. (1976): An introduction to sedimentology. 408 pp., 162 Figs., London–New York–San Francisco: Academic Press

(RS) Shepard, F. P. (1963): Submarine geology. 2nd ed., 557 pp., 222 Figs., New York–London: Harper & Row
Classic description of marine geology. Chapters: methods and instrumentation for exploring the ocean floor; Ocean waves and associated currents; Origin and history of the continental shelves; Continental slopes; Submarine canyons, deep channels and other marine valleys; Coral and other organogenic reefs; Deep ocean floor topography; Deep ocean deposits and stratigraphy; Suboceanic layers and origin of basins; Mineralogy and geochemistry of marine sedimentation; Recent sediments and the interpretation of Ancient sediments.

(S) Shirley, M. L., Ragsdale, J. A. (1966): Deltas in their geologic framework. Houston Geol. Soc., 251 pp., Houston
Results of a symposium about Recent and Ancient deltaic sedimentation. The Mississippi, the Rhine, the Godavari (India). Ancient examples from the Paleozoic and Tertiary in the USA.

(S) Shrock, R. P. (1948): Sequence in layered rocks. 507 pp., 397 Figs., New York–Toronto–London: McGraw Hill
Basic description of the "top-and-bottom" criteria of sediments, plutonites, and metamorphites.

(RS) Stanley, D. J. (ed.) (1973): The Mediterranean Sea: A Natural Sedimentation Laboratory. 765 pp., Stroudsburg: Dowden, Hutchinson and Ross

(S) Stanley, D. J., Kelling, G. (eds.) (1978): Sedimentation in submarine canyons, fans and trenches. 395 p., Stroudsburg: Dowden, Hutchinson and Ross

(E) Tait, R. V. (1971): Meeresökologie. Das Meer als Umwelt. dtv. Wiss. Reihe **4091,** 305 pp., Stuttgart: Thieme
Good introduction to the biotopes and ecofactors of marine environments. Chapters: The oceans; Marine plankton; Measuring and collecting; Physical and chemical parameters of the marine environment; Organic productivity; The ocean floor; The coastline; Ocean fisheries.

(S) Talwani, M., Hay, W., Ryan, W. B. F. (eds., 1979): Deep drilling results in the Atlantic Ocean: Continental margins and the paleoenvironment. Maurice Ewing Series **3**, Amer. Geophys. Union, 437 p., Washington
A collection of 20 important and comprehensive papers dealing with the sedimentary evolution of the Atlantic Ocean since the Late Jurassic, based on the results of Deep Sea Drilling.

(S) Tucker, M. E. (1981): Sedimentary Petrology. An Introduction. 252 pp., Oxford: Blackwell Scientific Publications
A very stimulating introduction to the petrology of sedimentary rocks, written with undergraduate students in mind.

(S) Vossmerbäumer, H. (1976): Allgemeine Geologie. Ein Kompendium. 277 pp., 116 Figs., Stuttgart: Schweizerbart
Excellent brief description of general geological criteria, in particular of sedimentary rocks. Many references.

(C) Wilson, J. L. (1975): Carbonate facies in geologic history. 471 pp., 30 Pls., 183 Figs., Berlin–Heidelberg–New York: Springer
Very important and stimulating description of carbonate sedimentation in geologic time. By comparing well-studied facies environments, theoretical models are set up which cover the development and typification of carbonate sedimentation in shallow-water environments. Special emphasis is placed on the interpretation of microfacies criteria (lists "Standard Microfacies Types" for nine standard Facies Belts). Chapters: Principles of carbonate sedimentation; Stratigraphy of carbonate deposits; Outline of carbonate petrography; The advent of frame-builders in the Middle Paleozoic; The Lower Carboniferous Waulsortian facies; Pennsylvanian-Lower Permian shelf margin facies; Late Paleozoic terrigenous/carbonate shelf cycles; (Permo-Triassic) "buildups" and Late Triassic ecologic reefs; Reef trends and basin deposits in the Late Jurassic facies of Europe and the Middle East; Shoaling upward shelf cycles and shelf dolomitization; The rise of rudists; The Middle Cretaceous facies in Mexico and the Middle East.

(E) Ziegler, B. (1972): Einführung in die Paläobiologie. Teil 1. Allgemeine Paläontologie. 245 pp., 249 Figs., Stuttgart: Schweizerbart
Clear and graphic introduction to the question of fossilization. Classification and taxonomy; Theory of evolution; Biostratigraphy; Modes of life; Ecology and biogeography of Recent and Ancient organisms.

1.4.2 Journals

A survey of the important geoscientific journals is found in Vossmerbäumer (1976: 56–58). The following journals are very important for *sedimentological studies and facies analyses of carbonate rocks.*

Arbeiten aus dem Geol.-Paläont. Institut der TU Stuttgart
Bulletin of the American Association of Petroleum Geologists (Tulsa/USA; very important!)
Bulletin of the Geological Society of America (New York/USA)
Bulletin of Marine Science
Bulletin of the State Geological Survey of Kansas (Lawrence/USA)
Chemical Geology (Amsterdam: Elsevier, 1966 ff.)
Contributions to Mineralogy and Petrology (Heidelberg)
Contributions to Sedimentology (Stuttgart: Schweizerbart, 1973 ff.)
Coral Reefs (Berlin–Heidelberg–New York: Springer, 1982 ff.)
Deep-Sea Research (Oxford, 1953 ff.; very important bibliography!)
Developments in Sedimentology (Amsterdam: Elsevier, 1964 ff., important series!)
Earth-Science Review (Amsterdam: Elsevier, 1966 ff.)
Facies (Erlangen, Germany, 1979 ff., devoted especially to facies analysis)
Fortschritte der Mineralogie (Stuttgart: Schweizerbart)
Geochimica et Cosmochimica Acta (Oxford; important!)
Göttinger Arbeiten zur Geologie und Paläontologie

Joides Initial Reports of the Deep Sea Drilling Project (up to 1980 about 50 vols), Washington; very important!)
Journal of Geology (Chicago/USA; important!)
Journal of Sedimentary Petrology (Tulsa/USA; most important journal!)
Marine Geology (Amsterdam: Elsevier, 1960 ff; important!)
„Meteor“ Forschungsergebnisse, Reihe A: Allgemeines, Physik und Chemie des Meeres (1966 ff.); Reihe C: Geologie und Physik (1968 ff.); Reihe D: Biologie (1967 ff.), Berlin–Stuttgart
Sedimentary Geology (Amsterdam: Elsevier, 1967 ff; important!)
Sedimentology (Oxford, 1962 ff; very important!)
Senckenbergiana maritima (Frankfurt a.M.)

For the *identification of fossils in thin-sections* important articles can be found in the following journals:

Acta Geologica et Geographica Universitatis Comenianae, Geologica (Bratislava/Czechoslovakia)
Acta Palaeontologica Polonica (Warsaw/Poland)
Archive des Sciences (Geneva/Switzerland)
Annales Univ. Bucaresti
Biomineralisation, Forschungsberichte, Akad. Wiss. Lit. Mainz
Bolletino Societa geologica Italia
Bolletino della Società dei Naturalisti, Napoli
Bulgarian Academy of Sciences, Bulletin of the Geological Institute, Ser. Paleontology; Paleontology, Stratigraphy and Lithology (Sofia/Bulgaria)
Bulletin Canadian Petroleum Geologists (Calgary/Canada)
Bulletin Cent. Rech. Explor.-Prod. Elf-Aquitaine (Pau/France)
Bulletion Soc. géol. France (Paris/France)
Dare Seama a le Sedintelor (Bucuresti/Romania)
Eclogae geol. Helvetiae (Basel/Switzerland)
Földtany Közlöny (Budapest/Hungary)
Geobios (Lyon/France)
Geologia Romana (Roma/Italy)
Geologija (Ljubljana/Yugoslavia)
Geološki Glasnik (Sarajevo, Titograd/Yugoslavia)
Geološki Vjesnik (Zagreb/Yugoslavia)
Geologický Zbornik, Geol. Carpathica, Slovenska akademia vied (Bratislava/CSSR)
Giornale di Geologia (Bologna/Italia)
Journal of Palaeontology (Tulsa, Oklahoma)
Lethaia (Oslo/Norway)
Mem. Inst. Geol. Geophys. Bucuresti (Bucuresti/Romania)
Micropaleontology (New York/USA)
Mitteilungen Bayer. Staatssammlung Paläont. hist. Geol. (München/Germany)
Mitteilungen Ges. Geologie- und Bergbaustudenten Österreich (Wien/Austria)
Notes Labor. Paléont. Univ. Genève (Genève/Switzerland)
Palaeontologia Jugoslavica (Zagreb/Yugoslavia)
Paleogeography, Paleoclimatology, Paleoecology (Amsterdam/Netherlands)
Palaeontology (London/England)
Prirodoslovna Istraživanja, Acta Geol. Jugosl. akad. znanosti umjetnosti (Zagreb/Yugoslavia)
Quarterly Colorado School Mines (Golden, Colorado/USA)
Rivista Espanola Micropaleontologia (Madrid/Spain)
Revue de Micropaléontologie (Paris/France)
Rivista Italiana di Stratigrafia e Paleontologia (Milano/Italy)
Studii si Cercetari de Geologie, Geofizika, Geografie, Ser. Geol. (Bucuresti/Romania)
Vesnik zavod geol. geofiz. istraživanja (Beograd/Yugoslavia)

2 Recent Carbonate Sedimentation

About 50% of Recent ocean floors are covered by carbonate sediments (Lisitzin, 1971). Areas of profound carbonate sedimentation are the shallow seas of the tropics and subtropics between 30 °N and 30 °S (Fig. 4) and the deep-sea bottoms. Approximately 1/3 of the deep-sea floors are covered by carbonate muds.

Excellent surveys of marine shallow-water carbonates are found in Bathurst (1971; 2nd ed., 1975) and in Milliman (1974). Marine deep-water carbonates are described in Hsü and Jenkyns (1974), Milliman (1974) and in particular in the very important *Initial Reports of the Deep Sea Drilling Project (Joides).* For non-marine carbonates in lacustrine and terrestrial environments individual papers must be referred to (see Sect. 2.4.2), since modern monographs of the latest information are scarce (e.g., Matter and Tucker, 1978); Pia (1933), however, is still very valuable reading.

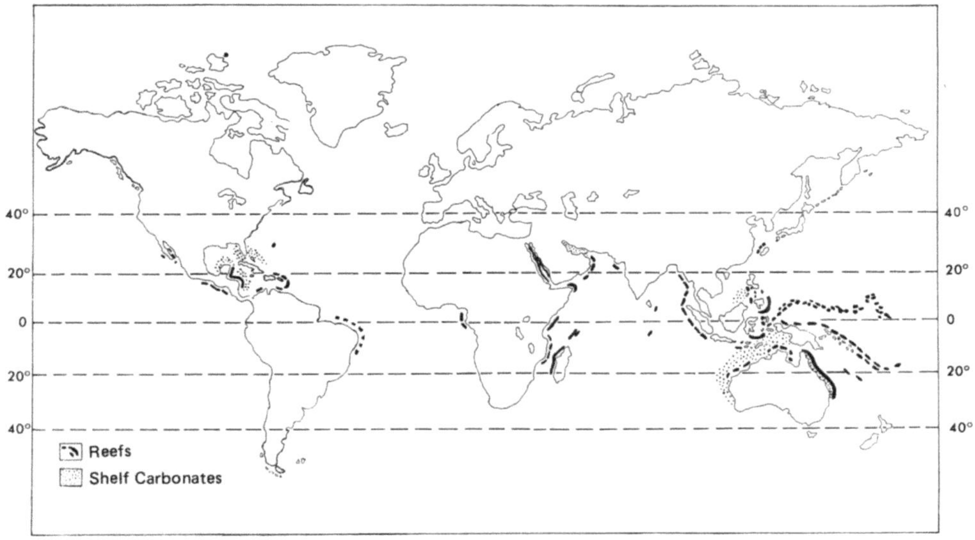

Fig. 4. Carbonate deposition in Recent shallow marine environments (shelf carbonates and reefs) is concentrated chiefly in the regions north and south of the equator, between 30 °N and 30 °S. Shelf carbonates, however, are not always restricted to these latitudes and, in fact, are accumulating today near western Ireland and western Scotland as well as between 32 °S and 40 °S, off southern Australia. After Wilson (1975)

2.1 Carbonate Mineralogy

According to Milliman (1974) and Folk (1974), the following parameters are important for the formation of the carbonate minerals listed in Table 1:

a) *Aragonite:* Mg in solution; high water temperatures (between 20 ° and 30 °C); high pH; presence of certain organic compounds; presence of Sr and possibly Ba and Pb.
b) *Calcite:* very low concentrations or absence of Mg; low water temperatures around 10 °C); low pH; presence of SO_4; presence of Na_2CO_3 and $(NH_4)_2CO_3$; presence of certain organic compounds.
c) *Dolomite:* high Mg/Ca ratio (> 5); high water temperatures (generally > 30 °C); ample input of solvents or high solution transfer.

The morphology of $CaCO_3$ crystals appears to be controlled essentially by the rate of crystallization and the Mg and Na content in the waters (see Fig. 7, p. 64).

Recent marine carbonates consist for the most part of carbonate skeletons or their organic detritus. Organisms are very important for the production of calcareous mud made up of micron-sized crystals as well as for the formation of larger particles, such as peloids, aggregate grains, and ooids (see Sect. 4.1.3). "Inorganic" carbonate formation seems to be possible only under special conditions, that is, when no organisms are available whose organic compounds (particularly asparagine, Mitterer, 1968, 1971) serve as centers of nucleation for the formation of shells (biomineralization). In contrast, in non-marine environments inorganic carbonate precipitation seems to be very common (CO_2 removal due to

Table 1. Petrographic properties of rock-forming carbonate minerals

	Aragonite	Calcite (Low-Mg calcite)	Mg-calcite (High-Mg calcite)	Dolomite
Formula	$CaCO_3$	$CaCO_3$	$CaCO_3$	$CaMg(CO_3)_2$
Crystal system	rhombic	trigonal	trigonal	trigonal
Common trace elements	Sr, Ba, Pb, K	Mg, Fe, Mn, Zn, Cu		Fe, Mn, Zn, Cu
Mol% $MgCO_3$	–	< 4	> 4 to > 20	40–50
Double refraction index	0.155	0.172		0.177
Specific gravity	2.94	2.72		2.86
Hardness	3.5–4	3		3.5–4
Crystal habit	usually acicular (fibrous) micritic	often isometric (sparry calcite), micritic, very seldom acicular	micritic, often acicular (fibrous)	often isometric (sparry dolomite), micritic, very seldom acicular
Occurrence	dominant in shallow-marine environment	dominant in deep-marine environment, common in freshwater environment	dominant in shallow-marine environment	chiefly in the shallowest marine environments

Table 2. Skeletal mineralogy of Recent calcareous organisms and algae. Rare or not yet confirmed results in parentheses. Modified after Lowenstam (1963), Milliman (1974), Flajs (1976), Pentecost (1980). Non-carbonate minerals not included

	Aragonite	Calcite	Mg-calcite	Calcite + Mg-calcite	Aragonite + Calcite
Algae					
Blue-Green Algae (Cyanobacteria)	×	×	×		
Chrysophyta: Coccolithophorida	(×)	×			
Rhodophyta (red algae): Corallinaceae	×		×	(×)	
Chlorophyta (green algae):					
Codiaceae	×				
Dasycladaceae	×				
Charophyta		×			
Calciodinoflagellata		×			
Foraminifera	×	×	×	×	
Sponges					
Calcispongia	×	×	×		
Sclerospongia	×		×		
Coelenterata					
Hydrozoa	×	×			
Octocorallia	×				
Scleractinia	×				
Bryozoa	×		×		×
Brachiopoda		×			
Mollusca					
Polyplacophora	×				
Scaphopoda	×				
Gastropoda		×			(×)
Prosobranchia	×				
Pteropoda	×				
Pelecypoda	×	×			×
Cephalopoda	×		(×)		
Polychaeta					
Serpulidae	×	×			
Arthropoda					
Cirripedia		×			(×)
Decapoda		×			
Ostracoda		×	(×)		
Echinodermata			×		
Chordata					
Tunicata	×				
Vertebrata					
Fish Otolites	×				

temperature and pressure changes or assimilation: e.g., tufa formation; concentration due to evaporation of pore waters; e.g., caliche formation; mixing of various water zones, etc).

Various groups of Recent organisms have different carbonate minerals in their shells (Table 2). The formation of Mg-calcite skeletons is influenced by the water temperature, the salinity, and the phylogenetic level of the organisms (Chave, 1954; Lowenstam, 1964).

How representative Recent mineralogical composition is for Ancient carbonate environments is being discussed at the moment; mineralogical differences seem to appear within larger groups (primary calcite skeleton in Paleozoic rugose corals, aragonite skeletons in Mesozoic to Recent hexacorals?; Sandberg, 1974). In contrast to their Recent analogs, Ancient ooids and perhaps also Ancient calcilutites seem to have been composed primarily of calcite rather than of aragonite. The oldest aragonitic sediments (skeletal grains excluded) seem to have come from the Tertiary (Loreau, 1979). The mineralogy of modern carbonate environments may not be representative of Ancient environments! Several authors have suggested that the sedimentary and early diagenetic environment of the Lower Paleozoic seas contained less magnesium than present-day sea water (Folk, 1974; Lindström 1979).

2.2 Carbonate Sedimentation in Marine Environments

2.2.1 Principles of Oceanographic Classification

Fig. 5 and Table 3 show the possible vertical and horizontal classifications of marine environments. The international classification systems, which do not coincide with each other completely, are based on biological, oceanographical, and topographical criteria.

Carbonates occur in: (1) eulittoral and supralittoral environments (principally in tropical and subtropical environments); (2) sublittoral environments (in water depths approximately between 20 and 200 m, shelf and shelf margin); and (3) in the

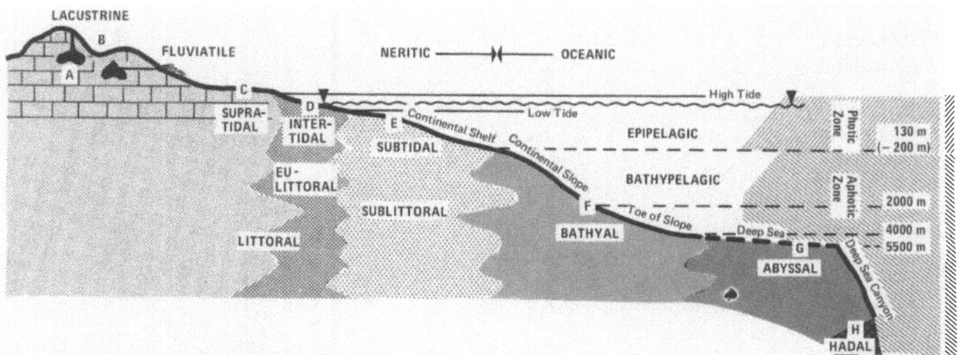

Fig. 5. Basic oceanographic classifications

Table 3. Principles of oceanographic classification. For the littoral system see Hedgpeth (1957) and Schneider (1976); the vertical classification of the "deep-sea" is in need of further refinement

	Riedl (1963)	English terminology	French terminology	German terminology		
			Adlittoral	Supralitoral (Sprühwasser-, Spritzwasserbereich)	Littoral System	Shallow Sea
	Supralitoral (Zone I)	Supratidal	Supralittoral			
Extreme high water spring tides						
Mean high water					Tidal Zones	
Mean sea level	Eulitoral (Zone II)	Intertidal	Sup. Medio- inf. Littoral	Eulitoral		
Extreme low water spring tides						
	Sublitoral (Zone III)	Subtidal	Infralittoral	Sublitoral (Unterwasserbereich)		
Lower boundary of the photic zone (100 – 200 m, very different) or upper edge of the continental margin		Bathyal	Epibathyal Mesobathyal	Bathyal		Deep Sea
about 4000 m						
(1500 – 5000 m)		Abyssal	Infrabathyal	Abyssal		
about > 5500 m		Hadal	Hadal	Hadal		

deep sea (down to the calcite compensation depth between 4000 and 5000 m; primarily planktonic sedimentation).

Carbonate sedimentation in shallow-water environments is strongly controlled by water temperature and salinity (see Sect. 10.3.4). "Warm-water carbonates" can be distinguished from "temperate-water carbonates". Warm-water carbonates develop on the subtropical carbonate shelves, which include two major categories: 1) protected shelf lagoons like the Bahamas, Florida, Belize, Batabano/Cuba, and the Great Barrier reef area; and 2) the open shelves like the Yucatan, western Florida, the eastern Gulf of Mexico, western North Atlantic, and Northern Australia. Protected shelf lagoons have an average depth of about 10 m, and are enclosed by reefs, shoals, and islands, thus producing low-energy environments. Wind-generated waves and tides are the most effective physical factors in these lagoons. Open-shelves are inclined seawards toward a shelf-break, which today lies between 140 and 230 m. The wave zone is spread over the shelf floor; oceanic and tidal currents are also active. Five major environmental zones can be recognized on the warm-water shelves (Sellwood, 1978): supratidal zone, shore zone and intertidal zone, marine platform (subtidal), reef belt, and upper shelf slope. The "temperate-water carbonates" may be locally abundant (e.g. Mannin Bay, western Ireland), but still have not been studied in detail. The lithological spectrum is characterized by hard substrates (beachrock etc.), bioclastic sands consisting of debris of mollusks, foraminifera, echinoderms, bryozoans, and red algal clasts; black muds, and red algal oncoids. Milliman (1974) differentiates the following environments of carbonate sedimentation: Eulittoral, sublittoral and deep-sea.

2.2.2 Eulittoral Shallow-Water Sedimentation (Plate 1)

A. Environments with coral reefs. Important environmental factors: Water temperature (optimum for reef corals 23–27 °C); water depth (hermatypic corals are limited to certain depths due to symbiosis with zooxanthelles [assimilating algae], below 30–40 m the number of species decreases markedly); salinity (optimum for reef corals between 30‰ and 40‰); water movement (higher water energy increases the supply of food and oxygen). References on the ecology of Recent reefs: Stoddart and Yonge (1971), Endean and Jones (1972), and particularly Schuhmacher (1976); of prime importance are the *Proceedings of the Third International Coral Reef Symposium* Miami (1977). The *reef zones* can be distinguished according to ecological and morphological criteria:

a) *Fore-reef or reef front:* the windward side, lying between the margin of the reef flat and the lower limits of coral growth at about 40 m. Characterized by special communities (e.g., Land and Goreau, 1970); a "spur and groove system" – closely spaced, steep-sided linear highs and lows that extend perpendicularly seaward from the reef flat (see Shinn, 1963); and sedimentation of reef debris.

b) *Reef flat or reef core:* frequently a combination of (1) and (2):

(1) the windward side, often with wave-breaking algal structures ("Lithothamnion ridges" in the Pacific reefs) and characteristic organic associations, (2) less differentiated leeward side. The reef core is composed of the reef framework (primarily

hermatypic reef corals bound by red algae) and is intergraded with medium to coarse grained bioclasts (corals, green algae, mollusks, foraminifera); the ratio of framework to bioclast material lies between 1:4 and 1:10.

B. Lagoon environments. Extensive shallow-marine environments, which are bordered on at least one side by reefs or islands or the mainland. The following lagoons can be differentiated according to the degree of salinity:

a) *Normal lagoons:* Rapid exchange of water with the adjacent ocean; often small local reefs (patch reefs). Example: Eniwetok atoll and Bikini atoll, Pacific (Emery et al., 1954).

b) *Brackish water lagoons:* A high influx of fresh water lowers the salinity to 10‰–15‰; in dry months the salinity rises to 50‰. These extreme differences and the quiet water conditions prevent reef organisms from settling. The shells of mollusks, foraminifera, and green algae provide the primary source of carbonate sediments. Example: Florida Bay (Ginsburg, 1964; Stockman et al., 1967).

c) *Hypersaline lagoons:* caused by restricted water circulation combined with high evaporation rates; salinities can exceed 45‰ (Bahamas) and 70‰ (Shark Bay, Australia). The communities are associated with distinct types of sediment. Mollusks, bryozoans, foraminifera, green algae. Examples: Great Bahama Bank (Cloud, 1962; Newell et al., 1959; Purdy, 1963); nearshore areas of the Persian Gulf (Kendall and Skipwith, 1968, 1969; Kinsman, 1964; Purser, 1973), Shark Bay (Logan and Cebluski, 1970), and Laguna Madre/Texas (Dalrymple, 1965).

Large stretches of marine grasses (*Thalassia, Posidonia, Cymodocea*) act in lagoon areas as sediment trappers and stabilizers, by slowing currents. They also serve as a substrate for special epibiontic organisms (e.g., *Melobesia,* gastropods) which live symbiotically with assimilating algae and together produce carbonate mud. Examples: Shark Bay (Davies, 1970); Florida (Turmel and Swanson, 1976); Bahamas (Ginsburg and Lowenstam, 1958; Scoffin, 1970; Land, 1970); Seychelles (Taylor and Lewis, 1970).

Algal mats made up of blue-green algae, green algae, and red algae likewise serve as sediment binders (Bathurst, 1967; Scoffin, 1970; Neumann et al., 1970), as a food source, and as a preferential environment for early diagenesis (Bathurst, 1967). See Sect. 5.3.12.4.

Lagoon sediments generally display greater textural and compositional variation than do most reef sediments. A wide range of compositions of carbonate sands (see Tables in Milliman, 1974: 182–84) and carbonate mud (see Sect. 4.1.2) is present.

Recent carbonate lagoons are different with respect to shape, sediments, and water energy (Jordan, 1978). Lime muds with minor skeletal grains are found in low-energy coastal lagoons. Back-reef lagoons are characterized by abundant skeletal sands and deposited within a moderate to low energy environment. Interreef lagoons with moderate energy contain skeletal sands as well as lime muds. Low-energy shelf lagoons are differentiated into those with predominantly skeletal material or those with lime muds. Atoll lagoons exhibit moderate to low water energy; the sediments range from muddy skeletal sands to pure lime mud.

C. Carbonate mud banks and carbonate cays

a) *Mud banks and tidal flats.* These sedimentary environments are subjected to periodic fluctuations of sea level or intermittent marine flooding and drying due to subaerial exposure; they are characterized by polygonal desiccation crack patterns, open-space structures (birdseyes, see Sect. 4.2.3.5), crinkly-laminate algal mats and mm-laminites (Sect. 4.2.3.3) as well as by stromatolites and typical feeding and protective burrows. Other criteria for tidal flats and also some shallow subtidal areas are large-scale erosion channels of various depths (1 to about 10 m). Sabkhas (salt-encrusted coastal flats) are typical for arid tidal and supratidal zones. Ginsburg (1975) gives an excellent survey of Recent and Ancient tidal sedimentation. Important observations stem from the Persian Gulf (Kinsman, 1969) and from the Caribbean (Shinn et al., 1969).

b) *Sand cays:* The composition of calcareous sand bars, whose long axes parallel the reef front, reflects the composition of the marine source (reef flat); they are stabilized by plant growth in high-energy environments. The beach sands of the Caribbean reefs consist of green algae (*Halimeda*), corals, foraminifera, mollusks, and coralline algae (e.g., Alacran reef, Folk and Robles, 1964) as well as ooids, peloids, and aggregate grains. Benthonic foraminifera and coralline algae dominate Pacific beach sands.

D. Other tropical and subtropical carbonates

a) *Vermetid reefs:* Encrusting gastropods belonging to the family Vermetidae flourish on hard substrates (Keen, 1971) and build structures similar to reefs in the clear agitated warm water of tidal environments. Examples: coasts of Florida; edges of the Bermuda platform; rocky shorelines in the Mediterranean, "trottoir" (Pérès, 1967).

b) *Serpulid reefs:* In the subtidal environments of the hypersaline Baffin Bay (southeastern Texas Gulf coast), Recent serpulids and balanids have built reef crusts up to 0.5 m thick (Andrews, 1964).

→

Plate 1. Recent Carbonate Particles

1. Aggregate grains (grapestones) and miliolid foraminifera. Berry Islands, Great Bahama Bank. Thin-section. Scale is 1 mm

2. Ooids, predominantly as superficial ooids with only a few laminae and a nucleus formed by large peloids. The ooids are partly micritized, therefore the lamination has been destroyed. Such micritized ooids are sometimes called "bahamite grains" (see Sect. 4.1.3.2); they occur chiefly in not too agitated environments. The particle, which deviates greatly from the rounded form, can be referred to aggregate grains (botryoidal lump after Illing, 1954, see Section 4.1.3.3). Cat Cay, Great Bahama Bank. Thin-section. Scale is 0.5 mm

3. Large aggregate grain. Various sizes of large rounded micritic peloids are bonded by aragonitic micrite cement. Berry Islands, Great Bahama Bank. Thin-section Scale is 0.5 mm

4. Extensively bored and micritized aggregate grains (lumps after Milliman, 1974; see Section 4.1.3.3). Berry Islands, Great Bahama Bank. Thin-section. Scale is 0.5 mm

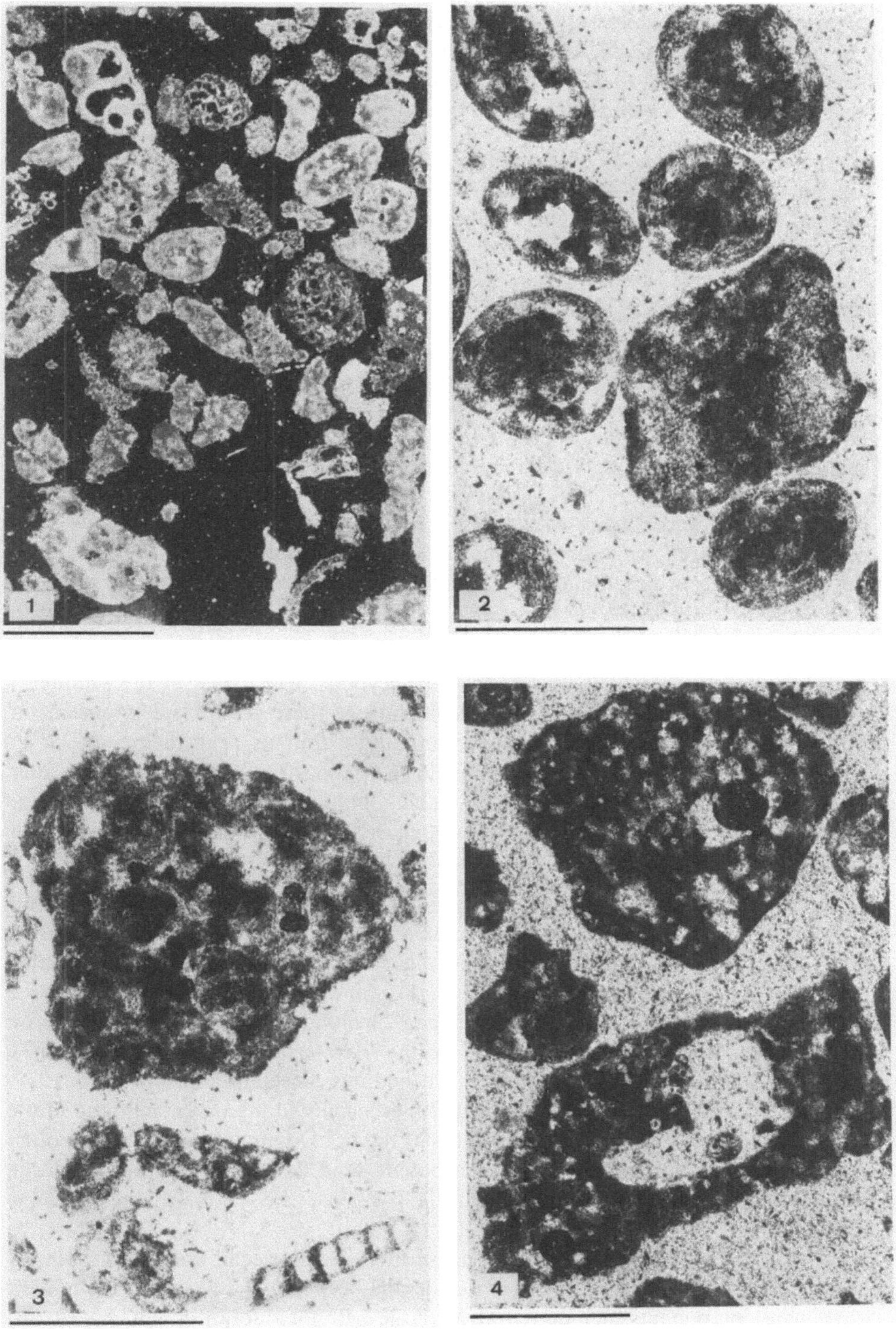
1
2
3
4

c) *Oyster reefs:* In shallow water (tidal ranges to 30 m) oysters can build extensive structures which serve as a habitat for gastropods, balanids and serpulids. They prefer slightly brackish environments such as estuaries (e.g., off the coasts of Texas and Georgia – Galtsoff, 1964; Wiedemann, 1972). Comparable fossil reef structures are the *Placunopsis* bioherms of Southern Germany (Bachmann, 1979).

E. Non-tropical carbonates of temperate and polar climatic regions. Although detrital sedimentation generally dominates in these environments, carbonate sediments can also be deposited in the higher latitudes of the eulittoral zone of the coastal environment, if large numbers of organisms with calcareous shells are available (mollusks, balanids, coralline algae, foraminifera). Examples: Mannin Bay, Ireland (Bosence, 1977), Alaska (Hoskin and Nelson, 1969).

2.2.3 Sublittoral Carbonate Sedimentation

In depths between about 20–200 m (= depth range of most Recent continental-shelves and upper slopes) calcareous sands consisting of bioclasts of mollusks, calcareous algae, echinoderms, foraminifera and calcareous lithoclasts as well as planktonic foraminifera and coccolithophorids are found. Many Recent shelf sediments are admixed with subfossil material (Milliman and Emery, 1968). In sublittoral carbonate sedimentation the following sediments are differentiated:

A. Sediments on the inner shelf. Characterized by greater sediment accumulation, more silt, and relatively fine-grained low-carbonate sediments. Carbonate-producing organisms are filter-feeders (e.g., pelecypods, which live in the sediment) and bulk detritus-feeders (e.g., echinoids). Both of these groups are important in sandy areas; foraminifera dominate in muddy sediments (Sarntheim, 1971). In nearshore high-energy environments organic associations similar to those in the outer shelf can be found (articulated coralline algae, bryozoans, ophiuroids). Examples: Gulf of Manfreddonia, Adriatic Sea – Oeltzschner and Sigl, 1970; the English Channel – Boillot, 1975.

B. Sediments on the outer shelf. Frequently more abundant in carbonates and plankton than the inner shelf environment; the biofraction is more diversified (mollusks, echinoids, bryozoans, coralline algae, balanids, sponges, tunicates, corals, benthonic and planktonic foraminifera). Prominent components are encrusting organisms, such as coralline algae, bryozoans, balanids, and corals which settle on hard substrates ("coralligenous facies" on the Mediterranean coasts, Pérès, 1967). Coralline algae appear as rhodolites (algal nodules, Bosellini and Ginsburg, 1971; Adey and MacIntyre, 1973; Milliman, 1977) especially in the outer shelf and in the upper part of the shelf slope (water depths to 100 m). Dominant organisms are often bryozoans which can build extensive carbonate sand flats as well as reefs, with the aid of red algae (e.g., Southern Australia – Wass et al., 1970; Mediterranean – Caulet, 1972).

a) *Algal ridges:* Small banks rising 1 to 5 m above the ocean floor occur in many outer shelves and upper slopes in water depths between 40 and 180 m. Dominant carbonate components include red algae, large benthonic foraminifera, bryozoans, and corals. Frequently these elevated topographic features serve as a substrate for deep water corals, balanids, and serpulids. Examples are known in the Gulf of Mexico (Ludwick and Walton, 1957; Logan, 1969), in the eastern coastal shelves of North and South America, and in the shelves of West Africa and Western Australia. These shallow-water reefs probably represent Holocene "relict" structures (MacIntyre and Milliman, 1970).

b) *Shelf oolites:* The ooid sands which are widely distributed on the shelf and on the upper shelf margin in water depths to 160 m must also be interpreted as Holocene and Pleistocene "relict" structures. Examples: North African Mediterranean coast (Fabricius et al., 1970 a, 1970 b), Gulf of Mexico (Gould and Steward, 1955), south-eastern coast of the USA (Pilkey et al., 1966), Persian Gulf (Houbolt, 1957), Arabian Sea (v. Stackelberg, 1970).

2.2.4 Deep-Sea Carbonates

Carbonates occurring in water depths greater than 200 m are generally classified as deep-sea carbonates. These carbonates are deposited in the bathyal and abyssal regions (see Fig. 5). The *bathyal region* corresponds to the submarine slope between the outer continental shelf and the continental rise. The upper limit is defined by the shelf break (most frequently at about 130 m); its lower limit is only loosely defined (1000–3000 m; often 2000 m). The bathyal zone lies below wave base and is largely aphotic; sedimentation is affected by density currents flowing down the continental slope within depressions and by currents flowing parallel or oblique to the base of the continental slope. The *abyssal region* covers an area of about 76% of the ocean bottom. The upper boundary may coincide with the 4 °C isotherm which occurs at 2000 m in the Atlantic Ocean, and between 1000 m and 1500 m in the Indian and Pacific Oceans. Because of variations in water masses and bottom currents, no constant bathyal/abyssal boundary can be established (Menzies, George, and Rowe, 1973). The lower limit of the abyssal region is about 6000 m. The region is characterized by lack of sunlight, tremendous hydrostatic pressure, very slight variations in salinity, and rather uniform water temperatures between +4 °C and –1 °C. The fauna consists of carnivores and detritus feeders; echinoderms are most common. An ample water circulation is provided by bottom currents, turbidity flows and physical differences in the water masses. No carbonates are deposited within the *hadal province,* the deepest region of the deep sea, because of the dissolution effects of the calcite compensation zone, which also strongly affects the composition of bathyal and abyssal sediments. The hadal region is characterized by long and narrow, V-shaped depressions beneath the general level of the deep-sea bottom. These deep-sea trenches are related to island arcs or are adjacent to active continental margins. The depth of the hadal zone is below 6000 m, down to more than 10,000 m in the Mariana Trench.

The selective dissolution of planktonic tests is influenced by: (1) mineralogy – aragonite is more soluble than calcite and disappears from the sediment at shallower depths than calcite; the tests composed of high magnesium calcite also disappear more rapidly than those made of calcite; (2) size – smaller particles tend to dissolve more rapidly than larger ones because a larger surface area is exposed to the corrosive medium; (3) organic coatings retard the solution of calcium carbonate particles in undersaturated waters; (4) cold bottom currents increase the intensity of dissolution; (5) burrowing organisms may contribute to enhanced dissolution; (6) ultrastructure of the skeletal elements (Berger, 1970; McIntyre and McIntyre, 1971; Schneidermann, 1973).

Selective dissolution allows the compensation zone to be subdivided into three major zones of dissolution (Hsü and Andrews, in Maxwell et al., 1970; Adelseck et al., 1973; Schneidermann, 1973): (1) Mesolytic Zone – lowest part of the compensation zone, immediately overlying the CCD and extending 500 m above it in the equatorial Atlantic; characterized by low $CaCO_3$ content, absence of planktonic foraminifera, and a coccolith assemblage of low diversity composed of solution-resistant species (predominantly isolated distal shields of placoliths); (2) Oligolytic Zone – ranging in depth from 5500 m at the lower boundary to 4400–4500 m at the upper boundary; characterized by a dissolved, fragmented assemblage of planktonic foraminifera and a calcareous nannofossil assemblage composed of a few corroded non-resistant coccoliths and abundant, well-preserved solution-resistant coccoliths; (3) Eolytic Zone – upper part of the compensation zone, separated from the CCD by 1500 m; characterized by a high $CaCO_3$ content and a diversified coccolith assemblage which shows initial dissolution. Considerable vertical fluctuation in the CCD must be taken into account for the Tertiary and the Cretaceous (see Sect. 10.3.8).

The *topography* of the bathyal slope as well as of the abyssal plains is strongly differentiated by submarine canyons, submarine fans, and by swells. Some of the topographic features are related to present-day sea-floor spreading, others are not (aseismic ridges, volcanic seamounts and guyots, and oceanic plateaus). Many seamounts now lie in depths of several hundred to several thousands of meters. Non-depositional conditions seem to characterize the flanks and sometimes also the tops of the seamounts, favoring the lithification of carbonate by the precipitation of Mg-calcite cements (Milliman and J. Müller, 1979), the replacement of calcareous material by phosphate (Heezen et al., 1973), and also the formation of ferro-manganese nodules and crusts. Many Pacific guyots show a section of Lower Cretaceous carbonates, deposited within a shallow-water environment above submarine volcanics, grading up to deep-water limestones of Upper Cretaceous and Tertiary age (Heezen et al., 1973). Oceanic plateaus are vast areas of relatively thick elevated ocean crust that commonly rise above the abyssal sea floor to within 2–3 km of the surface. These platforms are covered by thick sedimentary blankets; examples are the Magellan Rise (Winterer et al., 1973), the Maniki Plateau (Winterer et al., 1974), the Otonong Java Plateau (Klein, 1975), and the Shatsky Rise (Moberly and Larson, 1975) in the Pacific. The abyssal plains are covered by calcareous sediments above the calcite compensation depth, below it by red clay, radiolarian ooze and diatomaceous ooze. The flat plains with low reliefs may exhibit abyssal hills (up to 250 m high). Of special interest for the interpretation of Ancient continental margins are carbonate platforms interspersed with deeper basins (e.g., Bahama-Blake Plateau; Pourtales Terrace); these areas are generally characterized by shallow-water sedimentation, but may also include smaller deeper basins with pelagic sedimentation.

Deep-Sea Sediments are formed by pelagic sedimentation (deposition of fine-grained biogenous, predominantly planktonic particles settling through the water at slow rates of sedimentation, by the deposition of land-derived clay minerals (forming "red clay", "brown clay", etc.), by the sedimentation of slump sediments from the slope and of turbidites derived from shallow-water environments, and by ice-transport of sediments. Another important mode for the formation of deep-sea limestones is the precipitation of carbonate cements in non-depositional areas or in areas associated with igneous activity (Thompson, 1972; Milliman and J. Müller, 1979). Mg-calcite as well as calcite occur as cements, the latter predominantly below the boundary between intermediate and cold deep-water masses (700–1200 m, Tongue of the Ocean, Bahamas). Below this level magnesian calcite and aragonite cements are restricted to semi-enclosed seas with exceptionally warm waters (W. Schlager and James, 1978).

Following the definitions used by Murray and Renard (1891), true pelagic deposits (e.g., Globigerina ooze, radiolarian ooze) are often distinguished from "terrigenous deposits" (various muds). This corresponds somewhat to the distinction between "eupelagic" sediments (with little or no terrigenous or neritic influence) and "hemipelagic" sediments (with a substantial proportion of land- or shelf-derived material, deposited at relatively high rates, typically on continental slopes and rises as well as on abyssal plains). In consideration of the results of the Deep Sea Drilling Project, Berger (1974) has proposed a new *classification of deep-sea pelagic sediments,* including pelagic deposits, hemipelagic deposits, and various lithified sediments as well:

I. Pelagic deposits (oozes and clays)
$<25\%$ of the fraction $>5\,\mu m$ is of terrigenous, volcanogenic, and or neritic origin. Median grain size $<5\,\mu m$ (except authigenic minerals and pelagic organisms)

A. Pelagic clays. $<30\%$ $CaCO_3$ and siliceous organisms
1) 1–10% $CaCO_3$ – slightly calcareous clay
2) 10–30% $CaCO_3$ – very calcareous clay (marl)
3) 1–10% siliceous organisms – slightly siliceous clay
4) 10–30% siliceous organisms – very siliceous clay

B. Oozes. $>30\%$ $CaCO_3$ or siliceous organisms
1) $>30\%$ $CaCO_3$ to $<2/3$ $CaCO_3$ – marl ooze;
$>2/3$ $CaCO_3$ – chalk ooze
2) $<30\%$ $CaCO_3$, $>30\%$ siliceous organisms – diatom or radiolarian ooze

II. Hemipelagic deposits (muds)
$>25\%$ of the fraction $>5\,\mu m$ is of terrigenous, volcanogenic, and/or neritic origin. Median grain size $>5\,\mu m$ (except authigenic minerals and pelagic organisms).

A. Calcareous muds. $>30\%$ $CaCO_3$
1) $<2/3$ $CaCO_3$ – marl mud;
$>2/3$ $CaCO_3$ – chalk mud
2) $>30\%$ skeletal $CaCO_3$ – foram∼, nanno∼, coquina∼calcareous mud

B. Terrigenous muds. < 30% $CaCO_3$, quartz, feldspar, mica dominant (quartzose, arkosic, micaceous)
C. Volcanogenic muds. < 30% $CaCO_3$. Ash, palagonite, etc. dominant

III. Pelagic and/or hemipelagic deposits
1) Dolomite-sapropelite cycles
2) Black (carbonaceous) clay and mud – sapropelites
3) Silicified claystones and mudstones – chert
4) Limestone

Calcareous oozes consist of coccoliths and foraminifera, and in places, of "pteropods" as well as noncarbonate material:

a) Coccolith ooze (see Plates 8/1–3) consists of mass accumulations of calcitic tests of coccolithophorids (planktonic algae, Chrysophyta). The tests (coccospheres, diameter 10–100 µm) consist of tiny variously-shaped plates (coccoliths, diameter about 5–10 µm). The distribution within the oozes is controlled by the productivity in the upper water column, by the composition of the algal assemblages dependent on the water temperature (McIntyre and Be, 1967; McIntyre and McIntyre, 1971) and the salinity, and also by the sedimentation rate, which is very slow for isolated coccoliths but fast for coccoliths enclosed in fecal pellets of fishes and invertebrates. Coccoliths may comprise more than 50% of the carbonate fraction; during the Mesozoic and the Tertiary the percentage was sometimes even higher (Bramlette, 1958). Differential solution modifies the composition of the assemblages (A. McIntyre and R. McIntyre, 1971; Hill, 1975).

b) Planktonic foraminifera (Globigerina) *oozes* are the most conspicuous contributors to biogenous deep-sea sedimentation. The ooze consists generally of more than 30% calcitic tests of planktonic foraminifera together with coccoliths, and some benthic foraminifera. The composition of the fauna depends on the geographical latitude, the fertility, the water temperature and the differential solution of the species (Sliter et al., 1975). Differential dissolution generally merely modifies the original assemblages. Recent Globigerina oozes are found commonly between about 1600 m and 4500 m. Planktonic foraminiferal ooze and coccolith ooze are deposited in shallow-water environments, too (Kling, 1975; Scholle and Kling, 1972).

c) Pteropod ooze is composed of aragonitic shells of heteropod and euthecosome pelagic gastropods living in warm water zones in a depth of about 100 m. The deposition of the shells is limited to a maximum depth of about 3500 m because of the solubility of aragonite; main concentrations have been found between 1500 and 3000 m (Chen, 1964; Melguen and Thiede, 1974). Pteropod oozes often contain well-preserved spined and thin-walled planktonic foraminifera. Pteropod oozes are known from the Persian Gulf, from the Bermuda-Bahama slope, from the southwestern Atlantic, and the Indian Ocean.

These pelagic oozes are often redeposited. The winnowing of the sediments may produce lag deposits and assemblages with tests of selected sizes (Berger and von Rad, 1972). Vertical mixing several centimeters deep within the sediment may be caused by bioturbation.

The *distribution of pelagic sediments* is related to depth, latitude, temperature, fertility of the surface waters, ocean currents, distance from land, amount of organic matter in the sediments, and associated CO_2 production and oxidation state as well as the activity of burrowing organisms (Berger, 1974).

Shallower, carbonate-rich parts of the ocean are separated by the *calcite compensation depth* (CCD) from deeper, carbonate-poor deposits (Riedel and Funnel, 1964). At this boundary, the rate of supply and the rate of dissolution are approximately compensated. The upper limit of the compensation zone is the level below which sea water is undersaturated with respect to $CaCO_3$ ("saturation depth"; Pytkowitz, 1970). The level with increased solution rates of planktonic foraminifera and of coccoliths is called the lysocline (Berger, 1968), generally defined as the facies boundary between well-preserved and poorly-preserved calcareous assemblages. A clear distinction should be made between the sedimentologically defined "foraminiferal lysocline" or "coccolith lysocline" (Berger, 1973; Roth and Berger, 1975) and the "hydrographic lysocline" (Peterson, 1966), which corresponds to the level in the water column with a drastic increase in dissolution rates (Berger, 1970).

Aragonite dissolves at much shallower depths than calcite, usually between 1000 to 3500 meters (Friedman, 1965; Li et al., 1969). The "aragonite compensation depth" and "aragonite lysocline" (derived from the dissolution patterns of pteropods) lies higher than the calcite compensation depth (see Fig. 6). The CCD lies at a depth of about 4500 m in the Pacific, and at about 5500 m in the Atlantic.

The position and thickness of the CCD and the rate of $CaCO_3$ within it is a result of an interplay of the following parameters: (1) surface productivity; a high rate of production of $CaCO_3$ depresses the CCD (Arrhenius, 1963); (2) increase of $CaCO_3$ dissolution with increasing depth (pressure); (3) amount of CO_2 dissolved in the sea water is a function of the amount of organic debris oxidized in deep waters (sediments rich in organic carbon deposited in rather shallow depths of 300–600 m may also be affected by solution; West African continental slope); (4) temperature; solubility of CO_2 increases with decreasing temperature, increasing the rate at which $CaCO_3$ is dissolved; (5) physical and chemical properties of the deposited $CaCO_3$ particles; and (6) quantity of non-calcareous material supplied to the bottom; dilution by terrigenous sediments and siliceous biogenic particles (radiolaria, diatoms) decreases the $CaCO_3$ content of a given sediment.

Carbonate turbidites: Bioclastic and lithoclastic shallow-water carbonates can be transported by slumps and turbidity flows into bathyal and abyssal depths. Recent examples are known from the surrounding areas of the Bahamas (Bornhold and Pickey, 1971; Pickey and Rukker, 1966; Schlager and Chermak, 1979; Schlager, 1980), from the Gulf of Mexico (Davies, 1968), from the Hispaniola-Caicos Basin (Bennets and Pickey, 1976; Ditty et al., 1977) as well as from the Fiji Islands (Houtz et al., 1962). The carbonate material derives from reefs or carbonates platforms bordering the turbidite basin. Turbidites may also be composed of pelagic carbonates, resedimented from submarine rises in adjacent basins (van Andel and Komak, 1969). Rapidly emplaced carbonate turbidites are only slightly affected by the undersaturated waters of the deep sea (Hesse, 1975); see Sect. 10.2.4.2

Deep-water coral reefs: "Coral coppices", composed of low-diversity ahermatypic hexacorals, may form large reef structures in deep marine water, down to 850 m. Main growth occurs between 200 and 400 m, at water temperatures between 4 °C and 10 °C. The coral coppices form beds up to 60 m thick and reefs extending over several km², whose coarser detrital fraction consists of coral fragments (60–90% of the total sediment); the fine fraction consists of mollusks and echinoderm fragments (inhabitants of the reef), planktonic foraminifera and sometimes pteropods. Examples are known from many places, e.g., the offshore area of Norway (Teichert, 1958), along the continental slope of France, in the Blake Plateau (Stetson et al., 1962), Rockall Banks, Northeast Atlantic (Scoffin et al., 1980), Bahamas (Mullins et al., 1981) and off the coast of New Zealand (Squires, 1965). Similar deeper-water coral reefs may extend as far back as the Triassic (Stanley, 1979).

2.3 Lacustrine and Terrestrial Carbonate Sedimentation
(Plate 2)

The more important non-marine carbonate sediments found in Ancient limestones can be subdivided according to origin and formational environment:

1. Lacustrine carbonates: deposition in lakes
2. Terrestrial carbonates: deposition on land
   A. Freshwater limestones
      a) Tufa: porous } Calcareous sinter
      b) Travertine: dense, banded } Calcareous sinter
   B. Caliche
   C. Cave carbonates (speleothem carbonates)
   D. Aeolian carbonates

2.3.1 Lacustrine Carbonates

Carbonates appear in freshwater and salt-water lakes as
(A) inorganic precipitations (survey in Müller et al., 1972),
(B) algal sediments (see Pia, 1933), and
(C) shell accumulations.

A. Inorganic precipitation: Three factors are responsible for the formation of primary inorganic carbonate minerals in freshwater environments:

a) Loss or removal of CO_2 as a result of a change in pressure or temperature, or the contributions of assimilating water plants or phytoplankton. Particularly in humid environments (e.g., Plitvitzer lakes, Yugoslavia; Lake Balaton, Hungary).

b) Evaporitic concentration. Particularly in arid environments (e.g., Tuz Gölü, Anatolia, Turkey).

c) Mixture of water bodies with various pH values (river and lake water).

According to the models developed by Müller et al. (1972) Mg/Ca ratios of lake water regulate the formation of various mineral phases: Mg/Ca ratio <2 – precipitation of calcite (Example: Lake Constance); Mg/Ca ratio 2–7 – formation of high Mg-calcite (Lake Balaton); Mg/Ca ratio 7–12 – formation of high Mg-calcite and dolomite by transformation from high Mg-calcite (Lake Neusiedl, Austria); Mg/Ca ratio high, up to >12 – formation of aragonite (Van Gölü).

Carbonate deposition in lakes occurs under several different sets of circumstances: Relatively shallow lakes in arid and semi-arid settings deposit carbonate chiefly by evaporation. The resulting limestones are almost always impure, containing significant quantities of clay, and are usually associated with stromatolitic structures, dolomite, and other evaporitic minerals such as gypsum, halite, and in some cases, bitter salts. This type of depositional environment is illustrated by the Eocene Green River Formation in Wyoming, Colorado, and Utah (Eugster and Hardie, 1975; Surdam and Wolfbauer, 1975). A second type of lacustrine carbonate occurs in temperate regions: These lakes contain progradational marl sequences which result from carbonate formation in shallow water, mainly by biochemical precipitation associated with the photosynthetic activity of macrophytes (chiefly *Chara*). Examples are Littlefield Lake in Central Michigan (Murphy and Wilkinson, 1980), and also some Alpine lakes (see J. Schneider, 1980).

B. Carbonatic algal sediments (Plate 44) may be formed in lacustrine environments by various processes:

a) Biocorrosion of carbonate substrates (e.g., limestone coasts; see J. Schneider, 1977; or mollusk shells, see Philippon and Plaziat, 1975) by cyanophyceans, chlorophyceans, rhodophyceans, fungi and lichens. These organisms corrode the rock by etching processes and produce silt-sized sediments.

b) Formation of non-marine stromatolites by the interaction of sediment-binding algae (mainly filamentous and coccoid blue-greens) and sedimentation of fine-grained carbonates or terrigenous particles. Lacustrine stromatolites were already known from the Precambrian but did not become important till the late Mesozoic and Cenozoic. Examples have been described by Freytet and Plaziat (1965), Link et al. (1978), A. Schäfer and K. R. G. Stapf (1978), Smoot (1978), Surdam and Wray (1976) and Tucker (1978). A common feature of many lacustrine algal bioherms is the veneer formed on the algal tufa and adjacent sediments by laminated carbonate sinter during emergence phases due to a temporary fall in lake level (see Riding, 1979). Recent freshwater stromatolites are predominantly formed by trapping of carbonate sediment by blue-green algae and also mosses, with subsequent cementation by precipitated $CaCO_3$ (Eggleston and Dean, 1976).

Of special interest are blue-green algal bioherms formed within a brackish environment (D. K. Richter et al., 1979). The composition of the algal flora is similar to that of lacustrine stromatolites; a brackish environment can only be inferred from the coexistence of blue-green algae, brackish water pelecypods, and marine coralline algae.

c) Oncoids are formed by cyanophycean and non-skeletal chlorophycean algae, which circumcrust shells (e.g., gastropods) or rock fragments. Shape and size seem to depend upon water energy and the type of the adjacent sediment. Oncoids from Lake Constance ("Schnegglisteine") are 1–30 cm in diameter and show both a hard and smooth as well as a soft and spongy surface. Their shape is discoidal; large oncoids commonly have a slightly convex surface exposed to the open water and a conical subsurface embedded in the sandy substrate. The organization of the laminae and also the fabric seen in thin-sections result from the growth of algae and the nature of their filaments. Oncoid microstructure can only be related with difficulty to the algae which formed them, since algal filaments decay quickly. But it seems possible to attribute special algal fabrics to particular algal groups (see A. Schäfer and Stapf, 1979): A spongy microstructure is produced by *Phormidium, Calothrix* and/or *Dichothrix;* a dense microstructure is formed by *Schizothrix,* and a fan-like microstructure may be produced by *Rivularia.* These microstructures often occur together in one oncoid. Many oncoids are overturned by wave activity or by the activity of fishes.

d) "Lake Chalk" (Seekreide), a very fine-grained carbonate sediment (grainsizes approximately 10 microns), is formed in shallow, warm waters within protected

→

Plate 2. Lacustrine and Terrestrial Carbonates

1. Cave carbonates. Cross-section of stalactites which have grown together shows a crustose structure. The lamella are discontinuously deposited and terraced. Due to various rates of growth, cavity relicts have remained intact. Layers of carbonate were deposited around the outside surface of the three sinter tubes (seen as black circles). The internal structure of the crusts corresponds to that of caliche-crusts. Zoolithen cave near Burggaillenreuth, Northern Franconian Alb, Germany. Thin-section. Scale is 10 mm

2. Ancient lacustrine algal carbonates. Extracellular micritic calcite was precipitated around the filaments of the algae. See thin-section photomicrographs in Plate 47. Miocene fresh-water limestone: Hainsfarth quarry, Ries, Germany. SEM photomicrograph. Scale is 1 mm

3. Travertine, banded calcareous sinter. Annual "inorganically" precipitated, mm-thin concentric layers with fibrous bundles. See Reiff (1955). Interglacial ("Böttinger Marmor"). Bad Cannstatt near Stuttgart, Germany. Thin-section. Scale is 10 mm

4. Caliche. The fabric is characterized by particles formed in situ, whose morphological features are reminiscent of peloids, oncoids, and aggregate grains (see Section 2.3.2.2). Also typical – even for Ancient caliches – is the "brecciation" indicated by vertical and horizontal calcite veins. Fuerteventura, Canary Islands. Thin-section. Scale is 10 mm

5. Caliche. Crusts with calcitic "particles" formed in situ, which contain distinct pisoids (see Table 4). Fuerteventura, Canary Islands. Thin-section. Scale is 10 mm

6. Ancient caliche with finely laminated, dark layers, which have been transformed into pisoids with indistinct boundaries. In contrast to the stromatolites formed by algal activity, the laminar structures fill in the depressions in the microrelief. Such fabrics, together with the in situ brecciation and the sparseness of fossils, are good indicators of subaerial exposure and a period of non-deposition. Middle Triassic (Wetterstein limestone, Ladinian): near Complonhütte/Reißkofel mountain, Gailtal Alps, Carinthia, Austria. Thin-section. Scale is 10 mm

7. Caliche pisoids, formed at the top of lacustrine algal stromatolites. Altenglan Limestones, Rotliegendes, Lower Permian: Tiefenbach, Saarpfalz, Germany. Scale 1 mm

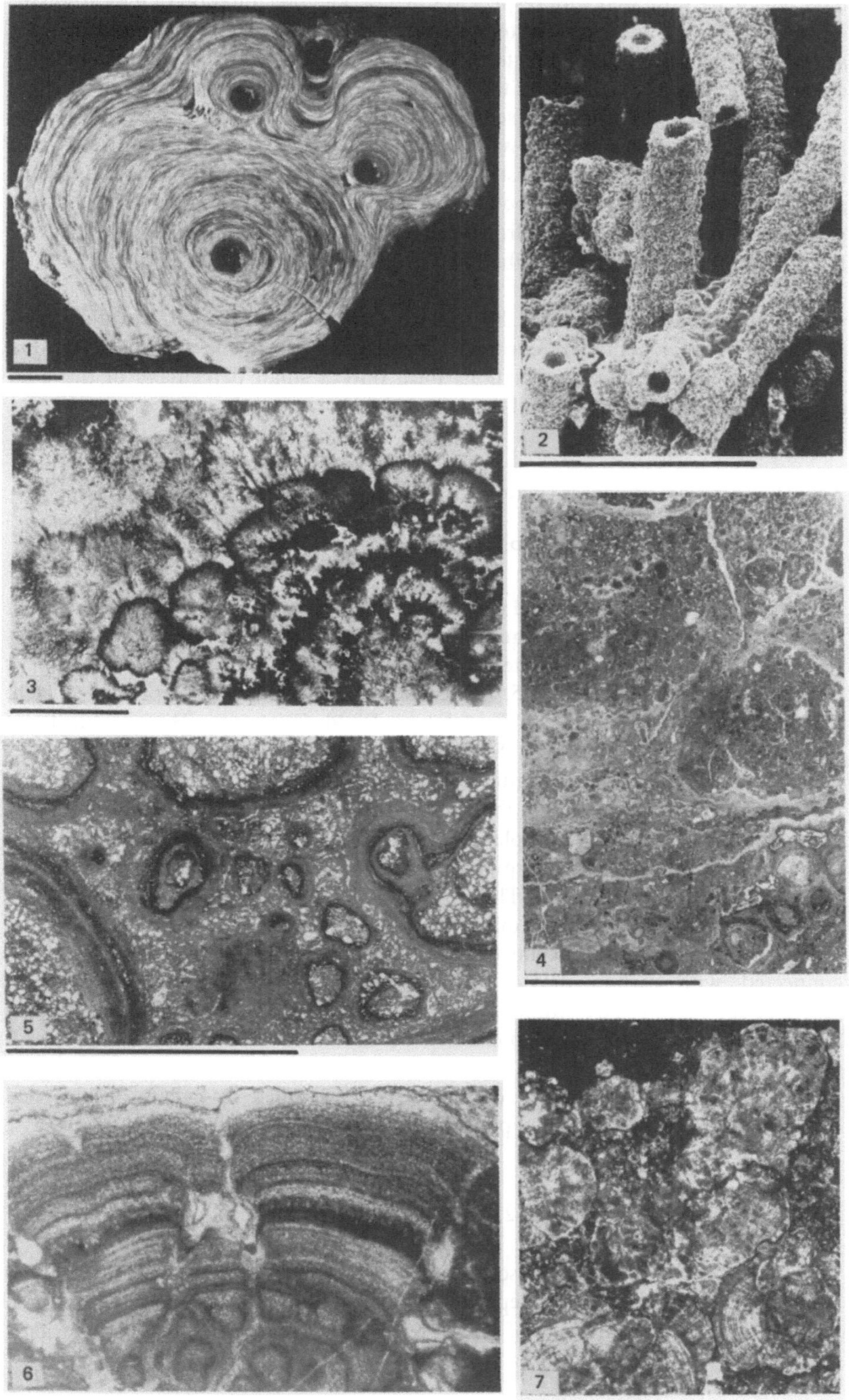
1
2
3
4
5
6
7

bays which are saturated with respect to $CaCO_3$. The precipitation of the carbonate is caused by an increase in temperature, evaporation and the activities of assimilating plants. Tiny fragments of shells, oncoids and calcareous crusts may also contribute to the accumulation of the lake chalk (see A. Schäfer, 1973).

e) Charophycean oogonia (see Sect. 5.3.12) may contribute greatly to the formation of lacustrine carbonates. Recent charophytes inhabit fresh and brackishwater environments; most species grow submerged in fresh standing water and upon sandy or muddy bottom. They often form subaquatic meadows; gyrogonites may be transported from their habitats and deposited in deeper parts of lakes. Examples have been described by Pia (1933).

C. *Accumulations of shells* may be common in lakes. Gastropod and pelecypod shells are swept together by currents within zones parallel to the coast (see Pia 1933). In general shell accumulations form only a small percentage of the total carbonate sediment deposited in lakes.

2.3.2 Terrestrial Carbonates (Plate 2)

2.3.2.1 Freshwater Limestones (Plate 2/3)

In wetter humid areas laminated carbonates made up of calcite occur at the mouth of springs, less often in streams and in lakes. These carbonates are formed either as (a) dense, banded calcareous sinter (= travertine, Plate 2/3, e.g., Bad Cannstatt near Stuttgart, Germany – Reiff, 1955; Leine Valley near Göttingen, Germany – Savelli and Wedepohl, 1969; Bad Laer, Teuteburger Wald, Germany – Hiltermann, 1977) or as (b) porous, loose tufas (e.g., Swabian Alb, Germany – Irion and Müller, 1968). Whereas inorganic origin dominates in travertines (CO_2 loss due to higher temperature near the mouth of the spring; sulfate solutions), tufas are formed by precipitation of calcite onto plants (usually algae and mosses). The role of assimilation during $CaCO_3$ precipitation is still a matter of discussion (see Pia, 1933; Stirn, 1964; Grüninger, 1965). The assimilation rates of Recent blue-green algae encountered in the deposition of carbonate tufa are low (Pentecost, 1978); only 1–5% of the calcite of tufa seems to have been precipitated by an equilibrium shift connected with carbon dioxide uptake by the algae. Photomicrographs of thin-sections as well as descriptions of the tufa fabrics categorized according to algal and fungal groups have been published by Irion and Müller (1968) and Smoot (1978).

A special type of terrestrial carbonate is represented by calcitic pisoids and concretions formed in spring-fed pools on playa surfaces (see Sect. 4.1.3.6).

2.3.2.2 Caliche (Calcrete) (Plate 2/4–7)

Calcium hydrogen carbonate dissolved in pore water rises to the surface by capillary action and upon evaporation deposits cryptocrystalline calcareous crusts on or in the surface. The occurrence of these terrestrial calcareous crusts is restricted

Table 4. Differences between sedimentary and diagenetic globoids

Characteristics	Caliche-peloids	Peloids	Caliche-ooids	Ooids	Caliche-pisoids	Oncoids
Shape	Irregularly rounded	Generally rounded	Spherical or oval	Generally spherical	Spherical to oval	Elliptical; cross-section irregular, oval shape, lobed outline
Structure	Cryptocrystalline calcite, sometimes sepiolite or opal	Cryptocrystalline calcite	Finely laminated envelopes of indistinct concentric layers of non-oriented cryptocrystalline calcite and tangentially arranged calcite needles. Thickness of envelopes to 0.05 mm, coatings thinner on corners and edges of singular nuclei	Well-defined concentrically arranged layers, sometimes radial structures	Envelopes not uniformly thick but layers well-defined. Alternation of thin light-colored layers (0.05 mm) and brownish layers (0.02 – 0.5 mm). Often upward growing. Repeated in situ brecciation	Non-concentric deposition micrite envelopes with or without biogenic structures
Nuclei	–	–	Usually rock fragments, pisoid fragments, or peloids	Bioclasts, ooids, peloids	Usually rock fragments or pisoid fragments	Usually bioclasts
Fabric	Generally missing, sometimes faint concentric zoning	–	No regular grain fabric	Regular grain fabric	Generally no regular fabric	Depositional fabric, caused by organisms
Size	0.02–8.0 mm, often 0.5–1.5 mm, or 0.02–0.10 mm	0.02 to > 1 mm	to 2 mm	to 2 mm	2 mm to 5 cm	mm- to cm-sized
Lithofacies	Especially in upper part of caliche crusts	Bioclasts, often micritic matrix	Caliche-pisoids, caliche-crusts; no fossils	Bioclasts, microfossils; often sparry cement	Caliche-ooids, caliche-peloids, caliche-crusts, angular rock fragments, breccia, no fossils	Bioclasts, coated grains (cortoids), ooids

primarily to arid or semiarid regions, where the annual relative humidity is about 200–600 mm and evaporation exceeds the annual rainfall. Exceptions are known in humid regions with an annual precipitation of over 150 mm and periods of intense evaporation (Müller and Hoffmeister, 1968; James, 1972).

Good descriptions of genesis, facies criteria and fabrics are found in Chapman (1974, Saudi Arabia); Degens and Rutte (1960, Greece); Goudie (1973, important survey of "duricrusts"); James (1972, survey of facies criteria); H. Kulke (1974, Algeria); Krejci-Graf (1960, Macronesia); Nägele (1962, "Albstein"); Nagtegaal (1969); Read (1976), Reeves (1970); Knox (1977; South Africa, description of profiles), Rutte (1958, 1960); A. Schäfer (1975); Siesser (1973); and Werner (1971). A general survey is found in Reeves (1976).

Caliche formations (Plate 2) are *characterized* by: a) Laminated crusts (Plate 2/5) made of micritic or microsparry calcite (rarely aragonite and Mg-calcite, Scholle and Kinsman, 1974) incorporating lithoclasts, fragments of caliches, rarely skeletal grains (e.g., snails or root elements). These "particles" formed in situ are called caliche-pisoids, caliche-ooids (= "diagenetic ooids" in Siesser, 1973; "vadose pisoids" in Dunham, 1969) or caliche-pellets (see Plates 2/4, 2/6). The definitions and the differences to marine globoids are listed in Table 4.

b) Laminar calcretes: a few centimeters thick, partially "brecciated," made of planar or wavy lamella consisting of stratose structures which are delimited by non-laminated, brownish envelopes. The grain sizes are more or less eqǔidimensional; the lamination is due to Fe-pigmentation or organic substances. The thickness of the lamella is 0.02–0.20 mm. Frequently the laminar structures fill in depressions in microreliefs (in contrast to stromatolites!). Such structures may be confused with physico-chemical coniatolites formed in supratidal regions and with stromatolites formed by algae (see Table 5). Krumbein and Giele (1979) have described "subaerial stromatolites" formed within calcretes in the Borrega Desert, California, and the Sinai Desert. These laminated carbonate crusts originate by carbonate deposition, caused principally, by the cyanobacterium *Pleurocapsa.*

c) Radial-granular or coarse-granular calcite or fibrous aragonitic "meniscus-cements" are described by James, 1972, and Scholle and Kinsman, 1974.

d) Displacive calcites (Watt, 1979) provide evidence of highly supersaturated precipitating solutions

e) Displacement of SiO_2 by calcite. Frequently microcrystalline quartz is found closely intergrown with fine-grained calcite (Nagtegaal, 1969).

f) "Tepee" structures (diameter in cm to m range) with steep sides (Watts, 1977); fenestral fabrics and small tubes, 5–20 microns wide, lined with sparite (algae or rhizoids?).

Laminar calcretes originating from successive cycles of lichen telogenesis, incipient pedogenesis and diagenesis (Klappa, 1979) are characterized by millimeter thick laminae with varying amounts of organic substance, algal filaments, fungal hyphae, calcite spherulites and decimicron-sized calcite grains arranged in surface parallel layers.

Caliche profiles may have thicknesses between < 1 m to more than 50 m (Brown, 1956). Often the top of the profile is characterized by pisoidal or "oncoidal" grainstones, whereas the lower part of the profile consists of laminated caliches or secondary mudstones.

Table 5. Differences between supratidal "marine" tufa (physico-chemically formed "coniatolites"), biogenetically formed stromatolites and non-marine tufas, particularly cave carbonates. Vadose fabrics do not necessarily indicate "non-marine" conditions! After Purser and Loreau (1973)

Characteristics	Coniatolites (vadose)	Stromatolites	Non-marine tufa, Cave sediments (vadose)
Contact to substrate	Often discordant; particles in substrate truncated and bored	Adapted to the substrate	
Micromorphology	Dripstone types, marginal stalactite structures	Layered structures dominant	Dripstone types + layered structures
Shrinkage cracks	Absent	Present	Absent
Fenestral fabrics	Absent	Common	Absent
Inclusions of detrital material	Only in small depressions	Often as continuous layers	Limestone fragments
Mineralogical composition	Very fine-grained or fibrous aragonite	Calcite	Larger calcite crystals
Microborings	Very common	Sometimes	Rare
Occurrence	On marine substrate (beach-rock, rubble bottom)	Preferentially in tidal zone	Terrestrial caves

The identification of *Ancient caliche formations* is extremely important for facies interpretation, because here indications of emergence and desiccation, extended periods of subaerial exposure and non-deposition are present (see Plate 2/6).

2.3.2.3 Cave Carbonates (Speleothems) (Plate 2/1)

The carbonate sinters precipitated from aqueous solutions are made up of carbonate minerals (calcite, aragonite, dolomite, hydromagnesite, magnesite, huntite, etc.); the carbonate sinters form structures which are variously described as stalagmites, stalactites (Plate 2/1), cave popcorn, cave pearls, etc., and are important for facies interpretations of carbonate rocks because they indicate:

A break in marine sedimentation (examples in Dunham, 1969).

Possible confusion may occur between cave pearls or pisoids and marine ooids as well as caliche-pisoids (see Table 4) and between cave sinters and stromatolites (differences explained in Thraikill, 1976).

The identification of Ancient speleothems appears to be possible by considering the types of carbonate cements. Folk and Assereto (1976) have shown that Ancient cave carbonates, even after the inversion of aragonite to calcite, are characterized by alternating phases of short fibrous calcite crystals and long palisade crystals. Examples have been found in the Permian of New Mexico and the Ladinian of Lombardy, Italy:

The Middle Triassic Calcare Rosso in northern Lombardy is an excellent example of the great variety of vadose cement types formed in paleospeleothems with fluctuating chemical environments (Assereto and Folk, 1980). Special criteria are aragonite "raggioni" formed by vadose seepage mainly on roofs of cavities and festooned cellular crusts interpreted as paleoflowstone. These giant paleo-aragonite rays form continuous crusts consisting of enormous fibrous crystals which line the walls of tens-of-centimeters to meter-sized cavities (formed by tepeeization). Raggioni are typically 20–40 mm long (up to 80 and 250 mm) and about 1 mm wide (maximum 4 mm). Ray boundaries are very sharply defined. Two types of terminations of the raggioni occur: squared ends and feathery terminations. Raggioni are believed to have grown directly as free-standing individual crystals, which were surrounded by a floatstone matrix after the completion of growth. These cave cements are also known from the Holocene Carlsbad speleothem, New Mexico (Folk and Assereto, 1976).

Paleoflowstone is characterized by 5–10 cm thick crusts with vague laminae with a scalloped appearance. The crinkled laminae are about 0.02–0.5 mm thick and composed of equant 2–4 μm calcite crystals. Between the laminae occur lenticules filled with a mosaic of alveolar fans composed of radiating, convex downward growing crystals. These paleoflowstones are believed to have been formed in the fissures of a tepee-deformed supratidal sediment after flooding by marine brines.

2.3.2.4 Eolian Carbonates (Plate 4/5,6)

Recent carbonate "eolianites" develop from wind-blown beach sands (Bretz, 1960; Land et al., 1967; MacKenzie, 1964 – Bermudas; Ball, 1967 – Bahamas). 95% of the land surface of Bermuda consists of Pleistocene carbonate dunes which build lobate bodies of sand. Cross-bedding structures, grain-size parameters like skewness and sorting (see Sect. 4.2.1), and possibly grain roundness (see Bigarella, 1972) can be used as facies criteria. Other criteria are remains of paleosols, the presence of terrestrial fossils (land snails, pollen), and the presence of rhizoconcretions in the paleosols. The carbonate particles (frequently ooids and skeletal grains) are bonded with calcitic granular cement B (Plate 4/5, 6).

2.4 References: Recent Carbonate Sedimentation

2.4.1 References: Marine Carbonates

Adelseck, C. G., Geehan, G. W., Roth, P. H. (1973): Experimental evidence for the selective dissolution and overgrowth of calcareous nannofossils during diagenesis. Bull. Geol. Soc. Amer. **84,** 2755–2762, New York

Adey, W. H., MacIntyre, I. G. (1973): Crustose coralline algae: a re-evaluation in the geological sciences. Bull. Geol. Soc. Amer. **84,** 883–903, New York

Adolphe, J. P., Billy, C. (1974): Biosynthèse de calcite par une association bactérienne aérobic. C. R. Acad. Sci. Paris **278,** Ser. D., 2873–2875, Paris

Andel, Tj. H. van, Komar, P. D. (1969): Ponded sediments of the Mid-Atlantic Ridge between 22° and 23° North latitude. Geol. Soc. Amer. Bull. **80,** 1163–1190, New York

Andrews, P. B. (1964): Serpulid reefs, Baffin Bay, Southeast Texas. In: Scott, A. J. (ed.), Depositional Environments South Central Texas Coast, 101–120, Gulf Coast Ass. Geol. Sci., Field Trip Guide Book

Arrhenius, G. (1963): Pelagic sediments, in Hill, M. N. (ed.): The Sea, vol. 3, 655–727, New York: Interscience

Bachmann, G. H. (1979): Bioherme der Muschel Placunopsis ostracina v. Schlotheim und ihre Diagenese. Neues Jahrbuch Geol. Paläont. Abh. **158**/3, 381–407, 17 Figs., Stuttgart

Bathurst, R. G. C. (1967): Subtidal gelatinous mat, sand stabilizer and food, Great Bahama Bank. J. Geol. **75,** 736–738, Chicago

Bathurst, R. G. C. (1971): see 1.4.1

Bennetts, K. R. W. and Pilkey, O. H. (1976): Characteristics of three turbidites, Hispaniola – Caicos Basin. Geol. Soc. Amer. Bull. **87,** 1291–1300, 13 Figs., New York

Berger, W. H. (1967): Foraminifera ooze: solution at depths. Science **156,** 383–385, Washington

Berger, W. H. (1968): Planctonic Foraminifera: selective solution and paleoclimatic interpretation. Deep-Sea Res. **15,** 32–43, Oxford

Berger, W. H. (1970): Planctonic Foraminifera: selective solution and the lysocline. Marine Geol. **8,** 111–138, Amsterdam

Berger, W. H. (1972): Deep-sea carbonates: dissolution facies and age-depth constancy. Science **236,** 392–395, Washington

Berger, W. H. (1973): Deep-sea carbonates: evidence for a coccolith lysocline. Deep-Sea Res. **20,** 917–921, Oxford

Berger, W. H. (1974): Deep-sea sedimentation, in Burke, C. A. and Drake, C. L. (eds.): The geology of continental margins, 213–241, New York: Springer

Berger, W. H. (1975): Dissolution of deep-sea carbonates: an Introduction. Cushman Found. Foram. Res. Spec. Publ. **13,** 7–10, 6 Figs., Lawrence

Berger, W. H. (1975): Deep-Sea Carbonates: Dissolution profiles from foraminiferal preservation. Cushman Found. Foram. Res. Spec. Publ. **13,** 82–86, 4 Figs., Lawrence

Berger, W. H. (1978): Sedimentation of deep-sea carbonate: maps and models of variations and fluctuations. J. Foram. Res. **8**/4, 286–302, 19 Figs., Lawrence

Berger, W. H. (1979): Impact of deep-sea drilling on paleo-oceanography. In, Talwani, M. and Ryan, W. B. F. (eds.): Deep sea drilling results in the Atlantic Ocean: Continental margins and paleoenvironment, 297–314, 12 Figs., Washington: Amer. Geophys. Union

Berger, W. H. and Rad, U. von (1972): Cretaceous and Cenozoic Sediments from the Atlantic Ocean. Initial Reports Deep Sea Drilling Project **14,** 787–954, 48 Pls., 59 Figs., Washington

Boillot, G. (1965): Organic gradients in the study of neritic deposits of biological origin: the example of the western English Channel. Marine Geol. **3,** 359–367, Amsterdam

Bornhold, B. D., Pilkey, O. H. (1971): Bioclastic turbidite sedimentation in Columbus Basin, Bahamas. Bull. Geol. Soc. Amer. **82,** 1341–1354, New York

Bosence, D. (1977): Ecological studies on two carbonate sediment-producing Algae. In: Flügel, E. (ed.): Fossil Algae, Recent Results and Developments, 270–278, 5 Figs., Berlin – Heidelberg – New York: Springer

Bosellini, A., Ginsburg, R. N. (1971): Form and internal structure of recent algal nodules (Rhodolites) from Bermuda. J. Geol. **79,** 669–682, Chicago

Bourrouilh, F. (1977): Les particules du Pleistocene bahamien et de sa croute: sédimentologie et diagenese. Sed. Geol. **17,** 247–284, Amsterdam

Bramlette, M.N. (1958): Significance of coccolithophorids in calcium carbonate deposition. Bull. Geol. Soc. Amer. **69,** 121–126, New York

Broecker, W. S., Takahashi, T. (1966): Calcium precipitation of the Bahama banks. J. geophys. Res. **71**/6, 1575–1602

Broecker, W. S., Takahashi, T. (1978): see 3.9

Carrol, J. J., Greenfield, L. J., Johnson, R. F. (1965): The mechanism of calcium and magnesium uptake from sea water by a marine bacterium. J. Cell. Comp. Physiol. **66,** 109–118

Caulet, P. J. (1972): Recent biogenic calcareous sedimentation on the Algerian continental shelf. In: Stanley, D. J. (ed.): The Mediterranean Sea. A sedimentation laboratory. 261–277, 7 Figs., 7 Tab.
Chave, K. E. (1962): Factors influencing the mineralogy of carbonate sediments. Limn. Oceanography **7**/2, 218–223, 7 Tab., 2 Figs., Lawrence
Chave, K. E. (1954): Aspects of the biochemistry of magnesium. J. Geol. **62,** 266–283, 587–599, Chicago
Chen, C. (1964): Pteropod ooze from Bermuda Pedestal. Science **144,** No. 3614, 60–62, Washington
Clarke, M. W. H., Keij, A. J. (1973): Organisms as producers of carbonate sediment and indicators of environment in southern Persian Gulf. In: Purser, B. H.: The Persian Gulf, 33–56, 7 Pls., 6 Figs., Berlin – Heidelberg – New York: Springer
Cloud, P. E. Jr. (1962): Carbonate deposition west of Andros Islands, Bahamas. U.S. Geol. Surv. Prof. Paper **350,** 138 pp., Washington
Dalrymple, D. W. (1965): Calcium carbonate deposition associated with blue-green algal mats, Baffin Bay, Texas. Publ. Inst. Marine Sci. **10,** 187–200, Port Aranas
Davies, D. K. (1968): Carbonate turbidites, Gulf of Mexico. J. Sed. Petrol. **38,** 1100–1109, Tulsa
Davies, G. R. (1970): Carbonate bank sedimentation, eastern Shark Bay, Western Australia. In: Logan et al. (eds.): Carbonate Sedimentation and Environments, Shark Bay, Western Australia. Amer. Ass. Petrol. Geol. Mem. **13,** 85–168, 33 Figs., Tulsa
Drew, G. H. (1911): The action of some denitrifying bacteria in tropical and temperate seas, and the bacterial precipitation of calcium carbonate in the sea. J. Marine Biol. Ass. U.K. **9,** 142–155
Easton, W. (1966): Underwater effect of Hurricane Betsy on some Bahamian reefs. Geol. Soc, Amer. Program Annual Meetings 1966, San Francisco, **59**
Emery, K. O. (1960): The Sea of Southern California. 366 pp., 248 Figs., New York – London: Wiley
Emery, K. O., Tracey, J. I., Ladd, H. S. (1954): Geology of Bikini and nearby Atolls. Part I. Geology. U.S. Geol. Surv. Prof. Paper **260,** A, 265 pp., 64 Pls., Washington
Endean, R., Jones, O. A. (1972): Biology and Geology of Coral Reefs. New York: Academic Press
Fabricius, F. H., Berdau, D., Munnich, K. O. (1970 a): Early Holocene ooids in modern littoral sands reworked from a coastal terrace southern Tunesia. Science **169,** 757–760, Washington
Fabricius, F., von Rad, U., Hesse, R., Ott, W. (1970 b): Die Oberflächensedimente der Straße von Otranto (Mittelmeer). Geol. Rdsch. **60**/1, 164–192, 7 Figs., 4 Tab., Stuttgart
Flajs, G. (1977): Die Ultrastrukturen des Kalkalgenskeletts. Palaeontographica **160 B,** 69–128, 17 Pls., 16 Figs., 1 Tab., Stuttgart
Folk, R. L. (1974): The natural history of crystalline calcium carbonate: Effect of magnesium content and salinity. J. Sed. Petrol. **44**/1, 40–53, 9 Figs., Tulsa
Folk, R. L., Robles, R. (1964): Carbonate sands of Isla Perez, Alacran Reef Complex, Yucatan. J. Geol. **72,** 255–292, Chicago
Friedman, G. M. (1965): Occurrence and stability relationships of aragonite, high-magnesian calcite and low-magnesian calcite under deep-sea conditions. Bull. Geol. Soc. Amer. **76,** 1191–1196, New York
Friedman, G. M. (1968): Geology and geochemistry of reefs, carbonate sediments and waters, Gulf of Aquaba (Elat), Red Sea. J. Sed. Petrol. **38**/3, 895–919, 32 Figs., Tulsa
Fuchs, Th. (1905): Über Pteropoden- und Globigerinenschlamm in Lagunen von Korallen-inseln. Verh. geol. Reichsanstalt Wien
Galtsoff, P. (1964): The American oyster *Crassostrea virginica* Gmelin. Fishery Bull. Fish and Wildlife Serv. **64,** 480 pp.
Garret, P., Smith, D. L., Wilson, A. O., Patriquin, D. (1971): Physiography, Ecology, and Sediments of two Bermuda patch reefs. J. Geol. **79,** 647–668, 21 Figs., Chicago
Gebelein, C. D. (1977): Dynamics of recent carbonate sedimentation and ecology. Cape Sable, Florida. Intern. Sed. Petrograph. Ser. **16,** 120 pp., 52 Photos, Leiden: Brill

Gevirtz, J. L., Friedman, G. M (1966): Deep-sea carbonate sediments of the Red Sea and their implications on marine lithification. J. Sed. Petrol. **36**/1, 143–151, 13 Figs., 1 Tab., Tulsa
Ginsburg, R. N. (1964): South Florida carbonate sediments. Ann. Meeting Geol. Soc. Amer. Guidebook, Field Trip No. 1, 72 pp.
Ginsburg, R. N. (ed.) (1975): see 1.4.1
Ginsburg, R. N., Lowenstam, H. A. (1958): The influence of marine bottom communities on the depositional environments of sediments. J. Geol. **66**, 310–318, Chicago
Ginsburg, R. N., Schroeder, J. (1973): Growth and submarine fossilization of algal cup reefs, Bermuda. Sedimentology **20**, 575–614, 26 Figs., 2 Tab., Amsterdam
Gould, H. R., Steward, R. H. (1955): Continental terrace sediments in the northeastern Gulf of Mexico. In: Finding Ancient shorelines. Soc. Econ. Paleont. Min. Spec. Publ. **5**, 2–19 Tulsa
Greenfield, L. J. (1963): Metabolism and concentration of calcium and magnesium and precipitation of calcium carbonate by a marine bacterium. Ann. New York Acad. Sci. **109**, 23–45, New York
Heath, R. R. (1969): Carbonate sedimentation in the Abyssal Equatorial Pacific during the past 50 million years. Bull. Geol. Soc. Amer. **80**, 689–694, 6 Figs., New York
Hedgpeth, J. W. (1957): Classification of marine environments. In: Hedgpeth, J. W. (ed.): Treatise Marine Ecol. Paleoecol. 1, Ecology, Mem. Geol. Soc. Amer. **67**/1, 17–27, 5 Figs., New York
Heezen, B. C., Matthews, J. L., Catalano, R., Natland, J., Coogan, A., Tharp, M., Rawson, M. (1973): Western Pacific guyots. Initial Report Deep Sea Drilling Project, **20**, 653–723, 3 Pls., Washington
Hesse, R. (1975): Turbiditic and non-turbiditic mudstones of Cretaceous flysch sections of the East Alps and other basins. Sed. **22**, 387–416, Amsterdam
Hill, M. E. (1975): Selective dissolution of mid-Cretaceous (Cenomanian) calcareous nannofossils. Micropaleont. **21**/2, 227–235, 2 Pls., New York
Hine, A. C., Wilber, R. J., Neumann, A. C. (1981): Carbonate sand bodies along contrasting shallow bank margins facing open seaways in Northern Bahamas. Amer. Ass. Petrol. Geol. Bull. **65**, 261–290, 24 Figs., 2 Tabs., Tulsa
Honjo, S. (1976): Coccoliths: production, transportation and sedimentation. Marine Micropaleont. **1**/1, 65–80, 2 Pls., Amsterdam
Hoskin, Ch., Nelson, R. V. Jr. (1969): Modern marine carbonate sediments, Alexander Archipelago, Alaska. J. Sed. Petrol. **39**/2, 581–590, 6 Figs., Tulsa
Houbolt, J. J. H. C. (1957): Surface sediments of the Persian Gulf near the Qatar Peninsula. Proefschrift, Univ. Utrecht, 113 pp., 31 Figs. on 14 Pls., 28 Figs., 8 supps., den Haag: Mouton & Co.
Houtz, R. E., Wellman, H. W. (1962): Turbidity current at Kadavu Passage, Fiji. Geol. Mag. **99**/1, 57–62, 1 Fig., London
Hsü, K. J., Jenkyns, H. C. (1974): see 1.4.1
Illing, L. V. (1954): Bahaman calcareous sands. Amer. Ass. Petrol. Geol. Bull. **38**, 1–95, Tulsa
JOIDES (Joint Oceanographic Institutions for Deep Earth Sampling): Initial reports of the deep-sea drilling project. Vol. 1 ff., Washington
Keen, A. M. (1961): A proposed reclassification of the gastropod family Vermetidae. Bull. Brit. Mus. Zool. **7**, 183–213, London
Kendall, C. G., Skipwith, P. A. d'E. (1968): Recent algal mats of a Persian Gulf lagoon. J. Sed. Petrol. **38**, 1040–1058, Tulsa
Kendall, C. G., Skipwith, P. A. d'E. (1969): Holocene shallow-water carbonate and evaporite sediments of Khor al Baxam, Abu Dhabi, Southwestern Persian Gulf. Bull. Amer. Ass. Petrol. Geol. **53**, 841–869, Tulsa
Kinsman, D. J. J. (1964): The recent carbonate sediments near Halat El Bahrani, Trucial Coast, Persian Gulf. Developments in Sedimentology **1**, 185–192, Amsterdam
Kinsman, D. J. J. (1969): Modes of formation, sedimentary associations and diagnostic features of shallow-water and supratidal evaporites. Bull. Amer. Ass. Petrol. Geol. **53**, 830–840, Tulsa
Klein, G. de Vries (1975): Resedimented pelagic carbonate and volcaniclastic sediments and sedimentary structures in Leg 30 DSDP Cores from the Western Equatorial Pacific. Geology **3**/1, 38–42, Boulder

Kling, S. A. (1975): A lagoonal coccolithophore flora from Belize (British-Honduras). Micropaleont. **21**/1, 1–13, New York
Lalou, C. (1957): Studies of bacterial precipitation of carbonates on sea water, J. Sed. Petrol. **27**, 190–195, Tulsa
Land, L. S. (1970): Carbonate mud: production by epibiotic growth on *Thalassia testudinum*. J. Sed. Petrol. **40**, 1361–1363, Tulsa
Land, L. S. Goreau, T. F. (1970): Submarine lithification of Jamaican reefs. J. Sed. Petrol. **40**, 457–462, Tulsa
Lewis, M. S. (1969): Sedimentary environments and unconsolidated carbonate sediments of the fringing coral reefs of Mahé, Seychelles. Marine Geol. **7**/2, 95–127, 14 Figs., Amsterdam
Li, Y. H., Takahashi, T., Broecker, W. S. (1969): Degree of saturation of $CaCO_3$ in the oceans. J. Geophys. Res. **74**, 5507–5525
Lisitzin, A. P. (1971): Distribution of carbonate microfossils in suspension and bottom sediments. Micropaleont. Oceans, 197–218, 14 Figs., Cambridge
Logan, B. W. (1969): Coral reefs and banks, Yucatan shelf, Mexico. In: Logan, B. W. et al. (eds.): Carbonate Sediments and Reefs, Yucatan Shelf, Mexico. Amer. Petrol Geol. Mem. **11**, 129–198, Tulsa
Logan , B. W., Davies, G. R., Read, J. F., Cebulski, D. E. (1970): Carbonate sedimentation and environments, Shark Bay, Western Australia. Amer. Ass. Petrol. Geol. Mem. **13**, 223 pp., Tulsa
Lowenstam, H. A. (1963): Biological problems relating to the composition and diagenesis of sediments. In: Donnelly, T. W. (ed.): The Earth Sciences, 137–195, Chicago: Univ. Chicago Press
Lowenstam, H. A. (1964): Coexisting calcite and aragonites from skeletal carbonates of marine organisms and their strontium and magnesium contents. In: Recent Researches in the Field of Hydrosphere, Atmosphere, and Nuclear Geochemistry, 303–404, Tolyo: Marzen Co.
Lucas, G. (1973): Facies carbonates des mers profondes. Bull. Centre Rech. Pau-SNPA **7**/1, 193–206, Pau
Ludwick, J. C., Walton W. R. (1957): Shelf-edge, calcareous prominences in northeastern Gulf of Mexico. Bull. Amer. Ass. Petrol. Geol. **41**/9, 2054–2101, 19 Figs., Tulsa
MacIntyre, I. G., Milliman, J. D. (1970): Physiographic features on the outer shelf and outer slope, Atlantic continental margin, southeastern United States. Bull. Geol. Soc. Amer. **81**, 2577–2598, New York
McIntyre, A., Be, A. W. H. (1967): Modern coccolithophorids of the Atlantic Ocean. – I. Placoliths and cyrtoliths. Deep-Sea Res. **14**, 561–597, 12 Pls., 17 Figs., 10 Tab., Amsterdam
McIntyre, A., McIntyre, R. (1971): Coccolith concentrations and differential solution in oceanic sediments. In: Funnel, B. M., Riedl, W. R. (eds.): The Micropaleontology of the Oceans, 252–261, Cambridge: Cambridge Press
McKee, E. D. (1959): Storm sediments on a Pacific Atoll. J. Sed. Petrol. **29**/3, 354–364, 5 Figs., Tulsa
Maiklem, W. R. (1968): The Capricorn reef complex, Great Barrier Reef, Australia. J. Sed. Petrol. **38**/3, 785–798, 15 Figs., Tulsa
Matthews, R. K. (1966): Genesis of recent mud in southern British Honduras, J. Sed. Petrol. **36**, 428–454, Tulsa
Maxwell, W. G. H. (1968): Atlas of the Great Barrier Reef. 258 pp., 166 Figs., Amsterdam – London – New York: Elsevier
Maxwell, W. G. H., Day, R. W., Fleming, P. J. G. (1961): Carbonate sedimentation on the Heron Island Reef, Great Barrier Reef. J. Sed. Petrol. **31**/2, 215–230, 8 Figs., Tulsa
Maxwell, W. G. H., Jell, J. S., McKellar, R. G. (1964): Differentiation of carbonate sediments in the Heron Island Reef. J. Sed. Petrol. **34**/2, 294–308, 12 Figs., Tulsa
Maxwell, W. G. H., Swinchatt, J. P. (1970): Great Barrier Reef: Regional variation in a terrigenous-carbonate province. Bull. Geol. Soc. Amer. **81**/3, 691–724, 21 Figs., New York
Melguen, M., Thiede, J. (1974): Facies distribution and dissolution depths of surface sediment components from the Vema Channel and the Rio Grande rise (Southwest Atlantic Ocean). Marine Geol. **17**, 341–353, 4 Figs., Amsterdam

Milliman, J. D. (1967): Carbonate sedimentation on Hogsty reef. A Bahamian Atoll. J. Sed. Petrol. **37**/2 658–678, 19 Figs., Tulsa
Milliman, J. D. (1974): see 1.4.1
Milliman, J. D. (1977): Role of calcareous algae in Atlantic continental margin sedimentation. In: Flügel, E. (ed.): Fossil Algae, Recent Results and Developments. 232–247, 7 Figs., Berlin – Heidelberg – New York: Springer
Milliman, J. D., Emery, K. O. (1968): Sea levels during the past 25,000 years. Science **162,** 1121–1123, Washington
Milliman, J. D., Müller, J. (1973): Precipitation and lithification of magnesian calcite in the deep-sea sediments of the eastern Mediterranean Sea. Sed. **20,** 29–46, Amsterdam
Mitterer, R. M. (1968): Amino acid composition of organic matrix in calcareous oolites. Science **162,** 1498–1499, Washington
Mitterer, R. M. (1971): Comparative amino acid composition of calcified and noncalcified polychaete worm tubes. Comp. Biochem. Physiol. **38** B, 405–409
Moberly, R. M., Larson, R. L. (1975): Mesozoic magnetic anomalies, ocean plateaus, and seamount chains in the north-western Pacific Ocean. Initial Report Deep Sea Drilling Project **32,** 945–957, Washington
Müller, G., Blaschke, R. (1969): Zur Entstehung des Tiefsee-Kalkschlammes im Schwarzen Meer. Naturwissenschaften **56,** 561–562, Heidelberg
Müller, G., Müller, J. (1967): Mineralogisch-sedimentpetrographische und chemische Untersuchungen an einem Bank-Sediment (Cross-Bank) der Florida Bay, USA. N. Jb. Min. Petrograph. Abh. **106,** 257–286, Stuttgart
Mullins, H. T., Newton, C. R., Heath, K., Vanburen, H. M. (1981): Modern deep-water coral mounds north of Little Bahama Bank: Criteria for recognition of deep-water coral bioherms in the rock record. J. Sed. Petrol. **51**/3, 999–1013, 7 Figs., Tulsa
Multer, H. G. (1969): Field Guide to some carbonate environments: Florida Keys and Western Bahamas. Miami Geol. Soc., Miami
Murray, J., Renard, A. F. (1891): Report on deep-sea deposits based on the specimens collected during the voyage of H.M.S. Challenger in the years 1872 to 1876. Rep. Voyage Challenger, 525 p., London: Longmans
Neumann, A. C., Gebelein, C. D., Scoffin, T. P. (1970): Composition, structure and erodability of subtidal mats, Abaco Bahamas. J. Sed. Petrol. **40,** 274–297, Tulsa
Newell, N. D., Imbrie, J., Purdy, E. G., Thurber, D. L. (1959): Organism communities and bottom facies, Great Bahama Bank. Bull. Amer. Mus. Nat. Hist. **117,** 177–228
Oeltzschner, H., Sigl, W. (1970): Sedimentologische Untersuchungen im Golf von Manfredonia (Südadria). Geol. Rdsch. **60,** 133–144, Stuttgart
Pérès, J. M. (1967): Les biocenoses benthiques dans le système phytal. Rec. Trav. Sta. Marine End. Bull. **42,** 3–113
Peterson, M. N. A. (1966): Calcite: rates of dissolution in a verticale profile in the central Pacific. Science **154,** 1542–1544, Washington
Pia, J. (1933): see 1.4.1
Pilkey, O. H., Blackwelder, B. W., Doyle, L. J., Estes, E., Terlecky, P. M. (1969): Aspects of carbonate sedimentation on the Atlantic Continental Shelf off the Southern United States. J. Sed. Petrol. **39**/2, 744–768, 15 Figs., Tulsa
Pilkey, O. H., Rucker, J. B. (1966): Mineralogy of Tongue of the Ocean's sediments. J. Marine Res. **24,** 276–285
Pilkey, O. H., Schnitker, D., Pevear, D. R. (1966): Oolites on the Georgia continental shelf edge. J. Sed. Petrol. **36,** 462–467, Tulsa
Purdy, E. G. (1963 a): Recent calcium carbonate facies on the Great Bahama Bank. 1. Petrography and reaction groups. J. Geol. **71**/3, 334–355, 5 Figs., 2 Tab., Pls. 1–5, Chicago
Purdy, E. G. (1963 b): Recent calcium carbonate facies of the Great Bahama Bank. 2. Sedimentary facies. J. Geol. **71**/4, 472–497, 4 Figs., 3 Tab., 1 Pl., Chicago
Purdy, E. G. (1974): Reef configurations: Cause and effect. In: Laporte. L. F. (Ed.): Reefs in Time and Space. Soc. Econ. Paleont. Min., Spec. Publ. **18,** 9–76, 43 Figs., Tulsa
Purser, B. H. (ed.) (1973): The Persian Gulf, Holocene Carbonate Sedimentation and Diagenesis in a Shallow Epicontinental Sea. 471 pp., 7 Pls., 250 Figs., 3 maps, Berlin – Heidelberg – New York: Springer

Pytkowitz, R. M. (1970): On the carbonate compensation depth in the Pacific Ocean. Geochim. Cosmochim. Acta **34,** 836–839
Riedel, W. R., Funnel, B. M. (1964): Tertiary sediment cores and microfossils from the Pacific ocean floor. Quart. Geol. Soc. London **120,** 305–368, London
Riedl, R. (1963): Probleme und Methoden der Erforschung des litoralen Benthos. Zool. Anz. **26,** Suppl., 505–567
Roth, P. H., Berger, W. H. (1975): Distribution and dissolution of coccoliths in the south and central Pacific. Cushman Found. Foram. Res. Spec. Paper **13,** 87–113, 3 Pls., 24 Figs., Lawrence
Sandberg, P. A. (1975): see 4.1.7.2
Sarntheim, M. (1969): The plankton-benthos ratio of mollusks in the Recent sediments of the Persian Gulf. Proc. 1st Internat. Conf. Planktonic Mikrofoss. Geneva 2, 594–598, 4 Figs., Leiden: Brill
Sarntheim, M. (1971): Oberflächensedimente im Persischen Golf und Golf Oman. II. Quantitative Komponentenanalyse der Grobfraktion. „Meteor" Forsch. Ergebn. C, **5,** 113 pp. Berlin: Borntraeger
Saunders, C. E., Schneidermann, N. (1973): Carbonate sedimentation on the Inner Shelf Isla Magueyes, Puerto Rico, Water Resources Research Inst. Univ. of Puerto Rico, 1–77, 19 Figs., Puerto Rico
Schäfer, W. (1967): Biofazies-Bereiche im subfossilen Korallenriff Sarso (Rotes Meer). Senck. lethaea **48,** 107–133, Frankfurt a.M.
Schlager, W., James, N. P. (1978): Low-magnesian calcite limestones forming at the deep-sea floor, Tongue of the Ocean. Bahamas. Sed. **25,** 675–702, 19 Figs., Amsterdam
Schlanger, S. O., Douglas, R. G. (1974): Pelagic ooze-chalk-limestone transition and its implication for marine stratigraphy. In: Hsü, K. J., Jenkyns, H. C.: Pelagic Sediments: On Land and Under the Sea, 117–148, 12 Figs., Oxford: Blackwell Sci. Publ.
Schneider, J. (1976): Biological and inorganic factors in the destruction of limestone coasts. Contrib. Sed. **6,** 112 pp., 11 Pls., 25 Figs., 19 Tab., Stuttgart: Schweizerbart
Schneidermann, N. (1973): Pelagic limestones of the Central Caribbean, Leg. 15. Initial Rep. Deep Sea Drilling Project **15,** 773–793, Pls. 1–8, Washington
Schneidermann, N. (1973): Deposition of coccoliths in the compensation zone of the Atlantic Ocean. Gulf Coast Sect. Soc. Econ. Paleont. Min., Proc. Symp. Calcareous Nannofossils, 140–151, 1 Fig., Houston
Scholle, P. A., Kling, St. A. (1972): Southern British Honduras: lagoonal coccolith ooze. J. Sed. Petrol. **42**/1, 195–204, 5 Figs., Tulsa
Schroeder, J., Zankl, H., Ginsburg, R. N. (1973): The dynamics of reef formation: Bahama reefs. 2nd Intern. Coral Reef Symp. Great Barrier Reef, Australia (Abstracts)
Schumacher, H. (1976): see 1.4.1
Scoffin, T. P. (1970): The trapping and binding of subtidal carbonate sediments by marine vegetation in Bimini lagoon, Bahamas. J. Sed. Petrol. **40,** 249–273, Tulsa
Seibold, E. (1962): Untersuchungen zur Kalkfällung und Kalklösung am Westrand der Great Bahama Bank. Sedimentology **1**/1, 50–74, 3 Tab., 8 Figs., Amsterdam – New York
Shepard, F. P. (1963): Submarine Geology, 2nd Ed., 557 pp., New York – Tokyo: Harper
Shepard, F. P. (1970): Lagoonal topography of Caroline and Marshall Islands. Bull. Soc. Geol. Soc. Amer. **81,** 1905–1914, New York
Shinn, E. A. (1963): Spur and groove formation on the Florida reef tract. J. Sed. Petrol. **33,** 291–303, Tulsa
Shinn, E. A., Lloyd, R. M., Ginsburg, R. N. (1969): Anatomy of a modern carbonate Tidal Flat, Andros Island, Bahamas. J. Sed. Petrol. **39**/3, 1202–1228, 37 Figs., Tulsa
Sliter, W. V., Be, W. H., Berger, H. (eds.) (1975): Dissolution of deep-sea carbonates. Spec. Publ. Cushman Foram. Res. **13,** 159 pp., 20 Pls., Lawrence
Stackelberg, V. U. v. (1970): Faziesverteilung in Sedimente des indisch-pakistanischen Kontinentalrandes (Arabisches Meer). Geol. Rdsch. **60,** 268–270, Stuttgart
Stanley, G. D. (1979): Paleoecology, structure, and distribution of Triassic coral buildups in western North America. Univ. Kansas Paleont. Contrib., **65,** 1–58, Lawrence
Stetson, T. R., Squires, D. F., Pratt, R. M. (1962): Coral Banks occurring in deep water on the Blake Plateau. Amer. Mus. Novitates, No. 2114, 39 pp.

Stieglitz, R. D. (1972): Scanning electron microscopy of the fine fraction of recent carbonate sediments from Bimini, Bahamas. J. Sed. Petrol. **42,** 211–226, Tulsa
Stieglitz, R. D. (1973): Carbonate needles: additional organic sources. Bull. Geol. Soc. Amer. **84,** 927–930, 1 Fig., New York
Stockmann, K. W., Ginsburg, R. N., Shinn, E. A. (1967): The production of lime mud by algae in South Florida. J. Sed. Petrol. **37,** 633–648, Tulsa
Stoddart, D. R., Yonge, M. (eds.) (1971): Regional variation in Indian Ocean coral reefs. Symp. Zool. Soc. London **28,** 584 pp, London
Storr, J. F. (1964): Ecology and oceanography of the coral-reef tract, Abaco Island, Bahamas, Geol. Soc. Amer. Spec. Paper **79,** 98 pp., New York
Suess, E. (1970): Interaction of organic compound with calcium carbonate. 1. Association phenomena and geochemical implications. Geochim. Cosmochim. Acta **34,** 157–168, Oxford
Swinchatt, J. P. (1965): Significance of constituent composition, texture and skeletal breakdown in some recent carbonate sediments. J. Sed. Petrol. **35,** 71–90, Tulsa
Taft, W., Harbaugh, J. W. (1964): Modern carbonate sediments of Southern Florida, Bahamas, and Espiritu Santo Island, Baja California: a comparison of their mineralogy and chemistry. Stanford Univ. Publ. Sci. **8**/2, 133 pp.
Taylor, J. D., Lewis, M. S. (1970): The flora, fauna and sediments of the marine grass beds of Mahe, Seychelles, J. Nat. Hist. **4,** 199–220
Teichert, C. (1958): Cold- and deep-water coral banks. Bull. Amer. Assoc. Petrol. Geol. **42**/5, 1064–1082, 3 Figs., New York
Turmel, R. J., Swanson, R. G. (1976): The development of Rodriguez Bank, a Holocene mudbank in the Florida reef tract. J. Sed. Petrol. **46**/3, 497–518, 23 Figs., Tulsa
Wagner, C. W., Togt, C. van der (1973): Holocene sediment types and their distribution in the southern Persian Gulf. In: Purser, B. H. (ed.): The Persian Gulf. 123–155, 20 Figs., 1 Tab., Berlin – Heidelberg – New York: Springer
Walther, J. (1910): Die Sedimente der Taubenbank im Golf von Neapel. Preuss. Akad. Wiss. Jg. 1910, Phys.-math. Kl., Anhang, Abh. III, 49 pp., 2 Pls., Berlin
Walter, M. R. (ed.) (1976): Stromatolites. Dev. Sediment **20,** 790 pp., Amsterdam – Oxford – New York: Elsevier
Wass, R. E., Conolly, J. R., MacIntyre, R. J. (1970): Bryozoan carbonate sand continuous along southern Australia. Marine Geol. **9,** 63–73, Amsterdam
Wiedemann, H. U. (1972): Shell deposits and shell preservation in Quaternary and Tertiary estuarine sediments in Georgia, USA. Sed. Geol. **7,** 103–115, Amsterdam
Wiens, H. J. (1962): Atoll Environments and Ecology, 532 pp., 88 Pls., 93 Figs., 27 Tab., New Haven – London: Yale Univ. Press
Wilson, J. L. (1976): see 1.4.1
Winterer, E. L., Ewing, J. I. et al. (eds.) (1973): Initial reports deep sea drilling project **17,** 930 pp. Washington: U. S. Government Printing Office
Zankl, H., Schroeder, J. H. (1972 a): Ecology, sedimentology and diagenesis of recent and fossil reefs. Geol. Rdsch. **61**/2, 480–483, Stuttgart
Zankl, H., Schroeder, J. H. (1972 b): Interaction of genetic process in Holocene reefs off North Eleuthera Island, Bahamas, Geol. Rdsch. **61**/2, 520–541, Stuttgart

2.4.2 References: Lacustrine and Terrestrial Carbonates

Assereto, R., Folk, R. L. (1980): Diagenetic fabrics of aragonite, calcite, and dolomite in an Ancient peritidal – spelean environment: Triassic calcare rosso, Lombardia, Italy. J. Sed. Petrol. **50**/2, 371–394, 17 Figs., Tulsa
Ball, M. M. (1967): Carbonate sand bodies of Florida and the Bahamas. J. Sed. Petrol. **37**/2, 556–561, Tulsa
Bigarella, J. J. (1972): Eolian environments; their characteristics, recognition, and importance. Spec. Publ. Soc. Econ. Paleont. Min. **16,** 12–61, 38 Figs., Tulsa
Bretz, J. H. (1960): Bermuda: a partly drowned, late mature, Pleistocene karst. Bull. Geol. Soc. Amer. **71,** 1729–1754, New York

Brown, C. N. (1956): The origin of caliche on the north-eastern Llano Estacado, Texas. J. Geol. **64,** 1–15, Chicago

Chappman, R. W. (1974): Calcareous duricrusts in Al-Hasa, Saudi-Arabia. Bull. Geol. Soc. Amer. **85,** 119–130, New York

Degens, E. T., Rutte, E. (1960): Geochemische Untersuchungen eines Kalkkrustenprofils von Altkorinth (Griechenland). N. Jb. Geol. Paläont. Mh. **1960,** 263–276, Stuttgart

Dunham, R. J. (1969): Vadose pisolite in the Capitan reef (Permian), New Mexico and Texas. Soc. Econ. Paleont. Min., Spec. Publ. **14,** 182–191, Tulsa

Eggleston, J. R., Dean, W. E. (1976): Freshwater stromatolitic bioherms in Green Lake, New York: In: Walter, M. R. (ed.): Stromatolites, 479–488, 8 Figs., Amsterdam – Oxford – New York: Elsevier

Esteban, M. C. (1974): Caliche textures and "Microcodium". Boll. Soc. Geol. Ital. **92,** 105–125, 13 Figs., Rome

Eugster, H. P., Hardie, L. A. (1975): Sedimentation in an Ancient playa-lake complex: the Wilkins Peat Member of the Green River Formation of Wyoming. Bull. Geol. Soc. Amer. **86,** 319–334, New York

Feth, J. H. (1964): Review and annotated bibliography of Ancient lake deposits (Precambrian to Pleistocene) in the Western United States. Bull. U.S. Geol. Surv. **1080,** 119 pp., Washington

Folk, R. L., Assereto, R. (1976): Comparative fabrics of length-slow and length-fast calcite and calcitized aragonite in a Holocene speleothem, Carlsbad caverns, New Mexico. J. Sed. Petrol. **46**/3, 486–496, 11 Figs., Tulsa

Freytet, P., Plaziat, J. C. (1965): Importance des constructions algaires dues à des Cyanophycées dans les formations continentales du Crétacé supérieur et de l'Eocène du Languedoc. Bull. Soc. géol. France **7,** 679–694, Paris

Golubić, S. (1976): The relationship between blue-green algae and carbonate deposits. In: Carr, N. G., Whitton, B. A. (eds.): The biology of Blue-Green Algae, 434–472, Oxford: Blackwell Sci. Publications

Goudie, A. (1973): Duricrusts in tropical and subtropical landscapes. 174 pp. Oxford: Clarendon

Grüninger, W. (1965): Rezente Kalktuffbildung im Bereich der Uracher Wasserfälle. Abh. Karst- und Höhlenkunde, Reihe E, Botanik 2, 113 pp., München

Hardie, L. A., Smoot, J. P., Eugster, H. P. (1978): Saline lakes and their deposits: a sedimentological approach. Spec. Publ. Int. Ass. Sed. **2,** 7–41, 8 Figs., Oxford

Hay, R. L., Reeder, R. J. (1978): Calcretes of Olduvai Gorge and the Nodolanya Beds of Northern Tanzania. Sed. **25,** 649–673, 13 Figs., Oxford

Hay, R. L., Wiggins, B. (1980): Pellets, ooids, sepiolite and silica in three calcretes of the southwestern United States. Sed. **27,** 559–576, 4 Tab., 8 Figs., Oxford

Hiltermann, H. (1977): Die Sinterkalke (Travertine) von Bad Laer am Teutoburger Wald. Natur und Heimat **37,** 77–80, 2 Figs., Münster

Holland, H. D., Kirspu, T., Huebner, S., Oxburgh, U. (1964): On some aspects of the chemical evolution of cave water. J. Geol. **72**/1, 36–67, Chicago

Irion, G., Müller, G. (1968): Mineralogy, petrology and chemical composition of some calcareous tufa from the Schwäbische Alb, Germany. In: Müller, G., Friedmann, G. M. (eds.): Recent developments in Carbonate Sedimentology in Central Europe, 157–171, 16 Figs., Berlin – Heidelberg – New York: Springer

James, N. P. (1972): Holocene and Pleistocene calcareous crusts (Caliche) Profiles: Criteria for subaerial exposure. J. Sed. Petrol. **42**/4, 817–836, 12 Figs., Tulsa

Kann, E. (1941): Krustensteine in Seen. Arch. Hydrobiol. **37,** 504–532, Wien

Klappa, C. F. (1978): Biolithogenesis of Microcodium: elucidation. Sed. **25,** 489–522, 10 Figs., Oxford

Klappa, C. F. (1979): Lichen Stromatolites: criterion for subaerial exposure and a mechanism for the formation of laminar calcretes (Caliche). J. Sed. Petrol. **49**/2, 387–400, 5 Figs., Tulsa

Knox, G. J. (1977): Caliche profile formation, Saldanha Bay (South Africa). Sed. **24**/5, 657–674, 9 Figs., Oxford

Krejci-Graf, K. (1960): Zur Geologie der Makronesen. 4. Krustenkalke. Z. dtsch. Geol. Ges. **112**, 37–61, Hannover

Krumbein, W. E., Giele, C. (1979): Calcification in a coccoid cyanobacterium associated with the formation of desert stromatolites. Sed. **26**, 593–604, 9 Figs., Oxford

Kulke, H. (1974): Zur Geologie und Mineralogie der Kalk- und Gipskrusten Algeriens. Geol. Rundschau **63**/3, 970–998, 7 Figs., Stuttgart

Link, M. H., Isborne, R. H., Awramik, S. M. (1978): Lacustrine stromatolites and associated sediments of the Pliocene Ridge Route formation, Ridge Basin, California. J. Sed. Petrol. **48**/1, 143–158, 8 Figs., Tulsa

MacKenzie, F. T. (1964): Bermuda Pleistocene Eolianites and Paleowinds. Sed. **3**/1, 52–64, 6 Figs., Amsterdam

Marzo, M., Esteban, M., Pomar, L. (1974): Presencia de Caliche Fósil en el Buntsandstein del valle del Congost (Provincia de Barcelona). Acta Geol. Hispan. **9**/2, 33–36, 5 Figs., Madrid

Matter, A., Tucker, M. E. (eds.) (1978): Modern and Ancient lake sediments. Spec. Publ. Int. Ass. Sed. **2**, 288 p., Oxford

Müller, G., Irion, G., Förstner, U. (1972): Formation and diagenesis of inorganic Ca-Mg carbonates in the lacustrine environments. Naturwissenschaften **59**, 158–164, 4 Figs. Heidelberg

Müller, G., Wagner, F. (1978): Holocene carbonate evolution in Lake Balaton (Hungary): a response to climate and impact of man. Spec. Publ. Int. Ass. Sed. **2**, 57–81, 11 Figs., Oxford

Murphy, D. H., Wilkinson, B. H. (1980): Carbonate deposition and facies distribution in a central Michigan marl lake. Sed. **27**/2, 123–136, Oxford

Multer, H. G., Hoffmeister, J. E. (1968): Subaerial laminated crusts of the Florida Keys. Bull. Soc. Geol. Amer. **79**, 183–192, New York

Murray, J. W. (1954): The deposition of calcite and aragonite in caves. J. Geol. 481–492, Chicago

Nägele, E. (1962): Zur Petrographie und Entstehung des Albsteins. N. Jb. Geol. Paläont. Abh. **115**, 44–120, Pls. 3–7, 10 Figs., Stuttgart

Nagtegaal, P. J. C. (1969): Microtextures in recent and fossil caliche. Leidse Geol. Medel. **42**, 131–142, Leiden

Netterberg, F. (1978): Distinguishing fossil from contemporary calcretes. Abstracts, Int. Congr. Sed. Jerusalem, **2**, 463–464, Jerusalem

Pentecost, A. (1978): Blue-green algae and freshwater carbonate deposits. Proc. Roy. Soc. London, B **200**, 43–62, 1 Plate, 11 Figs., 10 Tab., London

Philippon, J., Plaziat, J.-Cl. (1975): Roles respectifs de la corrosion et des Cryptogames perforantes dans la destruction de coquilles de Mollusques des mangroves. Consequences sur la fossilisation. Comptes Rendus Acad. Sci. Paris **281**, D, 617–620, 3 Figs., Paris

Pia, J. (1933): Die rezenten Kalksteine. Min. Petrograph. Mitt., N. F. Erg. Bd., 420 pp. 4 Pls., 22 Figs., Leipzig

Picard, M. D., High, L. R. J. (1972): Criteria for recognizing lacustrine rocks. In: Rigby, J. K., Hamblin, W. K. (eds.): Recognition of Ancient Sedimentary Environments, Soc. Econ. Paleont. Min. Spec. Publ. **16**, 108–145, 24 Figs., 12 Tab., Tulsa

Pomar, L., Esteban, M., Llimona, X., Fontarnau, R. (1975): Accion de Liquenes, Algas y Hongos en la Telodiagénesis de las Rocas Carbonatadas de la zona litoral prelitoral Catalana. Investigaciones Geol. Univ. Barcelona **30**, 83–117, 10 Plates, Barcelona

Purser, B. H., Loreau, J.-P. (1973): Aragonitic, supratidal encrustations on the Trucial Coast, Persian Gulf. In: Purser, B. H. (ed.): The Persian Gulf, 343–376, 20 Figs., Berlin – Heidelberg – New York: Springer

Read, J. F. (1976): Calcretes and their distinction from stromatolites. In: Walter, M. R. (ed.): Stromatolites, 55–71, 3 Pls., 1 Fig., Amsterdam – Oxford – New York: Elsevier

Reeves, C. C., Jr. (1968): Introduction to paleolimnology. Dev. Sed. **11**, 228 pp., 124 Figs., Amsterdam – London – New York: Elsevier

Reeves, C. C. Jr. (1970): Origin, classification, and geologic history of caliche on the southern High Plains, Texas and eastern New Mexico. J. Geol. **78**, 352–362, Chicago

Reeves, C. C. Jr. (1976): Caliche. Lubbock (Estacado Books)

Reiff, W. (1955): Über den pleistozänen Sauerwasserkalk von Stuttgart – Münster – Bad Cannstatt. Jber. Mitt. oberrhein. geol. Ver., N.F. **37**, 56–91

Richter, D. K., Herforth, A., Ott, E. (1979): Pleistozäne, brackische Blaugrünalgenriffe mit Rivularia haematites auf der Perachorahalbinsel bei Korinth (Griechenland). Neues Jahrbuch Geol. Paläont. Abhandlungen **159**/1, 14–40, 9 Figs., Stuttgart

Riding, R. (1979): Origin and diagenesis of lacustrine algal bioherms at the margin of the Ries crater, Upper Miocene, southern Germany. Sed. **26,** 645–680, 41 Figs., Oxford

Robbin, D. M., Stipp, J. J. (1979): Depositional rate of laminated soilstone crusts, Florida Keys. J. Sed. Petrol. **49**/1, 175–180, 6 Figs., Tulsa

Rutte, E. (1955): Der Albstein in der miozänen Molasse Südwestdeutschlands. Z. deutsch. geol. Ges. **105**/3. 360–383, 1 Plate, 3 Figs., Hannover

Rutte, E. (1958): Kalkkrusten in Spanien. N. Jb. Geol. Paläont. Abh. **106,** 52–138, Stuttgart

Rutte, E. (1960): Kalkkrusten im östlichen Mittelmeergebiet. Z. dtsch. geol. Ges. **112,** 81–90, Hannover

Savelli, C., Wedepohl, K. H. (1969): Geochemische Untersuchungen an Sinterkalken (Travertinen). Contrib. Min. Petrol. **21,** 238–256, 1 Fig., 7 Tab. Heidelberg

Schäfer, A. (1972): Petrographische und stratigraphische Untersuchungen an den rezenten Sedimenten des Untersees/Bodensee. N. Jb. Min. Abh. **117,** 117–142, 4 maps., Pls. 1–2, 3 Figs., Stuttgart

Schäfer, A. (1973): Zur Entstehung von Seekreide-Untersuchungen am Untersee. N. Jb. Geol. Paläontol. Mh. **1973,** 216–230, 6 Figs., Stuttgart

Schäfer, A. (1975): Kalkkrusten auf der Insel Christiani, Santorin-Gruppe, Ägäis (Griechenland). N. Jb. Paläontol. Abh. **150**/1, 1–18, 8 Figs., 1 Tab., Stuttgart

Schäfer, A., Stapf, K. (1972): Calcite "whitings" im Bodensee-Untersee. Natur u. Mus. **102,** 307–311, 3 Figs., Frankfurt/M.

Schäfer, A., Stapf, K. R. G. (1978): Permian Saar-Nahe Basin and Recent Lake Constance (Germany): two environments of lacustrine algal carbonates. Spec. Publ. Int. Ass. Sed. **2,** 83–107, 22 Figs., Oxford

Schmidle, W. (1910): Postglaziale Ablagerungen im nordwestdeutschen Bodenseegebiet. N. Jb. Min. (II), **1910,** 104–122, Stuttgart

Schneider, J. (1977): Carbonate construction and decomposition by epilithic and endolithic micro-organisms in salt- and freshwater. In: Flügel, E. (ed.): Fossil Algae. Recent Results and Developments, 248–260, 3 Pls., 4 Figs., Berlin – Heidelberg – New York: Springer

Schöttle, M., Müller, G. (1968): Recent carbonate sedimentation in the Gnadensee (Lake Constance), Germany. In: Müller, G. Friedman, G. M. (eds.): Recent Developments in Carbonate Sedimentology in Central Europe, 148–156, 5 Figs., Berlin – Heidelberg – New York: Springer

Scholle, P. A., Kinsman, D. J. J. (1974): Aragonitic and high-Mg calcite caliche from the Persian Gulf – a modern analog for the Permian of Texas and New Mexico. J. Sed. Petrol. **44,** 904–916, Tulsa

Schröder, H. G., Schneider, J. (1979): Sedimentgeologische Untersuchungen zum Karbonatkreislauf und zur Sedimentationsgeschichte des Attersees. Arb. Lab. Weyregg **3,** 229–242, 4 Figs., 1 Tab., Weyregg/Austria

Siesser, W. G. (1973): Diagenetically formed ooids and intraclasts in South African calcretes. Sed. **20,** 539–551, Oxford

Smoot, J. P. (1978): Origin of the carbonate sediments in the Wilkins Peak Member of the lacustrine Green River Formation (Eocene), Wyoming, U.S.A. Spec. Publ. Int. Ass. Sed. **2,** 109–127, 16 Figs., Oxford

Steel, J. (1974): Cornstone (Fossil Caliche) – its origin, stratigraphic, and sedimentological importance in the New Red Sandstone, Western Scotland. J. Geol. **82,** 351–369, 13 Figs., Chicago

Stirn, A. (1964): Kalktuffvorkommen und Kalktufftypen der Schwäbischen Alb. Abh. Karst- u. Höhlenkunde, Reihe E, Botanik **1,** 92 pp., München

Surdam, R. C., Wolfbauer, C. A. (1975): Green River Formation, Wyoming: a playa lake complex. Bull. Geol. Soc. Amer. **86,** 335–345, New York

Surdam, R. C., Wray, J. L. (1976): Lacustrine stromatolites, Eocene Green River Formation, Wyoming. In: Walter, M. R. (ed.): Stromatolites. Dev. Sed. **20,** 535–541, Amsterdam: Elsevier

Thrailkill, J. (1976): Speleothems. In: Walter, M. R.: Stromatolites, 73–86, 16 Figs. Amsterdam – Oxford – New York: Elsevier

Tietz, G. F. (1979): Untersuchungen zur Genese und Mineralverteilung rezenter Karbonate in Dolomithöhlen Frankens unter besonderer Berücksichtigung jahreszeitlicher Konzentrationsänderungen in Höhlenwässern. Habilitationsschrift, Univ. Erlangen-Nürnberg, 139 p., 17 Pls., Erlangen

Trimmel, H. (ed.) (1965): Speläologisches Fachwörterbuch (Fachwörterbuch der Karst- und Höhlenkunde). Landesver. Höhlenkunde Wien/Niederösterr. 109 pp., 20 Figs., Wien

Tucker, M. E. (1978): Triassic lacustrine sediments from South Wales: shore-zone, clastics, evaporites and carbonates. Spec. Publ. Int. Ass. Sed. **2,** 205–224, 17 Figs., Oxford

Walls, R. A., Harris, W. B., Nunan, W. E. (1975): Calcareous crust (caliche) profiles and early subaerial exposure of Carboniferous carbonates, northeastern Kentucky. Sed. **22,** 417–440, Amsterdam

Watts, N. L. (1977): Pseudo-Anticlines and other Structures in some Calcretes of Botswana and South Africa. Earth Surface Processes **2,** 63–74, 10 Figs.

Werner, D. (1971): Böden mit Kalkanreicherungs-Horizonten in NW-Argentinien. Ein Beitrag zur Genese der Kalkkrusten. Göttinger Bodenkundliche Ber. **19,** 167–181, Göttingen

Wolff, M., Füchtbauer, H. (1975): Die karbonatische Randfazies der tertiären Süßwasserseen des Nördlinger Ries und des Steinheimer Beckens. Geol. Jb., D. **14,** 3–53, 3 Pls., 7 Figs., Hannover

3 Carbonate Diagenesis

Because diagenetic structures related to lithification can be confused with primary depositional structures and textures, knowing the possible diagenetic processes is extremely important for the interpretation of microfacies characteristics. The diagenesis of carbonates and carbonate rocks includes all of the processes involving solution, cementation, lithification and alteration of the sediments during the interval between deposition and metamorphism. Despite great progress in the last two decades, the complex factors influencing the diagenesis of carbonate sediments have only been partially worked out; just in the last few years have sharper distinctions between vadose and phreatic diagenesis been made. How much basic concepts can change was illustrated by the discovery that submarine lithification is possible without freshwater input (Fischer and Garrison, 1967).

Important surveys on the diagenesis of carbonates and carbonate rocks are found in Bricker (1971), Bathurst (1970, 1971), Folk (1965, 1974), Friedman (1964 and 1975), Milliman (1974), Purdy (1968) as well as Chilingar et al. (1967).

3.1 Diagenetic Environments. Classification of Diagenetic Processes

The amount of alteration connected with diagenesis is dependent on the diagenetic environments (post-depositional environments, Purdy, 1968). Basically the following distinctions can be made:

a) Submarine diagenesis
   - In shallow seas
   - In deeper parts of the ocean
b) Meteoric diagenesis (= freshwater diagenesis)
   - Vadose environment: zone above the level of the groundwater table (see Figs. 6 and 7); diagenesis under the influence of rain water (meteoric-vadose) or of splash-waters on the coast (marine-vadose). Examples: subaerial limestone crusts and caliche, cave carbonates (see 2.3.2).
   - Phreatic environment: zone below or in the water table (meteoric-phreatic) or influenced by marine waters (marine-phreatic) (Figs. 6 and 7).

Freshwater diagenesis and freshwater cementation is characterized by a rapid and extensive change in (a) CO_2 concentration and isotope composition, (b) degree of $CaCO_3$ supersaturation, and (c) flow rates and intermixing of the water. The water flow, which is dependent on climate, topography and permeability, produces various diagenetic structures which partially overlap each other (see Matthews, 1967; Steinen and Matthews, 1973).

By considering the factors at work in the process of sedimentation as well as within the sediment itself, diagenetic processes can be classified as follows (Füchtbauer, 1974; Milliman, 1974; Folk, 1974):

I. Destructive diagenesis – carbonate degradation:
   by biological or mechanical erosion or by chemical dissolution
II. Constructive diagenesis – carbonate aggradation:
   A. Isochemical (without changing the chemical composition of the sediments):
      1. Cementation: Infilling of primary voids in or between particles, or of solution cavities by chemically precipitated cements. In this way the sediments are consolidated, compaction is diminished or prevented, and lithification is initiated.
         a) Early diagenetic cement formation ("cement A")
            – At the expense of metastable carbonate minerals (e.g., aragonite in skeletal grains)
            – By evaporation of pore water rich in carbonates, in the supratidal zone
            – On the ocean floor

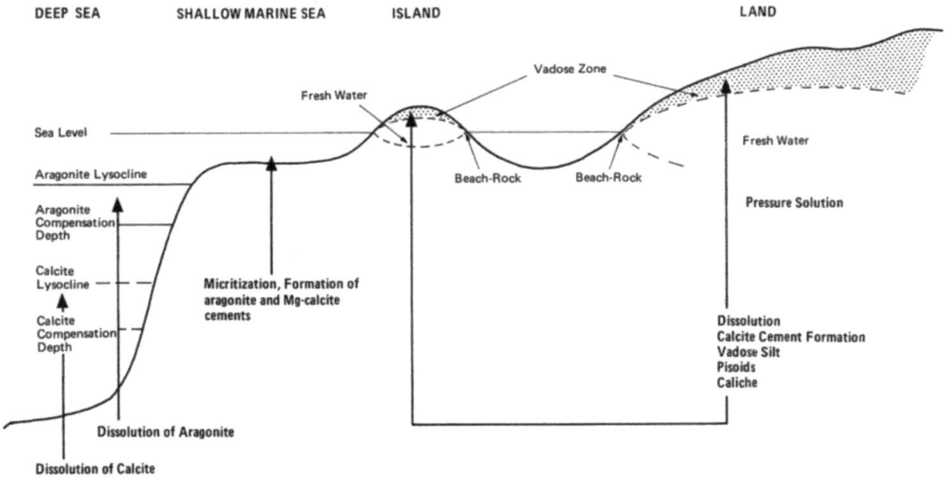

Fig. 6. Diagenetic environment of carbonate sediments. The diagenetic alterations take place in shallow seas (preferential aragonite and Mg-calcite cement formation); in deeper oceans [preferential dissolution at the aragonite compensation depth (ACD), aragonite lysocline (ALy) = level of first intense dissolution, calcite compensation depth (CCD), and calcite lysocline (CLy)]; under freshwater influence (meteoric diagenesis in the vadose zone or in the groundwater zone (phreatic zone); or by pressure solution in the sediment itself. Modified after Blatt et al. (1972)

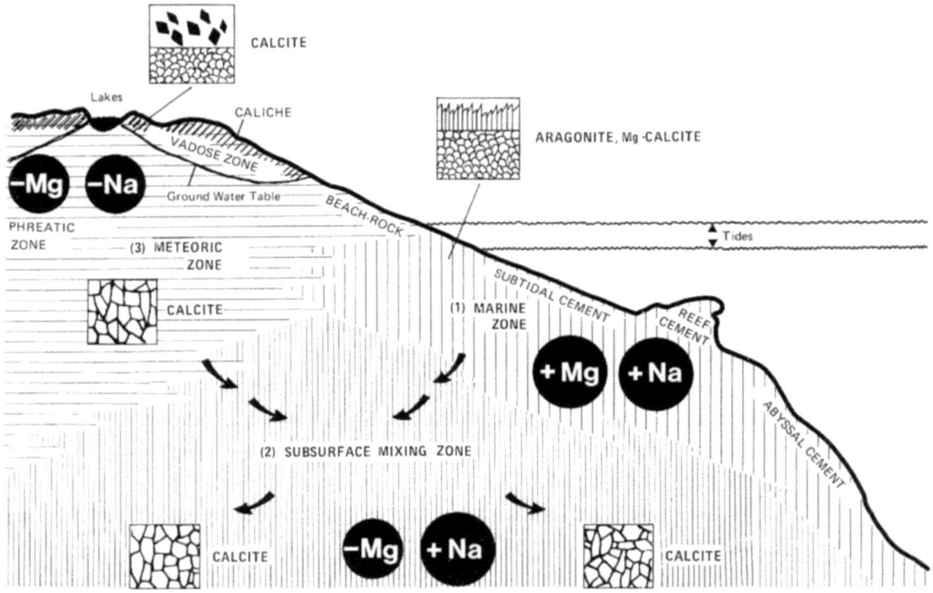

Fig. 7. Relations between diagenetic environments and formation of carbonate minerals (after R. L. Folk, 1974). *Zone 1,* high Na^+ and Mg^{2+} values = marine zone; aragonite or Mg-calcite cement with fibrous or micritic structures (e.g., beach-rock, subtidal, reefs) are formed. *Zone 2,* high Na^+ and low Mg^{2+} (removed by adsorption onto clay minerals and by dolomite formation) = subsurface mixing zone, calcitic sparite. Sparry calcite is formed in the phreatic zone; and micritic calcite in the vadose environment, sometimes as clearly defined rhombohedral crystals

b) Late diagenetic cement formation ("cement B")
Cement formed after sediment consolidation or after compaction; the necessary carbonate may be made available by pressure solution.

2. Neomorphism: This term was introduced by Folk (1965: 21) as a "comprehensive term of ignorance" for all mineral transformations in which the mineral either remains intact or is converted into a polymorphous mineral.
   a) Coalescive neomorphism: Larger crystals grow at the expense of smaller crystals ("aggrading" neomorphism, e.g., change of a lamellar calcitic brachiopod test into a test consisting of equant calcite crystals), or small crystals grow within a large crystal ("degrading" neomorphism, e.g., in the remains of echinoderms).
   b) Transformation (inversion) of aragonite to calcite by solution and in-situ precipitation in an aqueous environment.
      - Homoaxial transformation with conservation of the crystal shape (e.g., "spherulitic" microstructures in coelenterates; some radial structures of ooids (see 4.1.3.5); (calcite cement A)
      - Heteroaxial transformation: no optical or structural correspondence between aragonite and calcite crystals (e.g., in mollusk shells)

c) Recrystallization (Folk's terminology, 1965!): Growth of unstrained crystals at the expense of strained crystals of the same mineral, and increasing pressure and temperature conditions (metamorphism)

3. Dissolution processes in which carbonate is selectively dissolved, e.g., in clay seams (stylolite formation) and secondary porosity occurs

B. Allochemical (change in the chemical composition of the sediments):

1. Dolomitization
   a) Early diagenetic dolomitization in nonlithified sediment, in evaporitic environments with increased salinity
   b) Late diagenetic dolomitization, after consolidation of the sediment
2. Dedolomitization (recalcification): Dissolution of dolomite and replacement by calcite; usually with freshwater influence (Examples: Franconian Jurassic, Germany – Bausch, 1965; Italian Mesozoic – Mattavelli, 1966)
3. Dissolution of Mg-calcite and formation of calcite, whereby the texture of the organic hard parts or carbonate cements is preserved
4. Silicification (see Füchtbauer, 1974; R. C. L. Wilson, 1966)
5. Formation of authigenic minerals (see 4.1.5)

3.2 Carbonate Destruction

Biological erosion (see Plates 1/4, 25/6, 27/3, 31/7, 43/7) destroys carbonate particles and substrates and creates cavities. Organisms which burrow in sediments and bore into hard substrates are carbonate destroying organisms. Tables 6 and 12 give a survey of the significance of biological erosion.

Mechanical erosion forms fine and very fine grain fractions by fracturing and abrasion (see Fig. 30) and can even break up big reef blocks. The susceptibility to comminution is dependent on the microstructure of the tests of the organisms. In high-energy coastal areas mechanical erosion is generally slow but extensive.

Dissolution of $CaCO_3$ can arise in environments which are undersaturated with carbonates. Controlling parameters important for undersaturation include low water temperatures, high partial pressures of CO_2 or low pH, and increasing hydrostatic pressures. In tidal zones carbonate dissolution is strongly affected by diurnal and seasonal fluctuations in water temperature and in the O_2–CO_2 system; however, the significance of chemical dissolution for erosion appears here to be relatively low. Conversely, $CaCO_3$ dissolution is of major importance in the deeper oceans, where dissolution is directly proportional to increasing hydrostatic pressure in deeper waters, and is also influenced by the oxidation of organic material (see 2.2.4),

The following *factors* which may affect *resistance to dissolution* can be seen in *thin-sections:*

- Type of matrix between bioclasts. Possibilities: homogeneous or inhomogeneous matrix (e.g., homogeneous micrite or calcareous siltite, see 4.1.2); size of the micrite crystals; visible intercrystalline porosity or intergranular pores; matrix of calcite, dolomite, SiO_2, etc. These factors can affect the course of meteoric diagenesis.

Table 6. Carbonate construction and decomposition by epilithic and endolithic micro-organisms. After J. Schneider (1976)

	Processes		Significance in		
			Freshwater		Sea-water
Carbonate Construction	Trapping and binding of sediment particles	Sediment grains are mechanically deposited and bound by algal filaments to the surface of algal mats	Formation of travertine and lacustrine crusts	Formation of lake deposits, limestone crusts, stromatolites, and oncoids	Formation of supratidal, intertidal and subtidal algal mats (stromatolites, oncoids)
	Cementation by $CaCO_3$	Because of algal assimilation of CO_2 and HCO_3^{-1}, the pH between the algal filaments rises > 10; thus, CO_3^{-2} concentration rises, which may lead to inorganic $CaCO_3$ precipitation (non-oriented micrite)	Calcite, relatively common		Aragonite rare, because the presence of Mg^{2+} or organic substances precludes inorganic precipitation
	Encrustation with $CaCO_3$	Blue-green algae, green algae, and fungi can encrust their filaments and hyphens with micritic calcite, forming layered structures around the algal filaments	Calcite, only in freshwater		–
	Precipitation of $CaCO_3$ within the algal mucilage sheath	In calcified *Rivularia* colonies, calcite crystals precipitate onto trichomes embedded in the algal mucilage sheath	Calcite in lakes and rivers		–
Carbonate Decomposition	Biological corrosion by endolithic organisms	Boring blue-green, green, and red algae as well as fungi and lichens loosen up the substrate	In lacustrine and terrestrial environments as well as in marine environments		
	Preparation of the carbonate surface for inorganic solution	The loosened substrate can be dissolved more readily by rain water, leading to karren formations in terrestrial environments; sometimes dissolution by lichens	Particularly noticeable in terrestrial environments		
	Biological abrasion	The loosened substrate provides food for many microfossils, such as gastropods, echinoderms, fishes, etc.			Particularly noticeable in marine environments

- Organic substance in the bioclasts or in the matrix. The decay of organic substances can produce an acid milieu and thus a greater chance of dissolution.
- Mineralogy of the shell elements. The various solubilities of carbonate minerals lead to selective destruction and void formation.
- Size, shape, and orientation of the crystals. The higher surface/volume ratio of smaller elongated crystals increases the chance of dissolution.
- Micro- and ultrastructures of the shells. Shells with wall pores may be more susceptible to dissolution than homogeneous shells.

3.3 Cementation

A sufficient amount of dissolved carbonates is available for cement formation in shallow marine environments and in zones permeated with freshwater, depending on the various *stabilities of carbonate minerals:*

Submarine environment: Calcite > magnesian calcite ($\leqq$ 12 Mol% $MgCO_3$) > aragonite > very high Mg-calcite ($\geqq$ 12 Mol% $MgCO_3$)
Meteoric environment: Calcite > aragonite > High-Mg-calcite

According to Friedman (1975) and Wilson (1975) the following *cement formation processes* can be distinguished:

A. Meteoric diagenesis: Aragonite particles are dissolved in slightly acidic freshwater; thus, voids originate between the particles (as long as smaller particles or cements formed earlier can be dissolved) or in the particles (as long as micrite envelopes prevent the dissolution of the whole particle). During the dissolution of a great deal of material or of evaporites, collapse structures result, in which calcite is precipitated as cement in the voids or between the particles. Drusy cement or blocky cement is formed in the phreatic zone in this way; in the vadose zone the pore spaces are filled with air most of the time. Exceptions are known for grain-to-grain contacts, in which cements are formed from a thin film of water between the particles (meniscus cement, microstalactite cement, gravitative cement). In the vadose zone the pores generally remain open; the original structures remain intact. In contrast, in the phreatic zone carbonate-saturated or carbonate-supersaturated waters precipitate calcite cements which rapidly close the original pores. The original textures are essentially destroyed by neomorphism. Bioclasts, consisting of high-Mg calcite are an exception (e.g., coralline algae); they invert to calcite without destroying their original texture (D. K. Richter, 1979).

Vadose and phreatic diagenesis can be recognized from the different calcite mosaics. The phreatic calcite mosaic often consists of coarse crystals (up to a maximum size of several millimeters) which extend out from the coral skeletons into void filling cement ("cross-cutting" mosaic, after Pingitore, 1976). In contrast, the vadose mosaics are termed "fabric selective" since the boundaries of filled primary voids usually serve as crystal boundaries in the calcite mosaic. These two mosaics result from different modes of aragonite/calcite inversion in the vadose and phreatic zones. Vadose transformation takes place across micron-thick films of water, aragonite being dissolved on one side and calcite precipitated on the other. Phreatic diagenesis often involves the development of zones of chalky aragonite (several millimeters across) between fresh coral skeletons and the growing calcite mosaic. Extensive dissolution of the corals leaves a chalky friable framework and results in a calcite product with areas of secondary porosity, and calcite cements. Single calcite crystals may grow to large dimensions and extend beyond the coral skeleton into primary voids, producing the characteristic cross-cutting mosaic.

Meteoric diagenesis is indicated by: the above cement types as well as the existence of "vadose silt" (see 3.3.3), SiO_2 horizons formed within Ancient groundwater tables, subaerially formed dolomite "crusts," and the existence of dedolomitized zones. In geologically young sediments (Pliocene and Quaternary) the stages of the meteoric-vadose transformation of high-Mg calcite into stoichiometric calcite can be traced back by comparing the various amounts of Mg as well as Sr in the tests of echinids and corallinacean algae (D. K. Richter, 1979).

B. Marine diagenesis

1. *Shallow water:*

   a) Beach-rock (Plates 3, 4): Cementation (in coastal areas) in warm climates and in water supersaturated with carbonates, causing formation of aragonitic fibrous cement and micritic Mg-calcite cement.

Plate 3. Recent Carbonate Cements in Beach-Rocks

1. Aragonitic fibrous cement A and micritic cement B of Mg-calcite. A uniformly broad rim consisting of fibrous aragonite crystals growing normal to the pyroclastic particles (upper subtidal to lower intertidal zones). Cement A has been covered by the micritic Mg-calcite of cement B which was formed under vadose conditions (intertidal zone). Aragonite crystals have been partially dissolved by the action of fresh water. Fuerteventura, Canary Islands. SEM photomicrograph. Scale is 150 μm

2. Thin-section to Fig. 1. Fibrous cement A and micritic cement B. Scale is 500 μm

3. Micritic Mg-calcite cement A and whisker-like aragonite cement B. Due to the evaporation of vadose water in the tidal zones, broad micrite rims of cement A develop around carbonate and non-carbonate sand grains. This decrease in porosity, which is linked with cementation, leads to a decrease in the vadose conditions and to the appearance of phreatic micro-environments, in which whisker-like crystals of cement B are formed during continuous water saturation. Fuerteventura, Canary Islands. SEM photomicrograph. Scale is 25 μm

4. Thin-section to Fig. 3. Micrite cement A and fibrous cement B. Red algal bioclasts (coralline algae) are conspicuous among the carbonate particles. Scale is 500 μm

5. Meniscus cement. Concave cement "bridges" connect adjacent sediment grains. Because vadose water at the grain contacts remains for a fairly long time as a water film, grain-connecting cements can be precipitated here more readily than on the grain surfaces themselves. See Dunham (1969). Fuerteventura, Canary Islands. SEM photomicrograph. Scale is 150 μm

6. Thin-section to Fig. 5. Meniscus cement between corallinacean bioclasts and non-carbonatic lithoclasts. Open pores black. Scale is 500 μm

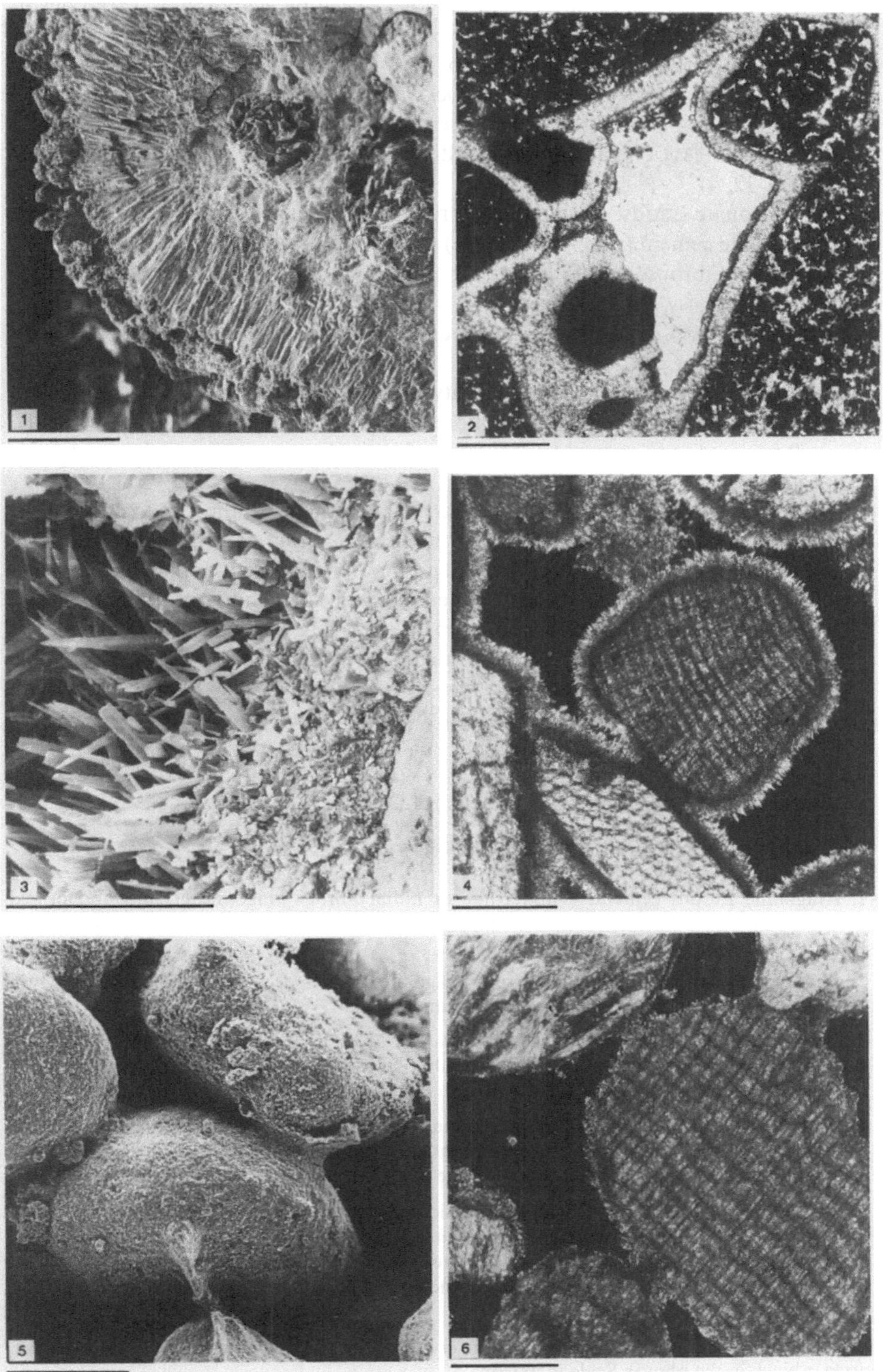
1
2
3
4
5
6

The precipitation of the cement may be caused by evaporation, variations in the chemistry of vadose waters, or by metabolic processes of microorganisms (blue-green algae; diatoms, Keupp, 1978).
It should be stressed that beachrocks are also formed in groundwater seeps along the top of beach zones in lake environments (Binkley et al., 1980).

b) Tidal zones: Cement formation initiated by blue-green algae (Pratt, 1979; see 5, 3, 12, 4)
c) Carbonate sands: Aragonitic fibrous cements or micritic Mg-cement or palisade cement (even-rim cement) grow around carbonate particles
d) Reefs: Fibrous aragonitic cement and micritic Mg-calcite cement as well as coarser drusy cement
e) Areas of non-deposition: Hardground formation by aragonite and Mg-calcite cementation
f) Environments of increased salinity: At temperatures over 30 °C and with hypersalinity (Dead Sea) formation of fibrous and micritic aragonite cements.
g) Restricted environments: Sulfate-reducing bacteria form calcite cements (Red Sea)

2. *Deeper Ocean:* Most of the deep-sea carbonate cements studied up to now consist of cryptocrystalline Mg-calcite. It can be supposed that cementation takes place preferentially during periods of non-deposition or restricted sedimentation.
The precipitation of Mg-calcite under open ocean conditions is temperature controlled; Mg-calcites with more than 10 mol% $MgCO_3$ tend to occur in waters less than 1000 m deep. Above a depth of 600 m the tests of planktonic foraminifera occurring in the deep fore-reef area of the Belize Reef are filled with Mg-calcite cement; below 600 m they are empty (James and Ginsburg, 1979).

Plate 4. Recent Carbonate Cements in Beach-Rocks and Eolianites

1. Aragonitic fibrous cement. Whisker-like or radial fibrous aragonite crystals are formed during continuous water saturation in the phreatic zone. Beachrock, Fuerteventura, Canary Islands. SEM photomicrograph. Scale is 10 μm

2. Thin-section to Fig. 1. The whisker-like aragonite crystals become longer along the grain contacts and in the narrow pores. Scale is 500 μm

3. Micritic cement. Single rhombohedral crystals of Mg-calcite grow adjacent to and through each other; often rosette-like aggregates result. Vadose beach-rock, Fuerteventura, Canary Islands. SEM photomicrograph. Scale is 10 μm

4. Thin-section to Fig. 3. Broad micrite cement rim around carbonate and non-carbonate sedimentary grains. A portion of the intergranular voids is totally infilled with micro-grained sedimentary particles and micrite cement. Scale is 500 μm

5. Calcitic blocky cement. Dogtooth calcitic blocky cements are formed in eolianites, preferentially in (meteoric) surface waters. In comparison to aragonite needles, crystals of calcitic blocky cements have radically different length/width ratios (approximately 3 : 1). Pleistocene dunes. Fuerteventura, Canary Islands. SEM photomicrograph. Scale is 10 μm

6. Thin-section to Fig. 5. Dogtooth cement growing on coralline algal bioclasts and on non-carbonate sedimentary grains. Broken crystals ("crystal silt") in the free pores (which appear black in the photo). Scale is 500 μm

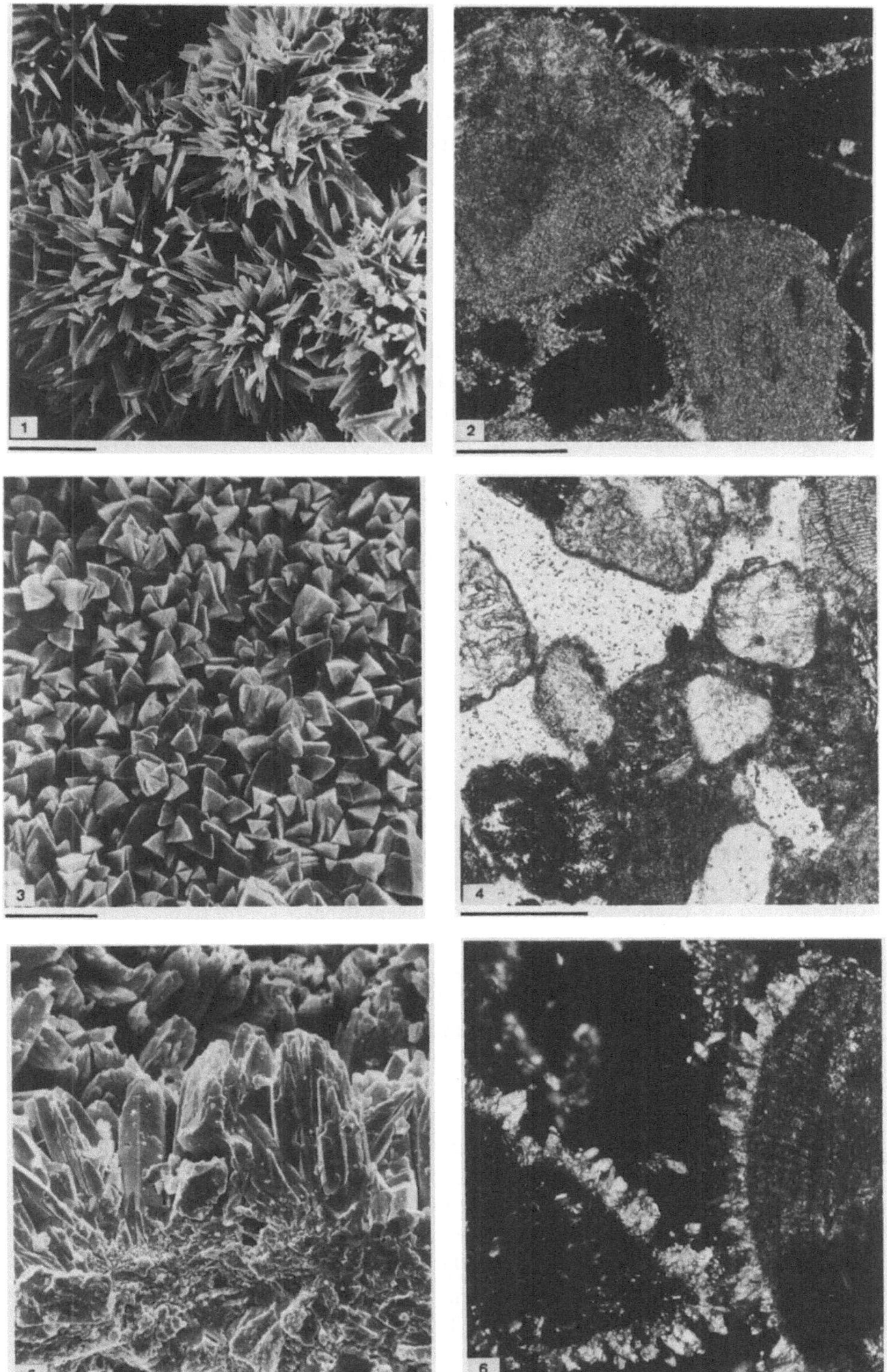
1
2
3
4
5
6

With decreasing water temperature and increasing depth, the magnesium content of calcite cement as well as the carbonate saturation of sea water both decrease, and the course of sea-floor diagenesis in deep-marine carbonates converges with that of fresh-water diagenesis. A cementation by primary low-magnesian calcite or a rapid conversion of high magnesian calcite to calcite on the sea-floor may be possible (Schlager and James, 1978).

3.3.1 Terminology of Cement Types (Figs. 8, 9)

Table 7 lists definitions and modes of genesis as well as synonyms for a few of the numerous carbonate cement types. Here it must be taken into account that some particles in thin-sections are too small to be resolved by binocular or petrographic microscopes; hence SEM studies are necessary (see Plates 3 and 4).

According to Milliman (1974) differences in the kind of cementation appear in intertidal and subtidal environments. In subtidal zones Milliman differentiates between shallow marine types (A – infilling of primary voids and biogenic bores with internal sediment, and later lithification; B – cementation of isolated particles by fibrous and micritic cements, typical for shelf environments) and five deeper water types.

3.3.2 Cementation Phases (Plates 3, 4, 5)

In many limestones two kinds of cements characterized by different fabric types can be seen, because cementation frequently occurs in successive diagenetic phases (e.g., marine cements and phreatic cements) (see Graf and Lamar, 1950):

Cement A (Plate 3/1–4). Fibrous cement, palisade-like overgrowth normal to the particles and growing outward into the pore space; crystals often brownish (submicroscopic contaminants). Recent observations: submarine aragonite or Mg-calcite cements invert to calcite cements poor in Mg, although they retain their isometric subhedral crystal shape and their orientation patterns. In addition to long fibrous crystals of aragonite cement and short fibrous crystals of Mg-calcite cement, cryptocrystalline (micritic) Mg-calcite cements can occur as well (e.g., Purser, 1969). Microstalactite cement A (aragonite and calcite) has been known up to now only in desiccated tidal flats in the Persian Gulf. Occurrence: Cement A is common in shallow marine environments, also appears, however, in beach-rocks and in deeper marine environments with restricted sedimentation (deep ocean ridges, guyots).

Theories concerning the origin of fibrous (or acicular) crystals seen in Ancient limestones consider not only replacement of marine acicular aragonite or high-Mg calcite but also void-filling calcite, vadose cements, or recrystallized micrite. Acicular cement-replacive mosaics may be composed of single crystals of fascicular optic calcite (Kendall, 1977) replacing a single precursing acicular bundle; radiaxial-fibrous crystals (Kendall and Tucker, 1973) in which individual crystals replace parts of adjacent acicular bundles; or radially-divergent ray and composite

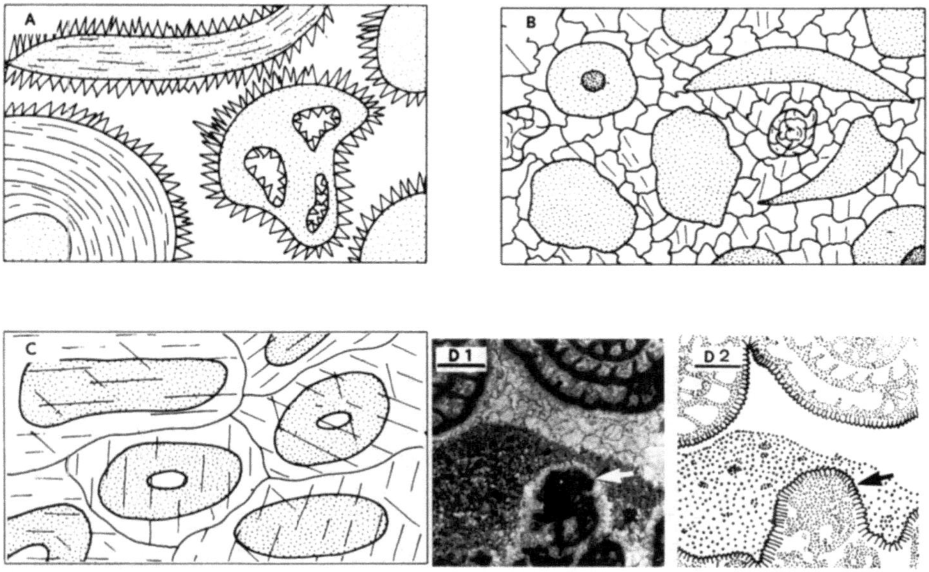

Fig. 8 A–D. Carbonate cement types. **A** Fibrous cements as overgrowth on bioclasts. This cement consists of crystals growing normal to the substrate, like "drusy cement". It corresponds to "cement A" (usually early diagenetic, see Sect. 3.3.2). The remaining intergranular pores are filled with "cement B" (generally late diagenetic). **B** Blocky cement or granular cement. **C** Syntaxial "rim cement" around the remains of echinoderms, in optical continuity with the original single crystals. **D1** and **D2** Vadose crystal silt in interstices bounded by fibrous cements. Legend: *Fine dots* – original micritic sediment; *parallel lines* – fibrous cements; *coarse dots* – vadose silt; *white area* – blocky cement. Permian of Texas. After Dunham (1969) and Blatt et al. (1977)

crystals (Mazzullo, 1980), which consist of straight-edged slender calcite fibers with square tips. Much uncertainty exists with regard to the precursor mineralogies of cements replaced by fibrous calcites. Most authors presuppose a primary marine aragonite or high-Mg calcite mineralogy; on the other hand, primary calcites are replaced by bladed to prismatic calcites in speleothems and Holocene fluvial pisoids as well as in laminated crusts (Braithwaite, 1979; Kendall and Boughton, 1977). It does not seem possible now to relate the replacement fabric to a corresponding precursor mineralogy and habitat. Aragonite cements occur in a variety of morphological habits (crusts with acicular crystals; fan druses or sphaerulites; botryoides; lath; micrite). Furthermore, aragonite and high-Mg calcite may both have similar crystal habits.

Ancient examples: common in reef limestones and calcirudites, e.g., Middle Triassic Wetterstein limestone of the Northern Alps, in fore-reef breccias of Devonian stromatoporoid-coral reefs. Giant radiaxial fibrous carbonate cements with crystal lengths up to several mm are known as "Großoolith" from the Middle Triassic limestones of the Northern Alps (Schmidegg, 1928; Brandner, 1978). These cements are found in solution cavities and have been attributed to phases of emersion (vadose zone).

Table 7. Carbonate cement types

Fabric Type	Author	Diagenetic cement fabrics		
	Important literature	Definition	Genesis	Synonyms
1. Fibrous (Pl. 3/1 – 2; 4/1 – 2; 5/1, 3, 7; 23/1)	Friedman, 1964, 1968; Gervitz and Friedman, 1966; Ancient: Purser, 1969; Zankl, 1971; Germann, 1968; Wolf, 1965; Mountjoy and Walls, 1977	Long axis of crystals ⊥ to boundaries of pore walls or particle surface; serrated crystal boundary in sections ⊥ to the long axis; crystals often brownish	Usually forms early diagenetic cement A (Graf and Lamar, 1950); Submarine: long-fibered aragonite cement; short-fibered Mg-calcite; inverted to calcite. Intertidal: calcite cement. Often in shallow seas, littoral to sublittoral, rarer in deep marine environment; in beach-rocks	Acicular; fibrous spatite; palisade cement; radial fibrous cement; fibrous fringing cement (W. Schneider, 1977)
2. Granular (Pl. 16/2)	Bathurst, 1958: 14	Anhedral to subhedral crystals, generally > 10 to 60 μm, usually no preferred orientation of crystals	Forms cement B; usually formed after lithification of sediment, usually after compaction in submarine environment; usually *after* formation of cement A; only isometric cement similar to cement B in subaerial environments; also in deep marine environments	Originally suggested for fabrics *between* particles, today often used as opposite to "fibrous", or as a general term; sometimes a synonym for drusy; blocky cement
3. Drusy (Pl. 53/3)	Bathurst, 1958: 14; Friedman, 1968	Anhedral to subhedral crystals, usually > 10 μm; crystal sizes increase from pore walls to center of cavities; crossed nicols uniform, extinction not undulose; crystal boundaries planar; more than 50% of crystals form 180° angle (enfacial junction) at the intersection of 3 crystals; often Fe-calcite	Forms (late diagenetic) cement B (Graf and Lamar, 1950). Infills the free space remaining after cementation in submarine environments. Also found in shallow and deep marine environments as well as in subaerial environments. Crystals grow on free surfaces in supersaturated solution. Drusy cement also found in vadose and phreatic zones	Drusy mosaic (Bathurst) originally suggested for cements in primary or secondary voids; drusy mosaic cement (W. Schneider, 1977)

4. Radiaxial fibrous cement (Pl. 5/2; 48/1)	Bathurst, 1959: 511; Bathurst, 1971; Kendall and Tucker, 1973; Bechstädt, 1974; W. Schneider, 1977	Cement crystals with curved twin lamella, convergent optic axes and divergent long axes in subcrystals; overgrowth ⊥ to pore walls, undulose extinction, nonplanar intercrystalline boundaries; inclusions	Controversial: submarine (Kendall and Tucker, 1973) or vadose zone (Bechstädt, 1974)	Open-space structures in reef limestones (e.g., “Großoolithe”, Triassic of the Alps) represent a special development of radiaxial fabrics (Brandner, 1978)
5. Micrite cement (Pl. 4/3 – 4)	Alexandersson, 1969; Ancient: Purser, 1969	Crystals < 4 µm or < 30 µm	Sometimes as cement A (as cryptocrystalline Mg-calcite), e.g., in beach-rocks; in reef cavities. Submarine: in environments with low rate of sedimentation, e.g., guyots	Submicrocrystalline cement (MacIntyre, 1967)
6. Syntaxial rim cement (Pl. 27/1)	Orme and Brown, 1963; Evamy and Shearman, 1965, 1969; Neugebauer and Ruhrmann, 1978; Burgess, 1979	Syntaxial growth of aragonite or calcite on an existing nucleus; well-documented for single crystals of echinoderm fragments, but also known from foraminiferal tests, mollusks, and also corals	Development of large optically continuous crystals on the single crystals of echinoderm fragments or on shell fragments. Indicating a freshwater subsurface or subaerial cementation, or an early marine cementation	Rim cement (Bathurst, 1958: 21); Syntaxial overgrowth; Syntaxial calcite cement (W. Schneider, 1977)
7. Meniscus cement (Pl. 3/5, 6)	Dunham, 1971	In intergranular pores, between but especially under particles	Consists of aragonite or Mg-calcite; originates in the vadose zone by carbonate precipitation from capillary films	
8. Dripstone cement (Pl. 44/4)	Purser, 1969: 224; Bernoulli and Wagner, 1971; Bechstädt, 1975	Thicker crusts of cement A on the roofs of cavities or under particles	After repeated emergence the drops remain hanging preferentially on roofs of cavities – vadose environment. Sporadically also in intertidal or supratidal zone (Purser and Loreau, 1973)	Microstalactitic druse, (Purser, 1969); gravitational cement, (Müller, 1971); Sinter or microsinter, (Sander, 1936)
9. Dog-tooth cement (Pl. 4/5 – 6; 5/4; 10)		Drusy or blocky cement with jagged dog-tooth corners	Occurs in meteoric environment; possibly in subtidal and intertidal zones	Dentate cement, scalenoedric calcite cement (W. Schneider, 1977)

Cement B (Plate 3/1–4). Fills in pore spaces still remaining after a previous cementation. Isometric drusy or blocky mosaics made up of equant anhedral crystals are formed. Often it can be observed that the crystal size increases towards the center of the void. In larger voids, elongated crystals may grow normal to the pore wall ("radial-fibrous cement"). Occurrence: Cement B is found in marine environments (shallow and deeper seas) as well as in meteoric zones. Chiefly calcitic blocky cements with isometric crystals are formed in the presence of fresh water, because there is no cement A in this kind of environment. In contrast to cement A, cement B is often composed of ferroan calcite.

3.3.3 Carbonate Cement Fabrics as Facies Criteria

Understanding of types and textures of cements is essential for the recognition of the major diagenetic environments and for the interpretation of porosity trends in carbonate rocks. Good summaries of diagenetic environments in carbonates have been published by Folk (1973, 1974), Matthews (1974), Bathurst (1976), Friedman and Sanders (1978), Moore (1979), and Longman (1980). Fig. 8 summarizes the characteristics of the different *diagenetic environments of the shallow subsurface* distinguished by Longman:

(1) Marine phreatic zone: refers to warm shallow (< 100 m) seas, with sediments or rocks in which all the pore spaces are filled with normal marine water. The marine phreatic environment may be divided into two zones: (a) Active marine phreatic zone in which water movement, in combination with other processes (e.g., photosynthesis, respiration of organisms, CO_2 degassing, bacteria), results in cementation, and (b) Stagnant marine phreatic zone in which water movement through the sediment is relatively slow and cementation does not occur.

Active marine phreatic zone: Cementation occurs where tides, waves and currents are capable of moving sea water into sediments; this works best at shelf margins where rigid substrates formed by organic binding or framing (e.g., reefs) may be present, and in some nearshore carbonate sand bars. The dominant cements precipitated in this environment are Mg-calcite and aragonite:

Mg-calcite cement forms isopachous crusts on the grains, consisting of micritic or steep-sided rhombic crystals. Marine aragonite cement is typically present as fibrous crystals in a variety of textures, which may be caused by the substrate on which the cementation takes place, the nature of the cavity, and the rate of crystallization. Commonly the crystals nucleate on grains and grow outward to form an isopachous fibrous fringe. Like Mg-calcite, marine aragonite may form in a variety of rock types (e.g., reefs, beach rocks, ooid shoals).

The main criterion of marine cements is their fibrous texture. Distinguishing ancient marine fibrous Mg-calcite from aragonite cement may be possible on the basis of crystal morphology: Orthorhombic aragonite commonly has square-tipped fibers (Assereto and Folk, 1976; Loucks and Folk, 1976). If this shape is preserved after neomorphism to calcite, there is evidence of a former aragonite mineralogy. This criterion has been applied, together with special growth patterns, to cements in Ancient reefs (e.g., Bechstädt, 1975; Mazzullo and Cys, 1977; Mountjoy and Walls, 1977; W. Schneider, 1977).

A Marine phreatic environment B Fresh water vadose environment C Freshwater phreatic environment

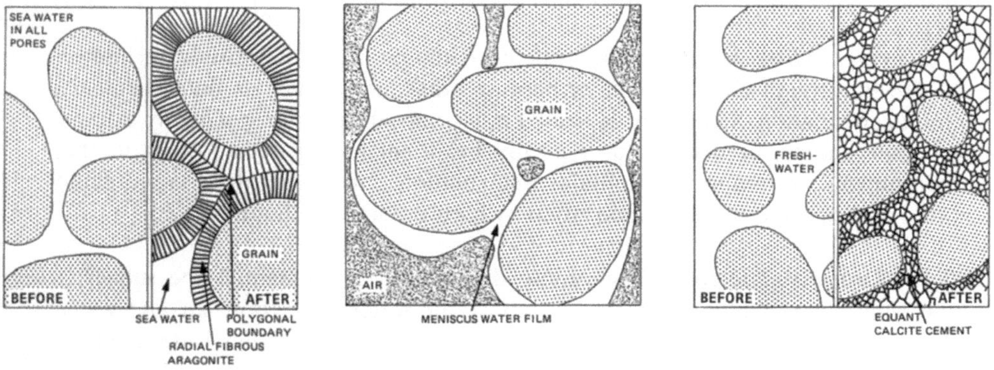

A Marine phreatic environment

Stagnant zone

Processes

1. Little or no water circulation through sediment
2. Bacterial (?) control on cementation
3. Water saturated with $CaCO_3$

Products

1. Little cementation except in skeletal micropores
2. No leaching
3. No alteration of grains
4. Micritization

Active zone

Processes

1. Water forced through sediments by waves, tides, or currents
2. All pores filled with sea water
3. No leaching in shallow marine environments

Products

1. Random aragonite needles
2. Isopachous fibrous aragonite
3. Botryoidal aragonite
4. Micrite Mg-calcite
5. Isopachous fibrous Mg-calcite
6. Mg-calcite pseudo-pellets
7. Polygonal boundaries between isopachous cements
8. Interbedded cements and sediments
9. Borings in cements
10. Most cementation in reefs or surf zones

B Fresh water vadose environment

Zone of solution

Processes

1. Solution by undersaturated meteoric water
2. Production of CO_2 in soil zone aiding solution

Products

1. Extensive solution
2. Preferential removal of aragonite if present
3. Formation of vugs in limestone

Zone of precipitation

Processes

1. Meniscus or pendant distribution of water
2. CO_2 loss of evaporation

Products

1. Minor cementation
2. Meniscus cements
3. Pendant cements
4. Equant calcite
5. Preservation of most porosity

C Freshwater phreatic environment

Zone of solution

Process

1. Solution by undersaturated meteoric water

Products

1. Development of moldic and/or vuggy porosity
2. Possible neomorphism of unstable grains

Stagnant zone

Processes

1. Little or no water movement
2. Water saturated with $CaCO_3$

Products

1. Little cementation
2. Stabilization of Mg-calcite and aragonite
3. Little or no leaching
4. Preservation of porosity
5. Neomorphism of aragonite grains with some preservation of textures

Active zone

Processes

1. Active water circulation
2. Some leaching of aragonite, leaching may be accompanied by calcite replacement
3. Rapid cementation

Products

1. Abundant equant calcite cement
2. Isopachous bladed calcite cement
3. Interlocking crystals
4. Crystals coarsen toward center of pores
5. Complete replacement of aragonite by equant calcite
6. Syntaxial overgrowths on echinoderms
7. Relatively low porosity

Fig. 9. Diagenetic effects in shallow-water carbonates. After Longman (1980)

Other characteristics useful in identifying Ancient marine phreatic cements include marine organisms (e.g., algal crusts) encrusting the cements, and internal sedimentation in the cements, following extensive boring. The syndepositional character of submarine cements may be evident from the existence of multiple cement generations encrusted by algae, the presence of skeletal debris and peloids between various layers of cements, and the coatings of isopachous fringes of short fibrous crystals on broken lithified algal crusts or on lithified reef material (Flügel, 1981).

Stagnant marine phreatic zone: Little cementation occurs because water moves too slowly through the sediments or because other factors such as fluid chemistry inhibit cementation. Sediments deposited in lagoons, deeper parts of gently sloping carbonate ramps, or distal talus typically begin their diagenetic history in this environment. A minor intragranular cementation caused by decaying organic matter may occur as well as micritization by boring algae and fungi.

(2) Freshwater vadose zone: The vadose diagenetic environment is the subaerial zone lying below the land surface and above the water table. Both air and water may be present in the pores. Because the zone is not water saturated, water tends to be held between grains by capillary forces or under grains as pendant drops. The cementation is strongly influenced by frequent wetting and drying. Water in this zone is generally meteoric (rain) water causing solution; production of CO_2 by microbial breakdown of organic material in the soil zone at the top of the vadose zone may also result in solution processes. Once vadose waters reach saturation, they begin precipitating calcite through evaporation or CO_2 degassing.

The vadose zone may be conveniently divided into two parts (Longman, 1980): (a) the soil zone or zone of solution, and (b) the capillary fringe zone or zone of precipitation. The boundary between these zones is gradational and may fluctuate considerably, particularly as a result of rainfall and depending on the climate.

Zone of solution: Any form of calcium carbonate may be dissolved in this zone at the top of the vadose zone. Aragonite is particularly susceptible to leaching and will tend to be removed preferentially along with calcite grains or cement. Solution of limestones may occur, resulting in an extensive karstification. In more moderate climates, solution of homogenous limestones may produce pitted outcrop surfaces and vugs. The only cementation that occurs in this zone is the formation of evaporitic and/or biologic caliche crusts at or near the air/rock interface.

Zone of precipitation: Cementation may occur in the lower part of the vadose zone; the cement morphology tends to reflect the distribution of the pore waters: If pore water is distributed between the grains in a meniscus fashion, the cement will also assume a meniscus texture (Dunham, 1971). If water is present as droplets on the underside of grains, a pendant cement will form (gravitative cement, Müller 1971; dripstone cements). Vadose cements are usually calcites in the form of very fine equant crystals; crystal terminations are rare; the cements are not isopachous. Freshwater vadose cements should not be confused with marine vadose cements which may also exhibit meniscus and pendant textures (marine dripstone cements, Purser and Loreau, 1973; Hanor, 1978). These cements are formed on marine beaches by fibrous aragonite, and micritic or fibrous Mg-calcite.

Meniscus cement may possibly be confused with large submarine syntaxial overgrowths developed on the lower surfaces of echinoderm fragments (Görür, 1979).

Vadose silts (crystal silt; Dunham, 1969) may give additional hints for the recognition of freshwater vadose environments and of subaerial exposure of the sediment. This diagenetic sediment is characterized by the following criteria (see Pl. 5/1, 3, 4, 6; Pl. 23/1):

- Mechanically deposited silt-sized sediment found in solution vugs, interstices, or primary vugs
- Crystal silt differs from ordinary internal sediment in its relationship to cement, in its occurrence in secondary voids, and in its texture
- Internal sedimentation generally postdates the precipitation of fibrous or drusy cements
- These cements show characteristics of internal erosion
- Vadose silt is generally overlain by blocky cements
- Vadose silts are composed of tiny single crystals of calcite (crystal silt) or of well-sorted, poorly defined crumbly, silt-sized aggregates of pellets, (pellet silt, about 50 microns in diameter)
- Recognizable skeletal particles are scarce or absent whithin the crystal silt or the pellet silt

"Vadose silt" is believed to have been formed inside preexisting crumbly carbonate rocks in the vadose zone of an island or coast during early emergence (Dunham, 1969).

Other criteria which may be of use for the recognition of a vadose diagenetic environment are *pisoids* (see Sect. 4.1.3.6). Vadose pisoids are characterized by

- Generally large grains, consisting of very fine concentric lamellae,
- Cores, consisting of broken pisoids,
- Corresponding boundaries of adjacent pisoids,
- Thickened lamellae in the lower part of the pisoids,
- Occurrence of pisoids within laminated calcareous crusts,
- Occurrence of vadose silt between pisoids,
- Strong brecciation of pisolithic limestones,
- No bioclasts,
- No sedimentary structures (e.g., cross-bedding).

(3) Freshwater phreatic zone: This zone lies between the vadose and the mixed marine phreatic-freshwater zones. All pore space is filled with meteoric water containing variable amounts of dissolved carbonate. The top of the phreatic zone is marked by the water table; the lower boundary is gradational with marine waters in areas proximal to the sea. In inland areas, the freshwater phreatic zone grades downward into a zone with subsurface fluids which may contain brines deriving from sediment compaction, clay dehydration, and other processes. The geometry of the freshwater meteoric zone is controlled by topography, rainfall, and the distribution of porosity and permeability in the rocks.

The saturation of water with respect to $CaCO_3$ may increase as water moves downward, therefore three zones of solution may be developed:

Undersaturated zone: If permeability pathways such as fractures, caves or sinkholes are available, meteoric water may pass through the vadose zone without becoming saturated with $CaCO_3$. In the freshwater meteoric zone the meteoric water will dissolve the sediment, producing a vuggy or moldic porosity which is similar to that of the vadose zone.

Active saturated zone: Extensive and rapid cementation is characteristic of the actively circulating part of the freshwater phreatic zone. The calcitic cement is distributed equally around all grains. The first cement may nucleate on crystals on the surfaces of grains and grow outward to produce bladed isopachous calcite rims. Equant calcite is more common than bladed calcite. Coarsening of calcite crystals toward pore centers may be a distinctive characteristic of early cementation in a freshwater phreatic environment (Loucks, 1977). Another characteristic for recognizing freshwater phreatic environments are syntaxial overgrowths on echinoderm fragments; these cements form most rapidly in the freshwater phreatic zone and generally surround the grains.

Examples of Ancient freshwater phreatic cements have been described by Land (1970), Jacka and Brand (1977), Loucks (1977), W. Schneider (1977) and Meyers (1978).

Stagnant freshwater phreatic zone: In the deeper parts of many freshwater phreatic environments and in many arid climates, water movement is very slow and little cementation occurs. Recrystallization of aragonite to calcite may be an important process.

One criterion which may distinguish phreatic from vadose cements is the *iron content of the cements* (estimated by staining with potassium ferricyanide): Calcites containing ferrous iron are probably precipitated in the phreatic zone where interstitial water has a low Eh. Vadose waters are normally oxidizing and most iron is present in the ferric form, which does not substitute for calcium. Changes of temperature, pressure and pH with depth may allow removal of ferrous ions from clay and facilitate the substitution of that iron into the calcite structure (Oldershaw and Scoffin, 1967). Burgess (1979) pointed out that iron-rich cements are also found in algal laminated sediments (due to algae and bacteria which concentrate iron during their life processes) and that the iron content of phreatic waters may be changed by climatic factors and the influence of vegetation.

(4) Mixing zone environment: The boundary between the marine phreatic and the freshwater phreatic environments is marked by brackish water (Runnels, 1969; Land, 1973a, 1973b). Mixing of sea water with freshwater can form a solution slightly undersaturated with calcite and oversaturated with dolomite. Dolomitization may occur in mixing zone environments with good water circulation (Badiozamani, 1973).

Interpreting *diagenetic sequences* is a very complex task because of the large number of diagenetic environments (see also Sect. 3.3) and also because several diagenetic environments generally influence the diagenetic history of a carbonate sediment. Table 8 summarizes the stages of an ideal diagenetic sequence to which a typical shallow-marine carbonate sediment might be exposed (Longman, 1980). The change of cement types across the water table may be used to identify paleo-water tables, especially in Tertiary and Pleistocene limestones (Halley and Harris, 1979; Buchbinder and Friedman, 1980).

Carbon and oxygen isotopes may be used to control the various diagenetic events documented in the different carbonate cement types (e.g., Hudson, 1975; Meyers and James, 1978; Scholle, 1979; Dickson and Coleman, 1980). An interesting example of the combined application of ^{18}O and ^{13}C isotope studies together with the investigation of rare elements and of cement types in thin-sections has been given by Mirsal (1978) and Mirsal and Zankl (1979), who studied the

Table 8. Typical stages of diagenesis in carbonate rocks (after Longman, 1980). Diagenesis of many carbonates starts in the shallow-water marine phreatic zone and ends with the stage 6. An uplift of the limestone into the freshwater vadose zone may result in additional solution by undersaturated meteoric water, producing some secondary porosity (stage 7). Stage 8 represents a new subsurface situation, characterized by a reduction of porosity

Stages	Environment	Characteristics	Porosity
1 Deposition	Marine phreatic, shallow-marine	Deposition of bioclasts (e.g., red algae, foraminifers, mollusks, echinoderms); the sediment contains both Mg-calcite and aragonite	45% ± 10%
2 Micritization and submarine cementation inside grains	Marine phreatic	Micritization of grains by boring microbiota (resulting in micrite envelopes). Minor cementation inside small pores of bioclasts. No alteration of grains, no leaching	40% ± 10%
3 Intergranular marine cementation	Active marine phreatic	Intergranular cementation by fibrous aragonite and Mg-calcite. The cement forms isopachous rims on all grains	20% ± 10%
4 Precipitation of sparry calcite cements	Change from marine phreatic to Freshwater phreatic	All pore space is filled with equant calcite, which coarsens towards pore centers. Bladed calcites may occur as well as syntaxial overgrowth on echinoderm fragments	10% ± 10%
5 Leaching of aragonite and some Mg-calcite to calcite	Freshwater phreatic undersaturated with aragonite	Aragonite and some Mg-calcite become leached, producing moldic porosity. Many Mg-calcite grains may neomorphose to normal calcite grains	20% ± 10%
6 Final recrystallization of unstable grains and micrite to calcite	Freshwater phreatic	Cementation of all moldic porosity by calcite. Grain shapes and textures may be still preserved. Micrite may recrystallize to microspar	5% ± 10%
7 Formation of vuggy porosity and recrystallization of micrite	Freshwater vadose	Vuggy porosity cuts across both grains and cements. Selective leaching of fossils may occur	10% ± 5%
8 Filling of vugs with sparry calcite	Freshwater phreatic	Vuggy porosity may be filled with equant calcite	5% ± 5%

genesis of void-filling cements in Devonian and Upper Triassic reefs. All parameters show a rhythmic fluctuation in successive cement layers. This reflects a similar change in the salinity conditions during cementation. The void-filling processes may have started in their early stages as a semi-closed system. Since the original mineralogy of the cements must have been influenced by these fluctuations in salinity, alternating layers of aragonite and calcite could have been formed. Inversion and grain growth are responsible for the formation of wider fibrous calcite cements within a restricted submarine environment. Blocky calcite represents the last stage of cementation (perhaps a few meters below the reef surface).

As shown by this and other examples the genetic interpretation of cement sequences should be based on a combined approach of thin-section studies (including staining techniques and cathodoluminescence, see Richter and Zinkernagel, 1981), geochemical investigations, and SEM studies which often enable diagenetic fabrics to be better understood (see Plates 3 and 4).

3.4 Sparite and Microsparite Terminology

The following names may be used to characterize carbonate groundmasses which consist of relatively large crystals and originate by pore-filling or by neomorphic inversions.

Plate 5. Ancient Carbonate Cements

1. Submarine and vadose cements. Algal fragments and aggregate grains have been the substrate for thin, isopachous crusts consisting of fibrous crystals (1). This submarine stage is followed by the development of bladed isopachous calcite rims (2) in the freshwater phreatic zone. Diagenesis within the freshwater vadose zone is indicated by vadose silt (3) filling up the remaining voids. Middle Permian: Velebit, Croatia.

2. Different types of sparry calcite crystals on the surface and in the interior of a shell fragment protected by a "micrite envelope". The shell was totally dissolved and replaced by late diagenetic sparite. Relicts of the original radial fibrous cements growing on the surface of the micritic rim are rare because of the strong recrystallization. Crystals with curved twin lamellae indicate the existence of radiaxial cement. Sparite fabric may be described using Folk's code system as follows: $1 = P_S \cdot E_4$; $2 = P \cdot B_5C_W$. Muschelkalk, Middle Triassic: Odenwald, Germany.

3. Cementation in a growth-framework built by sponges and algae. A thin cement crust (2) consisting of short fibrous crystals can be seen directly adjacent to the algal crusts (1). Up to five cement generations, consisting of radiaxial crystals (3), follow. These submarine cements may have been formed during alternating salinity conditions within a semi-closed cavity system. The remaining cavities are infilled with a silty sediment (4, possibly vadose silt), or with white drusy cement (5), indicating freshwater diagenesis. Middle Permian: Rocca de San Benedetto, Sosio valley, Sicily.

4. Freshwater cement and vadose silt. Aggregate grains (1) of platform carbonates act as substrate for dogtooth cement (2) which borders a cavity filled with vadose silt (3). Sulzfluh Limestone, Upper Jurassic: Weißfluh, Graubünden, Switzerland.

5. Recrystallized fibrous cements between bioclasts. Dasyclads (*Diplopora phanerospora* Pia) show micrite envelopes which were overgrown by sessile foraminifera. Cements consist of various generations (1, 2) of submarine, originally fibrous crystals forming isopachous rims; later generations are recrystallized. Bedded Dachstein Limestone, Upper Triassic: Hinterer Gosausee, Gosaukamm, Austria. – This Figure can be used for an exercise: Describe the sparite fabric using Folk's code system (see Sect. 3.4.2); answers in Sect. 12.

6. Vadose silt. Pelsparitic grainstone. Trogkofel Limestone, Lower Permian: Tarvisio, Italy.

7. Interaction of growth of algal crusts and syndepositional cementation. A calcisponge (?) is circumcrusted by spongiostromid algal crusts (1) which act as a substrate for fibrous cement (2) growing in cavities between algal crusts. The growth of multiple cements (3) belongs to a later stage. Middle Permian: Rocca di San Benedetto, Sosio valley, Sicily.

Scale for all Figures 2 mm

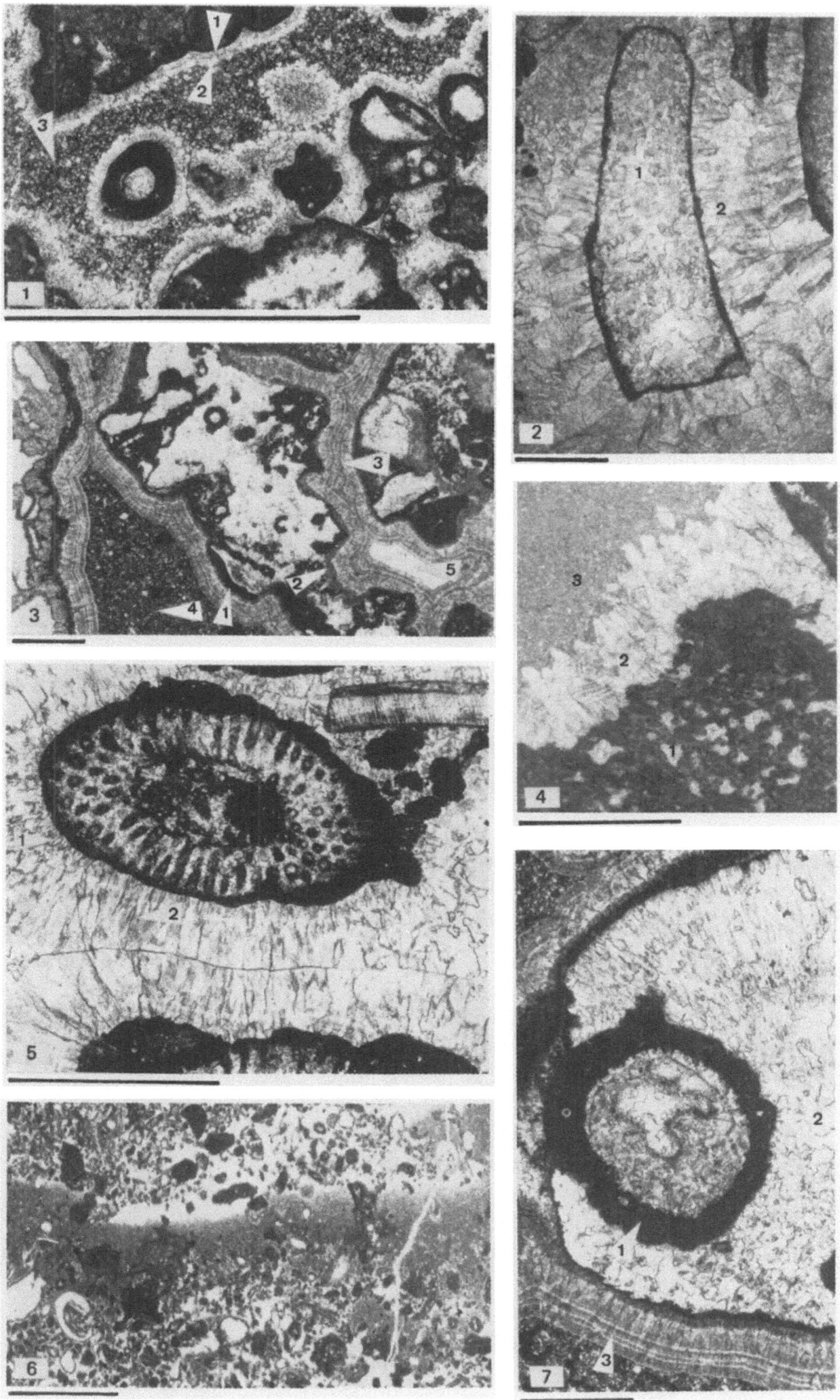
1
2
3
4
5
6
7

A. Sparite (Folk, 1959) (Plates 5/2; 33) characterized by crystal sizes generally larger than 10 microns and by light-colored translucent crystals.

Sparite is differentiated according to its origin:

a) *Orthosparite* (Wolf and Conolly, 1965; = eosparite, Nichols, 1967) originates as a cement in intergranular or intragranular pores
b) *Pseudosparite* (Folk, 1965; = neosparite, Nichols, 1967) formed by neomorphic processes (transformation, recrystallization).

B. Microsparite (Folk, 1959), characterized by crystal sizes generally less than 10 microns and over 4 microns, uniform and homogeneous crystal sizes and crystal shapes, and a relatively large amount of clay or organic substance between the crystals, in comparison to sparite.

Microsparite is often used to describe "recrystallized micrites", sometimes also – wrongly – to characterize inhomogeneous calcareous silty micrites. According to Folk (1974) microsparite originates from the recrystallization of micrite calcite, which is possible only after the removal of Mg^{++} ions (see Longman, 1977). The removal of Mg can be caused by a brackish environment, by freshwater input into carbonate beds, by the adsorption of Mg onto clay minerals, or by rain water and surficial weathering.

A variety of microspar, called "ant-egg spar" by Lindström (1979), consists of single, oblong calcite crystals (15–30 μm long, 1/3–2/3 μm thick) occurring in patches, which may be due to reworked palisade rim cements originally growing upon skeletal grains.

The term *spatite* introduced by Sander (1936) coincides only partially with Folk's definition of sparite. Spatite characterizes the deposition of carbonate minerals as cement growing normal to the surface of void spaces (essentially in the form of fibrous cements).

3.4.1 Differentiation of Orthosparite and Pseudosparite

Orthosparite (cement) is characterized by the following *criteria* (Bathurst, 1976):

1. Sparite between abraded and well-sorted particles which support each other mutually after sedimentation.
2. Two or more generations of sparite.
3. No relict structures as in neomorphic sparites.
4. Particles consisting of micrite (e.g., peloids not inverted into sparite).
5. Micrite envelopes around particles not inverted into sparite.
6. Mechanically deposited micrite unchanged.
7. Well-defined boundary between sparite and particles.
8. The margin of the sparry mosaic corresponds to an originally free surface.
9. Sparite surrounds small, partially filled void spaces.
10. Sparite in the "roof" of a small void space, whose bottom is filled with geopetal internal sediment.
11. Sparite crystals demonstrate an arrangement and shape, such as can be expected in pore-filling or in encrustation (stalactites, tufa).

12. Interfaces between the crystals are planar surfaces.
13. Crystals increase in size away from the overgrowth plane.
14. Optical axes of the crystals are preferentially oriented normal to the overgrowth plane.
15. Longer axes of the crystals are preferentially oriented normal to the overgrowth plane.
16. The sparry mosaic is characterized by a high percentage (up to more than 70%) of "enfacial junctions" (the angle of intersection of three intergranular crystal interfaces).
17. Characteristic neomorphic fabrics are absent.

Pseudosparite (neomorphic sparite) is characterized as follows:

1. Crystal size increases outward away from the pore wall.
2. Crystals generally larger than 6–10 microns, often 50–100 microns.
3. Crystals within the sparry mosaics of conspicuously different sizes.
4. Large tooth-like crystals grow normal to the surface of micritic areas.
5. Sparry particles are intersected by sparite crystals.
6. The majority of interfaces between the crystals are not planar but curved surfaces.
7. Sparry patches within the micrite.
8. Micritic relict structures within the sparry mosaic.
9. Low percentage of "enfacial junctions" (up to 3%).
10. Star-shaped, radial-fibrous structures.
11. Formation of clear chronological sequences or generation interrupted by cement phases.

The differentiation of orthosparite and pseudosparite can hardly ever be carried out to complete satisfaction (see Bathurst, 1976).

3.4.2 Descriptive Code for Orthosparite and Pseudosparite

According to Folk (1965) sparite can be categorized as follows (see also Fig. 10):

A. *Origin*
   a) *Direct precipitation in cavities*
      – Infilling of interparticle pore spaces (Symbol P)
      – Infilling of solution pores, e.g., in a snail shell (Smybol P_s)
   b) *Displacive precipitation:* Particles or micrite are displaced by fibrous calcite generally growing normal to a substrate (Symbol D)
   c) *Neomorphic inversion of Mg-calcite or aragonite* (Symbol N)
      – By inversion of aragonite (Symbol N_i)
      – By recrystallization of calcite (Symbol N_r)

Grain degrading (e.g., in echinoderms) can be designated by the Symbol N_d.

B. *Crystal shape*
   a) Equant, length: width ratio = 1.5 : 1 (E)
   b) Bladed, l : w = 1.5 : 1 – 6 : 1 (B)
   c) Fibrous, l : w = 6 : 1 (F)

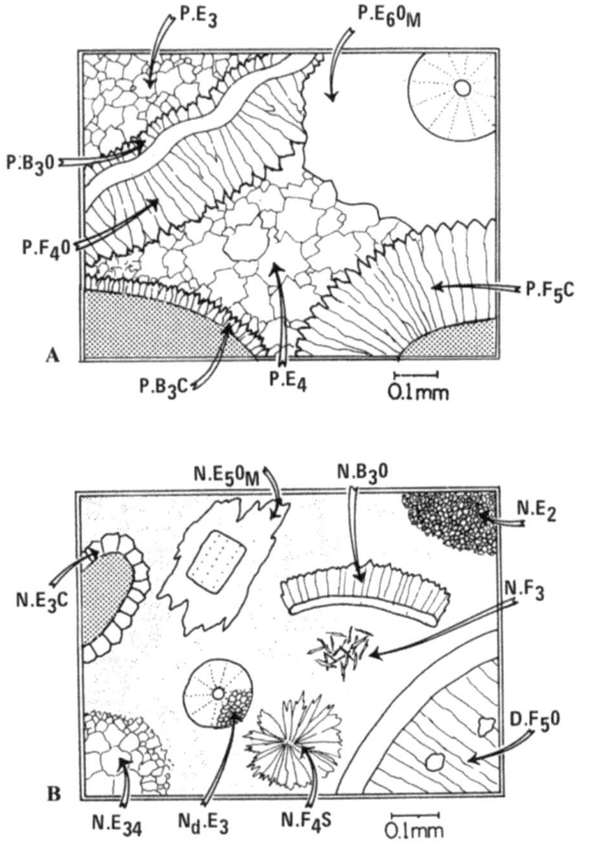

Fig. 10 A, B. Possible descriptions of **A** orthosparites originating by pore infilling and **B** pseudosparites originating by neomorphism. Examples: $P.F_5C = \mathrm{P}$ – pore filling; *F* – fibrous crystal habit; *5* – grain-size class 5; *C* – crusts. $P.E_4 = P$ – pore infilling; *E* – equant crystals; *4* – grain-size between 0.062 and 0.25 mm. $N.B_3O = N$ – neomorphism; *B* – bladed crystal habit; *3* – grain-size between 0.016 and 0.062 mm; *O* – overgrowth; $N.E_5O_M = N$ – neomorphism; E_5 – equant, coarse crystalline crystals; O_M – monocrystalline overgrowth on echinoderm fragments; $N_dE_3 = N_d$ – neomorphism with grain degrading; E_3 – equant, fine crystalline crystals. After Folk (1965)

C. *Crystal size* (Scale after Folk, 1959)

	Symbol
4.0 mm extremely coarsely crystalline	7
very coarsely crystalline	6
1.0	
coarsely crystalline	5
0.25	
medium crystalline	4
0.062	
finely crystalline	3
0.016	
very finely crystalline	2
0.004	
aphano crystalline	1
0.001 mm	

D. *Substrate*

a) *Overgrowth,* in optical continuity with the nucleus

– Perpendicular overgrowth (syntaxial on polycrystalline shells) Symbol O

- Overgrowth on monocrystalline skeletal grains, e.g. echinoderms — Symbol O_m
- Overgrowth with increasingly wider crystals — Symbol O_w

b) *Crusts,* not in optical continuity
- Perpendicular crusts — Symbol C
- Crusts with wider crystals at the periphery — Symbol C_w

c) *Spherulites* – radial aggregates of fibrous or bladed calcite — Symbol S

With the aid of this code as developed by Folk, an exact description of the sparite types is made possible. Examples in Fig. 10 and in Plate 5.

Friedman's (1965 a) precise terminology is useful for describing the *shape* as well as the *texture of sparry fabrics:*

1. *Crystal shape:* anhedral = without crystal planes; subhedral = with some crystal planes; euhedral = with well-developed crystal planes.
2. *Texture:*
   A. Equant = nearly equidimensional crystals
      a) xenotopic = fabric of anhedral crystals
      b) hypidiotopic = fabric of subhedral crystals
      c) idiotopic = fabric of euhedral crystals
   B. Non-equant = different sized crystals
      a) xenotopic
         - porphyrotopic = larger crystals in fine-grained matrix
         - poikilotopic = larger crystals incorporate smaller crystals
      b) hypidiotopic
         - same subgroups as in non-equant xenotopic!
      c) idiotopic

Difficulties may arise in the application of the descriptive terminology (Folk, 1965) used to classify diagenetic fabrics for limestones which have been affected by pressure solution and metamorphic processes. Coarse crystalline calcite in these rocks may represent unaltered precipitates, recrystallized precipitates, recrystallized precipitates enlarged by marginal replacement of the host rock, recrystallized host rock, or recrystallized skeletal grains. Often a multiple genesis occurs. Therefore, the descriptive scheme proposed by Logan and Semeniuk (1976) may be advisable.

3.5 Lithification of Lime Mud

Lithification of lime mud involves a change from solid carbonate phases within pore solutions, to a rock composed of low-Mg-calcite; this change is connected with a drastic reduction of the primary porosity to 2–3%. Assuming an often only slight compaction of carbonate muds, large amounts of carbonate are necessary for an intercrystalline and intergranular cementation, which together with neomorphic processes seems to be responsible for the lithification of oozes (see Bathurst, 1970). In general the following processes may be of importance for the lithification of lime muds:

- Dissolution of tiny crystals and of protruding corners of larger particles
- Transformation of aragonite to calcite
- Recrystallization of high-Mg-calcite
- Pressure-solution at grain contacts
- Syntaxial and epitaxial growth of cement on the surfaces of particles (e.g., nannofossils)

The relative importance of these processes depends on the mineralogy, the size and shape, the sedimentation rate of lime mud particles as well as the amount of clay minerals (which may retard cementation and recrystallization) and organic matter. Water depths and variable solution conditions in deeper marine environments (see Sect. 2.2.4) are other factors strongly modifying lithification of lime muds (see Scholle, 1977).

SEM studies of Ancient lime mudstones indicate that various diagenetic alterations of coccoliths and other nannofossils occur, which have been enlarged by the effect of pressure-solution at grain contacts and also at lithological boundaries (Flügel and Keupp, 1979). Diagenetic alterations of nannofossils include selective dissolution of smaller coccolith plates, prograding crystallization and a lateral coalescence of adjacent marginal coccolith plates, syntaxial growth of coccolith plates, epitaxial and non-epitaxial overgrowth on coccoliths, and grain diminution connected with pressure-solution. These features can be seen in lime mudstones originally deposited in shallow-water environments (e.g., Solnhofen lagoon, Upper Jurassic of Germany) as well as in bathyal environments (e.g., Oberalm basin, Upper Jurassic of the Northern Alps, Austria).

3.6 Compaction

Compaction refers to any process that decreases the bulk volume of rocks. This includes mechanical processes that decrease the bulk volume of single grains (grain deformation) or that cause closer packing of grains (re-orientation) and pressure solution which decreases the volumes of grains and of cement minerals. Research on compaction uses experimental approaches (e.g., Bhattacharyya and Friedman, 1979; Hugman and Friedman, 1979; Terzhagi, 1949; Hathaway and Robertson, 1961; Robertson, 1967; Fruth et al., 1966; Ebhardt, 1968), mathematical approaches (e.g., Rittenhouse, 1971; Coogan and Manus, 1975; Manus and Coogan, 1974), theoretical approaches (Bayer, 1978; DeBoer, 1977; Deelman 1975; Roll, 1974; Weyl, 1959), or observational approaches (e.g., Shinn et al., 1976; Coogan, 1970; Kahle, 1966; Meyers, 1980).

Many carbonate sediments undergo relatively little compaction or are only affected by compaction in a deep burial stage. Differences seem to exist in the possibility for a compaction of lime muds and of carbonate sands. Experiments indicate that carbonate sands need an overburden of up to 1000 m in order to become compacted whereas exponentially increasing pressures are necessary for the compaction of lime muds. The degree of compaction is strongly influenced by the cause of early cementation which may inhibit or diminish compaction.

Dolomitization may be another factor inhibiting compaction (see Coogan and Manus, 1975).

Estimating the degree of compaction of calcilutites or calcarenites is rather difficult because of the high variety of key variables which may influence the behaviour of the sediments. The compaction of calcarenites (carbonate sands) may be affected by the following factors:

- Amount of overburden
- Duration of burial stress
- Pore pressure
- Pore-water chemistry (the greater the degree of undersaturation with respect to $CaCO_3$, the less the amount of overburden needed to cause inter- and intragranular pressure solution)
- Clay content of the carbonates
- Pre-compaction diagenesis (early cementation and dolomitization)
- Grain size
- Grain architecture
- Grain mineralogy

Common criteria for the recognition of compaction in grainstones and packstones are:

- Breakage of grains
- Plastic bending and collapse
- Fragmentation of intact valves or single shells (e.g., ostracods, brachiopods)
- Collapsing and telescoping of shells
- Mashing of fragile skeletal grains between compaction-resistant grains (e.g., bryozoan fragments between echinoderms)
- Grain rotations (seen by rotated geopetal fabrics)
- Drag fabrics (due to re-orientation and stylolitization) with elongated grains showing alignment parallel or subparallel to the compaction-resistant grains
- Intergranular pressure solution at grain contacts (circumidenic planar micro-stylolites)
- Truncation of grains by adjacent grains
- Interpenetration of grains
- Densely packed grains (overpacking) forming a "mash" structure with almost no intergranular space

Many of these effects are also observed in barely compacted carbonate sands (e.g., breakage of grains, overclose packing). On the other hand the lack of deformed particles can not be used as good evidence of only weak compaction because fragile shells may resist high pressures if embedded within a plastic matrix (see Pray, 1960). Compression tests of prepared mixtures of varying proportions of ooids and lime mud show that many calcareous ooids remain undeformed; a linear relationship exists between an increasing proportion of lime mud and the proportion of undeformed ooids (Bhattacharyya and Friedman, 1979). Hence, ooids and shell fragments remain unbroken in micritic limestones, even though the original carbonate may have undergone significant compaction. This must be considered when the degree of compaction of limestones is estimated.

A few attempts have been made to quantify compaction in carbonate grainstones (Kahle, 1966; Coogan, 1970). The two principal measurable effects of the compaction of carbonate sands are (1) increase in grain volume percentage as compared to the original uncompacted sediment, and (2) increase in closeness of the grains as reflected in the increased number of grains which touch each other in a compacted sand. Increase in grain volume percentage can be directly measured in thin-sections of grainstones, using packing density indices as proposed by Kahn (1956) or Coogan (1970). Coogan's index is expressed by

$$\bar{x}\ PD\% = (gi/TL) \times 100$$

where $\bar{x}\ PD$ is the arithmetic mean of the separate measurements made on all traverses of the thin-section and TL is the total of all traverses; *gi* is the length of the grain intercept of the *i*th grain. Comparison of compacted oolitic grainstones with uncompacted oolitic grainstones shows that with increasing compaction the mean of the Packing Density shifts, the standard deviation is reduced and a strong kurtosis may develop.

The second feature, an increased number of grains in contact, is represented by the Packing Index proposed by Mason (1951) and used with modifications by Kahle (1966) and Coogan (1970) in measuring the compaction of oolitic and pelletoidal limestones. The rationale behind the use of the Packing Index is the idea that lime grainstones which are compacted should have more grains touching each other than those which are uncompacted. The Packing Index is more quickly determined than the Packing Density. It is less useful in an evaluation of rock compaction, however, because the Packing Index is affected more by the total number of grains counted, the shape of grains, their sorting, and the effects of cementation which may change grain contact configurations. Kahn's Packing Proximity Index (corresponding to Coogan's Packing Index) is expressed by

$$\bar{x}\ PI\% = GC/N \times 100,$$

Plate 6. Stylolites in Limestones

1. High columnar stylolites within a peloidal bioclastic wackestone. SEM photograph. Stylolite interfaces are partly covered with reactate calcite. Scale 1 mm

2. Irregularly anastomosing stylolite sets composed of low-amplitude and high-amplitude peaked stylolites. Bioclastic wackestone. Scale 2 mm

3. Anastomosing sets of high-amplitude peaked stylolites producing a fitted fabric (see Logan and Semeniuk, 1976). Bioclastic wackestone. Scale 2 mm

4. Columnar stylolites causing a stylobreccioid structure. Stylocumulate (material that has crystallized at stylolite interfaces under pressure solution conditions) consists of clay minerals and pyrite. Bioclastic wackestone with pelecypod shells. Scale 2 mm

5. Stylolitized foraminiferal wackestone with orbitolinid foraminifera (*Palaeodictyoconus* sp.), strongly affected by a post-calcite vein stylolitization. Scale 2 mm

6. High-amplitude peaked stylolites with residual clay films. Fine-grained bioclastic wackestone. Scale 2 mm

All samples from the Cretaceous of the Middle East

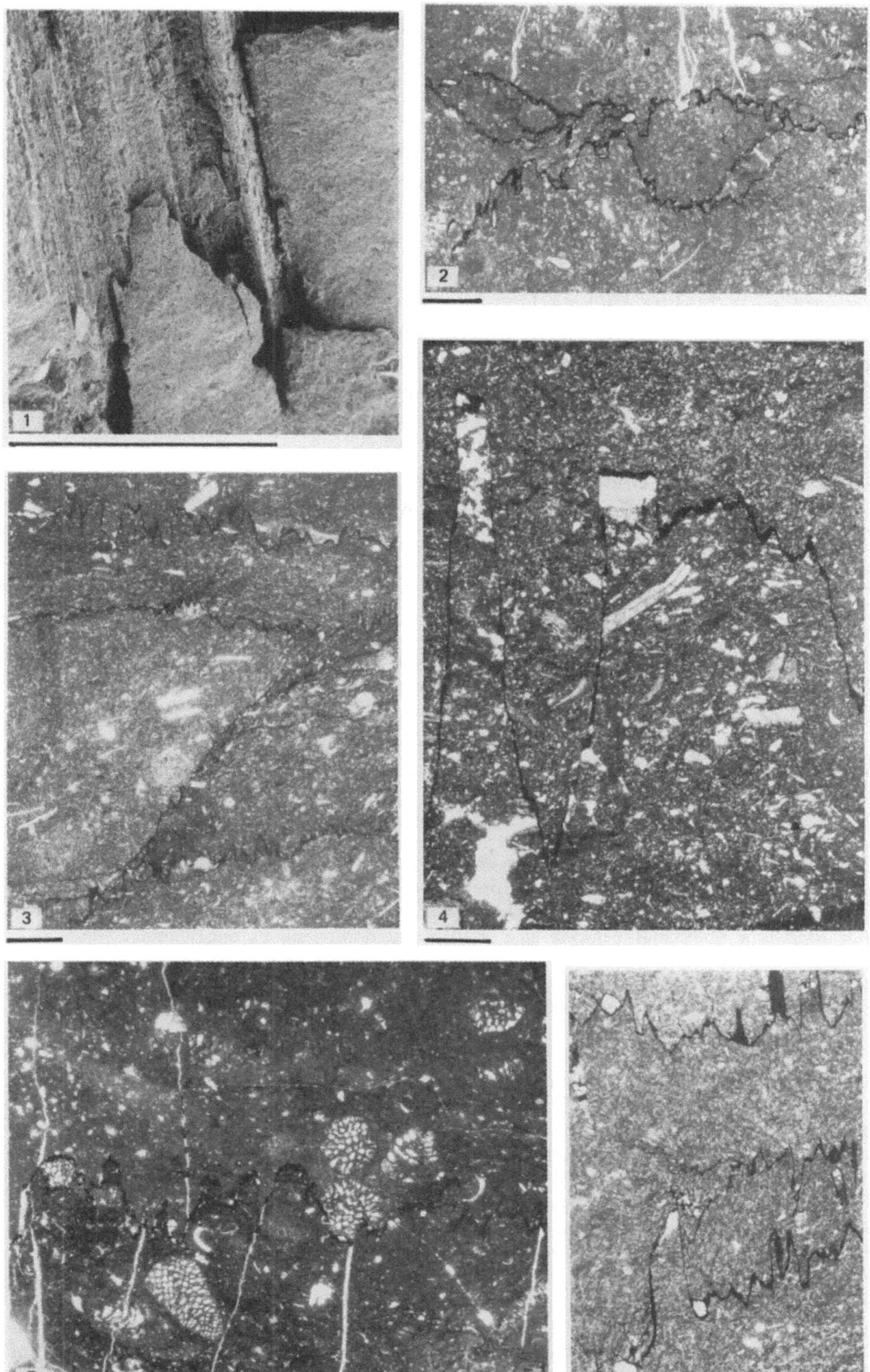
1
2
3
4
5
6

where *GC* is the number of grain-to-grain contacts along the line of traverse, and *N* is the number of grains along the same traverse. Values for repeated traverses are summed to derive the mean Packing Index.

A *Compaction Index* based on Packing Density was formulated by Coogan (1970) for oolitic grainstones. An increase in Packing Density reflects the movement of grains closer together, presumably as the result of increased overburden pressure. Thus, any increase in Packing Density may be thought of as an increase in the amount of compaction from some Packing Density value. This value may range from zero compaction for a particular carbonate sand to 100% Packing Density for a rock composed entirely of grains, i.e. one that is 100% compacted. Based on the concept of increasing Packing Density, the Compaction Index arbitrarily relates Packing Density or grain volume percentage in thin-section to the amount of compaction (Coogan and Manus, 1975). The diagram developed by Coogan and Manus for different kinds of carbonate sands (oolitic sands, grapestones, mixed skeletal sands, pelecypod sands, foraminiferal sands) is recommended in order to estimate the degree of compaction, but it must be refined with respect to the effect of grain shapes, pre-compaction solution effects, and a critical evaluation of the initial zero compaction index values as used by the authors.

3.7 Stylolites (Plate 6)

Experiments on the dry compaction of mixtures of mineral fragments with different hardness result in grain contacts identical to those generally attributed to the effect of stylolitization (Deelman, 1975).

Stylolites are thin zones of discontinuity within rocks. In thin-sections they have undulate to zig-zag sutures; in general they consist of conical to columnar

Fig. 11 A–D. Classification of stylolites in carbonate rocks. **A** – Configuration of pressure-solution surfaces; **B** – Classification of pressure-solution (stylolite) sets; **C** – Classification of pressure-solution fabrics; **D** – Classification of pressure-solution structures. *Iden-support:* Idens (an Iden is a body that behaves as a statistically homogeneous entity under physical or chemical conditions) form a self-supporting frame and are in contact at points. *Condensed fabric:* Idens are overclosely packed; grains are no longer in point contact but contact over a surface. *Fitted fabric:* Idens are in contact with neighbors all along their margins. *Stylocumulate/reactate support:* Larger idens are supported by finer-grained stylocumulate (material that passively accumulated as insoluble residue along a pressure-solution surface; composed of quartz, mica, dolomite and Fe-oxides, or of relatively more insoluble calcitic idens such as crinoid and brachiopod fragments). *Stylobedding:* A structure due to stylolitic parting. Stylobedding may be parallel to the original layering, but may also develop at various angles to the original sedimentary layering. *Stylolaminite:* Rock laminated due to the abundance of low-amplitude stylolites (e.g., quartzose stylolaminite, characterized by quartzose insoluble residue in the stylolite seams; calcareous or dolomitic stylolaminite). *Stylonodular limestone:* Lenses and nodules of limestone idens are separated by sheets and lenses of stylolaminite. *Stylomottled rock:* Carbonate rocks with irregularly mottled stylocumulate and reactate (material which has grown at pressure-solution interfaces under PT conditions of pressure solution; e.g., calcite, dolomite, or quartz); these develop where pressure solution was restricted to patches. *Stylobreccia:* Breccia in which fragments are bound by stylolites caused by fracturing accompanied by pressure solution. After Logan and Semeniuk (1976) projections with intervening depressions. These structures vary in size from microscopic sutured contacts (microstylolites) up to stylolites several meters in length. Aggregate stylolites with amplitudes greater than the grain diameter of the rock and intergranular stylolites with amplitudes smaller than the grain size of the host rock may be distinguished (Park and Schott, 1968). A detailed classification of intergranular pressure solution contacts has been published by Trurnit (1968).

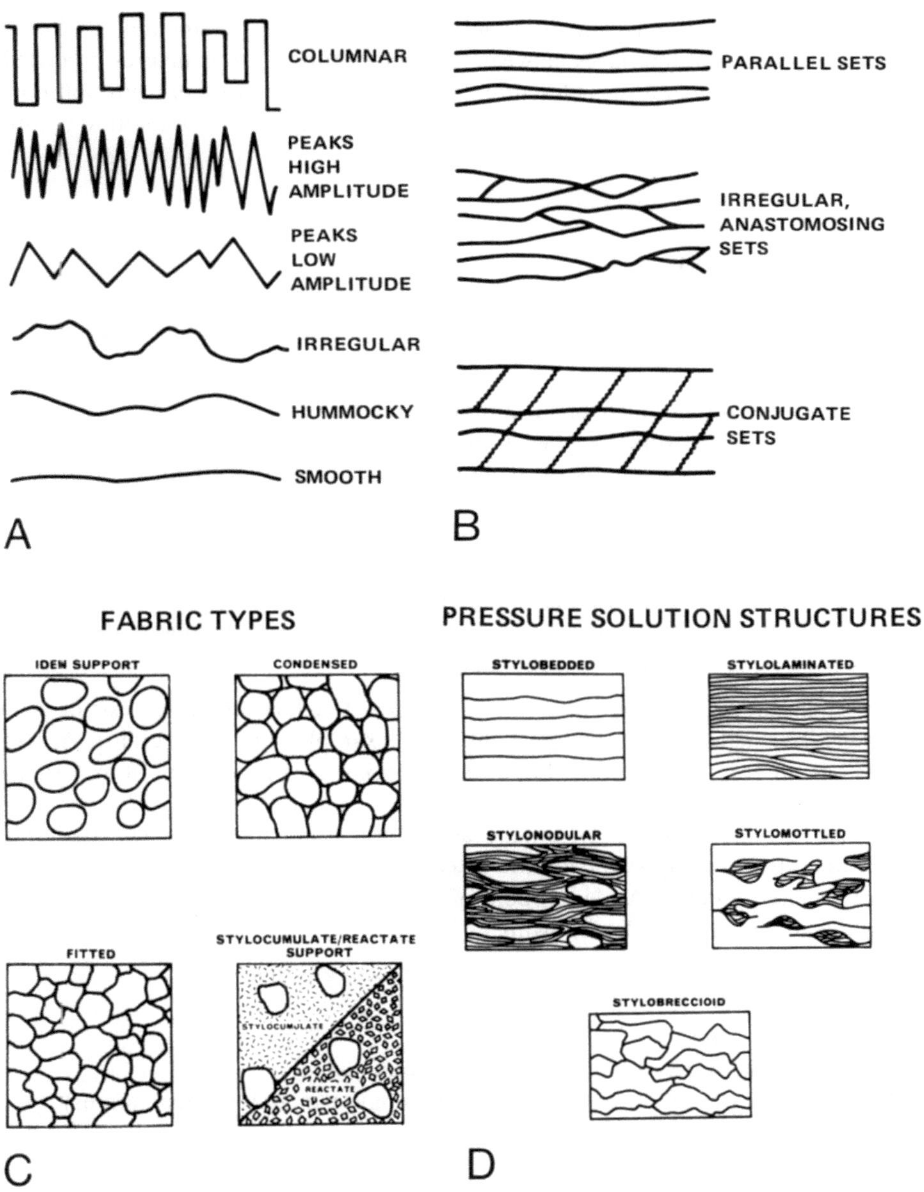

Stylolites are generally regarded as the result of pressure solution. This involves solution around points of contact between mineral grains in response to pressure (usually the weight of the overburden). Since certain minerals are more susceptible to pressure solution than others, various burial depths are necessary in order to produce a strong stylolitization. In limestones stylolitization may start just after the deposition of the grains, caused by the contact between adjacent grains (see Plate 6). Pressure solution is an important process for supplying carbonate for the formation of carbonate cements.

Pressure solution and stylolitization may change the original depositional textures and produce new diagenetic fabrics. In order to avoid confusion, it seems advisable to use an independent terminology (Logan and Semeniuk, 1976) to describe the types of interfaces between adjacent particles, the structures produced by pressure solution and shear fracture, the fabrics developed by pressure solution, and the resulting rock types. Fig. 11 and Plate 6 show the possibilities of this approach.

3.8 Calcite Veins

Many limestones display various systems of calcite filled veins and veinlets, which may be due to several processes:

- Crumbly fractures, which break around solid particles. These fractures form a part of a pore network; they were produced before or after compaction of the sediment. Crumbly fractures are generally transected by sharp-edged fractures.
- Sharp-edged fractures, which break through the particles. These fractures are produced during late diagenesis; their genesis is probably related to tectonic movements since they are often arranged in parallel sets.
- Short, thin, rather sharp-edged fractures without parallel sets, which rise vertically from particles. The origin of these fractures may be due to desiccation or dehydration (synaeresis) of carbonate mud.

Tectonically induced fractures can be important factors for the genesis of good permeability in carbonate rocks. Wellknown examples are the Tertiary Asmari limestone in Iran (McQuilcan, 1973), or Devonian reef limestones from Canada (Drummond, 1964).

Since ground water may obtain access to carbonate rocks via fractures, microfractures are important for the intensity of diagenetic processes within the phreatic environment. A description of microfractures and calcite-filled veins should take into account the following:

- Existence of simple (isolated) or several (transecting) veins
- Abundance (density) of the veins
- Width of the veins
- Position of the veins with respect to the bedding plane
- Mode and degree of the vein-fillings (mineralogy, cement type, generations)
- Association with stylolites
- Differences in the occurrence of veins in various beds

Experimentally deformed carbonate rocks indicate that the mechanical behavior and the origin of fracture is controlled by the amount of sparite, micrite and dolomite (Hugman and Friedman, 1979). Limestones with more than 50% micrite are relatively strong as compared to sparry limestones. In pure dolomites strength and ductility depend on the texture. Crystal size seems to be the most important factor affecting ultimate strength. The density of precursive microfractures increases with increasing confining pressure (or deeper burial). Very micritic rocks exhibit far fewer microfractures than crystalline carbonate rocks. Most microfractures in non-micritic particles of carbonate rocks are intragranular.

3.9 References: Carbonate Diagenesis

Adelseck, C. G. (1978): Dissolution of deep-sea carbonate: preliminary calibration of preservational and morphologic aspects. Deep-Sea Res. **25,** 1167–1185, Oxford

Adelseck, C. G. Jr., Geehan, G. W., Roth, P. H. (1973): Experimental evidence for the selective dissolution and growth of calcareous nannofossils during diagenesis. Geol. Soc. Amer. Bull. **84,** 2755–2762, New York

Alexandersson, T. (1972): Intragranular growth of marine aragonite and Mg-calcite: evidence of precipitation from supersaturated sea water. J. Sed. Petrol. **42,** 441–460, Tulsa

Alexandersson, T. (1973): Mediterranean beachrock cementation: Marine precipitation of Mg-Calcite. In: Stanley, D. J.: The Mediterranean Sea. A Natural Sedimentation Laboratory, Part. 5. Coastal and Shallow Water Sedimentation: Carbonate Sediments. 203–223, 15 Figs.

Alexandersson, T. (1974): Carbonate cementation in coralline algal nodules in the Skagerrak, North Sea: biochemical precipitation in undersaturated waters. J. Sed. Petrol. **44,** 7–26, Tulsa

Alexandersson, T. (1975): Etch patterns on calcareous sediment grains: petrographic evidence of marine dissolution of carbonate minerals. Science **189,** 47–48, Washington

Alexandersson, T. (1978): Petrographic saturometry in marine carbonate sediments. In: Johari, O. (ed.): Scanning Electron Microscopy, part 1, 503–511, SEM inc., AMF O'Hare

Andel, T. H. van (1975): Mesozoic/Cenozoic Calcite compensation depth and the global distribution of calcareous sediments. Earth and Planetary Sci. Letters **26**/2, 187–194

Assereto, R., Folk, R. L. (1976): Brick-like texture and radial rays in Triassic pisolites of Lombardy, Italy: a clue to distinguish ancient aragonitic pisolites. Sed. Geol. **16,** 205–222, Amsterdam

Assereto, R. L., Rizzini, A. (1975): Reworked ferroan dolomite grains in the Triassic "Oolite a Gasteropodi" of Camoniche Alps (Italy) as indicators of early diagenesis. N. Jb. Geol. Paläont. Abh. **148**/2, 215–232, 13 Figs., Stuttgart

Athy, L. F. (1930): Density, porosity and compaction of sedimentary rocks. Bull. Amer. Ass. Petrol. Geol. **14,** 1–24, Tulsa

Ayer, N. J. (1971): Statistical and Petrographic Comparison of Artificially and Naturally Compacted Carbonate Sediments. Thesis Univ. Illinois, Urbana, 92 p.

Badiozamani, K. (1973): The Dorag dolomitization model – application to the Middle Ordovician of Wisconsin. J. Sed. Petrol. **43,** 965–984, Tulsa

Badiozamani, K., MacKenzie, F. T., Thorstensen, D. C. (1977): Experimental carbonate cementation: salinity, temperature and vadose-phreatic effects. J. Sed. Petrol. **47,** 529–542, Tulsa

Baldwin, B. (1971): Ways of deciphering compacted sediments. J. Sed. Petrol. **41,** 293–301, Tulsa

Banner, F. T., Wood, G. V. (1964): Recrystallization in microfossiliferous limestones. Geol. J. **4**/1, 21–34, 6 Pls., London

Bathurst, R. G. C. (1958): Diagenetic fabrics in some British Dinantian limestones. Liverpool Manchester Geol. J. **2**/1, 11–36, Liverpool

Bathurst, R. G. C. (1959): The cavernous structure of some Mississippian Stromatactis reefs in Lancashire, England. J. Geology **67,** 506–521, Chicago
Bathurst, R. G. C. (1959): Diagenesis in Mississippian calcites and pseudobreccias. J. Sed. Petrol. **29,** 365–376, 6 Figs., Tulsa
Bathurst, R. G. C. (1970): Problems of lithification in carbonate rocks. Proc. Geol. Ass. **81**/3, 429–440, Leeds
Bathurst, R. G. C. (1971): see 1.4.1
Bathurst, R. G. C. (1976): Carbonate sediments and their diagenesis (2nd enlarged ed). 658 p., New York: Elsevier
Bausch, W. M. (1968): Clay content and calcite crystal size of limestones. Sedimentology **10,** 71–75, 2 Figs., Amsterdam
Bayer, U. (1978): Modelle instabiler Kompaktionsverläufe unter Sedimentation. Geol. Rundschau, **67**/3, 980–990, Stuttgart
Beales, F. W. (1965): Diagenesis in pelleted limestones. Soc. Econ. Paleont. Min. Spec. Publ. **13,** 49–70, 2 Pls., 5 Figs., Tulsa
Bechstädt, T. (1974): see 4.2.4.5
Bechstädt, T. (1975): Zyklische Sedimentation im erzführenden Wettersteinkalk von Bleiberg-Kreuth (Kärnten, Österreich). N. Jb. Geol. Paläont. Abh. **149**/1, 73–95, 8 Figs., Stuttgart
Benson, L. V., Achauer, C. W., Matthews, R. K. (1972): Electron microprobe analysis of magnesium and iron distribution in carbonate cements and recrystallized sediment grains from ancient carbonate rocks. J. Sed. Petrol. **42**/4, 803–811, 9 Figs., Tulsa
Berger, W. H. (1970): see p. 51
Berger, W. H. (1973): Deep-sea carbonate: evidence for a coccolith lysocline. Deep-Sea Res. **10,** 917–921
Berner, R. A. (1966): Chemical diagenesis of some modern carbonate sediments. Amer. J. Sci. **264,** 1–36, 4 Figs., Washington
Bernoulli, D., Wagner, C. W. (1971): Subaerial diagenesis and fossil caliche deposits in the Calcare Massicio Formation (Lower Jurassic, Central Apennines, Italy). N. Jb., Geol. Paläont. Abh. **138**/2, 135–149, Stuttgart
Bhattachyaryya, A., Friedman, G. M. (1979): Experimental compaction of ooids and lime mud and its implication for lithification during burial. J. Sed. Petrol. **49**/4, 1279–1286, 2 Figs., Tulsa
Binkley, K. L., Wilkinson, B. H., Owen, R. M. (1980): Vadose Beachrock Cementation along a Southwestern Michigan Marl Lake. J. Sed. Petrol. **50**/3, 953–962, 5 Figs., Tulsa
Blatt, H., et al. (1972): see 1.4.1
Bosellini, A., Rossi, D. (1974): Triassic carbonate buildups of the Dolomites, Northern Italy. Soc. Econ. Paleont. Min., Spec. Publ. **18,** 209–233, Tulsa
Braithwaite, C. J. R. (1979): Crystal textures of Recent fluvial pisolites and laminated crystalline crusts in Dyfed, South Wales. J. Sed. Petrol. **49,** 181–194, Tulsa
Brandner, R. (1978): Tektonisch kontrollierter Sedimentationsablauf im Ladin und Unterkarn der westlichen nördlichen Kalkalpen. Geol. Paläont. Mitteilungen Innsbruck **8,** 317–354, 22 Figs., Innsbruck
Bricker, O. P. (ed.) (1971): Carbonate Cements. John Hopkins Univ.-Stud. Geol. **19,** 376 pp., Baltimore–London: John Hopkins
Broecker, W. S., Takahashi, T. (1978): The relationship between lysocline depth and in situ carbonate ion concentration. Deep-Sea Res. **25,** 56–65, Oxford
Buchbinder, L. G., Friedman, G. M. (1980): Vadose, phreatic and marine diagenesis of Pleistocene-Holocene carbonates in a borehole: Mediterranean coast of Israel. J. Sed. Petrol. **50**/2, 395–407, 5 Figs., Tulsa
Burgess, C. J. (1979): The development of a Lower Jurassic carbonate tidal flat, Central High Atlas, Morocco, 2: Diagenetic history. J. Sed. Petrol. **49**/2, 413–428, 6 Figs., Tulsa
Bush, P. (1973): Some aspects of the diagenetic history of the Sabkha in Abu Dhabi, Persian Gulf. In: Purser, B. H.: The Persian Gulf, 395-407, 11 Figs., Berlin–Heidelberg–New York: Springer
Buxton, T. M., Sibley, D. F. (1981): Pressure solution features in a shallow buried limestone, J. Sed. Petrol. **51**/1, 19–26, 4 Figs., Tulsa

Chanda, S. K. (1967): Selective neomorphism and fabric discontinuities in limestones. J. Sed. Petrol. **37**/2, 688–690, 2 Figs., Tulsa

Chilingar, G. V., et al. (1967): see 1.4.1

Coogan, A. H. (1970): Measurements of compaction in oolithic grainstone. J. Sed. Petrol. **40**/3, 921–929, 3 Figs., Tulsa

Coogan, A. H., Manus, R. W. (1975): Compaction and diagenesis of carbonate sands. In: Chilingarian, G. V. and Wolf, K. A. (eds.): Compaction of coarse-grained sediments, vol. 1, 79–166, 43 Figs., Amsterdam: Elsevier

Cullis, C. G. (1904): The mineralogical changes observed in the cores of the Funafuti borings. In: Bonney, T. G. (ed.): The atoll of Funafuti. Roy. Soc. London, 392–420, London

Davies, P. J., Kinsey, D. W. (1973): Organic and inorganic factors in recent beach rock formation, Heron-Island, Great Barrier Reef. J. Sed. Petrol. **43**/1, 59–81, 9 Figs, Tulsa

DeBoer, R. B. (1977): On the thermodynamics of pressure-solution: interaction between chemical and mechanical forces. Geochim. Cosmochim. Acta **41**, 249–256, Oxford

Deelman, J. C. (1974a): Granulo-mechanical aspects of lithification, I. Neues Jb. Geol. Paläont. Abhandlungen **147**/1, 1–29, 15 Figs., Stuttgart

Deelman, J. C. (1974b): Granulo-mechanical aspects of lithification, II. Neues Jb. Geol. Paläont. Abhandlungen **147**/3, 237–268, 10 Figs., Stuttgart

Deelman, J. C. (1975): "Pressure solution" re-examined. Neues Jahrbuch Geol. Paläont. Abhandlungen **149**/3, 384–397, 2 Figs., 1 Table, Stuttgart

Deelman, J. C. (1976): Lithification analysis. Experimental observations. Geol. Rdsch. **64**/3, 1055–1078, 12 Figs., Stuttgart

Dickson, J. A. D., Coleman, M. L. (1980): Changes in carbon and oxygen isotope composition during limestone diagenesis. Sed. **27**, 107–118, Oxford

Drummond, J. M. (1964): An appraisal of fracture porosity. Bull. Canadian Petrol. Geol. **12**, 226–245, Calgary

Dunham, R. J. (1969): Early vadose silt in Townsend mound (reef), New Mexico. Soc. Econ. Paleont. Min. Spec, Publ. **14**, 139–182, 22 Figs., Tulsa

Dunham, R. J. (1969b): Vadose pisolite in the Capitan reef (Permian), New Mexico and Texas. Soc. Econ, Paleont. Min. Spec. Publ. **14**, 182–190, 22 Figs., Tulsa

Dunham, R. J. (1971): Meniscus Cement. In: Bricker, O. P. (ed.): Carbonate Cements, 297–302, 136 Figs., Baltimore–London: Harper

Dunnington, H. V. (1967): Aspects of diagenesis and shape change in stylolitic limestone reservoirs. 7th World Petrol. Congr., Mexico City, Proc. **2**, 339–352, Mexico City

Ebhardt, G. (1968): Experimentelle Untersuchungen zur Kompaktion karbonatischer Sedimente. Diss. Univ. Würzburg, 99 + VIII pp., 3 Pls., 9 Tab., 27 Figs., Würzburg

Evamy, B. D. (1969): The precipitational environment and correlation of some calcite cements deduced from artificial staining. J. Sed. Petrol. **39**/2, 787–821, 20 Figs., Tulsa

Evamy, B. D., Shearman, D. J. (1969): Early stages in development of overgrowth on echinoderm fragments in limestones. Sedimentology **12**, 3/4, 317–322, 1 Pl., 2 Figs., Amsterdam

Fairbridge, W. R. (1966): Diagenetic phases. Bull. Amer. Ass. Petrol. Geol. **50**, 3/1–2, 612–613, Tulsa

Fenninger, A. (1968): Das Kalzitgefüge der sparitischen Kalke des Plassen (Tithonium, nördliche Kalkalpen, Oberösterreich). Sedimentology **10**, 273–291, 9 Figs., 3 Tab., Amsterdam

Fischer, A. G., Garrison, R. E. (1967): Carbonate lithification on the sea floor, J. Geol. **75**/4, 488–496, 4 Pls., 1 Fig., Chicago

Fisher, W. L., Rodda, P. V. (1969): Edwards Formation (Lower Cretaceous), Texas: Dolomitization in a carbonate platform system. Amer. Ass. Petrol. Geol. Bull. **53**/1, 55–72, 14 Figs., 1 Tab., Tulsa

Flügel, E. (1981): Lower Permian Tubiphytes/Archaeolithoporella buildups in the Southern Alps (Austria and Italy). Soc. Econ. Paleont. Min. Spec, Publ. **30**, 143–160, 9 Figs., Tulsa

Flügel, E., Keupp, H. (1979): Coccolithen-Diagenese in Malm-Kalken (Solnhofen/Frankenalb; Oberalm/Salzburg). Geol. Rundschau **68**/3, 876–893. 3 Pls., 2 Tables, 1 Fig., Stuttgart

Flügel, H. W., Walitzi, E. (1966): Untersuchungen über Calcit-Regelung in Kalken mit Hilfe des Diffraktometer-Verfahrens. Österr. Akad. Wiss. Anz. math.-naturw. Kl. **1966**/3, 67–72, 1 Fig., Wien

Folk, R. L. (1959): see 6.4
Folk, R. L. (1962): Petrography and Origin of Silurian Rochester an McKenzies Shales, Morgan County, West Virginia. J. Sed. Petrol. **32**/3, 539–578, 12 Figs., 5 Pls., Tulsa
Folk, R. L. (1965): Some aspects of recrystallization in ancient limestones. Soc. Econ. Paleont. Min. Spec. Publ. **13,** 14–48, Tulsa
Folk, R. L. (1973): Carbonate petrography in the post-Sorbian age. In: Evolving concepts in sedimentology. John Hopkins Univ. Stud. Geol. **21,** 118–158, Baltimore
Folk, R. L. (1974): The natural history of crystalline calcium carbonate: Effect of magnesium content and salinity. J. Sed. Petrol. **44**/1, 40–53, 9 Figs., Tulsa
Folk, R. L., Assereto, R. (1976): Comparative fabrics of length-slow and length-fast calcite and calcitized aragonite in a Holocene Speleothem, Carlsbad Caverns, New Mexico. J. Sed. Petrol. **46**/3, 486–496, 11 Figs., Tulsa
Folk, R. L., Land, L. L. (1975): Mg/Ca ratio and salinity: Two controls over crystallization of dolomite. Bull. Amer. Ass. Petrol. Geol. **59**/1, 60–68, 2 Figs., Tulsa
Friedman, G. M. (1964): Early diagenesis and lithification in carbonate sediments. J. Sed. Petrol. **34**/4, 777–813, 53 Figs., Tulsa
Friedman, G. M. (1965a): Terminology of crystallization textures and fabrics in sedimentary rocks. J. Sed. Petrol. **35**/3, 643–655, 11 Figs., Tulsa
Friedman, G. M. (1965b): Occurrence and stability relationships of aragonite, high-magnesium calcite, and low-magnesium calcite under deep-sea conditions. Bull. Geol. Soc. Amer. **76**/10, 1191–1196, 1 Pl., 1 Fig., New York
Friedman, G. M. (1968): The fabric of carbonate cement and matrix and its dependence on the salinity of water. In: Müller, G., Friedman, G. M. (eds.): Recent Developments in Carbonate Sedimentology in Central Europe, 11–20, 8 Figs., New York
Friedman, G. M. (1975): The making and unmaking of limestones or the downs and ups of porosity. J. Sed. Petrol. **45**/2, 379–398, 22 Figs., Tulsa
Friedman, G. M., Amiel, A. J., Schneidermann, N. (1974): Submarine cementation in reefs: Example from the Red Sea. J. Sed. Petrol. **44**/3, 816–825, 18 Figs., Tulsa
Friedman, G. M., Kolesar, P. T. Jr. (1971): Fresh-water cements. In: Bricker, O. P. (ed.): Carbonate Cements, 122–126, Figs. 61–66, Baltimore–London: Hopkins
Friedman, G. M., Sanders, J. E. (1978): see 1.4.1
Fruth, L. S., Orme, G. R., Donath, F. A. (1966): Experimental compaction effects in carbonate sediments. J. Sed. Petrol. **36**/3, 747–754, 6 Figs., Tulsa
Füchtbauer, H. (1974): see 1.4
Gaither, A. (1953): A Study of porosity and grain relationships in experimental sands. J. Sed. Petrol. **23,** 180–195, Tulsa
Garrison, R. E., Kennedy, W. J. (1977): Origin of solution seams and flaser structure in Upper Cretaceous chalks of southern England. Sed. Geol. **19**/2, 107–138, 16 Figs., Amsterdam
Gavish, E., Friedman, G. M. (1969): Progressive diagenesis in Quaternary to Late Tertiary carbonate sediments: sequence and time scale. J. Sed. Petrol. **39**/3, 980–1006, 32 Figs., Tulsa
Germann, K. (1966): Ablauf und Ausmaß diagenetischer Veränderungen im Wettersteinkalk (alpine Mitteltrias). 122+XV pp., 8 Tab., 17 Pls., Diss. Univ. München
Germann, K. (1968): Diagenetic patterns in the Wettersteinkalk (Ladinian, Middle Trias), Northern Limestone Alps, Bavaria and Tyrol. J. Sed. Petrol. **38**/2, 490–500, 17 Figs., Tulsa
Gevirtz, J. L., Friedman, G. M. (1966): see 2.4
Ginsburg, R. N. (1957): Early diagenesis and lithification of shallow-water carbonate sediments in South Florida. In: Regional Aspects of Carbonate Deposition. S.E.P.M. Spec. Publ. **5,** 80–100, Tulsa
Ginsburg, R. N., James, N. P. (1976): Submarine botryoidal aragonite in Holocene reef limestones, Belize. Geol. **4,** 431–436, Boulder
Ginsburg, R. N., Marszalek, D. S., Schneidermann, N. (1971): Ultrastructure of carbonate cement in a Holocene algal reef of Bermuda. J. Sed. Petrol. **41**/2, 472–482, 13 Figs., Tulsa
Ginsburg, R. N., Schroeder, J. H. (1973): Growth and submarine fossilisation of algal cup reefs, Bermuda. Sedimentology **20,** 575–614, Oxford
Gökdag, H. (1974): Sedimentpetrographische und isotopenchemische (O^{18}, C^{13}) Untersuchungen im Dachsteinkalk (Obernor-Rät) der Nördlichen Kalkalpen. Diss. Univ. Marburg, 156 pp., 33 Pls., 2 encls., 10 Diagr., Marburg

Graf, D. L., Lamar, J. E. (1950): Petrology of Fredonia oolite in Southern Illinois. Bull. Amer. Ass. Petrol. Geol. **34,** 2318–2336, Tulsa

Griggs, D. T., Paterson, M. S., Heard, H. C., Turner, F. J. (1960): Annealing recrystallization in calcite crystals and aggregates. In: Griggs, D. T., David, Handin, J. (eds.): Rock Deformation. Mem. Geol. Soc. Amer. **79,** 21–37, New York

Grout, F. F. (1946): Microscopic character of vein carbonates. Econ. Geol. **41,** 475–502

Grover, G. Jr., Read, J. F. (1978): Fenestral and associated vadose diagenetic fabrics of tidal flat carbonates, Middle Ordovician, New Market Limestone, southwestern Virginia. J. Sed. Petrol. **48,** 453–474, Tulsa

Halley, R. B., Harris, P. M. (1979): Fresh-water cementation of a 1,000-year-old oolite. J. Sed. Petrol. **49**/3, 969–988, 14 Figs., Tulsa

Hanor, J. S. (1978): Precipitation of beach rock cements: mixing of marine and meteoric waters vs. CO_2 degassing. J. Sed. Petrol. **48,** 489–502, Tulsa

Harbaugh, J. W. (1961): Relative ages of visible crystalline calcite in late Paleozoic limestones. Kansas Geol. Surv., Bull. **152,** Reports of Studies, part 4, 91–126, Lawrence

Hathaway, J. C., Robertson, E. C. (1961): Microtexture of artificially consolidated aragonitic mud. US Geol. Surv. Prof. Paper **424,** 301–304, Washington

Hattin, D. E., Dodd, J. R. (1978): Holocene cementation of carbonate sediments in the Florida Keys. J. Sed. Petrol. **48**/1, 307–312, 5 Figs., Tulsa

Hesse, R., Butt, A. (1976): Paleobathymetry of Cretaceous turbidite basins of the East Alps relative to the calcite compensation level. J. Geol. **84**/5, 505–533, 12 Figs., Chicago

Hottinger, L., Levinson, G. (1978): Cementation of Reefs in the Gulf of Elat (Red Sea) by acervulinid foraminifera. Abstracts, Tenth Int. Cong. Sed., **1,** p. 315, Jerusalem

Hsü, K. J., Andrews, J. E. (1970): History of South Atlantic Basin. In: Maxwell, A. E. et al. (eds.): Initial Rep. Deep Sea Drilling Project **3,** 464–467, 1 Fig., Washington

Hugman, R. H. H., Friedman, M. (1979): Effects of texture and composition on mechanical behaviour of experimentally deformed carbonate rocks. Amer. Ass. Petrol. Geol. Bull. **63**/9, 1478–1489, 4 Figs., 6 Tables, Tulsa

Jacka, A. D., Brand, J. P. (1977): Biofacies and development and differential occlusion of porosity in a lower Cretaceous (Edwards) reef. J. Sed. Petrol. **47**/1, 366–381, 11 Figs., Tulsa

James, N. P., Ginsburg, R. N. (1979): The seaward margin of Belize barrier and atoll reefs. Spec. Publ. Int. Ass. Sed. **3,** 191 p., Oxford

James, N. P., Ginsburg, R. N., Marszalek, D. S., Choquette, Ph. W. (1976): Facies and fabric specifity of early subsea cements in shallow Belize (British Honduras) Reefs. J. Sed. Petrol. **46**/3, 523–544, 15 Figs., Tulsa

Jeans, C. V. (1980): Early submarine lithification in the Red Chalk and Lower Chalk of eastern England: a bacterial control model and its implication. Proc. Yorkshire Geol. Soc. **43**/1–2, 81–157, 44 Figs., 12 Tabs., London

Jenkyns, H. C. (1974): Origin of red nodular limestones (Ammonitico Rosso, Knollenkalke) in the Mediterranean Jurassic: a diagenetic model. Spec. Publ. Int. Ass. Sed. **1,** 249–271, 8 Figs., Oxford

Kaisin, F. Jr. (1956): Le rôle de la substitution dans l'altération méteorique des roches sédimentaires, specialement des calcaires. Univ. Louvain, Inst. Geol. Mém. **20,** 47–164

Kahle, C. F. (1966): Some observations on compaction and consolidation in ancient oolites. Compass **44,** 19–29

Kahn, J. S. (1956): Analysis and distribution of the properties of packing in sandsize sediments. I. On the measurement of packing in sandstones. J. Geol. **64,** 385–395, Chicago

Kendall, A. C. (1976): Origin of fibrous calcite cements that apparently replace foraminiferal tests. J. Sed. Petrol. **46**/3, 545–547, Fig. 1 A, 1 B, Tulsa

Kendall, A. C. (1977): Fascicular-optic calcite: a replacement of bundled acicular carbonate cements. J. Sed. Petrol. **47,** 1056–1062, Tulsa

Kendall, A. C., Broughton, P. L. (1978): Origin of fabrics in speleothems composed of columnar calcite crystals. J. Sed. Petrol. **48,** 519–538, Tulsa

Kendall, A. C., Tucker, M. E. (1973): Radiaxial fibrous calcite: a replacement after acicular carbonate. Sedimentology **20,** 365–389, Amsterdam

Keupp, H. (1979): Biogen beeinflußte Beachrock-Zementation auf Raratonga, Cook-Inseln. Natur und Mensch, Jahresmitteilungen Naturhist. Ges. Nürnberg, **1978,** 35–40, 8 Figs., Nürnberg

Krebs, W. (1969): Early void-filling cementation in Devonian fore-reef limestones (Germany), Sedimentology **12,** 279–299, Amsterdam

Lahann, R. W. (1978): A chemical model for calcite growth and morphology control. J. Sed. Petrol. **48**/1, 337–344, 4 Figs., Tulsa

Land, L. S. (1967): Diagenesis of skeletal carbonates. J. Sed. Petrol. **37,** 914–930, Tulsa

Land, L. S. (1970): Phreatic versus vadose meteoric diagenesis of limestones: evidence from a fossil water table. Sedimentology **14,** 175–185, 6 Figs., Amsterdam

Land, L. S. (1973a): Contemporaneous dolomitization of middle Pleistocene reefs by meteoric water, North Jamaica. Bull. Marine Sci. **23,** 64–92

Land, L. S. (1973b): Holocene meteoric dolomitization of Pleistocene limestones, north Jamaica. Sed. **20,** 411–422, Oxford

Land, L. S., Goreau, F. T. (1970): Submarine Lithification of Jamaican reefs. J. Sed. Petrol. **40,** 457–462, Tulsa

Land, L. S., Mackenzie, F. T., Gould, S. J. (1967): Pleistocene history of Bermuda. Geol. Soc. Amer. Bull. **78,** 993–1006, New York

Lindström, M. (1979): Früher Kalzitzement in skandinavischen Orthocerenkalken: Beziehungen zum Glaukonit. Geol. Rundschau **68**/3, 952–964, 8 Figs., Stuttgart

Lippmann, F. (1973): Sedimentary Carbonate Minerals. 228 p., Berlin: Springer

Logan, B. W., Semeniuk, V. (1976): Dynamic metamorphism; processes and products in Devonian carbonate rocks. Canning Basin, Western Australia. Geol. Soc. Australia Spec. Publ. **6,** 138 p., 97 Figs., 11 Tables, Sidney

Longman, M. W. (1980): Carbonate diagenetic textures from nearshore diagenetic environments. Amer. Ass. Petrol. Geol. Bull. **64**/4, 461–487, 18 Figs., Tulsa

Loucks, R. G. (1977): Porosity development and distribution in shoal-water carbonate complexes – subsurface Pearsall Formation (Lower Cretaceous), south Texas. Texas Univ. Bur. Econ. Geol. Rep. Inv. **89,** 97–126, Austin

Loucks, R. G., Folk, R. L. (1976): Fanlike rays of former aragonite in Permian Capitan reef pisolite. J. Sed. Petrol. **46,** 483–485, 5 Figs., Tulsa

Lucia, F. J. (1968): Recent sediments and diagenesis of South Bonaire, Netherlands Antilles. J. Sed. Petrol. **38**/3, 845–858, 23 Figs., Tulsa

Manus, R. W., Coogan, A. H. (1974): Bulk volume reduction and pressure solution derived cement. J. Sed. Petrol. **44,** 466–471, Tulsa

Mapstone, N. B. (1975): Diagenetic history of a North Sea chalk. Sedimentology **22,** 601–614, Oxford

MacIntyre, I. G. (1967): Submerged coral reefs, west coast of Barbados, West Indies. Canadian J. Earth Sci. **4,** 461–474, Ottawa

MacIntyre, I. G., Mountjoy, E. W., d'Anglejan, B. F. (1968): An occurrence of submarine cementation of carbonate sediments off the West Coast of Barbados, W. I. – J. Sed. Petrol. **38**/2, 660–664, 5 Figs., Tulsa

Marschner, H. (1968): Relationship between carbonate grain size and non-carbonate content in carbonate sedimentary rocks. In: Müller, G., Friedman, G. M. (eds.): Recent Developments in Carbonate Sedimentology in Central Europe, 55–57, 2 Figs., Berlin–Heidelberg–New York: Springer

Masson, P. H. (1951): Measurement of grain packing in sandstone. Amer. Ass. Petrol. Geol. Annual Meeting, Abstract, St. Louis

Mattavelli, L. (1966): Osservazioni petrografiche sull sostituzione della dolomite con la calcite (dedolomitizzazione) in alcune facies carbonate italiane. Atti Soc. Ital. Sci., Nat. Mus. Civ. Storia Nat. Milano **105**/3, 294–316, 16 Figs., Milano

Matter, A., Douglas, R. G., Perch-Nielsen, K. (1975): Fossil preservation, geochemistry and diagenesis of pelagic carbonates from Shatsky Rise, northwest Pacific. In: Larson, R. L., Moberly, R., et al. (eds.): Initial Reports of the Deep Sea Drilling Project **32,** 891–921, 7 Pls., 5 Figs., Washington

Matthews, R. K. (1967): Diagenetic fabrics in biosparites from the Pleistocene of Barbados, West Indies. J. Sed. Petrol. **37**/4, 1147–1235, 10 Figs., Tulsa

Matthews, R. K. (1968): Carbonate diagenesis: Equilibration of sedimentary mineralogy to the subaerial environment: Coral Cap of Barbados, West Indies. J. Sed. Petrol. **38**/4, 1110–1119, 1 Fig., Tulsa

Matthews, R. K. (1974): A process approach to diagenesis of reefs and reef associated limestones. Soc. Econ. Paleont. Min. Spec. Publ. **18**, 234–256, 18 Figs., Tulsa

Maxwell, W. G. H. (1962): Lithification of carbonate sediments in the Heron Island Reef, Great Barrier Reef. J. Geol. Soc. Australia **8**, 217–238, Adelaide

Mazzullo, S. J. (1981): Facies and burial diagenesis of a carbonate reservoir: Chapman Deep (Atoka) Field, Deleware Basin, Texas. Amer. Ass. Petrol. Geol. Bull. **65**, 850–865, 15 Figs., Tulsa

Mazzullo, S. J. (1980): Calcite pseudospar replacive of marine acicular aragonite, and its implications for aragonite cement diagenesis. J. Sed. Petrol. **50**/2, 409–422, 7 Figs., Tulsa

Mazzullo, S. J., Cys, J. M. (1977): Submarine cements in Permian boundstones and reef-associated rocks, Guadelupe Mountains, west Texas and southeastern New Mexico. Soc. Econ, Paleont. Min. Permian Basin Sect., Guidebook **77**-16/1, 151–200

Meyers, W. J. (1978): Carbonate cements: their regional distribution and interpretation in Mississippian limestones of southwestern New Mexico. Sed. **25**, 371–400, Oxford

Meyers, W. J. (1980): Compaction in Mississippian skeletal limestones, southwestern New Mexico. J. Sed. Petrol. **50**/2, 457–474, 8 Figs., Tulsa

Milliman, J. D. (1971): Carbonate lithification in the deep sea. In: Bricker, O. P. (ed.): Carbonate cements, 95–102, Figs. 49–50, Baltimore–London: Hopkins

Milliman, J. D. (1974): see 1.4.1

Mirsal, I. (1978): Zementmineralisation in fossilen Korallenriffen. Eine petrographische und geochemische Faktorenanalyse. Diss. Univ. Marburg/Lahn, 120 p., 29 Figs., Marburg

Mirsal, I. A., Zankl, H. (1979): Petrography and geochemistry of carbonate void-filling cements in fossil reefs. Geol. Rundschau **68**/3, 920–951, 21 Figs., 4 Tables, Stuttgart

Mišik, M. (1966): Tentative microscopical classification of veinlets in limestones. Geol. Sborn. Slov. Akad. Vied Bratislava **17**, 337–344, Bratislava

Mišik, M. (1968): Some aspects of diagenetic recrystallization in limestones. XXXIII Intern. Geol. Cong. Rep. Sect. **8**, 129–136, 15 Figs., Prague

Mountjoy, E. W., Walls, R. A. (1977): Some examples of early cements from Devonian buildups of Alberta, Proc. 3rd Internat. Coral Reef Symp. Miami **2**, 155–161, 5 Figs., Miami

Müller, G. (1971): "Gravitational" cement: An indicator for the vadose zone of the subaerial diagenetic environment. In: Bricker, O. P. (ed.): Carbonate Cements, 301–302, Fig. 137, Baltimore–London: Hopkins

Müller, G., Tietz, G. (1975): Transformation of carbonate sands into limestone and dolostone: Fuerteventura, Canary Island, Spain. 9th Int. Congr. Sed. Nice **7**, 143–148, 7 Figs., Nice

Neugebauer, J. (1973): The diagenetic problem of chalk. The role of pressure solution and pore fluid. N. Jb. Geol. Paläont. Abh. **143**/2, 223–245, 8 Figs., Stuttgart

Neugebauer, J. (1974): Some aspects of the cementation in chalk. Int. Ass. Sed. Spec. Publ. **1**, 149–176, Oxford

Neugebauer, J. (1975): Fossil-Diagenese in der Schreibkreide: Coccolithen. Neues Jahrbuch Geol. Paläont. Monatshefte **1975**/8, 489–502, 6 Figs., Stuttgart

Neugebauer, J. (1975): Foraminiferen-Diagenese in der Schreibkreide. Neues Jahrbuch Geol. Paläont. Abhandlungen **150**/2, 182–206, 8 Figs., Stuttgart

Neugebauer, J. (1979): Fossil-Diagenese in der Schreibkreide: Echinodermen. Clausthaler Geol. Abhandlungen **30**, 198–229, 14 Figs., Clausthal-Zellerfeld

Newell, N. D. (1955): Depositional fabric in Permian reef limestones. J. Geol. **63**, 301–309, 2 Pls., Chicago

Nichols, R. A. H. (1967): The ". . . Sparite" Complex: Eosparite v. Neosparite. J. Sed. Petrol. **37**/4, 1247–1248, Tulsa

Oldershaw, A. E., Scoffin, T. P. (1967): The source of ferroan and non-ferroan calcite cements in the Halkin and Wenlock limestones. Geol. J. **5**/2, 309–320, 2 Figs., 2 Tab., Liverpool

Orme, G. R., Brown, W. W. M. (1963): Diagenetic fabrics in Avonian limestones of Derbyshire and North Wales. Proc. Yorkshire Geol. Soc. **34**, 51–66, Pls. 7–13, 1 Tab., Leeds

Otte, C. Jr., Parks, J. M. Jr. (1963): Fabric studies of Virgil and Wolfcamp Bioherms, New Mexico. J. Geol. **71**/3, 380–396, 5 Figs., 3 Pls., 2 Tab.. Chicago

Park, W. C., Schot, E. H. (1968): Stylolites: Their nature and origin. J. Sed. Petrol. **38,** 175–191, Tulsa

Pia, J. (1933): Kohlensäure und Kalk, Die Binnengewässer **13,** 17 Figs., 58 Tab., 5 supps., 183 pp., Stuttgart

Pingitore, N. E. (1970): Diagenesis and porosity modification in *Acropora palmata,* Pleistocene of Barbados, West Indies. J. Sed. Petrol. **40**/2, 712–721, 9 Figs., Tulsa

Pop, G. (1976): Origin of some Mesozoic basinal limestones from the Resita zone (South Carpathians). Anuarul Inst. Geol. Geofiz. **48,** 57–96, 12 Pls., Bucaresti

Pratt, B. R. (1979): Early cementation and lithification in intertidal cryptalgal structures, Boca Jewfish, Bonaire, Netherland Antilles. J. Sed. Petrol. **49**/2, 379–386, 4 Figs., Tulsa

Pray, L. C. (1960): Compaction in calcilutites, an abstract. Bull. Geol. Ass. Amer. **71,** p. 1946, New York

Purdy, E. G. (1968): Carbonate diagenesis: an environmental survey. Geol. Romana **7,** 183–228, 6 Pls., 10 Figs., Rome

Purser, B. H. (1969): Syn-sedimentary marine lithification of Middle Jurassic limestones in the Paris Basin. Sedimentology **12,** 205–230, Oxford

Rad, U. v. (1974): Great Meteor and Josephine Seamounts (Eastern North Atlantic): composition and origin of bioclastic sands, carbonate and pyroclastic rocks. „Meteor" Forsch.-Ergebn. C. **19,** 1–61, 10 Pls., 26 Figs., 7 Tab., Berlin–Stuttgart

Richter, D. K. (1976): Gravitativer Meniskuszement in einem holozänen Oolith bei Neapolis (Süd-Peleponnes, Griechenland). N. Jb. Geol. Paläont. Abh. **151**/2, 192–223, Stuttgart

Richter, D. K. (1979): Die Stufen der meteorisch-vadosen Umwandlung von Mg-Calcit in Calcit in rezenten bis pliozänen Biogenen Griechenlands. Neues Jb. Geol. Paläont. Abhandlungen **158,** 277–333, Stuttgart

Richter, D. K., Zinkernagel, U. (1981): Zur Anwendung der Kathodolumineszenz in der Karbonatpetrographie. Geol. Rundschau **70**/3, 1276–1302, 4 Pls., 2 Figs., 1 Tab., Stuttgart

Rittenhouse, G. (1971): Pore-space reduction by solution and cementation. Bull. Amer. Ass. Petrol. Geol. **55,** 80–91, Tulsa

Robertson, E. C. (1967): Laboratory consolidation of carbonate sediment. In: Richards, A. F. (ed.): Marine Geotechnique, 118–127, 5 Figs., Urbana: Univ. Illinois

Robinson, R. B. (1967): Diagenesis and porosity development in recent and Pleistocene Oolites from Southern Florida and the Bahamas. J. Sed. Petrol. **37**/2, 355–364, 9 Figs., Tulsa

Roll, A. (1974): Langfristige Reduktion der Mächtigkeit von Sedimentgesteinen und ihre Auswirkung – eine Übersicht. Geol. Jahrbuch, A **14,** 76 p., 44 Figs., 3 Tables, Hannover

Runnels, D. D. (1969): Diagenesis, chemical sediments, and the mixing of natural waters. J. Sed. Petrol. **39,** 1188–1201, Tulsa

Schmidegg, O. (1928): Über geregelte Wachstumsgefüge. Jahrbuch geol. Bundesanstalt **78,** 1–52, Wien

Sander, B. (1936): see 1.4.1

Schlager, W. (1974): Preservation of cephalopod skeletons and carbonate dissolution on ancient Tethyan sea floors. In: Hsü, K. J., Jenkyns, H. C. (eds.): Pelagic Sediments, on Land and Under the Sea. Spec. Publ. Inter. Ass. Sediment **1,** 37–58, Oxford

Schloz, W. (1972): Zur Bildungsgeschichte der Oolithenbank (Hetangium) in Baden-Württemberg. Arb. Inst. Geol. Paläont. Univ. Stuttgart **67,** 101–212, Stuttgart

Schmalz, R. F. (1967): Kinetics and diagenesis of carbonate sediments. J. Sed. Petrol. **37**/1, 60–67, 2 Figs., 1 Tab., Tulsa

Schneider, J. (1976): Biological and inorganic factors in the destruction of limestone coasts. Contrib. Sed. **6,** 112 pp., 11 Pls., 25 Figs., 19 Tab., Stuttgart: Schweizerbart

Schneider, W. (1973): Einige Beobachtungen zur Diagenese in den devonischen Karbonatkomplexen des ostrheinischen Schiefergebirges unter besonderer Berücksichtigung der Quarzbildung. N. Jb. Geol. Paläont. Mh. Stuttgart **1973**/4, 231–257, 6 Tab., 16 Figs., Stuttgart

Schneider, W. (1977): Diagenese devonischer Karbonatkomplexe Mitteleuropas. Geol. Jahrbuch, ser. D. **21,** 107 p., Hannover

Schneidermann, N. (1973): Pelagic limestones of the Central Caribbean, Leg 15. In: Saunders, E. N. T. (ed.): Initial Report of the Deep Sea Drilling Project, XV, 773–793, Pls. 1–8, Washington

Schneidermann, N., Sandberg, P. S., Wunder, S. J. (1972): Recognition of early cementation of aragonitic skeletal carbonates. Nature, Physical Science **249,** 88–89, 3 Figs., London

Scholle, P. A. (1971): Diagenesis of deep-water carbonate turbidites, Upper Cretaceous Monte Antola Flysch, Northern Apennines, Italy. J. Sed. Petrol. **41,** 233–250, 9 Figs., Tulsa

Scholle, P. A. (1974): Diagenesis of Upper Cretaceous chalks from England, Northern Ireland and the North Sea. Int. Ass. Sed. Spec. Publ. **1,** 177–220, Oxford

Scholle, P. A. (1977): Chalk diagenesis and its relation to petroleum exploration: Oil from chalks, a modern miracle. Amer. Ass. Petrol. Geol. Bull. **61**/7, 982–1009, 13 Figs., Tulsa

Scholtgen, J. I. (1972): Beach Rocks. Zentralbl. Geol. Paläont. I, **1971**/9–10, 655–672, Stuttgart

Scholz, R. (1973): Zur Sedimentologie und Kompaktion der Schreibkreide von Lägerdorf in SW-Holstein. Neues Jahrbuch Min. Abhandlungen **118**/2, 111–133, 10 Figs., 5 Tables, Stuttgart

Schroeder, J. H. (1972): Calcified filaments of an endolithic algae in recent Bermuda reefs. N. Jb Geol. Paläont. Mh. **1972**/1, 16–33, 18 Figs., Stuttgart

Schroeder, J. H. (1972): Fabrics and sequences of submarine carbonate cements in Holocene Bermuda cup reefs. Geol. Rdsch. **61**/2, 708–730, 12 Figs., Stuttgart

Schroeder, J. (1973): Submarine and vadose cements in Pleistocene Bermuda Reef Rock. Sed. Geology **10,** 179–204, 9 Figs., 1 Tab., Amsterdam

Septfontaine, M. (1976): Microfacies et diagenese de quelques niveaus jurassique des Prealpes medianes du Chablais occidentales (Haute-Savoie, France). Eclogae geol. Helvet., **69**/1, 29–61, 2 Pls., 5 Figs., Basel

Seyfried, H. (1978): Der subbetische Jura von Murcia (Südost-Spanien). Geol. Jb., B **29,** 3–201, 28 Pls., 36 Figs., Hannover

Shinn, E., Halley, R. B., Hudson, J. J., Lidz, B. H. (1976): Limestone compaction: an enigma. Geol. **5,** 21–25, Boulder

Sibley, D. F., Murray, R. C. (1972): Marine diagenesis of carbonate sediment, Bonaire Netherlands Antilles. J. Sed. Petrol. **42**/1, 168–178, 7 Figs., Tulsa

Sliter, W. V., Be, A. W. H., Berger, W. H. (1975/76): Dissolution of Deep-Sea Carbonate. Cushman Found. Foram. Res. Spec. Publ. **13**

Stauffer, K. W. (1962): Quantitative petrographic study of Paleozoic carbonate rocks. Caballo Mountains, New Mexico. J. Sed. Petrol. **32**/3, 357–396, 1–38, Tulsa

Stehli, F. G., Hower, J. (1961): Mineralogy and early diagenesis of carbonate sediments. J. Sed. Petrol. **31,** 358–371, 11 Figs., Tulsa

Steinen, R. P. (1974): Phreatic and vadose diagenetic modification of Pleistocene limestone: petrographic observations from subsurface of Barbados, West Indies. Amer. Ass. Petrol. Geol. Bull. **58,** 1008–1024, Tulsa

Steinen, R. P. (1978): On the diagenesis of lime mud: Scanning electron microscopic observations of subsurface material from Barbados, W. I. – J. Sed. Petrol. **48**/4, 1139–1148, 7 Figs., Tulsa

Steinen, R. P., Matthews, R. K. (1973): Phreatic vs. vadose diagenesis: stratigraphy and mineralogy of a cored borehole on Barbados, W. I. – J. Sed. Petrol. **43**/4, 1012–1020, 2 Figs., Tulsa

Terzaghi, R. (1940): Compaction of lime mud as a cause of secondary structure. J. Sed. Petrol. **10,** 78–90, 5 Figs., 2 Tab., Tulsa

Thorstenson, P. C., MacKenzie, F. T., Ristuet, B. L. (1972): Experimental vadose and phreatic cementation of skeletal carbonate sand. J. Sed. Petrol. **42**/1, 162–167, 5 Figs., Tulsa

Trurnit, P. (1968): Pressure solution phenomena in detrital rocks. Sed. Geol. **2,** 89–114, Amsterdam

Tucker, M. E. (1973): Sedimentology and diagenesis of Devonian pelagic limestones (Cephalopodenkalk) and associated sediments of the Rhenohercynian geosyncline, West Germany. N. Jb. Geol. Paläont. Abh. **142**/3, 320–350, 24 Figs., 1 Tab., Stuttgart

Tucker, M. E., Kendall, A. C. (1973): The diagenesis and low-grade metamorphism of Devonian styliolinid-rich pelagic carbonates from West Germany: possible analogues of recent pteropod ooze. J. Sed. Petrol. **43,** 672–683, 17 Figs., Tulsa

Violanti, D., Premoli Silva, I., Cita, M. B., Kersey, D., Hsü, K. J. (1979): Quantitative characterization of carbonate dissolution facies of the Atlantic Tertiary sediments. An Attempt. Riv. Ital. Paleont. Strat. **85**/2, 517–548, Pls. 51–52, 12 Figs., 5 Tables, Milano

Wardlaw, N. C. (1962): Aspects of diagenesis in some Irish Carboniferous limestones. J. Sed. Petrol. **32,** 776–780, Tulsa

Weynschenk, R. (1949): Beiträge zur Geologie und Petrographie des Sonnwendgebirges (Tirol), besonders der Hornsteinbreccien. Schlern-Schriften **59,** 66 pp., Pls. 1–15, 7 Figs., 3 Supps., Innsbruck

Weyl, P. K. (1959): Pressure solution and the force of crystallization – a phenomenological theory. J. Geophys. Res. **64,** 2001–2025

Wikinson, B. H., Landing, E. (1978): "Eggshell diagenesis" and primary radial fabric in calcite ooids. J. Sed. Petrol. **48**/4, 1129, 4 Figs., Tulsa

Wilson, J. L. (1975): see 1.4.1

Wilson, R. C. L. (1966): Silica diagenesis in Upper Jurassic limestones of southern England. J. Sed. Petrol. **36**/4, 1036–1049, 5 Figs., 1 Tab., Tulsa

Winder, C. G. (1968): Carbonate diagenesis by burrowing organisms. XXIII Intern. Geol. Congr., Rep. Sect. **8,** 173–183, 2 Pls., Prague

Winland, H. D. (1969): Stability of calcium carbonate polymorphs in warm shallow seawater. J. Sed. Petrol. **39**/4, 1579–1587, 3 Figs., Tulsa

Wolf, K. H. (1965): see 4.1.7.2

Wolf, K. H., Conolly, J. R. (1965): Petrogenesis and paleoenvironment of limestone lenses in Upper Devonian Red Beds of New South Wales. Palaeo.-3, **1,** 69–111, 18 Pls., 4 Tab., 6 Figs., Amsterdam

Wolfe, M. J. (1968): Lithification of a carbonate mud: Senonian chalk in Northern Ireland. Sed. Geol. **2,** 263–290, 16 Figs., Amsterdam

Wong, P. K., Oldershaw, A. (1981): Burial cementation in the Devonian, Kaybob Reef Complex, Alberta, Canada. J. Sed. Petrol. **51**/2, 507–520, 11 Figs., Tulsa

Zankl, H. (1969): Structural and textural evidence of early lithification in fine-grained carbonate rocks. Sedimentology **12,** 241–256, 7 Figs., Amsterdam

Zankl, H. (1971): A model of sedimentation and diagenesis in a Triassic reef. In: Bricker, O. P. (ed.): Carbonate Cements, 189–192, Figs. 98–101, Baltimore–London: Hopkins

Zorn, H. (1970): Frühdiagenetische Vorgänge im Salvatore-Dolomit (Mitteltrias, Kanton Tessin, Schweiz). Z. deutsch. geol. Ges. **121,** Jg. 1969, 163–177, 4 Pls., Hannover

4 Microfacies Characteristics

Microfacies characteristics as seen in thin-sections include: the types of "groundmass", the different kinds of particle categories, depositional textures, and solution textures. All these criteria can lead to information about the sedimentological and diagenetic development of limestones. Also included are, of course, textural and structural criteria which can be represented by frequency analyses, grain size analyses, or morphometrical characteristics.

4.1 "Groundmass" and Particles

Groundmass designates the primary matrix (broadly defined as micrite) as well as sparite formed by cementation or neomorphic processes. The criteria for describing orthosparite and pseudosparite have already been discussed (see Sects. 3.4.1 and 3.4.2).

The term *particles* pertains to synsedimentary carbonate grains or to the elements of carbonate rocks transported into the sedimentation area. To avoid confusion, we will also include in the term particles elements with similar diagenetic origins, such as caliche-pisoids. The term particles essentially corresponds to Folk's "allochems" (1959) or the term grains as used in the English literature.

4.1.1 **Recent Carbonate Grains** (Plate 1)

Theoretically it is possible to distinguish between organic and inorganic carbonate particles in Recent marine environments. However, because many "inorganic" particles such as ooids originate directly or indirectly as a result of interaction with organisms, Illing (1954) speaks of "skeletal components" and "non-skeletal components"; all those particles which do not stem from organic hardparts belong to the latter category.

A comparison of Tables 9 and 13 shows that more terms are used for the description of particles of limestone rocks than for the study of Recent carbonate sediments, where generally only skeletal grains, peloids, ooids, lumps or grapestones, and lithoclasts are distinguished.

The origin and occurrence of Recent carbonate particles will be discussed in Sects. 4.1.2 and 4.1.3 (see also Plate 1). The significance of Recent groups of organisms for carbonate sedimentation follows from Sect. 2.2 and Tables 6 and 10.

Table 9. "Non-biogenic" carbonate particles in Recent seas

	Shape	Size	Mineralogical composition	Organic interaction Organic substances	Distribution
Pellets	Oblong, cylindrical, spherical; surface dull or smooth and glossy; white color	Length < 0.2–3 mm, usually 0.2–1 mm; diameter 0.1–0.5 mm	Aragonite or Mg-calcite; usually cryptocrystalline	Fecal pellets contain a relatively large amount of organic substance	Dominant in subtidal and lower intertidal zones off low energy coasts; reef flats and lagoons, often adjoining a zone with aggregate grains; common in deep seas but preserved less often than in shallow waters
Ooids	Spherical or oval, nucleus with coating of one or more laminae made of crystals with certain orientation patterns (tangential or radially oriented c-axis); pearly luster; cream color	Diameter 0.06–1 mm, often 0.15–0.60 mm; Laminae thickness 1–3 µm	Aragonite; in non-marine area Mg-calcite and aragonite	Relatively large amount of organic substance; probable algal activity during formation	Restricted to tropical and subtropical, normal marine or hypersaline areas; generally high-energy shallow water (tides and wave action); usually in depths < 2 m (but to 15 m); ooids with few laminae occur preferentially in low-energy environments
Aggregate grains	Irregularly shaped aggregate of two or more strongly micritized particles bonded by cryptocrystalline matrix or cement	Larger diameter about 0.5–1.5 mm	Aragonite	Large amount of organic substance, often with algal encrustation and boring algae	Restricted to shallow water areas with slow current flow and low wave energy
Lithoclasts	Sharply delimited calcareous fragments	< 0.5 mm to several cm	Synsedimentary lithoclasts: aragonite, Mg-calcite; Postsedimentary lithoclasts: calcite	—	Relatively near subaerial or submarine source areas

Table 10. Sedimentological significance of Recent organisms with carbonate hard parts. B = boring; F = frame building; A = accumulation of scattered hard parts; ST = sediment trapping; C = cementing and encrusting; + no special sedimentological significance; ( ) occasional sedimentological significance

	Deep Sea	Sublittoral	Intertidal	Supratidal	Limnic-Lacustrine
Cyanophyta (blue-green algae)		B, ST	B, ST	B, ST	B, ST
Chrysophyta: Coccolithophorida	A	A			
Chlorophyta (green algae)		A	A		
Charophyta					A
Rhodophyta (red algae)	(F)	A, C, F	A, C, F		
Foraminifera	A	A, C	A, C		
Porifera	+, (A)	B	B		
Hydrozoa		F, C			
Octocorallia	A	A			
Scleractinia	(F)	F, ST	F		
Bryozoa		A, F, C			
Brachiopoda		+			
Amphineura		+	+		
Scaphopoda	(A)	(A)			
Lamellibranchiata	+	A, B	A, B		+
Gastropoda		A	A	A	A
Pteropoda	A	A			
Cephalopoda	+	+			
Annelida		A, C	A, C		
Cirripedia		A, C	A, C, B		
Ostracoda	(A)	(A)	(A)	A	A
Decapoda	+	+	+		+
Crinoidea	+	+			
Ophiuroidea	+	(A)			
Echinoidea	+	+	+		
Holothuroidea	+	+	+		
Tunicata		A	A		
Fishes Otolites	+	+	+		

Basically, the mutual interaction between organisms and sedimentation, as summarized in Table 10 is as follows:

a) Mechanically and/or chemically boring organisms (Tables 6 and 12) which contribute to the disintegration and destruction of the carbonate substrate and calcareous shells. The end products of their boring are silt- or micron-sized carbonate particles.

b) Sediment trapping and baffling organisms which bind mechanically deposited carbonate grains with their organic surfaces (e.g., algal mats) or in which fine-grained sediments are deposited between the parts of the organism (e.g., between the branches of a coral colony) in local quiet water environments.

c) Sediment forming organisms, which contribute to rock formation by accumulation of hardparts (e.g., coccolithophorids, foraminifera (see Hallok, 1981), mollusks, crinoids).

d) Organisms encrusted on a hard substrate, which contribute to the origin of solid biogenic structures as well as to cement formation in voids (e.g., coralline algae, Alexandersson, 1977).

e) Frame building organisms whose growth forms favor the formation of larger biogenic structures (e.g., reef corals).

These various interactions aid in the classification of carbonate rocks (e.g., Dunham, 1962, see Sect. 6.2.2).

4.1.2 Micrite (Plates 7, 8)

Taking into account the bimodal grain-size distribution observed in most fine-grained limestones, Folk (1959) differentiated between a microcrystalline, very fine grained "matrix" (microcrystalline calcite, abbreviated: micrite) and sand-sized carbonate grains, which will be designated subsequently (Sect. 4.1.3) as "particles". Frequently a broader matrix definition is advisable: "the part of the rock in which all the allochemical and terrigenous sediment particles are embedded."

In general, micrite is understood to be the very fine grained matrix of carbonate rocks or the micro-grained basic elements of abiogenic and biogenic particles. In the field or in hand specimens micritic limestones are commonly designated as calcilutite, calcareous pelite, mudstone or cryptite.

Subcategories are distinguished according to grain sizes:

- *Minimicrite* (Folk, 1974): grain sizes between < 1 and 1 micron (Plate 8/6),
- *Micrite* (Folk, 1959): grain sizes between 1 to 4 microns or often < 3.5 microns, i.e., between 2–3 microns.
- *Microsparite* (Folk, 1959): grain sizes between 5 to 10 microns, often 5–6 microns.

Minimicrite and micrite are characterized in thin-sections by grain fabrics consisting of anhedral to subhedral calcite crystals with planar and curved grain

Plate 7. Micrite in Solnhofen Limestone, Jurassic

1. Fracture planes of "Fäule" (marls) with spherical void spaces due to coccoid blue-green algae. Lower Tithonian (Solnhofen limestones): Schernfeld near Eichstätt, Southern Franconian Alb, Germany. SEM photomicrographs

2. Surfaces of "Flinzen" (limestone). FE-4 Maxberg/Solnhofen, FE-5 Langenaltheim. SEM photomicrographs

Despite considerable neomorphic alteration, the sizes of the micrite crystals have a maximum of less than 2 microns and are therefore, within the range of Folk's definition of micrite. Solnhofen limestone is regarded as a typical example of the original definition of micrite. The majority of fine-grained limestones in thin-sections which are called "micrite" are actually composed of crystals larger than 5 microns. (From Keupp, 1977)

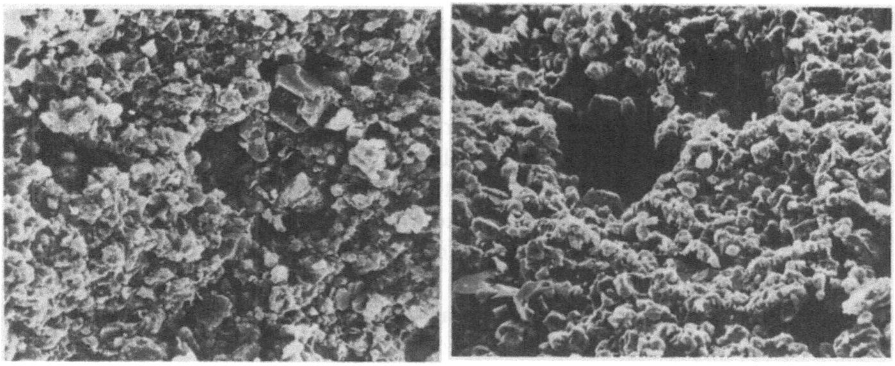

FE-1 5µ S70u-11

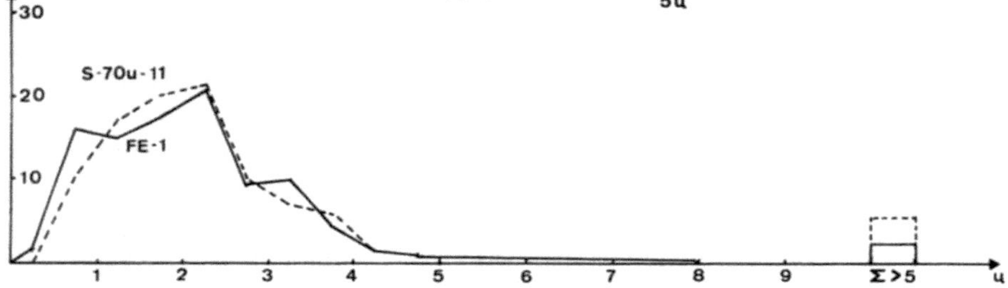

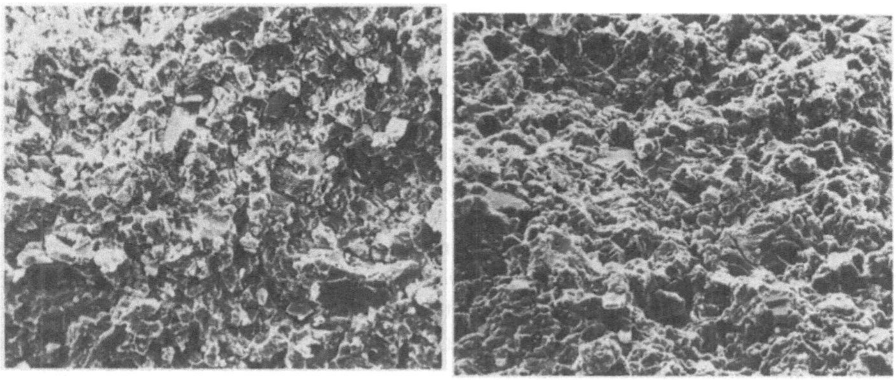

FE-4 5µ FE-5

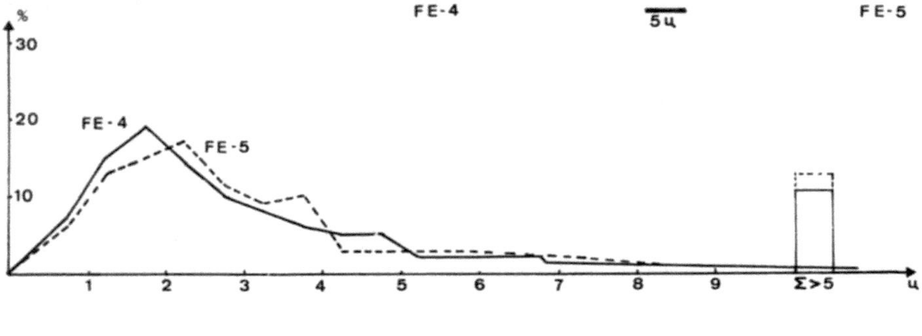

contacts (see E. Flügel et al., 1968); microsparite is characterized by more or less equant subhedral and euhedral calcite crystals (see Folk, 1965, 1974). In correspondence to the model of Folk, carbonate mud consisting of aragonite and Mg-calcite inverts to calcite, whereby the liberated Mg^{2+}, which forms a "cage" around the calcite crystals, prevents a growth larger than 2–3 microns. After Mg^{2+} removal (e.g., freshwater saturation, Longman, 1977) the calcite crystals can continue growing monomorphically to microsparite (see Sect. 3.4) and to pseudosparite (see Sect. 3.4.1). Mg-calcite with large amounts of $MgCO_3$ is severely limited in its possibilities for growth, and thus can form only minimicrite. These minimicrites appear black in transmitted light, and white in direct light. Minimicrite is found in porcelaneous foraminifera, coralline algae, blue-green algae (see Plate 8/6) and in peloids derived from fecal pellets.

Because the above grain-size ranges can only, at the most, be estimated in 30–50 micron thick MF thin-sections, many authors have suggested other upper limits for micrite grain sizes: 10 µm (Elf-Aquitaine, 1975), 20 µm (Dunham, 1962), 30 µm (Leighton and Pendexter, 1962; Schneidermann, 1973). Bossellini (1964) sensibly differentiates micrite I (grain sizes < 4 microns) and micrite II (4–30 microns). Other authors such as Müller-Jungbluth and Toschek (1969) attempt a more detailed grain-size differentiation (aphanite, cryptite, lutite, etc.).

Studies based on electron microscopic grain-size measurements (Folk, 1965; E. Flügel et al., 1968; Gerthofferova and Mišik, 1971; Minoura, 1975; Keupp, 1977) show that most of the micritic limestones possess a clear maximum between 1 and 2 microns. This maximum is also present in the Solnhofen Plattenkalk (Plate 7) designated by Folk as typical micrite.

A more extensive categorization of micrites is only possible in electron microscope studies of limestone fracture surfaces or of polished and etched surfaces (Borza and Harmann, 1970; Fischer et al., 1967; E. Flügel, 1967; Shoji and Folk, 1964).

The name micrite should be reserved for thin-section and SEM studies in line with Folk's definition. In hand specimens fine grained limestones with grains of clay and silt size can be designated as mudstone, calcilutite, or calcisiltite.

In taking into account the modes of genesis of micrites (Table 11), Wolf (1965 a) and Wolf and Conolly (1965) differentiate between "primary" *orthomicrite*, with more or less autochthonous or allochthonous carbonate mud particles (automicrite and allomicrite), and "secondary" *pseudomicrite*, which are the result of neomorphic processes. Honjo (1969) drew attention to the fundamentally important difference between "*nannoagorites*" (Plate 8/1–5) and "*orthomicrites*": The former consist of more than 50% calcareous nannofossils and their particles, exhibit solution structures, and contain relatively little cement. Orthomicrite in Honjo's terms consists of subhedral, polygonal clay-sized calcite grains which meet at interfaces. The term "nannomicrite" (Farinacci, 1968) corresponds to nannoagorite. Mišik (1970: 680) calls micro-grained calcite in organisms skeletons, e.g., within agglutinated foraminifera, "orgmicrite."

Modes of Origin of Micrite (Table 11)

1. An exclusively *chemical precipitation* in conjunction with high salinity and temperature extremes was postulated for carbonate mud in the hypersaline lagoons

Table 11. Interpretation and nomenclature of micritic limestones

<table>
<tr><th colspan="2">Modes of origin for carbonate mud or micrite</th><th colspan="2">Nomenclature of micrite</th><th></th></tr>
<tr><td>Change of water chemism by</td><td>1) temperature increase, salinity variation, and water energy
2) bacterial action and decomposition of organic substances
3) assimilation by plants</td><td>Automicrite</td><td rowspan="2">Orthomicrite</td><td rowspan="3">Primary micrite</td></tr>
<tr><td rowspan="2">Supply of carbonate particles as</td><td rowspan="2">4) submicroscopic calcareous algal fragments
5) disintegrated hard parts of invertebrates
6) borings of endolithic algae, fungi, and invertebrates
7) fine detritus from bioerosion
8) fine detritus from abrasion
9) tests of micro-organisms
10) nanno-organisms</td><td rowspan="2">Allomicrite</td></tr>
<tr><td>Nannomicrite (nannoagorite)</td></tr>
<tr><td>Diagenetic changes by</td><td>11) grain degrading as a result of recrystallization
12) cryptocrystalline cement
13) recrystallized peloids
14) caliche</td><td colspan="2">Pseudomicrite</td><td>Secondary micrite</td></tr>
</table>

of the Bahama Bank (Cloud, 1962). Mineralogical and isotopic chemical data (Broeker and Takahashi, 1966; Husseini and Matthews, 1972) raised doubts about this interpretation (see also Lowenstam, 1966). An inorganic $CaCO_3$ precipitation in a marine environment seems to be possible only then when organisms are present which do not need carbonates for building their skeletons. This may be the case particularly in deeper oceans (see Milliman et al., 1969). In shallow environments chemical precipitation seems to be restricted to protected lagoons, sabkhas, or to lakes.

To what extent the aragonite "clouds," which appear suddenly as "whitings" in warm carbonate-supersaturated shallow seas, are exclusively due to inorganic factors, such as temperature increase, is controversial. In the Bahamas and the Persian Gulf precipitation is caused by CO_2 removal by assimilating algae. Whitings are sometimes due to the activity of fishes living near the bottom (Neumann and Land, 1975).

2. *Bacterial activity and decomposition of organic substances:* The question of bacterial precipitation of calcium carbonate in the sea is still a controversial issue (Alexandersson, 1978). The role of bacteria during carbonate precipitation in the ocean, as discussed earlier by Drew (1913) and Bavendamm (1932), has been confirmed by direct observation (Krumbein and Cohen, 1977; Oppenheimer, 1961) as well as by experimental studies (Adolphe and Billy, 1974; Greenfield, 1963; Lalou, 1957; McCunn, 1972; Deelman, 1975). Trichet (1972) describes the influence of bacteria in decomposed cyanophycean algae on carbonate formation; Maurin and Noel (1977) have drawn attention to similar origins for Devonian micrites.

Plate 8. Micrite in SEM Photomicrographs – Nannomicrite and Algal Micrite

1. Coccospheres of *Ellipsagelosphaera britannica* (Stradner). Lower Tithonian (Lower Solnhofen shale) from Solnhofen, Southern Franconian Alb, Germany. ×4150

2. Coccoliths and partially disintegrated coccospheres of *Cyclagelosphaera margereli* Noel. Uppermost Kimmeridgian of Painten, Southern Franconian Alb, Germany. ×4650

3. Coccolithophorids [including *Cribrosphaerella ehrenbergi* (Arkhangelsky)] are the main constituent of soft chalks. Løgstør, Denmark. ×3750

4. In addition to coccolithophorids, nannoconids (planktonic micro-fossils of controversial systematic position) are of stratigraphic and rockforming importance in Tithonian to Lower Cretaceous pelagic micrites. *Conusphaera mexicana* Trejo. Upper Jurassic (Tithonian, Oberalm limestones): Puch near Hallein, Salzburg, Austria. ×8300

5. In micritic deep water and shallow water limestones, "calcispheres" sometimes appear in rock forming abundance (see Plate 20/6, 7); they are generically defined by their characteristic wall structure. These hollow spheres, probably derived from planktonic algae, are distinctly larger than coccoliths. In the picture *Pithonella gustafsoni* Bolli from Solnhofen limestones in the Southern Franconian Alb, Germany. ×1700

6. In contrast to the "nannomicrites" or "nannoagorites" seen in Figs. 1–5, "algal micrites" feature conspicuously small and equant micrite crystals. This is particularly true for "minimicrites" formed by blue-green algae (see Sect. 4.1.2). SEM photomicrograph of micrite envelopes around filamentous Cyanophytes, with crystals growing within the algal mucus. Lower Tithonian. Kapfelberg near Kelheim, Southern Franconian Alb, Germany. ×1400

Scale is 5 μm

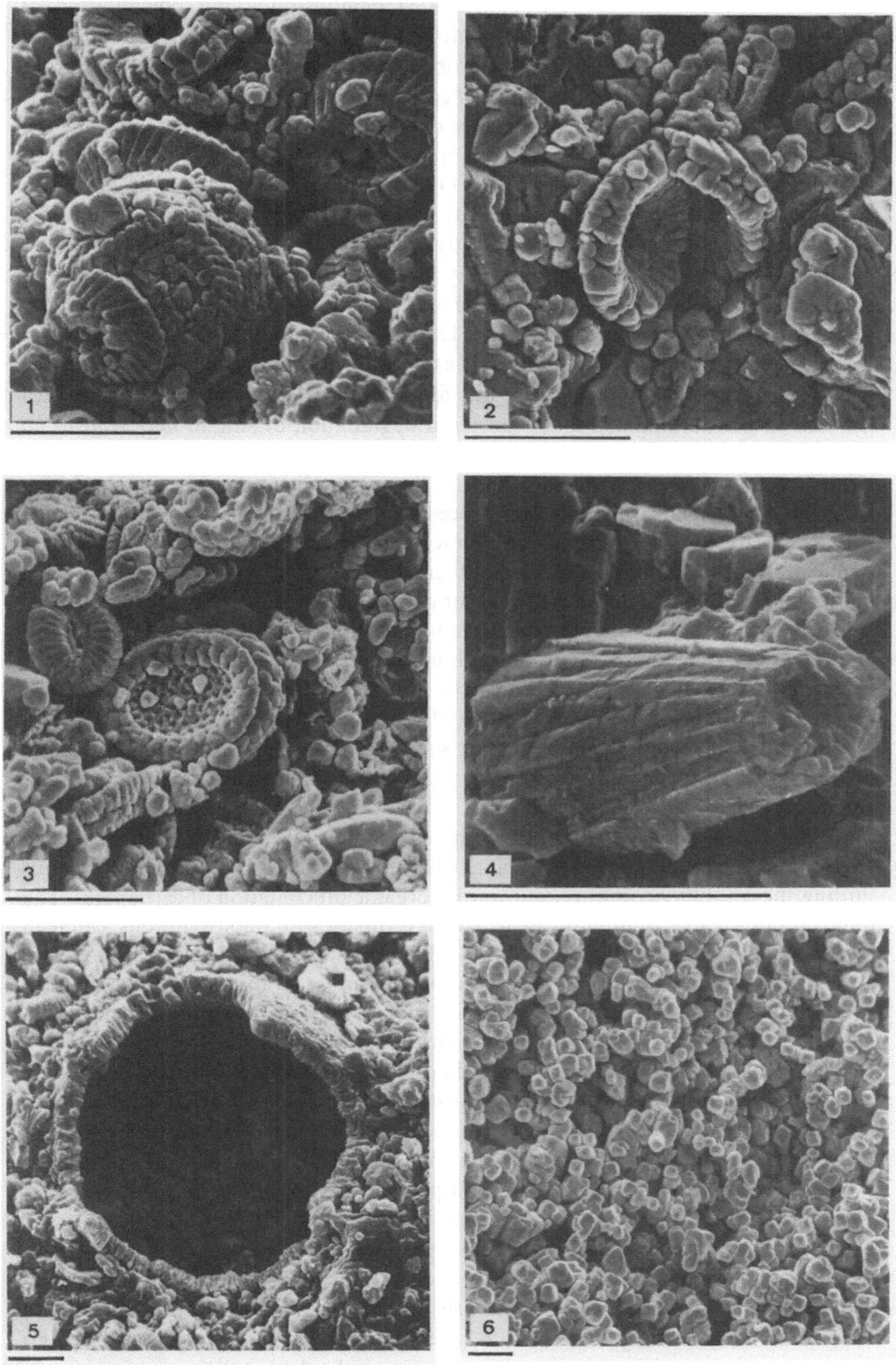
1
2
3
4
5
6

Laboratory experiments by Deelman (1975) demonstrated the precipitation of aragonite through the activity of sulfate-reducing bacteria. The metabolism of marine yeasts is another type of microbial activity capable of mobilizing and reprecipitating considerable quantities of carbonate material. However, very little is known about the quantitative significance of bacterial micrite formation, despite the fact that the origin of Ancient micrite is frequently interpreted by this theory (e.g., Hadding, 1958; Salomon, 1914). Cloud (1962) estimates the amount of carbonate mud sediment formed by bacterial activity to be less than 5% in the area around the Bahama Banks (further discussion in Purdy, 1963: 485).

Basinal black micritic limestones may be formed in deep restricted basins as a result of the bacterial attack of the sulfate ions in solution, which is then followed by the reaction of sulfide with bicarbonate. Evaporitic sulfate minerals can originate in deep barred-basins (see Friedman, 1972 a), but they do not accumulate due to the bacterial reduction that results in the formation of low-magnesian calcite. An increase in pH accompanying the bacterial degradation of sulfate may lead to an additional precipitation of high-magnesian calcite (which is then diagenetically altered to low-magnesian calcite). Basinal limestones in the rock record resulting from these biochemical degradation processes are rich in H_2S; they are blackened with FeS during the sulfate-replacement process. Such basinal limestones (analogous to the sediments of the southern Red Sea) are usually rich in organic matter and may constitute source rocks for petroleum.

The bacterial decomposition of organic substances leads to the formation of ammonia and CO_2 in carbonate-rich solutions. Ammonium carbonate is degraded by bacterial and algal activity, and calcite and aragonite are formed, as shown in experiments by Berner (1971), McCunn (1972), and even as early as Philippi (1907). In this way the decomposition of organic substances can also induce the production of very fine grained carbonates (e.g., "Verwesungsfällungskalk" in the Upper Jurassic sponge reefs, Fritz, 1958).

3. *Plant assimilation, algal activity:* Photosynthetic algae and bacteria change the environmental conditions by reducing CO_2 to organic carbon compounds, thus shifting the solubility equilibrium toward inorganic precipitation of carbonates. The removal of CO_2 by assimilating algae, which is linked with photosynthesis, can lead to the precipitation of aragonite and Mg-calcite (Monty, 1965); bacterial interaction is very probable here, too. This is valid in particular for Cyanophyceae, which form stromatolites in the intertidal and supratidal zones. This mode of origin is of special importance for micrites formed in tidal zones.

4. *Submicroscopic algal fragments:* The bioclastic sedimentation of aragonitic needles, which form the basic element of Codiacean skeletons (*Penicillus*; *Halimeda*, see Plate 28/4), and of brown algae (e.g., *Padina*) has been described for Recent back-reef and bank areas of the Caribbean and in lagoons in the Pacific (Lowenstam, 1955; Lowenstam and Epstein, 1957; Stockman et al., 1967; Matthews, 1966; Neumann, 1965; Neumann and Land, 1975; Stieglitz, 1973).

In situ measurements of the carbonate production rates of algae from a Recent shallow subtropical lagoon show that production rates during one year on one m^2 lie between about 30 and 50 g calcium carbonate for *Penicillus* and *Halimeda,* and 240 g for *Padina*. The growth rates measured suggest that the algae renew their standing stock approximately every four to six weeks during their growing season

(Wefer, 1980). Neumann and Land have shown that in shallow water environments an excess of aragonite needles occurs, which is then transported in suspension to deeper basins (see Fig. 12). A morphological differentiation of "inorganic" and "organic" aragonite needles has not been possible up to now; better criteria are provided by the $^{18}O/^{16}O$ ratio.

5. *Disintegrated hardparts of invertebrates:* As early as 1879 Sorby indicated that after the death of the organism, the shells of various mollusks and corals break up into micron-sized crystallites ("Sorby principle"). Experiments (Purdy, 1963; Chave, 1964; Folk and Robles, 1964) as well as studies of the fine-fraction of Recent carbonate sediments (Matthews, 1966; Stieglitz, 1972; Futterer, 1979) furnish evidence for the great importance of this mode of origin. Here, too, bacteria are probably at work in the further diminuation of the scattered skeletal elements. The degradation of the hardparts takes place in the context of mechanical or biological erosion (mode of origin 7). The manner in which degradation of hardparts takes place is controlled by various factors – abrasion and fracturing in high-energy environments, biological erosion (boring, Table 12) in low-energy environments (Swinchatt, 1965).

The disintegration of skeletal particles (pelecypods, brachiopods, barnacles, coralline algae) into their microscopic structural elements, needles, fibers, flakes, and granules by a natural maceration of the organic parts produces large amounts of lime muds on the sea bottoms in old neritic seas (e.g. the North Sea; Alexandersson, 1979; Fitzgerald et al., 1979). Because of this a great number of potential calcareous fossils are lost from the sediments.

The grain sizes resulting after the destruction of Recent organisms are shown in Fig. 30. As shown by Fütterer (1977), grains smaller than 20 μm can no longer be identified by SEM. A better method, which allows the identification of the main groups of carbonate producers in the silt and clay fraction, is to study the stable biopolymers within the organic tissue (Böhm et al., 1980).

6. *Borings of endolithic algae, fungi and invertebrates:* The boring activity of plant and animal organisms mentioned earlier (Table 6) produces clay- and silt-sized boring products. Of fundamental importance is the role of thallophytes during the formation of micrite envelopes or cortoids (see Sect. 4.1.3.7).

Marine endolithic micro-organisms and grazing organisms are responsible for the erosion in the littoral zone of limestone coasts in the Northern Adriatic (Torunski, 1979). Endolithic micro-organisms, mainly cyanophytes, corrode the rock surface, grazing gastropods and sea urchins rasp away rock particles while grazing on the microflora. Products of these processes are 10–30% dissolved carbonate and 70–90% granular carbonate; shape and size of the eroded particles are determined by the boring pattern of the endolithic micro-organisms, which produce silt to fine sand-sized sediments rather than micritic sediments. This fine-grained material represents 3–25% of the near-shore sedimentation.

7. *Fine detritus from bioerosion:* This mode of origin, which is closely associated with mode 6, occurs only then when organisms searching for food mechanically rasp away the carbonate shells of animals or carbonate substrates with algal growth. This is true in particular of fishes (Gygi, 1975; Frydl and Stearn, 1978), crabs, mollusks, and echinoderms which digest sediment and then excrete it in the form of flakes or relatively compact fecal pellets (see Sect. 4.1.3.2; Folk and Robles, 1974).

Table 12. Bioerosion by boring organisms (after Golubič et al., 1975; Warme, 1975; Boekschoten, 1966; Bromley, 1968; Menzis, 1957; Wetzel, 1938; Torunski, 1979). Carbonate destruction is not only caused by the boring of organisms but also by feeding activity (e.g. fishes which eat mollusks or corals). Herbivorous organisms (e.g. snails which graze on algal mats) and organisms which look for nutrients in sediments (e.g. holothurians, various arthropods) also contribute substantially to carbonate destruction and bioturbation of carbonate mud

Groups of Organisms	Examples	Biotope
A. Fungi and Bacteria	Bacterial attack of organic substances leads to the disintegration of the skeletal grains in the microstructural units. The distribution of boring fungi, in contrast to boring algae, is not dependent on light but is dependent on the amount of organic substance found in the substrate inhabited. The borings, generally independent of the crystal fabric, have diameters < 4 µm, often 1–2 µm. Involved in the Recent formation of micrite envelopes around bioclasts and the micritization of particles, see also B.	Marine and non-marine; shallow and deep seas
B. Algae	Blue-green algae, eucaryotic green-algae and, rarely, red algae represent boring organisms in carbonates and mollusk shells. Depth range is dependent on the intensity and quality of light, because of the role it plays in photosynthesis. Exceptions are known for some blue-green algae (see 4.1.3.7). Species-specific boring patterns are produced. Diameter of the borings < 1 µm to about 100 µ. Involved in the Recent formation of micrite envelopes and micritization.	Preferentially in shallow sea areas down to the lower boundary of the photic zone; often in intertidal and upper subtidal zones. Primary involvement in the bioerosional disintegration of calcareous crusts. Characteristic spatial distribution patterns. Endolithic microborers are also found down to 160 cm below the sediment/water interface in the sediment. They may contribute to the porosity development during early sediment burial.
C. Sponges	Chemico-mechanical boring primarily by species of the desmosponge family Clionidae. Carbonate sediment is removed during the process of boring. Examples: *Cliona lampa* (subtidal zone, Bermudas) removed 6–7 kg/m^2 of carbonate substrate in 100 days; up to 5% of the carbonate sediment in the deeper fore-reef area off the coast of Jamaica is derived from the boring products of *Cliona*.	Preferentially in deeper intertidal and subtidal zones (to depths of several 100 m). Borings in shells and limestones (hard grounds). Destruction of Recent coral reefs.
D. Bryozoans	Elongated and branched borings, forming net-like patterns especially in shells; diameters 100–200 µm.	Nearshore littoral environment; also in brackish waters

E. Mollusks	Boring pelecypods: Recent Mytilidae (with *Lithophaga*) and Gastrochaenidae in carbonates and hard parts. Club-shaped, mechanical, less often chemical borings; diameter in the cm range. In various water depths. Examples: reef area of California: 30–50 m (Plates 12/2; 43/7).	In various marine environments. Boring pelecypods especially in shallow seas
	Boring snails: Naticids. Bore with the help of radula teeth. Examples: *Murex, Cassis; Natica* (diameter of borings 0.7–2 mm). Pelecypods are the major victims of these predators.	Marine, littoral environment, sandy and rocky coasts
F. Worms	Sipunculids and polychaetes; destruction of coral colonies	Marine, particularly in shallow water areas, tidal zones and reef biotopes
G. Arthropods	Cirripedia with *Lithotrya* and Acrothoracicaceae, shallot-shaped borings	Particularly coastal and tidal environments
H. Echinoderms	Chiefly regular echinoids; coastal breakdown (1 cm/year); destruction of coral reefs. Biological erosion and sediment production by grazing sea urchins on the Adriatic coast. In the subtidal zone the erosion rate is 0.96 mm/year (depth 0–1 m) to 0.02 mm/year (depth 9–10 m), corresponding to an erosion rate of 7.5 kg/year/1 m shoreline and 0.15 kg/year/1 m, respectively.	Preferentially in littoral and sublittoral zone; breakdown of tropical carbonate coasts

8. *Fine detritus from abrasion:* In high-energy environments (e.g., breakers) the finest detritus can be abraded, generally with the aid of bioerosion. Lindholm (1969: 270) suggested the name "calcisiltite" for a fine-grained carbonate sediment with grain sizes between 5 and 15 microns, poor sorting, partially oriented grains, and quartz and clay particles, which result from the interaction of abrasion, bioerosion, and organic decomposition. Micritic limestones which originate in such a manner are common, because the combination of abrasion and submarine erosion contributes greatly to the destruction and degradation of limestone coasts (e.g., Adriatic coast, J. Schneider, 1977). Abrasion of limestones on the coasts of the Persian Gulf and Puerto Rico have also been closely studied. As shown by Pilkey and Noble (1966) and Kukal and Saadallah (1973), considerable amounts of eroded calcite and dolomite particles are brought into the Persian Gulf by rivers as well as by aeolian sedimentation.

Calcisiltites and microsparites may sometimes be confused, but the latter have equant grain sizes (see Sect. 3.4).

9. *Tests of micro-organisms:* Foraminifera and pteropod tests contribute greatly to the formation of Recent deepwater carbonates (Sect. 2.2.4). However, the micron-sized elements of these sediments consist primarily of nanno-organisms. Planktonic foraminiferal and nannoplankton limestones are not necessarily restricted to great water depths, as shown by the spatial association of these pelagic limestones with shallow-water reef complexes and with beds of reef-derived, redeposited material (Tertiary, Guam; Garrison et al., 1975).

10. The distribution of *nanno-organism* tests (chiefly coccolithophorids) is determined by the food supply (dependent on temperature and lighting conditions), by the sedimentation rate, bottom currents (redistribution of sediments) and dissolution conditions (see Keupp, 1978). Coccoliths have been known since the Jurassic as a very common element of deep and shallow water limestones (E. Flügel and Franz, 1968; Keupp, 1977). Many Upper Jurassic and Lower Cretaceous pelagic limestones of the Mediterranean region (e.g., *Calpionella* ls.) are quantitatively composed of phytoplanctonic elements like *Nannoconus* (see Noel, 1978). This abundance of coccoliths led to suggestions that such micrites should be named nannomicrites, nannoagorites, or "ultramicrites" in order to distinguish them from the other micrites formed by chemico-biochemical and bioclastic factors (see Plate 8/1–5).

The modes of origin sketched below for "pseudomicrite" correspond to changes in the course of diagenesis:

11. *Grain diminuation by recrystallization* is seen in Recent *Lithothamnia* (e.g., Wolf, 1965b) and Devonian algal limestones. More recent observations of Alexandersson (1977) point out that grain degrading essentially involves an early cement formation within the algal framework. The recrystallization of calcarenites can also lead to the formation of calcilutites (Example: Lower Carboniferous in Ireland, Wardlaw, 1962), where sparite is dissolved and micrite precipitated ("sparmicritisation," Kahle, 1977: 415).

12. *Cryptocrystalline cement* (Plates 3/1–4; 4/3–4): Many carbonate cements, in particular those made of Mg-calcite, are composed of micrite-sized crystals (see Sect. 3.3.1). Distinguishing between diagenetic micrites and micrites already present as a depositional fabric is difficult in thin-sections.

13. *Recrystallized peloids, "clotted structure"*: Many micritic limestones exhibit a "clotted fabric," which Cayeux (1935: 271) calls "structure grumeleuse." This designates a micrite (Plate 50) which tends to crumble into subrounded elements, which then appear to be partially or totally embedded in a somewhat coarser crystal mosaic (sometimes microsparite-sized). The "clots," which are usually 50–150 microns in size, generally exhibit only very indistinct boundaries; some of the "particles" have the appearance of being fused. There are two theories for this, namely, that the matrix (micrite) recrystallizes (Schwarzacher, 1961) or that mud particles disintegrate early and "fuse" during compaction; at present the latter theory is preferred (e.g., Illing, 1954; Bachmann, 1973). It must therefore be taken into account that micrite may only correspond to a "pseudomatrix" (Dickinson, 1970) which is formed by the deformation of weak detrital grains in connection with compaction and pressure-solution. Fabricius (1966) points out the possible significance of endobiontic organisms and detritus-feeders for the origin of the clotted structure.

14. *Caliche-micrite* (Sect. 2.3.2.2; Plate 2/4–7)

It must be emphasized that the possibilities for interaction as outlined here involve only schematic causal complexes. In Recent shallow seas often several genetic factors work together (particularly modes 2 and 3 as well as 4–7). Calcareous mud sedimentation in deep-seas is controlled by modes 9 and 10 and sometimes by inorganic precipitation.

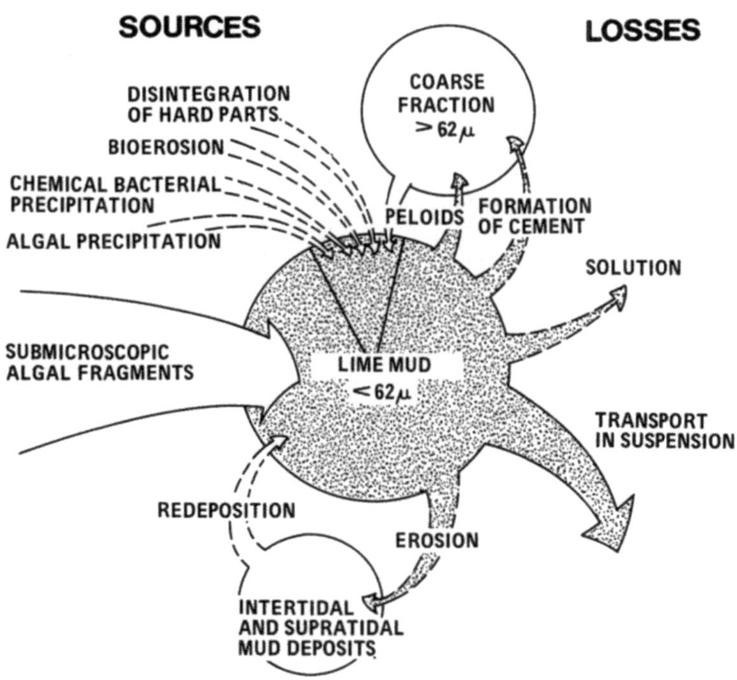

Fig. 12. Calcareous mud budget in the Bight of Abaco, Bahamas. Simplified after Neumann and Land (1975). Details in text

Figure 12 shows the calcareous mud budget of a deep marine lagoon, roughly 7 meters deep, in the Bight of Abaco (Bahamas), which illustrates the many factors active during sedimentation production and removal. The supply of bioclasts and a chemico-bacterial precipitation are both minimally significant. The mass of carbonate mud material is made available by submicroscopic algal fragments. Although some of the calcareous mud is lost during the formation of the cement, during the formation of peloids, and by erosion and removal in the intertidal and supratidal zones, so much carbonate mud is formed in this way that removal of the fine grained sediments takes place in the surrounding deeper basins. The fine mud transported in suspension forms mud mounds ("lithoherms," Neumann et al., 1972).

Enormous difficulties are encountered when Ancient micrites are genetically interpreted; these difficulties can, however, be surmounted by the combined use of microscopic and submicroscopic methods of analysis. Experimental investigations are also of primary importance for elucidating the problem of chemical and biochemical precipitation of fine-grained carbonates (see Lippmann, 1973; Ricketts, 1980). After SEM investigations, "algal micrites" (mode 3, possibly 2) appear to be characterized by conspicuously equant and homogeneous fabrics (minimicrites, see Plate 8/6). In general one must assume that not all the known Recent modes of origin can also be assumed for Ancient micrites. This holds true even for mode 4, which is so very significant in the Recent period; mode 10 is somewhat of an exception (nannomicrite, not until the Mesozoic?). On the whole, though, calcareous algae, which are morphologically and systematically comparable to those Recent green algae which provide aragonite needles, are not known until the Upper Triassic (*Boueina*), and more abundantly not until the Lower Cretaceous.

4.1.3 Limestone Particles (Plate 9)

Particles are defined as mechanically deposited grains which originate before sedimentation. Categories of particles used during microfacies studies are listed in Table 13. The basic categories are skeletal grains, micritic round bodies (peloids), aggregate grains, oncoids, ooids, cortoids, and resediments (intraclasts and extraclasts).

Discussions of definitions and origins are found in Bosellini (1964), Catalov (1970), E. Flügel and Kirchmayer (1962), Kendall (1969), Kerkmann (1966), Monty (1963), Skupin (1970), Wilson (1967), Wolf (1965 a), and Elf-Aquitaine (1975).

4.1.3.1 Skeletal Grains

A distinction between non-fragmented and fragmented skeletal grains appears to be just as meaningful as the common subdivision of bioclasts in microfacies studies (Grabau, 1904) according to:

a) size: siltite – up to about 20 microns; fine and coarse arenite – up to 2 mm; rudite – larger than 2 mm,

b) roundness (see 4.2.2), and c) "circumcrusts" (=cortoids, see 4.1.3.7 and Plate 10).

Six types of skeletons can be differentiated on the basis of their resistance to fragmentation, with reference to Ginsburg (1956) and Wilson (1975):

1. Needles embedded in organic tissue yield scattered particles. Examples: sponges, octocorals, holothurians, tunicates, green algae (e.g., *Penicillus*).
2. Segmented skeletons yield arenitic particles. Examples: crinoids, articulated red algae (e.g., *Melobesia*), green algae (e.g., *Halimeda*; Dasycladaceae).
3. Branched skeletons yield non-equant fragment sizes. Examples: corals (*Acropora*; "*Disphyllum*" in the Devonian, "*Thecosmilia*" in the Mesozoic), bryozoans, calcisponges, red algae.
4. Chambered skeletons are in general relatively resistant. Examples: gastropods, serpulids, brachiopods, some pelecypods and foraminifera.
5. Encrusted skeletons are usually resistant to mechanical breakup. Examples: sessile foraminifera, stromatoporoids, corals, bryozoans, serpulids, red algae.
6. Massive skeletons are most resistant to breakup because of their size. Examples: coral colonies; coralline algae.

The fragmentation of hard parts is controlled by the destruction of the organic tissue and microstructural types (see Force, 1969), primary porosity of the skeletons, mode of life (endo- or epibiota), boring organisms, existence of predators, mechanical breakdown of the skeletons in a high-energy environment, transportation and isolation of the hardparts, and dissolution.

The average size of the organic remains more frequently found in thin-sections is seen in Fig. 13. In Sect. 5 a few of the important "thin-section fossils" will be discussed in more detail.

4.1.3.2 Peloids (Plates 1/3; 9/1; 10; 31/1; 37/2)

Peloids are commonly recorded as major contributors to intra-reef sedimentation in Recent and Ancient environments, particularly those peloids accompanied by structureless micrite. Pelleted and unpelleted micrites between reef-building organisms often show distinct intergradations. The origin of peloids found in micritic reef limestones may be explained by the prolific boring activity of clionid sponges or molluscs, by fecal pelleting of invertebrates, or by a clotting of intra-reef Mg-calcite precipitates (Pedley, 1979). Because "pellet" is often considered a synonym for "fecal pellet" in the English literature, McKee and Gutschick (1969) have suggested the name "peloid" for all particles of various origin which consist of cryptocrystalline carbonate.

Refer to Table 13 under *Synonyms* to the term "*pseudooid*": The original definition (Bornemann, 1886: 227) was applied to worn carbonate rock and mineral fragments which the author observed in thin-sections of "Muschelkalk". Later the definition was expanded and now designates – especially in the French literature – well-rounded, non-laminated, structureless micritic particles of various origin. The term should no longer be used now (despite the recommendations of E. Flügel and Kirchmayer, 1962), because it leads to false associations (pseudo-"ooids"), is unsuitable for compound words and, moreover, coincides with the term peloid. Adequate German terms for pellet are: "Pelitkörner" and "Pille" (less satisfactory).

Continued p. 130

Plate 9. Limestone Particles in Thin-Sections ►

1. Recent ooids. Relatively thin concentric laminae around nuclei consisting of peloids and gastropod shells. Distinct tangential structures. White Banks, Great Bahama Bank. Scale is 0.5 mm

2. Ooids in various stages of compaction (see Section 3.4), which are recognizable as ooids with pitting marks -1, distorted ooids -2, and cracked ooids -3. The carbonate cement is partially made up of meniscus cement -4 (indication of vadose diagenesis). In addition to ooids a few cortoids -5 appear. The occurrence of large normal ooids and small superficial ooids corresponds to Situation 4 of the Carozzi model (see Section 4.1.3.5, Fig. 19). As in many oolites, benthonic organisms, which avoid unstable bottoms, are absent in this sample. Middle Jurassic: Central Iran. Scale is 0.5 mm

3. Radial fibrous ooids. Normal ooids -1 and broken and regenerated ooids -2 ("hiatus ooids") in micritic matrix with areas of sparite (birdseyes -3, see Section 4.2.3.5). Shape and structure of the ooids indicate, along with an autochthonous occurrence in low-energy micrite environments with few organisms, a relatively restricted sedimentation area with increased salinity (see Table 18). Lower Cretaceous: S Cusano Mutri, Southern Apennines, Italy. The sample comes from the cyclic limestone sequences described by Simone (1974) and identified as a tidal facies. Scale is 0.5 mm

4. Calcareous ooids in arenaceous carbonate matrix. Original thin-section of Kalkowsky (1908). The ooids of the Bunter (sandstone), which were dealt with more recently by Usdowski (1962) and Haage (1970), are characterized by combinations of various shapes. In the picture spherical normal ooids as well as cerebroid compound ooids (German "Ooidviellinge") are characterized by cone structures. Rogenstein, Lower Bunter sandstone: Sangerhausen near Bad Harzburg, Germany. Photo H. Zankl. Scale is 0.5 mm

5. Pisoid. In comparison to ooids, striking features are the larger size and the very distinct but discontinuous and terraced lamellae structure. Pisoids can be distinguished from biogenically formed oncoids by the extremely fine lamination and the absence of organic structures. In addition, pisoids frequently occur in crust limestones derived from caliche which have very few or no fossils (see Plate 2). Middle Triassic (Wetterstein limestone): Reißkofel/Gailtal Alps, Carinthia, Austria. Scale is 5 mm

6. Tangentially structured, large normal ooids with distinctly abraded laminae together with small ooids, some radially structured. Conspicuously poor sorting. These are transported ooid sands deposited on the margin of an oolite bar (Franke et al., 1975). Lower Carboniferous (Ilpen oolite, Tournaisian): Heiligenhaus, Kettwig, Germany. Scale is 1 mm

7. Aggregate grain characterized by very irregular outline, distinct invaginations and protuberances, "inclusions", and peripheral overgrowth on organisms (sessile foraminifera at the top), and a relatively large amount of micritic matrix. These are "bio-lumps" according to the definition given in Section 4.1.3.3. The two cement types A and B are clearly recognizable in the sparry "groundmass". Middle Permian: Velebit, Croatia, Yugoslavia. Scale is 1 mm

8. Algal oncoids with distinct nuclei consisting of shells and irregular concentric biogenic coatings of cyanophycean algae (*Girvanella*). As shown by the indentations on the surface of the oncoid, made by fusulinid foraminifera tests, the algal oncoids were not yet lithified when they were embedded. For proper orientation turn picture 90° clockwise. Lower Permian (Grenzland formation): Rattendorf Sattel, Carnic Alps, Austria. Scale is 5 mm

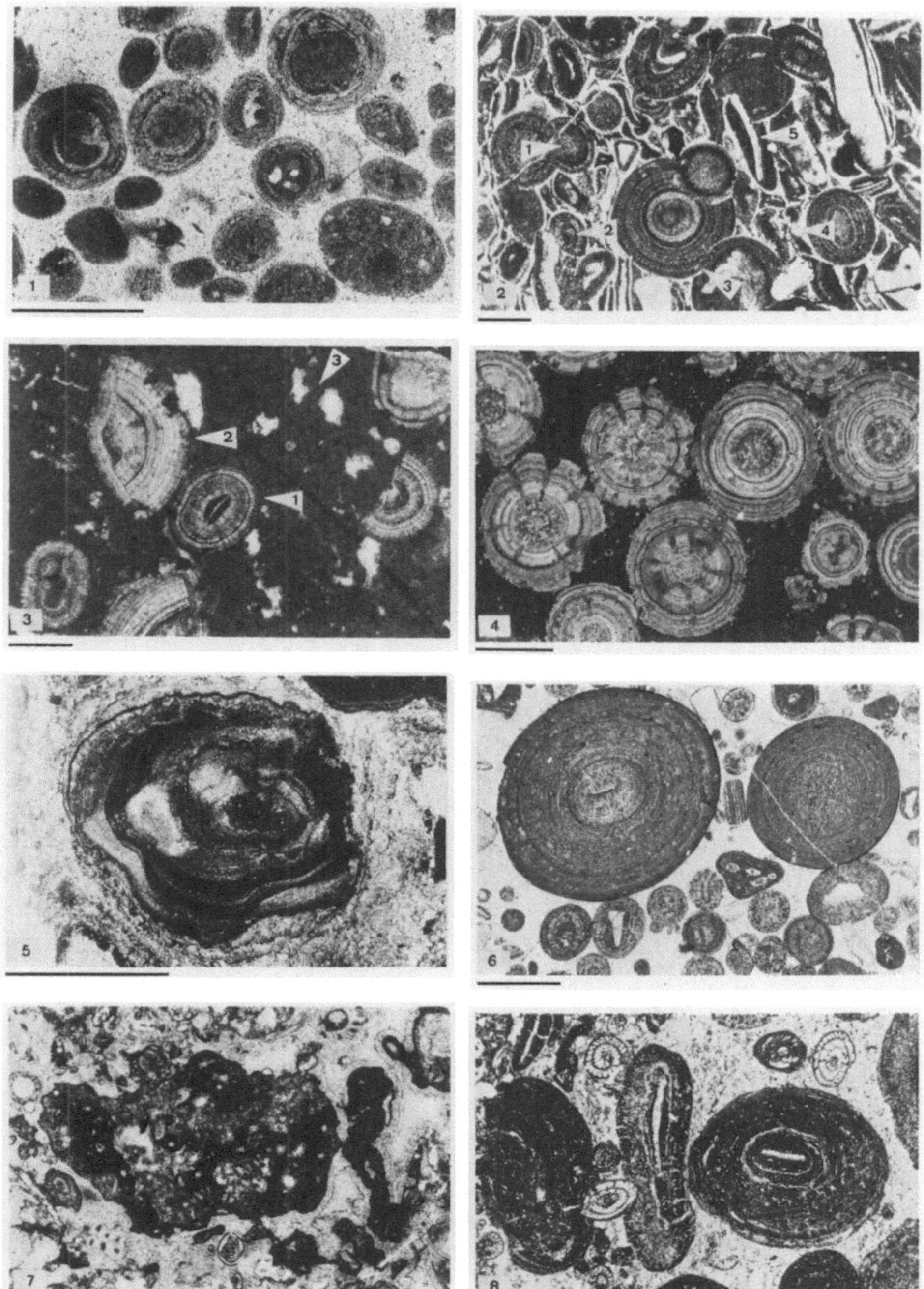

Table 13. Microfacies particles in carbonate rocks. For the most part the categories have a descriptive character, whereas the subcategories have a genetic character. Ooids and oncoids are listed together as "coated grains". Peloids, oncoids, ooids and aggregate grains may all be called globoids. The intraclasts and extraclasts, which can not always be clearly distinguished from one another, are listed under the lithoclasts or limeclasts. Algal oncoids and foraminifera-(algal) oncoids belong to the skeletal oncoids, as defined by Riding (1977). * = first definition

Category	Definition	Modes of Genesis			Discussion
		Subcategories	Origin	Characteristics	
Skeletal grains	Hard parts of organisms appearing in thin-sections	Biomorpha Syn: biosomata Monty (1963)		Non-fragmented hard parts (e.g. bivalved pelecypod shells), almost whole fossils (e.g., coral colony)	
		Bioclasts	Biological, mechanical and chemical destruction. Allochthonous or parautochthonous deposition	Hard part fragments (e.g., crinoid ossicles) and broken, partially rounded skeletal grains (= sphaeroclasts), arrondids (Paulsen, 1964: 9); size: sand- and silt-sized	
Peloids (= pellets) Syn: Pseudooid Bornemann (1886), chondroid Monty (1963), cryptolithe Catalov (1970), Schlammkörner Fabricius (1966)	Micritic, subrounded particles, usually without any internal structure. Outline usually elliptical or circular. McKee and Gutschick (1969) *	Fecal pellets – Pl. 37/2	Excrement of sediment and plankton feeders (e.g., crustaceans, holothurians, gastropods, fishes), often produced by endobiontic organisms (peloids together with bioturbate structures!)	Elongated micritic particles: often conspicuously black due to much organic matter, often in clusters, "good sorting". Some exhibit sieve-like structure (cross-section); (e.g., anomuran coprolites; size: 0.02–0.50 mm. Ratio width/breadth = 1/2 to 1/4. See Pl. 18/13)	Brown (1964) Brönnimann (1976) Bandel (1974) Skupin (1969, 1970)
		Algal peloids – Pl. 39/1; 44/5; 49/1; 50/3	Spherulitic carbonate encrustations around blue-green algae or decomposition products of various algae	Subrounded micritic particles, sometimes indistinct algal structures still visible (filamentous algae), often together with calcareous algae. Size: 0.05–0.20 mm	Friedmann et al. (1973) Wolf (1965 b)

		Pseudopeloids Syn: "small intraclasts" – Pl. 31/1; 37/2	Reworking and rounding off of consolidated calcareous mud or aggregates of aragonite and calcite crystals. Partially coincides with very small intraclasts	Subrounded micritic particles, very finegrained	Fahraeus et al. (1974) *
		Bahamite peloids	Bioclasts or ooids totally micritized by boring algae, fungi, etc.	Subrounded and angular micritic particles often together with bioclasts or cortoids micritized to various degrees	Beales (1958) * Purdy (1963) Gygi (1969)
		Pelletoids	Recrystallized particles, e.g., rounded off bioclasts (= arrondids)	Subrounded micritic particles generally without internal structures	Blatt et al. (1972: 422) * Paulsen (1964)
Aggregate grains Syn: "Kornaggregatklumpen", "Traubenklumpen" Sarntheim (1965), botroidi Sestini (1964), bothrolite, (chondrolite Monty) (1963), colloclasts Sander (1967), tuberoid Fritz (1968: 77) "Bahamite" auct. – Pl. 1/1	Irregular lobular aggregates of particles agglutinated by sparitic or micritic cement or by algae. Particles: generally only a few strongly micritized ooids, bioclasts, and peloids. Less cement than particles. The transitions between the categories distinguished here are flexible	Grapestones Syn: bothryodal lumps Wilson (1967); composite grains – Pl. 1/1; 33; 38/1	Cementation of hard carbonate particles in areas with very low rate of sedimentation, good water circulation and changing levels of agitation. The intergranular cement formation is strengthened by synsedimentary algal encrustations	Grape-like clusters of particles generally cemented together by sparite. Often together with micritized particles and algae. Size: ~ 0.5–2 mm	Illing (1954: 30) * Sudgen (1963) Roberts and Moore (1971) Winland and Metthews (1974)
		Algal lumps Syn: Algenklümpchen Pümpin (1965) – Pl. 9/7	Similar to grapestones, however, more interaction of organisms. Often in protected environments (lagoons etc). Micritic particles with algal encrustations found together with inhomogeneous micritic fabrics can be called algal intraclasts	The particles of aggregate grains are agglutinated by algae and/or sessile foraminifera (e.g., *Bacanella, Lithocodium*). The single aggregate grains are often bound together by algal encrustations. Filamentous algae dominate (e.g. *Cayeuxia, Zonotrichites*)	

Table 13. (continued)

Category	Definition	Modes of Genesis			Discussion
		Subcategories	Origin	Characteristics	
		Lumps, Syn: amorphous lumps Wilson (1967); cryptocrystalline grains Purdy (1963) – Pl. 1/4; 31/6; 38/1; 42/9; 47/1	Similar to grapestones, but > 50% micrite as bonding agent. Many of these are micritized and recrystallized grapestones	Particles bonded by micritic cement	
		Caliche lumps – Pl. 2/4	Cementation within the vadose zone	Similar to marine lumps, but usually found together with pisoids, laminated crusts, etc. (see 2.3.2.2). No scattered particles, generally in sediments without fossils	
Oncoids Syn: Biopisolith Catalov (1970), Mumies Steinmann (1880), "algal biscuit"!	Irregularly formed particles with nonconcentric, partially overlapping micritic laminae, derived from biogenic deposition around a nucleus (e.g., organisms) by algae, foraminifera, etc. mm- to cm-sized. A. Heim (1916)*	Algal oncoids – Pl. 9/8; Pl. 10; 39/2; 42/1	Biogenic coating by algae around a solid nucleus; often cm-sized	The algae responsible for the formation of oncoids are preserved as fossils. Algal groups: a) Cyanophyceae (e.g., porostromate algae such as *Girvanella*, microstromatolites), b) Rhodophyceae (e.g., Corallinaceae), which form "rhodolites" – Pl. 29/2	Gasche (1956) Babic (1969) Ihrion and Müller (1968) Riding (1977)
		Foraminifera – (algal) oncoids – Pl. 29/4; 42/1	Biogenic coating by sessile foraminifera (and algae) around a solid nucleus. Often cm-sized	The tests of the foraminifera grow upon micritic oncoid laminae. Examples: *"Sphaerocodium"* from the Carnian, some *"Osagia"* from the Late Paleozoic	Johnson (1946) Peryt (1977)

		Micritic oncoids Syn: Einfachonkoid Müller-Jungbluth and Toschek (1969); circumcrusts Wolf (1965 b) – Pl. 10; 30/6	Biogenic coating by algae (and bacteria?); possibly also thick micrite envelopes formed by boring algae, if so, then indistinct boundary between nucleus and "envelope"	Micritic coating of a solid nucleus, generally without distinct biogenic structures	
Ooids	Regularly formed generally spherical to elliptical particles with a nucleus surrounded by concentric laminae sometimes displaying radial structures. Kalkowsky (1908) *	Superficial ooids (Einfach-Ooide) Syn: Proto-Ooid, Rinden-Ooid Kremer (1953); Mantel-Ooid Schmidt (1961) – Pl. 1/2; 32/2	In relatively low-energy environments or in areas with an excess of large nuclei, whose diameters approximate the critical size of marine ooids (generally < 2 mm). Formed by an interaction of inorganic and organic processes	One or only very few laminae of various thicknesses around a large nucleus	Illing (1954) *
		Ooids (Normal-Ooide) Syn: true ooids Wilson (1967) – Pl. 9/1–4, 6; 31/3; 32/2; 47/4	In high energy shallow waters supersaturated with $CaCO_3$. Inorganically formed with interaction of organisms (blue-green algae). Ooids having radial structures appear to prefer waters deviating in salinity from normal marine conditions	Generally many concentric laminae around a nucleus, diameter usually between 0.20 and 0.60 mm. Upper limit often 1 mm. "Ooid-aggregates" (Ooidbeutel, etc.) often occur. Sometimes deformed by synsedimentary fragmentation or compaction or tectonic activity (distorted oolites)	Bausch (1963) Bornemann (1886) Carozzi (1963) Usdowski (1963) Fabricius (1967) Schloz (1972) Haage (1970) Loreau and Purser (1973)
		Micritic ooids – Pl. 47/4	a) "Pelagic ooids", formed by blue-green algae, which may trap planktonic particles or b) micritized normal ooids	Concentric laminae of micrite, diameter between 0.20 and 0.40 mm	Jenkyns (1972) Purser (1980)

Table 13. (continued)

Category	Definition	Modes of Genesis			Discussion
		Subcategories	Origin	Characteristics	
Pisoids Syn: diagenetic ooids Siesser (1973) – Pl. 2/6, 7; 9/5; 14/1, 14/3	Conspicuously regular concentric micritic laminae, large particles (generally > 2 mm) often broken up. Sometimes vadose silt in the outer laminae; nuclei usually rock and pisoid fragments. Repeated in situ brecciation during growth and formation of caliche crusts		Inorganic formation with fresh-water influx (vadose zone), sometimes caliche formations (subaerial, arid climate), sometimes cave pearls		Kirchmayer (1964) Donahue (1965) Gradzinsky and Radomski (1967) Dunham (1969) Siesser (1973) Adeleye (1975)
Cortoids (= grains with micrite envelopes), "coated grains" – Pl. 5/2; Pl. 9/2; Pl. 10; Pl. 11/3; 47/1	Rounded bioclasts, lithoclasts, and peloids with thin micrite envelopes, "boundaries" between "nucleus" and "envelope" indistinct; tubes in the periphery of the nuclei either empty or filled with micrite (diameter often about 5 µm)		Boring Cyanophyceae, Chlorophyceae and fungi produce tubules filled with micrite in bioclasts, etc. The repetition of this process leads to "micritization" (destructive genesis). Constructive genesis is caused by early cementation of filamentous algae.		Bathurst (1966) * Swinchatt (1969) Loreau (1970a) Rooney and Perkins (1972) Sibley and Murray (1972) Friedman et al. (1971) Golubić et al. (1975)
Intraclasts Syn: Endoclast Monty (1963)	Synsedimentary resediments often with inclusions of other particles, mm- to cm-sized, generally poorly sorted and not well-rounded < 0.2 mm, often hardly distinguishable from peloids. Folk (1959) *.	Plasticlast (= colloclast) Sander (1967)	Reworking of weakly consolidated lime mud by waves or currents disturbing the bottom ("mud pebbles")	Micritic particles usually with inclusions, generally < 0.5 mm. Composition is the same as the matrix	Folk (1959) *
		Algal Intraclast Pl. 11/2; 47/3	Reworking of lime mud covered by algal films	Only slightly separated micritic particles of various sizes, with porostromate and spongiostromid algal structures	

		Autoclast (= protointraclast Bosellini, 1964)	Early diagenetic movement of sediment, together with submarine or subaerial dehydration of the sediment	Separated pieces of sediment whose boundaries still fit, as in jigsaw puzzles	Sander (1936) * Bosellini (1964)
		Pseudo-intraclast	Sediment burrowing by endobiontic organisms	Bioturbated, sometimes separated pieces of micrite	Wobber (1965: 685) *
		Caliche-intraclasts – Pl. 2/4	"Pieces" of caliche formations, diagenetic structures	Together with crusts, pisoids, etc. see 2.3.2.2	
Extraclasts Syn: Alloclast Folk (1959), Lithoclast s. str. Folk (1959), exoclast Monty (1963); gravelles; calclithite fragments – Pl. 11/1; 31/5; 48/1;	Postsedimentary resediments, carbonate and noncarbonate rock fragments, often angular; particles (e.g., fossils) contained in the fragments often truncated. No compaction features. Often polymict. mm- to cm-sized Folk (1959)*		Erosional debris of harder (earlier) rocks, which were eroded by subaerial or submarine processes		Wolf (1965) *
Terrigenous particles	e.g., clay minerals, terrigenous quartz		Minerals carried into the sedimentary basin		
Authigenic minerals	e.g., glauconite, authigenic quartz, pyrite		Formation of new minerals within the sediment		

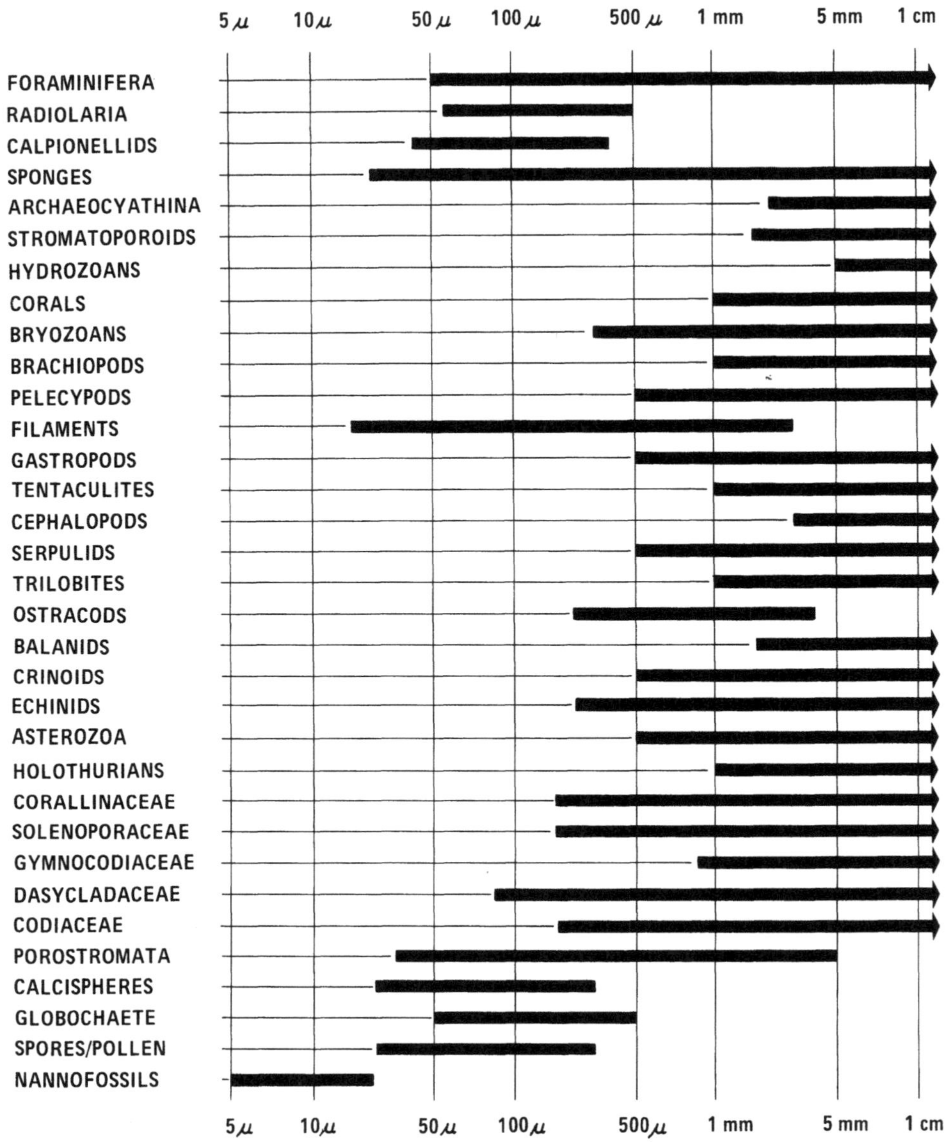

Fig. 13. The average sizes of skeletal grains in thin-sections

Peloids are generally smaller than all other kinds of particles. Recent peloids, which frequently are fecal pellets, often have a diameter of less than 1 mm (see Fig. 15). Ancient peloids have similar dimensions (see Fig. 14). It is quite noticeable that the calcareous peloids – independent of origin or geological age – are rarely larger than 0.5 mm, very often lying within a range of 0.05 to 0.20 mm.

Of all the modes of origin listed in Table 13, it is easiest to recognize peloids derived from fecal pellets (mode 1) and reworked carbonate mud particles (mode 3).

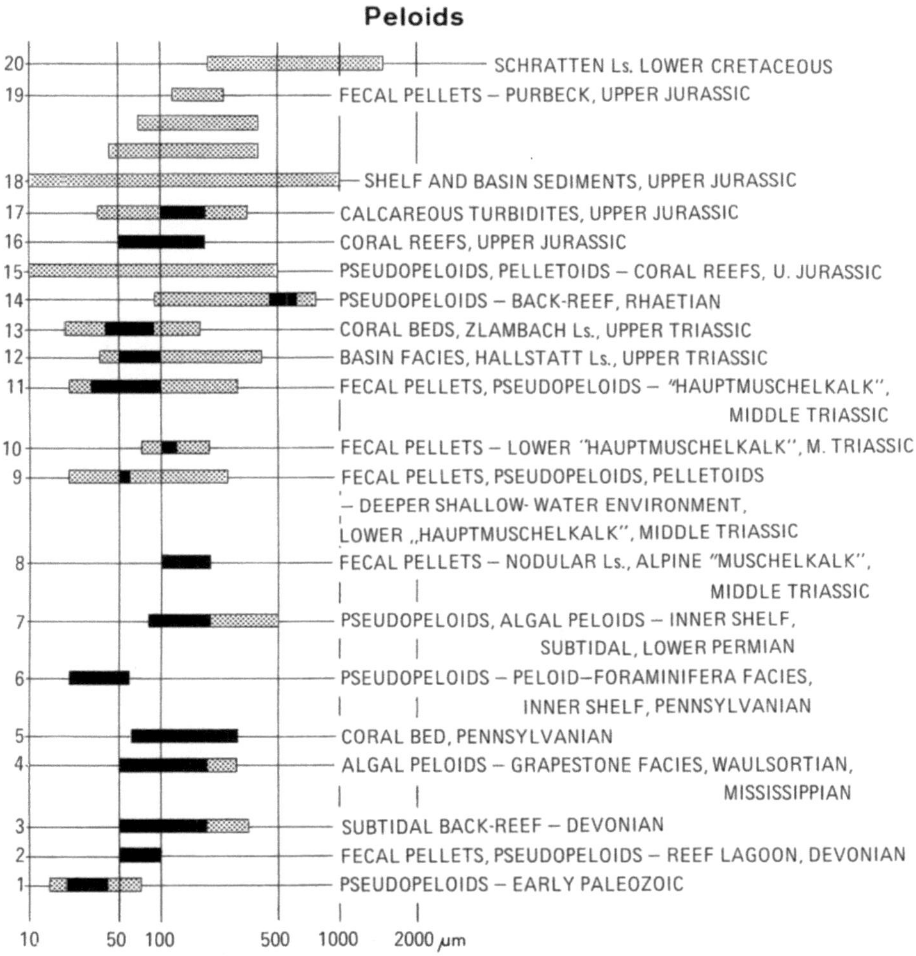

Fig. 14. Sizes of Ancient peloids, compiled after various sources, including the author's own observations. *Black bars* represent commonly occurring values

Discussion of Peloids

1. *Fecal pellets* (small coprolites) are characterized as follows

- Spherical, ovoid, or rod-shaped outline
- Length: width ratio of 2 : 1
- Well-rounded
- Often thin dark rim (organic substance) helps define outer boundaries of pellets in a micritic matrix
- Extremely fine-grained (minimicrite, see 4.1.2); sometimes, however, aggregates of various-sized crystals occur too
- Gregarious occurrence

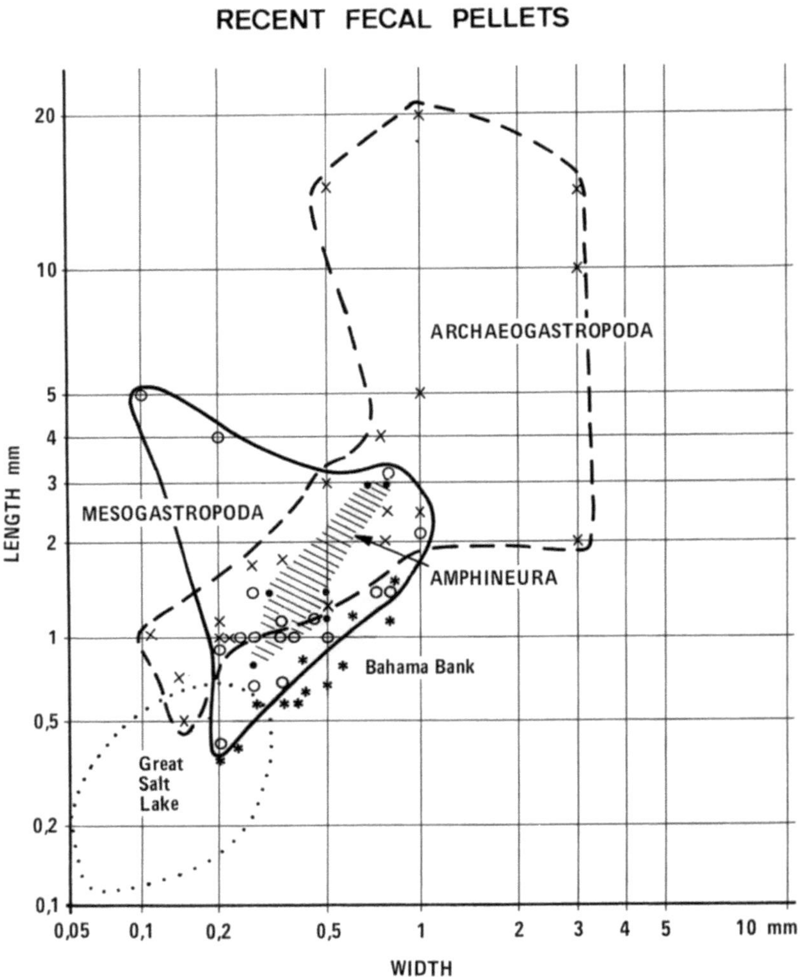

Fig. 15. Sizes of Recent fecal pellets. Compiled after Bandel (1974), Bathurst (1971), Cloud (1962), Eardley (1939), and others

- Good "sorting"
- Often concentrated in isolated nests in small burrows
- Large amount of organic substance
- Often altered to glauconite, pyrite, or phosphorite
- Sometimes with sieve-like, lamella-like, and ring-like internal structures:

These structured fecal pellets are derived from crustaceans (thalassinids, decapods). Fecal pellets with a length of up to 10 mm may be produced by the crab *Callianassa.* A more exact classification of these microfossils, which are quite common in the Mesozoic, seems meaningful because of the stratigraphic information they provide (see Senowbari-Daryan, 1978).

Fecal pellets are especially abundant in micritic limestones. Producers of these particles are: brachiopods, amphineura, gastropods, pelecypods, worms, ostracods, copepods, anomuran decepods, echinoderms, and tunicates. In the straits of Florida more than 100 species are involved in the production of fecal pellets. Fecal pellets are produced by detritus-feeders and herbivorous animals as well as by plankton-feeders such as copepods and fishes. The fecal pellets of nektonic animals contribute substantially to a more rapid sedimentation of coccolithophorids (Hattin, 1975; Honjo, 1976).

Fecal pellets are a dominant constituent of Recent subtidal and shallow intertidal coastal areas, where they occur preferentially in low-energy zones. Although they generally have a soft consistency, fecal pellets can be transported without destruction even into high-energy environments if early intergranular cementation has taken place (Wetzel, 1937). A rapid cementation of fecal pellets is possible particularly in lower-energy shallow-water areas supersaturated with $CaCO_3$. However, fecal pellets are also often found in deep-sea sediments (see Standard Microfacies Type 2).

A good example of how important these particles are for the make-up of the sediment can be seen in the region of the Great Bahama Banks, west of Andros Island; here the "pellet-mud facies" stretches over an area of about 10,000 km^2 and consists of peloids which comprise up to 75% of the total sand fraction (up to 38% of the total sediment). This is a shallow area (water depth 2–6 m) with minimal water energy and increased salinity ($>40‰$). Despite the fact that the number of species and form diversity is low, most of these peloids are fecal pellets.

2. *Algal peloids* occur together with other algal structures (stromatolitic crusts, oncoids) and with calcareous algal bioclasts. They can originate as

- Spherulitic carbonate encrustations around coccoid blue-green algae, particularly within algal mats (Recent: Bahamas – Monty, 1967; Gebelein, 1969; hypersaline coastal ponds – Friedman et al., 1973; lacustrine environment – Scholl and Taft, 1964. Ancient: lacustrine – Buchbinder et al., 1974). The size of the spherulitic peloids is 150–500 microns.
- Organic detritus of various algae: Peloids originate from bioclasts of green, red, and blue-green algae after boring, transport and recrystallization. Wolf (1965 b) described the manifold possibilities for destruction of algal fragments, as illustrated in a Lower Devonian algal reef.

In contrast to fecal pellets, algal peloids are characterized by poor "sorting," irregular shapes, and diverse sizes (20 to 2000 microns). Often they are found together with aggregate grains; some of the peloids probably are derived solely from the breakup of these aggregate grains.

According to SEM photomicrographs, algal peloids as well as fecal pellets consist of equigranular micrite grains between 0.5 to about 2 microns in size.

Ancient examples of algal peloids have been found in Permian Zechstein limestones (Füchtbauer, 1964), in Carboniferous and Triassic lagoonal and shelf limestones (e.g., Meizer, 1971), and in deeper water carbonates.

3. *Pseudopeloids:* Fahraeus et al. (1974: 27) have described a Lower Paleozoic carbonate sequence characterized by alternating layers of limestone breccia and micritic limestones. Found in the micrites are areas of numerous peloids arranged

sometimes in fine lenticular layers; these are called "pseudopellets" and have been interpreted as rounded and sorted carbonate mud reworked by water currents. These particles correspond, then, to "small intraclasts" (Wilson, 1967). Indications of such a mode of origin are:

- Hydraulic equivalence of pseudopeloids and terrigenous particles, commonly found together
- Similar composition of micrite in the peloids and in the matrix
- Very irregular contact between the areas with and without peloids (interpreted as bottom reworking)
- Preferential occurrence of peloids in lenses (indication of currents) or in laminae
- Occurrence together with sorted terrigenous particles
- Absence of fossils and bioturbation (hence, no fecal pellets)

The size of these pseudopeloids lies between about 12 and 70 microns, usually at 30 microns.

An origin as a "miniresediment" appears to be quite probable particularly for particles that originate in deeper shelf environments (see Wilson, 1969). However, such peloids are also likely to occur in somewhat higher energy back-reef environments (e.g., in Upper Rhaetian reefs, see Fabricius, 1966).

In many cases it is only possible to distinguish small intraclasts (plasticlasts) by size, where the conventional limit can be regarded as about 0.20 mm. Similarly, it is likely to be very difficult to clearly distinguish "lithic pellets" and worn, structureless bioclasts of pseudopeloids. When the nongenetic collective term pelletoids is not used, the term "arrondids" can be substituted (Paulsen, 1964).

A good example of pseudopeloids was described by Carss and Carozzi (1965) from Devonian calcarenites, whose peloids are derived from periodic reworking of sediment in extremely shallow water environments with restricted sedimentation.

4. *Mud peloids:* "Mud aggregates" (Purdy, 1963, Bahamas) are found in Recent protected shallow marine environments; these were interpreted by Illing (1954) as inorganic aggregate formations, by other authors as micritized bioclasts. Fabricius (1966) understands "mud grains" to be fecal pellets as well as redeposited peloids or reworked micrite particles.

5. *Bahamite peloids:* With progressive micritization (see 4.1.3.7) of bioclasts, rounded particles develop, which can be called peloids because of their lack of original fine structures. A large number of the peloids in the Bahama Banks which are not derived from fecal pellets probably correspond to such micritized particles: This mode of origin of peloids must be reckoned with particularly in environments in which organisms producing fecal pellets are rare.

During a study of Lower Paleozoic limestones rich in peloids, Beales (1958) introduced the name "bahamite" for limestones which are similar in particle composition, and probably in origin of the carbonate sediments, to the Bahama Bank. The name bahamite grain or Bahamite (Gygi, 1969) characterizes peloids that, according to Bathurst (1971: 389), sometimes derive from micritic bioclasts. Depending on the source material, bahamite peloids are generally larger than algal peloids or pseudopeloids and are conspicuously well rounded. They may possibly be confused with pelletoids which originate by recrystallization. Generally, though, the occurrence together with cortoids and peloids should be an indication of bahamite peloids.

6. *Pelletoids:* This term was suggested by Blatt et al. (1972: 422) for peloids which originate by recrystallization of bioclasts and ooids. Often a strong micritization precedes the recrystallization. If the source material consists of "well-rounded bioclastic and lithoclastic grains of all sizes transported in agitated water", Paulsen (1964: 13) calls them arrondids. This term coincides with the original term pseudooid.

Tuberoids (Fritz, 1958) are to be considered as special kinds of peloids (Plate 50/3). This name was suggested in connection with the petrographic description of sponge limestones in the Upper Jurassic of Southern Germany. The author understood these to be nodular and rounded, various-sized, dark-colored micritic particles, which derive from carbonate precipitation around remains of organisms; precipitation is induced by the decay of the organic matter in siliceous sponges and algal crusts ("Verwesungsfällungskalk"). Thin-section and SEM analyses (Meyer, 1976; Behr and Behr, 1976) have shown that some of these particles correspond to algal fragments (Rivulariaceae). Accordingly, then, these tuberoids (see also 5.3.4) should be classified as algal peloids.

Matyja (1978) has proposed the name "putroid" for limestone particles which were formed by the decay and subsequent calcification of organic matter.

Beales (1971) pointed out that carbonate mud with peloids exhibits a special sequence of cementation in which neomorphic microsparite or sparite is formed between micritized and microsparry particles, thus forming a "pelsparitic" structure which does not correspond to a primary depositional fabric.

The relationships between peloids and a "clotted structure" were outlined in Sect. 4.1.2.

4.1.3.3 Aggregate Grains (Plates 1/1, 3, 4; 5/1; 9/7; 31/6; 38/1; 47/1)

Carbonate aggregates were first observed in the Bahama Banks (Illing, 1954; Purdy, 1963; Bathurst, 1971; Windland and Matthews, 1974; Roberts and Moore, 1971; Fabricius, 1977; Cros, 1979), then in the Persian Gulf (Sudgen, 1963; Taylor and Illing, 1969) and also in the Gulf of Batabano south of Cuba (Bathurst, 1971). These slightly rounded, strongly micritized particles are tightly bonded to each other by micritic aragonite cement or Mg-calcite cement. Illing (1954: 30) called these irregularly formed grains "lumps". He differentiated three morphological types: (a) "grapestones" – grapelike aggregates with protruding particles, (b) "botryoidal lumps" (Plate 1/2) – grapestones with thin oolitic envelopes, (c) "encrusted lumps" – smoother surfaces than grapestones, often with internal voids. The second type (b) corresponds to superficial ooids with aggregate grains as nuclei. In Purdy's definition (1963) particular emphasis was placed upon internal voids. Besides grapestones, Purdy describes other types (friable aggregates, organic aggregates), which are differentiated by their different consistencies and organic interactions (encrusting foraminifera, red algae, serpulids) as well as by various degrees of cementation. Milliman (1967, 1974) distinguishes aggregate grains with little matrix (corresponding to grapestones) and "lumps" with more than 50% matrix.

In Table 13 the following groups have been distinguished, with reference to Milliman and by taking account of the cementation of particles by algae in Ancient aggregate grains:

- *Grapestones*, Plate 1/1; 33; 38/1
- *Algal lumps* (or "bio-lumps", when the biogenic agglutination is due to foraminifera, serpulids, etc.), Plate 9/7
- *Lumps* (micrite-rich; often probably micritized grapestones), Plates 1/4, 31/6, 38/1, 42/9, 47/1.

Because overlapping exists between these groups, only the grapestones and lumps, on the one hand, and the algal or bio-lumps, on the other hand, appear to be useful categories.

Aggregate grains occur in situ through the agglutination of adjacent carbonate grains. According to Windland et al. (1974) peloids, ooids, and mollusk fragments are first agglutinated by encrusting organisms (miliolid and ophthalmid foraminifera, etc.) and by Mg-calcite formed in the mucilage sheaths of Cyanophyceae living on the substrate. Because the aggregate grains are often covered by subtidal algal mats and thus are stationary over longer periods of time, the precipitation of micrite cement by boring algae increases. During a second stage the internal voids are infilled by fibrous aragonitic cement. Cros (1979) emphasizes the interrelationships between biogenic encrustations, sediment trapping of loose silt-material between the individual particles of the aggregate grains, and external as well as internal cements (partly induced by organisms, e.g., diatoms).

According to Fabricius (1977) grapestones may be regarded as products of a biomineralization by boring thallophytes, probably blue-green algae.

The formation of aggregate grains occurs generally in subtidal and intertidal shallow-water environments with restricted circulation (to about 10 m water depth). High-energy particles in aggregate grains (ooids) may indicate a change of water turbulence (strong to weak) or a general change in the sedimentary conditions (example in Winland et al., 1974). Recent algal lumps form in the shallow brackish water of the tidal zone (Monty, 1967: B 83). In hypersaline coastal ponds of the Black Sea, cement formation occurs between the particles of the aggregate grains by means of a subaquatic bacterial degradation of calcium sulphate, which stems from earlier gypsum concretions (Baltres, 1975). Grapestones, consisting of peloids, ooids and oncoids, cemented by fibrous aragonite or cryptocrystalline Mg-calcite, are also known from algal mats of sea-marginal hypersaline pools (e.g., Solar Lake, Red Sea Coast; Friedman, 1978).

The size of the aggregate grains varies between about 0.5 and several mm, but usually lies around 1 mm.

In the Bahama Banks the percentage of aggregate grains in the total amount of carbonate particles reaches 60%–80%. An increased number of aggregate grains occur in sediments which are poor in carbonate mud. This observation has also been confirmed in Ancient calcarenites (e.g., Middle Triassic Marmolata limestone, Southern Tyrol; bedded Dachstein limestone in the Upper Triassic, Gesäuse, Austria; Upper Jurassic Sulzfluh limestone, Vorarlberg, Austria). Because the origin of aggregate grains is limited to environments with very restricted or absent carbonate mud sedimentation, the conspicuously small amount of non-carbonate insoluble residues in rocks of this facies type is understandable (see Sect. 9.3).

4.1.3.4 Oncoids (Plates 9/8; 29/2; 39/2; 42/1)

Irregularly formed carbonate particles which derive from biogenic and mechanical deposition around shells, etc. are known in Recent high or low energy, normal marine environments (deep intertidal, shallow subtidal); these nodules occur more often in hypersaline areas (Great Salt Lake, Utah; Persian Gulf; Red Sea; Shark Bay, Australia) and in fresh water ("Schnegglisteine" in Lake Constance, Germany; Würm Lake and Lake Starnberg, Bavaria, Germany; Bahamas; in rivers).

A. Heim (1916: 566) suggested the name oncoid for these spheroidal lobulate particles and considered the following properties to be particularly characteristic: "Non-concentric succession of more or less concentric laminae; form not necessarily round but rather lobulate; limestone dense, with or without inclusions; boundary sharply defined or indistinct." Pia (1927: 37) called them "Oncolithi" – "nodules lying loose in the rock . . . without any distinct organic structures, which often coat a tiny foreign body" – and regarded them as a subgroup of the blue-green algae Spongiostromata. Although Heim attributed the formation of oncoids to the "aggressive activity of bacterial colonies," Pia's theory of carbonate precipitation by algae is today the widely accepted explanation. In comparison to the original definition, the term oncoid has been broadly expanded today to include particles formed as biogenic accretions by recognizable algae (Cyanophyceae, Rhodophyceae) as well as particles with more or less non-concentric micrite accretions coating a nucleus. The term "Mumie" (Steinmann, 1880: 152; Gasche, 1956; Pümpin, 1965) coincides with oncoid – "total coating of shell fragments by other organisms, partially retaining the form of the enclosed body". Occasionally the name is still used to describe relatively large oncoids.

The particles called micritic oncoids in Table 13 as well as some of the algal oncoids belong, genetically speaking, to the stromatolites (Sect. 5.3.12.4), because "skeletal oncoids" (Riding, 1977) or cyanophycean oncoids consist of laminated, biosedimentary structures formed by microscopic algae (and bacteria). The biogenic accretions are produced by (a) trapping and agglutinating fine particles of sediment to the surface of the mucilaginous sheath of algal filaments and (b) extracellular precipitation of $CaCO_3$ in the mucilaginous sheath of algae without building a true skeleton. Oncoids correspond to "type ss" (spheroidal stromatolites) in the morphological stromatolite classification (Logan et al., 1964).

Oncoids appear together with layered stromatolites as well as, more commonly, with algae, aggregate grains, and cortoids, either as single particles or in groups of particles. Here it must be noted that oncoids appear to represent a morphological and ecological adaptation of encrusting organisms to soft bottoms and/or low-energy environments, whereas true stromatolites are generally associated with hard grounds and/or relatively high-energy environments (Monty, 1974; Peryt, 1975).

Composition of Oncoids

Oncoids generally consist of very fine grained micrite (often minimicrite) layers and of silt- and sand-sized detrital grains, whereby the latter are quantitatively more important in larger oncoids than in micro-oncoids. The grain size and mineralogical composition of non-carbonate material usually corresponds to the non-carbonate minerals in the vicinity of the oncoids. Apparently only quartz is taken up by the oncoids in smaller amounts than are available in the sedimentary environment, as illustrated in Table 14 by an example from the marine Lower Permian of the Carnic Alps, Austria.

Fe-hydroxide layers occur, in addition to the carbonate layers – particularly when sedimentation lags or is interrupted (Szulczewski, 1967).

Limonite nodules, known as "snuff-boxes" in the Middle Jurassic of England, represent a special type of ferruginous oncoids (Gatral et al., 1972).

Table 14. Mineralogical composition of algal oncoids and micritic matrix from Upper Pseudoschwagerina limestones (Lower Permian, Carnic Alps)

	Oncoids (7 samples)	Matrix (4 samples)
Calcite	> 50%	> 50%
Dolomite	< 10%	< 25%
Quartz	10–25%	25–50%
Illite	absent	very rare, 10–25%
Chlorite	10–25%	10–25%
Kaolinite	very rare, < 10%	rare, < 10%
Sericite	< 25%	< 25%
Goethite	present	present
Pyrite	present	present

The individual layers may be wavy or crinkly. They are usually oriented slightly off-center around a nucleus (bioclasts, lithoclasts, ooids, etc.).

The material for the nucleus usually comes from the immediate vicinity (see e.g., Kutek and Radwânski, 1965). Often oncoids already present, form the nuclei for new oncoids ("double oncoids" or "compound oncoids"; "Algenbeutel" – Kerkmann, 1966).

In contrast to Recent oncoids, Ancient oncoids seem to have been more strongly calcified and rapidly lithified. This is indicated by frequent borings as well as often very diverse epibiontic communities (see Plate 42/1).

Difficulties are encountered in the biogenic interpretation of various nodular structures described from Precambrian and Cambrian limestones, and are interpreted as blue-green algal oncoids or catagraphs (Zhuravleva 1964, 1979). These structures include concentrically laminated nodules, radially structured nodules (*Sterosphaeroides* group; Reitlinger, 1959; and *Radiosus* group; Zhuravleva, 1964), vesicular nodules (*Vesicularites* group; Reitlinger, 1959), clotted and lumpy structures like forms of the *Nubecularites* group (Reitlinger, 1969). Some of these structures are believed to represent carbonate particles formed by the activities of different blue-green algae, because of morphological similarities (e.g., between *Vesicularites* and some Chroccoccales, or *Asterophoides* and some Rivulariaceae), and also because of a corresponding isotopic composition of the graphitic coatings on the oncoids. Other structures like *Nubecularites* and *Hieroglyphites* seem to be aggregate grains or ooids and pisoids only.

Oncoid Shapes

The oncoid shape is dependent on

- The presence or absence of a well-developed nucleus. Generally the shape of the nucleus is more or less outlined. If the nucleus is absent or is very small, Cyanophyceae can form oncoids by concentric growth.
- The growth forms of algal thalli, determined by the bifurcation pattern of Cyanophyceae filaments (Golubić and Fischer, 1975).
- The substrate; deep invaginations in the oncoids may be caused by the contours of an uneven substrate (Ginsburg, 1960).
- Effect of abrasion during the transport from shallow to deeper environments

According to Logan et al. (1964), and Radwânski and Szulczewski (1966), three morphological groups of oncoids can be differentiated (Fig. 16):

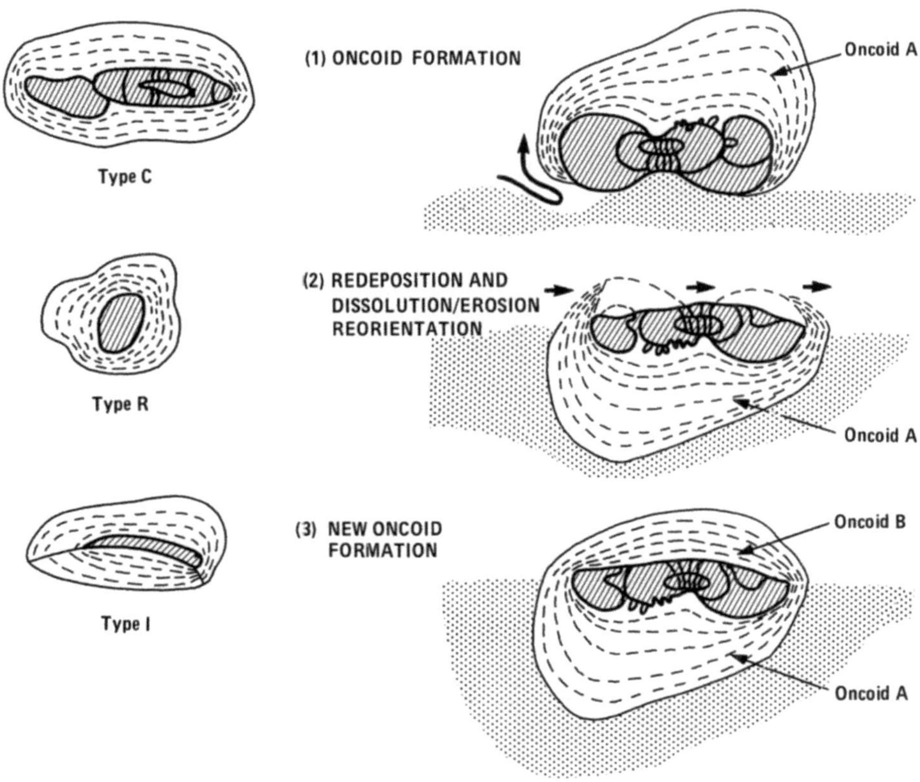

Fig. 16. Oncoid types. Shells (here, ammonite fragments) serve as nuclei. *Type C* represents a normal oncoid; the shape of the oncoid is dependent on the shape of the nucleus. In *Type R* dome-shaped structures occur, which are dependent on the various rates of growth. *Type I* (inverted) demonstrates interruptions in the development of the oncoid followed by inversion, dissolution, and erosion of the surface. A second oncoid then grows in the opposite direction on the surface of the first. Modified after Radwânski and Szulszewski (1966)

1. Type C – Layers form concentrically around a nucleus; the oncoid shape is dependent on the shape of the nucleus
2. Type R – Oncoids with dome-shaped layers growing at different rates, depending on the conditions.
3. Type I – The normal growth of the oncoid is interrupted on one side; the particle is inverted, followed by dissolution and erosion. A second oncoid grows on top of the eroded surface opposite the first oncoid.

Type C indicates the existence of uninterrupted algal growth and periods of sedimentation lag; Type I indicates synsedimentary reworking and erosion. Similar asymmetrical oncoids develop even then when a certain size and, hence, a certain weight is attained. The oncoids then become stationary and grow preferentially only in one direction (similar to Type R); after reorientation they can again grow in all directions (Example: Lake Constance/Gnadensee, Germany, Schöttle and Müller, 1968).

Classification of Oncoids

Oncoids may be grouped according to origin, size, and shape.

A. *Main genetic groups* (see Table 13):
   1. *Algal oncoids,* formed from biogenic accretion of
      a) Cyanophyceae, rarely Chlorophyceae. Examples: Recent – Golubić and Fischer, 1975; Ancient – *Rothpletzella* oncoids in the Silurian and Devonian, Flügel and Wolf, 1969, Plate 28/2; *Girvanella* oncoids in the Carboniferous and Permian, Toomey, 1974; Obrhel, 1977,
      b) Rhodophytes. Recent oncoids formed by Corallinaceae are called "rhodolites" (Bosellini and Ginsberg, 1969; Plate 29/2). Examples: Tertiary (Orszag-Sperber, 1977; Early Tertiary of the Mediterranean region and the Miocene of the Vienna Basin [Leitha limestones]); Upper Paleozoic (Toomey, 1975, oncoids from ancestral Corallinaceae).
      The form and the internal structures of Recent rhodoids appear to be controlled by water energy (Montaggioni, 1979). High-energy rhodoids are characterized by globular and laminar thalli, high algal diversity, encrusting foraminifera, and a skeletal nucleus. In quiet waters columnar rhodoids are formed without a nucleus by only one or just a few red algal species.
   2. Zoogenic *foraminiferal-algal* (serpulid algal, etc.)-*oncoids:* Biogenic coatings of encrusting animals around a nucleus, frequently intergrown with algae. Catalov (1970: 157) speaks of "zoophytic oncoids". Examples: Ancient – *Osagia* oncoids in the Upper Paleozoic (Johnson, 1946; Henbest, 1963); Zechstein (Peryt and Peryt, 1975); Alpine Triassic "*Sphaerocodium*" of the Raibl formation (Babić, 1969; Peryt, 1977).
   3. *Micritic oncoids,* derived from
      a) Encrustation by unfossilized algae (similar to 1 a),
      b) Agglutination of micrite grains to the surface of algal felt ("Mikrithüllen" – Bachmann, 1973: 41),

c) Increased micritization by boring algae, fungi, etc.
   This type corresponds to the "circumcrusts" of Wolf (1965) and somewhat to the "micro-oncoids" of Dragastan (1969). Micritic oncoids which occur gregariously and are not reworked indicate deeper water environments (e.g., oncoids in the Zechstein basin (Füchtbauer, 1968). Examples: Zechstein (Bausch and Wiontzek, 1961; Füchtbauer, 1964, 1968; Quester, 1964; Kerkmann, 1966; Peryt, 1977); Altenglan limestone in the non-marine Rotliegend (Stapf, 1970).

B. *Other genetic classifications:*
Maslov (1960) classifies Paleozoic oncoids as

- Morphotype *Osagia*, characterized by easily recognizable concentric laminated structures around a nucleus
- Morphotype *Nubecularites*, characterized by poorly recognizable laminated structures, which occur in clusters.

A similar classification is used by Reitlinger (1959) and Zhuravleva (1964) for Precambrian and Lower Cambrian oncoids and "catagraphs" (non-laminated nodules with clotted and lumpy structures, and obscured layering; Maslov, 1953). At present 17 groups of oncolites are distinguished according to external morphology and internal structures (Zhuravleva, 1979).

Dragastan (1969) combines genetic and morphological elements:
a) spherical-concentric oncoids (= type ss, after Logan et al. 1964)
   - Compound oncoids (subtype "accolé")
   - Fused oncoids (subtype "amalgamé"), correspond to type I of Logan et al. (1964),
b) ellipsoidal oncoids
   - Styliform oncoids, sometimes growing normal to the substrate (subtype "columnoidal"), correspond to microstromatolites,
   - Oncoids with sessile foraminifera (*Nubecularia*) and very different shapes

C. *Size:*
Maslov (1960) differentiates micro-oncoids and macro-oncoids according to size. Kutek and Radwânski (1965) speak of micro-oncoids (< 0.20 cm, usually without any fine structures), piso-oncoids (0.20 – 1.00 cm), and macro-oncoids (> 1.00 cm). Sometimes cm-sized "Mumien" (mumies) are called "large oncoids" ("Groß-onkoide" – Busse, 1968).

After attaining a certain size, the oncoid growth may come to a halt (Schöttle and Müller, 1968). In addition to inorganic factors (supply of nuclei, water energy, etc.), the size of the oncoids also seems to be strongly determined by the various abilities of the algae to bind sediment. This is shown in a comparison of *Archaeolithoporella* oncoids, *Girvanella* oncoids, and *Archaeolithophyllum* oncoids from the Southern Alpine and Turkish Permian (Fig. 17).

D. Classification according to *shapes* has already been mentioned (see "Oncoid shapes").

Difficulties may arise when trying to *distinguish oncoids from other round particles*, for instance:

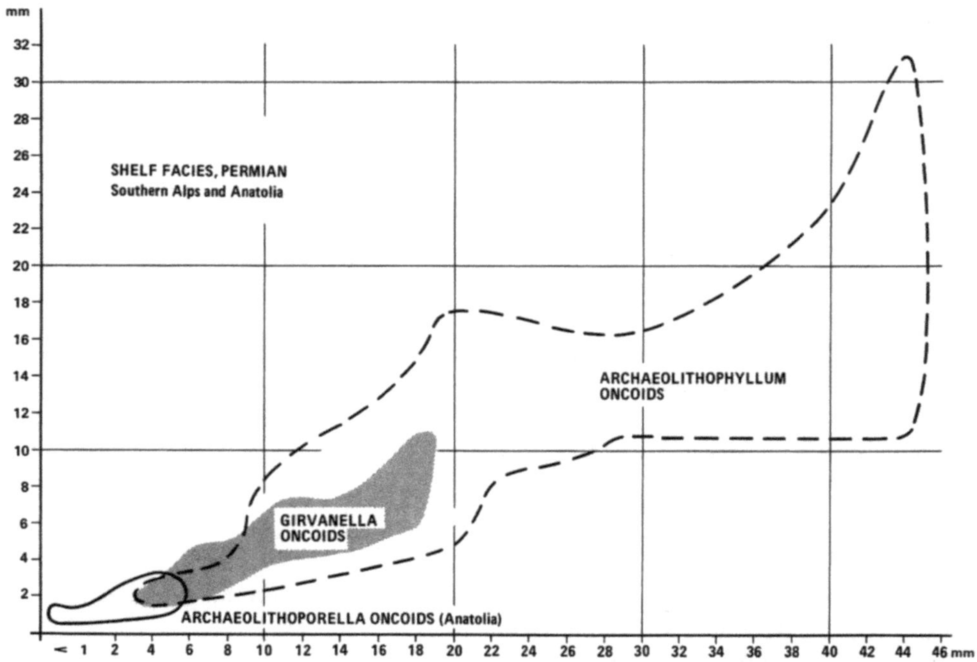

Fig. 17. The size of algal oncoids from the Lower Permian (*Girvanella* and *Archaeolithophyllum* oncoids, Rattendorf formation of the Carnic Alps, Austria) and from the Middle Permian of Western Anatolia, Turkey, (*Archaeolithoporella* oncoids) depends on the genera and species which form oncoids. A total of more than 600 data

- *Oncoids and ooids* (see Fabricius, 1967: 150), a problem whenever larger ooids deviate in their shape from a normal spherical or ellipsoidal form and micritic oncoids are present, which frequently exhibit a relatively well-developed concentric structure.
- *Oncoids and aggregate grains*, particularly algal lumps. One must then take a closer look at whether the algae build structures by accretion around a nucleus or between particles.
- *Oncoids and "coniatoids"* (Purser and Loreau, 1973). Inorganic encrustations made up of uniformly thick coatings of aragonitic lamellae occur in the supratidal zone of the Persian Gulf. These coniatoids are distinguished from oncoids as follows: (1) the microstructure of the lamellar envelopes is very dense; (2) detrital grains incorporated within the lamellae are missing; (3) the predominant core material of coniatoids are rock fragments. In contrast oncoids exhibit small or no hard cores at all (see Purser, 1980). Coniatoids are often associated with supratidal coniatolite crusts (see Table 5). Fossil coniatoids have been described by Wright (1981).

Notes regarding a comparison of Recent and Ancient oncoids

It is debatable whether the soft non-fossilized algal nodules which Ginsburg (1960), Gebelein (1969), and Monty (1965, 1967) as well as Pratt (1979) have described in shallow, brackish or normal marine environments (Florida, the Bahamas, Bermuda, the Dutch West Indies) can serve as Recent analogues for well-lithified calcareous oncoids known since the Precambrian. In the opinion of Monty (1972), Recent freshwater oncoids can be better compared to Ancient (marine) oncoids than to Recent marine oncoids, in regard to morphology, structure, and microstructure. This is very likely due to the displacement of blue-green algae, also living in marine environments until the Tertiary, by Rhodophyceae (Monty, 1973). Hence, it is important for a facies interpretation of Ancient oncoids to compare also Recent freshwater oncoids (see also 2.3.1) when making comparisons. The significance of such comparisons can be clearly seen in Tables 15 and 16.

Peryt (1981) emphasizes the importance of an evolution of oncoid types during geological times: During the Phanerozoic, porostromate and spongiostromate oncoids existed in different environmental settings. From the Cambrian to the Jurassic, porostromate oncoids flourished in somewhat deeper marine subtidal environments, probably with fairly slow sedimentation rates. Spongiostromate oncoids occurred in lacustrine and, when associated with stromatolites, in transitional continental-marine environments. Starting with the Jurassic, spongiostromate oncoids replaced porostromate oncoids and became the only type of oncoid in shallow-water marine settings, continuing to flourish in lacustrine environments. With the Eocene, spongiostromate oncoids were replaced in marine environments by rhodoids. The occurrence of spongiostromate oncoids is now limited to continental settings or to a few extreme transitional marine-continental environments.

Table 15. Comparison of Recent and Ancient oncoids

	Recent		Ancient
	Fresh-water	Marine	Marine
Environmental factors	Low water energy, periodic turbulence possible, soft mud bottom	Restricted water flow, low rate of sedimentation	
Environment	Shores of lakes, areas at and near mouths of rivers	Tidal zones and upper subtidal area, water depth to about 3 m	a) Protected quiet water areas in lagoons or backreefs, b) Relict surfaces of regressions or tongues of transgressions
Origin	Interaction of periodic growth of Cyanophyceae, calcite formation, and carbonate solution; allochthonous particles are incorporated into the layers	Diurnal rhythm of mechanical deposition and growth of Cyanophyceae. Thick layers of sedimentary grains agglutinated by organic material are separated by very thin algal films	Comparable to Recent fresh-water oncoids

Table 16. Oncoids – Characteristics and environments

	Recent-Marine	Recent-Hypersaline	Recent-Freshwater	Ancient-Marine	Ancient-Freshwater
Environment	1. Lower tidal zone, upper subtidal zone 2. Subtidal, in carpets of marine grasses	3. Algal mats in coastal ponds, salinity 42‰	4. Sand bars in rivers 5. Lake Constance (Gnadensee) 6. Rivers	7. Nearshore shallow water environments 8. Shallow water environment 9. Shallow platforms, possibly with freshwater influx	10. Lakes Rivers
Water Depth	1. 1.5–3 m 2. 0.5–3 m	3. to 5 m	4. 0.50–1.5 m 5. to 3 m 6. –	7. – 8. very few m 9. shallow	10. –
Water Energy	1. Sluggish current 2. Sluggish current 1–11 cm/sec	3. Low	4. – 5. Low 6. –	7. Current present	10. Different shapes of oncoids in turbulent and quiet environments
Shape and Structure of Oncoids	1. Spheroidal structures (ss type) 2. Sometimes partially "rooted", domed, up to 90% consists of thick (~ 1 mm) layers of mechanically deposited sedimentary grains < 250 μ; 10% consists of Cyanophyceae filaments	3. Various shapes, generally subrounded, alternating layers of cryptocrystalline and fibrous calcite	4. Flattened ellipsoids, concentric layers of various species of Cyanophyceae which are interrupted during growth by erosion and carbonate dissolution 5. Flattened ellipsoids, porous and nonporous micritic layers of calcite with detrital sedimentary material 6. Spherical and ellipsoidal, porous and compact layers (1–8 mm thick) with numerous micrite layers containing algae and fungi	7. Ellipsoids of micrite layers with and without *Girvanella* 8. Dependent on the nucleus, bioclasts embedded in concentric micrite layers 9. Ellipsoidal, spherical micrite layers surrounding gastropods/ *Girvanella*-oncoids	10. Ellipsoidal, flat, with algal layers consisting primarily of Rivulariaceae
Size of Oncoids	1. cm-sized 2. cm-sized (width 0.5–4 cm, height 0.25–3 cm), annual growth about 30 mm	3. 1–2 mm	4. 5–25 cm 5. 1–30 cm 6. 0.05–13 cm, annual growth 2.2 mm	7. 3–20 × 2–10 mm 8. < 0.2–4.5 cm 9. to 60 mm	10. to 4 cm

4.1.3.5 Ooids (Plates 1/2; 9/1–4, 6; 31/3; 32/2; 47/4)

Calcareous ooids are among the limestone particles most important for microfacies interpretations, but are also the most controversial in regard to origin.

A. Definition: Calcareous ooids have various shapes; they are usually more or less spherical or ellipsoidal carbonate particles with uniform, concentric laminae coating a nucleus; in thin-sections tangential and/or radial structures can be recognized in the laminae. Under crossed nicols an interference cross may appear (Bausch, 1963).

The name was suggested by Kalkowsky (1908: 72) "for the spherical components of the oolites . . ., for the single grain". In contrast to the English nomenclature, "ooid" should be used for the particles and "oolite" for the rock consisting of ooids (see Teichert, 1970).

B. Classification and description of ooids

1. Ooids can be differentiated according to the *number of laminae*
   a) *Normal ooids* or *ooids* with several laminae (Plates 9/2, 3, 6; 31/3; 32/2),
   b) *Superficial ooids* (Carozzi, 1957, 1960: 238) with one or only very few laminae (Plate 1/2). Carozzi restricts the term to ooids with a single lamina. German synonyms are "Rindenooid", "Mantelooid," and "Proto-ooid." Superficial ooids are not exclusively derived from the early stages of ooid formation (Carozzi, 1960). Recrystallization and dissolution of the ooid nucleus as well as some of the inner laminae may produce similar patterns in thin-sections ("Oolithoids" – Krech, 1911: 82). The genesis of extremely thin "oolitic films" on carbonate sand grains in low-energy environments was described in the Bahama Banks by Bathurst (1967). Such films can not be fossilized, but do possibly illustrate incipient ooid formation.
2. Ooids can be differentiated according to *size* – especially in the English literature
   a) *Ooids*, smaller than 2 mm,
   b) *Pisoids* (pisolites), larger than 2 mm. See Sect. 4.1.3.6,
3. Ooids can be differentiated according to *shape*
   a) *Single ooids:* spherical (the sphericity value may decrease with increasing diameter, particularly for terrestrial ooids), ovular, rod-shaped or lens-shaped, sometimes flattened, cerebroid (determined by cone structures).
   b) *Compound ooids* (complex ooids, Plate 9/4) develop as
      - Multiple ooids ("Ooidviellinge"): Intergrowths of two or more ooids. Normal ooid growth is prevented at the surface of contact to the adjacent ooid; hence, the ooid grows preferentially in one direction. Genetically speaking, ooids with cone structures also belong to this category (see Usdowski, 1962).
      - Polyooids: Several complete ooids serve as nuclei for a large new ooid. Polyooids ("Ooidbeutel") can be indicators of breaks in deposition.

Generally the shape of the ooids is determined by the size and shape of the nuclei as well as the thickness of the laminae.

4. Depending on whether ooids are *complete* or *incomplete,* they can be differentiated as follows:
   a) *Whole ooids:* normal ooids or superficial ooids without a change in the shape of the structure of the laminae.
   b) *Broken or abraded ooids:* due to mechanical breakup by agitated water (see Plate 9/3).
   c) *Half-moon ooids* (Carozzi, 1963; "Hemiooide"-Kalkowsky, 1908: 85; some distorted ooids, Carozzi): formed by leaching and collapse of the inner laminae, which then accumulate as a geopetal fabric on the bottom of the ooid. The residual internal cavity then fills with sparite or dolomite. In hypersaline environments salts are leached out; in supratidal environments, carbonates (Castellarin and Sartoni, 1973; Radwânski and Birkenmayer, 1977).
   d) *Cracked ooids* (Plate 9/2): originate from mechanical stress and pressure solution (see Radwânski and Birkenmajer, 1977).
   e) *Pitted ooids* (Plate 9/2): originate from pressure solution (see Radwânski, 1965).
   f) *Distorted ooids* (Carozzi, 1961a; "verformte Ooide" – Bachmann, 1973; "oolithes confluants" – Mišik, 1968; "spastoliths" – Pettijohn, 1957; "snouted ooids" – Radwânski and Birkenmajer, 1977): ooids are bent into an s-shape by sedimentary boudinage; distortion is due to compaction. See Plate 9/2 and Conley (1977).

Broken and regenerated ooids ("Hiatusooide" – Berg, 1944) are preferentially associated with a hypersaline environment, as are ooids with radial structures (see below): Middle Carboniferous of Canada (Shenk, 1975), Keuper of Poland (Labecki and Radwânski, 1967), Neocomian in the Dinaric Alps of Yugoslavia (Radoičić, 1960), Miocene of Italy (Roda, 1965), Recent Great Salt Lake in Utah (Carozzi, 1961b), see also Radwânski and Birkenmajer (1977), and Plate 9/3 and Plate 35. Such imperfect ooids which have been caused by sedimentary and diagenetic processes are to be kept separate from *deformed ooids*; these are due to the tectonic stress of oolitic limestones and may be indicators of the deformation style (Exampes and discussion in Cloos, 1947; Nissen, 1964; and Badoux, 1970). Another cause of imperfect ooids is an early compaction: Since the prelithification compaction of ooids in carbonate sediments results in inhomogeneous deformation of the grains, the coexistence of deformed and undeformed ooids provides a signal for an overburden compaction of unlithified sediment. Post-lithification deformation will produce a homogeneous deformation fabric with a very high percentage of deformed ooids (Bhattacharyya and Friedman, 1979).

C. Microstructures and ultrastructures. Basically the following microstructures can be recognized in thin-sections of the laminae of Ancient ooids:

1. *Tangential structures* (Plates 9/1, 9/6) characterized by light-colored lamellae with tangentially oriented crystals and darker cryptocrystalline lamellae. Both types of lamellae exhibit a distinctly concentric orientation. Under crossed nicols a bar cross can be seen readily.

2. *Radial structures* (Plate 9/3, 6) characterized by lighter lamellae with radially oriented fibrous crystals and by darker cryptocrystalline layers. A bar cross is very occasionally seen in the outer layers.
3. *Micritic ooids* characterized by dark cryptocrystalline layers with no preferred crystal orientation.

These types with different micro- and ultrastructures correspond to the subtypes of ooids differentiated by Purser (1980): "ooid type alpha" is characterized by concentric laminations of various thicknesses (normal ooids and superficial ooids) composed of tangentially oriented crystals; ooids with radial structures are included in the groups "ooid type beta" (generally irregularly shaped), and "ooid type gamma" (with asymmetrically developed envelopes, e.g., ooids of the protected coastal lagoons of Texas; Frishman, 1969).

As shown in SEM analyses of Recent ooids (Fabricius and Klingele, 1970; Loreau, 1970a; Loreau and Purser, 1973), the tangential structures as well as the radial structures are to be regarded as primary structures; apparently, then, there is a connection between the microstructures or ultrastructures, on the one hand, and the environment, on the other (Tables 17 and 18). Ooids with primary, purely concentric textures or radial concentric textures seem to be related to a change in the mode of transport from suspension to bed load when the ooids reach a critical size (0.06 mm; Heller et al., 1980).

In contrast to ooids with tangential structures, which originate in high-energy environments (but also in hypersaline waters; Garber et al., 1981), "quiet water ooids" in both marine marginal zones and non-marine environments (hypersaline coastal regions; terrestrial salt lakes, cave pearls) are characterized by a radial-fibrous structure. This observation has also been confirmed by the study of Ancient ooids: ooids with radial structures are frequently found in limestones formed in environments with salinities deviating from normal marine values; this must be assumed on the basis of the greatly reduced diversity of organisms, the occurrence of Characeae and Ostracoda as well as broken and regenerated ooids. A few examples have been listed in Table 19.

Not all radially structured ooids must necessarily be derived from primary radial accretion (see e.g., Hahne et al., 1968; Loreau, 1979): Ooids composed essentially of radially arranged large calcite crystals, which transect only faintly developed concentric laminations, are probably the result of diagenetic alteration and recrystallization. These "spheroids" (or "spherulites") have been described in terrestrial environments (cave pearls; Hahne et al., 1968), lacustrine settings (Great Salt Lake; Eardly, 1938), but also in protected lagoonal reef environments (Maxwell et al., 1964). On the other hand the ever mounting number of observations of Recent ooids seem to argue in favor of the fact that the radial structure represents an original feature even for ooids consisting of calcite (see e.g., Simone, 1974; Poncet, 1976, and Plates 9/3, 9/6).

It should, however, be questioned whether or not the structural comparison of Recent and Ancient ooids is admissible, particularly in regard to original mineralogical structures. Sandberg (1975) assumes that in contrast to Recent shallow-water ooids, Paleozoic ooids consist primarily of calcite, and that aragonite ooids have existed only since about the Tertiary. Sandberg's suggestion is supported by Wilkinson (1979), who stresses the fact that aragonite inversion results in sparry

Table 17. Relationships between ooid structure and environment in the Persian Gulf. After Loreau and Purser (1973)

<table>
<tr><td>Water Energy</td><td>Very Low</td><td>Low</td><td colspan="2">Moderate to high</td><td>High to Very High</td></tr>
<tr><td>Morphology of the Environment</td><td>Sabkha flats, well protected deflations behind the coastline</td><td>Shallow depressions between megaripples</td><td>Shallow depressions between sedimentary sills</td><td>On shoals</td><td>Prodelta, seaward</td></tr>
<tr><td>Microstructures in Thin-sections</td><td>Radial structure</td><td>Tangential structure</td><td>Tangential structure</td><td>Tangential structure</td><td>Tangential structure</td></tr>
<tr><td rowspan="2">Ultrastructure of peripheral ooid laminae</td><td rowspan="2">Radial</td><td colspan="2">Loose to more or less dense felts of aragonite needles</td><td rowspan="2">Dense felts of rod-shaped aragonite crystals with tangential orientation</td><td rowspan="2">Loose felts of rod-shaped aragonite crystals</td></tr>
<tr><td>sporadic subhedral nannocrystals with radial orientation</td><td>more or less tangentially oriented or nannocrystals</td></tr>
<tr><td></td><td>Transitional environment, supratidal/intertidal</td><td>Lower intertidal zone</td><td colspan="3">Intertidal and shallow subtidal zones</td></tr>
</table>

Table 18. Ooids with different microstructures probably originate in environments with various levels of water energy and salinity

	"Tangential Structure"			"Radial Structure"		
	Locality	Environment	Mineralogy	Locality	Environment	Mineralogy
Recent Calcareous Ooids	Bahama Banks	Marine, Platform margin	Aragonite	Bimini lagoon Bahamas, very thin ooid film	Protected areals of marine grasses	Aragonite
	Persian Gulf	Marine, Tidal zone, protected coastal region	Aragonite	Persian Gulf	Evaporitic tidal flats	Aragonite
	Gulf of Suez	Marine, Coastal lagoons	Aragonite	Gulf of Aqaba	Hypersaline coastal pond	Aragonite
				Laguna Madre, Texas	Hypersaline	Mg-calcite
				Salt Lake, Utah	Hypersaline	Calcite or aragonite
				Cave pearls Styria, Austria; Ruhr district, Germany	Caves and mines, Freshwater	Calcite
				Texas, Caliche-ooids	Vadose pisoids terrestrial	Calcite
Shape of Ooids	Regular			Regular, but frequently irregular deviations from the spherical or oval form, sometimes asymmetrical		
Size of Ooids	Usually 0.15 mm to 1.5 mm			Various sizes, to 5 mm		
Proportion of laminae to total diameter	5 – 75%			75 – 80%; thickness of lamellar envelope not dependent from nucleus size		
Water Energy	High Energy			Low Energy		

Table 19. Occurrence and facies of radially structured Ancient calcareous ooids

Stratigraphic Range	Locality	Paleoenvironment	Lithology	Fossil Content	Author
Mississippian	Alberta, Canada	Shallow water lagoons	Oosparite	Low diversity, foraminifera, calcispheres	Carozzi (1961b)
Zechstein, Permian	Poland	Subtidal	Oosparite	Stromatolites, sessile foraminifera	Piatkowski (1977)
Bunter	Harz margin, N. Germany	Coastal areas	"Rogensteine" Carbonate sandstone	absent	Usdowski (1962) Haage (1970)
Keuper	Holy Cross mountains, Poland	Hypersaline lagoons with anhydrite	Carbonates and dolomite with ooids	Thin-shelled pelecypods; Characeae, agglutinated foraminifera	Labecki and Radwánski (1967)
Upper Rhaetian	Northwestern Calcareous Alps	Back-reef facies	Oosparite	Biogenic content to 13 vol%, algae, foraminifera	Fabricius (1967)
Lias	Geiselstein, Ammergau Alps	Rapidly subsiding shallow water platform	Oosparite	Biogenic content 3 vol%, echinoderms	Fabricius (1967)
Upper Jurassic	Holy Cross mountains, Poland	Marginal facies	Oosparite	Very rare fragments of shells	Radwánski (1965)
Upper Jurassic/ Lower Cretaceous	Bahama Banks	Carbonate platform	Oopelsparite	Filamentous Cyanophyceae, benthonic foraminifera	d'Argenio et al. (1975)
Lower Cretaceous	S. Apennines, Italy	Platform, hypersaline lagoon	Oosparite and biopelmicrite	Filamentous Cyanophyceae, benthonic foraminifera	Simone (1974)

calcite with gross obliteration of primary crystal fabric, and finds no evidence that recrystallization of tangential aragonite gives rise to a secondary calcitic radial fabric in ooid cortices. A similar result was obtained in SEM studies of Jurassic calcareous ooids (Purser, 1979), supporting the hypothesis, first ventured by Sorby (1879), that the radial structure and the calcitic mineralogy of Ancient ooids correspond to the original composition.
The discussion of the primary mineralogy of Ancient calcareous ooids is further complicated by mineralogical and textural variations described in Holocene ooids from the Baffin Bay, Texas (Land et al., 1979). Radial coatings are Mg-calcite, whereas the tangential and micritic coatings are aragonite. This may be caused by differences in the rate of nucleation and precipitation, degree of agitation, and in the presence of certain types of organic compounds.

Micritic ooids with very fine micritic layered structures, often unrecognizable may be due to the following *factors*:

a) total micritization of tangentially structured calcareous ooids by boring algae or fungi (see Margolis and Rex, 1971).
b) formation of anhedral nannocrystals whose uniform growth on the nuclear surface is controlled by organic films (Loreau, 1973). The process of the primary formation of micrite laminae may be aided by the mucilage sheaths of endophytes (Mitterer, 1968; Trichet, 1969). According to Cros (1979), the cortex of micritic ooids may be formed by fragmentation and dissolution of adherent nannoparticles together with a renewed precipitation of nannocrystals. This process does not seem to be influenced by biological perforation.
c) mechanical deposition of pelagic particles (essentially coccoliths) in the mucilage sheaths of Cyanophyceae which have bored into the nucleus. Such an origin was postulated by Jenkyns (1972) for the micrite rings (0.20–0.40 mm) of "pelagic ooids". In accordance with the definitions given in Table 13, these particles, which have been described in Jurassic seamount sediments (Western Sicily, Hungary, Poland, San Vigilio-oolites of the Southern Alps; Tithonian-Hauterivian of Morocco, see W. Schlager, 1980) are actually small, very uniformly formed oncoids. However, because this kind of biogenic structure can only be detected with an electron scanning microscope, just "pelagic ooids" will be discussed here.
   Hesse (1973) conjectured a similar mode of origin for Tertiary ooids in Pacific seamount sediments.
d) recrystallization. Bolliger and Burri (1970) pointed out that micritic ooids occur preferentially in sediments without clay and that radially fibrous ooids occur preferentially in sedimentary cycles of marls and limestones. The radially fibrous ooids may have been protected by the clay from recrystallization (see Marschner, 1968).

Purser (1980, p. 30) calls cryptocrystalline ooids with faint or missing concentric or radial structures "bahamithes" (see Plate 1/2). This definition of "bahamite grain" is different from that used in the study of Ancient carbonates (see Sect. 4.1.3.2).

D. Size and sorting. The diameter of Recent ooids is generally less than 1 mm (see Fig. 18), often around and below 0.50 mm. Ancient ooids have a broader spectrum

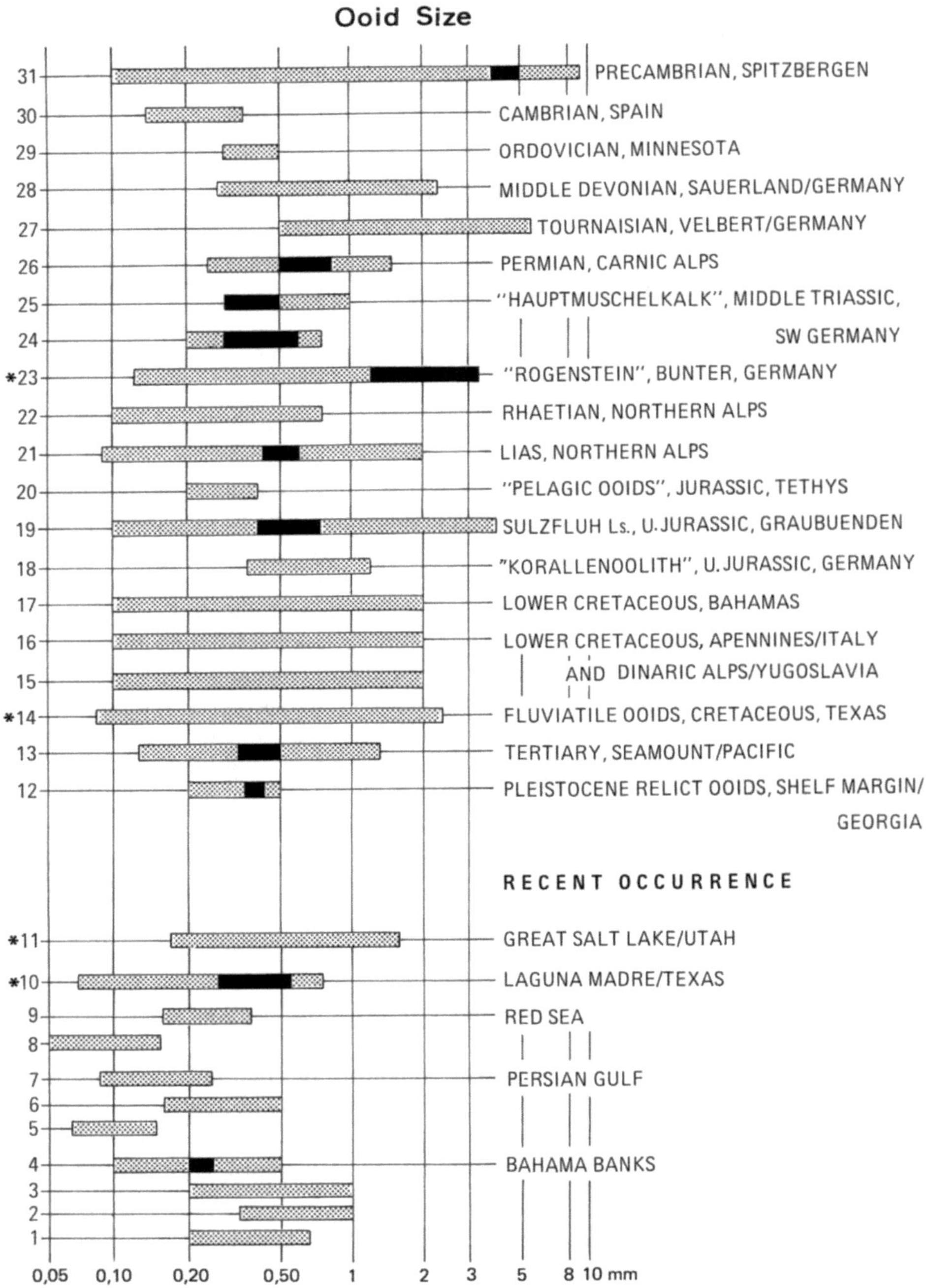

Fig. 18. Diameters of Recent and Ancient calcareous ooids from the literature and the author's own observations. * = hypersaline or fluviatile occurrence

of grain sizes, with values frequently between 0.10 and 2 mm. There are no size differences between marine and "non-marine" ooids.

Very large ooids/pisoids (up to about 10 mm in diameter) and complex ooids (up to 15 mm) have been described from various Precambrian dolostones (see Radwánski and Birkenmajer, 1977). The enormous size is assumed to have resulted from a profuse carbonate precipitation, possibly of primary dolomite. This may be the result of different chemistry and precipitation conditions in the Precambrian epeiric seas. Due to the absence of skeletal organisms, the cores of these pisoids are made up of other ooids. This sometimes results in the formation of complex and compound pisoids, which indicate multiple redepositions of the ooids within shallow subtidal, intertidal and supratidal environments.

When comparing Recent ooids to Ancient ooids, size and sorting can provide information about the paleobathymetry als well as indicate whether the ooids are autochthonous or allochthonous:

1. *Water depth*: Marine ooids originate in the tropics and subtropics in high-energy shallow-water environments influenced by wave action or tidal currents, whereby both the percentage of ooids in the total sediment fraction and the thickness of the ooid laminae increase with increasing water depth. The majority of ooids originate in depths down to 2 m. According to Newell et al. (1960), the correlation of the net concentration of ooids to water depth is important for interpreting Ancient oolite limestones (see Fabricius, 1967; Ott, 1969): 0–7 m – more than 70 vol. % ooids, 2–7 m – 70–40 vol. % ooids, 2–15 m – 40–20 vol. % ooids.

2. The question of whether ooids are autochthonous or allochthonous is an interesting one when *ooids occur in micritic limestones* (generally regarded as low-energy formations). Listed below are several possible interpretations:

a) *allochthonous ooids*: These ooids derive from neighboring high-energy environments (Example: Bahama Banks, Purdy, 1963) or have worked their way out of earlier ooid sediments (Example: ooids on the North African coast derived from redeposited Pleistocene coastal dunes, Fabricius et al., 1970, "relict ooids"). After transport, shallow-water ooids may also be deposited at great depths (e.g., in the Great Bahama Canyon at 1800 m, Andrews et al., 1970).

b) *autochthonous ooids*: These ooids originate in low-energy environments, e.g. in protected bays and lagoons or at greater depths (see "Micritic ooids"). According to Lacey and Carozzi (1967) evidence of redeposition is given by (1) size, sorting, and skewness of ooids and non-ooids, (2) ratio of ooid content to mean diameter, and (3) ratio of laminae thickness or number of laminae to nucleus size. With regard to the number of laminae it is also necessary to take into consideration abrasion during transport or even during ooid formation.
Criteria for autochthonous ooids are:

- Greater deviation from the spheroidal or ellipsoidal shape,
- Laminae thickness less than the size of the nucleus,
- Lamellae of the laminae often indistinctly formed, often truncated,
- Generally very few laminae,
- Radial structure common,
- Nuclei not always recognizable (or present),

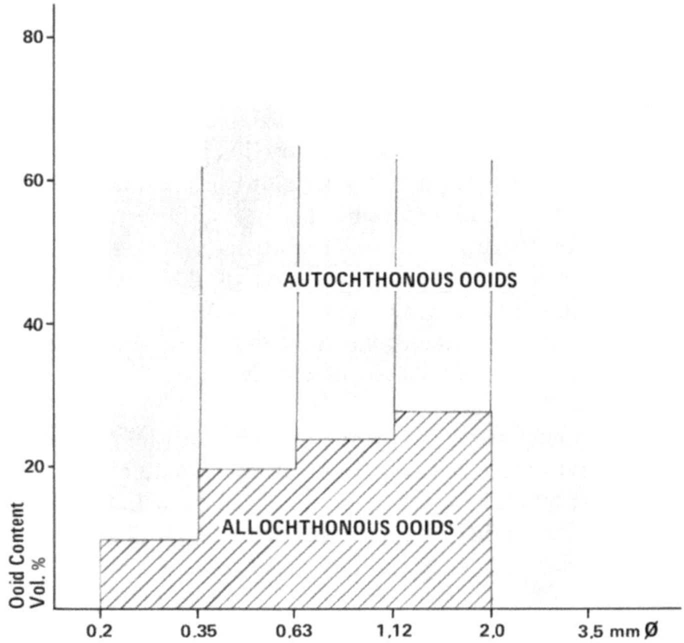

Fig. 19. Relationship between ooid content and average ooid diameter in autochthonous ooids (Rogenstein) and allochthonous ooids (calcareous sandstone with oolites) in the Lower Bunter. Modified after Usdowski (1962)

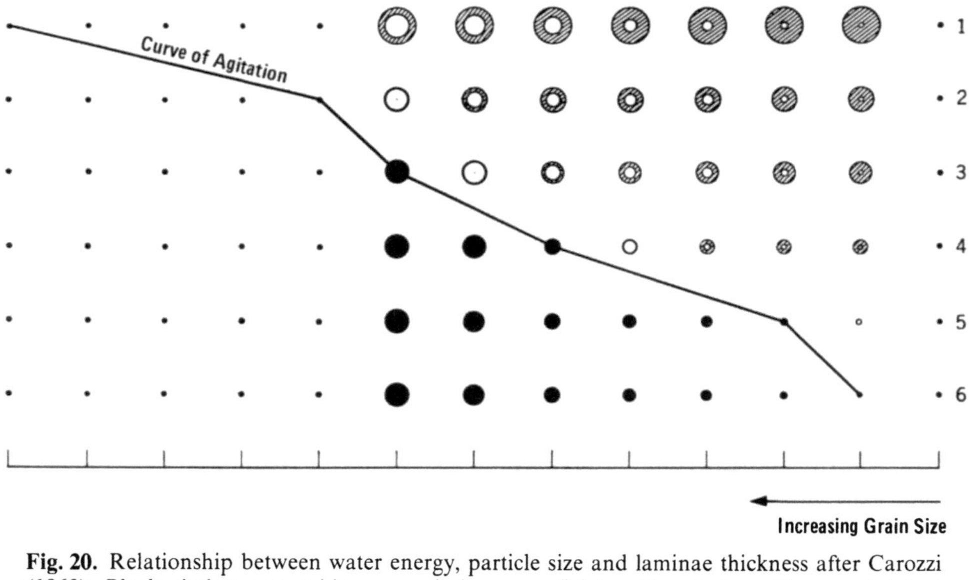

Fig. 20. Relationship between water energy, particle size and laminae thickness after Carozzi (1960). Black circles: non-ooids; open circles: superficial ooids; circles with cross-hatched signature: normal ooides

- Size of the ooids very small (about 0.05–0.15 mm),
- Poor sorting,
- Often relatively large amount of insoluble residues.

According to Sass et al. (1972), the small size of "quiet water ooids" is due to their origin in a suspension (larger ooids arise from saltation and rolling in currents along the bottom of a stream).

As illustrated in Fig. 19 by ooids in the calcareous oolitic facies of the Northern German Lower Bunter, allochthonous ooids differ from autochthonous ooids in that they are fewer in number and smaller in diameter. This is because continued transport leads to greater destruction of the larger ooids, whereas the smaller ones are left behind intact.

In measuring ooid sizes in thin-section, the effect of cutting introduces errors, which, however, can be eliminated by using the Fabricius method (1967).

Carozzi (1957, 1960) developed a *model* (Fig. 20) for *correlating the size of the nucleus, the thickness of the laminae, and the local water energy*; this is described by six situations:

Situations 1 and 2: The local water energy is greater than the maximum current intensity. Ooid formation includes all the particles available until the diameter of the largest ooid and the diameter of the largest movable grain are the same (or nearly the same). Oolitic sediments originate, in which the constant maximum diameter of the ooids is related to the intensity of the water agitation.

Situation 3: The local water energy is too weak to keep the largest particles in the turbulent zone. Smaller particles, however, serve as nuclei for ooids. A sediment of large non-ooids and many small ooids originates, of which the largest of the small particles are superficial ooids, and the others normal ooids.

Situation 4: The local water energy is lower than the maximum current intensity. A sediment originates in which large superficial ooids and smaller normal ooids as well as non-ooids of various sizes are represented.

Situation 5: The local water energy is somewhat greater than the water energy needed to move the smallest grain. The smallest grains become superficial ooids, larger particles remain unchanged.

Situation 6: The local water movement is equal to or less than the water energy needed to move the smallest grain. Ooid formation does not take place.

This theory assumes that ooids can no longer be moved by currents after attaining a critical size, thus resulting in a cessation of growth (criticism by Bathurst, 1967: 451).

E. Occurrence and origin of ooids. Recent calcareous ooids and pisoids are known in the following environments (see Fig. 21):

1. *Terrestrial*: (a) *caliche-ooids* (diagenetic ooids); Examples in Siesser (1973), morphological differences in comparison to marine ooids, see Table 4, (b) *cave pearls*; Examples in Donajue (1965, 1969), Gradzinski and Radomski (1967, 1976), Hahne et al. (1968), Kirchmayer (1964), (c) *Pisoids* formed in *shallow pools on a playa surface* (Risacher and Eugster, 1979) which active hot springs (20°–75° C) flow over; these pisoids (a few millimeters to 20 cm in diameter) are structurally very similar to cave pearls, but they are formed in an evaporitic

setting. The pisoids are found together with simple and arborescent concretions, sinter terraces and rimstone dams.

2. *Fluviatile*: Example in McGannon (1975, Cretaceous of Texas). The microstructure of fluviatile pisoids resembles cave pearls. Fluviatile pisoids are associated with laminated crystalline crusts consisting of acicular fibres growing radially from terminations of older prismatic crystals (Braithwaite, 1979).
3. *Lacustrine*: (a) *fresh-water lakes*; Examples in Carozzi (1964, Triassic of Virginia), Picard and High (1968, Eocene of Utah), (b) *salt lakes*; Example Great Salt Lake, Utah (Carozzi, 1964; Kahle, 1974; Sandberg, 1975).
4. *Marginal marine, hypersaline*: Example Laguna Madre, Texas (Freeman, 1962; Rusnak, 1960). The proportion of ooids is hardly more than 30%; characteristic features are asymmetric ooids, and associations of normal, superficial, and compound ooids; Gulf of Aqaba, Red Sea (Friedman et al., 1973).
5. *Marine*: The following *factors* appear to be important for *ooid formation in Recent shallow-sea environments*:
   - Presence of algae, bacteria, or organic substances (see Mitterer, 1971; Shearman et al., 1970; Trichet, 1968; Suess and Fuetterer, 1972; Hesse, 1973; Fabricius, 1977).
   - Water supersaturated with $CaCO_3$
   - At least occasional, relatively strong water movement
   - Supply of potential nuclei
   - Few organisms which remove carbonate from the water
   - Warm water, generally above 20 °C
   - Normal or increased salinity
   - Very shallow water, generally less than 2 m
   - Relatively constant environmental factors (duration of growth for "Recent" ooids 100 to 1000 years!)

The characteristics and environments of some Recent and Ancient calcareous ooids are listed in Table 20 and Fig. 20. When comparing Recent ooids from the

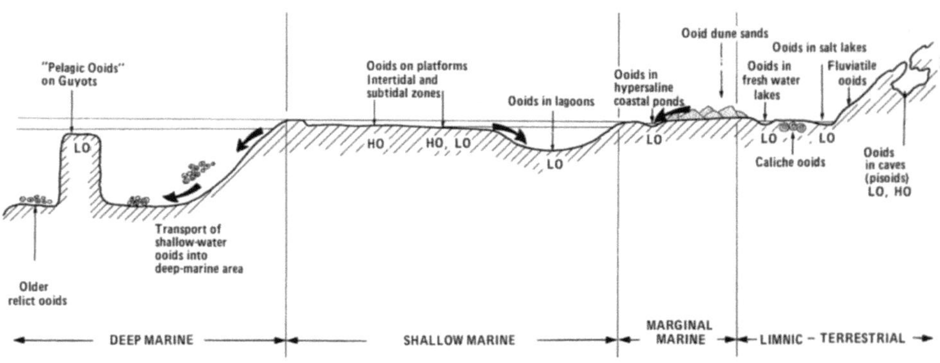

Fig. 21. Calcareous ooids originate in various marine and non-marine environments. High energy ooids (*HO*) of agitated water and low energy ooids (*LO*) of relatively quiet water are differentiated according to microstructure, grain size, and association with certain organisms

Table 20. Characteristics of Recent and Ancient calcareous ooids. Data after Newell et al. (1960), Purdy (1963), Fabricius (1967), W. Ott (1969) and author's own observations

	Recent	Ancient (Northern Alps)		
	Bahama Banks	Upper Rhaetian	Lias	Upper Jurassic
Average ooid content	60–66%	22.6%	53.5%	50–62%
Average median values (diameter in mm)	0.25–0.42	0.31	0.47	0.54
Average sorting coefficient	1.09–1.34	1.27	1.16	0.95
Laminae thickness (LT) Nucleus thickness (NT)	LT = NT (oolite wall, bank) LT = NT (bank margin)	Usually LT < NT	Usually LT < NT	LT = NT
Number of laminae	Shallow water: many laminae + small nucleus Deeper water: few laminae + larger nucleus	A few laminae	Usually many laminae (to 10)	A few to several laminae (to 10)
Nucleus	Bioclasts, peloids	Various bioclasts, peloids	Echinoderms, peloids	Bioclasts, "older" ooids
Laminar structure	Tangential	Radial	Radial	Tangential
Single layers	Sometimes very thin (1–3 μ)	Often relatively thick	Various thicknesses	Various thicknesses
Ooid form	Determined by nucleus shape; spherical	Determined by nucleus shape	Usually a perfect sphere	Determined by nucleus shape
"Porosity" or cement content	ca. 16–19%	25%	21%	17%
Average skeletal grain content	1.7% (pure oolite facies) up to 10.5%	13%	3%	9%
Bedding structures	Very pronounced	rare	rare	rare

Bahama Banks to Ancient ooids from the Upper Rhaetian back-reef environments, the Lias of the Northern Alps and the platform sediments of the Upper Jurassic Sulzfluh limestones, it is immediately apparent that despite similarities in the frequency and sorting of the ooids, as well as in the scarcity of organisms, distinct differences in morphology do exist.

The depositional and diagenetic development of Holocene ooid accumulations of a Bahamian ooid shoal is described by P. M. Harris (1979). The facies pattern, that is, a belt of oolithic grainstone bordering a belt of oolithic packstone which becomes increasingly muddy with depth, is quite common in the geological record.

F. Diagenesis of ooids. Ancient ooids are often micritized, leached, and recrystallized. As Shearman et al. (1970) have shown, subaerial exposure of ooids most likely leads to destruction of the protective organic substances. Many times mechanical abrasion and the formation of "pitting" structures, related to grain-to-grain dissolution, signal the beginning of diagenetic changes (Radwánski, 1965). Description of the diagenetic processes involved for ooids are found in Fabricius et al. (1970), Hesse (1973), Kahle (1974), Sandberg (1975).

Ooids are of special interest in recognizing lithification during burial and the degree of compaction in micritic limestones. Compression tests of Recent ooids and carbonate muds (Bhattacharyya and Friedman, 1979) indicate a linear relationship between an increasing proportion of lime mud and the proportion of undeformed ooids; as the proportion of lime mud increases, that of broken ooids decreases. Compaction-deformation of unlithified sediments, as opposed to rock deformation, seems to result in inhomogenous deformation of carbonate particles; therefore coexistence of deformed and undeformed ooids (peloids, or shells) may provide a signal that the sediments were compacted before they were converted into limestones. Generally, deformation of components is less likely to occur in mudstones and wackestones than in grainstones.

4.1.3.6 Pisoids (Plates 2/6, 7; 9/5; 14/1, 3)

Pisoids should be separated from the structurally similar ooids because of their remarkable size (generally 2–10 mm diameter) and their "non-marine" genesis. A distinction between pisoids, ooids and algal oncoids (algal pisolites of American authors) is important for the interpretation of sedimentary environments.

In cross-sections pisoids are generally made up of densely spaced, very regular concentric laminae. A nuclear grain may serve as an initial site for the accretion of the laminae. The nuclei are often fragments of other pisoids.

The shape of the pisoids depends on environmental conditions: pisoids formed in agitated waters (e.g., in caves or in geysers) exhibit shapes in which original irregularities of the nucleus are rounded by the accretion of concentric laminae; in quiet-water pisoids irregularities of the nucleus are increased by the addition of new laminae, thus producing irregular lobate forms.

The mineralogical composition varies, but commonly consists of calcite or aragonite. Occurrences of bauxite, gibbsite, hematite, limonite, silica and phosphorite are also known (see Carozzi, 1960). Phosphatic, ferruginous, and siliceous pisoids are formed by the replacement of primary calcareous pisoids or by the redeposition of non-calcareous pisoids.

Pisoids are formed in terrestrial, lacustrine and vadose-marine environments (see Sect. 4.1.3.5). Important terrestrial environments are calcareous hardpans (caliche) and lateritic crusts (duricrusts), caves and mines, or geysers and hot spring deposits. Lacustrine pisoids are known from freshwater lakes, salt lakes, and from small rivers. Nearshore hypersaline environments are where "vadose-marine pisoids" are formed (Esteban, 1976; Esteban and Pray, 1976; Peryt and Piatkowski, 1977).

Three modes of origin have been suggested for pisoids (Donahue, 1969; Esteban, 1976): (1) Chemical precipitation from agitated solutions: the envelope of the concentric laminae grows by chemical precipitation and is abraded by contacts with neighbouring grains, resulting in nearly spherical or ovoid shapes with continued growth of concentric laminae; examples are cave pearls (see Sect. 2.3.2.3) or pisoids formed in hot springs (e.g., Mammoth Hot Springs, Yellowstone; Czechoslowakia), (2) Chemical or biochemical precipitation from non-agitated solutions: due to the lack of abrasion an irregular shape develops. This holds true for fluviatile pisoids (Braithwaite, 1979) and also for pisoids, formed during small eustatic fluctuations by an interaction between vadose diagenesis and hypersaline immersions (Esteban and Pray, 1978); (3) During soil formation, especially in humid tropical or subtropical climates. These caliche pisoids are characterized by composite nuclei, asymmetric "dripstone" pisoids, occurrence of pisoids in reverse size-graded units, and by a fitted polygonal pisolite fabric. Caliche pisoids are not truly sedimentary particles, but form chemically by evaporation in a semiarid climate. The difference between these diagenetic particles and morphologically similar sedimentary particles is shown in Table 4.

Caliche pisoids and pisoids formed in hypersaline vadose environments are similar with respect to their microscopic fabrics; both types may display indistinct or distinct laminations consisting of five or more laminae; both types may show the "typical clotted peloid texture" believed to be diagnostic for caliche pisoids only (see Esteban, 1976; Chafetz and Butler, 1980). Caliche pisoids and hypersaline pisoids are also similar with respect to the downward elongation of laminae, to the development of fibrous radial and blocky cements with some interstitial sandy or silty matrix within the interstitial space, and also with regard to the association with fenestral fabrics in dolomite strata (these fabrics seem to be grain-supported in hypersaline environments, and mud-supported in paleosols). Hypersaline pisoids may be recognized by the regional setting of pisolite accumulations, the presence of (marine) fossils, the brickwork-like texture and the elongated crystal rays with square terminations (interpreted as relict aragonite needles; Assereto and Folk, 1976). The occurrence of well crystallized calcite and/or bladed sparry calcite with triangular-shaped cross-sections differentiates the caliche pisoids.

Vadose pisoids may originate in hypersaline environments (e.g., Scholle and Kinsman, 1974) or in subaerially exposed freshwater environments (e.g., James, 1972; Dunham, 1969). Vadose pisoids, believed to have been formed by capillary and percolating groundwaters, have been described by Bernoulli and Wagner (1971), Wardlaw and Reinson (1971), Burri et al. (1973), Bosellini and Rossi (1974), and Catalano et al. (1974). In some cases the interpretation of these examples is not at all unequivocal: this is shown by the controversy concerning the pisoids from the Lower Permian of the Capitan Reef Complex (Dunham, 1965; Kendall, 1969; Smith, 1974; Esteban and Pray, 1976) which have been regarded as of subaerial vadose origin, primarily marine but secondarily vadose origin, or formed in a depositional facies close to high intertidal situations in hypersaline environments.

Vadose pisoids may be associated with algal oncoids, indicating a rather complex depositional history. A partly algal and partly vadose origin of pisolithic Zechstein limestones of Northern Poland can be interpreted as the effect of transportation of algal oncoids from a subtidal zone of oncoid formation to supratidal flats where vadose deposits were formed, followed by a transport of the vadose pisoids and/or algal-vadose pisoids to the subtidal zone (Peryt and Piatkowski, 1977).

4.1.3.7 Coated Grains (Cortoids) (Plates 5/2; 10; 11/3; 47/1)

In accordance with their significance as facies indicators, bioclasts, peloids, ooids, and other particles with coatings of relatively thin micrite envelopes are put into a separate category and called "cortoids" (from "cortex" – Latin for bark). The micrite envelopes (Plate 5/2) appear black in transmitted light, white in direct light and are shown in SEM analyses (Loreau, 1970; Carbone et al., 1974) to consist of crystals 0.25 to 5 microns in size, usually about 2 microns.

The term cortoid does not coincide completely with the English term "coated grains" because the latter sometimes also includes oncoids and ooids. R. C. L. Wilson (1967) calls these cortoids "coated shells."

Micrite envelopes or cortoids should not be mistaken for superficial rim cements, which line conduit systems with a broad multilayered fringe (up to 90 μm

Plate 10. Microfacies of Shallow Water Limestones

The section, which was cut perpendicular to the bedding plane, shows the following genetically important MF characteristics: Cortoids –1 – almost all the larger bioclasts are coated by micrite envelopes. Particles with shells and corals as "nuclei" seem to be due to micritization by boring micro-organisms. In larger particles such as coral fragments –2, the thin micrite envelopes are coated with biogenic crusts of sessile organisms (algal oncoids or bio-oncoids). "Nuclei" for these oncoids or for the micrite envelopes can be Solenoporaceae –3, shells –4, and especially fragments of solitary corals (longitudinal section –5, cross-section –6). Micritized fragments of echinoderms –7 are rarely seen. Larger algal oncoids, sometimes micritic oncoids, are also rare –8. Foraminifera –9, gastropods –10, and sporadic dasycladaceans (*Heteroporella* –11) exhibit only insignificantly thin micrite envelopes. A fine arenitic pelsparite layer can be seen in the top half of the picture; the micritic peloids –12 are conspicuously small. The pores between the particles are infilled by cement A –13 and cement B –14, (very light-colored in the picture); dogtooth cement also occurs –15. These microfacies correspond to Standard Microfacies Type 11 after Wilson (environments of constant wave action, platform carbonates), see Section 9. Late Triassic (Dachstein limestone): Loser highway near Bad Aussee, Salzkammergut, Styria, Austria. Exercise: Classify the sample after a) Folk (1959 or 1962); b) Dunham (1962); c) Leighton and Pendexter (1962); d) Plumley et al. (1962). Answers in Section 12. Scale is 10 mm

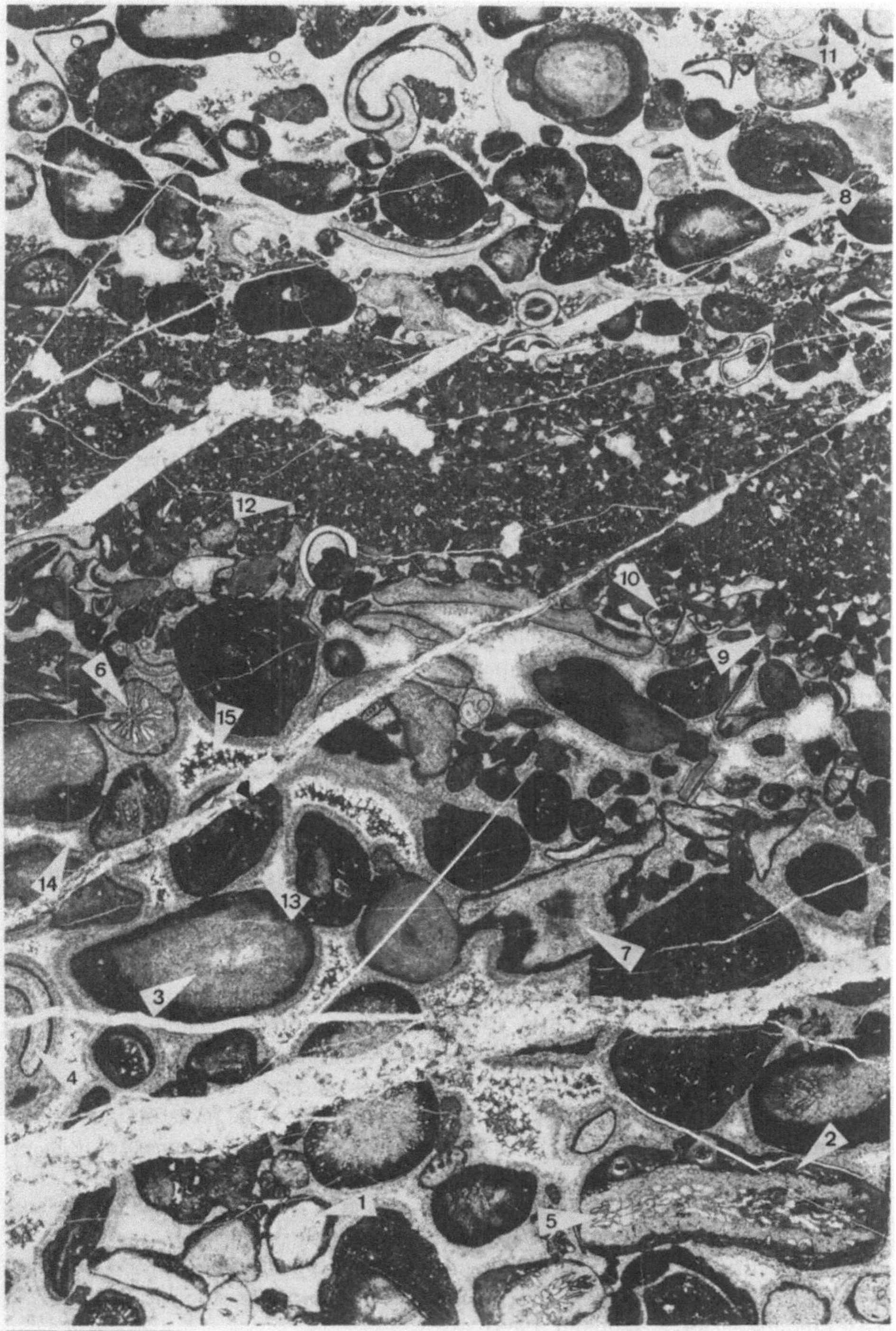
11
8
12
10
9
6
15
14
13
7
3
4
2
1
5

in thickness) of minimicrite and microsparite. In Recent reefs these rims are composed of magnesium calcite (James et al., 1976). Similar to Recent counterparts the superficial rim cements also had a profound binding effect on Ancient reef-detrital sediments (see Pedley, 1979).

Origin of Cortoids

1. *Micritization* (Bathurst, 1966; cryptocrystallization, Milliman, 1974: 274): Micro-organisms which have attacked and bored into the surface of bioclasts and other particles produce, especially in tidal zones, tiny little voids which are infilled by (bacterially precipitated?) micrite after the death of the organisms. If this process stretches over a longer period of time, a micritic replacement structure arises, which is centripetally rather than centrifugally oriented (e.g., micritic oncoids, see 4.1.3.4). The micrite envelopes of Recent particles consist of aragonite or Mg-calcite (Buchbinder and Friedman, 1970; Winland, 1968); in general a large amount of organic substance is present.

Organisms which have been proved to be borers are: ?bacteria, photosynthesizing Cyanophyceae, eukaryotic green and red algae as well as heterotrophic fungi (Bathurst, 1966; Friedman et al., 1971; Golubić et al., 1975; Rooney and Perkins, 1972). The size of the bores fluctuates between < 1 micron and 50 microns. In thin-sections there is usually an irregular boundary between bored bioclasts and micrite envelopes. Distinct boundaries may originate then when biogenic circumcrusts of carbonate particles (e.g., Corallinaceae) have been micritized by boring algae (Sibley and Murray, 1972). The bores frequently resemble straight tubes running perpendicular to the surface of the particle (diameter about 5–30 microns), rounded cone-shaped indentations, or shallow depressions.

The complete process of micritization consists in loosening the surface, abrading and rounding off the particles, and, later, formation of micritic peloids (see 4.1.3.2); see Illing (1954), Purdy (1968), and Kendall and Skipwith (1969). Intensive micritization may also occur within caliche crusts in the meteoric zone (James, 1972) resulting in a transformation of carbonate particles into particles without microstructures.

In Ancient cortoids a preferential boring of the original aragonitic shell parts has been found, whereas primary calcitic bioclasts, such as echinoderms, are only slightly or not at all micritized (Examples in Bachmann, 1973; Catalov, 1971).

Micritization, as described here, is probably temperature-dependent: According to Gunatilaka (1976) no cortoids originate from carbonate particles bored by thallophytes in cooler seas (Ireland). This is due to the fact that the micrite in the microborings does not originate by way of bacterial metabolic processes but rather as cement dependent on (higher) water temperatures and the supersaturation of $CaCO_3$.

2. Micrite envelopes of anhedral *nannocrystals* arise whose growth on the surface of the particles is *induced by organic films* (Loreau, 1970a, see 4.1.3.5).

3. Selective dissolution of fragments of organic shells due to the *different solubilities of aragonite and Mg-calcite* can lead to the formation of *residual micrite* (Alexandersson, 1972; Winland, 1968).

4. The *intergrowth of externally calcified filamentous algae* on the surface of carbonate particles leads to the formation of micrite envelopes which protect the particles from being destroyed (Kobluk and Risk, 1977). The thickness of these micrite envelopes, which are formed in about one year, may reach 500 microns. These "constructive" micrite envelopes are not easily distinguished from the "destructive" micrite envelopes described by Bathurst (1966). An important control on the development of constructive envelopes is the degree of agitation of the sediment substrate. Constructive envelopes as well as destructive envelopes will probably not form under agitated conditions (Alexandersson, 1972; Kobluk and Kahle, 1978).

5. *Decomposition of indigenous organic matter* and simultaneous *replacement by micrite* of both the original organic matter and the clay-sized carbonate particles (Purdy, 1968; Burgess, 1979). This would explain partially micritized grains in which the original fabric is still visible.

Micrite envelopes are relatively resistant to dissolution, but susceptible to dolomitization (see Buchbinder and Friedman, 1970).

The Value of Cortoids as Facies Indicators

The potential usefulness of endolithic algae in recognizing paleophotic zones, and as an aid in determining paleobathymetry has been stressed by many authors (e.g., Mägdefrau, 1937; Hessland, 1949; Golubić et al., 1975). Having observed that various abundances of cortoids appear at different water depths, some researchers (Perkins and Halsey, 1971; Rooney and Perkins, 1972; Swinchatt, 1969) use micritized carbonate bioclasts as paleobathymetric indicators of water depth, usually postulating water depths of less than 15–20 m. Friedman et al. (1971) pointed out the dangers inherent in such an interpretation. Basically the sources of error are these

- Boring organisms are not restricted to algae dependent on light and, hence, depth, but also include fungi, which are independent of light (down to depths of several thousand meters)
- Many endolithic algae demonstrate adaptation to the supply of light in deeper ocean regions (Golubić et al., 1975)
- Greater water energy generally prevents colonization by endolithic algae. Recent accumulations of boring algae are found mainly in protected, relatively low-energy environments.
- Micritization by boring algae is not the only possibility for forming micrite envelopes
- Cortoids can be transported

By considering the various morphologies of the borings at various water depths (Golubić et al., 1975) as well as the different shapes of algal and fungal borings (see Bathurst, 1971; Boekschoten, 1966; Porter and Zebrowski, 1937; Zeff and Perkins, 1979) a better evaluation of cortoids as bathymetric criteria should be possible in the future. Evidence of this are the constant distribution patterns in Ancient shelf and reef carbonates, where "coated bioclasts" occur preferentially in grainstones in the high-energy environment of shallow shoals (Standard Microfacies Type 11, see

8.2); coated bioclasts found in wackestones and packstones in the open shelf region have probably drifted in from neighboring high-energy environments (Standard Microfacies Type 10).

The application of microborings as bathymetric indicators generally hinges on depth zonations recognized from modern endoliths and the persistence of similar forms through geologic time. When endoliths from shallow-water environments are compared to deep-marine environments below the photic zone (Zeff and Perkins, 1979), some types of microborings together with a generally lower diversity seem to be indicative of microbial alterations in the deep-sea. A clear distinction of photosynthetic microborers and endolithic marine fungi as well as the recognition of the types restricted to deep-sea environments will yield valuable criteria for the paleobathymetrical interpretation of deep-sea carbonate buildups, displaced carbonates, and submarine hardgrounds. Differences in the direction of penetration of microborings between photic and aphotic endolithic organisms may be sufficient criteria for differentiating paleophotic zones in Ancient carbonate sequences. A three-fold bathymetrical subdivision has been found in the distributional patterns of microborings on the Puerto Rican shelf and slope: an upper photic zone assemblage (intertidal to 20 m) with endolithic blue-green algae oriented perpendicular to substrate surfaces; a lower photic zone assemblage (20–85 m) with endolithic green algae and red algae, and an aphotic zone (down to 530 m), characterized by the absence of algal borings and the presence of fungal microborings. Microborings in the lower photic zone and in the aphotic zone are oriented parallel to substrate surfaces and penetrate only slightly into substrate interiors (Budd and Perkins, 1980). Assuming that the endolithic microboring forms represent long-lived species of algae and fungi indigenous to tropical settings, relative depth values of Recent microborings are useful for the paleobathymetrical interpretation of Ancient carbonates.

4.1.3.8 Intraclasts and Extraclasts (Plate 11/1, 2; 31/5; 47/3; 48/1)

Intraclasts are reworked products of a substrate which has already been weakly consolidated within the basin of deposition. Reworking can take place by:

- Waves or currents disturbing the bottom (origin of mud pebbles)
- Organic activity on the surface of the sediment (e.g., grazing benthonic organisms, fishes living near the bottom)
  Intraclasts originating from water movement and organic activity on the sea floor are called "plasticlasts" by Folk (1959).
- Intraclasts originating from organic activity in the sediment (burrowing endobiontic organisms, bioturbation). These are called "pseudointraclasts" by Wobber (1965).
- Early diagenetic changes in volume (e.g., submarine sediment dehydration, compaction; leaching of evaporitic layers) lead to the formation of "autoclasts" (Sander, 1936) or "protointraclasts" (Bosellini, 1964), whose outer boundaries, though separated, would fit into each other like the pieces of a jigsaw puzzle
- Desiccation and formation of "mud pebbles", particularly in the upper tidal and supratidal zones

- Local sliding of the weakly consolidated sediment
- Reworking of arenitic and micritic carbonates already consolidated by carbonate cement, e.g., by storm waves (reworked dolomite crusts in Middle Triassic Wetterstein limestone as indicators of early supratidal dolomitization, see K. Germann, 1969).

Two types of intraclasts are rather common in shallow-water limestones: angular, calcarenitic intraclasts – produced by the mechanical erosion of lithified beachrocks within the intertidal and supratidal zones, and mud intraclasts – formed by the dessication of supratidal, partly lithified and sometimes dolomitized calcareous mudstones.

The reworking of a substrate that has already been lithified – for instance, after a break in deposition or after a restricted rate of sedimentation – can lead to the origin of resediments which are sometimes difficult to distinguish from allochthonous erosional detritus (e.g. coastal material). Synsedimentary "intraclasts (Folk, 1959, 1962) and "extraclasts" brought into the basin from an outside source should be kept separate in nomenclature, if possible.

Extraclasts (Plate 31/5) are characterized as follows:

- Fossil content (stratigraphically "older" fossils)
- Truncation of fossils and other particles at the extraclast boundary
- No compaction features
- Different cements in the extraclasts (e.g., greater neomorphism) than in the rest of the particles
- Presence of recrystallized veins
- Oxidation boundaries
- Extraclasts are often either conspicuously angular or conspicuously round
- Carbonate extraclasts often occur together with non-carbonate (terrigenous) particles such as sandstone or chert fragments.

Submarine rock-falls at the edge of steep carbonate cliffs can produce limestone breccias which are characterized by poorly sorted, densely packed and non-graded extraclasts of rather homogeneous microfacies, and by the lack of matrix (intrasparites).

Black pebbles (cailloux noirs) are a special type of extraclast important for facies interpretations; they are known in Mesozoic and Recent carbonates (Barthel, 1974; Seyfried, 1980). These involve conspicuously black, cm-sized calcareous lithoclasts which originate by infiltration of organic substances into a porous limestone bed in the Florida Keys in the Recent. The organic substance is derived from a substrate overgrown with mangroves and xerophytes which superimposed the limestone bed. Erosion and redeposition may also transport the lithoclasts into various sedimentary environments near land (shallow-marine, freshwater ponds). Black pebbles are thus indicators of terrestrial plant growth near land, and of transgressions or fluctuations in sea level.

Another mode of genesis for the black pebbles has been discussed by Bechstädt (1975): Subtidal sediments are cemented within the vadose zone, followed by brecciation and resedimentation. Small fragments deposited in shallow depressions may be dissolved by organic acids; the black color is due to bitumina or pyrite formed in the reducing environment. Black pebbles or "black breccias" associated with grayish-green marls or marly dolomites can sometimes indicate cyclic emersions and karstification of lagoonal carbonate platforms, producing residual sediments (see Bechstädt, 1979).

Blackened limestone and caliche rubble is also produced in situ at the margin of hypersaline lakes (Yucatan peninsula, Ward et al., 1970). Blackened limestone in Ancient carbonate series may call attention to carbonate breaks (subaerial exposure adjacent to hypersaline waters).

Folk (1959) suggested the term intraclasts, intending "to embrace the entire spectrum of carbonate rock fragments, . . . regardless of degree of cohesion or time gap between deposition of the original layer of sediment and later reworking of it," whereas Chanda (1967) attempts to provide a nomenclature for the various ages of the reworked substrate of the sedimentary basin: "autolimeclast" refers to intraclasts which have originated from in situ reworking by diagenetic and sometimes tectonic processes (inhomogeneous breccia, solution breccia, etc.). The name "infralimeclast" designates fragments which have originated through reworking of pieces of sediment already consolidated in the basin (e.g., tidal sediments; lithified coral reef sediments, boulders of fore-reef rock; subsolution relicts; Plate 15/2). Such fine subdivisions are rarely made.

Carbonate petrologists often consider carbonate grains as essentially of intrabasinal origin and regard the presence of extrabasinal carbonate and non-carbonate particles in calcarenites as an "unlucky mistake". Zuffa (1980) has emphasized the importance of distinguishing extrabasinal and intrabasinal detritus within "hybrid arenites" (arenites containing both intrabasinal and extrabasinal components) in order to avoid unreliable reconstructions of paleobasins and paleosource areas. Four main groups of arenaceous grain types can be recognized: (1) non-carbonate extrabasinal, (2) carbonate extrabasinal, (3) non-carbonate intrabasinal, and (4) carbonate intrabasinal. Their three-dimensional configuration establishes a first-level classification of the main types of arenites. In addition two more groups (neovolcanic particles; carbonate particles of either extrabasinal or intrabasinal, i.e., undetermined origin, = "limeclasts") are required for a truly practical scheme.

Frequently the term *lithoclast* (Folk, 1962) or *limeclast* (Wolf, 1965) is used to characterize erosional detritus which has been brought in from an outside source (= extraclasts) as well as pieces of sediments which have been more or less synsedimentarily reworked (= intraclasts). Some authors restrict lithoclast, however, to resediments which have been brought in (extraclasts). When intraclasts derive from synsedimentary reworking of non-consolidated or only weakly consolidated calcareous mud sediments, these particles are often larger than peloids (however, an artificial division between these two particle categories is usually made at 0.20 or 0.50 mm); sorting is generally very poor.

Aggregate grains (see 4.1.3.3) do not belong to the intraclasts because of their totally different mode of origin.

The value of lithoclasts as facies indicators

Intraclasts can indicate the following:

- Relatively weak bottom current (plasticlasts, sometimes imbricate structures)
- Soft bottoms (pseudointraclasts)
- Very shallow intertidal and supratidal zones (algal intraclasts, mud pebbles, see Table 13 and Fagerstrom, 1967)
- Break in deposition (accumulations of intraclasts)
- Compaction, dissolution, leaching (inhomogeneous breccia or autoclasts)
- Dissolution on the ocean floor (subsolution breccia)
- Successive diagenesis (carbonate cements in arenitic intraclasts)

4.1.4 Terrigenous Particles

All non-carbonate minerals or rock fragments which have been brought into the depositional environment of carbonates are called terrigenous particles. Among the most important are

- Clay minerals
- Quartz
- Siliceous rocks (cherts, siliceous shale, sandstones)
- Argillaceous shale (rare)
- Pyroclastic particles

Theoretically, of course, a great number of non-carbonate terrigenous particles may occur (a list can be found in Charollais and Davaud, 1974); in practice, however, the ones listed above predominate. Descriptions can be carried out using the conventional petrographic methods, or analyses of insoluble residues, separately or together (see Sect. 9.1.1).

The amount of detrital quartz grains can be used as a criterion for the intensity of the water energy. The mean and maximum diameter should be considered with respect to the efficiency of winnowing processes. Roundness is another criterion important for the interpretation of transportation processes.

4.1.5 Authigenic Minerals

Minerals generated in place subsequent to the formation of the sediment are called authigenic minerals. General surveys of the more common authigenic minerals in carbonate rocks can be found in Carozzi (1960), Theodorovich (1961), and Füchtbauer (1974).

The following characteristics indicate an authigenic origin of minerals:

- Euhedral crystal form,
- Pseudomorphs after carbonate particles (e.g., ooids) and fossils
- Inclusions which correspond mineralogically to the matrix of the rock (e.g., carbonate inclusions within authigenic quartz crystals; D. K. Richter, 1971)
- Zones of inclusions arranged parallel to idiomorphic faces of euhedral crystals

- Cross-cutting relationships between crystal boundaries and host rock fabrics
- Crystal aggregates such as spherulites and radiated clusters
- Differences in crystal form and color as compared with detrital mineral grains
- Differences in the optical properties in comparison to detrital grains (especially for feldspars; Füchtbauer 1950)
- Abundant occurrence of minerals, which can not be interpreted by a sorting effect during transport

Among the more common authigenic minerals found in carbonate rocks are (Topkaya, 1950): quartz, glauconite, albite, K-feldspars, tourmaline, muscovite, zircone and – as an indicator of conditions similar to those for gyttia (see Fabricius, 1961) – framboidal pyrite aggregates. Of particular importance are those authigenic minerals which provide information about the course of burial diagenesis and about synsedimentary geochemical parameters (see A. Strasser, 1979; Buyce and Friedman, 1975). Interesting examples have been described by Grimm (1962), Nachsel (1966), and E. Franz (1967), who used authigenic quartz as an index mineral for an evaporite facies.

Micritic limestones represent a favorable environment for the development of authigenic K- and Na-feldspars, particularly where they contain impurities such as clay minerals or pyroclastics. Authigenic feldspars are characterized by small crystals (generally 0.2 mm) which occur in low concentrations (less than 2%); the occurrence of abundant authigenic feldspar may indicate a replacement of wind-transported volcanoclastics (Buyce and Friedman, 1975). Authigenic feldspars may form at low pressure and low temperature, after burial or even at the sediment-water interface. Both strongly alkaline and hypersaline environments have been correlated with authigenic feldspar formation which tends to be associated with evaporitic conditions (Füchtbauer, 1972). The dolomitization of illitic carbonate beds may also cause the formation of authigenic feldspar (Swett, 1968; Bausch, 1980).

An important source of authigenic minerals (albite, mica, chlorite, quartz, magnesite) in carbonate rocks is the interaction of saturated brines, rich in Si and Al, with dolomites, limestones, and silicates precipitated earlier; this may originate from crystallization during diapirism at temperatures of less than 300 °C (Kulke, 1976).

Authigenic tourmaline seems to be rather common in carbonate rocks within evaporite sequences.

4.1.6 Notes on the Differentiation of Particles

The various *particles categories* in Table 13 should be used essentially for a *description* of the particles recognizable in thin-sections of limestones. As long as a genetic interpretation of the particle is possible and meaningful, the *subcategories* can also be applied.

In practice *difficulties* are presented by:

a) Too small bioclasts: Because even Recent bioclasts less than 0.06 mm in size can not generally be determined by binocular microscopes or polarization

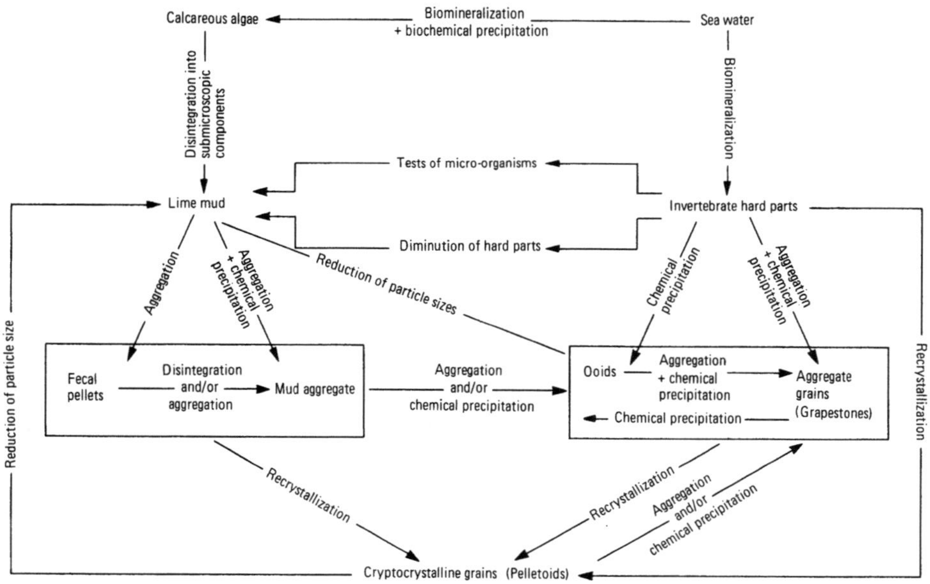

Fig. 22. Genetic interrelationships between particle categories. Modified after Kendall (1969)

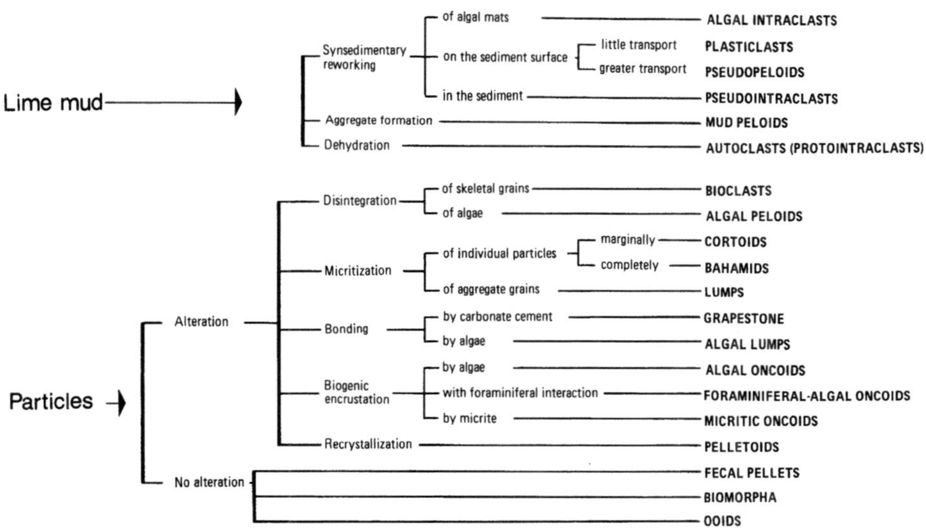

Fig. 23. Modes of origin of limestone particles

microscopes (see Feray et al., 1962; Matthews, 1966; Jindrich, 1969; Swinchatt, 1965), to determine quantitatively small bioclasts of uncertain systematic classification they should be considered as belonging to a separate category (e.g., "skeletal grains," in a very general sense).

b) Differentiating small intraclasts and peloids: For practical reasons it is recommended that a line be drawn between the groups at a particle diameter of about 0.20 mm, it must be remembered, however, that many peloids are products of reworking ("pseudopeloids" – see 4.1.3.2) and, hence, can be regarded genetically as intraclasts.

c) Differentiating grapestones and lumps: Often neither of these subcategories can be meaningfully separated from each other. In line with a suggestion by Milliman (1974), the increased micrite concentration in the lumps can be used to characterize these particles (see Sect. 4.1.3.3).

d) Differentiating intraclasts and extraclasts: Quite often the genetically neutral term lithoclast is more sensible, even if "intraclast" is more useful for building compound names for limestone types.

e) Differentiating sedimentary round particles and particles which are morphologically similar but which have originated diagenetically, e.g., caliche-ooids and caliche-peloids. In this case differences in fossil content (generally absent or reduced in caliche formations) and the types of carbonate cement, in addition to morphological details (see Table 4), can be of help.

Figures 22 and 23 summarize the genetic interrelationships between the particle categories.

4.1.7 References: Groundmass and Particles

4.1.7.1 References: Micrite

Adolphe, J. P., Billy, C. (1974): Biosynthèse de calcite par une association bactérienne aérobic. C. R. Acad. Sci. Paris **278** D, 2873–2875, Paris

Alexandersson, T. (1977): see 5.3.12.7

Alexandersson, T. (1978): Carbonate sediments – bacterial precipitation. In: Fairbridge, R. W. and Bourgeois, J. (eds.): The Encyclopedia of Sedimentology, **6,** 90–94, 2 Figs., Stroudsberg: Dowden, Hutchinson and Ross

Alexandersson, T. (1979): see 10.3.9

Bachmann (1973): see 4.1.7.2

Bathurst, R. G. C. (1966): see 4.1.7.2

Bathurst, R. G. C. (1971): see 1.4.1

Bavendamm, W. (1932): Die mikrobiologische Kalkfällung in der tropischen See. Arch. Mikrobiol. **3,** 205–276

Black, M. (1933): The precipitation of calcium carbonate on the Great Bahama Bank. Geol. Mag. **70,** 455–466

Berner, R. A. (1971): Bacterial processes effecting the precipitation of calcium carbonate in sediments. In: Bricker, O. P. (ed.): Carbonate Cements, 247–251, Baltimore–London: Hopkins

Böhm, L., Dawson, R., Liebezeit, G., Wefer, G. (1980): Suitability of monosaccharides as markers for particle identification in carbonate sediments. Sed., **27,** 167–177, 4 Figs., Oxford

Borza, K., Harman, M. (1970): Lithologisches und elektronenmikroskopisches Studium von Kalksteinen des Oberjuras und der Unterkreide. Geol. Zbornik, Geol. Carpathica **21**/1, 139–151, 19 Figs., Bratislava

Bosellini (1964): see 4.1.7.2
Brasier, M. D. (1975): An outline of history of seagrass communities. Palaeont. **18,** 681–702, London
Broecker, W. S., Takahashi, T. (1966): see 2.4.1
Bubela, B., Ferguson, J., Davies, P. J. (1975): Biological and abiological processes in a simulated sedimentary system. J. Geol. Soc. Austral., **22,** 135–144
Cayeux, L. (1935): Les roches sédimentaires de France. Roches carbonatées (calcium et dolomies). 436 pp., Paris: Masson
Chave, K. E. (1964): Skeletal durability and preservation. In: Imbrie, J., Newell, N. D. (eds.): Approaches in Paleoecology, 377–387, New York: Wiley
Cloud, P. E. (1962): Environment of calcium carbonate deposition west of Andros Island, Bahamas. Geol. Surv. Prof. Pap. **350,** 138 pp., 10 Tab., 46 Figs., Washington
Dalrymple, D. W. (1965): Calcium carbonate deposition associated with blue-green algal mats. Publ. Inst. Marine Sci. **10,** 187–200
Deelman, J. C. (1975): Two mechanisms of microbial carbonate precipitation. Naturwiss., **62**/10, 484–485, 1 Fig., Heidelberg
Dickinson, W. R. (1970): Interpreting detrital modes of graywacke and arkose. J. Sed. Petrol., **40,** 695–707, Tulsa
Drew, H. G. (1913): On the precipitation of calcium carbonate in the sea by marine bacteria, and on the action of denitrifying bacteria in tropical and temperate seas. J. Marine Biol. Ass. **9,** 479–524
Dunham, R. J. (1962): see 6.4
Elf-Aquitaine (1975): see 1.4.1
Fabricius, F. (1966): see 5.5
Farinacci, A. (1968): La tessitura della micrite nel calcare "Corniola" del Lias medio. Accad. Naz. Lincei, Red. Cl. Sci. Fis. Mat. Naturali, Fasc. 2, Ser. 8, **44,** 284–289, Pls. 1–4, Roma
Fischer, A. G., Honjo, S., Garrison, R. E. (1967): Electron Micrographs of Limestones and their Nannofossils. Monogr. Geol. Paleont. **1,** 141 pp., 94 Figs., Princeton
Fitzgerald, M. G., Parmenter, C. M., Milliman, J. D. (1979): Particulate calcium carbonate in New England shelf waters: result of shell degradation and resuspension. Sed. **26,** 853–857, 2 Figs., Oxford
Flügel, E. (1967): Elektronenmikroskopische Untersuchungen an mikritischen Kalken. Geol. Rdsch. **56,** 341–358, 2 Pls., 2 Figs., 1 Tab., Stuttgart
Flügel, E., Franz, H. E. (1968): Über die lithogenetische Bedeutung von Coccolithen in Flachwasser-Karbonaten. Eclogae geol. Helvet. **60,** 1–17, Basel
Flügel, E., Franz, H. E., Ott, W. F. (1968): Review on electron microscope studies of limestones. In: Müller, G., Friedman, G. M. (eds.): Recent Developments in Carbonate Sedimentology in Central Europe, 85–97, 2 Pls., 4 Figs., Berlin–Heidelberg–New York: Springer
Flügel, E., Keupp, H. (1979): see 3.9
Folk, R. L. (1959): see 6.4
Folk, R. L. (1965): Some aspects of recrystallization in ancient limestones. Spec. Publ. Soc, Econ. Paleont. Min. **13,** 14–48, 14 Figs., Tulsa
Folk, R. L. (1974): see 3.6
Folk, R. L., Robles, R. (1964): see 4.2.1.2
Friedman, G. M. (1972): Significance of Red Sea in problem of evaporites and basinal limestones. Amer. Ass. Petrol. Geol. Bull. **56,** 1072–1086, Tulsa
Fritz, G. (1958): Schwammstotzen, Tuberolithe und Schuttbrekzien im Weißen Jura der Schwäbischen Alb. Arb. geol. Inst. TH Stuttgart, N. F. **13,** 118 pp., 24 Figs., 5 Pls., Stuttgart
Frydl, P., Stearn, C. W. (1975): Rate of bioerosion of parrotfish in Barbados reef environments. J. Sed. Petrol. **48**/4, 1149–1158, 2 Figs., Tulsa
Fütterer, D. (1978): Die Feinfraktion (Silt) in marinen Sedimenten des ariden Klimabereiches. Quantitative Analysenmethoden, Herkunft und Verbreitung. Habilschrift, Univ. Kiel, 246 pp., separate vol. with 53 Pls., Kiel
Garrison, R. E., Schlanger, S. O., Wachs, D. (1975): Petrology and paleogeographic significance of Tertiary nannoplankton – foraminiferal limestones, Guam. Palaeo – 3, **17,** 49–54, 8 Figs., Amsterdam

Gerthofferova, H., Mišik, M. (1971): Electron microscope study of some limestones and dolomites of the Carpathian Mountains. Acta Geol. Geogr. Univ. Comenianae, Geol. **21,** 163–186, Pls. 20–40, 3 Figs., Bratislava

Golubić, S. (1969): Distribution, taxonomy and boring patterns of marine endolithic algae. Amer. Zool. **9,** 747–751, 6 Figs., Uthica

Greenfield, L. J. (1963): Metabolism and concentration of calcium and magnesium and precipitation of calcium carbonate by a marine bacterium. New York Acad. Sci. Ann. **109**/1, 23–45, New York

Gygi, R. A. (1975): *Sparisoma viride* (Bonnaterre), the Stoplight Parrotfish, a major sediment producer on coral reefs of Bermuda? Eclogae Geol. Helvet. **68**/2, 327–359, 11 Figs., 2 Tab., Basel

Hadding, A. (1958): Origin of the lithographic limestones. Kungl. fysiogr. Sallsk. I. Lund. Forhanl. **28**/4, 21–32, Stockholm

Hay, W. W., Wise, S. W., Stieglitz, R. D. (1970): Scanning electron microscope study of fine grain size biogenic carbonate particles. 1. Transact. Gulf Coast. Geol. Soc. **20,** 287–302, 23 Figs.

Hill, M. E. (1975): Selective dissolution of mid-Cretaceous (Cenomanian) calcareous nannofossils. Micropaleont. **21**/2, 227–235, 2 Pls., New York

Honjo, S. (1969): Study of fine grained carbonate matrix: Sedimentation and diagenesis of "micrite". Spec. Paper Paleont. Soc. Japan **14,** 67–82, Pls. 15–19, Tokyo

Husseini, S. I., Matthews, R. K. (1972): Distribution of high-magnesium calcite in lime muds of the Great Bahama Bank: Diagenetic implications. J. Sed. Petrol. **42**/1, 179–182, 7 Figs., Tulsa

Illing, L. V. (1954): see 2.4.1

Kahle, Ch. F. (1977): Origin of subaerial Holocene calcareous crusts: role of algae, fungi and sparmicritisation. Sedimentology **24**/3, 413–435, 7 Figs., Oxford

Keupp, H. (1977): Ultrafazies und Genese der Solnhofener Plattenkalke (Oberer Malm, Südliche Frankenalb). Abh. Naturhist. Ges. Nürnberg **37,** 128 pp., 30 Pls., 18 Figs., Nürnberg

Keupp, H. (1978): Das kalkige Nannoplankton der „Roten Mergel" (Tithon-Basis) in der Südlichen Frankenalb und ein assemblage – Vergleich mit anderen Proben des oberen Weißjura. Geol. Bl. NO – Bayern, **28**/2/3, 80–117, 11 Figs., 3 tables, Erlangen

Kitano, Y., Hood, D. W. (1962): Calcium carbonate crystal forms from sea water by inorganic processes. J. Oceanograph. Soc. Japan **18,** 35–39, Tokyo

Kitano, Y., Hood, D. W. (1965): The influence of organic material on the polymorphic crystallization of calcium carbonate. Geochim. Cosmochim. Acta **29,** 29–41, London

Kukal, Z., Saadallah, A. (1973): Aeolian admixtures in the sediments of the N. Persian Gulf. In: Purser, B. H. (ed.): The Persian Gulf, 115–122, Heidelberg: Springer

Krumbein, W. E., Cohen, C. (1977): Primary production, mat formation and lithification: Contribution oxygenic and facultative anoxygenic Cyanobacteria. In: Flügel, E. (ed.): Fossil Algae, 37–56, 19 Figs., Berlin–Heidelberg–New York: Springer

Laffitte, R., Noel, D. (1967): Sur la formation des calcaires lithographiques. C. R. Acad. Sci. Paris **264**/11, 379–382, 2 Pls., Paris

Lalou, C. (1957): Studies on bacterial precipitation of carbonates in sea water. J. Sed. Petrol. **27**/2, 190–195, 2 Figs., Tulsa

Land, L. S. (1970): Carbonate mud: production by epibiontic growth on *Thalassia testudinum.* J. Sed. Petrol. **40,** 1361–1363, Tulsa

Leighton, M. W., Pendexter, C. (1962): see 6.4

Lindholm, R. C. (1969): Carbonate petrology of the Onondaga Limestone (Middle Devonian), New York: A case for Calcisiltite. J. Sed. Petrol. **39**/1, 268–275, 5 Figs., Tulsa

Lippmann, F. (1973): see 3.9

Longman, M. W. (1977): Factors controlling the formation of microspar in the Bromide formation. J. Sed. Petrol. **47**/1, 347–350, 3 Figs., Tulsa

Loreau, J. P. (1972): Pétrographique à calcaires fins au microscope électronique à balayage: introduction à une classification des "micrites". C. R. Acad. Sci. Paris **274,** 810–813, 3 Pls., Paris

Lowenstam, H. A. (1955): Aragonite needles secreted by algae and some sedimentary implications. J. Sed. Petrol. **25,** 270–272, Tulsa

Lowenstam, H. A. (1966): Biologic problems relating to the composition and diagenesis of sediments. Earth Science, 137–195, 4 Pls., 13 Figs.
Lowenstam, H. A., Epstein, S. (1957): On the origin of sedimentary aragonite needles of the Great Bahama Bank. J. Geol. **65,** 364–375, Chicago
Matthews, R. K. (1966): Genesis of recent lime mud in southern British Honduras. J. Sed. Petrol. **36**/2, 428–454, 8 Figs., 5 Tab., Tulsa
Matyja, B. A. (1978): Early diagenetic structures in Oxfordian calcareous deposits. Przeglad Geol., **3 (299)**, 156–158, 4 Figs., Warszawa
Maurin, A. V., Noel, D. (1977): A possible bacterial origin for Famennian micrites. In: Flügel, E. (ed.): Fossil Algae, 136–142, 3 Figs., Berlin–Heidelberg–New York: Springer
May, J. A., Perkins, R. D. (1979): Endolithic infestation of carbonate substrates below the sediment – water interface. J. Sed. Petrol., **49**/2, 357–378, 4 Figs., Tulsa
McCunn, H. J. (1972): Calcite and aragonite phenomena precipitated by organic decay in high lime concentrate brines. J. Sed. Petrol. **42**/1, 150–154, 8 Figs., Tulsa
Milliman, J. D. (1974): see 1.4.1
Milliman, J. D., Ross, D. A., Teh-Lung Ku (1969): Precipitation and lithification of deep sea carbonates in the Red Sea. J. Sed. Petrol. **39**/2, 724–736, 14 Figs., Tulsa
Minoura, N. (1975): An electron microscope study of the Upper Pennsylvanian limestones of Kansas. J. Fac. Sci. Hokkaido Univ., Ser. IV, Geol. Min. **16**/2–3, 193–330, 30 Pls., 16 Figs., 3 Tab., Sapporo
Mišik, M. (1970): Verwischung der ursprünglichen Merkmale in kalkigen Sedimenten und Kalken bei der Dolomitisierung. Verh. Geol. Bundesanst. **1970**/4, 673–688, 2 Pls., 2 Figs., Wien
Molnia, B. F., Pilkey, O. H. (1972): Origin and distribution of calcareous fines on the Carolina continental shelf. Sedimentology **18,** 293–310, 13 Figs., Amsterdam
Monty, Cl. (1965): Recent algal stromatolites in the Windward Lagoon, Andros Island, Bahamas. Ann. Soc. Geol. Belgique **88**/5–6, 269–276, Bruxelles
Müller-Jungbluth, W., Toschek, P. H. (1969): see 6.4
Neumann, A. C. (1965): Processes of recent carbonate sedimentation in Harrington Sound, Bermuda. Bull. Marine Sci. **15**/4, 987–1035, 13 Figs.
Neumann, A. C., Keller, G. H., Kofoed, J. W. (1972): "Lithoherms" in the Straits of Florida. Geol. Soc. Amer., Abstr. Programs **6,** 611
Neumann, A. C., Land, L. S. (1975): Lime mud deposition and calcareous algae in the Bight of Abaco, Bahamas: A budget. J. Sed. Petrol. **45**/4, 763–786, 10 Figs., Tulsa
Noel, D. (1969): Etudes des Coccolithes in situ dans les roches: La notion de nannofacies. Proc. 1st Intern. Conf. Planktonic Microfoss. Geneva **2,** 486–491, 4 Pls., Leiden : Brill
Noel, D. (1978): Les calcaires à Nannoconus (dits Calcaires à Calpionelles) de l'Ermitage d'Esparron (Isère), au passage Tithonique – Berriasien. Note Lab. Paléont. Univ. Geneve, **13,** 109–113, 2 Pls., Geneva
Oppenheimer, C. H. (1961): Note on the formation of spherical aragonitic bodies in the presence of bacteria for the Bahama Bank. Geochim. Cosmochim. Acta **23,** 295–296, Oxford
Patriquin, D. G. (1972): Carbonate mud production by Epibionts on *Thalassia*: an estimate based on leaf growth rate data. J. Sed. Petrol. **42**/3, 687–689, 1 Fig., Tulsa
Philippi, F. (1907): Über die Dolomitbildung und chemische Abscheidung von Kalk in heutigen Meeren. Neues Jahrbuch Min. Geol. Paläont. **1907,** Festband, 397–445, Stuttgart
Pilkey, O. H., Nobles, D. (1966): Carbonate and clay mineralogy of the Persian Gulf. Deep Sea Research, **13,** 1–16, Oxford
Purdy, E. G. (1963): see 2.4.1
Ricketts, B. D. (1980): Experimental investigation of carbonate precipitation in hydrated silica gels. J. Sed. Petrol., **50**/3, 963–970, 2 Figs., Tulsa
Roth, P. H., Mullin, M. M., Berger, W. H. (1976): Coccolith sedimentation by fecal pellets: Laboratory experiments and field observations. Bull. Geol. Soc. Amer. **86,** 1079–1084, 3 Figs., New York
Salomon, W. (1914): Über die Bildung dichter Kalke. Geol. Rdsch. **5,** 478–480, Berlin

Schneider, J. (1977): Carbonate construction and decomposition by epilithic and endolithic micro-organisms in salt- and freshwater. In: Flügel, E. (ed.): Fossil Algae. Recent Results and Developments, 248–260, 3 Pls., 4 Figs, Berlin–Heidelberg–New York: Springer

Schneidermann, N. (1970): Genesis of some Cretaceous Carbonates in Israel. Israel J. Earth Sci. **19,** 97–115, 13 Pls., 5 Figs., Jerusalem

Schneidermann, N. (1973): see 3.9

Scholle, P. A., Kling, St. A. (1972): Southern British Honduras: Lagoonal coccolith ooze. J. Sed. Petrol. **42**/1, 195–204, 5 Figs., Tulsa

Schwarzacher, W. (1961): Petrology and structure of some Lower Carboniferous reefs in northwestern Ireland. Bull. Amer. Ass. Petrol. Geol. **45,** 1481–1503, Tulsa

Shoji, R., Folk, R. L. (1964): Surface morphology of some limestone types as revealed by electron microscopy. J. Sed. Petrol. **34**/1, 144–155, Pls. 1–5, Tulsa

Sorby, H. C. (1879): Origin of limestones. Proc. Quart. J. Geol. Soc. London **35,** 56–95, London

Steinmann, G. (1889): Über Schalen- und Kalksteinbildung. Ber. naturforsch. Ges. Freiburg **4,** 288–293, Freiburg

Stieglitz, R. D. (1972): Scanning electron microscopy of fine fraction of recent carbonate sediments from Bimini, Bahamas. J. Petrol. **42**/1, 211–226, 4 Figs., Tulsa

Stieglitz, R. D. (1973): Carbonate needles: additional organic sources. Geol. Soc. Amer. Bull. **84,** 927–930, 1 Fig., New York

Stockman, K. W., Ginsburg, R. N., Shinn, E. A. (1967): The production of lime mud by algae in South Florida. J. Sed. Petrol. **37**/2, 633–648, 14 Figs., Tulsa

Swinchatt, J. P. (1965): see 2.4.1

Torunski, H. (1979): Biological erosion and its significance for the morphogenesis of limestone coasts and for nearshore sedimentation (northern Adriatic). Senckenbergiana maritima **11,** 193–265, 31 Figs., 4 Pls., 4 tables, Frankfurt

Trichet, J. (1972): Etude du mécanisme de la nucléation du carbonate de calcium dans les dépots algaires. Lien avec la biogéochimie comparée de quelques tapes de dépots dus à des Cyanophycées (Polynésie française). C. R. Congr. Geol. Intern. Montreal, Sect. **7,** 631–638, Montreal

Wardlaw, N. C. (1962): Aspects of diagenesis in some Irish carboniferous limestones. J. Sed. Petrol. **32**/4, 776–780, Pl. 1, Tulsa

Wefer, G. (1980): Carbonate production by algae Halimeda, Penicillus and Padina. Nature **285,** no. 5763, 323–324, 2 Figs., London

Wefer, G., Böhm, L., Dawson, R., Liebezeit, G., Meyer, M. (1978): Can the analysis of monosaccharides aid in the identification of carbonate producers in the fine-sediment fraction? Abstracts, Int. Congr. Sed., **2,** 726–727, Jerusalem

Wells, J., Illing, L. V. (1963): Present-day precipitation of calcium carbonate in the Persian Gulf. In: van Straaten, L. M. J. U. (ed.): Deltaic and Shallow Marine Environments. Dev. Sed. **1,** 429–435, Amsterdam: Elsevier

Wolf, K. H. (1965a): Petrogenesis and paleoenvironment of Devonian algal limestones of New South Wales. Sedimentology **4**/1–2, 113–178, Amsterdam

Wolf, K. H. (1965b): "Grain-diminuation of algal colonies to micrite". J. Sed. Petrol. **35**/2, 420–427, 7 Figs., Tulsa

Wolf, K. H. (1978): Limestones. In: Fairbridge, R. W. and Bourgeois, J.: The Encyclopedia of Sedimentology, 434–446, 14 Figs. 3 Pls., Stroudsburg: Dowden, Hutchinson and Ross

Wood, A. (1941): "Algal dust" and the finer-grained varieties of carboniferous limestones. Geol. Mag. **78,** 192–200, London

4.1.7.2 References: Limestone Particles

Adeleye, D. F. (1975): Deviation of fragmentary oolites and pisolites from desiccation cracks. J. Sed. Petrol. **45**/4, 794–798, 3 Figs., Tulsa

Alexandersson, T. (1972): Micritization of carbonate particles: processes of precipitation and dissolution in modern shallow-water sediments. Bull. Geol. Inst. Univ. Uppsala, N. S. **3,** 201–236, Uppsala

Andrews, J. E., Shepard, F. P., Hurley, R. J. (1970): Great Bahama Canyon. Geol. Soc. Amer. Bull. **81,** 1061–1078, New York

Argenio, B. d', Castro, P. de, Emiliani, C., Simone, L. (1975): Bahamian and Apenninic limestones of identical lithofacies and age. Bull. Amer. Ass. Petrol. Geol. **86,** 524–530, 6 Figs., Tulsa

Assereto, R., Folk, R. D. (1976): see Sect. 3.9

Babić, L. (1969): *Sphaerocodium* or onkoids from the Upper Triassic Dolomite of Western Yougoslavia. Geol. Vjesnik, Zagreb **23,** 11–19, 1 Pl., 1 Fig., Zagreb

Bachmann, G. H. (1973): Die karbonatischen Bestandteile des Oberen Muschelkalkes (Mittlere Trias) in Südwest-Deutschland und ihre Diagenese. Arb. Inst. Geol. Paläont. Univ. Stuttgart. N.F. **68,** 1–99, 67 Figs., Stuttgart

Badoux, H. (1970): Les oolites déformées du Vélar (massif de Morcles). Eclogae geol. Helvet. **63**/2, 539–548, 19 Figs., Basel

Baltres, A. (1975): Lithification of some modern carbonate sediments in a hypersaline lake adjacent to the Black Sea. Sed. Geol. **13,** 291–302, 4 Figs., Amsterdam

Bandel, K. (1974): Fecal pellets of Amphineura and Prosobranchia (mollusca) from the Caribbean Coast of Columbia, South America. Senck. marit. **6**/1, 14 Figs., Frankfurt/M.

Barthel, K. W. (1974): Black pebbles, fossil and recent, on and near coral islands. Proc. Soc. Intern. Coral Reef Symp. **2,** 395–399, 5 Figs., Brisbane

Bates, R. L., Jackson, J. A. (eds.) (1980): Glossary of Geology. 749 pp., Falls Church, Virginia: American Geol. Institute

Bathurst, R. G. C. (1966): Boring algae, micrite envelopes, and lithification of molluscan biosparites. Geol. J. **5**/1, 15–32, 3 Pls., Liverpool

Bathurst, R. G. C. (1967): Oolitic films on low energy carbonate sand grains, Bimini Lagoon, Bahamas. Marine Geol. **5**/2, 89–109, 14 Figs., 2 Tab., Amsterdam

Bathurst, R. G. C. (1968): Precipitation of ooids and other aragonite fabrics in warm seas. In: Müller, G., Friedman, G. M. (eds.): Recent Developments in Carbonate Sedimentology in Central Europe, 1–10, Berlin–Heidelberg–New York: Springer

Bathurst, R. G. C. (1971): see 1.4.1

Bausch, W. M. (1963): Die Oolithe des Malms. Fortschr. Geol. Rheinland-Westf. **10,** 343–346, Krefeld

Bausch, W. M. (1980): Tonmineralprovinzen in Malmkalken. Erlanger Forschungen, B, Naturwiss. Medizin **8,** 78 pp., 33 Figs., Erlangen

Bausch, W., Wiontzek, H. (1961): Petrographische Untersuchungen am Hauptdolomit von Rehden. Erdöl, Kohle **14,** 686–692, Hamburg

Beales, F. W. (1958): Ancient sediments of Bahaman type. Bull. Amer. Ass. Petrol. Geol. **42**/8, 1845–1880, Tulsa

Beales, F. (1971): Cementation in ancient pelletted limestones. In: Bricker, O. P. (ed.): Carbonate Cements, 215–224, Figs. 112–113, Baltimore–London: Hopkins

Bechstädt, T. (1979): The Lead-Zinc deposit of Bleiberg-Kreuth (Carinthia, Austria): Palinspastic situation, paleogeography and ore mineralisation. Verhandlungen Geol. Bundesanstalt **1978**/3, 221–235, 8 Figs., Wien

Behr, K., Behr, H.-J. (1976): Cyanophyten aus oberjurassischen Algen-Schwamm-Riffen. Lethaia **9,** 283–292, 8 Figs., Oslo

Bender, F. (1951): Fossile Pilze aus dem Eisenoolithhorizont des Lias α in Württemberg. Palaeontographica, B, **91,** 152–158, Pls. 22–24, 1 Fig., Stuttgart

Berg, G. (1944): Vergleichende Petrographie oolithischer Eisenerze. Arch. Lagerstättenforsch. **76,** Berlin

Bernoulli, D., Wagner, C. W. (1971): Subaerial diagenesis and fossil caliche deposits in the Calcare Massico Formation (Lower Jurassic, Central Apennines, Italy). N. Jb. Geol. Paläont. Abh. **138**/2, 125–149, Stuttgart

Bhattacharyya, A., Friedman, G. M. (1979): see 3.9

Blatt, H., Middleton, G., Murray, R. (1972): see 1.4.1

Boekschoten, G. J. (1966): Shell borings of sessile epibiontic organisms from the Dutch coast. Palaeo-3 **2,** 233–379, Amsterdam

Bolliger, W., Burri, P. (1970): Sedimentologie von Schelf-Carbonaten und Beckenablagerungen im Oxfordien des zentralen Schweizer Jura. Beitr. Geol. Karte Schweiz, N.F. **140,** 1–96, 15 Pls., 37 Figs., 4 Tab., Bern

Bornemann, J. G. (1886): Beiträge zur Kenntnis des Muschelkalks, insbesondere der Schichtenfolge und der Gesteine des unteren Muschelkalkes in Thüringen. Jb. preuss. geol. Landesanst. **1885,** 267–321, Berlin

Bosellini, A. (1964): Sul significato genetico e ambientale di alcuni tipi di rocce calcaree in base alle piu recenti classificazioni. Mem. Mus. Storia Nat. Venezia Tridentina, Anno 27–28, 1964–65, **15,** Fasc. **2,** 1–58, 8 Pls., 6 Figs., Trento

Bosellini, A. (1966): Protointraclasts: Texture of some Werfenian (Lower Triassic) limestones of the Dolomites (Northeastern Italy). Sedimentology **6,** 333–337, 2 Figs., Amsterdam

Bosellini, A., Ginsburg, R. N. (1971): Form and internal structure of recent algal nodules (rhodolites from Bermuda). J. Geol. **79,** 669–682, Chicago

Bosellini, A., Rossi, D. (1974): Triassic carbonate buildups of the Dolomites, northern Italy. Soc. Econ. Paleont. Min. Spec. Publ. **18,** 209–233, Tulsa

Braithwaite, C. J. R. (1979): Crystal textures of recent fluviatile pisolites and laminated crystalline crusts in Dyfed, South Wales. J. Sed. Petrol. **49**/1, 181–194, 5 Figs., Tulsa

Brönnimann, P. (1976): Revision of the lectotype of *Favreina salevensis* (Pareas) (Crustacea, Decapoda) and description of favreine form-species from the Jurassic and Cretaceous of Scotland, Portugal, Yougoslavia and Pakistan. Paläont. Z. **50,** 40–56, Pls. 5–7, 7 Figs., Stuttgart

Brown, P. R. (1964): Petrography and origin of some Upper Jurassic beds from Dorset, England. J. Sed. Petrol. **34,** 254–269, Tulsa

Buchbinder, B., Begin, Z. B., Friedman, G. M. (1974): Pleistocene algal Tufa of Lake Lisan, Dead Sea Area, Israel. Israel J. Earth Sci. **23,** 131–138, 13 Figs., Jerusalem

Buchbinder, B., Friedman, G. M. (1970): Selective Dolomitization of Micritic Envelopes: a possible clue to original mineralogy. J. Sed. Petrol. **40,** 514–517, 2 Figs., Tulsa

Budd, D. A., Perkins, R. D. (1980): Bathymetric zonation and paleoecological significance of microborings in Puerto Rican Shelf and Slope Sediments. J. Sed. Petrol. **50**/3, 881–994, 9 Figs., Tulsa

Burgess, C. J. (1979): see 3.9

Burri, P., DuDresnay, R., Wagner, C. W. (1973): Tepee structures and associated diagenetic features in intertidal carbonate sands. Sed. Geol. **9,** 221–228, Amsterdam

Busse, R. (1968): „Mumien“, Algenumwachsungen im Hauptrogenstein am Oberrhein. Aufschluß **19**/12, 346–348, 4 Figs., Heidelberg

Carbone, F., Civitelli, G. (1974): SEM non dispersive detector analysis and interpretation of microstructures in coated grains from central Italy. Geol. Romana **13,** 45–62, 34 Figs., Rom

Carozzi, A. (1956): An intraformational conglomerate by mixed sedimentation in the Upper Cretaceous of the Roc-De-Chère., Autochthonous Chains of High Savoy, France. J. Sed. Petrol. **26**/3, 253–257, 1 Fig., Tulsa

Carozzi, A. V. (1957): Contribution de l'étude des propriétes géométriques des oolithes – l'exemple du Grand Lac Salé, Utah, USA. Inst. National Genève Bull. **58,** 1–52, Geneva

Carozzi, A. V. (1960): see 1.4.1

Carozzi, A. V. (1961a): Distorted oolites and pseudoolites. J. Sed. Petrol. **31**/2, 262–274, Figs. 1–13, Tulsa

Carozzi, A. V. (1961b): Oolithes remaniées, brisées et régénerées dans le Mississippien des Chaines frontales, Alberta, Central Canada. Arch. Sci. **14**/2, 281–996, 8 Figs., Geneva

Carozzi, A. V. (1962): Cerebroid oolites. Trans. Illinois State Acad. Sci. **55**/3–4, 238–249

Carozzi, A. V. (1963): Half-Moon Oolites. J. Sed. Petrol. **33**/3, 633–645, 9 Figs., Tulsa

Carozzi, A. V. (1964): Complex ooids from Triassic Lake Deposits, Virginia. Amer. Sci. **262**/2, 231–241, 3 Pls., Washington

Carss, B., Carozzi, A. V. (1965): Petrology of Upper Devonian pelletoidal limestones, Arrow Canyon Range, Clark County, Nevada. Sedimentology **4**/3, 197–222, Amsterdam

Castellarin, A., Sartoti, R. (1973): Desiccation shrinkage and leaching vugs in the calcari grigi Infraliassi tidal flat (S. Massenza and Loppio, Trento, Italy). Eclogae geol. Helvet. **66**/2, 339–343, Basel

Catalano, R., d'Argenio, B., Lo Cicero, G. (1974): I ciclotemi triassici di Capo Rama (Monti di Palermo). Geol. Romana **13**, 125–145, Roma

Catalov, G. (1970): Fabric parts of limestones, Bulgarian Acad. Sci. Comm. Geol. Bull. Geol. Inst. Ser. Strat. Lithology **19**, 137–172, 4 Pls, 4 Figs., Sofia

Catalov, G. (1971): Centripetal replacement, micritization and the part played by boring algae in some Triassic limestones in the Fore-Balkan. C.R. Acad. bulgare Sci. **24**/2, 349–352, 8 Figs., Sofia

Chafetz, H. S., Butler, J. C. (1980): Petrology of recent caliche pisolites, spherulites and speleothem deposits from central Texas. Sed. **27**, 497–518, 17 Figs., Oxford

Chanda, S. K. (1967): Petrogenesis of the calcareous constituents of the Lameta Group around Jabalpur, M. P., India. J. Sed. Petrol. **37**/2, 425–437, 20 Figs., Tulsa

Chanda, S. K., Bhattacharyya, A., Sakar, S. (1977): Deformation of ooids by compaction: implications for lithification. Geol. Soc. Amer. Bull. **88**, 1577–1585, Oxford

Charollais, J., Davaud, E. (1974): see 7.3.2

Cherry, R. D., Higgo, J. J. W., Fowler, S. W. (1978): Zooplankton fecal pellets and element residence time in the ocean. Nature **274**, no. 5668, 246–248, London

Choquette, P. W. (1955): A petrographic study of the "State College" Siliceous oolite. J. Geology **63**, 337–347, 1 Pl. ., 3 Figs., 1 Tab., Chicago

Cloos, E. (1947): Oolite deformation in the South Mountain Fold, Maryland. Bull. Geol. Soc. Amer. **58**, 843–918, New York

Conley, C. D. (1977): Origin of distorted ooliths and pisoliths. J. Sed. Petrol. **47**/2, 554–564, 11 Figs., Tulsa

Cook, H. E., Enos, P. (eds.) (1977): see 1.4.1

Cros, P. G. (1979): Genèse d'oolithes et de grapestones, plate-forme des Bahamas (Joulters Cays, Grand Banc). Bull. Centre Rech. Explor.-Prod. Elf-Aquitaine **3**/1, 63–139, 15 Plates, 9 Figs., Pau

Dahanayake, K. (1977): Classification of oncoids from the Upper Jurassic carbonates of the French Jura. Sed. Geol. **18**/4, 337–353, 9 Figs., Amsterdam

Davies, P. J., Bubela, B., Ferguson, J. (1978): The formation of ooids. Sed. **25**, 703–730, 19 Figs., Oxford

Davies, R. A. (1966): Quiet water oolites from the Ordovician of Minnesota. J. Sed. Petrol. **36**/3, 813–818, 5 Figs., Tulsa

Donahue, J. (1965): Laboratory growth of pisolite grains. J. Sed. Petrol. **35**/1, 251–256, 6 Figs., Tulsa

Donahue, J. (1969): Genesis of oolite and pisolite grains: an energy index. J. Sed. Petrol. **39**/4, 1399–1411, Menasha

Dragastan, O. (1969): "Micro-oncolithes" dans le Jurassique supérieur de la vallée du Bicaz (Carpates orientales, Roumaine). Bull. Soc. géol. France (7), **11**, 655–659, Pls., 21–22, Paris

Dunham, R. J. (1969): see 2.4.2

Eardley, A. J. (1938): Sediments of the Great Salt Lake Utah. Bull. Amer. Ass. Petrol. Geol. **22**/10, 1305–1411, Tulsa

Ernst, H. (1978): Zu Bathymetrie und Sedimentstrukturen der Schreibkreide von Lägerdorf/ Holstein (Coniac-Santon): Eine quantitative Analyse der Foraminiferen-Fauna. Mitt. Geol.-Paläont. Inst. Univ. Hamburg **48**, 53–78, 8 Pls., 2 Figs., Hamburg

Esteban, M. C. (1976): Vadose pisolite and caliche. Bull. Amer. Ass. Petrol. Geol. **60**, 2048–2057, 4 Figs., Tulsa

Esteban, M., Pray, L. C. (1976): Non-vadose origin of pisolitic facies, Capitan reef complex (Permian), Guadalupe Mountains, New Mexico and West Texas (abstract). Bull. Amer. Ass. Petrol. Geol. **60**, 607, Tulsa

Esteban, M., Pray, L. C. (1978): Origin of the Pisolite Facies of the Shelf Crest. In: Hileman, M. E., Mazzullo, S. J. (eds.): Upper Guadelupian Facies, Permian Reef Complex, Guadelupe Mountains, New Mexico and West Texas. Soc. Econ. Paleont. Min. Permian Basin, 1977 Field Conf. Guidebook, 479–483, Midland

Fabricius, F. H. (1966): see 5.5

Fabricius, F. H. (1967): Die Rät- und Lias-Oolithe der nordwestlichen Kalkalpen. Geol. Rdsch. **56**/1, 140–170, 10 Figs., 2 Pls., 2 Tab., Stuttgart
Fabricius, F. H. (1977): Origin of marine ooids and grapestones. Contrib. Sed. **7,** 113 p., 18 Pls., 6 Figs., 1 Tab., Stuttgart
Fabricius, F. H., Berdau, H. D., Münnich, K. O. (1970): Early Holocene ooids in modern littoral sands. Reworked from a coastal terrace, Southern Tunesia. Science **169,** 757–760, 2 Figs., Washington
Fabricius, F. H., Klingele, H. (1970): Ultrastrukturen von Ooiden und Oolithen: Zur Genese und Diagenese quartärer Flachwasserkarbonate des Mittelmeeres. Verh. Geol. Bundesanst. **1970**/4, 594–617, 1 Fig., 4 Pls., Wien
Fagerstrom, J. A. (1967): Development, floatation and transportation of mud crusts – neglected factors in sedimentology. J. Sed. Petrol. **37**/1, 73–79, 7 Figs., Tulsa
Fahraeus, L. E., Slatt, R. M., Nowlan, G. S. (1974): Origin of carbonate pseudopellets. J. Sed. Petrol. **44**/1, 27–29, 2 Figs., Tulsa
Feray, E., et al. (1962): see 6.4
Ferguson, J., Buebla, B., Davies, P. J. (1978): Synthesis and possible mechanisms of formation of radial carbonate ooids. Chem. Geol. **22**/4, 285–308, 4 Figs., Amsterdam
Flügel, E., Kirchmayer, M. (1962): Zur Terminologie der Ooide und Pseudooide. N. Jb. Geol. Paläont. Mh. **1962**/3, 113–123, 2 Figs., Stuttgart
Flügel, E., Wolf, K. H. (1969): „Sphaerocodien" (Algen) aus dem Devon von Deutschland, Marokko und Australien. N. Jb. Geol. Paläont. Mh. **1969**/2, 88–103, 4 Figs., Stuttgart
Folk, R. L. (1959): see 6.4
Folk, R. L. (1962): see 6.4
Force, L. M. (1969): Calcium carbonate size distribution on the W. Florida Shelf, and experimental studies on the microarchitectural control of skeletal breakdown. J. Sed. Petrol. **39,** 902–934, Tulsa
Franke, W., Eder, W., Engel, W. (1975): Sedimentology of a Lower Carboniferous shelf-margin (Velbert Anticline, Rheinisches Schiefergebirge, W-Germany). N. Jb. Geol. Paläont. Abh. **150**/3, 314–353, 16 Figs., 1 Tab., Stuttgart
Freeman, T. (1962): Quiet water oolites from Laguna Madre, Texas. J. Sed. Petrol. **32**/3, 475–483, 10 Figs., Tulsa
Friedman, G. M. (1978): "Solar Lake": A sea-marginal pond of the Red Sea (Gulf of Aqaba or Elat) in which algal mats generate carbonate particles and laminites. In: Krumbein, W. (Ed.): Environmental biogeochemistry and geomicrobiology, vol. 1, The Aquatic Environment, 227–235, 4 Figs., Ann Arbor: Ann Arbor Science
Friedman, G. M., Amiel, A. J., Braun, M., Miller, D. S. (1973): Generation of carbonate particles and laminites in algal mats, examples from sea-marginal hypersaline pools, Gulf of Aqaba, Red Sea. Bull. Amer. Ass. Petrol. Geol. **57**/3, 541–558, Tulsa
Friedman, G. M., Gebelein, C. D., Sanders, J. E. (1971): Micritic envelopes of carbonate grains are not exclusively of photosynthetic algal origin. Sedimentology **16,** 89–96, Amsterdam
Frishman, S. A. (1969): Geochemistry of oolites, Baffin Bay, Texas. Thesis Univ. Texas, 82 p., Austin
Fritz, G. (1958): see 5.3.4
Füchtbauer, H. (1964): see 9.4
Füchtbauer, H. (1968): Carbonate sedimentation and subsidence in the Zechstein Basin (Northern Germany). Recent Developments in Carbonate Sedimentology in Central Europe, 196–204, 4 Figs., Berlin–Heidelberg–New York: Springer
Garber, R. A., Friedman, G. M., Nissenbaum, A. (1981): Concentric aragonitic ooids from the Dead Sea. J. Sed. Petrol. **51**/2, 455–458, 3 Figs., Tulsa
Gasche, E. (1956): Über die Entstehung der Mumien und übrigen Kalkknollen aus dem Sequan des Berner Jura. In: Ziegler, P. A.: Geologische Beschreibung des Blattes Courtelary (Berner Jura) und zur Stratigraphie des Sequanien im zentralen Schweizer Jura. Beitr. Geol. Karte Schweiz, N. F. **102,** 3 pp., Basel
Gatral, M., Jenkyns, H. C., Parsons, C. F. (1972): Limonite concretions from the European Jurassic, with particular reference to the "snuff-boxes" of southern England. Sedimentology **18,** 79–103, 12 Figs., Amsterdam

Gebelein, C. D. (1969): see p. 346
Germann, K. (1969): Reworked dolomite crusts in the Wettersteinkalk (Ladinian, Alpine Triassic) as indicators of early supratidal dolomitization and lithification. Sedimentology **12**, 257–277, Amsterdam
Ginsburg, R. N. (1956): see p. 202
Ginsburg, R. N. (1960): Ancient analogues of recent stromatolites. Inst. Geol. Congr. XXI, Sess. Norden, part **22**, 26–35, Copenhagen
Glazek, J. (1965): Wspolczesne onkolity w potokach polnocnego weetnamu i Tatr Polskich. Rocz. polsk. Tow. geol. **35**/3, 221–242, 4 Pls., Figs., Tab. (pol. with engl. summary).
Golubić, St., Brent, G., Lecampion, Th. (1970): Scanning electron microscopy of endolithic algae and fungi using multipurpose casting-embedding technique. Lethaia **3**, 203–209, 5 Figs., Oslo
Golubić, St., Fischer, A. G. (1975): Ecology of calcareous nodules forming in little Connestage creek near Lancaster, Pennsylvanian. Verh. Intern. Verein. Limnol. **19**, 2315–2323, 9 Figs., Stuttgart
Golubić, St., Perkins, R. D., Lukas, K. J. (1975): Boring microorganisms and microborings in carbonate substrates. In: Frey, R. W. (ed): The Study of Trace Fossils, 229–259, 10 Figs., Berlin–Heidelberg–New York: Springer
Grabau, A. W. (1904): On the classification of sedimentary rocks. Amer. Geol. **33**, 228–247
Grabau, A. W. (1920): A Textbook of Geology, part I, General Geology, 83 pp., Boston: Heath
Gradzinski, R., Radomski, A. (1967): Pisoliths from Cuban caves. Ann. Soc. geol. Pologne **37**/2, 243–265, 4 Figs., 3 Tab., 6 Pls., Krakow
Gradzinski, R., Radomski, A. (1976): Cave ooids in a tertiary karst shaft at Pogorzyce, southern Poland. Acta geol. Polon. **26**/3, 395–403, Warszawa
Gunatilaka, A. (1976): Thallophyte boring and micritization within skeletal sands from Connemara, Western Ireland. J. Sed. Petrol. **46**/3, 548–554, Figs. 1–3C
Gygi, R. A. (1969): Zur Stratigraphie der Oxford-Stufe (oberes Jura-System) der Nordschweiz und des süddeutschen Grenzgebietes. Beitr. Geol. Karte der Schweiz, N.F. **136**, 123 pp., 11 Figs., 9 Tab., 13 Pls., 6 supps., Bern
Haage, R. (1970): Beitrag zur Genese der Ooide von Kalkoolithen am Beispiel des Unteren Buntsandsteines im Süden der DDR. Geol. **19**/1, 106–112, 7 Figs., Berlin
Hahne, C., Kirchmayer, M., Ottemann, J. (1968): „Höhlenperlen" (Cave Pearls), besonders aus Bergwerken des Ruhrgebietes. Modellfälle zum Studium diagenetischer Vorgänge an Einzelooiden. N. Jb. Paläont. Abh. **130**/1, 1–46, Pls. 1–8, 25 Diagr., Stuttgart
Hallock, P. (1981): Production of carbonate sediments by selected large benthic foraminifera on two Pacific coral reefs. J. Sed. Petrol. **51**/2, 467–474, 6 Tabs., Tulsa
Häntzschel, W. (1939): Schlick-Gerölle und Muschel-Klappen als Strömungsmarken im Wattenmeer. Natur und Volk **69**, 412–417, 5 Figs., Frankfurt/M.
Harris, P. M. (1979): Facies anatomy and diagenesis of a Bahamian ooid shoal. Sedimenta **7**, 163 p., 57 Figs., Miami
Harris, P. M., Halley, R. B., Lukas, K. J. (1979): Endolithic microborings and their preservation in Holocene-Pleistocene (Bahama-Florida) ooids. Geol. **7**, 216–220, 4 Figs., Boulder
Hattin, D. E. (1975): Petrology and origin of fecal pellets in Upper Cretaceous Strata of Kansas and Saskatchewan. J. Sed. Petrol. **45**/3, 686–696, 19 Figs., Tulsa
Heim, A. (1916): Monographie der Churfürsten-Mattstock-Gruppe. 3. Teil: Lithogenesis. Beitr. Geol. Karte Schweiz, N.F. **20**, Basel
Heller, P. L., Komar, P. D., Pevear, D. R. (1980): Transport processes in ooid genesis. J. Sed. Petrol. **50**/3, 943–952, 4 Figs., Tulsa
Henbest, L. G. (1963): Biology, mineralogy and diagenesis of some typical late paleozoic sedentary foraminifera and algal-foraminiferal colonies. Cushman Found. Foram. Res. Spec. Publ. **6**, 1–44, 7 Tab., Bridgewater
Herrmann, A. (1956): Der Zechstein am südwestlichen Harzrand. Geol. Jb. **72**, 1–72, 4 Pls., 14 Figs., 1 Tab., Hannover
Hesse, R. (1973): Diagenesis of a seamount oolite from the West Pacific, Leg 20, DSDP. In: Heezen, B. C., MacGregor, J. D. (eds.): Initial Reports of the Deep Sea Drilling Project **20**, 363–387, 9 Pls., 4 Figs., Washington: US Gov. Printing Office

Hessland, I. (1949): Investigation of the Lower Ordovician of the Siljan district, Sweden. II. Lower Ordovician penetrative and enveloping algae from the Siljan district. Bull. Geol. Inst. Univ. Uppsala **33,** 409–424, Uppsala

Honjo, S. (1976): see 2.4.1

Ihrion, G., Müller, G. (1968): see 2.4.2

Illing, L. V. (1954): Bahaman calcareous sands. Amer. Ass. Petrol. Geol. Bull. **38,** 1–95, Tulsa

James, N. P. (1972): see 2.4.2

Jenkyns, H. C. (1972): Pelagic "oolites" from the Tethyan Jurassic. J. Geol. **80,** 21–33, 8 Figs., Chicago

Jindrich, V. (1969): see 4.2.1.2

Johnson (1946): Lime-secreting algae from the Pennsylvanian and Permian of Kansas. Bull. Geol. Soc. Amer. **57,** 1087–1120, New York

Jones, F. G., Wilkinson, B. H. (1978): Structure and growth of lacustrine pisoliths from recent Michigan Marl Lakes. J. Sed. Petrol. **48**/4, 1103–1110, 3 Figs., Tulsa

Jones, H. A. (1965): Ferruginous oolites and pisolites. J. Sed. Petrol. **35**/4, 838–845, 10 Figs., Tulsa

Kahle, C. F. (1974): Ooids from Great Salt Lake, Utah as an analogue for the genesis and diagenesis of ooids in marine limestones. J. Sed. Petrol. **44**/1, 30–39, Tulsa

Kalkowsky, E. (1908): Oolith and Stromatolith im norddeutschen Buntsandstein. Z. deutsch. geol. Ges. **60,** 68–125, Berlin

Kendall, Ch. G. St. C. (1969): An environmental re-interpretation of the Permian evaporite/carbonate shelf sediments of the Guadelupe Mountains. Geol. Soc. Amer. Bull. **80,** 2503–2526, 9 Pls., 12 Figs., New York

Kendall, C. G. St. C., Skipwith, P. A. d'E. (1969): Holocene shallow-water carbonate and evaporite sediments of Khor als Bazam, Abu Dhabi, Southwest Persian Gulf. Bull. Amer. Ass. Petrol. Geol. **53,** 841–869, Tulsa

Kerkmann, K. (1966): Über „Oolithe" und „Stromatolithe" und die Beteiligung von Algen an der Kalksteinbildung. Wiss. Z. Hochschule Architektur u. Bauwesen **13**/3, 293–302, 19 Figs., Weimar

Kirchmayer, M. (1964): Höhlenperlen (cave pearls) aus Bergwerken, Sitzungsber. Österr. Akad. Wiss. Math.-naturw. Kl. **173**/5–7, 309–349, 18 Figs., 6 Tab., Wien

Klement, K. W., Toomey, D. F. (1967): Role of the blue-green algae *Girvanella* in skeletal grain destruction and lime-mud formation in the lower Ordovician of West Texas. J. Sed. Petrol. **37**/4, 1045–1051, 7 Figs., Tulsa

Kobluk, D. R. (1977): Calcification of filaments of boring and cavity-dwelling algae, and the construction of micrite envelopes. In: Romas, R. C. (Ed.): Geobotany, 195–207, 7 Figs., New York

Kobluk, D. R. Kahle, Ch. F. (1977): Bibliography of the endolithic (boring) algae and fungi and related geological processes. Bull. Canadian Petrol. Geol. **25**/1, 208–223, Calgary

Kobluk, D. R. Kahle, Ch. F. (1978): Geologic significance of boring and cavity-dwelling marine algae. Bull. Canadian Petrol. Geol. **26**/3, 362–379, Calgary

Kobluk, D. R., Risk, M. J. (1977): Calcification of exposed filaments of endolithic algae, micrite envelope formation and sediment production. J. Sed. Petrol. **47**/2, 517–528, Tulsa (see also: Bull. Amer. Ass. Petrol. Geol. **61**/7, 1977)

Köhler, E. E. (1977): see 4.1.7.3

Kohlmeyer, J. (1969): The role of marine fungi in the penetration of calcareous substances. Amer. Zool. **9,** 741–746

Kornicker, L. S., Purdy, E. G. (1957): A Bahamian fecal pellet sediment. J. Sed. Petrol. **27,** 126–128, Tulsa

Krech, K. (1911): Beitrag zur Kenntnis der oolithischen Gesteine des Muschelkalkes um Jena. Jb. Preuß. geol. Landesanst., Jg. 1909, 59–133, Berlin

Kutek, J., Radwanski, A. (1965): Upper Jurassic onkolites of the Holy Cross Mts. (Central Poland). Bull. Acad. polon. Sci. Ser. Sci. geol. geogr. **13**/2, 155–160, 6 Pls., Warszawa

Labecki, J., Radwanski, A. (1967): Broken ooids in lagoonal Keuper deposits of the Western Margin of the Holy Cross Mts. Bull. Acad. Polonaise Sci. **15**/2, 93–99, Warszawa

Lacey, J. E., Carozzi, A. V. (1967): Critères de distinction entre oolithes autochthones et allochthones. Application au calcaire de Sainte-Genévieve (Viséen) de l'Illinois (USA). Bull. Centre Rech. Pau-SNAP **1/2,** 279–313, 4 Pls., 13 Figs., Pau
Land, L. S., Behrens, E. W., Frishman, S. A. (1979): The ooids of Baffin Bay, Texas. J. Sed. Petrol. **49**/4, 1269–1278, 3 Figs., Tulsa
Logan et al. (1964): see 5.3.12.7
Loreau, J. P. (1969): Ultrastructure et diagenèses de oolithes marines anciennes (jurassiques). C.R. Acad. Sci. Paris **269,** 819–822, 2 Pls., Paris
Loreau, J. P. (1970a): Ultrastructure de la phase carbonatée des oolithes marines actuelles. C.R. Acad. Sci. Paris **271,** 816–819, 3 Pls., Paris
Loreau, J. P. (1970b): Contribution à l'étude des calcaires héterogènes par l'emploi simultane de la microscopie phototonique et de la microscopies électronique à balayage, Problème particulier de la micritisation. J. Microscopie **9**/6, 727–734, 4 Pls., Paris
Loreau, J. P. (1973): Nouvelles observations sur la genèse et la signification des oolithes. Sci. Terre **18**/3, 213–244, 3 Pls., 3 Figs., 1 Tab., Nancy
Loreau, J.-P. (1979): Nature calcitique initiale et diagenèse de oolithes Jurassiques du Bassin de Paris. Ass. Sed. français, Publ. spec. **1,** 417–429, 16 Figs., Paris
Loreau, J. P., Purser, B. H. (1973): Distribution and ultrastructure of Holocene ooids in the Persian Gulf. In: Purser, B. H. (ed.): The Persian Gulf, 279–328, 20 Figs., Berlin–Heidelberg–New York: Springer
Mägdefrau, K. (1937): Lebensspuren fossiler „Bohr"-Organismen. Beiträge naturkundl. Forschung SW-Deutschland **2**/1, 54–67, Pls. 4–6, Karlsruhe
Margolis, S., Rex, R. W. (1971): Endolithic algae and micrite envelope formation in Bahaman oolites as revealed by scanning electron microscopy. Geol. Soc. Amer. Bull. **82,** 843–852, 20 Figs., New York
Marschner, H. (1968): Relationship between carbonate grain size and non-carbonate content in carbonate sedimentary rocks. In: Friedman, G. M., Müller, G. (eds.): Recent Developments in Carbonate Sedimentology in Central Europe, 55–57, 2 Figs., Berlin–Heidelberg–New York: Springer
Maslov, V. P. (1953): Printsipy nomenklatury i sistematike stromatolitov. Izvest. Akad. Nauk SSSR, ser. geol., **1953**/4, Moskva
Maslov, V. P. (1960): Stromatoliti (ikh genezikh, metod izuchenya, syach s faunvami i geologicheskoe izachenie na premere ordovika russkoi platformy. Trudy geol. inst. Akad. nauk SSSR **41,** 1–188, Moskva
Matthews, R. K. (1966): see 2.4.1
Maxwell, W. G. H., Jell, J. S. and McKellar, R. G. (1964): see 2.4.1
Mazzulo, S. J. (1977): Shrunken (geopetal) ooids: evidence of origin unrelated to carbonate-evaporite diagenesis. J. Sed. Petrol. **47**/1, 392–397, 5 Figs., Tulsa
McCannon, D. E. Jr. (1975): Primary fluviatil oolites. J. Sed. Petrol. **45**/3, 719–727, 11 Figs., Tulsa
McKee, E. D., Gutschick, R. C. (1969): History of Redwall limestone of northern Arizona. Geol. Soc. Amer. Mem. **114,** 1–26, New York
Meijer, J. J. de (1971): Carbonate petrology of algal limestones (Lois-Ciguera Formation, Upper Carboniferous, Léon, Spain). Leidse Geol. Meded. **47**/1, 1–97, 21 Plates, Leiden
Meyer, R. (1976): Mikrofazielle Untersuchungen in Schwamm-Biohermen und -Biostromen des Malm Epsilon (Ober-Kimmeridge) und obersten Malm Delta der Frankenalb. Geol. Bl. NO-Bayern **25,** 149–177, 14 Figs., Erlangen
Milliman, J. D. (1967): see 2.4.1
Milliman, J. D. (1974): see 1.4.1
Milliman, J. D., Barretto, H. T. (1975): Relict magnesian calcite oolite and subsidence of the Amazon shelf. Sedimentology **22,** 137–145, 4 Figs., Oxford
Mitterer, R. M. (1968): see 2.4.1
Mitterer, R. M. (1971): see 2.4.1
Mišik, M. (1968): Traces of submarine slumping and evidences of hypersaline environment in the Middle Triassic of the West Carpathians core mountains. Geol. Zborn. **19**/1, Slov. Akad. Vied, 205–224, Bratislava

Monaghan, P. H., Melba, L. L. (1956): The origin of calcareous ooliths. J. Sed. Petrol. **26,** 111–118, Tulsa

Montaggioni, L. F. (1979): Environmental significance of rhodolites from the Mascarene reef province, Western Indian Ocean. Bull. Centre Rech. Expora.-Prod. Elf-Aquitaine **3**/2, 713–723, 3 Plates, Pau

Monty, Cl. (1963): Bases d'une nomenclature des roches calcaires marines. Ann. Soc. Géol. Belgique **86,** 1962–1963, Mém. 2, B 87–B 122, 1 Pl., Bruxelles

Monty, Cl. (1965). Recent algal stromatolites in the Windward Lagoon, Andros Island, Bahamas. Ann. Soc. Geol. Belgique Bull. **86**/6, B 269–B 276, Bruxelles

Monty, Cl. (1967): Distribution and structure of recent stromatolitic algal mats. Eastern Andros Island, Bahamas. Ann. Soc. Géol. Belgique **90**/3, 55–100, 19 Pls., 13 Figs., Bruxelles

Monty, Cl. (1972): Recent algal stromatolitic deposits, Andros Island, Bahamas, preliminary report. Geol. Rdsch. **61**/2, 742–783, 32 Figs., Stuttgart

Monty, Cl. (1973): Precambrian background and phanerozoic history of stromatolitic communities, an overview. Ann. Soc, Geol. Belgique **96**/3, 585–624, Bruxelles

Monty, Cl. (1977): see 5.3.12.7

Müller-Jungbluth, W. U., Toschek, P. H. (1969): Karbonatsedimentologische Arbeitsgrundlagen (Begriffe, Erläuterungen, Hinweise). Veröff. Univ. Innsbruck 8, Alpenländ. Studien **4,** 32 pp., 3 Pls., Innsbruck

Neugebauer, J. (1978): Micritization of crinoids by diagenetic solution. Sed. **25,** 267–283, Oxford

Newell, N. D., Purdy, E. G., Imbrie, J. (1960): Bahamian oolitic sand. J. Geol. **68**/5, 481–497, 3 Figs., 4 Pls., Chicago

Nissen, H. U. (1964): Calcite fabric analysis of deformed oolites from the South Mountain Fold, Maryland. Amer. J. Sci. **262,** 892–903, 4 Figs.

Obrhel, J. (1977): Onkolithe und Algen von Unterkarbon in Mähren. Vestn. Ustred. ustavu geol. **52,** 29–34, 4 Pls., Praha

Orzag-Sperber, F., Poignant, A.-F., Poisson, A. (1977): Paleogeographic significance of rhodolites: some examples from the Miocene of France and Turkey. In: Flügel, E. (ed): Fossil Algae. Recent Results and Developments, 286–294, 4 Pls., Berlin–Heidelberg–New York: Springer

Ott, W. (1969): Zur Geologie des Sulzfluh-Kalkes (Malm) in Graubünden und Vorarlberg. Unveröff. Diss. TH Darmstadt, 187 pp., 10 Pls., Darmstadt (typescript)

Paulsen, S. (1964): Aufbau und Petrographie des Riffkomplexes von Arnegg im höheren Weißen Jura der Schwäbischen Alb (Württemberg). Arb. Geol. Paläont. Inst. TH Stuttgart, N.F. **42,** 98 pp., 22 Pls., 20 Figs., Stuttgart

Pedley, H. M. (1979): Miocene bioherms and associated structures in the coralline limestone of the Maltese Islands: their lithification and paleoenvironment. Sed. **26,** 577–591, 5 Figs., Oxford

Perkins, R. D., Halsey, S. D. (1971): Geologic significance of microboring fungi and algae in Carolina shelf sediments. J. Sed. Petrol. **41,** 843–853. Tulsa

Perkins, R. D., Tsentas, C. I. (1976): Microbial infestation of carbonate substrates plated on the St. Croix shelf, West Indies. Geol. Soc. Amer. Bull. **87,** 1615–1628, 46 Figs., New York

Peryt, T. M. (1975): Significance of stromatolites for the environmental interpretation of the Buntsandstein (Lower Triassic) rocks. Geol. Rdsch. **64**/1, 143–158, 10 Figs., Stuttgart

Peryt, T. M. (1977): Environmental significance of foraminiferal-algal oncolites. In: Flügel, E. (ed.): Fossil Algae. Recent Results and Developments, 61–65, 2 Figs., Berlin–Heidelberg–New York: Springer

Peryt, T. M. (1981): Phanerozoic oncoids – an overview. Facies **4,** 197–214, Plates 10–11, 4 Tables, Erlangen

Peryt, T. M., Peryt, D. (1975): Association of sessile tubular foraminifera and schizophytic algae. Geol. Mag. **112,** 612–614, London

Peryt, T. M., Piatkowski, T. S. (1977): Algal-vadose pisoliths in the Zechstein Limestone (Upper Permian) of Northern Poland. Sed. Geol. **19,** 275–286, 9 Figs., Amsterdam

Pettijohn, F. J. (1957): see 1.4.1

Pia, J. (1927): Thallophyta. In: Hirmer, M. (ed.): Handbuch der Paläobotanik **1,** 31–136, München–Berlin: Oldenbourg

Piatkowski, T. S. (1977): Radial ooids from the Zechstein limestone of the Peri-Baltic syneclise (Northern Poland). Kwart. Geol. **21**/4, 757–768, 4 Pls., 4 Figs., Warsaw

Picard, M. D., High, L. R. (1968): Sedimentary cycles in the Green River Formation (Eocene), Unita Basin, Utah. J. Sed. Petrol. **38,** 378–383, Tulsa

Pomar, L., Esteban, M., Llimona, X., Fontarnau, R. (1975): Acción de Liquenes, Algas y Hongos en la Telodiagenesis de las Rocas Carbonatadas de la zona litoral prelitoral Catalana. Inst. Invest. Geol. Univ. Barcelona **30,** 83–117, 10 Pls., Barcelona

Poncet, A. (1976): Faciès carbonates de barrière-récif dans l'Eodévonien du Nord-Est du massif armoricain (Cotentin). Bull. B.R.G.M., sér. 2/I, **1** 49–68, 2 Pls., 7 Figs., Paris

Porter, C. L., Zebrowski, G. (1937): Lime-loving molds from Australian sands. Mycologia **29,** 252–257

Pratt, P. R. (1979): see 3.9

Pümpin, V. F. (1965): Riffsedimentologische Untersuchungen im Rauracien von St. Ursanne und Umgebung (Zentraler Schweiz. Jura). Eclogae geol. Helv. **58**/2, 799–876, 1 Pl., 36 Figs., Basel

Purdy, E. G. (1963): Recent calcium carbonate facies of the Great Bahama-Bank. 1. Petrography and reaction groups. J. Geol. **71**/3, 334–355, 5 Figs., Pls. 1–5, Chicago

Purdy, E. G. (1968): see 3.9

Purser, B. H. (1980): see 1.4

Purser, B. H., Loreau, J. P. (1973): Aragonites, supratidal encrustations on the Trucial Coast, Persian Gulf. In: Purser, B. H. (ed.): The Persian Gulf, 343–376, 20 Figs., Berlin–Heidelberg–New York: Springer

Quester, H. (1964): Petrographie des erdgashöffigen Hauptdolomits im Zechstein 2 zwischen Weser und Ems. Z. dtsch. geol. Ges. **114,** Jg. 1962, 461–483, Hannover

Radoičić, R. (1960): see 5.5

Radwánski, A. (1965): Pitting processes in clastic and oolitic sediments. Ann. Soc. Geol. Pologne **35**/2, 179–210, Taf. 11–20, 4 Figs., Krakow

Radwánski, A., Birkenmajer, K. (1977): Oolitic/pisolitic dolostones from the Late Precambrian of south Spitsbergen: their sedimentary environment and diagenesis. Acta geol. Polonica **27**/1, 1–39, 20 Figs., Warszawa

Radwánski, A., Szulczewski, M. (1966): Jurassic stromatolites of the Villany Mountains (Southern Hungary). Ann. Univ. Sci. Budapest Rolando Eötvös, Sect. Geol. **9,** 87–107, 6 Pls., Budapest

Reitlinger, E. A. (1959): Atlas mikroskopicheskikh organicheskikh ostatkov i problematikh drevnikh tolsch sibiri. Trudy geol. in-ta Akad. Nauk SSSR, **25**

Riding, R. (1977): Skeletal stromatolites. In: Flügel, E. (ed.): Fossil Algae. Recent Results and Developments, 57–60, 1 Fig., Berlin–Heidelberg–New York: Springer

Risacher, F., Eugster, H. P. (1979): Holocene pisoliths and encrustations associated with spring-fed surface pools, Paston Grandes, Bolivia. Sed. **26,** 353–370, 7 Figs., Oxford

Roberts, H. H., Moore, Cl. H. Jr. (1971): Recently cemented aggregates (grapestones), Grand Cayman Island, B.W.I. In: Bricker, O.P. (ed.): Carbonate Cements, 88–90, Figs. 46–47, Baltimore–London: Hopkins

Roda, C. (1965): Livelli a struttura grumosa e livelli ad ooliti rotte e rigenerate nel calcare miocenio del M. Alpi (Potenza). Geol. Romana **4,** 181–220, Rome

Rooney, W. S. Jr., Perkins, R. D. (1972): Distribution and geologic significance of microboring organisms within sediments of the Arlington Reef Complex, Australia. Bull. Geol. Soc. Amer. **83,** 1139–1150, 16 Figs., New York

Rusnak, G. A. (1960): Some observation of recent oolites. J. Sed. Petrol. **30**/3, 471–480, 7 Figs., Tulsa

Sandberg, P. A. (1975): New interpretation of great Salt Lake ooids and of ancient non-skeletal carbonate mineralogy. Sedimentology **22,** 497–537, Oxford

Sander, B. (1936): see 1.4.1

Sander, N. J. (1967): see 6.4

Sarntheim, M. (1965): Sedimentologische Profilreihen aus den mitteltriadischen Karbonatgesteinen der Kalkalpen nördlich und südlich von Innsbruck. Verh. Geol. Bundesanst. **1965,** 1/2, 19–161, 1 Pl., Wien

Sass, E., Weiler, Y., Katz, A. (1972): Recent sedimentation and oolite formation in the Ras Matarma Lagoon, Gulf of Suez. In: Stanley, D. J. (ed.): The Mediterranean Sea: A Natural Sedimentation Laboratory, 279–292, 10 Figs., Stroudsburg: Dowden, Hutchinson and Ross

Schäfer, A., Stapf, K. R.G. (1978): Permian Saar-Nahe Basin and Recent Lake Constance (Germany): two environments of lacustrine algal carbonates. Spec. Publ. Int. Ass. Sed. **2,** 83–107, 22 Figs., Oxford

Schenk, P. (1975): Facies and phases of the Altamont Limestone, a Megacyclothem (Pennsylvanian), Iowa to Oklahoma. Bull. Geol. Soc. Amer. **78,** 1369–1384, 3 Pls., 5 Figs., New York

Schlager, W. (1980): Mesozoic calciturbidites in deep sea drilling project hole 416 A. Recognition of a drowned carbonate platform. Initial Rep. Deep Sea Drilling Project **50,** 733–749, 14 Figs., Washington

Schloz, W. (1972): Zur Bildungsgeschichte der Oolithenbank (Hettangium) in Baden-Württemberg. Arb. Inst. Geol. Paläont. Univ. Stuttgart, N.F. **67,** 101–212, 40 Figs., 8 Pls., 11 Tab., Stuttgart

Schmidt, V. (1969): Facies, diagenesis, and related reservoir properties in the Gigas Beds (Upper Jurassic), Northwestern Germany. Spec. Publ. Soc. Econ. Paleont. Min. **13,** 124–168, 32 Figs., Tulsa

Scholle, P. A., Kinsman, D. J. J. (1974): Aragonitic and high-Mg calcite from the Persian Gulf – a modern analog for the Permian of Texas and New Mexico. J. Sed. Petrol. **44,** 904–916, Tulsa

Schöttle, M., Müller, G. (1968): see 2.4.2

Scudeler Baccelle, L. (1976): Ovoidi carbonatici della piattaforma liassica veneta. Boll. Soc. Geol. Ital. **95,** 101–131, 1 Plate, 21 Figs., 2 Tab., Roma

Senowbari-Daryan, B. (1978): Anomuren-Koprolithen aus der Obertrias der Osterhorngruppe (Hintersee/Salzburg, Österreich). Ann. naturhist. Mus. Wien **82,** 99–107, 1 Plate, 3 Figs., Wien

Sestini, G. (1964): Introduzione allo studio e alla descrizione delle rocce carbonate. 25 S., Ciclostilato, Rapp. interno Centro Studi Geol. Appenn., C.N.R., Firenze

Seyfried, H. (1980): Über die Bildungsbereiche mediterraner Jurasedimente am Beispiel der Betischen Kordillere (Südost-Spanien). Geol. Rdsch. **69,** 149–178, 11 Figs., Stuttgart

Shearman, D. J., Twyman, J., Zand Karimi, M. (1970): The genesis and diagenesis of oolites. Proc. Geol. Ass. **81**/3, 561–575, 5 Figs., Leeds

Sibley, D. F., Murray, R. C. (1972): Marine diagenesis of carbonate sediment, Bonaire, Netherlands Antilles. J. Sed. Petrol. **42**/1, 168–178, 7 Figs., Tulsa

Siehl, A., Thein, J. (1978): Geochemische Trends in der Minette (Jura, Luxemburg/Lothringen). Geol. Rundschau **67**/3, 1052–1077, 10 Figs., 3 Tab., Stuttgart

Siesser, W. G. (1973): Diagenetically formed ooids and intraclasts in South African calcretes. Sedimentology **20,** 539–551, 12 Figs., Oxford

Simone, L. (1974): Genesi e significato ambientale degli ooidi a struttura fibrosa-raggiata di alcuni depositi mesozoichi dell'area Appennino-dinarica e delle Bahamas meridionali. Boll. Soc. Geol. Ital. **93,** 513–545, 26 Figs., Rom

Skupin, K. (1969): Ein Fecal-Pellet-Typ aus dem Unteren Hauptmuschelkalk (Trochitenkalk) des Neckar-Jagst-Kocher-Gebietes. N. Jb. Geol. Paläont. Mh. **1969**/6, 377–383, 2 Figs., Stuttgart

Skupin, K. (1970): Feinstratigraphische und mikrofazielle Untersuchungen im Unteren Hauptmuschelkalk (Trochitenkalk) des Neckar-Jagst-Kocher-Gebietes. Arb. Geol. Paläont. Inst. Univ. Stuttgart, N.F. **63,** 173 pp., 4 Pls., Stuttgart

Smith, D. B. (1974): Sedimentation of Upper Artesia (Guadalupian) cyclic shelf deposits of northern Guadalupe Mountains, New Mexico. Bull. Amer. Ass. Petrol. Geol. **58,** 1699–1730, Tulsa

Sorby, H. C. (1879): The structure and origin of limestones. Proc. Geol. Soc. London **35,** 39–95, Plates 1–18, London

Stapf, K. (1970): Lithologische Untersuchung der Altenglaner Schichten im saarpfälzischen Unterrotliegenden mit besonderer Berücksichtigung der Karbonatgesteine. unpubl., Diss. Univ. Mainz, 231 pp., 52 Figs., 26 supps., Mainz (typescript).
Steinmann, G. (1880): Die Mumien des Hauptrogensteins, N. Jb. Min. Geol. Paläont., Stuttgart
Stoffers, P., Summerhayes, C. P., Dominik, J. (1980): Recent pelletoidal carbonate sediments off Alexandria, Egypt. Marine Geol. **33,** M 1–M 8, 3 Figs., 2 Tab., Amsterdam
Strasser, A. (1979): see 7.3.2
Sudgen, W. (1963): Some aspects of sedimentation in the Persian Gulf. J. Sed. Petrol. **33**/2, 355–364, 4 Figs., Tulsa
Suess, E., Fütterer, D. (1972): Aragonitic ooids: experimental precipitation from sea-water in the presence of humid acid. Sedimentology **19,** 1/2, 129–139, 6 Figs., Amsterdam
Swinchatt, J. P. (1969): Algal boring: A possible depth indicator in carbonate rocks and sediments. Geol. Soc. Amer. Bull. **80**/7, 1391–1396, New York
Szulczewski, M. (1967): Stromatolitic structures within Middle Jurassic transgressive deposits at Wola Morawicka (southern margins of the Holy Cross Mountains, Central Poland). Ann. Soc. Geol. Pologne **37**/4, 515–528, Pls. 37–40, 5 Figs., Krakow
Taylor, J. M. C., Illing, L. V. (1969): Holocene intertidal calcium carbonate cementation, Qatar, Persian Gulf. Sedimentology **12,** 69–107, Amsterdam
Toomey, D. F. (1974): Algally coated grains from the Leavenworth limestone (M. Pennsylvanian, Midcontinent Region, USA). N. Jb. Geol. Paläont. Mh. **1974**/3, 175–191, Stuttgart
Toomey, D. F. (1975): Rhodoliths from the Upper Paleozoic of Kansas and the Recent. A comparison. N. Jb. Geol. Paläont. Mh. **1975**/4, 242–255, Stuttgart
Teichert, C. (1970): Oolite, oolith and ooid: Discussion. Bull. Amer. Ass. Petrol. Geol. **54**/9, 26, Tulsa
Trichet, J. (1969): Etude de la composition de fraction organique des oolités, comparaison avec celle des membranes des bactéries et des cyanophycées. C.R. Acad. Sci. (Paris) **267**/19, 1492–1494, Paris
Usdowski, H. E. (1962): Die Entstehung der kalkoolithischen Fazies des norddeutschen Unteren Buntsandsteins. Beitr. Min. Petr. **8**/3, 141–179, 25 Figs., Heidelberg
Usdowski, H. E. (1963): Der Rogenstein des norddeutschen Unteren Buntsandsteins, ein Kalkoolith des marinen Faziesbereiches. Fortschr. Geol. Rheinland-Westf. **10,** 337–342, 1 Fig., 1 Tab., Krefeld
Vernet, J. P., Ferrari, J. M. (1969): Ultrastructure des oolithes ferrugineuses du Callovien du Jura meridonal. II. C.R. Soc. Phys. Hist. nat. Genève **1969**/4, 1, 62–67
Ward, W. C., Folk, R. L., Wilson, J. L. (1970): Blackening of eolianite and caliche adjacent to saline lakes, Isla Mujieres, Quintana Roo, Mexico. J. Sed. Petrol. **40**/2, 548–555, 9 Figs., Tulsa
Wardlaw, N. C., Reinson, G. E. (1971): Carbonate and evaporite deposition and diagenesis, Middle Devonian Winnipegosis and Prairie Evaporite Formations of south-central Saskatchewan. Bull. Amer. Ass. Petrol. Geol. **55,** 1759–1786, Tulsa
Werner, F. (1959): Zur Kenntnis der Eisenoolithfazies des Braunjura: von Ostwürttemberg. Arb. Geol. Paläont. Inst. TH Stuttgart, N. F. **23,** 169 pp., 5 Pls., 54 Figs., 8 Tab., Stuttgart
Wetzel, W. (1937): Die koprogenen Beimischungen mariner Sedimente und ihre diagnostische und lithogenetische Bedeutung. N. Jb. Geol. Min. Paläont., supp. vol. **78** B, 109–122, Stuttgart
Wilckinson, B. H. (1979): Biomineralization, paleoceanography, and the evolution of calcareous marine organisms. Geol. **7,** 425–427, Boulder
Wilson, R. C. L. (1967): Particle nomenclature in carbonate sediments. N. Jb. Geol. Paläont. Mh. **1967/68,** 498–510, 4 Figs., Stuttgart
Wilson, J. L. (1969): Microfacies and sedimentary structures in "deeper water" lime mudstones. Spec. Pap. Soc. Econ. Paleont. Min. **14,** 4–19, Tulsa
Wilson, J. L. (1975): see 1.4.1
Winland, H. D. (1968): The role of high Mg calcite in the preservation of micrite envelopes and textural features of aragonite sediments. J. Sed. Petrol. **38**/4, 1320–1325, 7 Figs., Tulsa
Winland, H. D., Matthews, R. K. (1974): Origin and significance of grapestone, Bahama Islands. J. Sed. Petrol. **44**/3, 921–927, 8 Figs., Tulsa

Wobber, F. J. (1965): Sedimentology of the Lias (Lower Jurassic) of South Wales. J. Sed. Petrol. **35**/3, 638–703, 8 Figs., Tulsa
Wolf, K. H. (1965a): Petrogenesis and palaeoenvironment of Devonian algal limestones of New South Wales. Sedimentology **4**, 113–178, 39 Figs., Amsterdam
Wolf, K. H. (1965b): Gradational sedimentary products of calcareous algae. Sedimentology, **5**, 1–37, 25 Figs., 4 Tab., Amsterdam
Wright, V. P. (1981): Algal aragonite-encrusted pisoids from a Lower Carboniferous schizohaline lagoon. J. Sed. Petrol. **51**/2, 479–489, 6 Figs., Tulsa
Zeff, M. L., Perkins, R. D. (1979): Microbial alteration of Bahamian deep-sea carbonates. Sed. **26**, 175–201, 17 Figs., Oxford
Zhuravleva, Z. A. (1964): Onkolity i katagrafii Rifeya i nizhnego Kembriaya Sibiri i iz stratigraficheskoe znachenie. Akad. nauk SSSR, Geol. Inst., Trudy, **114**, 1–73, 24 Plates, 2 Figs., 4 Tables, Moskva
Zhuravleva, Z. A. (1979): The importance of oncolites for the Precambrian stratigraphy. Bull. Centre Rech. Explora-Prof. Elf-Aquitaine **3**/2, 881–885, Pau
Zuffa, G. G. (1980): Hybrid Arenites: Their composition and classification. J. Sed. Petrol. **50**/1, 21–30, 3 Figs., Tulsa

4.1.7.3 References: Authigenic Minerals in Carbonate Rocks

Baskin, Y. (1956): A study of authigenic feldspars. J. Geol. **64**, 132–155, Chicago
Bausch, W. M. (1980): see 4.1.7.2
Berner, R. A. (1967): Diagenesis of iron sulfides in recent marine sediments. In:Lauff, G. H. (ed.): Estuaries. Amer. Ass. Adv. Sci. Spec. Publ. **83**, 757 pp. Washington
Berner, R. A. (1969): The synthesis of framboidal pyrite. Econ. Geol. **64**, 383–384
Birch, G. F. (1979): Phosphatic rocks on the western margin of South Africa. J. Sed. Petrol. **49**/1, 93–110, 8 Figs., Tulsa
Burst, J. (1958): "Glauconite" pellets: their mineral nature and applications to stratigraphic interpretations. Amer. Ass. Petrol. Geol. Bull. **42**, 310–327, Tulsa
Buyce, M. R., Friedman, G. M. (1975): Significance of authigenic K-feldspar in Cambrian-Ordovician carbonate rocks of the Proto-Atlantic shelf in North America. J. Sed. Petrol. **45**, 808–821, Tulsa
Carozzi, A. V. (1951): Le comportement des feldspaths authigènes dans le Crétacé supérieur helvétique. Arch. Sci. Genève **4**, 134–137, Geneva
Carozzi, A. V. (1958): see 4.2.1.2
Carozzi, A. V. (1960): see 1.4.1
Catalov, G. (1970): Authigene Albites in the Triassic carbonate rocks (Teteven-Anticlinorium). Comptes Rendus Acad, bulgare Sci. **23**, 1275–1278, 2 Figs., Sofia
Fabricius, F. H. (1961): Die Strukturen des „Rogenpyrits" (Kössener Schichten, Rät) als Beitrag zum Problem der „Vererzten Bakterien". Geol. Rundschau **51**, 311–319, Plates 32–34, Stuttgart
Franz, E. (1967): Zur Frage der Genese authigener idiomorpher Quarze im Salinar. Z. angewandt. Geol. **13**/3, 157–159, Berlin
Friedman, G. M., Brown, M. (1975): Shoaling and tidal deposits that accumulated marginal to the Proto-Atlantic Ocean: the Tribes Hill Formation (Lower Ordovician) of the Mohawk Valley, New York. In: Ginsburg, R. N. (ed.): Tidal Deposits, 307–314, New York–Heidelberg–Berlin: Springer
Füchtbauer, H. (1948): Einige Beobachtungen an authigenen Albiten. Schweiz. Min. Petrograph. Mitteilungen **28**, 709–716, Bern
Füchtbauer, H. (1950): Die nichtkarbonatischen Bestandteile des Göttinger Muschelkalkes mit besonderer Berücksichtigung der Mineralneubildungen. Heidelberger Min. Petrograph. Beiträge **2**, 235–254, Heidelberg
Füchtbauer, H. (1956): Zur Entstehung und Optik authigener Feldspäte. Neues Jahrbuch Min. Monatshefte **1956**/1, 9–23, Stuttgart
Füchtbauer, H. (1974): see 1.4.1
Funk, H. (1975): The origin of authigenic quartz in the Helvetic Siliceous Limestone (Helvetischer Kieselkalk), Switzerland. Sed. **22**, 299–326, Oxford

Grimm, W.-D. (1962): Idiomorphe Quarze als Leitmineralien für salinare Fazies. Erdöl und Kohle **15,** 880–887, 10 Figs., Hamburg

Hein, J. R., Griggs, G. B. (1972): Distribution and scanning electron microscope (SEM) observations of authigenic pyrite from a Pacific deep-sea core. Deep-Sea Res. **19,** 133–138, Oxford

Honess, A. P., Jeffries, C. D. (1940): Authigenic albite from the Lowville limestone at Bellefonte, Pennsylvania. J. Sed. Petrol. **10,** 12–18, Tulsa

Kastner, M. (1971): Authigenic feldspars in carbonate rocks. Amer. Min. **56,** 1403–1442

Kastner, M., Waldbaum, D. R. (1968): Authigenic albite from Rhodos. Amer. Min. **53,** 1579–1602

Köhler, E. E. (1977): Zum Stand der Glaukonitforschung – eine Bibliographie. Zentralblatt Geol. Paläont. **1976**/1, 974–1017, Stuttgart

Kulke, H. (1974): Zur Geologie und Mineralogie der Kalk- und Gipskrusten Algeriens. Geol. Rundschau **63**/3, 970–998, 7 Figs., Stuttgart

Kulke, H. (1976): Diagenese, beginnende Metamorphose und Mineralneubildungen der Karbonat-, Ton- und Sandsteinfolgen im Trias-Salz des Diapirs Rocher de Sel de Drjelfa (Algerien). Geol. Jb., D **19,** 3–73, 4 Pls., 9 Figs., Hannover

Kuslansky, G. H., Friedman, G. M. (1981): Chertification of crinoids may yield a product resembling "dedolomite". J. Sed. Petrol. **51**/3, 795–798, 4 Figs., Tulsa

Land, L. S. (1979): Chert-Chalk Diagenesis: The Miocene Island slope of North Jamaica. J. Sed. Petrol. **49**/1, 223–232, 6 Figs., Tulsa

Lange, H., Sarntheim, M. (1970): Glaukonitkörner in rezenten Sedimenten des Persischen Golfes. Geol. Rundschau **60,** 256–264, Stuttgart

Lawrence, J. R., Kastner, M. (1975): $^{18}O/^{16}O$ of feldspars in carbonate rocks. Geochim. Cosmochim. Acta **39,** 97–102, Oxford

Lindström, M. (1980): Glauconite shrinkage and limestone cementation. J. Sed. Petrol. **50**/1, 133–138, 2 Figs., Tulsa

McRae, S. G. (1972): Glauconite. Earth-Sci. Rev. **8,** 397–440, Amsterdam

Mišik, M. (1963): Authigenic quartz and authigenic feldspars in Mesozoic limestones of West Carpathians. Geol. Sbornik **14**/2, 295–300, 3 Plates, Bratislava

Nachsel, G. (1969): Idiomorphe Quarze und Vertaubungen im Kaliflöz „Staßfurt" des Südharz-Kalireviers. Z. angewandt. Geol. **15**/8, 420–425, 7 Figs., Berlin

Odin, G. S. (1972): Observations on the structure of the glauconite vermicular pellets: a description of the genesis of these granulaes by neoformation. Sed. **19,** 285–294, Amsterdam

Odin, G. S. (1978): Nature, formation et significance des glauconies. Abstracts, Int. Congr. Sed. **2,** 478–479, Jerusalem

Odin, G. S., Matter, A. (1981): De glauconiarum origine. Sed. **28,** 611–641, 20 Figs., Oxford

Oehler, J. (1979): Deposition and diagenesis of biogenic silica. In: Trudinger, Swaine (eds.): Biological Factors in Mineral Cycling, 33 p., 2 Figs.

Perrenoud, J.-P. (1952): Etude du feldspath potassique contenu dans le „Pontiskalk" (Trias, Valais). Schweiz. Min. Petrograph. Mitteilungen **32,** 179–184

Peterson, M. N., Goldberg, E. D. (1962): Feldspar distribution in South Pacific pelagic sediments. J. Geophys. Res. **67,** 3477

Peterson, M. N., von der Borch, C. C. (1965): Chert: Modern inorganic deposition in a carbonate-precipitation locality. Science **149,** 1501–1503, Washington

Richter, D. K. (1971): Fazies- und Diagenesehinweise durch Einschlüsse in authigenen Quarzen. Neues Jahrbuch Geol. Paläont. Monatshefte **1971**/10, 604–622, Stuttgart

Riech, V. (1978): Zur Coelestinbildung im germanischen Muschelkalk Süddeutschlands. Geol. Jahrbuch D **29,** 3–77, 6 Plates, 21 Figs., 4 Tab., Hannover

Riech, V., von Rad, U. (1979): Silica diagenesis in the Atlantic Ocean: Diagenetic potential and transformations. In: Talwani, M., Hay, W., Ryan, W. B. F. (eds.): Deep drilling results in the Atlantic Ocean: Continental margins and paleoenvironment. Ewing Ser. **3,** Amer. Geophys. Union, 315–340, 3 Plates, 13 Figs., Washington

Schmidt, V. (1965): see 9.4

Schöner, H. (1960): Über die Verteilung und Neubildung der nichtkarbonatischen Mineralkomponenten der Oberkreide aus der Umgebung von Hannover. Beitr. Min. Petrographie **7,** 76–103, Heidelberg

Swett, K. (1968): Authigenic feldspars and cherts resulting from dolomitization of illitic limestones: a hypothesis. J. Sed. Petrol. **38,** 128–135, Tulsa
Theodorovich, G. I. (1961): Authigenic minerals in sedimentary rocks. 120 p. New York: Consultants Bureau
Topkaya, M. (1950): Recherches sur les silicates authigènes dans les roches sédimentaires. Bull. Lab. Géol. Min. Géophys. Mus. géol. Lausanne **97,** 1–132, Lausanne
Triplehorn, D. M. (1966): Morphology, internal structure and origin of glauconite pellets. Sed. **6,** 247–266, Amsterdam
Van Straaten, L. M. J. U. (1948): Note on the occurrence of authigenic feldspar in non-metamorphic sediments. Amer. J. Sci. **246,** 569–572, New Haven

4.2 Textural and Structural Criteria

The frequency, size, and roundness of particles as well as the fabrics formed from particles represent, in addition to the various types of particles, the most important microfacies criteria.

4.2.1 Grain-Size Analyses

4.2.1.1 Methods and Examples

Grain-size analyses serve to describe, classify, and genetically interpret sediments and sedimentary rocks. Initially, grain-size analyses were done primarily on siliceous clastic rocks, but in the last few years, repeated attempts have been made to apply these analyses to carbonates and carbonate rocks as well (calcarenites and calcirudites). These grain-size analyses should help to provide more information about the depositional environment of carbonates.

In the following paragraphs a few of the methods will be described for determining grain size, for graphically displaying grain-size distributions, and for evaluating the information obtained in facies analyses of grain-size parameters. The purpose of this Section is actually to encourage increased application of these methods in describing and interpreting calcarenites.

I. Determination of grain sizes

A. Definition of grains and grain sizes: A grain is a clastic particle with well-defined boundaries, which retains its individuality during transport and deposition.

Grain size definitions:

1. In *soft rocks* the grain size is defined by the *mesh size of a sieve* or by its nominal diameter, Wadell (1932) (= diameter of a theoretical sphere with the same volume as the grain).

2. In *hard rocks grain sizes in thin-sections* are defined by measuring the maximum apparent grain diameter (e.g., Friedman, 1958; Hötzl, 1966) or determining a circle with the same area as the grain (Hörnstein, 1960; apparatus: particle-size analyzer, Zeiss).

B. Standard grain-size classes; see Fig. 24. In the English literature the Wentworth classification system is used; in the German literature, the Atterberg classification system. To simplify computation Krumbein (1934) proposed a transformation of the logarithmic scales to arithmetic scales, thus, replacing the Wentworth grade scale with the *Phi scale* and the Atterberg scale with the *Zeta scale* (hardly ever used, see Walger, 1965). The Phi unit corresponds to the negative logarithm to the base 2 of the grain size measured in millimeters. A grain size of 1 mm corresponds to a Phi unit of zero. Smaller grain sizes have positive Phi units; larger grain sizes, negative Phi units. Advantages of the Phi scale: all class intervals are the same; the distribution of particle sizes can be plotted on arithmetic paper and the calculation of statistical parameters is simplified. Phi values can be obtained from conversion tables (Page, 1955; Müller, 1964; Griffith, 1968) or from nomograms (Friedman and Sanders, 1978).

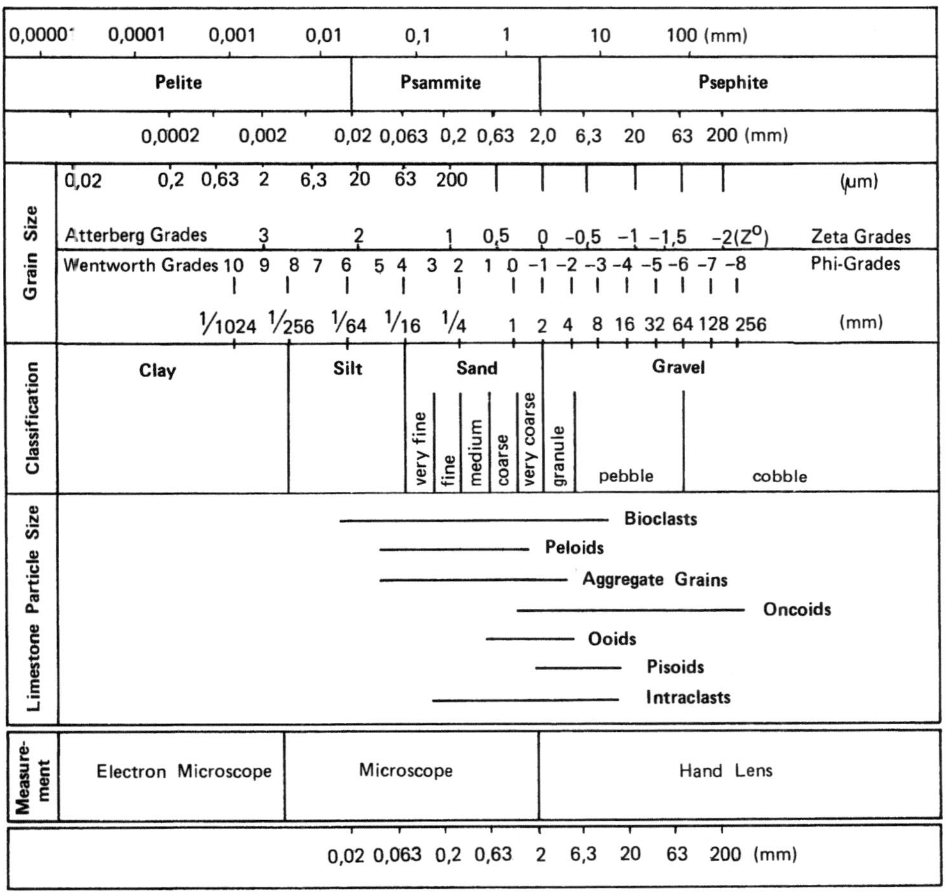

Fig. 24. Grain-size classifications of sediments or carbonate particles. Expanded after Vossmerbäumer (1976)

C. Grain-size determination in soft rocks: Grain-size diameters larger than 20 mm are measured directly. Grain-size diameters between 20 and 0.063 mm are determined by sieve analysis. Grain-size diameters less than 0.063 mm are measured by pipette sedimentation analysis.

D. Grain-size determination in hard rocks: For carbonate rocks this is usually only possible in thin-sections, peels, or polished sections. Grain sizes in thin-sections are a function of the mineralogical composition, the true grain size, grain shape, orientation of the grains, and packing of the grains.

1. *Methods.* The maximum apparent grain diameter is measured. For hand specimens the hand lens developed by Müller (1967) can be used; for thin-sections, an ocular with a millimeter scale can be used or the photomicrographs themselves can be measured.

The lower limits of measurement lie around 6 μm; particles less than 20 μm really should not be measured, because the particles often can not be properly identified and classified (see Plas, 1962). This has been confirmed experimentally by Ferey et al. (1962). Recent bioclasts of less than 60 μm frequently could no longer be distinguished and must, therefore, be put into a general group (e.g., bioclasts).

Sources of error:

a) Smaller grain diameters are measured than are really present, due to the random way the sample is cut.

b) When the point-counting method is used larger grains have a higher probability of being counted, thus, shifting the distribution curves towards coarser grain sizes.

c) The values can be higher or lower than the true values of the grains, depending on their shape and orientation.

a) and c) partially cancel each other out; values which are suspected of being too low can be eliminated by converting the numerical frequencies (thin-section data) to sieve frequencies (Friedman, 1958). The orientation effect can be diminished by measuring normal to the bedding plane. Source of error b) can be eliminated by "ribbon sampling" (Plas, 1962), in which all the grains lying within a band are measured. The band width should be at least as large as the maximum apparent grain diameter. According to Stauffer (1966), the values obtained from the band method cannot be compared directly with sieve data (the band method leads to numerical frequencies; point-counting and line methods, to volume frequencies!).

The number of measurements needed per thin-section. Ordinarily at least 300 grains are needed; calcarenites, however, require about 500 grains for accurate results. The statistical probability can be checked by referring to Fig. 38. The optimal sample size for thin-section size studies has been discussed by B. K. Sahu (1977).

2. *Conversion of thin-section data into sieve data.* Because most interpretations of grain-size distributions are based on the evaluation of unconsolidated sediment (sieve data), many authors feel thin-section data should be converted into sieve data. Here, though, the following problems should be considered:

a) Statistical comparisons show both of the methods to be separate and distinct. Burger and Skala (1976) showed that the calculation of sieve data from thin-section data could not be mathematically supported.

b) Empirical studies of sandstones (Friedman, 1958) found linear correlations between sieve data and thin-section data; theoretically, then, conversion formulas or conversion nomograms could possibly be used (after Friedman, in Müller, 1964). In reality conversion methods are rarely used. In fact, conversions of thin-section data for calcarenites (W. F. Ott, 1969) produced no significant differences in the parameters; see Fig. 29. Folk (1966) discusses the nature of this problem in detail.

II. Graphing grain-size distributions

In well-sorted sediments the distribution of grain sizes should correspond to a logarithmic distribution. If the log-histogram approximates a parabola, the Gaussian or "normal" distribution curve describes the empirical grain size distribution. In many cases, however, the log-histogram closely approximates a hyperbola (Bagnold and Barndorff-Nielsen, 1980). This hyperbolic distribution may be caused by mixtures of normal distributions or by an equilibrium established by two opposing tendencies.

A. Diagrams: a) The most usual method of graphic display is the cumulative frequency curve. Grain diameters are plotted along the abscissa in Phi units or in millimeters; the cumulative frequency percentages, along the ordinate. Plotting on probability paper is the preferred method; the grain-size distribution does not appear here as a curve but rather as a straight line, which actually consists of several straight line segments. The low and high percentages, thus, appear distinctly separated in the graph. Breaks, with different slopes, in the cumulative frequency curve indicate different sedimentation patterns (e.g., deposition in suspension or by rolling or sliding).

b) Histograms are clear and simple, but not very useful for genetic interpretations, see Walger (1965).

c) Frequency distribution curves represent the first derivative of the cumulative curves. This method is rarely used because the parameters are difficult to read accurately.

B. Interpretations of cumulative frequency curves: Breaks and discontinuous curves produced by plotting on logarithmic probability paper are interpreted as:

a) the results of mixing two or more overlapping lognormal sediment populations

b) the mixing of truncated lognormal populations; these correspond to the subpopulations produced by various modes of sediment transport (e.g., Visher, 1969; Sagoe and Visher, 1977).

III. Statistical grain-size parameters

Parameters are modes which serve to characterize grain-size distributions. They can be calculated from Phi values, which in turn have been read off a probability plot.

A. Calculation and interpretation of parameters:

1. *Average grain size:* the easiest to determine but the most inaccurate. This can be expressed by the following graphic measures:

a) *Median diameter:* M, Md, phi 50 – a projection of the 50% value of the ordinate onto the abscissa; characterizes only the central half of the population.

b) *Modal diameter:* Mo, M – a grain diameter corresponding to the highest point of a frequency distribution curve, with an arithmetically scaled ordinate; this is then the grain diameter which appears most frequently. However, it is not a very good measure of the average grain size.

c) *Graphic mean* (after Otto, 1939): M phi = (phi 16 + phi 84)/2. This formula neglects the central third of the distribution; hence, a better formula is Mz (after Folk and Ward, 1957) = (phi 16 + phi 50 + phi 84)/3.

The cumulative frequency curve is described by quartile measures read from the cumulative frequency curves on a logarithmic scale. In the graphic approach phi values are read from the curve for various percentiles. The 25th and 75th percentiles define the so-called quartile deviations. These quartiles represent the bulk of the particles, where Q_1 = 25th quartile, Q_2 = 50th quartile = Md, Q_3 = 75th quartile; percentile measures (P_1–P_{100}) represents the other percentages of the ordinate.

Information obtained from "average" grain sizes: The average value depends on the grain sizes of the material available and the amount of energy in the transport medium. No genetic interpretation of Md or Mz is possible for admixed distributions (e.g., coarser material washed into finer material). The Hjulström diagram (1939) attempts to provide a quantitative comparison of the grain sizes and the average current velocity. But care should be exercised in applying this diagram to Ancient sedimentary rocks.

The various average sizes of carbonate particles in different parts of the reef are used to describe reef complexes (Recent, Ginsburg, 1956); see also Fig. 31, p. 198.

2. *Sorting* (Fig. 25) shows the number and percentage of grain-size fractions represented in a grain mixture. The sorting coefficient is a measure of variability.

Formulas for calculation:

a) The inclusive graphic standard deviation, after Folk and Ward (1957), is the most accurate sorting standard and the one most often used:

$$\sigma_i = (\text{phi } 84 - \text{phi } 16)/4 + (\text{phi } 95 - \text{phi } 5)/6.6$$

This index includes the tails of the cumulative frequency curves, which are very important for interpretations.

b) A more simplified formula after Inman (1952) is:

$$\text{sigma phi} = (\text{phi } 84 - \text{phi } 16)/2$$

c) The easiest formulas to use are:

$$So = Q_3/Q_1 \quad \text{or} \quad \sqrt{Q_3/Q_1} \quad \text{(Trask coefficient)}$$

The sorting class is generally given after Füchtbauer (1959).

Information obtained from sorting: Sorting is dependent on the grain size of the material available for deposition, on the kind of sedimentation, on the current conditions (constant or periodic), and on the rate of sedimentation. A diagram of sorting versus mean size produces a sinusoidal curve (see Fig. 26) for Recent carbonate sands (Folk, 1962; Folk and Robles, 1964). This pronounced sinusoidal trend depends on the effectiveness of the surface waves, the force of erosion, and the

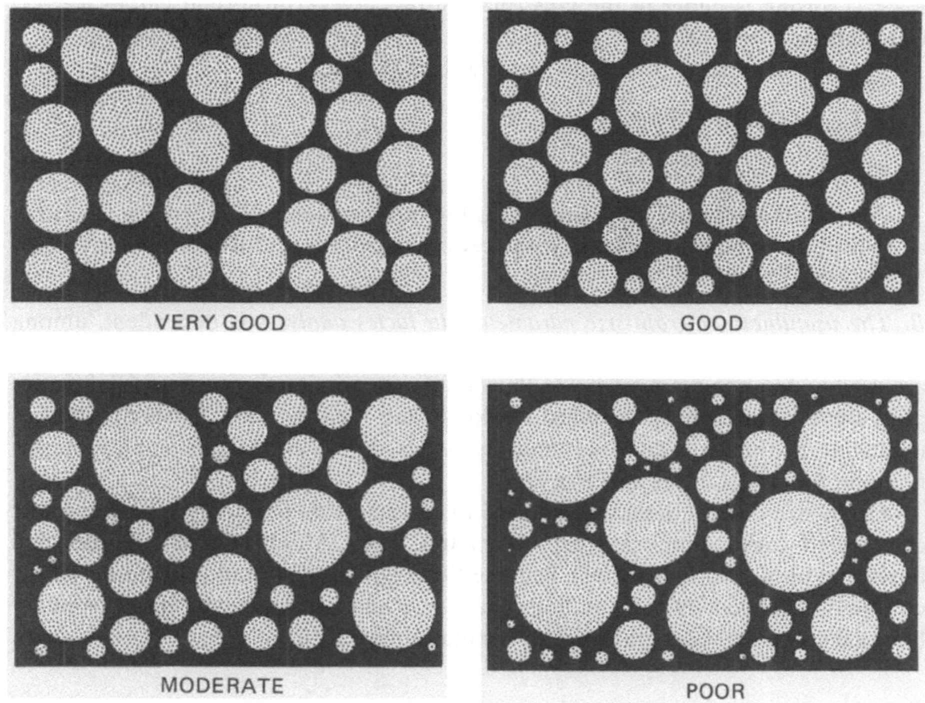

Fig. 25. Comparison charts for particle sorting. After Anstey and Chase (1979)

shapes of the bioclasts; high-energy environments produce a more obvious sinus pattern, whereas in lower-energy environments the trend is flatter.

3. *Skewness* (Sk) designates the asymmetry of the distribution curves. A tendency toward finer grain fractions is expressed by a negative log Sk; towards coarser grain fractions, by a positive log Sk.

a) Inclusive graphic skewness, after Folk and Ward (1957)

$$Sk_i = \frac{\text{phi } 16 + \text{phi } 84 - 2 \text{ phi } 50}{2\,(\text{phi } 84 - \text{phi } 16)} + \frac{\text{phi } 5 + \text{phi } 95 - 2 \text{ phi } 50}{2\,(\text{phi } 95 - \text{phi } 5)}$$

This formula was simplified by Warren (1974).

b) Another formula used is:

$$Sk = Q_1\, Q_3/(Md)^2$$

Information obtained from skewness: Sk is given an environment sensitive significance by many authors (e.g., Friedman, 1967; Folk and Robles, 1964; Hails et al., 1969). Dune sands often show a positive skewness; beach sands, a negative skewness (see Füchtbauer, 1970).

4. *Kurtosis* (K_G) characterizes the degree of peakedness, or flattening of the tails of the curve. If $K_G > 1$, sorting is better in the middle part (leptocurtic curves); if

$K_G < 1$, sorting is better in the tails (platycurtic curves). In normal curves $K_G = 1$. Calculation:

a) Graphic kurtosis, after Folk and Ward (1957)

$$K_G = (\text{phi } 95 - \text{phi } 5)/2.44\,(\text{phi } 75 - \text{phi } 25).$$

b) $K = Q_3 - Q_1/2\,(P_{90} - P_{10})$

Information obtained from kurtosis: Little information is obtained from kurtosis alone, but it can be used together with skewness (see Folk and Robles, 1964).

B. The usefulness of grain-size parameters in facies analyses is dependent, among other things, upon the Phi classes (whole Phi, 1/2 Phi, and 1/4 Phi values) which are used to describe grain sizes. 1/4 Phi values appear to be the most useful if scatter plots are to be interpreted genetically (Moiola and Weiser, 1968). A list of references indicating the importance of scatter plots (M_z vs Sk; M_z vs σ_i; Sk vs σ_i; Sk vs K_G) is found in Sect. 4.2.1.2; see also Fig. 26.

Grain-size distributions are also used to characterize the extent of the diagenetic changes in carbonate rocks (e.g., Germann, 1966, Middle Triassic Wetterstein limestone in the Northern Alps), because the original lognormal distribution of the crystals is changed by cementation, coalescive neomorphism, etc. and secondary bimodal and polymodal grain-size distributions may arise.

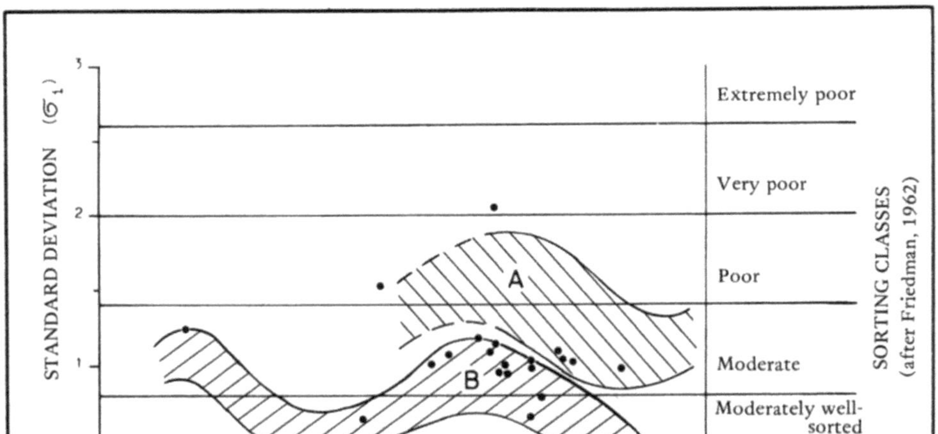

Fig. 26. Parameter diagram. Sorting (standard deviation) vs. average grain size (mean M_Z) of calcarenites in Sulzfluh limestones (Upper Jurassic). Sinusoidal trends. *A* (subtidal environment) and *B* (beach) after Folk and Robles (1964). The diagrams indicate beach deposition for calcarenites in Sulzfluh limestones. The line between "sorted" and "unsorted" is rather artificial, but it may be drawn between beach and subtidal sediments

IV. Clasticity index

Carozzi (1950, 1958) calls the maximum apparent grain diameter (out of at least 100 grains) the clasticity index, and considers it a measure of the (highest) water and transport energies (see Fig. 27). It is often used in microfacies analyses, but it must be used with caution. The degree of winnowing of fine material, and not necessarily the maximum size of the particles is characteristic for water energy!

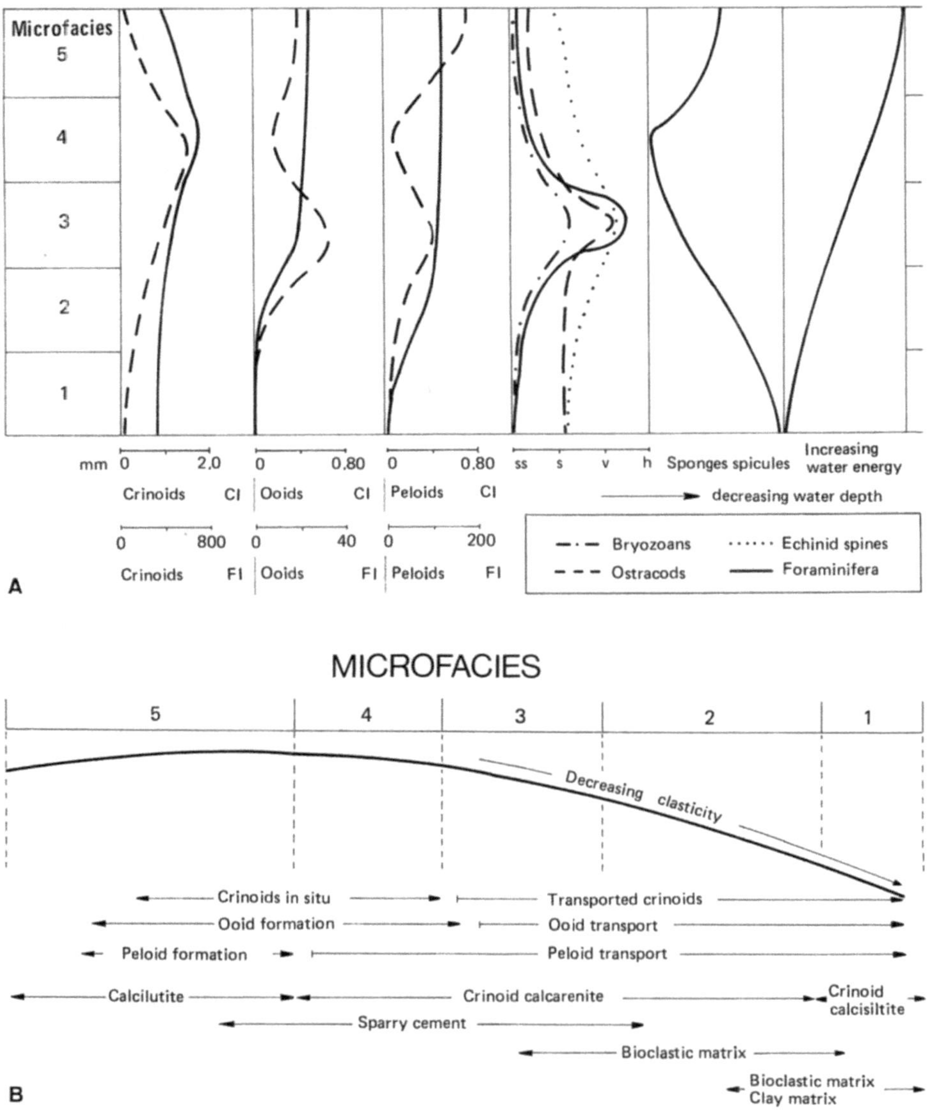

Fig. 27. Differentiation of microfacies types by particle size (*CI* = clasticity index) and analysis of number of particles (*FI* = frequency index) as illustrated by Lower Carboniferous crinoid limestones. Simplified after Walpole and Carozzi (1961)

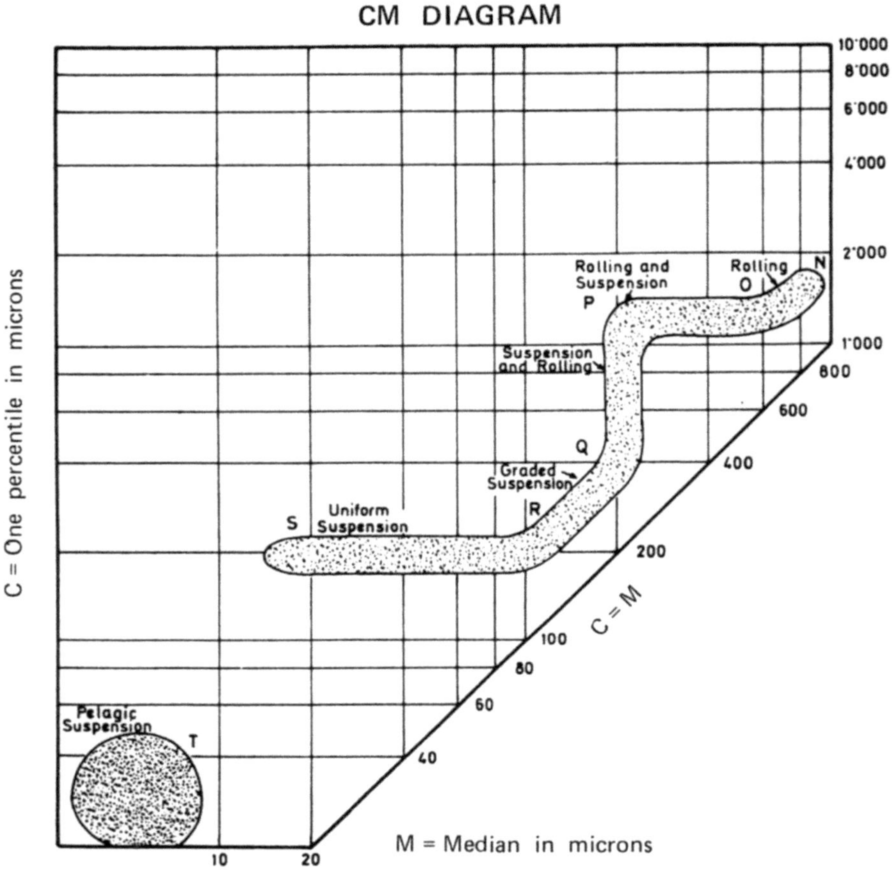

Fig. 28. CM patterns of sediments deposited in different environments. The one percentile (*C*) measures the coarsest grains in the deposit and the median (*M*) measures the fine grains in the tail. Discrimination between environments is made by examination of the patterns produced by plotting a number of samples from the same unit on CM or other Passega diagrams (see text). After Passega (1964)

V. Passega diagrams

Passega (1957) suggested a CM diagram to characterize transport and deposition mechanisms, in which the median M (in millimeters or Phi values) is plotted on the abscissa for each sample, and the 1st percentile of the grain-size distribution (in mm or Phi values, corresponding to the approximate value of the maximum grain size) is plotted on the ordinate. Phi values are plotted on simple graph paper; millimeter values on log log paper. Samples from genetically uniform depositional environments are represented by points forming CM patterns, whose shape and distance from the CM line reflect different modes of transportation and five depositional environments (see Fig. 29):

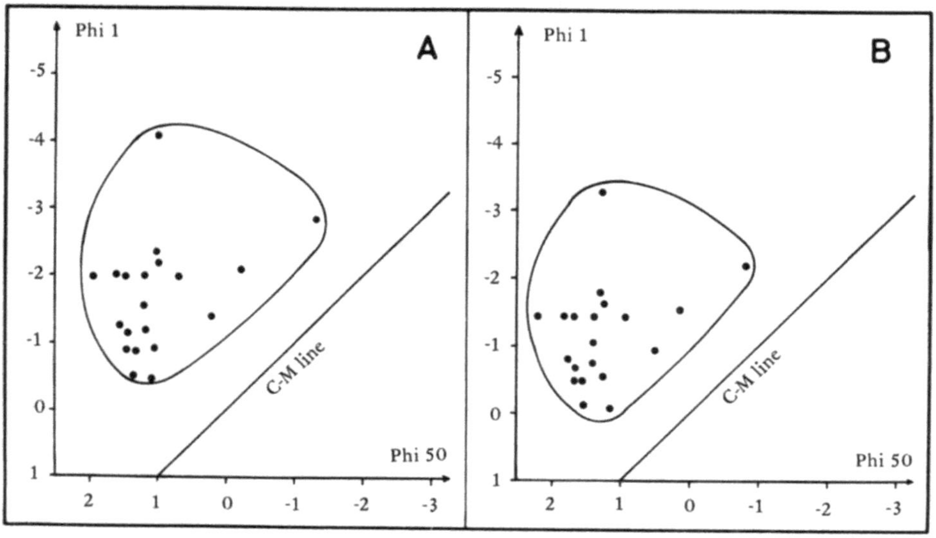

Fig. 29 A, B. Sulzfluh limestones, Upper Jurassic (see Sect. 11.4). Passega diagram. Characteristic point pattern for beach sediments. No substantial difference in the use of number frequencies (**A**) and weight % (**B**). Thin-section data converted into sieve data, after the Friedman method

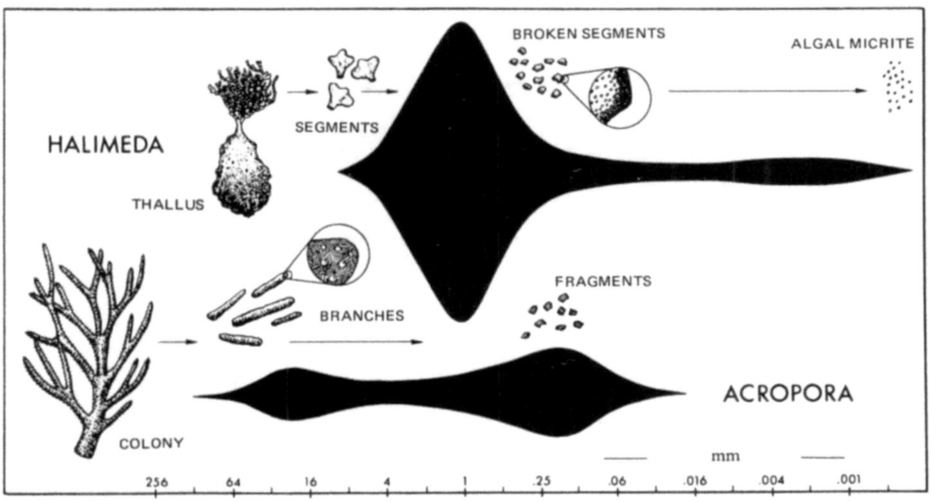

Fig. 30. Grain-size analyses of Ancient calcarenites must consider the fact that the disintegration of hard parts produces many different grain fractions. Example: Recent carbonate sands, Alacran reef, Yucatan (simplified after Folk and Robles, 1964). Codiacean algae disintegrate into micron-sized particles; corals, into sand- and gravel-sized bioclasts

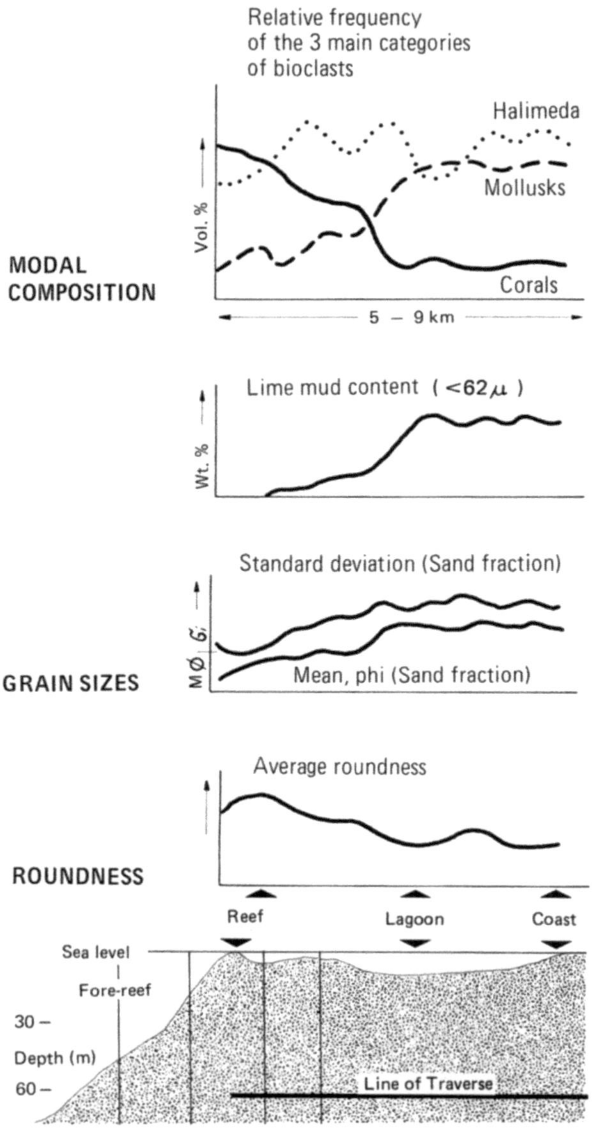

Fig. 31. Modal composition, grain sizes, and roundness of carbonate particles in a Recent reef-zone (Florida). Traverse length about 7.5 km. Simplified after Swinchatt (1965)

1. River sediments – normally L-shaped or boomerang-shaped
2. Tractive currents, marine bottom currents or on shelves – similar to Type 1, usually boomerang-shaped.
3. Turbidite pattern, straight and parallel to the CM line; the axes of the pattern lie on lines that are defined by C = 2.3 M to C = 4.2 M.
4. Beach sand pattern – club-shaped, with the head pointing downwards; minimum size of the median always near the 125 μm line; see Fig. 29.
5. Quiet water environment – spherical pattern some ways from the CM line.

The CM diagram is formed by a group of samples representing all the textures in the deposit and is composed of different segments (QR parallel to line C = M, and RS parallel to the M axis; Fig. 28) which appear to be representative of different transport mechanisms. Currents competent to support the coarsest suspension particles are represented by the segment QR; incompetent currents by the segment RS. The representation of grain size in these CM diagrams expresses, in a very simple fashion, the bottom distribution of depositional turbulence. When CM diagrams are used, it should be remembered that they are not sensitive to small environmental variations, but only define general hydraulic characteristics within a larger environment.

Passega (see p. 206) suggested other diagrams to characterize the various transport mechanisms, which permit a better separation of fine, very fine, and coarse fractions. The evaluation of these FM, LM, AM, and Cs or Cu diagrams provides paleobathymetric information (Passega, 1964) and can also be used for sorting parameters (Passega, 1972, 1977).

The Passega diagrams represent probably the most useful method for interpreting grain-size distributions. The method has been used for Recent sediments as well as for various sedimentary rocks. Calcarenites in particular have been studied in the Upper Jurassic of the Northern Alps by H. W. Flügel and Pölsler (1965, turbidites in basinal facies), Hötzl (1966, calcarenites and calcirudites), W. F. Ott (1969, near-coastal platform carbonates), and by Zacher (1973) in Lower Cretaceous shallow-water limestones. The interpretation of grain-size distributions of calcarenites must also consider the influence of bioturbation and of low-diversity biota which produce large amounts of equidimensional bioclasts (e.g., foraminifera).

As long as the various primary possibilities for the disintegration of organic hardparts (Fig. 30) are taken into account, grain-size analyses along facies traverses are good indicators, together with structural features, for classifying and interpreting depositional environments (Fig. 31).

4.2.1.2 Annotated Bibliography: Grain-Size Analyses

A portion of the vast amount of literature on the methods and interpretation of grain-size distributions in sediments and sedimentary rocks is listed below and coded as follows: M = methodologically important publications; F = articles about grain-size parameters as facies indicators; L = studies of grain-size distributions in limestones.

(M) Adams, J. (1977): Sieve size statistics from grain measurement. J. Geol. **85**/2, 209–227, 10 Figs., Chicago
Empirically determined and theoretically derived conversion formulas; important work.

(F, M) Allen, G. P., Castaing, P., Klingebiel, A. (1972): Distinction of elementary sand populations in the Gironde estuary (France) by R-mode factor analysis of grain-size data. Sedimentology **19**, 1/2, 21–35, 7 Figs., Amsterdam
A statistical analysis makes possible a good differentiation of four transport mechanisms dependent on grain size (less than 3 Phi – uniform suspension + bottom transport, greater than 5 Phi – surface creep). The CM curve reflects the amount of each group

Anstey, R. L., Chase, T. L. (1974): see 1.4.1

(M) Bagnold, R. A., Barndorf-Nielsen, O. (1980): The pattern of natural size distribution. Sed. **27**, 199–207, 10 Figs., Oxford

(L) Basan, P. R. (1973): Aspects of sedimentation and development of a carbonate bank in the Baracuda Keys, South Florida. J. Sed. Petrol. **43**/1, 42–54, 5 Figs., Tulsa

(L) Boccaletti, M., Ficcarelli, G., Manetti, P., Turi, A. (1969): Analisi stratigrafiche, sedimentologiche e petrografiche delle formazioni mesozoiche della Val di Lima (Prov. di Lucca). Mem. Soc. geol. Ital. **8**, 847–922, 49 Figs., Pisa. Passega diagrams are used to interpret the depositional environments of Mesozoic calcarenites.

(L) Braithwaite, C. J. R. (1973): Settling behaviour related to sieve analysis of skeletal sands. Sedimentology **20**, 251–262, 4 Figs., Oxford

(M, F) Buller, A. T., McManus, J. (1972): Simple metric sedimentary statistics used to recognize different environments. Sedimentology **18**, 1/2, 1–21, 14 Figs., Amsterdam
A comparison of over 800 grain-size analyses of Recent sediments from various depositional environments shows that the interpretation of statistical parameters is dependent upon the scales used (Phi scale for sediments with little grain-size variation, e.g. sands; arithmetic scale for sediments with larger grain-size variations). For the latter group the quartile and median values after Trask (QDa and Md in millimeters) as well as the skewness (Sk a) are most significant. Parameter diagrams (QDa : Md in mm) graphed on log-log paper allow eolian, fluviatile, coastal, and quiet-water environments to be differentiated.

(M) Burger, H., Skala, W. (1976): Comparison of sieve and thin-section technique by a Monte-Carlo-Model. Computers and Geosci. **2**, 123–139, Oxford

(M, F) Carozzi, A. V. (1958): Micro-mechanisms of sedimentation in the epicontinental environment. J. Sed. Petrol. **28**/2, 133–150, 13 Figs., Tulsa
The largest transported grain measurable under the microscope (from at least 100 sedimentary grains) is considered the measure of the (highest) water energy = clasticity index. Although applied relatively frequently in microfacies analyses, caution is necessary when this parameter is genetically interpreted.

(F) Chappell, J. (1967): Recognizing fossil strand lines from grain-size analysis. J. Sed. Petrol. **37**/1, 157–165, 10 Figs., Tulsa
Recent coastal sands exhibit negative skewness values; dune sands, positive skewness values. However, for Quaternary sands deviating results are produced.

(L) Chave, K. E. (1960): Carbonate skeletons to limestones: problems. Trans. N.Y. Acad. Sci. **23**, 14–24
Recent coastal sands show bimodal distributions. This can be explained by the abrasion of calcareous material which produces finer intermediate sizes.

(M) Coleman, C., Entsminger, L. (1977): Sieving vs. settling tube: a comparison of hydrodynamic and granulometric characteristics of beach and beach ridge sands. In: Tanner, W. F. (ed.): Coastal Sedimentology, 299–312, Tallahassee: Florida State University

(M) Correns, C. W. (1934): Grundsätzliches zur Darstellung der Korngrößenverteilung. Zbl. Min. etc., A **1934**, 11, 321–331, Stuttgart

(M, L) Davis, J. C., Conley, C. D. (1977): Relation between variability in grain size and area of samples on a carbonate sand beach. J. Sed. Petrol. **47**/1, 251–256, 6 Figs., Tulsa
The statistical check on the variability in grain size on a carbonate sand beach (Yucatan, Mexico), with the aid of Latin squares, shows that the range of variation is independent of the sample size and that this range can be characterized in thin-sections.

(M) Deelmann, J. C. (1972): Automatic grain size analysis by means of electromagnetic surface measurements. J. Sed. Petrol. **42**/3, 732–735, 10 Figs., Tulsa
Surface roughness of samples, thin-sections, sections of limestones, etc.

(M, F) Douglas, D. J. (1968): Grain-size-indices, classification and environment. Sedimentology **10**, 83–100, 12 Figs., 1 Tab., Amsterdam

(L) Early, Ch. F., Goodell, H. G. (1968): The sediments of Card Sound, Florida. J. Sed. Petrol. **38**/4, 985–999, 11 Figs., Tulsa
Shelf sediments. The size of carbonate particles decreases towards the sea, but increases near small coral reefs located parallel to the shoreline. Very few carbonates in the inner bay area, but much quartz sand with a wide spectrum of sorting.

(L) Emery, K. O. (1963): Marine geology of Guam. Geology and hydrology of Guam, Marin Island. Geol. Surv. Prof. Paper **403–B**, B 1–B 76, 65 Figs., 3 Tab., Washington
Grain-size analyses of Recent shallow-water reef carbonates.

(L, F) Fabricius, F. (1967): Die Rät- und Lias-Oolithe der nordwestlichen Kalkalpen. Geol. Rdsch. **56,** 140–170, 10 Figs., 1 Pl., 2 Tab., Stuttgart
When compared with Recent limestone ooids, grain-size parameters of Ancient ooids can provide information about the paleobathymetry and water energy, whereas other calcareous particles do not seem to be quite as suitable for these purposes. Parameters used were Md, So, Sk, and Kt.

Feray, D. E., et al. (1962): see 6.4

(L) Flood, P. G., Orme, G. R. (1977): A sedimentation model for platform reefs of the Great Barrier Reef, Australia. Proc. Third Internat. Coral Reef Symp. **2,** 111–117, 5 Figs., Miami
A sedimentation model is outlined for platform reefs. It is based on the premise that the grain-size distribution of bioclastic sediments is indicative of the hydrodynamic processes acting within the depositional environments. The grain-size distribution is subdivided into the weight percent corresponding to the three size populations: 0.5 mm and coarser, 0.125 mm and finer, and intermediate sizes. Triangular plots of the sediments reflect the dominant hydrodynamic processes in various parts of the reefs. The scheme is similar to the limestone Energy Index Classification, proposed by Plumley et al. (1962).

(L) Flügel, H. W., Pölsler, P. (1965): Lithogenetische Analyse der Barmstein-Kalkbank B_2 nordwestlich von St. Koloman bei Hallein (Tithonium, Salzburg). N. Jb. Geol. Paläont. **1965**/9, 513–527, 6 Figs., 2 Tab., Stuttgart
Grain-size parameters and the Passega diagram indicate the presence of turbidites in pelagic basinal limestones.

(L) Folk, R. L. (1962): Sorting in some carbonate beaches of Mexico. Trans. New York Acad. Sci. (2), **25**/2, 222–224, 11 Figs., 2 Tab., New York
The diagram "sorting: mean grain size" produces in terrigenous (siliceous) and calcareous sediments a sinusoidal distribution, which in calcareous sediments is dependent on the strength of the breakers and on the various degrees of disintegration of organic shells. The sorting of coastal sediments is independent of the grain shape and the wave energy. The latter influences the mean grain size. Surface samples exhibit a standard deviation of $\sigma_i < 0.70$; underwater samples, a $\sigma_i > 0.90$.

(L) Folk, R. L. (1966): A review of grain-size parameters. Sedimentology, **6,** 73–93, 1 Fig., 3 Tab., Amsterdam
Very important critical survey of the methods of grain-size representation and interpretation!

(L) Folk, R. L. (1967): Sand cays of Alacran Reef, Yucatan, Mexico. J. Geol. **75**/4, 412–437, 2 Pls., 12 Figs., Chicago

(M) Folk, R. L. (1968): Bimodal supermature sandstone: Product of the desert floor. 12th Intern. Geol. Congr. Prag **8,** 9–32, 7 Figs., Prag
Methodologically interesting study of the causes of a bimodal grain-size distribution and how this can be recognized.

(L) Folk, R. L., Robles, R. (1964): Carbonate sands of Isla Perez, Alcaran Reef Complex, Yucatan. J. Geol. **72**/3, 255–292, 4 Pls., Chicago
Important investigations of the distribution, composition, grain sizes and grain shapes of a Recent carbonate sand coast. 60% of the coastal sediments of Isla Perez are composed of the remains of Halimeda; 25% of coral fragments; 15% of foraminifera, etc. Six sedimentary populations classified according to origin, grain size, and grain shape as well as topographical position, are present: (a) Large blocky fragments of massive corals originate in high-energy zones, (b) Branches of coral which have broken off (*Acropora*) build ramp-like ridges in the carbonate sand or are arranged normal to the coastline under water, (c) The most common sediment is a sand consisting primarily of disc-like *Halimeda* fragments (Phi 0), (d) Well-sorted 2 Phi sands of coral debris are found on the coast in low-energy areas, (e) Gastropod fecal pellets occur in two shallow protected bays with little water movement, (f) Lime mud which originates from the abrasion of corals and *Halimeda* is found in subtidal sediments. Figure 30 of this book shows the various grain sizes during diminution of carbonate skeletal grains. The sorting of these coastal sediments also produces parameter diagrams with sinusoidal distributions. Submarine sediments of Isla Perez differ from coastal sediments because of large amounts of very fine grains and poorer sorting.

(M) Folk, R. L., Ward, W. C. (1957): Brazos River bar, a study in the significance of grain-size parameters. J. Sed. Petrol. **27,** 3–27, Tulsa
The authors suggest new formulas for calculating the grain-size parameters M, σ_i (inclusive graphic standard deviation), Sk_i (inclusive graphic skewness), and K_G (graphic kurtosis); in comparison with the commonly used Trask or Inman values, these new formulas consider a greater portion of the grain-size spectrum.

(F) Force, L. M. (1969): Calcium carbonate size distribution on the West Florida shelf and experimental studies on the microarchitectural control of skeletal breakdown. J. Sed. Petrol. **39,** 902–934, 21 Figs., Tulsa
The comparison of grain sizes of carbonate bioclasts and quartz in shallow near-coastal inner shelf environments with the results of tank experiments (abrasion studies on pelecypod shells) shows that *Chione, Strombus,* and *Aequipecten* disintegrate during mechanical stress into three grain fractions, according to the microstructure of the shell (layers – 500–250 μm, separate layers – 34–4 μm, and crystal aggregates – 0.50–0.125 μm). Waves and tidal action lead to a redistribution of destroyed skeletal grains: The particles which originate on the inner shelf in the wave zone and measure less than 500 μm are deposited on the outer shelf and in lagoons. Coarser particles are usually found in the surf zone.

(M) Friedman, G. M. (1958): Determination of sieve-size distribution from thin-section data for sedimentary petrological studies. J. Geol. **66**/4, 394–416, 9 Figs., Chicago

(M) Friedman, G. M. (1962): Comparison of moment measures for sieving and thin-section data in sedimentary petrological studies. J. Sed. Petrol. **32**/1, 15–25, 8 Figs., Tulsa
Correlation of mean sieve sizes with thin-section grain sizes. Conversion tables!

(M) Friedman, G. M. (1967): Dynamic processes and statistical parameters compared for size frequency distribution of beach and river sands. J. Sed. Petrol. **37,** 327–354, Tulsa
Grain-size parameter diagrams allow Recent coastal and river sediments to be distinguished. Important article (see also Friedman, 1961, J. Sed. Petrol. **31**!).

(M) Friedman, G. M. (1979a): Differences in size distribution of populations of particles among sands of various origins. Sed. **26,** 3–32, 31 Figs., Oxford
Discusses the interrelationships between textural parameters and depositional processes for beach, dune, and river sands, followed by the description of regional trends of size distribution in a beach-dune setting, a fluviatile point-bar setting, a continental shelf-slope setting, and a continental shelf setting.

(M) Friedman, G. M. (1979b): Differences in size distributions of populations of particles among sands of various origins: addendum to IAS Presidential Adress. Sed. **26,** 859–862, 2 Figs., Oxford
Includes plots of Inclusive Graphic Standard Deviation vs. Inclusive Graphic Skewness for beach and inland dune sands as well as for inland and near-shore dune sands.

(M) Füchtbauer, H. (1959): Zur Nomenklatur der Sedimentgesteine. Erdöl u. Kohle **12,** 605–613, 7 Figs., Hamburg
Description of the common grain-size classifications used in the German literature.

(M) Füchtbauer, H. (1974): see 1.4.1

(M, L) Germann, K. (1966): Ablauf und Ausmaß diagenetischer Veränderungen im Wettersteinkalk (alpine Mitteltrias). Diss. Univ. München, 122 pp., 17 Pls., 8 Tab., München
Grain-size curves and parameters of diagenetically changed carbonate rocks provide information about the degree of diagenesis when variations in the grain sizes of the matrix are considered.

(L, F) Ginsburg, R. N. (1956): Environmental relationships of grain size and constituent particles on some South Florida carbonate sediments. Bull. Amer. Ass. Petrol. Geol. **40,** 2384–2427, Tulsa
Important articles about the grain-size distribution in several traverses which run through a reef area. Various grain-size ranges and mean values in different environments can be seen.

(M) Gnaccolini, N. (1968): Caratteristiche granulometriche di alcuni campioni di sabbie di Spiaggia, di duna costeria, di fiume, di ambiente marino epineritico e di arenarie presumbilmente deposte da corrente di torbidita. Mem. Soc. Geol. Ital. **7,** 361–378, 12 Figs., Pisa

Important comparison of various methods used in facies differentiation (coasts, coastal dunes, river sands, marine shallow-water sands, turbidites). Differentiation of coastal sands, river sands, and turbidite sands, on the one hand, and coastal sands and river sands, on the other hand, is possible with the parameter diagram Sk_i or $So_s : \alpha_s$ (simple skewness measure). Sk : K_G (graphic kurtosis) permits the differentiation of coastal dunes and river sediments. Values and signs of skewness (Sk or α) have various meanings. The Passega diagram distinguishes between normally deposited marine sands and turbidite sands. The samples come from the Southern Adriatic Sea.

(M) Griffiths, J. C. (1967): see 1.4.1

(F) Hails, J. R., Moyt, H. J. (1969): The significance and limitations of statistical parameters for distinguishing ancient and modern sedimentary environments of the Lower Georgia coastal plain. J. Sed. Petrol. **39**/2, 559–580, 10, Tulsa
Recent and Pleistocene coastal sediments are compared with respect to grain-size parameters. The skewness values and signs prove to be especially useful.

(M) Harrell, J. A., Eriksson, K. A. (1979): Empirical Conversion Equations for Thin-Section and Sieve Derived Size Distribution Parameters. J. Sed. Petrol. **49**/1, 273–280, 4 Figs., Tulsa
Using size analysis of Ancient sandstones, thin-section-to-sieve and sieve-to-thin-section empirical conversion equations have been derived, via linear regression analysis, for several graphical size distribution statistics. Conversions of high accuracy are possible only for the median and graphic mean.

(M) Hörnstein, A. (1960): A method and a set of apparatus for mineralogic-granulometric analysis with a microscope. Bull. Geol. Inst. Uppsala **38,** 105–137, Uppsala

(M) Hjulström, F. (1939): Transportation of detritus by running water. In: Trask, P. D. (ed.): Recent marine sediments. Amer. Ass. Petrol. Geol. **23,** 5–31, 4 Figs., Tulsa
Experiments show the representation in "Hjulström diagrams" of the causal relationships between sediment deposition, transport, and erosion in currents. The transport phase corresponds to an equilibrium state between the rate of flow and the supply of grain sizes. If the grain size remains unchanged, increasing rates of flow cause erosion; decreasing rates of flow, sediment deposition. This diagram is used by a few authors to ascertain current speeds from the frequency of certain grain sizes.

(L) Hoskin, Ch. M., Nelson, R. V. Jr. (1971): Size modes in biologic carbonate sediment, Southeastern Alaska. J. Sed. Petrol. **41**/4, 1026–1037, 8 Figs., Tulsa
Skeletal endobiontic organisms in the subtidal zones, and balanids in the tidal zones are disintegrated by mechanical and biological erosion, which produces pelecypod and balanid bioclasts with sizes between 12 mm and 0.1 mm. The grain size distributions and parameters are of no value for facies interpretations.

(L, F) Hötzl, H. (1966): Zur Kenntnis des Tressenstein-Kalkes (Ober-Jura, Nördliche Kalkalpen). N. Jb. Geol. Paläont. Abh. **123**/3, 281–310, Pls. 32–36, 10 Figs., 4 Tab., Stuttgart
Calcarenites and calcirudites were analyzed in respect to their grain-size parameters (So, Sk, Kg) and interpreted from a Passega diagram. Important article for grain-size analyses based upon thin-sections.

(M) Inman, D. L. (1952): Measures for describing the size distribution of sediments. J. Sed. Petrol. **22,** 125–145, Tulsa
Formulas for calculating grain-size parameters. See Isphording (1972).

(M) Isphording, W. C. (1972): Analysis of variance applied to measures of central tendency and dispersion in sediments. J. Sed. Petrol. **42**/1, 107–121, 8 Figs., Tulsa
Critical work about the dependency of the grain-size distribution on various sieve data and on various methods of calculation. Sieve intervals of 0.5 and 1.0 Phi were compared with grain-size parameters which were calculated according to Davis and Ehrlich (1970, central-moment measure), Folk and Ward (1957) and Inman (1952). Result: In calculating the median value and the standard deviation of coastal and river sediments, no differences were found in the three methods at intervals of 0.5 Phi; in calculating the mean diameters, very slight differences were found in the three methods at 1.0 Phi intervals. When calculating the standard deviation of coastal and river sediments on the basis of 1.0 Phi data, the results become less accurate, regardless of the method of calculation used. Conspicuous differences occur during the calculation of skewness and kurtosis when using the central-moment measure method and the other two methods. Although the Inman

formula leads to results for mean diameters, standard deviations, and skewness which are statistically comparable to those values of Folk and Ward, the latter method is the preferred one. Kurtosis, according to Davis and Ehrlich, must be calculated with 0.5 Phi values.

(M, L) Jaanusson, V. (1952): Untersuchungen über die Korngröße der ordovizischen Kalksteine. Geol. Fören. Stockholm. Föranhandl. **74,**/2, No. 469, 121–130, 3 Figs., Stockholm
Measurement of the grain diameter according to the Rosival method permits the differentiation of "sandy fractions" (diameter larger than 0.1 mm) and "groundmass" (diameter less than 0.1 mm).

(L) Jindrich, Vl. (1969): Recent carbonate sedimentation by tidal channels in the Lower Florida Keys. J. Sed. Petrol. **39**/2, 1–553, 14 Figs., Tulsa
Important studies of sediment mixing in a tidal channel in a Recent carbonate sedimentary environment overlying Pleistocene carbonates whose rock fragments are admixed with Recent bioclasts. The channel sediments are characterized by anomalous size/roundness ratios and by negative skewness distributions.

(M) Kellerhals, R., Shaw, J., Arora, V. K. (1973): On Grain Size from Thin-Sections. J. Geol. **83,** 79–96, 7 Figs., 2 Tab., Chicago
Important critical summary of the methods proposed for microscopic grain-size analysis. A numerical method is used to obtain grain-size distributions of major and minor axes of ellipses produced by random cuts through ellipsoids of various shapes. The sieve curve is located between the cumulative distributions of these major and minor axes. The empirical results of Friedman (1958) are substantiated theoretically. Grid sampling combined with measurement of the apparent major axis, and the use of number frequencies provide an efficient procedure for obtaining approximate sieve curves.

(F, M) Klovan, J. E. (1965): The use of factor analysis in determining depositional environments from grain-size distribution. J. Sed. Petrol. **36,** 115–125, Tulsa
Factor analysis of Recent sediments in Barataria Bay, Louisiana, shows that the sediment distribution is dependent on the wind-wave energy, the current energy, and the gravitational energy. The advantages of factor analysis lie in the fact that the total spectrum of grain sizes is considered, that for the most part no artifical statistical description is necessary, and that groupings which correspond to natural conditions are made possible.

(F, M) Koldijk, W. S. (1968): On environment-sensitive grain-size parameters. Sedimentology **107,** 57–69, 8 Figs., Amsterdam
Tests the usefulness of grain-size parameters (after Folk and Ward; Friedman) for distinguishing coastal and river sands. Results: Folk's inclusive standard deviation and skewness are useful. Computer programs have been developed for parameters of Folk and Ward, Friedman and Trask.

(M) Köster, E. (1964): Granulometrische und morphometrische Meßmethoden an Mineralkörnern, Steinen und sonstigen Stoffen. 336 pp., 109 Figs., 68 Tab., Stuttgart: Enke
Survey of methods of grain-size measurements and the description of grain shapes.

(M) Krumbein, W. C. (1935): Thin-section mechanical analysis of indurated sediments. J. Geol. **43,** 482–496, Chicago

(M) Krumbein, W. C. (1938): Korngrößeneinteilung und statistische Analyse. N. Jb. Min. Geol. Pal., Abh. **73,** 137–150, Stuttgart
Description of the Phi grade scale.

(L, F) Lacey, J. E., Carozzi, A. V. (1967): Critères de distinction entre oolithes autochthones et allochthones. Application du calcaires de Saint-Géneviève (Viséen de l'Illinois, USA). Bull. Centre Rech. Pau SNPA **1, 2,** 279–313, 4 Pls., 14 Figs., Pau
Differentiation of autochthonous and allochthonous ooids with the aid of grain-size distributions (skewness).

(L, F) Lewis, M. S. (1969): Sedimentary environments and unconsolidated carbonate sediments of the fringing coral reef of Mahé, Seychelles. Marine Geol. **1,** 95–127, 14 Figs., Amsterdam
Grain-size parameters (sorting, skewness, etc.) exhibit no clear correlations with certain parts of reefs, but certainly do with the "availability" of the skeletal grains and with different areas of turbulence.

(F, L) Lynts, G. W. (1966): Relationships of sediment-size distribution to ecologic factors in Buttonwood Sound, Florida Bay. J. Sed. Petrol. **36**/1, 66–74, 8 Figs., Tulsa

The factor-vector analysis of ecological factors (water depth, temperature, salinity, pH and Eh) in 74 samples from Buttonwood Bay, which is scarcely 3 m deep, shows that no linear correlation exists between the ecological factors and the grain sizes. The grain size is determined by the distribution of the marine grass *Thalassia,* which traps fine-grained sediment.

(L) Maiklem, W. R. (1968): Some hydraulic properties of bioclastic carbonate grains. Sedimentology **10,** 101–109, 4 Figs., Amsterdam
Shape and density of bioclastic carbonate grains decisively influence grain-size distribution. Important work!

(L) Mallik, T. K. (1976): Grain-size variation in the Kavaratti Lagoon sediments, Lakshadweep, Arabian Sea. Marine Geol. **20,** 57–75, 10 Figs., Amsterdam
Important work about the grain-size distribution in a Recent atoll lagoon. The fraction 2–1 mm consists of corals and *Halimeda*; the fraction 1–0.5 mm of foraminifera; the fraction 0.5–0.25 mm of corals, and the fraction < 0.25 mm of various bioclasts.

(L) Maxwell, W. G. H., Day, R. W., Fleming, P. J. G. (1961): Carbonate sedimentation on the Heron Island Reef, Great Barrier Reef. J. Sed. Petrol. **31**/2, 215–230, 9 Figs., Tulsa
Grain-size parameters, roundness, and composition of carbonate sands on the reef platform permit the differentiation of facies zones.

(L, M) Meischner, K. D. (1964): Allodapische Kalke, Turbidite in riffnahen Sedimentationsbecken. Dev. Sed. **3,** 156–191, 3 Pls., 5 Figs., Amsterdam: Elsevier
Discussion of the Hjulström diagram in connection with the sedimentological characterization of allodapic limestones (turbidites).

(M) Miller, M. C., Komar, P. D. (1977): The development of sediment threshold curves for unusual environments (Mars) and for inadaequately studied materials (Foram sands). Sed. **24,** 709–721, Oxford
Generalized curves for grain threshold under liquid flow and under gas flow, based on laboratory data, are utilized to generate a series of threshold curves for aeolian and water flow threshold on the planet Mars, and the threshold of foraminiferal sands in the deep-sea.

(M, F) Moiola, R. J., Weiser, D. (1968): Textural parameters: an evaluation. J. Sed. Petrol. **38**/1, 45–53, 8 Figs., Tulsa
Important work about the information provided by grain-size parameter diagrams of Recent sediments. Those parameters prove useful which have been calculated on the basis of 0.25 Phi sieve data. 0.5 Phi and 1.0 Phi sieve data are only useful in a few cases. The "mean diameter versus skewness" permits a distinction to be made between coasts, coastal sands, and inland dune sands; the "mean diameter versus standard deviation" permits the differentiation of coastal and river sands, and river and coastal dune sands.

(M) Müller, G. (1964): see 1.4.1

(F) Müller, G. (1966): Grain size, carbonate content, and carbonate mineralogy of recent sediments of the Indian Ocean off the Eastern Coast of Somalia. Naturwissenschaften **53**/32, 547–550, 3 Figs., Berlin
Grain sizes decrease towards the sea, the ratios of clay/silt/sand permit the recognition of sedimentation zones running parallel to the coast.

(M) Müller, G. (1967): A pocket microscope for grain-size measurement in the field. J. Sed. Petrol. **37,** 703–704, 2 Figs., Tulsa
Description of the "Müller hand lens" which makes grain size measurement possible directly in the field.

(M) Münzer, H., Schneiderhöhn, P. (1953): Das Sehnenschnittverfahren. Heidelbg. Beitr. Min. Petrogr. **3,** 456–472, Heidelberg

(M) Neumann, R. (1963): Die Auswertung von Korngrößenverteilungen durch Häufigkeitsanalysen. N. Jb. Geol. Paläont. Mh. **1963**/9, 492–501, 2 Tab., 2 Figs., Stuttgart
Frequency analyses after Daeves and Beckel evaluate the role grain-size factors play in the total distribution

(L, F) Ott, W. F. (1969): Zur Geologie des Sulzfluhkalkes in Graubünden und Vorarlberg. Unpubl. Diss. TH Darmstadt, 187 pp., 10 Pls., Darmstadt (typescript)
Detailed description of the methods of grain-size analysis of carbonate rocks as illustrated by Upper Jurassic Sulzfluh limestones. The parameter diagrams (M vs Sk) and the Passega

diagram provide the most information about the genesis of calcarenites. Important comparison of various methods of measurement!

(M) Otto, G. H. (1938): see 1.3.

(M) Page, H. G. (1955): Phi-millimeter conversion table. J. Sed. Petrol. **25,** 285–292, Tulsa
Conversion tables of millimeter values into Phi values.

(M) Passega, R. (1957): Texture as characteristic of clastic deposition. Bull. Amer. Ass. Petrol. Geol. **41**/9, 1952–1984, 17 Figs., Tulsa
Important work. The CM diagram (Md vs 1% percentile) characterizes the kind of material transport, the conditions for deposition, and the sedimentary environment. If the samples originate from genetically homogeneous environments, the clustering of points determined by Md and percentile produce characteristic patterns. The following types of sediments can be distinguished according to the CM pattern and according to the distance from the CM line: (1) River sediments, CM pattern L-shaped or boomerang-shaped; (2) Tractive currents, e.g. tidal currents, pattern L-shaped; (3) Turbidity currents, CM pattern a straight line parallel to the CM line; (4) Beach sand sediments, CM pattern club-shaped, far from the CM line. The method suggested by Passega has been repeatedly, and successfully, used to characterize Recent and Ancient depositional environments! (See Fig. 28).

(M) Passega, R. (1964): Grain size representation by CM patterns as a geological tool. J. Sed. Petrol. **34**/4, 830–847, 11 Figs., Tulsa
In investigations of more than 10,000 sediment samples from various depositional environments, the author has refined the genetic information provided by the CM diagram. A subdivision of the CM pattern into various segments and the separate analysis of the segments permit the recognition of different transport mechanisms within a sediment. By considering the largest grain transported in a uniform suspension (Cs or Cu) statements can be made about the water depth. Paleobathymetric profiles and charts are also included.

(M, F) Passega, R. (1972): Sediment sorting related to basin mobility and environment. Bull. Amer. Petrol. Geol. **56**/12, 2440–2450, 3 Figs., 1 Tab., Tulsa

(M) Passega, R. (1977): Significance of CM diagrams of sediments deposited by suspensions. Sed. **24,** 723–733, 3 Figs., Oxford
Discusses the possibilities of facies interpretation using CM diagrams. CM diagrams also reflect two main types of currents occurring in sedimentary environments (bottom turbulence competent to support the coarsest particles in suspension; bottom turbulence incompetent to support the coarsest particles in suspension). Principal characters of CM patterns can be schematically represented by a bar diagram which gives significant information about the bottom currents and the distribution of bottom turbulence.

(M) Passega, R., Byramjee, R. (1969): Grain-size image of clastic deposits. Sedimentology **13,** 233–252, 5 Figs., Amsterdam
In addition to the CM diagram, which characterizes the coarse fraction of samples, new diagrams have been suggested to characterize the fine and very fine fractions, namely, FM, LM, and AM diagrams (F, L, and A = weight % of the grains smaller than 125, 31, and 4 μm, resp.).

(L, M) Pilkey, O. H., Morton, R. W., Lutenauer, J. (1967): The carbonate fraction of beach and dune sands. Sedimentology **8,** 311–327, 5 Figs., 5 Tab., Amsterdam
Studies of grain sizes of the carbonate fraction and acid insoluble residues show that similar factors are at work in grain-size distributions.

(M) Plas, L. van der (1962): Preliminary note on the granulometric analysis of sedimentary rocks. Sedimentology **1,** 145–157, 1 Fig., Amsterdam
A method of grain-size measurement in thin-sections; the point-counting method and the line method are rejected in favor of the "ribbon sampling method." Here all the grains are measured which are recognizable within the width of a band larger than 20 μ. The band width must be just as large as or larger than the maximum apparent grain diameter. This method leads to "numerical frequencies" which are not comparable with the volume frequencies obtained from sieve data.

(M) Rizzini, A. (1968): Sedimentological representation of grain sizes. Mem. Soc. Geol. Ital. **7,** 65–69, 19 Figs., Rome

(M) Sagoe, K. M., Visher, G. S. (1977): Population breaks in grain-size distributions of sand – a theoretical model. J. Sed. Petrol. **47**/1, 285–310, 6 Figs., Tulsa

(M) Sahu, B. K. (1974): Probabilistic solution to thin-section size measures. Indian J. Earth Sci. **1**/1, 22–36, 1 Fig., 4 Tab.

(M) Sahu, B. K. (1977): Optimal sample size for thin-section size studies. Sed. Geol. **19**/3, 185–196, 4 Tab., Amsterdam

(L, F) Sarntheim, M. (1970): Sedimentologische Merkmale für die Untergrenze der Welleneinwirkung im Persischen Golf. Geol. Rundschau **59**/2, 649–666, 7 Figs., Stuttgart
Although the grain-size distribution of Recent marl provides no information about the lower depths affected by the forces of waves; in calcareous sediments on "surfaces" and near islands a clear decrease in grain sizes with increasing water depth is noticeable; the wave base is first reached at a water depth of 50 m.

(M, L) Sarntheim, M. (1971): Oberflächensedimente im Persischen Golf und Golf von Oman. II. Quantitative Komponentenanalyse der Grobfraktion. Meteor. Forsch.-Ergebnisse C 5, 1, **113,** 4 Pls., 45 Figs., 8 Tab., Berlin – Stuttgart: Borntraeger
Important work about the relationships between grain sizes and grain composition in fractions larger than 63 μm.

(M) Sevon, W. D. (1968): First and second degree regression correlation on the size analysis statistical parameters of Trask, Imman, Folk and Ward, and Friedman. J. Sed. Petrol. **38**/1, 238–240, 2 Tab., Tulsa
Good correlation of the methods used by the authors for mean and standard deviation values; conversely, individual authors end up with substantially different results in statistical comparisons of the values for kurtosis and skewness!

(M) Sindowski, K. W. (1958): Die synoptische Methode der Kornkurven-Vergleiche zur Ausdeutung fossiler Sedimentationsräume. Geol. Jb. **73,** 235–275, 68 Figs., Hannover
As a result of about 5000 grain-size analyses of sands and sandstones, ten typical grain curves are produced which are supposed to characterize certain depositional environments.

(M) Stauffer, P. H. (1966): Thin-section analysis: a further note. Sedimentology **7**/3, 261–263, Amsterdam

(M) Swan, D., Clague, J. J., Luternauer, J. L. (1979): Grain-size statistics II: Evaluation of grouped moment measures. J. Sed. Petrol. **49**/2, 487–500, 6 Figs. Tulsa

(L) Swinchatt, J. P. (1965): see 2.4.1

(M) Tanner, W. F. (1964): Modification of sediment size distributions. J. Sed. Petrol **34**/1, 156–164, 8 Figs., Tulsa
Because of the interaction of several transport mechanisms, the primary filtering of certain grain sizes as well as the inability of the method to measure finer grain sizes, zig-zagging grain-size curves result which do not correspond to the normal distribution scheme found in the Hjulström diagram. This applies particularly to shallow water moved by waves.

(M, F) Tanner, W. F. (1966): The surf "break": key to paleogeography? Sedimentology **7**/3, 203–210, 3 Figs., Amsterdam
The grain-size curves of Recent coastal sediments which originate in environments of weak to moderate surfs are characterized by a break (at about 1.5 Phi); this is caused by winnowing of certain fractions. This surf break can be used as a facies criteria for recognizing older coastlines.

(M, L) Usdowski, H.-E. (1962): Die Entstehung der kalkoolithischen Fazies des norddeutschen Unteren Buntsandsteins. Beitr. Min. Petrogr. **8,** 141–179, 25 Figs., Heidelberg
Sieve analyses and thin-section analyses (method after Schneiderhöhn and Münzer) are compared for leached ooids. Curves and parameters (So, Sk) are almost identical.

(M, F) Valia, H. S., Cameron, B. (1977): Skewness as a paleoenvironmental indicator. J. Paleont. **47**/2, 784–793, 13 Figs., Tulsa
Miocene, Delaware. Negative and positive skewness together with lithological and textural features are good indicators of conditions of deposition.

(M) Vandenberghe, N. (1975): An evaluation of CM patterns for grain-size studies of fine-grained sediments. Sedimentology **22,** 615–622, 4 Figs., Oxford
The analysis of argillaceous sediments with the aid of CM, AM, and LM diagrams shows that the pattern of a "uniform suspension" does not necessarily provide useful information but can be explained by an abundance of clay in the sediment or by a change in the supply of clay and silt.

(M, F) Visher, G. S. (1969): Grain size distribution and depositional processes. J. Sed. Petrol. **39**/3, 1074–1106, 21 Figs., Tulsa
Important work about the possibilities of a genetic interpretation of grain-size curves; critical assessment of the various methods (see Sagoe and Visher, 1977!).
(M) Wadell, H. (1932): Volume, shape and roundness of rock particles. J. Geol. **40**, 443–451, Chicago
(M, F) Walger, E. (1961): Die Korngrößenverteilung von Einzellagen sandiger Sedimente und ihre genetische Bedeutung. Geol. Rdsch. **51**/2, 484–507, 8 Figs., Stuttgart
Considerable ranges of variation in grain-size distributions occur even in single layers of sandy sediments. The interpretation of grain-size analyses should be based on the recognition of "sedimentation units".
(M) Walger, E. (1965): Zur Darstellung von Korngrößenverteilungen. Geol. Rdsch. **54**/2, 976–1002, 11 Figs., Stuttgart
Grain size data must absolutely always be represented in Phi grades or Zeta grades (the Wentworth or Atterberg scale, resp.), if a genetic interpretation is being used.
(M) Warren, G. (1974): Simplified form of the Folk-Ward Skewness parameter. J. Sed. Petrol. **44**/1, 259, Tulsa
Simplification of the formula suggested by Folk and Ward (1957) for calculating the skewness and asymmetry of the grain-size distribution. The following formula can, thus, be used: $Sk_i = (\text{phi } 84 - \text{phi } 50)/(\text{phi } 84 - \text{phi } 16) - (\text{phi } 50 - \text{phi } 5)/(\text{phi } 95 - \text{phi } 5)$.
(L) Zacher, W. (1973): Das Helvetikum zwischen Rhein und Iller (Allgäu – Vorarlberg). Geotekton. Forsch. **44**, 1–74, 4 Pls., 12 Figs., Stuttgart

4.2.2 Morphometric Criteria

The form of sedimentary particles is described in general by the shape, sphericity, and angularity as well as roundness and surface structures. Studies of grain shape strive for information about the duration and extent of mechanical forces during transport.

4.2.2.1 Methods and Examples

The shape of sedimentary particles may provide information about the kind and duration of sediment transport. This holds true in particular for pebbles and for siliceous sand grains (see Füchtbauer, 1970; Pettijohn et al., 1972). Particle shape can be described by form, roundness and surface texture. Although many more parameters have been proposed in order to characterize the shape, none has been universally accepted (see Barrett, 1980). The form, i.e., the overall shape of the particle, is generally defined by elongation and flatness, angularity, sphericity, or roundness. The last two aspects, however, are the most commonly used and are defined as follows: *Sphericity* = the approximation of the shape to a sphere; *Roundness* = a change in shape, with increasing loss of sharp corners and edges on the particle.

In Sect. 4.2.2.2 a few articles are listed which suggest methods for determining parameters of roundness and sphericity. Many current methods of estimation, which are simpler and less precise versions of the methods of Wadell (1935) and Krumbein (1941), make use of comparison charts to rapidly determine morphometric criteria. The most popular ones are the charts of Russel and Taylor

(1937), Pettijohn (1957), Krumbein and Sloss (1963), and Powers (1953); see Figs. 32 and 33. Lees (1964) does not accept the standard charts of Krumbein and Sloss and suggests a scheme based on the "degree of angularity."

Carbonate particles cannot be directly compared with quartz grains or pebbles, with regard to morphometric characteristics, because the particles vary in their ability to resist mechanical and chemical influences and because they have different initial shapes. However, because a change in shape may also be observed in bioclasts and carbonate lithoclasts, to some extent in intraclasts, descriptions of *roundness and sphericity of transported carbonate particles* can be quite helpful:

The chart of Pilkey et al. (1967), which is based on empirical observations (see Fig. 33), is recommended for describing the roundness of carbonate bioclasts (especially shell fragments). Here Class 1 comprises particles without rounding; Classes 2 and 3 are characterized by increasing roundness; and Class 4 includes fairly well rounded to very well rounded particles. Another comparison scheme for the roundness and sorting of bioclasts and intraclasts, developed by Catalov (1967) during a study of Triassic limestones, also produces good results.

Folk (1962) pointed out that the average roundness in calcarenites is dependent on the following factors:

a) primary roundness of the skeletal grains (e.g., foraminifera),
b) the microstructure,
c) more rapid rounding of larger particles than of smaller particles,
d) rate of organic productivity; (even in high-energy environments not all of the bioclasts are rounded if there is a very high organic productivity!),
e) force of the breakers,
f) duration of high-energy conditions.

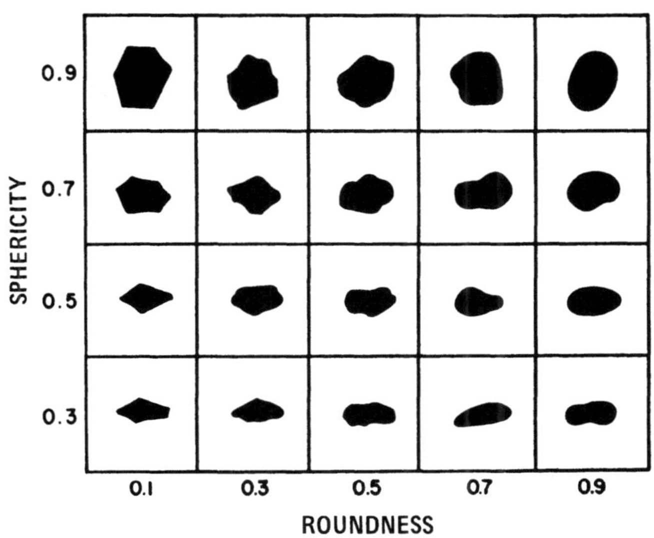

Fig. 32. Comparison chart for roundness and sphericity. After Krumbein and Sloss (1963). This set of pebble images is based on the Wadell key for roundness. It is necessary to describe at least 100 particles, at the same time taking into consideration the various grain-size classes

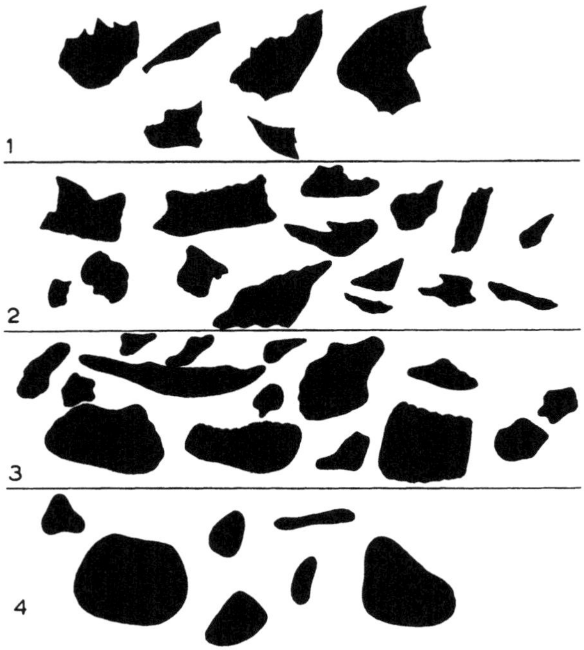

Fig. 33. Comparison chart for describing the roundness of carbonate bioclasts. After Pilkey et al. (1967)

The relationship between winnowing, sorting, and rounding is shown in Fig. 61.

Studies on the rounding of Recent carbonate bioclasts can be found in Folk and Robles (1964, Alacran reef complex, Yucatan), Pilkey et al. (1967, coastal and dune sands, Atlantic coast, USA), Maxwell et al. (1961, barrier reef, Australia), and Swinchatt (1965, Florida). Chave (1964, in Imbrie and Newell) described experiments on the durability and preservation of Recent skeletal organisms.

In Ancient environments the determination of the degree of rounding is especially important for the energy-index classification (Plumley et al., 1962), based on the various water energies at the time of sedimentation; see Catalov (1972) and Section 6.2.4.

4.2.2.2 References: Morphometric Criteria

Abbott, P. L., Peterson, G. L. (1978): Effects of abrasion durability on conglomerate clast populations: examples from Cretaceous and Eocene conglomerates of the San Diego Area, California. J. Sed. Petrol. **48** 1, 31–42, 10 Figs., Tulsa

Barrett, P. J. (1980): The shape of rock particles, a critical review. Sed. **27,** 291–303, 6 Figs., 5 Tab., Oxford

Beal, M. A., Shepard, F. P. (1956): A use of roundness to determine depositional environments. J. Sed. Petrol. **26,** 46–60, Tulsa

Behrens, M. (1977): Zur Stereometrie von Geröllen. Mitteilungen Geol. Paläont. Inst. Univ. Hamburg **47,** 1–124, 29 Figs., Hamburg

Cailleux, A. (1962): Morphoskopische Analyse und ihre Bedeutung für die Paläoklimatologie. Geol. Rdsch. **40,** 11–19, Stuttgart

Catacosinos, P. A. (1965): Tables for the determination of sphericity and shape of rock particles. J. Sed. Petrol. **35**/2, 354–365, 1 Fig., 11 Tab., Tulsa

Catalov, G. A. (1972): An attempt at energy index (EI) analysis of the Upper Anisian. Ladinian and Carnian carbonate rocks in the Teteven Anticlinorium (Bulgaria). Sed. Geol. **8,** 159–175, 5 Figs., 4 Pls., Amsterdam

Chave, K. E. (1964): Skeletal durability and preservation. In: Imbrie, J., Newell, N. D. (eds.): Approaches to Paleoecology, 377–387, New York: Wiley

Czarnecka, E., Gillott, J. E. (1980): Roughness of limestone and quartzite pebbles by the modified Fourier method. J. Sed. Petrol. **50**/3, 857–868, 6 Figs., Tulsa

Dobkins, J., Folk, R. L. (1970): Shape development on Tahiti-nui. J. Sed. Petrol. **40,** 1167–1203, Tulsa

Emerich, G. H., Wobber, F. J. (1963): A rapid visual method for estimating sedimentary parameter. J. Sed. Petrol. **33**/4, 831–843, 4 Figs., 5 Tab., Tulsa

Flemming, N. C. (1965): Form and function of sedimentary particles. J. Sed. Petrol. **35**/2, 381–390, 4 Figs., Tulsa

Folk, R. L. (1962): Spectral subdivision of limestone types. Mem. Amer. Ass. Petrol. Geol. **1,** 62–84, 7 Figs., Tulsa

Folk, R. L., Robles, R. (1964): see 4.2.1.2

Füchtbauer, H. (1970): see 1.4.1

Köster, E. (1964): Granulometrische und morphometrische Meßmethoden an Mineralkörnern, Steinen und sonstigen Stoffen 336 pp., 109 Figs., 68 Tab., Stuttgart: Enke

Krumbein, W. C. (1941 a): Measurement and geological significance of shape and roundness of sedimentary particles. J. Sed. Petrol. **11,** 64–72, Tulsa

Krumbein, W. C. (1941 b): The effects of abrasion on size, shape and roundness of rock particles. J. Geol. **49,** 482–520, Chicago

Krumbein, W. C., Sloss, L. L. (1963): see 1.4.1

Kuenen, Ph. H. (1964): Experimental abrasion: 6. Surface action. Sedimentology **3,** 92–143, 1 Fig., 6 Tab., Amsterdam

Lees, G. (1964): A new method for determining the angularity of particles. Sedimentology **3**/1, 2–21, 11 Figs., Amsterdam

Lüttig, G. (1962): Geröllmorphometrie des Zechsteinkonglomerates im Schacht Rossenray. 1. Fortschr. Geol. Nordrhein-Westf. **6,** 385–390, Krefeld

Maxwell, W. G. H., Day, R. W., Fleming, P. J. G. (1961): see 2.4.1

Mills, H. H. (1979): Downstream rounding of pebbles – a quantitative review. J. Sed. Petrol. **49**/1, 295–302, 1 Fig., Tulsa

Mrazek, J. (1975): Korngestaltsuntersuchungen am Körnerpräparat und im Dünnschliff. Z. angew. Geol. **21,** 280–284, Berlin

Patro, B., Sahu, B. K. (1977): Discriminant analysis of sphericity and roundness data of clastic quartz grains in rivers, beaches and dunes. Sed. Geol. **19,** 301–311, 1 Fig., 1 Tab., Amsterdam

Pettijohn, F. J. (1957): see 1.4.1

Pettijohn, F. J., et al. (1972): see 1.4.1

Pilkey, O. H., Morton, R. W., Lutenauer, J. (1967): The carbonate fraction of beach and dune sands. Sedimentology **8,** 311–327, 4 Tab., 5 Figs., Amsterdam

Plumley et al. (1962): see 6.4

Powers, M. C. (1953): A new roundness scale for sedimentary particles. J. Sed. Petrol. **23,** 117–119, Tulsa

Russel, R. D., Taylor, R. E. (1937): Roundness and shape of Mississippi river sands. J. Geol. **45,** 225–267, Chicago

Sames, C. W. (1969): Morphometric data of some recent pebble associations and their application to ancient deposits. J. Sed. Petrol. **36**/1, 126–142, 6 Figs., Tulsa

Shepard, F. P., Young, R. (1951): Distinguishing between beach and dune sands. J. Sed. Petrol. **31,** 196–214, Tulsa

Smalley, I. J. (1967): The presentation of subjective shape and roundness data. Sedimentology **8,** 35–38, 1 Fig., Amsterdam

Smith, N. S. (1972): Flume experiments on the durability of mud clasts. J. Sed. Petrol. **42**/2, 378–383, 6 Figs., Tulsa

Sneed, E. D., Folk, R. L. (1958): Pebbles in the lower Colorado River, Texas, a study in particle morphogenesis. J. Geol. **66,** 114–150, Chicago

Swinchatt, J. P. (1965): see 2.4.1
Wadell, H. (1933): Sphericity and roundness of rock particles. J. Geol. **41,** 310–331, Chicago
Wadell, H. (1935): Volume, shape and roundness of quartz particles. J. Geol. **43,** 250–280, Chicago
Waskom, J. D. (1958): Roundness as an indicator of environment along the coast of Panhandle, Florida. J. Sed. Petrol. **28**/3, 351–360, 9 Figs., Tulsa
Whalley, W. B. (1972): The description and measurement of sedimentary particles and concept of form. J. Sed. Petrol. **42**/4, 961–965, Tulsa

4.2.3 Fabrics

The general term fabrics, introduced by Sander (1951), includes "textures" as well as "structures". Among the textural characteristics important for microfacies analyses are, in addition to grain sizes, shapes, and grain roundness (see Sects. 4.2.1, 4.2.2) particle-to-particle relationships expressed in terms of packing and grain orientation as well as open-space structures and geopetal fabrics. Structural characteristics appear in thin-sections under the microscope as lamination, nodular structures, as genetically different discontinuities, and as sedimentary fissure fillings.

4.2.3.1 Orientation and Packing of Particles (Plate 11)

Current transport produces certain orientation patterns even in carbonate particles which are interesting for the discussion of paleocurrents (example in Stauffer, 1962). Generally, in conditions of constant current energies, a smaller carbonate particle will be transported in a manner different from a quartz grain available at the same spot, because quartz has a specific gravity of 2.65; calcite and aragonite, on the other hand, have a density of 2.72 and 2.95, respectively. However, as the density of carbonate grains is reduced by high intraparticle porosity and by gas formation associated with the decay of organic substances, carbonate particles may exhibit lower densities than water in certain circumstances and will be transported, then, in suspension.

Plate 11. Sedimentary Structures; Orientation of Particles

1. Cross-bedded calcareous sandstone, part of a turbidite sequence, overlain by a densely packed limestone conglomerate, composed of polymict clasts. Clasts contain shallow-water as well as slope elements, indicating an intrabasinal origin of the conglomerate. Late Devonian: Ferragut Vell, Menorca

2. Tempestite, storm deposit. The sediment consists of reworked bioclasts of various sizes, deposited within distinct layers. These layers are separated by laminated lime mud. Along with some overturned bioclasts there are different kinds of intraclasts, indicating a rapid erosion. Upper Triassic: Panormide Platform, northwestern Sicily

3. Imbrication structures, indicating current energy. Direction of currents is from right to left. Particles are micritized shells, fusulinids, and echinoderms. Open shelf environment. Lower Permian (bedded Trogkofel Limestone): Forni Avoltri, Carnia, Italy

Scale for all Figures 10 mm

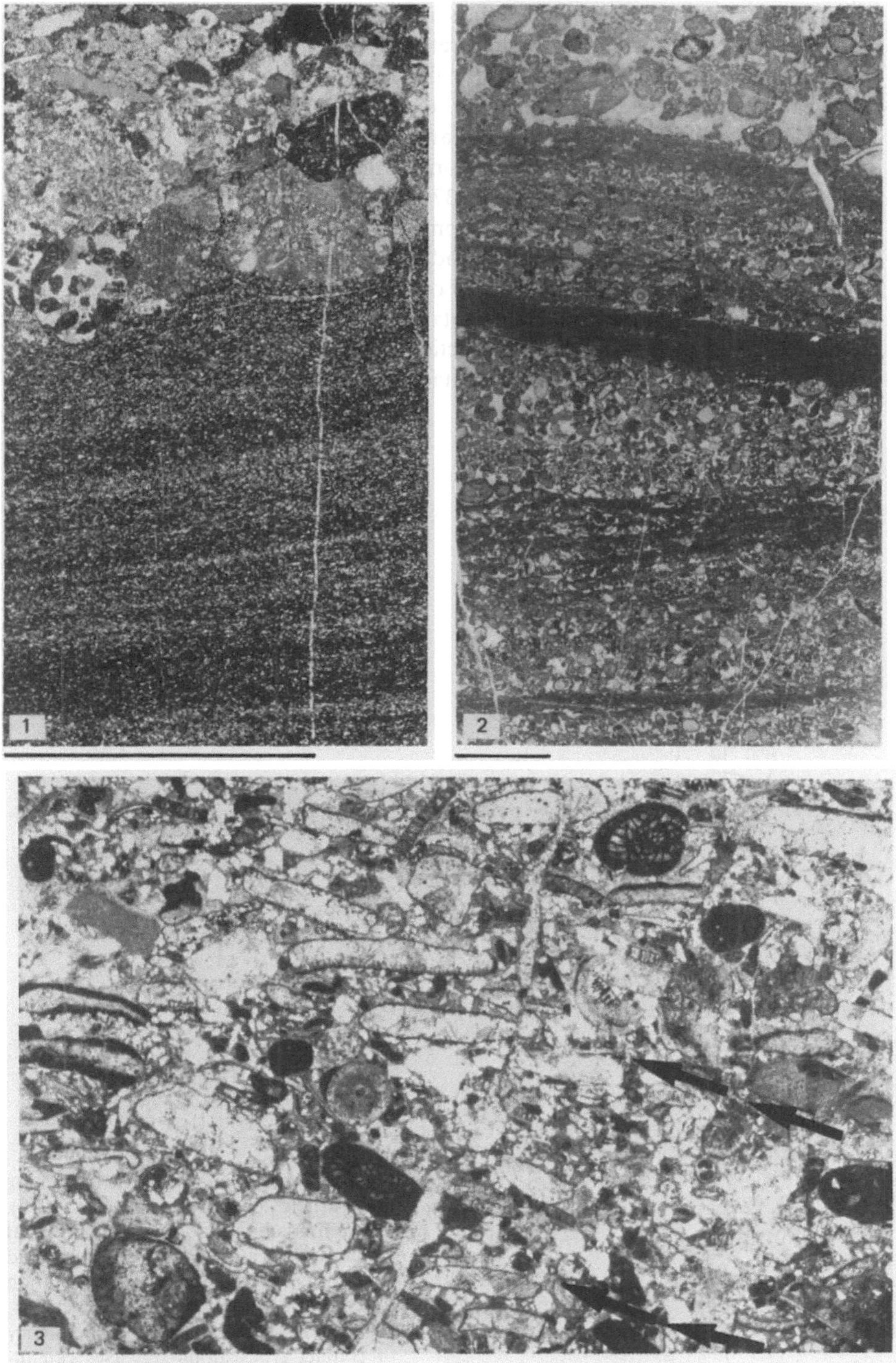
1
2
3

a) *Orientation*: Orientation patterns which can be followed over a larger area are important for microfacies analyses. Basically, the same difficulties in interpreting these patterns occur as in the interpretation of quartz grain orientation (see Vossmerbäumer, 1976). Up until now only a few studies have been available in which calcareous particles have been used to determine paleocurrents (e.g., Stauffer, 1962). Bioclasts (crinoid columnals, Schwarzacher, 1961; fusulinids, King, 1948) and intraclasts are particularly suitable for orientation measurements. Details can be found in Potter and Pettijohn (1963: 37–40). Lindholm (1980) interpreted the orientation pattern of the long axes of elongate peritidal intraclasts as an alignment transverse to flow, thus indicating the trend of the shore-line.

b) *Packing*: Packing is the spacing or density pattern of the particles. This characteristic, which provides important information about porosity and permeability can be described in thin-sections by Taylor's method (1950), in which the frequency and kind of particle contacts are described (see Fig. 34). Various

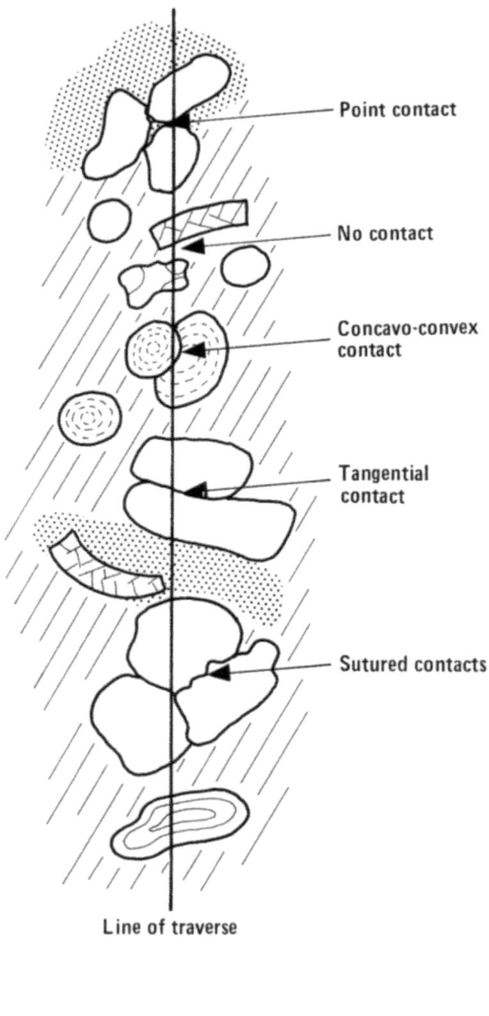

Fig. 34. Description of particle packing and particle contacts after Taylor (1950). Differentiated according to tangential contacts, concavo-convex contacts, and sutured contacts caused by dissolution (often with microstylolites). When expressed in percentages, the frequency of contact types along traverses can indicate the type of packing

packing proximities (Kahn 1956; Pettijohn et al. 1972) provide reference values for individual samples; see Sect. 3.6.

Another measure of packing is the packing index (German – "Offenheitsgrad", Sander, 1951), which is the quotient of the total number of particles and the number of grain-to-grain contacts along a traverse in a thin-section.

The identification of packing types is essential for understanding the limestone classifications (see 6.2.2) of Dunham (1962) as well as for interpreting compaction structures (see 3.6). Measurement errors were discussed by Harrell (1981).

4.2.3.2 Graded Bedding (Plate 52/2)

Bio- and lithoclastic grain- and rudstones often show sedimentary units grading from the coarsest material at the base to the finest at the top. These units may vary from a few mm to 1 m and more in thickness. Graded beds are most commonly found at the base of turbidites formed within deeper depositional areas, but it should be taken into account that waning currents (e.g. ebb tides), the activity of burrowing organisms (Rhoads and Stanley, 1965), beds of seagrass (Wanless, 1981) or the effects of hurricanes may also produce this fabric.

Reverse (or inverse) grading with the finest material at the base and the coarsest at the top is not uncommon; therefore, some caution is necessary when using graded bedding as a top-and-bottom criterion. Reverse order of grading is a typical feature of high-concentration gravity flow deposits (grain- and debris-flow deposits), but is found also in beach or in base-surge deposits.

4.2.3.3 Lamination (Plate 14/1–3)

The existence of lamination assumes that the size or the kind of carbonate materials deposited alternate over a period of time. This alternation can be caused by variations in the kinds and frequencies of sessile organisms binding sediment, or by fluctuating water energies. Recent laminated carbonate sediments originate particularly in environments with little or no bioturbation and with low species diversities. Such areas are located in shallow seas, especially in tidal and supratidal zones. The following types can be distinguished according to the arrangement of the lamination:

1. *Laminites* (Plate 14/2), in which more or less uniformly structured laminations are separated by even thinner parallel laminae (e.g. bituminous films in the Hauptdolomit of the Northern Alpine Triassic; interruptions in growth of algal mats).

2. *Rhythmites,* in which cyclic alternation of the laminae sequences can be observed.

"Planar-laminar" bedding (PLF fabrics) and irregular laminoid fabrics (e.g. fenestral fabrics, see 4.2.3.5) can be distinguished from one another by the uniformity of the lamination.

Because the formation of lamination fabrics in carbonates is very often associated with the reciprocal relationship between non-skeletal algal/bacterial communities and sedimentation, a general survey based on Recent observations of *cryptalgal fabrics* (Aitken, 1967) and a discussion of modes of origin (after Monty, 1976) is given in Tables 22 and 23 (p. 224, 225). See Plate 14/3.

4.2.3.4 Bioturbation Fabrics (Plates 12; 15/3)

Various burrowing organisms are active in the homogenization of fine-grained carbonate sediments deposited in intertidal and subtidal environments. Endobiontic mollusks, crustaceans, and annelids as well as rhizoids of plants (e.g., *Thalassia*) produce irregularly formed structures and also contribute to an alteration of the original grain-size distribution, because finer sedimentary particles are removed. The intensity of bioturbation is influenced by the distributional patterns, which are controlled by the stability of the substrate, and by the predominance of endobiota in shallow-water environments.

The distribution of bioturbation and shelly fauna can be used to describe a biofacies model for sedimentation in euxinic basins with a tripartite layering in their water columns (Byers, 1979): a mixed surface layer, well oxygenated, about 50 m deep, represents an aerobic environment characterized by shelly fauna and bioturbation. Between 50 and 150 m oxygen decreases rapidly with depth; in this dysaerobic zone bioturbation persists but shelly epifauna is missing. In the deepest stagnant zone below 150 m no bioturbation and no shelly fauna occur owing to anaerobic conditions; fine muds settle from suspension and produce a laminated nonfossiliferous sediment.

Bioturbation in micritic limestones is often characterized by a mottled structure and differences in color and grain sizes.

4.2.3.5 Open-Space Structures and Fenestral Fabrics (Plates 13; 14; 45/2)

Many micritic, pelmicritic, and arenitic limestones exhibit evenly or irregularly distributed voids of synsedimentary to post-sedimentary origin. These voids are filled with mechanically deposited internal sediments (e.g., micrite, siltite, or peloids) and/or chemically deposited cement (sparite). Such voids – not to be confused with sparry interparticle interstices or solution pores (see Krebs, 1969;

Plate 12. Bioturbation and Boring in Limestones

1. Burrowed, bioclastic wackestone/packstone. Bioturbation is indicated by the arrangement of bioclasts in a circular pattern and by a different packing of the particles. Skeletal grains are trilobite fragments and echinoderms. Ordovician: Öland island, Sweden

2. Strongly bored coral. Borings with remains of lithophagous boring pelecypods. Upper Jurassic (Kimmeridgian): Lower Saxony, Germany

3. Punctate brachiopod shell with sediment filled boring. Fine-grained peloidal sediment with trilobite fragments. Ordovician: Öland island, Sweden

Scale for all Figures 2 mm

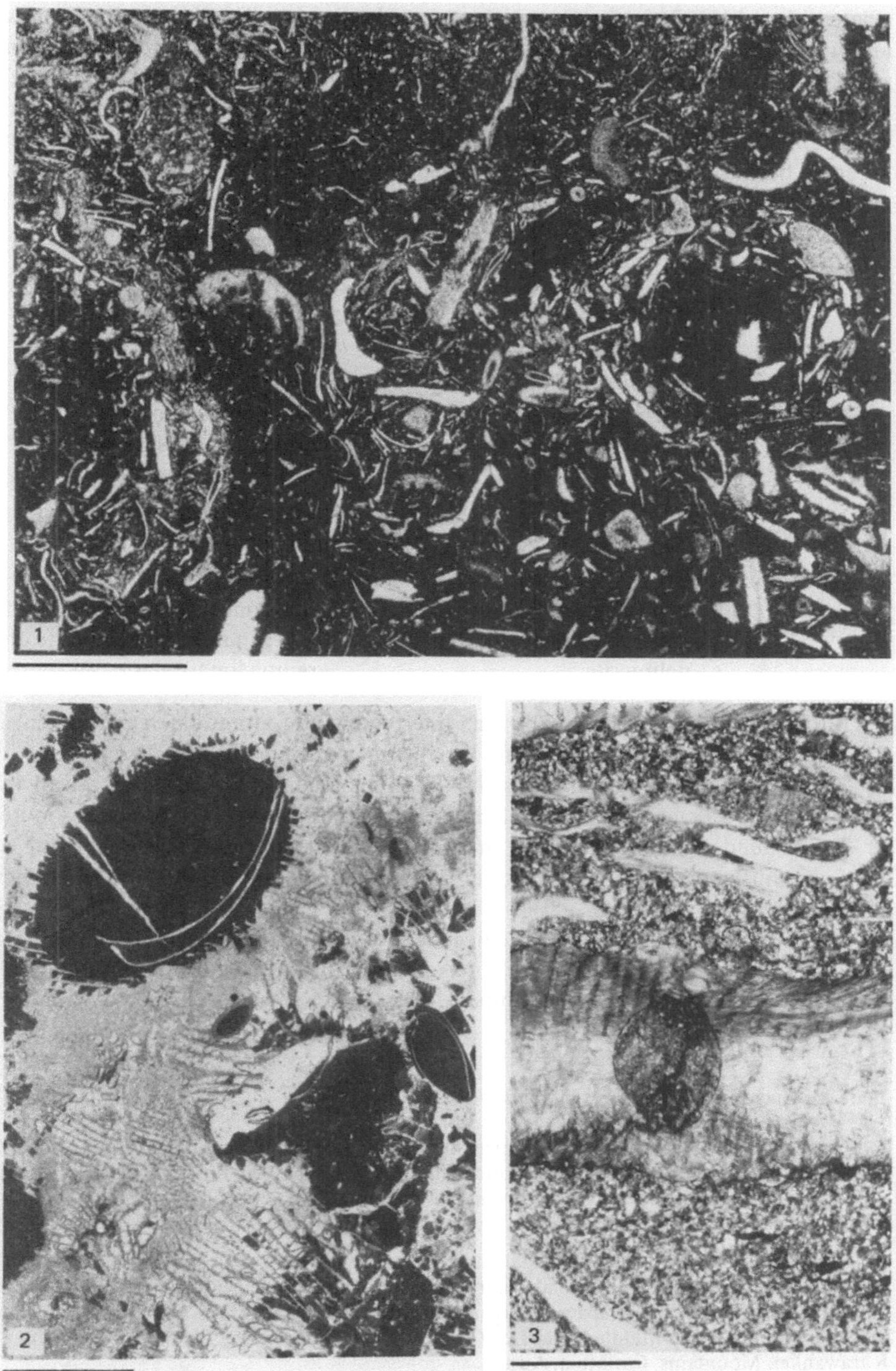
1
2
3

Scoffin, 1972) – were first described by Dupont (1882) with the name *Stromatactis* (Plates 13/1, 14/1–2) in the Devonian and in the Belgian Waulsortian limestones, and were interpreted as hydrozoans. Ham (1954) introduced the name *birdseyes* for small sparry-filled voids in the Ordovician limestones of Oklahoma. Folk (1959) suggested the name dismicrite (=disturbed micrite) for micrites which contain sparry-filled voids of various origin. This name, however, has generally not been accepted because it is far too broadly defined.

It is recommended to distinguish between birdseyes (German: "Sparitaugen") and Stromatactis (see Wolf, 1965; Deelman, 1972; Bechstädt, 1974) using the following criteria:

	Birdseyes	Stromatactis
Shape	Spherical, oval, elongated, irregular	Maze-like; relatively flat base; irregular digitate roof
Void Filling	Sparite, often two generations, no internal sediment	(Often planar) internal sediment at the bottom and sparite at the top
Size	Small, often 1–3 mm	Larger than birdseyes, several mm to cm
Orientation	Irregularly distributed or parallel to the bedding plane; in micrite, microsparite, or pelmicrite.	Long planar areas of sparite often concordant to bedding plane; in micrite, pelmicrite or arenitic limestones

Birdseyes occur preferentially in intertidal and supratidal environments, Stromatactis in the intertidal zone and in subtidal environments.

Misuse of the term Stromatactis should be avoided. Stromatactis is basically a mass of sparry calcite embedded in a lithified carbonate mud. The structure is more or less extended subhorizontally, has a finely digitate upper surface and a predominantly smooth undulose lower surface, and is linked with other similar masses in a reticular pattern.

The more common processes active in the genesis of voids are compiled in Table 21. Of special interest for the interpretation of the growth of Paleozoic mud mounds is the suggestion by Bathurst (1980), that the origin is linked with the

Plate 13. Open-Space Structures in Shallow-Water Limestones

1. Fenestral fabric, Stromatactis type. Open spaces are filled by micrograined, silty carbonate and drusy cements, forming geopetal fabrics. Internal filling may be due to meteoric (vadose) processes; this is indicated by the occurrence of common "black pebbles" (see Sect. 4.1.3.8). Upper Triassic (Hauptdolomit): Gaissau near Hallein, Salzburg, Austria

2. Open-space structure within algal mats. Two types can be recognized: primary, interlamellar interspaces; and secondary, irregularly distributed voids, caused by solution and enlargement of primary interspaces. Larger voids are filled with fibrous and blocky cements (cements A and B). Middle Triassic (Wetterstein Limestone, Ladinian): Hoch-Obir, Karawanken Mountains, Carinthia, Austria

Scale for both Figures 10 mm

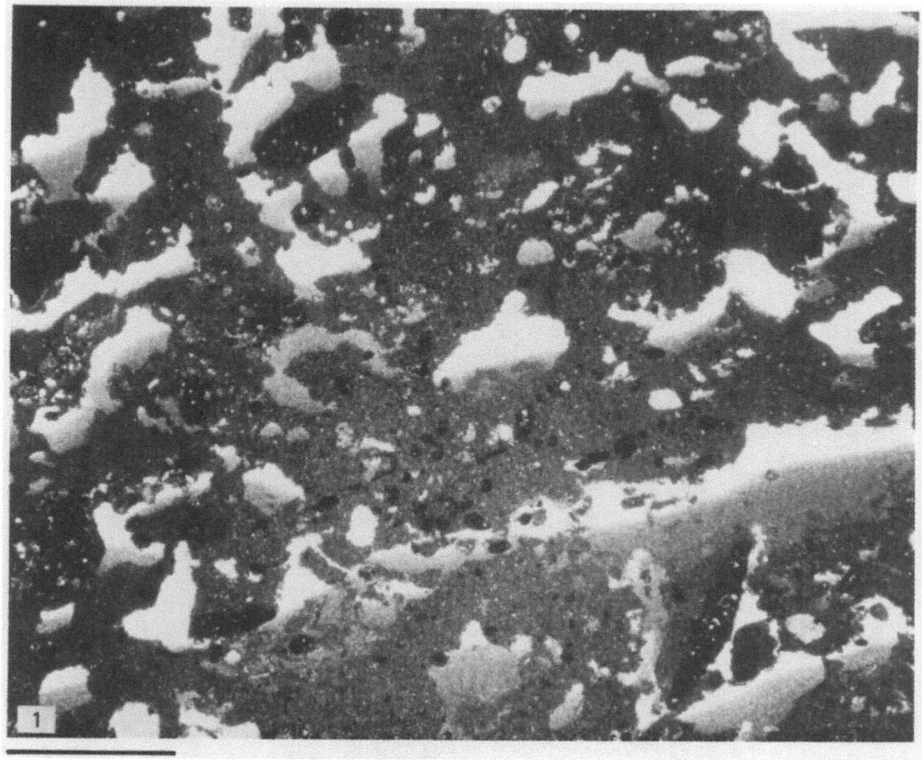
1

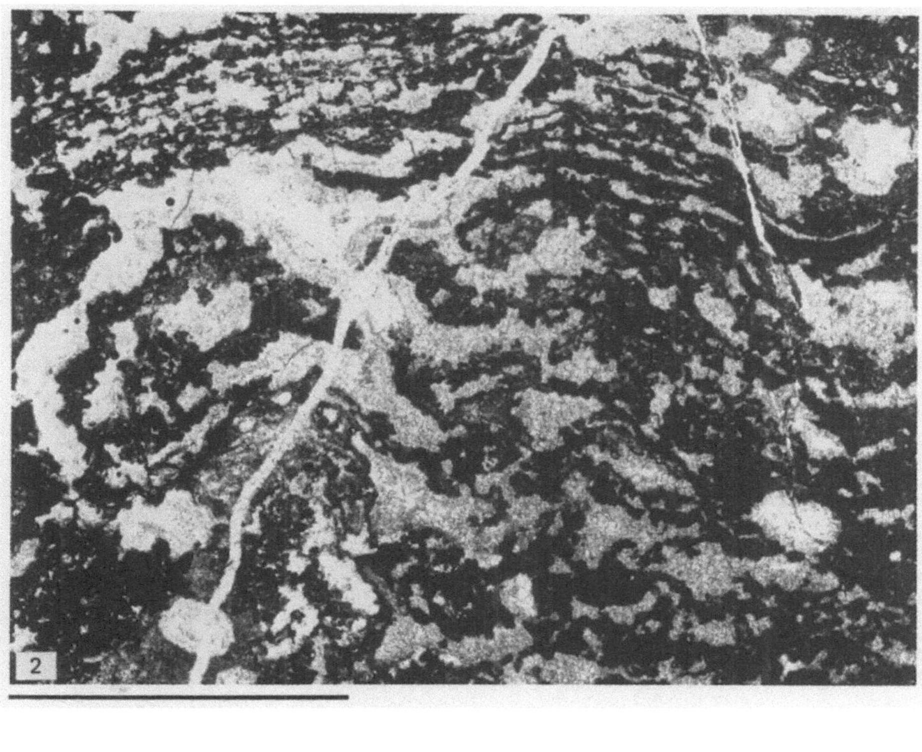
2

Table 21. Modes of origin of open-space structures in limestones, known as birdseyes and Stromatactis. Recent modes of origin are marked with an*. Wolf (1965) and Zorn (1971) call them "orthosparitaugen" (primary void fillings) and "pseudosparitaugen" (postsedimentary to epigenetic patches of recrystallization). Modes of origin 1, 5, and 6 are probably for birdseyes concordant to the bedding plane. Irregularly distributed structures can originate by modes 2, 3, and 8

Origin of Void	Birdseyes		Stromatactis	
	Process	Author	Process	Author
Void retains the shape of organism no longer preserved	1. Interspaces within algal mats*	Ham (1952) Philcox (1964) Deelman (1972)	10. Recrystallized algae	Prentice (1950)
			11. Recrystallized reef builders, e.g., bryozoans	Lowenstam (1950); Carozzi and Textoris (1963)
			12. Cavity formed after decay of soft-bodied organism in the sediment	Bathurst (1959) Ott and Parks (1963)
Indirect organic interaction	2. Gas bubbles of decaying organisms*	Korn (1932) Hodgson (1958) Cloud (1960) Shinn (1968b)		
	3. Bioturbation by endobiontic organisms	Fruth and Scherreiks (1975)	13. Bioturbation*	Tucker (1973); Bechstädt and Mostler (1974)
			14. Burrows, e.g., of crustaceans*	Shinn (1968a)
	4. Cavities supported by algae, erosional forms of receding tidal waters*	Wolf (1965)		

Without organic interaction	5. Shrinkage pores during desiccation in tidal zones	Fischer (1964)		
	6. Inclusion of air during subaerial exposure and reinundation in tidal zones*	Burri (1970) Deelman (1972)		
	7. Leaching of anhydrite*			
	8. Different behaviors of capillary and non-capillary sediment pores in comparison to water	Deelman (1972)	15. Slumping in mud, e.g., on reef flanks	Schwarzacher (1961)
			16. Different compaction during dehydration of carbonates poor in clay *; water escape structures	Heckel (1972) Bechstädt (1974) Lowe (1975)
			17. Subaerial solution, formation of void spaces, infilling with vadose silt	Semeniuk (1971)
			18. Cement (and sediment) filling of a system of cavities which developed between submarine-cemented crusts on a carbonate mud	Bathurst ((1980)
	9. Late diagenetic, selective recrystallization ("Pseudosparitaugen")	Wolf (1965) Zorn (1971)	19. Late diagenetic, selective recrystallization	Wolf (1965)
			20. Pressure-solution	Logan and Semeniuk (1975) Discussion: Bathurst (1980)

LAMINAR FENESTRAL FABRICS

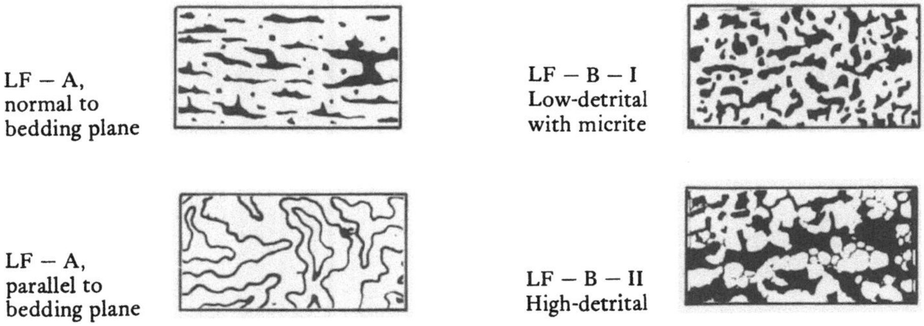

Fig. 35. LF-A and LF-B fabrics. *Dark areas* = sparite in the voids; *white areas* = micritic matrix. After Müller Jungbluth and Toschek (1969)

cementation and subsequent deformation and reworking of a succession of submarine crusts; this results in the development of cavities in the intercalated, less cemented carbonate muds.

It is to be expected that the voids which have occurred in shallow subtidal and intertidal environments will be reworked later in vadose conditions. Spar-filled cavities may also be due to percolation of gas connected with the intercalation of airborne volcanic dust (tephra); see Lindström (1979).

Tebbutt et al. (1965) suggested the name *fenestra* or window for these voids: ... "a primary or penecontemporaneous gap in rock framework, larger than grain-supported interstices." These gaps represent voids which are still open or filled with sediment or cement, and which have no apparent support in the framework of primary grains composing the sediment. The orientation of fenestrae can be concordant to stratification, cross-bedded, or irregularly formed. The term *laminoid-fenestral fabric* (German – "LF-Gefüge") corresponds to the elongate

Plate 14. Open-Space Structures

1. Laminoid fenestral fabric. LF-A fabric after Tebbutt et al. (1965). Open-space structures – *Stromatactis* – occur in pisoid layers sorted according to particle sizes; they are characterized by horizontal planar cavities. The flat bottoms of the cavities are generally filled with internal micritic and pisoidic sediment; the irregular digitate upper surfaces, with sparry calcite cement. Late Triassic (Dachstein limestone, ?Upper Carnian): Neśza quarry, NW Budapest, Hungary. Scale is 10 mm

2. Algal laminites with mm micritic and pelmicritic layers, interbedded with *Stromatactis* cavities. In the lower right hand corner of the picture are sparry-filled vertical "threads", which are probably due to Cyanophytes. Late Triassic (Hauptdolomit, Norian): Gaissau, Salzburg, Austria. Scale is 10 mm

3. Cryptalgal fabrics with voids which are due to shrinkage cracks formed in the supratidal zone. Notice at the top the broken and reworked pisoids and laminar caliche crusts (formation in meteoric environment). Detailed description in Catalano et al. (1974). Late Triassic: Capo di Rama, Monti di Palermo, Sicily. Scale is 10 mm

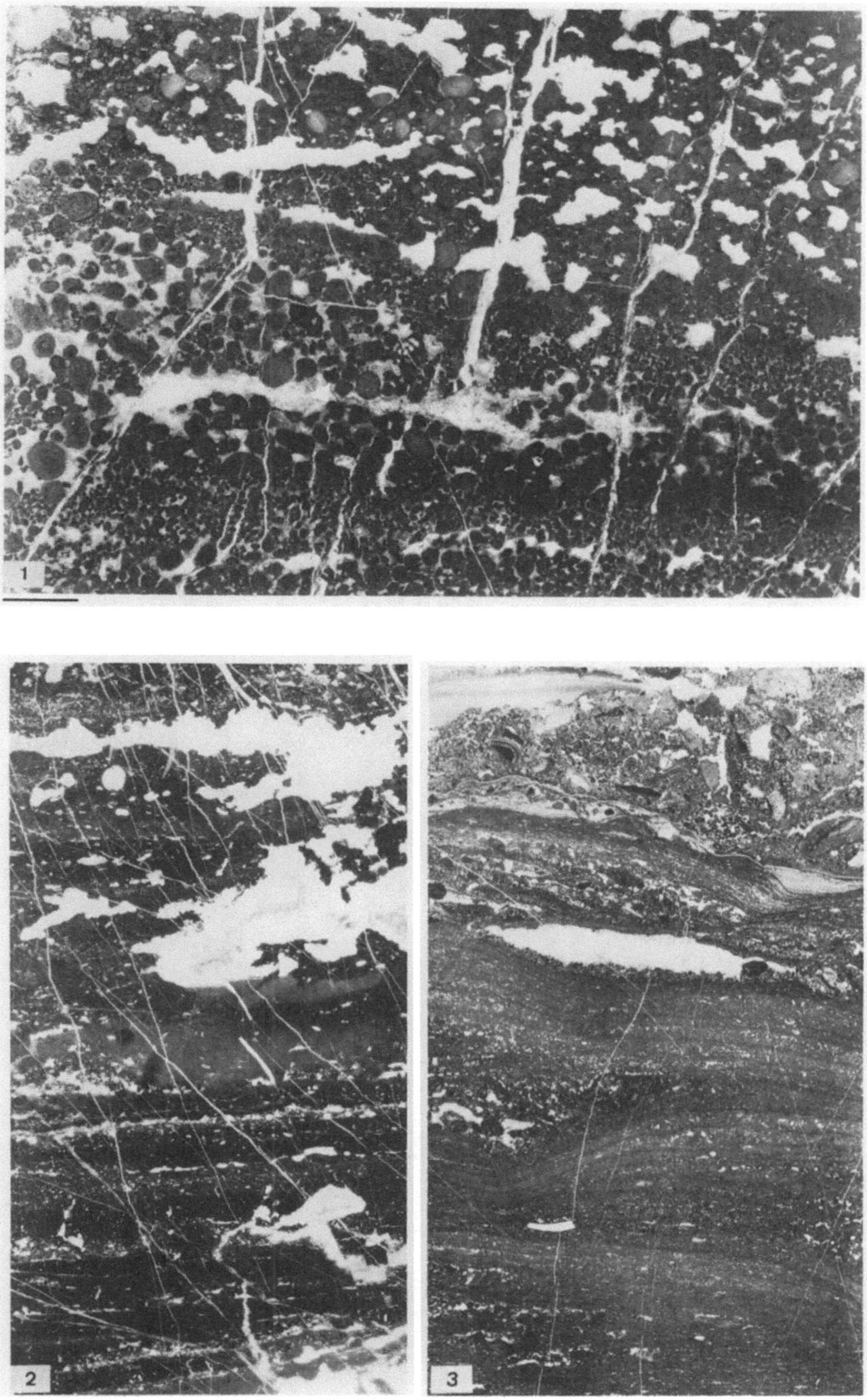

Table 22. Nomenclature of "cryptalgal fabrics" (Monty, 1976)

Cryptalgal fabrics (= biosedimentary structures which are a result of the activity of non-calcified algae and bacteria, which bind sediment and/or precipitate carbonate)	Layered fabrics	*Laminated fabrics* (distinct lamination with discontinuities)	
		Laminoid fabrics (without laminae; more or less horizontal structures, e.g. fenestra, produce an irregular layered fabric)	*Laminoid-fenestral fabric*
			Laminoid-boundstone fabric
			Lenticular-laminoid fabric
	Non-layered fabrics	*Thrombolitic fabrics* (Aitken, 1967). cm-sized micritic and sparitic areas together with algal filaments	
		Massive fabrics (no preferred orientation of particles)	
		Radial fabrics (radially growing algal filaments, e.g. recent *Scytonema*, Ancient *Ortonella*)	

Table 23. Modes of origin of "cryptalgal fabrics" after Recent observations by Monty (1976)

Layered fabrics

1. Laminae formation due to diurnal differences in algal growth
2. Alternation between preferred vertical and preferred horizontal growth within the thalli
3. Various growth rates of algae species in an algal mat
4. Algal growth inside the algal mats
5. Periodic calcification of Cyanophyceae mats
6. Precipitation of carbonate crystals with different habits
7. Periodic enrichment of elements (e.g., Fe, Mn)
8. Periodic influx of detrital particles
9. Periodic cementation
10. Different diagenetic behaviors of algal mats (e.g., preferential formation of dolomite)
11. Alignment of detrital particles and agglutination by algae
12. Separation of superposed algal mats due to gas bubble formation or shrinkage and dessication; formation of fenestral fabrics

Non-layered fabrics

Thrombolitic fabrics

1. Oxidation of dead colonies of algae
2. Intergrowing of colonies of algae
3. Internal dissolution of algal colonies

Massive fabrics

1. Trapping and agglutination of detrital particles
2. Predominance of sediment accumulation as opposed to algal activity
3. Algae agglutinate sediment deposited earlier
4. Mats of diatoms
5. Diagenetic phenomena

stratification mentioned above: thin carbonate sand or silt particles are separated by long, horizontal, irregularly formed fenestrae (see Plate 14/1). Two types can be differentiated (Fig. 35):

a) LF-A fabrics (Plate 14/1): with distinct laminoid fenestral fabrics (riddled with holes, like Swiss cheese)
b) LF-B fabrics: with irregular laminoidal fabrics. These are subdivided into
   - LF-B I: "low-detrital"; voids embedded in micrites
   - LF-B II: "high-detrital"; voids embedded in a sediment with numerous particles

LF-A and LF-B structures are due to cryptalgal fabrics, to shrinkage, or to the formation of gas and water bubbles in the sediment (see Tab. 23). The terms birdseyes and Stromatactis should serve to describe the shape of the voids; the name *Loferite* (Fischer, 1964) can be used to describe carbonate rocks abundant in fenestral fabrics. Mineralized lichens and roots of plants produce structures in caliche crusts which may be confused with marine open-space structures (Klappa, 1978). The name *keystone vugs* is used for mm- to cm-sized open cavities within upper intertidal calcarenites produced by the degassing of the sediments (Emery, 1945).

Open-space structures in reef limestones: Fenestral fabrics should not be confused with voids of different size found within the framework of Recent and fossil reefs (Schwarz, 1927; Newell, 1955; Krebs, 1971; Chevalier, 1973; Scoffin and Garret, 1974; Bonem, 1977; Mirsal, 1978; Brandner, 1978). These voids result from the selective, non-synchronous growth of the reef-building organisms, from the activity of boring organisms, from episodic storms, and also from synsedimentary and postsedimentary solution effects. The voids are often filled by several generations of marginal fibrous calcite (Cement A), followed by white blocky calcite (Cement B) or central infillings of fine-grained micritic or silty sediments. The fibrous cements can be formed during rhythmically fluctuating salinity conditions (see Sect. 3.3.3), whereas the blocky cement is formed in deep burial conditions under the influence of meteoric waters.

4.2.3.6 Nodular Structures (Plate 53/1, 2)

Nodular limestones are very common in some deeper-water carbonate sequences (e.g., Garrison and Fischer, 1969; Jenkyns, 1974; Tucker, 1974; Kukal, 1975). The theories about the origin of nodular limestones include:

- Differential submarine cementation and subsequent resedimentation by gravity flow processes (Hopkins, 1977; Seyfried, 1980; Weber, 1965). Alternation of periods of sedimentation (associated with the formation of soft bottoms) and of non-sedimentation (associated with the formation of hard bottoms and solution hard bottoms). A slumping of these variously lithified layers may produce the nodular texture (Seyfried, 1980).
- In situ submarine cementation during early diagenesis (Garrison and Fischer, 1969; Jenkyns, 1974)
- Subsolution during deposition at the sedimentary interface (Hollmann, 1962)
- Late burial diagenesis (Dvorak, 1972; Garrison and Kennedy, 1977), sometimes in connection with pressure-solution
- Combination of early lithification and burial diagenesis (compaction and pressure-solution; Tucker, 1974)
- Structural boudinage and differential compaction (Born, 1921; Nichols, 1966) as well as a mechanical rearrangement of particles in sediment mixtures (Deelman, 1974)
- Concretional growth due to diagenetic differentiation of carbonate and argillaceous material (Hildebrand, 1929; Schindewolf, 1925; Kukal, 1975)
- Microbial decomposition of organic matter (Gründer and Rösler, 1965)

There are relatively few reports of nodular calcareous units from modern deeper-water environments (e.g., Müller and Fabricius, 1974: high-Mg calcite nodules in micritic matrix from depths of 2680 to 3336 m in the Mediterranean Sea; Mullins et al., 1980: Bahamian slopes, 333–1280 m). The Bahamian nodules appear to have been formed essentially in situ by submarine cementation; a combination of bottom currents, burrowing and chemical processes are the major controls in the formation of these nodules.

4.2.3.7 Discontinuity Structures (Plate 15/1–5)

Often bedding surfaces can be observed in thin-sections and hand specimens which represent interruptions in normal sedimentation or periods of non-deposition; these are called discontinuities, erosion or corrosion surfaces, whereby the nomenclature of Heim (1924, 1958) illustrates the most probable mode of genesis. Discontinuities are related to the following processes:

1. *Subsolution:* Carbonate dissolution at the bottom of the ocean in an environment of cool ocean currents and, commonly, at great depths (e.g., Hollmann, 1962, 1964). W. Schlager (1974) and other authors see in subsolution not a purely inorganic process but rather an interaction of different cementation processes, organic corrosion, and winnowing of unconsolidated sediment during times of non-deposition.

2. *Omission:* Non-deposition or a break in sedimentation. Omission surfaces are often called *hardgrounds,* because during times of non-deposition increased cementation can lead to lithification of the ocean floor (general survey in Fürsich and Wendt, 1976).

In many cases a combined process of carbonate dissolution and removal of fresh sediment and dissolution residues by currents may account for the formation of discontinuity surfaces seen in the geological record.

Hardgrounds are formed in a wide variety of water depths. The majority lie under a few tens of meters, others have been found in deeper waters, down to 3000 m. Paleogeographic setting indicates that hardgrounds are usually regressive phenomena, typically formed on seamounts or submerged rises in the pelagic environment or on positive carbonate shelf areas (Bromley, 1978; Dravies, 1979).

Hardgrounds may represent enormously long intervals of interrupted sedimentation; intervals of between 25,000 and 125,000 years have been postulated as the range of the hardground formation for the Ordovician orthoceratite limestone of Sweden (Lindström, 1979).

Hardgrounds are quite common in chalk sediments but often somewhat blurred in limestone sequences. Deformed components may serve as signals for the presence of hardgrounds or emersion surfaces, because overburden stresses may be locally amplified, thus exceeding the limit required to cause deformation of components earlier than usual. Examples are described by Sakar et al. (1980).

3. *Dereption:* Removal by bottom currents of sediment deposited in the ocean depths. Relatively slow bottom currents are responsible for keeping the sea bottom free of fresh deposits, as shown in a Recent oceanic hardground in the Eastern Pacific (Malfait and Van Andel, 1980).

4. *Condensation:* Sedimentation lag. Thin carbonate banks originate which may contain several paleontological zones or "ghost fauna."

Characteristics of discontinuities (hardgrounds) in thin-sections are:

1. Conspicuous differences in color in micrite zones running parallel to the bedding; caused by Fe and Mn enrichment (thin seam or thicker crusts),
2. Fe-Mn crusts, sometimes with sessile micro-organisms (e.g., "foraminifera reefs"; see Plates 15/1 and 15/5; serpulids),

Plate 15. Microfacies Features of Deeper Marine Limestones ►

1. Discontinuities. The fragments of echinoderms in the bottom half of the picture have been dissolved on a discontinuity plane formed as a hardground and subsequently truncated. Above these are "stromatolitic limonite crusts" whose genesis is uncertain (see Section 4.2.4.5). The top half of the picture shows (red) biomicrite with sponge spicules, shell fragments, and the cross-section of an echinoid spine. Lower Lias (Adnet limestone): Leis quarry near Adnet, Salzburg, Austria. Scale is 2 mm

2. Subsolution fragment. Sedimentary particles (red biomicrite) dissolved at the bottom of the sea form an open fabric whose voids are filled with internal sediment (bottom) and chemically deposited sparite (top). Geopetal fabrics appear as graded grains –1 and –2 "sand traps" (German "Sandfang" = sand-sized particle at the bottom of a small cavity; see Section 4.2.3.9). Lias (red nodular limestone): Kiefer quarry, Adnet, Salzburg, Austria. Scale is 2 mm

3. Discontinuities? The fine arenitic biomicrites in the lower half of the picture are covered by dark red argillaceous micrites, which penetrate deep into the microrelief of the substrate. Because individual skeletal grains have penetrated into the lower region as well and the whorled alignment of the shell fragments –1 indicates bioturbation, the new deposition of sediment probably occurred without the formation of a hardground. Accumulations of glauconites in the biomicrite –2 indicate breaks in sedimentation. Lower Ordovician (Limbata limestone, Arenigian): Gettlinge quarry, Öland, Sweden. Scale is 2 mm

4. Limonitic cauliflower structure in a solution cavity below the discontinuity plane. The thin-section displays three types of sediment: –1 biomicrite with foraminifera (top left), echinoid spine (below that) and shell fragment, –2 dark red argillaceous micrite with a thin basal manganese crust, and –3 very fine arenitic biomicrite with ostracods and filaments. Cauliflower structures originate by migration of colloidal Fe-solution. Late Triassic (Hallstatt limestone, Norian): Hallstatt, Upper Austria. Scale is 2 mm

5. Fe/Mn crusts with a column of superimposed sessile foraminifera (*Tolypammina?;* "foraminifera reefs" after Wendt, 1969). The foraminifera –1 have grown on thin limonite crusts –2. Sessile arenaceous foraminifera, not dependent on the water depth or the submarine topography, are abundant in Recent manganese nodules and crusts (see Wendt, 1974). Late Triassic (Hallstatt limestone, Carnian): Feuerkogel near Bad Aussee, Styria, Austria. Scale is 2 mm

6. Filaments. The very thin calcitic skeletal grains, which often occur abundantly, probably represent larval stages of mollusks (see Section 5.3.13). Limestones with filaments can be classified as deposits in subtidal and bathyal environments. Late Triassic (Pötschen limestone, Norian): Pötschen pass, ESE Bad Goisern, Upper Austria. Scale is 2 mm

7. Submarine fissure fillings. In the picture from right to left can be seen: –1 "source rock" of the fissures, which is an echinoderm-bearing pelsparite (Upper Rhaetian limestone); –2 red echinoderm biomicrite; –3 red micrite, almost barren of fossils; –4 pseudosparitic calcite vein; –5 echinoderm sparite with densely packed echinoderm fragments, made coarser by formation of rim cement. This is a vertical fissure system repeatedly ruptured during several phases and filled with sediments of various ages; it can be explained by the destruction of the Late Triassic platform carbonates and the subsidence of the Jurassic sedimentation area (see Section 4.2.3.8). Triassic/Lias (fissure filling Lower Lias): Kirchen quarry, Adnet, Salzburg, Austria. Scale is 2.5 mm

3. Irregular serrate sutures running parallel to the bedding planes; the sutures are caused by karren consisting of solution pits of varying shapes and sizes (millimeters to meters),
4. Truncated calcareous bioclasts (e.g., ammonites) which have been dissolved along a planar surface,
5. Small truncated voids,
6. Accumulations of glauconite below the discontinuity surfaces and glauconite cements upon the surface of the hardgrounds (Plate 15/3)
7. Borings, frequently boring algae (diameter between 0–6 microns),
8. Tiny little pebbles,
9. Subsolution fragments (Plates 15/2),
10. "Stromatolitic" structures (e.g., cauliflower structures), in limonite crusts (Plate 15/4). These structures should not be confused with thin unfossiliferous, enamel-like limonite crusts, which turn to hematite because of diagenetic dehydration and may show shrinkage structures (Seyfried, 1978). Limonite crusts can also be formed during emersion phases!
11. Mixed fauna.

The *karren* development on a discontinuity surface is the result of various chemical reactions, precipitation, and climatic factors. It is influenced by the lithology and texture of the limestone, the slope of the limestone surface (preferential development of "kamenitzas" on surfaces that are horizontal or only slightly inclined; see Sweeting, 1966), and the presence or absence of vegetation and/or till cover (chemical agressivity of the water moving over the surface is increased by these factors). Kamenitzas are solution pits with smooth and rounded sides, several millimeters to several decimeters (or even meters) in diameter, and several millimeters in depth. These structures are known in all climatic regions and have also been found on discontinuity surfaces of Ancient limestones (Kobluk et al., 1977).

"Stromatolitic limonite crusts", as shown in Fig. 36, appear relatively abundantly in discontinuity surfaces. The modes of origin of these crusts have been summarized by Seyfried (1978):

1. Early migration and precipitation of Fe/Mn minerals to the ocean floor.
   According to geochemical studies (Fe, Mn, Co, Ni, Cr, Pb, Ti) by Drittenbass (1979), a hydrogenous origin by inorganic or biochemical fixation of metallic compounds from seawater is possible for Jurassic ferromanganese crusts and nodules exhibiting laminated, stromatolite-like, crenulated structures with common sessile foraminifera.
2. Inorganic formations, primarily by migration of colloidal Fe-solutions which have originated diagenetically (cauliflower structures).
   Another possibility is the rhythmic precipitation of sulphides, later on oxidizing to goethite and being dehydrated to hematite (Lindström, 1979).
3. Limonitization of organic structures (e.g., Jenkyns, 1970; Tietz, 1975) or inorganic Fe-precipitation on top of an organogenic substrate (e.g., algal mats, Fürsich, 1971).

Limonite crusts are very often deposited on a solution hard bottom (Fabricius, 1968), which is characterized by boring algae, overgrowths of sessile foraminifera, serpulids, etc. and usually by a strong mineralization as well.

The question as to whether genuine *stromatolites* occur in condensation horizons is still being debated, especially in regard to Mesozoic deposits. However, evidence seems to be mounting in favor of the third explanation listed above (e.g., Krystyn, 1971; Jenkyns, 1974; Bernoulli and Jenkyns, 1974; Wendt, 1970, 1971):

a) The frequently observed communities of foraminifera, serpulids, bryozoans, and crinoids indicate water depths of about 50–100 m,
b) The lamellibranches which occur in submarine fissures indicate relatively shallow water,
c) The occurrence of masses of *Discohelix* indicates water depths of 100 m or more. Algal activity during the formation of crusts is also plausible at these depths (see 5.3.12.3). Difficulties arise when assumptions are made of water depths ranging from about 200 m to more than 1000 m (Germann, 1971; Garrison and Fischer, 1969).

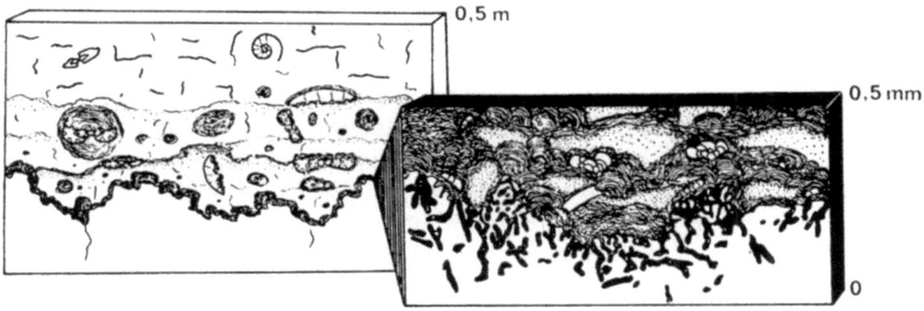

Fig. 36. Discontinuity planes in a Jurassic condensation horizon. The Fe/Mn impregnated zone displays microborings at its base. Above this, "stromatolitic" cauliflower structures have developed, on top of which sessile foraminifera have grown. After Wendt (1970)

"Deep-water stromatolites", formed by cyanophyta and bacteria, can be caused by the existence of submarine hydrothermal activity. Siliceous and ferromanganese stromatolites may be the result of such processes (Montenat et al., 1978; Walter et al., 1972).

The identification of impermeable discontinuity layers like submarine hardgrounds (but also caliche layers or beachrocks) is important in petroleum exploration and recovery. Secondary recovery can be seriously impeded by impermeable layers within a generally permeable reservoir (Videtich and Matthews, 1980).

Another type of discontinuity surface is the subaerially formed *emersion surface,* characterized by karstification (see above), the formation of paleosols, and by the infilling of pockets, channels and lenses by younger sediments. These emersion surfaces often develop at stratigraphic unconformities. The deposition of ores (e.g.,

Pb-Zn-Fe sulfides) is common below and/or along and above emersion surfaces. One such example, interesting in economic terms, is the "Metalliferous" of southwestern Sardinia, a Cambrian carbonate complex, unconformably overlain by Ordovician conglomerates. Other examples are known from the Devonian-Carboniferous disconformity in the Eastern Alps and from Triassic disconformities in the Southern Alps (see Assereto et al., 1976).

4.2.3.8 Fissure Fillings (Neptunian Dikes) (Plates 15/7; 16/1)

Because of tensile stress during the genesis of large-scale horst and graben structures on the margins of large sills and subsiding troughs, fissures originate parallel and oblique to the bedding planes (German – "S- und Q-Spalten"); these fissures are characterized in the Alpine Mediterranean Mesozoic by special accumulations of fossils and by microfacies types frequently different from the surrounding rock. Impressive examples of this have been described in the Triassic and the Jurassic of the Alps and of Sicily by Wendt (1970, 1971), Schöll and Wendt (1971) and Jurgan (1971); in Devonian cephalopod limestones, by Tucker (1973). The two types of fissures differ distinctly from each other in facies, condensed fauna, dimensions and filling mechanisms. Because the dimensions of the fissures extend downwards into the centimeter and millimeter range, reference to these structures is also appropriate for microfacies work (see Plate 15/7). Often these fissures are lined with cement tapestries and several generations of sediment fillings caused by repeated rupture. Typical for the fissure fillings are monotonous concentrations of particular fossils (brachiopods, ammonites, gastropods) and the absence of Fe/Mn crusts and resediments as well as the lack of graded bedding.

The microfacies of the fissure fillings is often characterized by red micrites and biomicrites with foraminifera, small juvenile gastropods and ammonites, as well as crinoids or brachiopods. Lithoclasts, derived from the boundaries of the fissures, may be common.

Plate 16. Facies Interfingering in Thin-Sections

1. Fissure fillings. In normal sediment –1 with filaments, radiolarians, and Globochaete –2, fissures –3 have ruptured and have been completely filled by fine arenitic sediment with juvenile ammonites and pelecypod shells. At the bottom of the picture a cavern –4 filled with pelecypod shells (*Halobia* coquina) has been intersected. If such sedimentary structures are not noticed, the biostratigraphic zonation and the recognition of ghost fauna is very difficult. Examples of these difficulties are found in Hallstatt limestones in the Alpine Mediterranean region (see Krystyn et al., 1971). Late Triassic (Hallstatt limestone): Siriuskogel in Bad Ischl, Upper Austria. Scale is 1 mm

2. Sedimentary deposits of red micrite with ammonites –1, echinoderms –2, foraminifera –3, and Globochaete in gray pelsparitic Dachstein limestones of the fore-reef environment –4. Such "interfingering" indicates the relatively close proximity of shallow water and basinal sediments. Late Triassic (Norian): Southern foot of the Hohes Brett, Göll Massif, Berchtesgaden Alps, Germany. Scale is 1 mm

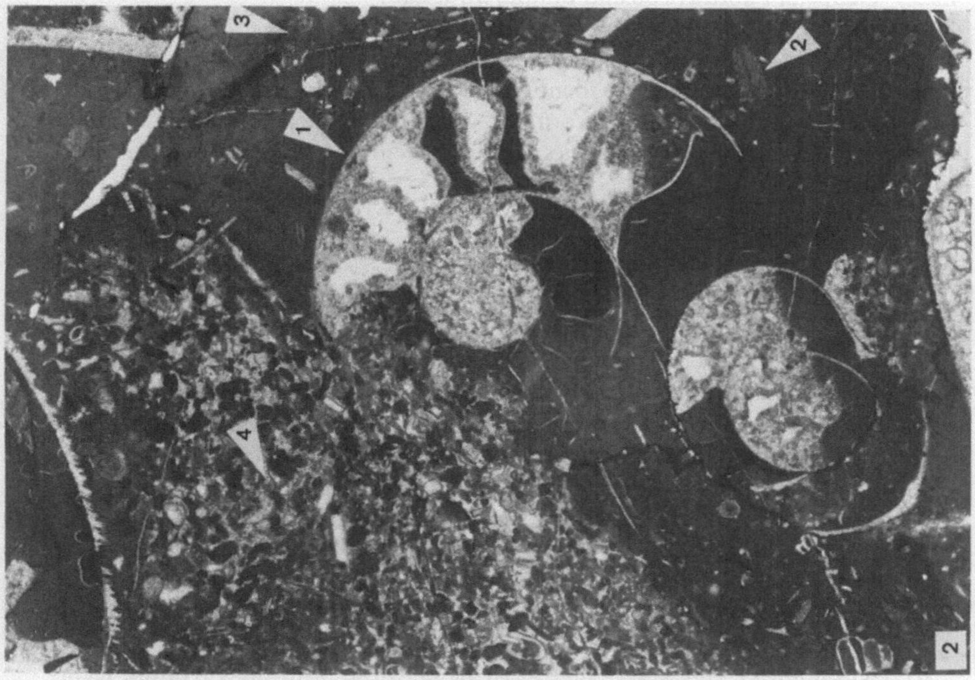

4.2.3.9 Geopetal Fabrics (Top and Bottom Structure) (Plates 5/6; 10; 11/1, 3; 13/1, 2; 15/1; 25/1–3, 6, 7; 35; 52/2)

Some of the criteria essential for identifying top and bottom structures can be very clearly seen in thin-sections. In accordance with the classification of Sander (1936: 56) the following geopetal fabrics, which occur most commonly, can be differentiated as

1. Mechanical deposition
   a) Sand-sized particles trapped in bottom reliefs or at the bottom of small cavities, e.g., in fossils. (German – "Sandfang"; see Plate 15/2).
   b) Grain grading (coarse grains at the bottom, fine grains at the top; see Plate 15/2). Frequently reverse graded bedding arises (fine sediment at the bottom, coarse grains at the top), particularly in limestones characterized by fenestral fabrics. Such fabrics can be caused diagenetically (e.g., caliche sediments) or by the Bernoulli effect (Fisher and Mattinson, 1968). Reverse grading can be attributed either to kinetic sieving (small grains fall through the interstices between larger particles when agitated, thus displacing the larger particles upwards) or to dispersive pressure between grains arising from intergrain impacts.
   c) Polar deposition of particles on bedding planes of discontinuities (see Plate 52/2).
   d) Deposition of thin saucer-shaped shells or valves with the curvature facing upwards.
   e) Truncated cross-bedding (Harbaugh, 1959).
2. Chemical deposition
   a) Cavity with sparry calcite "roof" and bottom filled with internal sediment or sand-sized particles (German – "fossile Wasserwaage"; see Plates 15/2; 19/1; Plate 35). Some restrictions are necessary with regard to the interpretation of top and bottom structures in depositional paleoslopes using intraskeletal geopetal fabrics (Wieszorek, 1979). An early lithification of internal sediments within the shells of ammonites or other mollusks and brachiopods may cause reversed geopetal fabrics if the shells have been redeposited. Redeposition is possible even in a low-energy environment as a consequence of the burrowing activity of organisms (Clifton, 1971).
   b) Open-space structures such as Stromatactis (Plate 13/1; 14/1, 2).
   c) Laminoid fenestral fabrics with garland-like shapes hanging from the roof (see Fig. 35).
   d) Partially dissolved ooids and half-moon ooids, whose inner laminae have collapsed and collected at the bottom.
3. Biogenic deposition
   a) Algal crusts aligned in the same direction (light orientation!).
   b) Sessile organisms in situ (e.g., stromatoporoid colonies, solitary corals).

Examples evaluating geopetal fabrics in carbonate rocks are found in the general surveys by Fiege (1938) and Shrock (1948), as well as in the papers of Schwarzacher (1948), Fuganti (1964), Hückel and Jacobshagen (1962), and Schulz (1967).

Geopetal fabrics, especially the partial infillings of shells, are also widely used as indicators of depositional paleoslopes (e.g., D. Richter, 1968).

4.2.4 References: Limestone Fabrics

4.2.4.1 References: Orientation and Packing of Particles

Allen, J. R. L. (1962): Petrology, origin and deposition of the highest Lower old Red Sandstone of Shropshire, England. J. Sed. Petrol. **32,** 657–697, Tulsa

Clifton, H. E. (1971): Orientation of empty pelecypod shells and shell fragments in quiet water. J. Sed. Petrol. **41**/3, 671–682, Tulsa

Dunham, R. J. (1962): see 6.4

Futterer, E. (1978): Studien über die Einregelung, Anlagerung und Einbettung biogener Hartteile im Strömungskanal. Neues Jb. Geol. Paläont. Abhandlungen **156**/1, 87–131, 8 Figs., 3 Tab., Stuttgart

Harrell, J. (1981): Measurement errors in the thin-section analysis of grain-packing. J. Sed. Petrol. **51**/2, 674–676, 3 Figs., Tulsa

Kahn, J. S. (1956): The analysis and distribution of packing in sand-size-sediments. J. Geol. **64,** 385–395, Chicago

King, P. H. (1948): Geology of the southern Guadalupe Mountains, Texas, US Geol. Surv. Prof. Paper **215,** 183 pp., Washington

Lindholm, R. C. (1980): Intraclast orientation in Cambro-Ordovician Limestones in western Maryland. J. Sed. Petrol. **50**/4, 1205–1212, 5 Figs., Tulsa

Pettijohn, F. J., et al. (1972): see 1.4.1

Potter, P. E., Pettijohn, F. J. (1963): see 1.4.1

Rhoads, D. C., Stanley, D. J. (1965): Biogenic graded bedding. J. Sed. Petrol. **35,** 956–963, Tulsa

Sander, B. (1951): see 4.3.2

Schwarzacher, W. (1961): Petrology and structure of some Lower Carboniferous Reefs in Northwestern Ireland. Bull. Amer. Ass. Petrol. Geol. **45**/9, 1481–1503, 17 Figs., 2 Tab., Tulsa

Smalley, I. G. (1964): Representation of packing in a clastic sediment. Amer. J. Sci. **262,** 242–248, Washington

Stauffer, K.' (1962): Quantitative petrographic study of Paleozoic carbonate rocks Caballo Mountains, New Mexico. J. Sed. Petrol. **32**/3, 357–396, 38 Figs., Tulsa

Taylor, J. M. (1950): Pore space reduction in sandstones. Amer. Ass. Petrol. Geol. Bull. **34,** 701–716 Tulsa

Vinopal, R. J., Coogan, A. H. (1978): Effect of particle shape on the packing of carbonate sands and gravels. J. Sed. Petrol. **48**/1, 7–24, 17 Figs., Tulsa

Vossmerbäumer, H. (1976): see 1.4.1

4.2.4.2 References: Graded Bedding

Rhoads, D. C., Stanley, D. J. (1965): see Orientation and Packing

Wanless, H. R. (1981): Fining-upwards sedimentary sequences generated in seagrass beds. J. Sed. Petrol. **51**/2, 445–454, 7 Figs., Tulsa

4.2.4.3 References: Lamination

Aitken, J. D. (1967): see 5.3.12.7

Monty, Cl. (1976): The origin and development of cryptalgal fabrics. In: Walter, M. R. (ed.): Stromatolites. Dev. Sed. **20,** 193–249, 36 Figs., Amsterdam – Oxford – New York: Elsevier

4.2.4.4 References: Bioturbation

Byers, Ch. W. (1977): Biofacies patterns in euxinic basins: a general model. Soc. Econ. Paleont. Min. Spec. Publ. **25,** 5–27, 8 Figs., Tulsa

Frey, R. W. (1973): Concepts in the study of biogenic sedimentary structures, J. Sed. Petrol. **43,** 6–19, Tulsa
Reineck, H.-E. (1967): Parameter von Schichtung und Bioturbation. Geol. Rundschau **56,** 420–438, Stuttgart

4.2.4.5 References: Open-Space Structures

Bathurst, R. G. C. (1959): The cavernous structure of some Mississippian Stromatactis reefs in Lancashire, England, J. Geol. **67,** 506–521, Chicago
Bathurst, R. G. C. (1980): Stromatactis – origin related to submarine-cemented crusts in Paleozoic mud mounds. Geol. **8,** 131–134, 3 Figs., Boulder
Bechstädt, Th. (1974): Sind Stromatactics und radiaxial-fibröser Calcit Faziesindikatoren? N. Jb. Geol. Paläont. Mh. **174**/11, 643–663, 17 Figs., Stuttgart
Bechstädt, Th., Mostler, H. (1974): Mikrofazies und Mikrofauna mitteltriadischer Beckensedimente der Nördlichen Kalkalpen Tirols. Geol. Paläont. Mitt. Innsbruck **4,** 5/6, 1–74, Innsbruck
Bolliger, W., Burri, P. (1970): see 4.1.7.1
Bonem, R. M. (1977): Comparison of cavities and cryptic biota in modern reefs with those developed in Lower Pennsylvanian (Morrowan) bioherms. Proc. Third Internat. Coral Reef Symp., Miami, **1,** 75–80, 4 Figs., Miami
Brandner, R. (1978): see 3.9
Carannante, G., Ferreti, V., Simone, L. (1975): La cavità paleocarsiche cretaciche di Dragoni (Campania). Boll. Soc. Naturalisti Napoli **83,** Jg. 1974, 1–12, 6 Figs., Napoli
Chevalier, J. P. (1973): Geomorphology and geology of coral reefs in French Polynesia. In: Jones, O. A., Endean, R. (eds.): Biology and geology of coral reefs **1,** 113–141, New York: Academic Press
Cloud, P. E. Jr. (1960): Gas as sedimentary and diagenetic agent. Amer. J. Sci. **258** A, Bradley-Vol., 35–45
Deelman, J. C. (1972): On mechanisms causing birdseye structures. N. Jb. Geol. Paläont. Mh. **1972**/12, 582–595, 7 Figs., Stuttgart
Dunham, R. J. (1970): Keystone vugs in Carbonate Beach Deposits. Bull. Amer. Ass. Petrol. **54**/5, 845, Tulsa
Dupont, E. (1882): Les Iles coralliennes de Roly et de Philippeville. Bull. Mus. Roy. Hist. Nat. **1,** 89–160, Pls. 7–8, Bruxelles
Emery, K. O. (1945): Entrapment of air in beach sand. J. Sed. Petrol. **15,** 39–49, Tulsa
Fischer, A. G. (1964): The Lofer Cyclothems of the Alpine Triassic. Kansas Geol. Surv. Bull. **169,** 107–149, 38 Figs., Lawrence/Kansas
Folk, R. A. (1959): Practical petrographic classification of limestones. Bull. Amer. Ass. Petrol. Geol. **43,** 1–38, Tulsa
Fruth, I., Scherreiks, R. (1975): Facies and geochemical correlation in the Upper Hauptdolomit (Norian) of the eastern Lechtaler Alps. Sediment, Geol. **13,** 27–45, Amsterdam
Gnoli, M., Jaanusson, V., Leone, F., Serpagli, E. (1981): A Lower Devonian stromatactis-bearing carbonate mound from southern Sardinia. N. Jb. Geol. Paläont. Mh. **1981**/6, 339–345, 5 Figs., Stuttgart
Ham, W. E. (1952): Algal origin of the "Birdseyes" limestone in the McLish Formation. Oklahoma Acad. Sci., Proc. **33,** 200–203
Heckel, P. H. (1972): Possible inorganic origin for Stromatactis in calcilutite mounds in the Tully Limestone, Devonian of New York. J. Sed. Petrol. **42**/1, 7–18, 6 Figs., Tulsa
Hodgson, W. A. (1958): On the origin of calcite-filled cavities in the British Dinantian limestones. Eclogae geol. Helvet. **51,** 649–656, Basel
Illing, L. V. (1959): Deposition and diagenesis of some Upper Paleozoic carbonate sediments in western Canada. 5th World Petrol. Congr. Proc. Paper 2, Sect. **I,** 23–50, New York
Korn, H. (1932): Entgasungsspuren in Sedimenten. Chemie der Erde **7,** 1–23, 6 Figs., Jena
Krebs, W. (1969): Early void-filling cementation in Devonian fore-reef limestones (Germany). Sedimentology **12,** 279–299, Amsterdam

Krebs, W. (1971): Devonian reef limestones in the eastern Rhenish Schiefergebirge. In: Müller, G., Friedman, G. M. (eds.): Sedimentology of central parts of Central Europe, 45–81, Frankfurt: Kramer
Lindström, M. (1979 a): see 4.2.4.7
Lowenstam, H. A. (1950): Niagaran reefs of the Great Lake area. J. Geol. **58,** 430–487, Chicago
Müller-Jungbluth, W. V., Toschek, P. H. (1969): see 6.4
Newell, N. D. (1955): Depositional fabric in Permian reef limestones. J. Geol. **63,** 301–309, Chicago
Otte, C. Jr., Parks, J. M. Jr. (1963): Fabric studies of Virgil and Wolfcamp bioherms, New Mexico. J. Geol. **71**/3, 380–396, 2 Tab., 5 Figs., 3 Pls., Chicago
Philcox, M. E. (1963): Banded calcite mudstone in the lower Carboniferous "Reef" knolls of the Dublin Basin, Ireland. J. Sed. Petrol. **33**/4, 904–913, 11 Figs., Tulsa
Prentice, J. E. (1950): The carboniferous limestone of the Manifold valley region, North Staffordshire. Geol. Soc. London Quart. **106,** 171–209, London
Sander, B. (1936): see 1.4.1
Schwarz, A. (1927): Wachstum, Absterben und Diagenese eines paläozoischen Korallenriffes. Nach Untersuchungen an Lahnmarmor-Platten von Wirbelau. Senckenbergiana lethaea **9**/2, 49–64, Frankfurt
Schwarzacher, W. (1961): Petrology and structure of some Lower Carboniferous Reefs in Northwestern Ireland. Bull. Amer. Ass. Petrol. Geol. **45,** 1481–1503, 17 Figs., 2 Tab., Tulsa
Scoffin, T. P. (1972): Cavities in the reefs of the Wenlock Limestone (Mid-Silurian) of Shropshire, England. Geol. Rdsch. **61**/2, 565–578, 12 Figs., Stuttgart
Scoffin, T., Garret, P. (1974): Processes in the formation and preservation of internal structures in Bermuda patch reefs. Proc. Second Internat. Coral Reef Symp. **2,** 429–448, Brisbane
Semeniuk, V. (1971): Subaerial leaching in the limestones of the Bowan Park Group (Ordovician) of Central Western New South Wales. J. Sed. Petrol. **41**/4, 939–950, 9 Figs., Tulsa
Shinn, E. A. (1968 a): Burrowing in recent lime sediments of Florida and the Bahamas. J. Paleont. **42**/4, 879–894, Pls. 109–112, 17 Figs., Menasha
Shinn, E. A. (1968 b): Practical significance of Birdseyes structures in carbonate rocks. J. Sed. Petrol. **38**/1, 215–223, 13 Figs., Tulsa
Tebbutt, G. E., Conley, C. D., Boyd, D. W. (1965): Lithogenesis of a carbonate rock fabric. Contrib. Geol. **4**/1, 1–13, 1 Pl., Laramie
Textoris, D. A., Carozzi, A. V. (1964): Petrography and evolution of Niagaran (Silurian) reefs. Indiana. Bull. Amer. Ass. Petrol. Geol. **48**/4, 397–426, 24 Figs., Tulsa
Tucker, M. E. (1973): Sedimentology and diagenesis of Devonian pelagic limestones (Cephalopodenkalke) and associated sediments of the Rhenohercynian Geosyncline, West Germany. N. Jb. Geol. Paläont. Abh. **142**/3, 320–350, Stuttgart
Wolf, K. H. (1965): Littoral environment, indicated by open-space structures in algal limestones. Palaeo-3, **1,** 183–223, Amsterdam
Zorn, H. (1971): Paläontologische, stratigraphische und sedimentologische Untersuchungen des Salvatoredolomits der Tessiner Kalkalpen, Schweiz. Paläont. Abh. **91,** 90 pp., Basel

4.2.4.6 References: Nodular Structures

Born, A. (1921): Zum Thema Kramenzelkalk. Geol. Rundschau **12,** 343–345, Stuttgart
Deelman, J. C. (1974): see 3.9
Dvorak, J. (1972): Shallow-water character of nodular limestones and their paleogeographic interpretation. Neues Jahrbuch Geol. Paläont. Monatshefte **1972,** 509–511, Stuttgart
Garrison, R. E., Fischer, A. G. (1969): Deep-water limestones and radiolarites of the Alpine Jurassic. Soc. Econ. Paleont. Min. Spec. Publ. **14,** 20–56, Tulsa
Garrison, R. E., Kennedy, W. J. (1977): Origin of solution seams and flaser structure in Upper Cretaceous chalks of southern England. Sed. Geol. **19,** 107–137, Amsterdam
Gründel, J., Rösler, H. (1963): Zur Entstehung der oberdevonischen Kalkknollengesteine Thüringens. Geologie **12**/9, 1009–1038, Berlin

Hildebrand, E. (1928): Über die chemisch-physikalischen Bedingungen der Knollenkalkbildung. Z. deutsch. geol. Ges. **80**/3, 308–342, Hannover
Hopkins, J. C. (1977): Production of foreslope breccia by differential submarine cementation and downslope displacement of carbonate sands, Miette and Ancient Wall buildups, Devonian, Canada. Soc. Econ. Paleont. Min. Spec. Publ. **25**, 155–170, Tulsa
Kukal, Z. (1975): On the origin of nodular limestone. California Min. Geol. **20**, 359–368
Mullins, H. T., Neumann, A. C., Wilber, R. J., Boardman, M. R. (1980): Nodular carbonate sediment on Bahamian slopes: possible precursors to nodular limestones. J. Sed. Petrol. **50**/1, 117–131, 13 Figs., Tulsa
Nichols, R. A. (1966): Petrology of irregular nodular beds, Lower Carboniferous, Anglesey, North Wales. Geol. Mag. **103**, 477–486, London
Nobel, J. P., Howells, K. D. M. (1974): Early lithification of the nodular limestones in the Silurian of New Brunswick. Sed. **21**, 587–609, Oxford
Schindewolf, O. H. (1925): Einige Bemerkungen zur Entstehung der oberdevonischen Kramenzelgesteine. Zentralblatt Min. Geol. Paläont., B, **1925**, 405–411, Stuttgart
Seyfried, H. (1980): see 4.1.7.2
Tucker, M. E. (1974): Sedimentology of Paleozoic pelagic limestone: the Devonian Griotte (Southern France) and Cephalopodenkalk (Germany). Spec. Publ. Int. Ass. Sed. **1**, 71–92, Oxford
Weber, P. (1965): Bildung und Regelung von Kalkknollengefügen. Untersuchungen im Oberdevon des Rheinischen Schiefergebirges. Decheniana **118**/1, 55–84, Bonn

4.2.4.7 References: Discontinuities and Fissure Fillings

Al-Gailani, M. B. (1980): Geochemical identification of unconformities using semi-quantitative x-ray fluorescence analysis. J. Sed. Petrol. **50**/4, 1261–1270, 4 Figs., Tulsa
Assereto, R., Brigo, L., Brusca, C., Omenetto, P., Zuffardi, P. (1976): Italian Ore/Mineral Deposits Related to Emersion Surfaces – A Summary. Mineralia Deposita **11**/2, 170–179, Berlin – Heidelberg – New York
Bernoulli, D., Jenkyns, H. C. (1974): Alpine, Mediterranean, and central Atlantic Mesozoic facies in relation to the early evolution of the Tethys. Spec. Publ. Soc. Econ. Paleont. Min. **19**, 128–160, Tulsa
Bromley, R. G. (1968): Burrows and borings in hardgrounds. Medded. Dansk. Geol. Förens. **18**, 247–250, Kopenhagen
Bromley, R. G. (1975): Trace fossils at omission surfaces. In: Frey, R. W. (ed.): The Study of Trace Fossils, 399–428, New York: Springer
Bromley, R. G. (1978): Hardground diagenesis. In: Fairbridge, R. W., Bourgeois, J. (eds.): The Encyclopedia of Sedimentology, 397–400, Stroudsburg: Dowden, Hutchinson and Ross
Dravis, J. (1979): Rapid and widespread generation of Recent oolitic hardgrounds on a high-energy Bahamian platform, Eleuthera Bank. J. Sed. Petrol. **49**/1, 195–208, 12 Figs., Tulsa
Drittenbass, W. (1979): Sedimentologie und Geochemie von Eisen-Mangan-führenden Knollen und Krusten im Jura der Trento-Zone (östliche Südalpen, Norditalien). Eclogae geol. Helvet. **72**/2, 313–345, 10 Figs., Basel
Fabricius, F. H. (1968): Calcareous sea bottoms of the Raetian and Lower Jurassic sea from the west part of the Northern Calcareous Alps. In: Müller, G., Friedman, G. M. (eds.): Recent Developments in Carbonate Sedimentology in Central Europe, 240–249, 1 Pl., 2 Figs., Berlin – Heidelberg – New York: Springer
Fürsich, F. (1971): Hartgründe und Kondensation im Dogger von Calvados. N. Jb. Geol. Paläont. Abh. **138**, 313–342, Stuttgart
Fürsich, F. T., (1979): Genesis, environments, and ecology of Jurassic hardgrounds. Neues Jb. Geol. Paläont. Abhandlungen **158**/1, 1–63, Stuttgart
Fürsich, F. T., Wendt, J. (1976): Hartgründe und Kondensations-Lagerstätten. Zbl. Geol. Paläont. Teil II **1976**, 5/6, 238–245, 4 Figs., Stuttgart
Garrison, R. E., Fischer, A. G. (1969): Deep-water limestones and radiolarites of the Alpine Jurassic. Spec. Publ. Soc. Econ. Paleont. Min. **14**, 20–56, Tulsa

Germann, K. (1971): Mangan-Eisen-führende Knollen und Krusten in jurassischen Rotkalken der Nördlichen Kalkalpen. N. Jb. Geol. Paläont. Mh. **1971,** 133–156, Stuttgart
Heim, A. (1924): Über submarine Denudation und chemische Sedimente. Geol. Rdsch. **15,** 1–47, Stuttgart
Heim, A. (1958): Oceanic sedimentation and submarine discontinuities. Eclogae geol. Helvet. **51,** 642–649, Basel
Hollmann, R. (1962): Über Subsolution und die „Knollenkalke" des Calcare Ammonitico Rosso Superiore in Monte Baldo (Malm, Norditalien). N. Jb. Geol. Paläont. Mh. **1962**/4, 163–179, 8 Figs., Stuttgart
Hollmann, R. (1964): Subsolutions-Fragmente. N. Jb. Geol. Paläont. Abh. **119**/1, 22–82, Pls. 7–10, 7 Figs., Stuttgart
Jaanusson, V. (1961): Discontinuity surfaces in limestones. Bull Geol. Inst. Univ. Uppsala **40,** 221–241, 2 Figs., Uppsala
Jenkyns, H. C. (1971): The genesis of condensed sequences in the Tethyan Jurassic. Lethaia **4,** 327–352, 11 Figs., Oslo
Jenkyns, H. C. (1974): see 3.9
Jurgan, H. (1969): Sedimentologie des Lias der Berchtesgadener Kalkalpen. Geol. Rdsch. **58,** 464–501, Stuttgart
Kennedy, W. J., Garrison, R. E. (1975): see p. 522
Kobluk, D. R., Pemberton, S. G., Karolyi, M., Risk, M. J. (1977): The Silurian-Devonian disconformity in southern Ontario. Bull. Canadian Petrol. Geol. **25**/6, 1157–1186, 23 Figs., Calgary
Krystyn, L (1971): Stratigraphie, Fauna und Fazies der Klaus-Schichten (Aalenium-Oxford) in den östlichen Nordalpen. Verh. geol. Bundesanst. **1971**/3, 486–509, Wien
Krystyn, L., Schäffer, G., Schlager, W. (1971): Über die Fossil-Lagerstätten in den triadischen Hallstätter Kalken der Ostalpen. Neues Jahrbuch Geol. Paläont. Abhandlungen **117**/2, 284–304, Stuttgart
Lindström, M. (1979 a): Calcitized tephra, sedimentary sills and micro-vents, in Lower Ordovician pelagic-type limestone, Sweden. J. Sed. Petrol. **49**/1, 233–244, 7 Figs., Tulsa
Lindström, M. (1979 b): Diagenesis of Lower Ordovician hardgrounds in Sweden. Geologica Palaeontologica **13,** 9–30, 3 Plates, 5 Figs., Marburg
Malfait, B. T., van Andel, T. H. (1980): A modern oceanic hardground on the Carnegie Ridge in the eastern Equatorial Pacific. Sed. **27,** 467–496, 25 Figs., Oxford
Massari, F., Medizza, F. (1973): Stratigrafia e paleogeografia del Campaniano-Maastrichtiano nelle Alpi Meridionale (con particolare riguardo agli Hard Grounds della Scaglia rossa Veneta). Memm. Ist. Geol. Min. Univ. Padova, **28,** 1–62, 4 Plates, 18 Figs., Padova
Montenat, C., Ott'd'Estevou, P., Vachard, D. (1978): Stromatolithes silico-métalliques du Messinien d'Espagne méridionale. Bull. Soc. géol. France, sér. 7, **20**/5, 773–777, 1 Pl., 1 Fig., Paris
Palmer, T. J., Fürsich, F. T. (1974): The ecology of a Middle Jurassic hardground and crevice fauna. Paleontology **17,** 507–524, London
Purser, B. H. (1969): Synsedimentary marine lithification of Middle Jurassic limestones in the Paris Basin. Sed. **12,** 105–230, Amsterdam
Rose, P. R. (1970): Stratigraphic interpretation of submarine versus subaerial discontinuity surfaces: An example from the Cretaceous of Texas. Geol. Soc. Amer. Bull. **81,** 2787–2798, New York
Sakar, S., Bhattacharyya, A., Chanda, S. K. (1980): Recognition of hardgrounds and emersion surfaces: a new criterion. J. Sed. Petrol. **50**/1, 83–89, 7 Figs., Tulsa
Schlager, W. (1969): Das Zusammenwirken von Sedimentation und Bruchtektonik in den triadischen Hallstätterkalken der Ostalpen. Geol. Rdsch. **59,** 289–308, Stuttgart
Schlager, W. (1974): Preservation of cephalopod skeletons and carbonate dissolution on ancient Tethyan sea floor. Spec. Publ. Intern. Ass. Sed. **1,** 49–70, 10 Figs., Oxford
Schöll, W. U., Wendt, J. (1971): Obertriadische und jurassische Spaltenfüllungen im Steinernen Meer (Nördliche Kalkalpen). N. Jb. Paläont. Abh. **139**/1, 82–98, 10 Figs., Stuttgart
Seyfried, H. (1978): see 3.9
Seyfried, H. (1981): Genesis of hardgrounds in the Tethyan Jurassic. Abstract 2nd Europ. Meeting Int. Ass. Sed., Bologna, 4 pp., 1 Fig., Bologna

Sweeting, M. M. (1966): The weathering of limestones, with particular reference to the Carboniferous limestones of northern England. In: Dury, G. H. (ed.): Essays in Geomorphology, 177–210, London: Heinemann

Tietz, G.-F. (1975): Petrographische und geochemische Untersuchungen an einer kondensierten Kalkbank des unteren Silurs der Karnischen Alpen (Österreich). Verhandlungen Geol. Bundesanstalt **1976**/2, 207–242, Wien

Tucker, M. E. (1973): Sedimentology and diagenesis of Devonian pelagic limestones (Cephalopodenkalk) and associated sediments of the Rhenohercynian Geosyncline, West Germany. N. Jb. Paläont. Abh. **142**/3, 320–350, 24 Figs., 1 Tab., Stuttgart

Videtich, P. E., Matthew, R. K. (1980): Origin of discontinuity surfaces in limestones: isotopic and petrographic data, Pleistocene of Barbados, West Indies. J. Sed. Petrol. **50**/3, 971–980, 4 Figs., Tulsa

Voigt, E. (1959): Die ökologische Bedeutung der Hartgründe (hardgrounds) in der oberen Kreide. Paläont. Zeitschrift **33**, 129–147, Stuttgart

Walter, M. R., Bault, J., Brock, T. D. (1972): Siliceous algal and bacterial stromatolites in hot springs and geyser effluents of Yellowstone National Park. Science **178**, 402–505, Washington

Wendt, J. (1969): Foraminiferen-„Riffe“ im karnischen Hallstätter Kalk des Feuerkogel (Steiermark, Österreich). Paläont. Z. **43**, 177–193, Pls. 21–22, Stuttgart

Wendt, J. (1970): Stratigraphische Kondensation in triadischen und jurassischen Cephalopodenkalken der Tethys. N. Jb. Geol. Paläont. Mh. **1970**, 433–449, Stuttgart

Wendt, J. (1971): Genese und Fauna sedimentärer Spaltenfüllungen im mediterranen Jura. Paläontograph. A **136**, 122–192, Pls. 15–18, Stuttgart

Wendt, J. (1974): Encrusting organisms in deep-sea manganese nodules. Spec. Publ. Intern. Ass. Sed. **1**, 437–447, Oxford

Wendt, J. (1976): Submarine Spaltenfüllungen. Zbl. Geol. Paläont. Teil II **1976**, 5/6, 245–251, 3 Figs., Stuttgart

Wiedenmayer, F. (1963): Obere Trias bis mittlerer Lias zwischen Saltrio und Tremona (Lombardische Alpen) – Die Wechselbeziehungen zwischen Stratigraphie, Sedimentologie und syngenetischer Tektonik. Eclogae geol. Helvetiae **56**/2, 529–640, Basel

4.2.4.8 References: Geopetal Fabrics

Carozzi, A. V. (1963): Half-moon oolites. J. Sed. Petrol. **33**, 633–645, Tulsa

Cullison, J. S. (1938): Origin of composite and incomplete internal moulds and their possible use as criteria of structure. Bull. Geol. Soc. America **49**/6, 981–988, New York

Deelman, J. C., deCoo, J. C. M. (1976): Experimental analysis of floored interstices in clastic carbonate sediments. Sed. Geol. **16**, 1–13, Amsterdam

Fay, M. (1976): Riffnahe Resedimente im Raum Kelheim: Lithologie, Genese und stratigraphische Bemerkungen. N. Jb. Geol. Paläont. Abh. **152**/1, 51–74, 5 Figs., Stuttgart

Fiege, K. (1937): Zyklische Sedimentation. Abh. preuß. geol. Landesanst., N. F., **177**, Berlin

Fisher, R. V., Mattinson, J. M. (1968): Wheeler George turbidite-conglomerate series, California: inverse grading. J. Sed. Petrol. **38**, 1013–1023, Tulsa

Fuganti, A. (1964): Strutture geopete in calcari mesozoici del Trentino e del Veronese (Alpi orientali). Atti Acc. Roveretana Agiati, ser. 6, **4**, fasc. B 5–9, 5 Pls., Trento

Harbaugh, J. W. (1959): Small scale cross-lamination in limestones. J. Sed. Petrol. **29**/1, 30–37, 6 Figs., Tulsa

Hückel, B., Jacobshagen, V. (1962): Geopetale Sedimentgefüge im Hauptdolomit und ihre Bedeutung für die tektonische Analyse der Nördlichen Kalkalpen. Z. dtsch. geol. Ges. **113**, 2/3, Jg. 1961, 305–310, Taf. 5–6, 1 Fig., Stuttgart

Koch, E. (1959): „Sandfang“. Ein Hinweis auf die Lagerungsverhältnisse. N. Jb. Geol. Mh. 443–450, 7 Figs., Stuttgart

Richter, D. (1968): Nur teilweise mit Sediment gefüllte Brachiopodenschalen als „fossile Wasserwaagen“ – ein Hinweis auf Lagerungsverhältnisse massiger Kalkabfolgen. Neues Jahrbuch Geol. Paläont. Monatshefte **1968**/1, 32–37, Stuttgart

Sander, B. (1936): see 1.4.1

Schulz, O. (1967): Die synsedimentäre Mineralparagenese im oberen Wettersteinkalk der Pb-Zn-Lagerstätte Bleiberg-Kreuth (Kärnten). Tschermaks min. petrograph. Mitt. **12**, 2/3, 230–289, 16 Figs., Wien
Schwarzacher, W. (1948): Sedimentpetrographische Untersuchungen kalkalpiner Gesteine. Hallstätter Kalke von Hallstatt und Ischl. Jb. geol. Bundesanst. **91**, Jg. 1946, 1–48, 15 Figs., Wien
Seilacher, A. (1973): Biostratinomy: the sedimentology of biologically standarized particles. In: Ginsburg, R. N. (ed.): Evolving concepts in sedimentology, 159–177, Baltimore: John Hopkins Univ. Press
Shrock, R. R. (1948): Sequence in Layered Rocks. 507 pp. New York – London: McGraw Hill
Wieczorek, J. (1979): Geopetal structures as indicators of top and bottom. Ann. Soc. Geol. Pologne **49**, 215–221, Plates 1–2, 2 Figs., Warsaw

4.3 Frequency Analysis (Constituent Analysis)

Limestone classifications and microfacies categorization call for a quantification of the features recognizable in thin-sections.

Knowing the percentage composition (=modal analysis) – subdivided according to matrix, sparite, and particles – is important for

a) Arranging samples in limestone classifications,
b) Following the facies development over time and space,
c) Information about the "ostracomass" (Jaanusson, 1972), that is, the total mass of skeletal material produced at a certain time in a certain area.

4.3.1 Methods and Examples

I. Methods

A. Lower limits of the measurable particles: Because small peloids, bioclasts, and intraclasts can no longer be differentiated below a minimum size, and because grouping Recent skeletal grains is impossible for sizes less than 0.06 mm and only about 60% successful for sizes between 0.06 and 0.125 mm (Feray et al., 1962), many authors classify all particles less than 0.10 mm as matrix (e.g., Jaanusson, 1972). Thereby, the "overestimation error," which is dependent on the thickness of the thin-section, is also reduced (see Fig. 39, p. 246).

B. Dual measurements: Skeletal grains can be registered with frequency analysis measurements, particularly with the point-counting method, as

- Particles with voids; the voids are counted as constituents of the particles (grain-bulk, Dunham, 1962; gives vol%).
- Particles without voids; the voids are not counted (grain-solid, Dunham, 1962; gives wt%).

Figure 37 shows the two possibilities of making measurements. In order to guarantee the comparability of the measurements, it is necessary to specify the

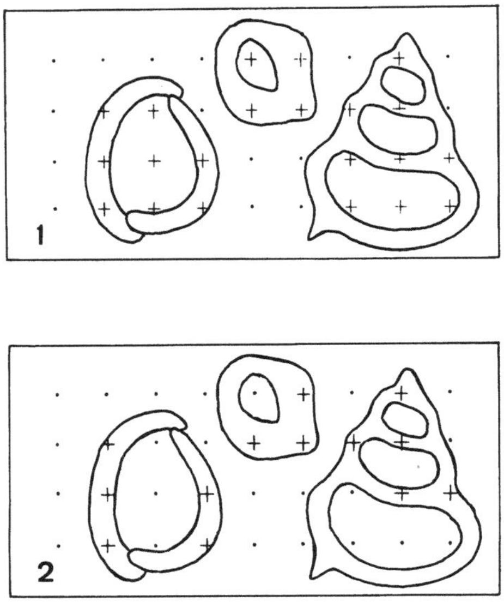

Fig. 37. Dual measurements. *1* grain bulk; the voids in the particles or the micrite and sparite in the particles are all counted as constituents of the particles. The points counted are indicated by crosses. *2* grain solid; the voids are not counted as constituents of the particles but rather are regarded separately as matrix or sparite. After Jaanusson (1972)

method of measurement each time. For microfacies analyses the "grain-solid" definition is commonly used. If the amount of matrix is especially significant or the percentage of cement should be emphasized, then dual measurements (grain-solid and grain-bulk) are to be recommended. R.C.L. Wilson (1965) pointed out the problems that arise during the quantification of "composite" particles (e.g., ooids with quartz nuclei, oncoids with shells as nuclei, bioclasts with voids).

C. Methods of measurement: Basically there are two methods of frequency analyses – a quantitative method and a semi-quantitative one.

1. Quantitative Methods

Of all the methods suggested for determining the amount of sediment in a sample (see Sarntheim, 1971), the following can be used in thin-sections:

a) *Point-counting* ("Punktzählverfahren, Punktfeld-Methode" after Sander, 1951): This method was developed by Glagolew (1933) and improved by Chayes (1956). Descriptions in German of the method are found in Lamprecht (1954) and Müller (1964). The thin-section is completely covered with a point grid. Every point which lies on a particle or on matrix or sparite is counted (see Fig. 37). Usually special cross-webbed oculars are used together with a specially adapted stage micrometer (spacing intervals of 0.2 and 0.3 mm). The use of point-counting devices (Emerson, 1958) as well as STATITEST counting instruments cuts the time needed enormously. Almost completely automated counting methods have been suggested by Harvey and Steinmetz (1971, image analysis, QUANTIMED), Fisher (1971, QUANTIMED) and Conley and Davis (1973, IDECS).

The number of points is statistically important information. This number is dependent upon the size and the distribution of the particles. The mean deviation

(Chayes, 1956: 39) decreases with an increasing number of points and increases with an increasing percentage of particles, that is up to 50%, whereupon the mean deviation again drops. Infrequently occurring particles (less than 10%) are subject to slightly higher errors than more frequently occurring particles (see Table 24).

The number of points needed to attain a certain reliability can be controlled with the aid of a chart published by Plas and Tobi (1965); Fig. 38; see also

Table 24. Average probable error of the point-counting method in percentages, with numbers of points counted (n) and various percentages of one particle. The standard deviation after Chayes (1956) is used, where $\sigma = 100 \cdot \sqrt{p(1-p)/n}$

Number of points counted (n)	1%	10%	20%	30%	40%	50%	60%
100	1	3	4	4.6	4.9	5	4.9
500	0.4	1.3	1.8	2.1	2.2	2.2	2.2
1 000	0.1	1.0	1.3	1.5	1.6	1.6	1.6
3 000	0.02	0.5	0.7	0.8	0.9	0.9	0.9

Bandemer (1965). Example: Out of 500 points counted, 140 points (=28%, P) correspond to ooids. When these values are plotted on the chart, the intersection of the 500 value and the 28% value leads to a standard deviation of $2\,\sigma = 4\%$ (absolute error). This means that the ooid content of the samples lies between $28 \pm 4\% = 24\%$ and 32% by volume (relative to a confidence interval of 95%). If this error is acceptable, the counting can then be brought to a halt. Otherwise the total number of points must be raised. In order to keep sample errors small, several hundred (at least 300 to 500) points must be counted; for poorly sorted and variously sized particles, as many as 1000–3000. Empirical studies of the number of points needed for counting calcarenites have been conducted by Purdy (1963), Harbaugh and Demirmen (1964) and Schenk (1967).

The point-counting method works with percentage by volume (for grain-bulk measurements) because it is the amount of space that the single particle groups or matrix and cement types take up in the sediment that is being determined.

b) *Micrometric method after Rosiwal (1898):* The intercept distances which are taken up by particles on parallel lines are measured in the thin-section. The spacing between two lines should be selected in such a way that large particles are not intersected twice. The LEITZ integrating stage is ordinarily operated by four or six tiny rods on screws. This method is rarely used today in microfacies analyses.

c) *Grain number analysis* (counting all particles within a unit area): Sarntheim (1971) showed that only by counting particles according to fractions of specific grain sizes could reproducible results be obtained. This method is generally limited to grain-size classes of 0.063–2.00 mm – relatively common in limestones. The method is not applicable to rudite fractions, because very often too few particles larger than 2 mm are available for counting.

This method became well known in heavy minerals analyses as the Fleet method (see Galehouse, 1969). As in the point-counting method, the number of the

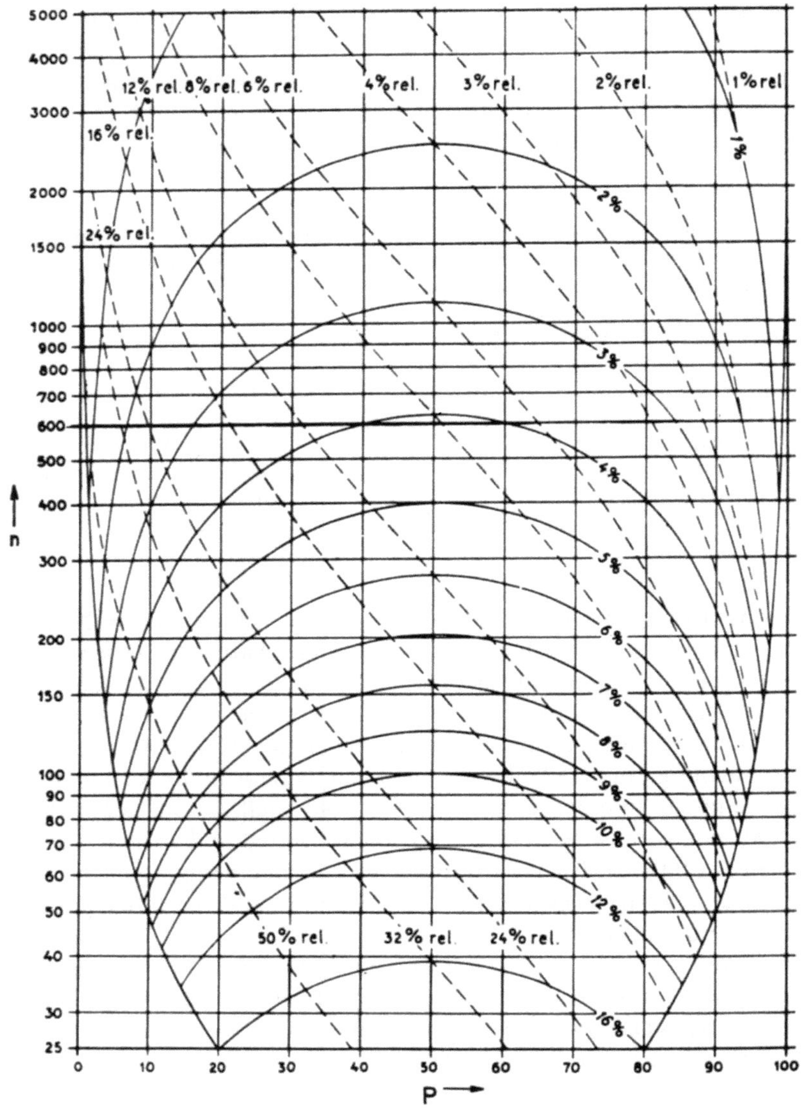

Fig. 38. Estimation of the statistical probability after the point-counting method. Key: $n=$ total number of points counted; $P=$percentage of particle (e.g., ooids) frequency in the thin-section. The probability can be expressed by the standard deviation, thus enabling the true content of the particle (p) to be estimated as vol.%. *Dark curved lines,* absolute error; *Dotted lines,* relative error. After Plas and Tobi (1965)

particles to be counted can be controlled with the help of the chart in Plas and Tobi (Fig. 38).

Carozzi (1950, 1958) called the absolute number of a certain particle or of all particles within a square area on a thin-section the *frequency index:* The width of the square to be measured should be at least ten times as large as the average particle diameter. The determination of the frequency index follows the *ribbon method,* in which all the particles lying within the band width to be measured are included (see Plas, 1962). This method leads to a determination of numerical percentages when the band width is considerably larger than the largest grain to be counted. In order to obtain volume percentages, the *line method* should be introduced: All those particles are counted which are intersected by equally spaced lines laid upon the thin-section. The ribbon and the line methods yield number counts that are biased differently with respect to size, shape, orientation, and projection diameter of the grains relative to the sampling lines or the sampling area. Grains with greater projection diameters lying perpendicular to the sampling lines have a greater chance of being counted. Hence, number counts obtained by line or ribbon methods do not give the actual numbers of grains and may not yield true statistical criteria for the sample; converting these grain numbers into volume (weight) or area measurements produces no statistically significant results, either. In spite of these limitations, both methods are the most common counting techniques used because they are quick, simple and easy to carry out. The effectiveness of these counting methods can be increased by the theory of bias correction for irregular grains developed by Sahu (1977).

Carozzi's frequency index is often used for microfacies analyses (e.g., Carozzi, 1961; Carozzi and Frost, 1966; Walpole and Carozzi, 1961; Skupin, 1973). By combining it with the "clasticity index" (maximum size of a category of particles in thin-section, see 4.2.1.1), an attempt has often been made to describe the changes over time in the microfacies types (Example: Fig. 27). Basically, a statistical check should always be made on the results (Student's t-test).

d) *Method of Latin squares:* This statistical method is based on the possibility of quantitatively splitting the problem up into smaller units. It can be used to see to what extent information can be provided about the composition of the whole area (total thin-section, etc.) by counting just part of the area. More details and examples are found in Marsal (1967).

2. *Semi-Quantitative Methods (Estimation Methods)*

Estimating percentages of particles with or without the help of comparison charts leads to values for limestones, if they are not too inhomogeneous, that correspond nicely to the results of the point-counting method (see Fig. 43). Preceding the estimation, one must make sure that the particles are categorized and that the overestimation and underestimation errors dependent upon the particle diameters are taken into consideration (see Fig. 39).

The estimation can be improved by the *mosaic field method* (Sander, 1951): Mosaic fields of 1 cm² are marked out on peels at constant intervals. Each mosaic field is then subdivided into 10 equal columns and 10 rows. The 100 square units which result from this are then equivalent in area to the field of view in the microscope, which has been delimited by a square aperture. The estimation can then take place accurately at ±1% (Examples in Eltgen, 1969).

Comparison charts have been published by Folk (1951), Terry and Chilingar (1958), Fediaevsky (1963), Baccelle and Bosellini (1965, see Fig. 40), and Schäfer (1969, see Figs. 41, 42). In considering the interference factors found by Allen (1956), it is recommended that different comparison charts be used for different particles, and that wherever possible the background (light – "sparite", dark – "micrite"), which influences the comparison somewhat, should also be taken into account.

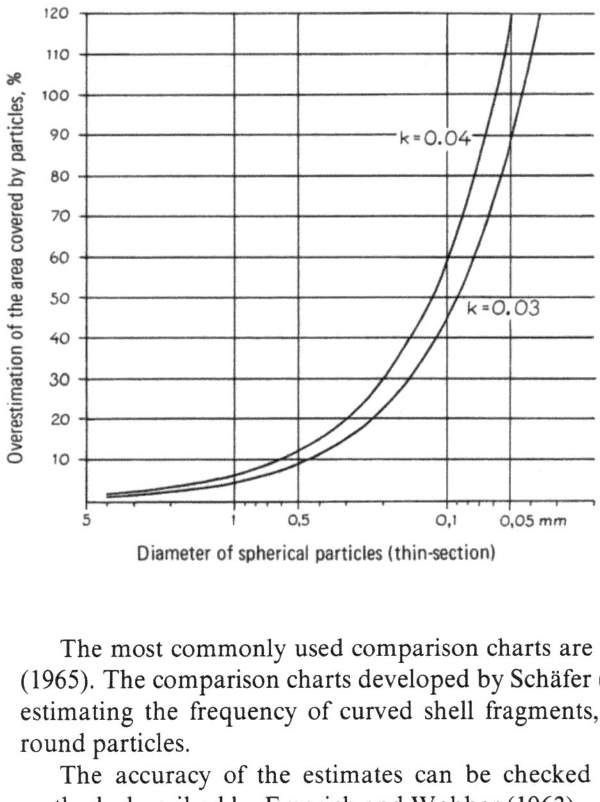

Fig. 39. Correlation between particle sizes and overestimation of particle sizes in thin-section analyses. k = thin-section thickness in millimeters. A particle with a diameter of 0.10 mm may be overestimated by as much as 45%, because, for example, in ooids the thin-section area is nearly always less than the true maximum diameter. After Jaanusson (1972)

The most commonly used comparison charts are those of Baccelle and Bosellini (1965). The comparison charts developed by Schäfer (Fig. 41, 42) serve as a basis for estimating the frequency of curved shell fragments, crinoid columnals, and small round particles.

The accuracy of the estimates can be checked with the aid of the statistical methods described by Emerich and Wobber (1963).

Comparison of Quantitative and Semi-Quantitative Methods

According to Dennison and Shea (1966), series of experiments to check the sources of error in the visual estimation of particle frequencies produced no differences in the results, regardless of whether comparison charts were used, whether or not the observers were qualified or experienced, or whether the background was light or dark. Consequently, one must assume that the semi-quantitative methods produce fairly comparable results, regardless of the kind of estimation. However, it would

Continued p. 259

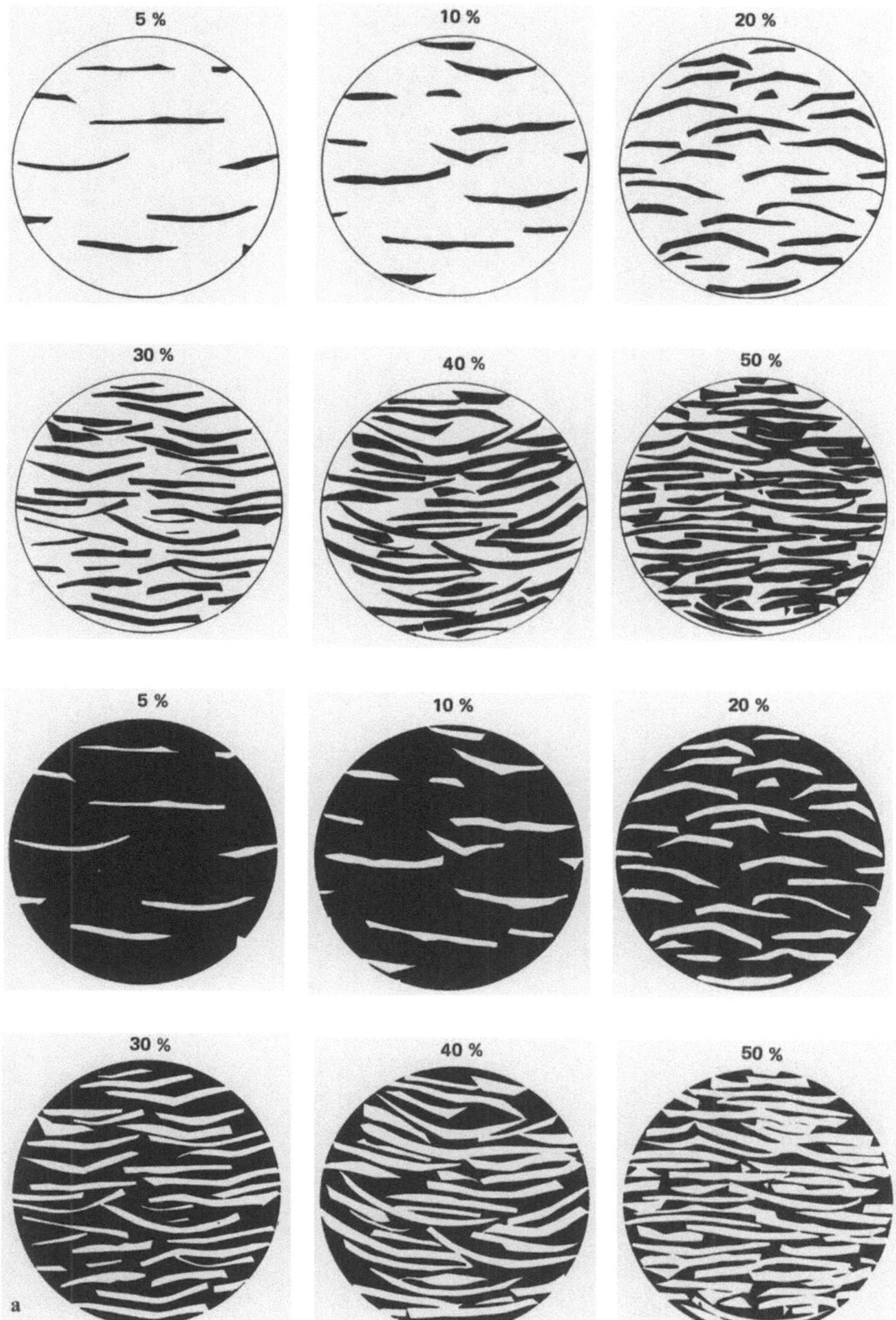

Fig. 40 a–i. Comparison charts for frequency estimations of bioclasts, peloids, ooids, oncoids, and lithoclasts. After Baccelle and Bosellini (1965)

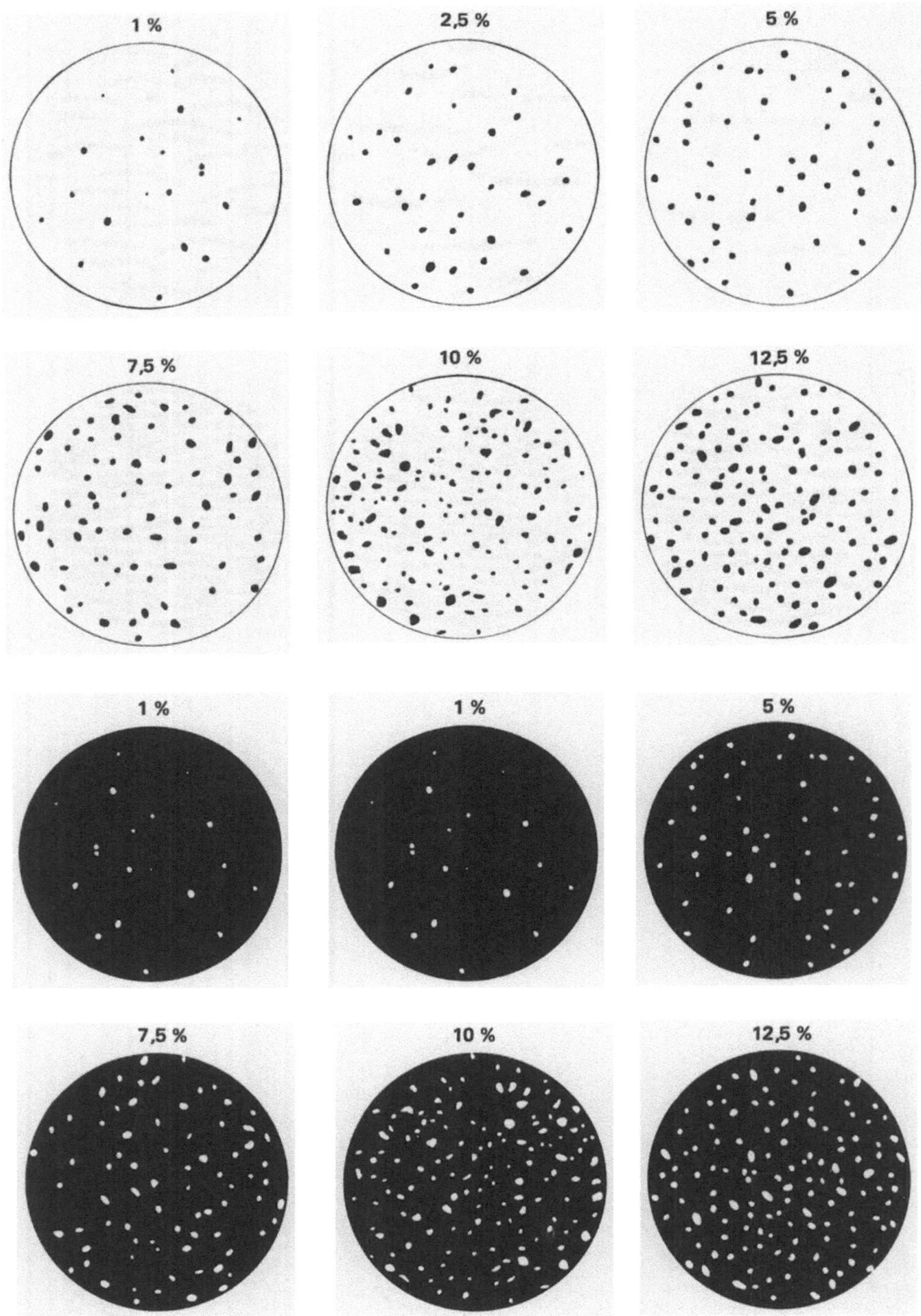

Fig. 40b

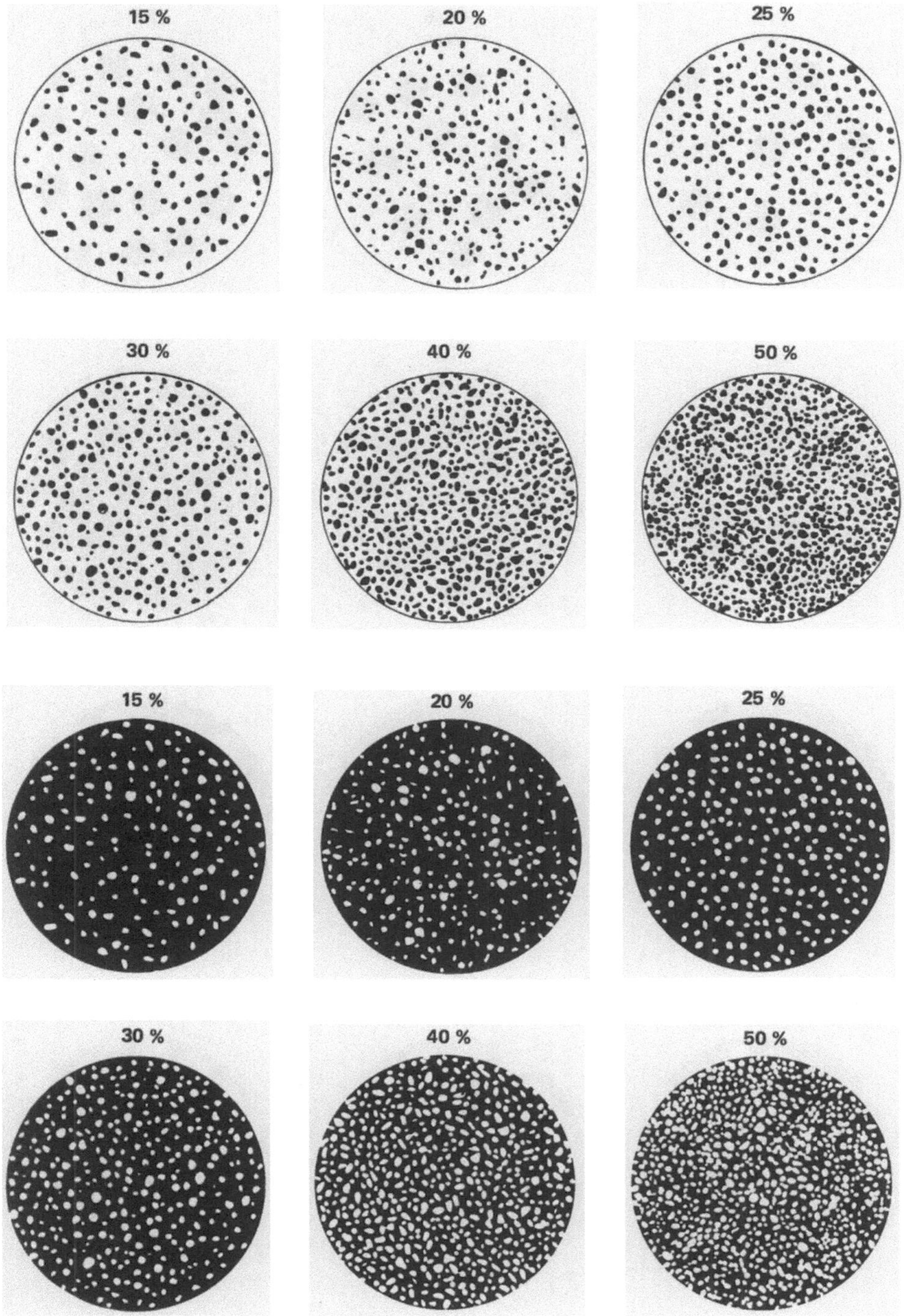

Fig. 40 c

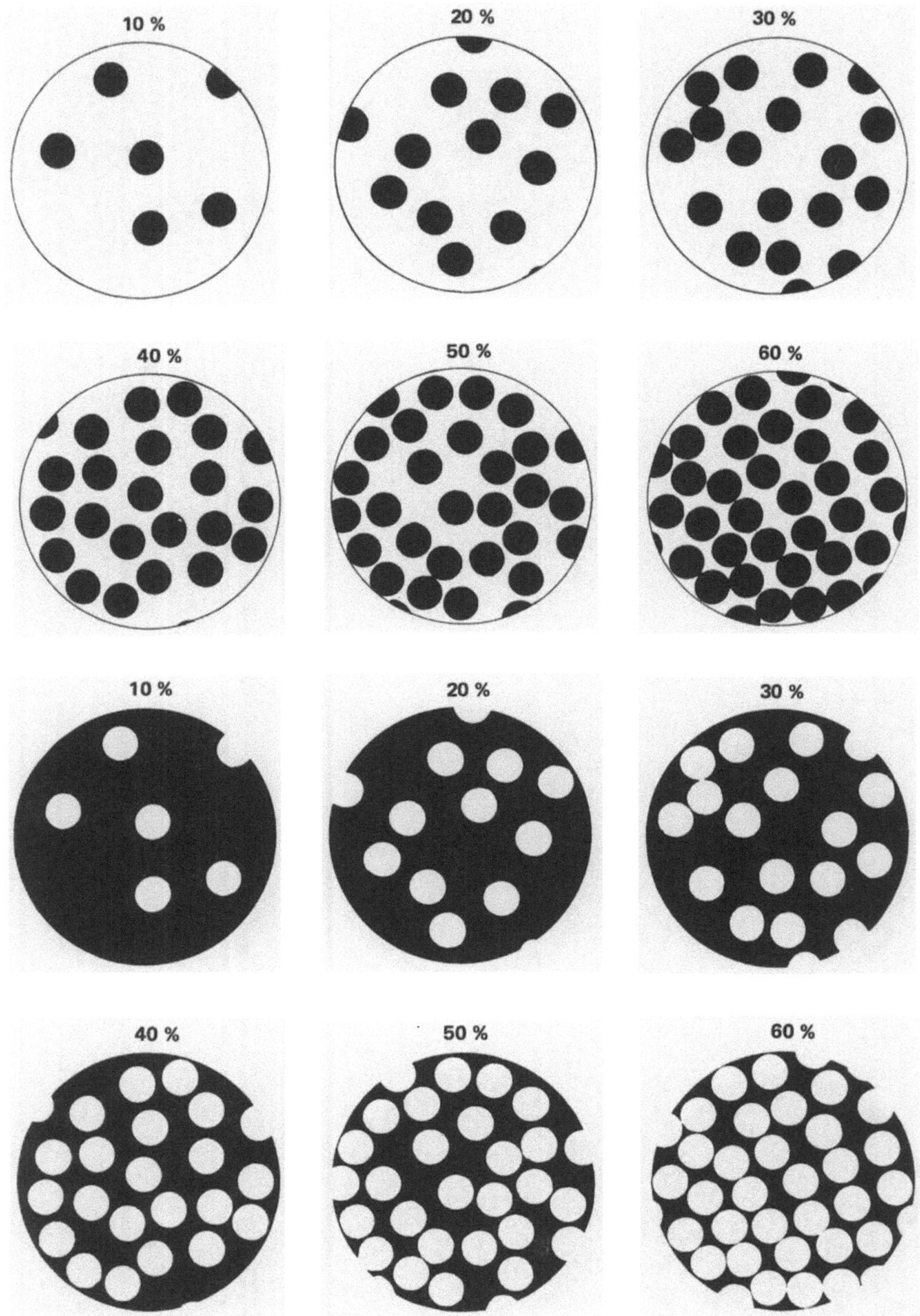

Fig. 40 d

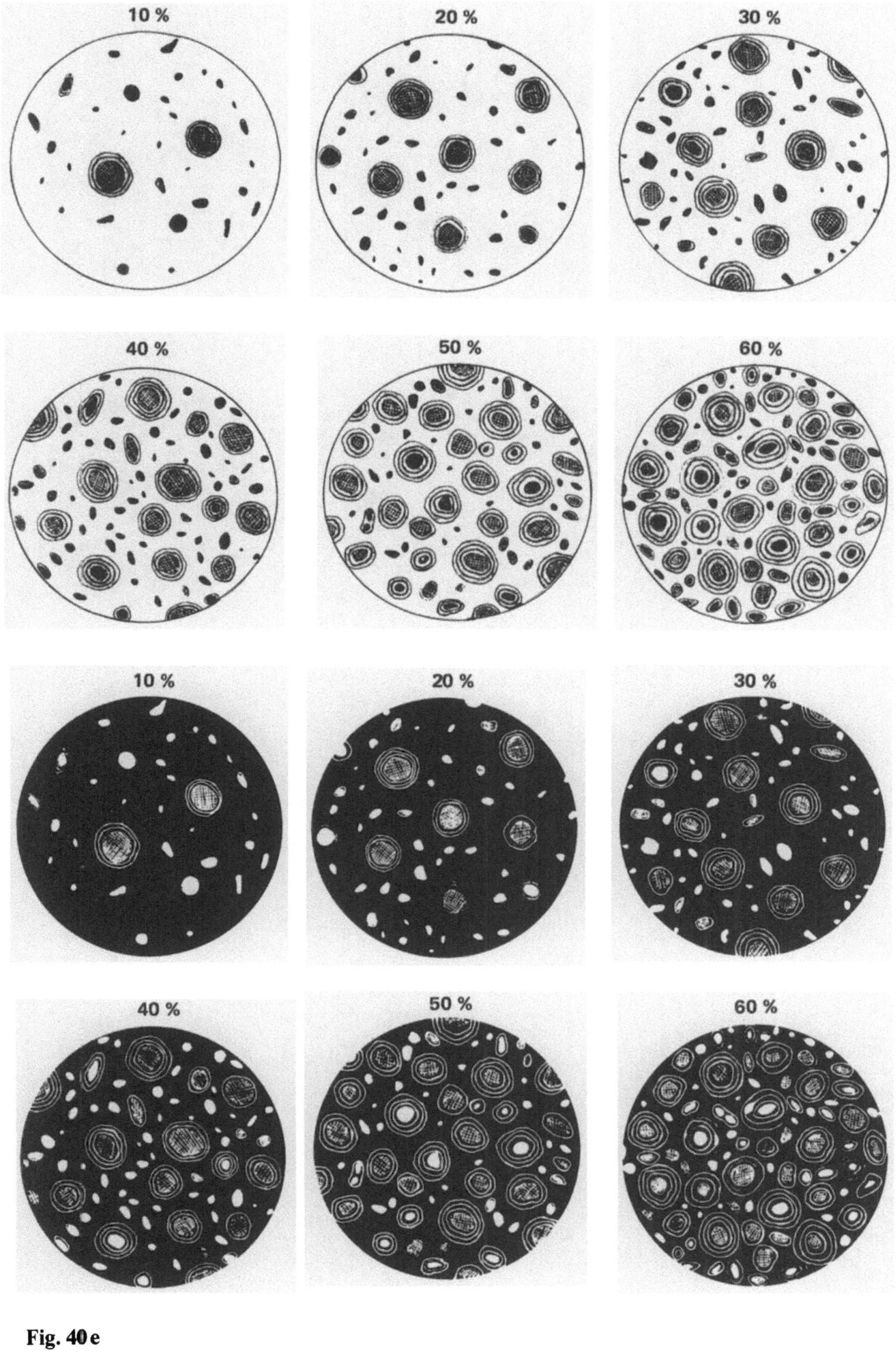

Fig. 40 e

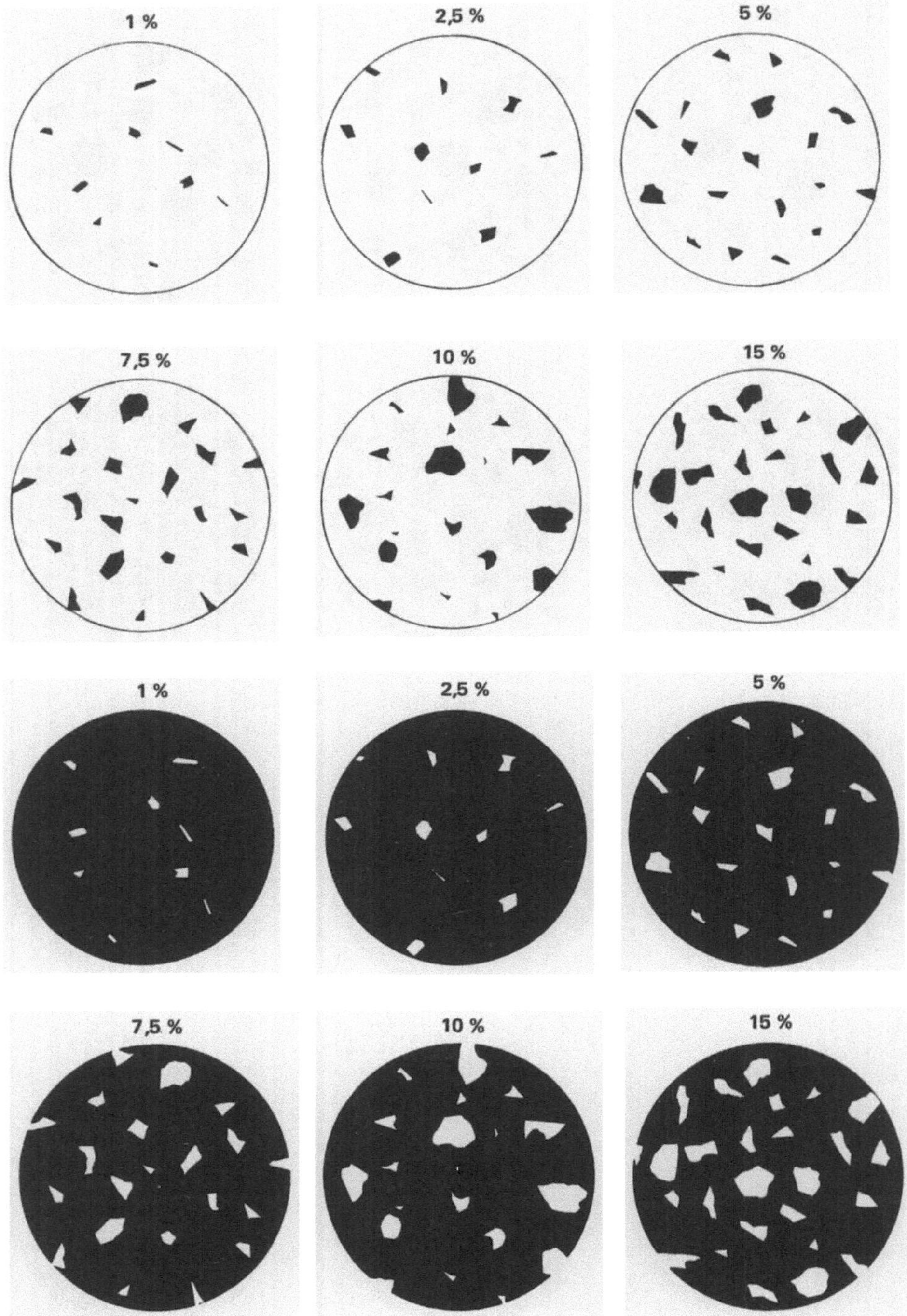

Fig. 40f

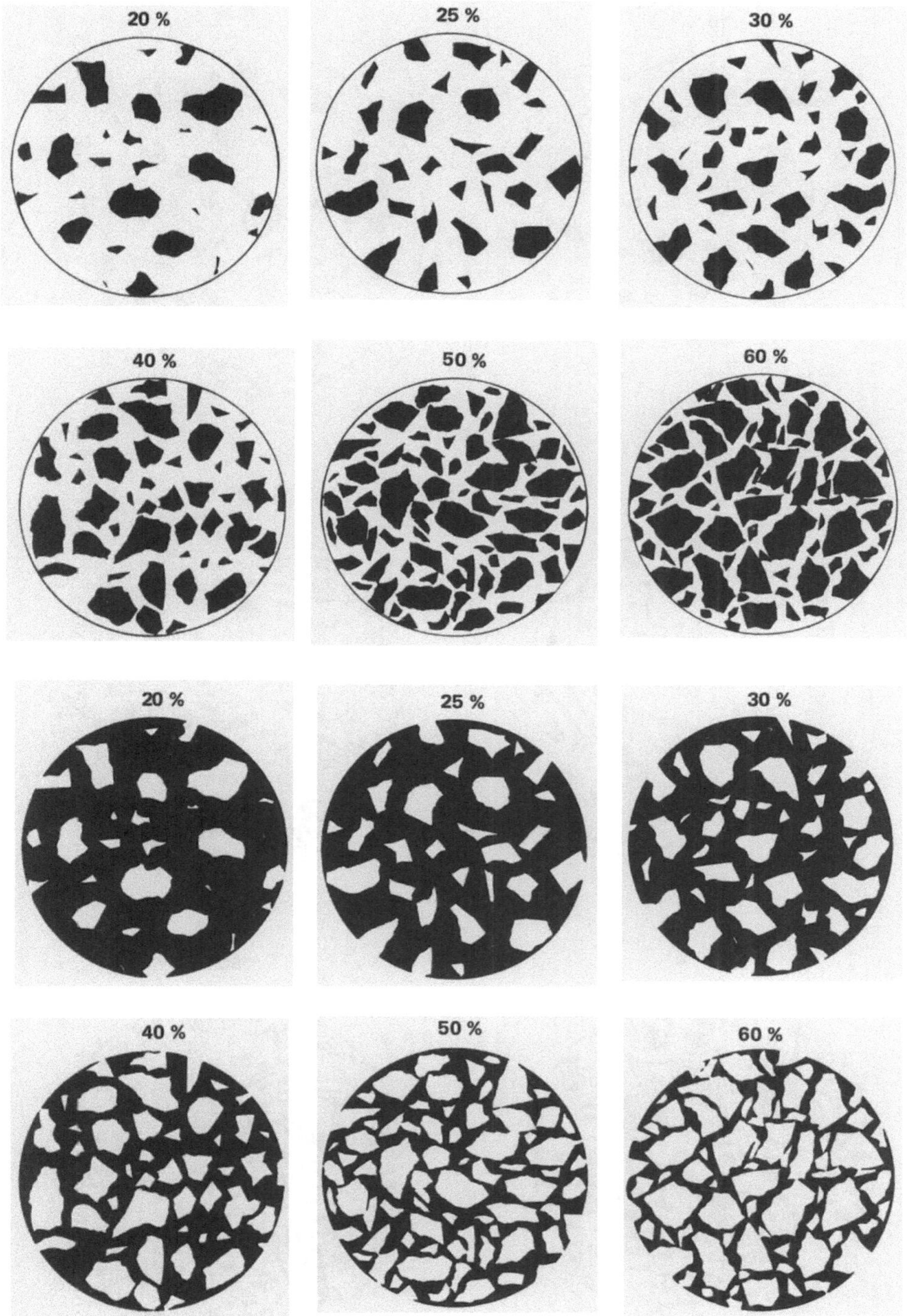

Fig. 40g

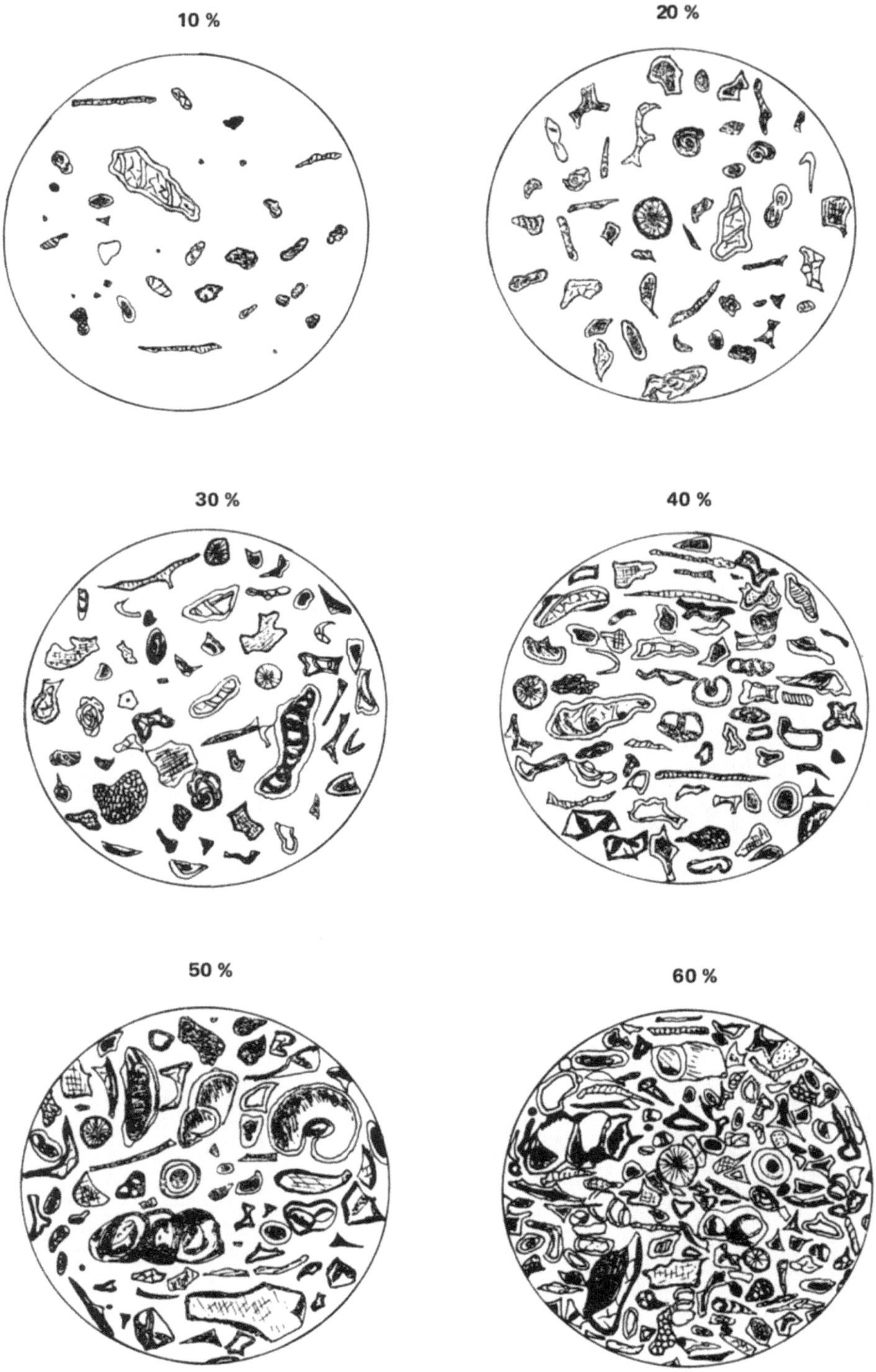

Fig. 40h

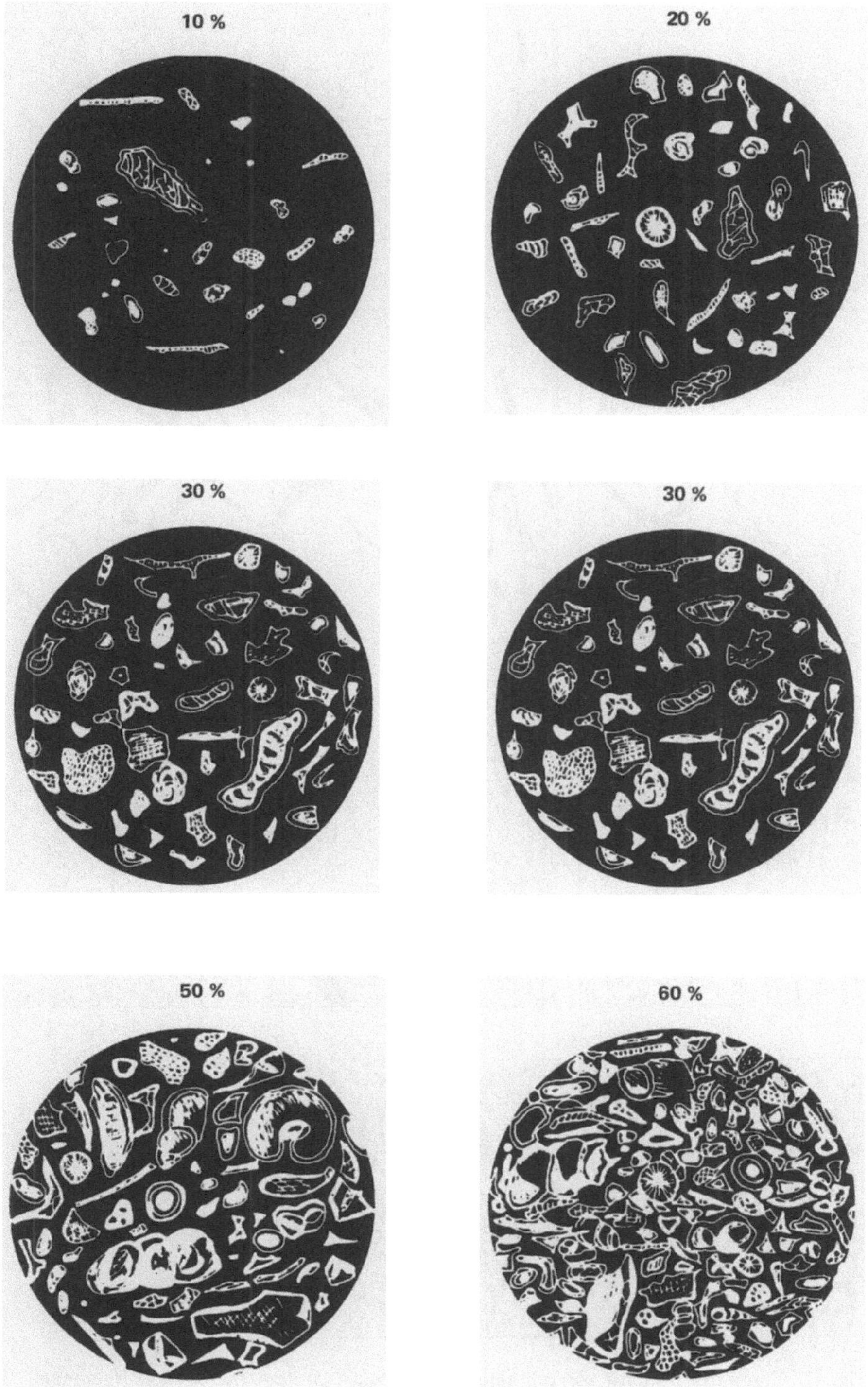

Fig. 40 i

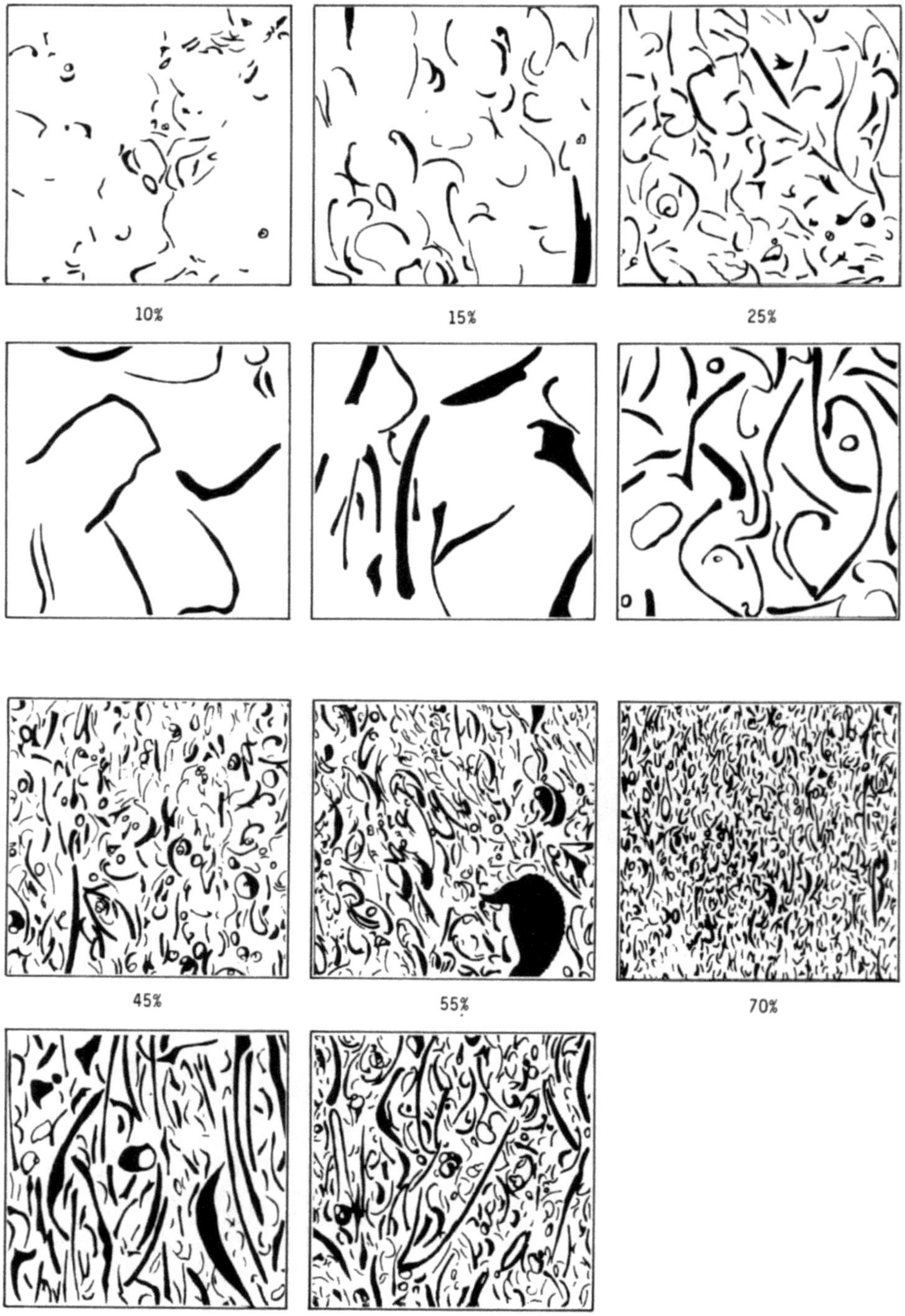

Fig. 41. Comparison charts for estimating the bioclast content (here, shell fragments). The percentage boundaries have been chosen such that they correspond to those used in the carbonate classification systems. After Schäfer (1969)

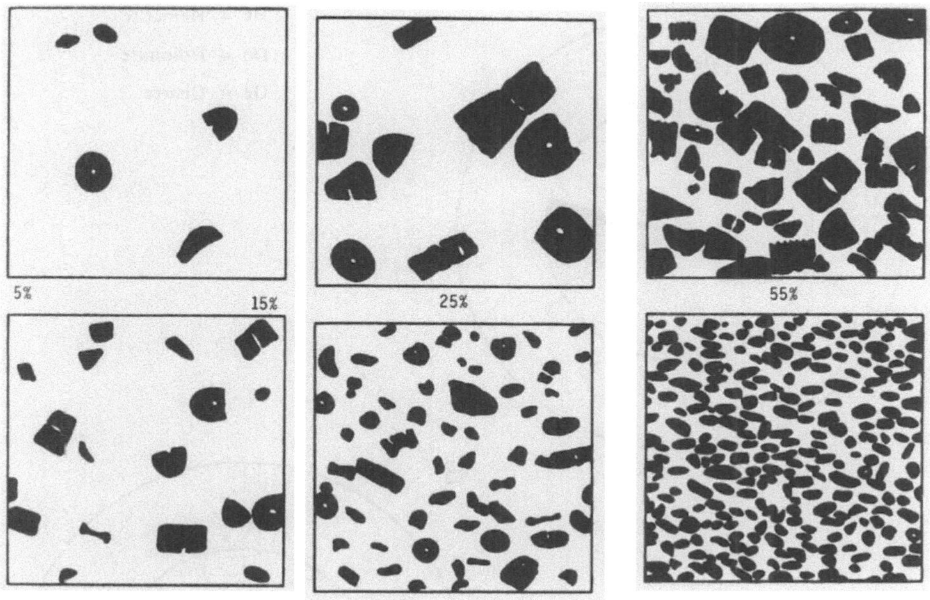

Fig. 42. Comparison charts for the frequency of crinoid fragments. After Schäfer (1969)

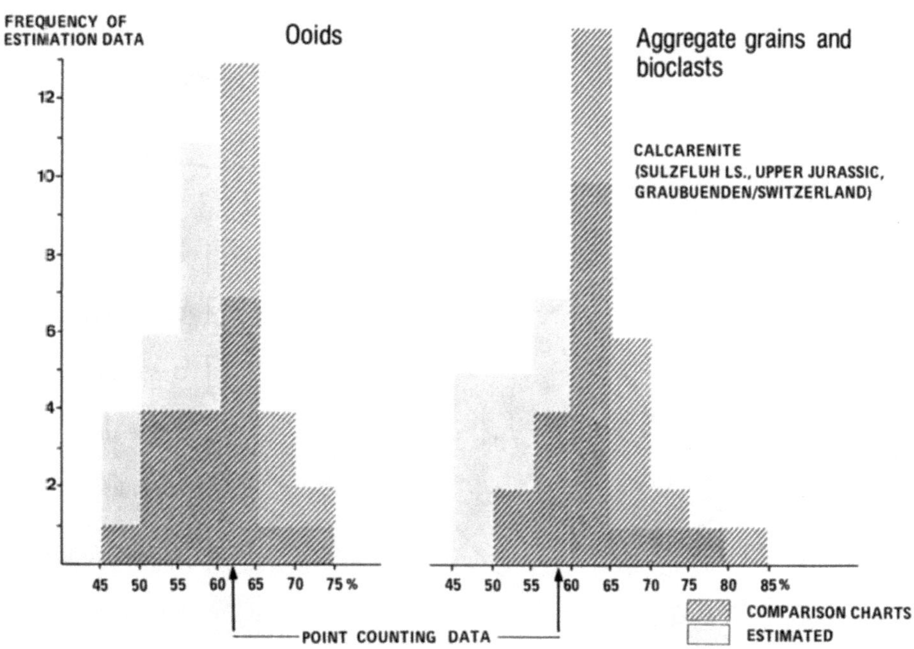

Fig. 43. Comparison of the point-counting method and estimation data (with and without the use of comparison charts). 30 people tested. Ooids in oosparites as well as aggregate grains + bioclasts in calcarenites. Platform carbonates (Sulzfluh limestone, Upper Jurassic: Graubuenden, Switzerland)

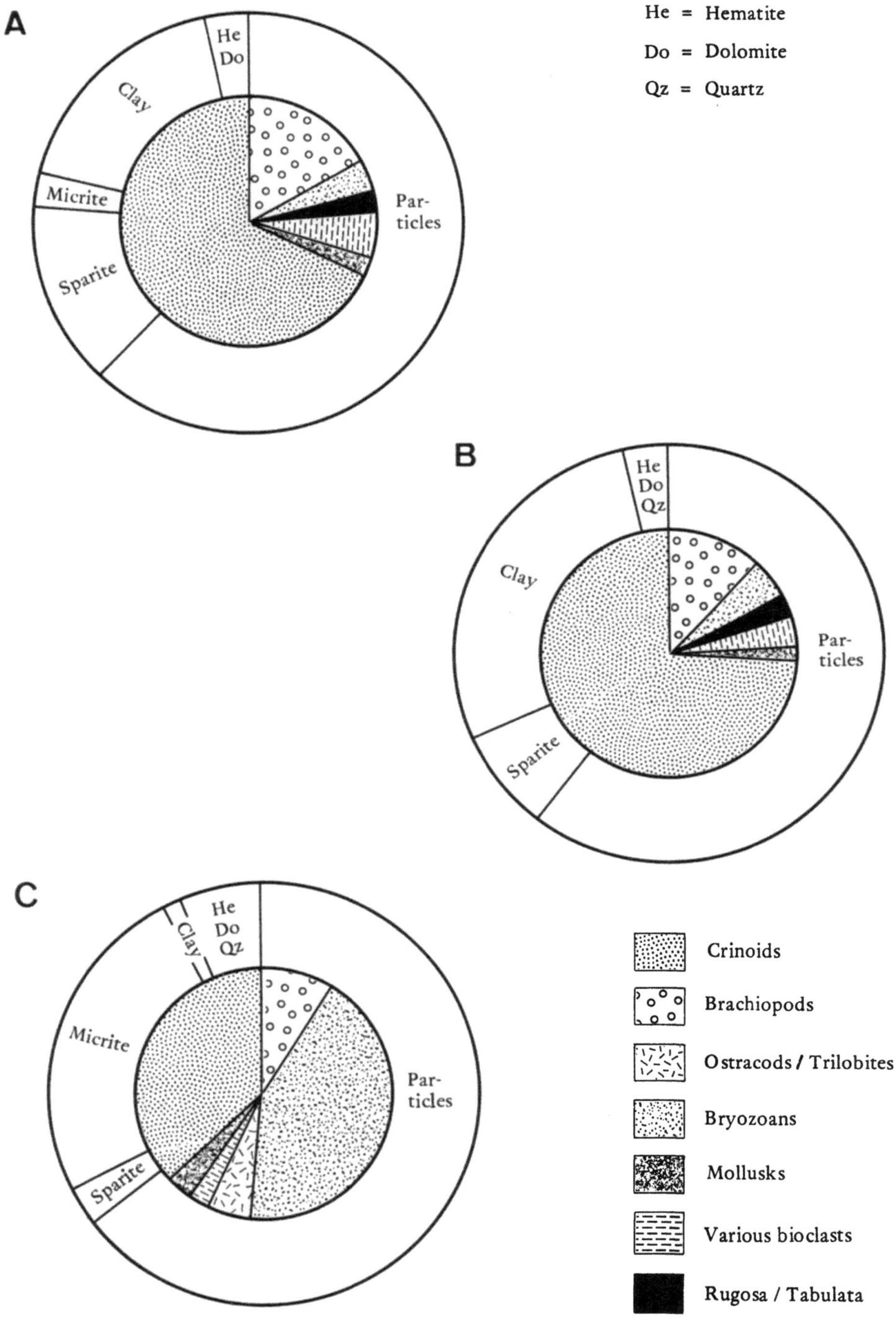

Fig. 44 A–C. Circle graphs of point-counting measurements of "detrital limestones". **A** bioclastic grainstone, **B** bioclastic packstone, **C** bioclastic wackestone. Lower Devonian (Lower Emsian): NW Spain. The detrital limestones are probably derived from storm deposits. *Outer ring,* modal composition; *Inner circle,* composition of the biofraction. After Stel (1975)

seem wise to carry out frequency analyses using thin-section photographs or subdividing the area taken up by the thin-section in a microfilm reader into several smaller squares which can be estimated separately.

Series of experiments carried out in the microfacies courses at Erlangen-Nuremberg University in West Germany produced favorable results in regard to the comparability of point-counting measurements and estimates (Fig. 43). Similar results have been found by A. Strasser (1979).

Generally, the estimates carried out without using comparison charts are somewhat lower than the estimates using the comparison charts of Baccelle and Bosellini. The majority of the estimates carried out by students and by post-graduates lie very close to the values obtained from the point-counting method.

Similar results were produced by a comparison of estimates and frequency values obtained from point-counting analyses (1000 points) of particles from Lower Carboniferous calcarenites (Mamet, 1964). However, very small particles such as foraminifera and calcispheres must be excluded; these are better described by the total number of individuals in a certain thin-section area. Similar methods were used by Hohenegger and Piller (1975) as well as by Schäfer and Senowbari-Daryan (1978) to characterize the frequencies of foraminifera in Upper Triassic reef limestones of the Northern Alps. Arnaud-Vanneau (1979) differentiated Cretaceous platform carbonates by using the frequency of microfossils (counted in an area of 3 cm^2) and the mean grain size of 100 particles; sorting is expressed by the total of the five smallest grains, divided by the total of the five largest grains, $\times$ 100.

Frequency ranges (about 1%–10%, 10%–25%, 25%–50%) make a more rapid estimate of characteristic frequencies possible and are often completely satisfactory for limestone classification (see Rich, 1964).

Method of microfacies elements frequency (Mišik, 1964): If one sets the total number of thin-section samples of a stratigraphic unit equal to 100%, then the "frequency" of a microfacies characteristic can be expressed by the number of thin-sections or percentage of the characteristic in the samples. Mišik regards the number as a reference value for the stability or homogeneity of organisms or groups of particles in geologic time.

II. Diagrams

In most cases graphing the results of frequency analyses is more preferable than listing them in a table. The following kinds of diagrams are useful:

a) *Circle graph:* Each particle or groundmass type is allotted to a part of a circle. The percentages are then presented within concentric rings of equal area, going from the middle with the lowest percentage to the outside with the highest percentage, in order to make the very low fractions stand out visually (see Eltgen, 1969). The method can be used to graph individual samples as well as to represent a survey of frequency analyses from a quarry or a profile; in such surveys the arithmetic means are graphed. Figure 44 shows a modification: The percentages of micrite, sparite, and non-biogenic particles are presented in an outer ring; the sectors of the inner circle are reserved for the composition of the biofraction.

b) *Coordinate system:* The sample interval is plotted along the abscissa, the percentages of microfacies characteristics along the ordinate. Subparallel or zigzag curves indicate mutual interaction between the features seen in the thin-section samples ("petrolog").

c) *Triangular diagram:* Each apex represents 100% of one of three constituents. The end members may be either particles/micrite/sparite or the most prominent types of particle (e.g., foraminifera/calcareous algae/aggregate grains).

III. Examples

Frequency analyses permit facies zones to be recognized and demarcated as long as enough material from traverses or profiles has been studied. Parts of Recent carbonate shallow-water and reef environments can be distinguished by the quantification of sediment composition as well as grain size and roundness (Ginsburg, 1956, Swinchatt, 1965, Florida; Purdy, 1963, Bahamas; Maxwell et al., 1964, Heron Island reef). Figure 45 shows the differentiation in the arenitic carbonate fraction expressed by particle frequency, as illustrated by a reef belt off the coast of Florida. Generally one can assume that frequency gradients reflect the distribution of communities of organisms. However, single biogenic particles may be over-represented in comparison to the biomass because they stand a better chance of becoming fossilized (e.g., red algae in Florida), whereas other sedimentologically and ecologically important elements may not even be encompassed by frequency analyses (e.g., marine grasses in Florida). An important basic requirement is that the relationships between modal analysis and grain sizes be taken into account: In Recent carbonates only the particles which are larger than

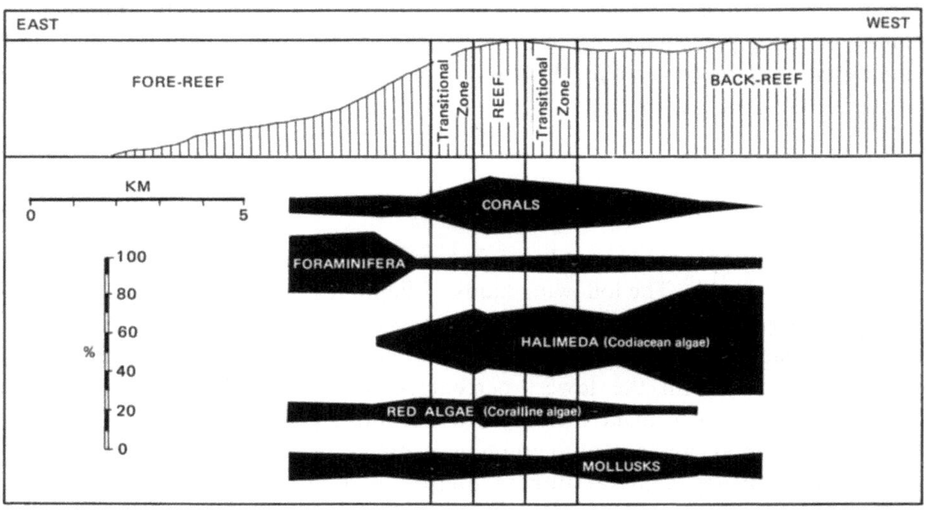

Fig. 45. Quantitative distribution of the most important groups of organisms in traverses perpendicular to the reef belt, Florida. Simplified after Ginsburg (1956)

0.16 mm appear to be more or less autochthonous, whereas smaller particles have often been transported off.

Extensive regional investigations of Ancient carbonate rocks, which have used frequency distributions have been done by: Jaanusson (1972, Ordovician: Sweden), Mamet (1964, Waulsortian facies, Lower Carboniferous: Belgium), Walpole and Carozzi (1961, Mississippian: Canada), Schenk (1967, Pennsylvanian: USA), Toomey (1966, Pennsylvanian: Kansas), and Stauffer (1962, Paleozoic: New Mexico).

4.3.2 References: Frequency Analysis

Allen, J. (1956): Estimation of percentages in thin-sections – considerations in visual psychology. J. Sed. Petrol. **26,** 160–161, Tulsa

Arnaud-Vanneau, A. (1979): Répartition de la microfaune dans les différents paléomilieux Urgoniens. Geobios, Mém. spec. **3,** 255–275, 2 Pls., 4 Figs., Lyon

Bacelle, L., Bosellini, A. (1965): Diagrammi per la stima visiva della composizione percentuale nelle rocce sedimentarie. Ann. Univ. Ferrara, N.S., Sez. IX, Sci. Geol. Paleont. **1**/3, 59–62, 24 Pls., Ferrara

Bandemer, H. (1965): Für die vorgegebene Genauigkeit notwendige Punktzahl beim Point-Counter-Verfahren. Z. angew. Geol. **11**/3, 147–148, Berlin

Baily, M. B. (1965): The sampling error in modal analysis. Amer. Min. **50,** 1/2, 196–211

Carozzi, A. V. (1950): Contribution a l'etude des rhythmes de sedimentation. Arch. Sci. **3, 17–40,** 95–146, Genf

Carozzi, A. V. (1958): Micro-mechanisms of sedimentation in the epicontinental environment. J. Sed. Petrol. **28**/2, 133–150, 13 Figs., Tulsa

Carozzi, A. V. (1961): Reef petrography in the Beaverhill Lake Formation, Upper Devonian, Swan Hills area, Alberta, Canada. J. Sed. Petrol. **31,** 497–513, Tulsa

Carozzi, A. V., Frost, S. H. (1966): Turbidites in dolomitized flank beds of Niagaran (Silurian) Reef, Lapel, Indiana. J. Sed. Petrol. **36**/2, 563–575, 10 Figs., Tulsa

Chayes, F. (1950): Measurement of intercept distances in thin-sections. Trans Amer. geophys. Union **31**/6, 860–872, Washington

Chayes, F. (1956): Petrographic Modal Analysis. 113 pp., 15 Tab., New York: Wiley

Conley, C. D., Davis, J. C. (1973): Carbonate petrography by pattern recognition. Bull. Amer. Ass. Petrol. Geol. **57**/2, 399–406, 3 Figs., Tulsa

Demirmen, F. (1971): Counting error in petrographic point-count analysis: a theoretical and experimental study. J. internat. Ass. math. Geol. **3**/1, 15–41

Dennison, J. M., Shea, J. H. (1966): Reliability of visual estimates of grain abundance. J. Sed. Petrol. **36**/1 81–89, 4 Figs., Tulsa

Dunham, R. J. (1962): see 6.4

Eltgen, H. (1969): see 7.3.2

Emerich, G. H., Wobber, F. H. (1963): A rapid visual method for estimating sedimentary parameters. J. Sed. Petrol. **33**/4, 831–843, 4 Figs., 5 Tab., Tulsa

Emerson, D. O. (1958): A stage for macropoint counting. Amer. Min. **43,** 1000–1003

Fediaevsky, A. (1963): Méthode d'étude quantitative des microfaciès calcaires. Rev. Micropal. **6**/3, 175–182

Feray, D. E., Heuer, E., Hewatt, W. G. (1962): Biological, genetic and utilitarian aspects of limestone classification. In: Ham, W. E. (ed.): Classification of Carbonate Rocks. Mem. Amer. Ass. Petrol., Geol. **1,** 20–32, Tulsa

Fisher, C. (1971): The new Quantimed 720. Microscope **19,** 1–20

Folk, R. L. (1951): A comparison chart for visual percentage estimation. J. Sed. Petrol. **21**/1, 32–33, 1 Fig., Tulsa

Fox, W. T., Brown, J. A. (1965): The use of time-trend analysis for environmental interpretation of limestones. J. Geol. **73,** 510–618, Chicago

Friedman, G. M. (1965): In defence of point counting analysis. A discussion. Sedimentology **4,** 247–253, Amsterdam

Galehouse, J. S. (1969): Counting grain mounts: number percentage vs. number frequency. J. Sed. Petrol. **93**/2, 812–815, 1 Fig., Tulsa

Ginsburg, R. N. (1956): Environmental relationships of grain size and constituent particles in some South Florida carbonate sediments. Bull. Amer. Ass. Petrol. Geol. **40,** 2384–2427, 10 Figs., Tulsa

Glagolew, A. A. (1933): On the geometrical methods of quantitative mineralogical analysis of rocks. Trans. Inst. Econ. Min. Moscow **59,** 1–47, Moscow

Harbaugh, J., Demirmen, F. (1964): see 7.3.2

Harvey, R. D., Steinmetz, J. C. (1971): Petrography of carbonate rocks by image analysis. Florida Dep. Natural Res. Spec. Publ. **17,** Bur. Geol., 161–170, 4 Figs., Tampa

Hennig, A. (1958): Kritische Betrachtungen zur Volumen- u. Oberflächenmessung in der Mikroskopie. Zeiss-Werkz. **6**/30, 76–86, Oberkochen

Hohenegger, J., Piller, W. (1975): Ökologie und systematische Stellung der Foraminiferen im gebankten Dachsteinkalk (Obertrias) des nördlichen Toten Gebirges (Oberösterreich). Paleo-3 **18,** 241–276, 8 Pls., 5 Figs., Amsterdam

Hörnstein, A. (1959): A method and a set apparatus for mineralogic-granulometric analysis with a microscope. Bull. geol. Inst. Univ. Uppsala **38,** 105–137

Jaanusson, V. (1972): Constituent analysis of an Ordovician limestone from Sweden. Lethaia **5,** 217–237, 11 Figs., Oslo

Lamprecht, J. (1954): Die Glagolewsche „Punktmethode" und ihre Anwendung. Geologie **3,** 1085–1094, Berlin

Mamet, B. (1964): Sédimentologie des Faciès "Marbres Noires" du Paléozoique franco-belge. Inst. Roy. Sci. Nat. Belgique, Mém. **151**

Marsal, D. (1967): see 7.3.2

Maxwell, W. G. H., Jell, J. S., McKellar, R. G. (1964): Differentiation of carbonate sediments in the Heron Island reefs. J. Sed. Petrol. **34,** 294–308, Tulsa

Mišik, M. (1964): Lithofazielle Studien im Lias der Großen Fatra im westlichen Teil der Niederen Tatra. Sbornik geol. vied., Zap. Karpaty **1,** Bratislawa

Müller, G. (1964): see 1.4.1

Payton, C. E. (1966): Petrology of the carbonate members of the Swope and Dennis Formation (Pennsylvanian), Missouri and Iowa. J. Sed. Petrol. **36,** 576–601, Tulsa

Plas, L. van der (1965): In defence of point counting analysis, a reply. Sedimentology **4,** 247–253, Amsterdam

Plas, L. van der, Tobi, A. C. (1965): A chart for judging the reliability of point counting results. Amer. J. Sci. **263,** 87–90, 1 Fig., New Haven

Purdy, E. G. (1963): see 2.4

Rich, M. (1964): Petrographic classification and method of description of carbonate rocks of the Bird Spring Group in Southern Nevada. J. Sed. Petrol. **34**/2, 365–378, Tulsa

Rosiwal, A. (1898): Über geometrische Gesteinsanalysen. Verh. geol. Reichsanst. **1898,** 143–175, Wien

Sander, B. (1951): Einführung in die Gefügekunde als Geologischer Körper, 2. Teil. Die Korngefügemerkmale. 409 pp., 4 Pls., 153 Figs., Wien – Innsbruck: Springer

Sarntheim, M. (1971): Oberflächensedimente im Persischen Golf von Oman. II. Quantitative Komponentenanalyse der Grobfraktion. „Meteor" Forschungsergebnisse, Reihe **C, 5,** 113 pp., Pls. 1–4, 8 Figs., Berlin – Stuttgart: Borntraeger

Sahu, B. K. (1974): Probabalistic solution to thin-section size analysis expressing size as intercepts and frequency as number or as measure. Indian J. Earth Sci. **1**/2, 168–176, 4 Tables

Schäfer, K. (1969): Vergleichs-Schaubilder zur Bestimmung des Allochemgehalts bioklastischer Karbonatgesteine. N. Jb. Geol. Paläont. Mh. **1969**/3, 173–184, 9 Figs., Stuttgart

Schäfer, P., Senowbari-Daryan, B. (1978): Die Häufigkeitsverteilung der Foraminiferen in drei oberräthischen Riff-Komplexen der Nördlichen Kalkalpen (Salzburg, Österreich). Verh. geol. Bundesanst. **1978**/2, 73–96, 2 Figs., 4 Pls., Wien

Schenk, P. E. (1967): Facies and phase of the Altamont Limestone and Megacyclothem (Pennsylvanian), Iowa to Oklahoma. Geol. Soc. Amer. Bull. **78,** 1369–1384, 5 Figs., 3 Pls., New York

Shepard, F. P., Moore, D. G. (1954): Sedimentary environments differentiated by coarse-fraction studies. Bull. Amer. Ass. Petrol. Geol. **38**/8, 1792–1802, Tulsa

Skupin, K. (1973): Stratigraphy and microfacies in the crinoidal limestones (Trochiten limestone, Triassic) of Southwest Germany. Sed. Geol. **9,** 1–19, 9 Figs., Amsterdam

Smosna, R., Warshauer, S. (1978): Fossil diversity in thin-sections. J. Sed. Petrol. **48**/1, 331–336, 2 Figs., Tulsa

Stanley, S. M. (1966): Paleoecology and diagenesis of Key Largo Limestone, Florida. Bull. Amer. Ass. Petrol. Geol. **50,** 1927–1947, Tulsa

Stauffer, K. W. (1962): Quantitative petrographic study of Paleozoic carbonate rocks, Caballo Mountains, New Mexico. J. Sed. Petrol. **32,** 357–396, Tulsa

Stel, J. H. (1975): The influence of hurricanes upon the quiet depositional conditions in the Lower Emsian Lavid Shales of Colle (NW Spain). Leidse Geol. Meded. **49**/3, 475–486, 14 Figs., Leiden

Strasser, A. (1979): see 7.3.2

Swinchatt, J. P. (1965): see 2.4

Terry, R. D., Chilingar, G. V. (1958): Summary of "Concerning some additional aids in studying sedimentary formations" by Shvetsov, M. S. J. Sed. Petrol. **25,** 229–234, Tulsa

Toomey, D. F. (1966): see 7.3.2

Walpole, R. L. Carozzi, A. V. (1961): Microfacies study of Rundle Group (Mississippian) of Front Ranges, Central Alberta, Canada. Bull. Amer. Ass. Petrol. Geol. **45**/11, 1810–1846, 1 Pl., 25 Figs., 6 Tab., Tulsa

Wilson, R. C. L. (1965): Dual classification for point-counting limestones. Nature **207,** No. 4999, 849–850, 3 Figs., London

5 Fossils in Thin-Sections

In this chapter the following questions will be discussed: which fossils can be expected to appear in thin-sections in various geological time units (Sect. 5.1); how, despite difficulties caused by random sectioning, groups may be determined or classified (5.2); and which special problems and interpretations apply to certain more abundant groups (5.3). The calcareous algae will be discussed in detail, because of their significance and the little attention they often receive. Section 5.4 is concerned with the "microfacies zones" which have been proposed for the Mesozoic of the Tethyan region.

Well-illustrated introductions to the wealth of organisms which appear in thin-sections are found in Cayeux (1935), Maslov (1937), Majewske (1969), and in particular, Horowitz and Potter (1971) as well as Lucas et al. (1976) – see Sect. 5.5, p. 360.

5.1 General Survey

Because of their different rates of evolution and various diversities, organisms appearing in thin-sections of limestones have different abundances in individual geological systems. Dissimilarities are also generated by ecological factors. Moreover, certain groups are unlikely to be recognized in thin-sections because of the minute sizes of their organic remains. And some organism groups are almost always underrepresented or even missing in thin-sections when compared with the residues of limestones; this is especially true for agglutinated foraminifera, juvenile tests of mollusks, many echinoderms, ostracods, and conodonts. Special examples of this are the microfossils found in basinal Triassic Hallstatt limestones. Whereas hundreds and thousands of remains of holothurians occur in acid residues, in thin-sections of the same samples generally only very few sclerites can be seen. The description of the fauna should never, therefore, be limited entirely to thin-sections or macrofauna.

Table 25 gives a cursory sketch of the organic groups appearing more abundantly in thin-sections of limestones. The large number of – sometimes stratigraphically important – "microproblematica" have been left out and are reserved for a later Section (5.3.13).

Table 25. Frequencies and stratigraphic significance of "fossils in thin-sections". + = rare in thin-sections; / = common in thin-sections; × = relatively abundant in thin-sections; ⊕ = need oriented sections for generic determinations

	Precambrian	Cambrian	Ordovician	Silurian	Devonian	Lower Carboniferous	Upper Carboniferous	Permian	Triassic	Lias	Middle Jurassic	Upper Jurassic	Lower Cretaceous	Upper Cretaceous	Early Tertiary	Late Tertiary
Foraminifera		—	—	—	/	×	×	×	/	/	/	—	—	—	—	—
Radiolaria		—	/	/	/	/	—	—	/	/	×	×	×	/	—	—
Calpionellids												—	—			
Other Tintinnids			+	+	+	+	- -	- -	+	- -	- -	/	—	- -	—	—
Archaeocyathina		⊕														
Siliceous sponges		—	—	—	—	—	—	—	—	—	—	×	—	—	—	—
Calcareous sponges		—	—	—	—	—	⊕	⊕	×	—	—	—	—	—	—	—
Sclerospongia								—	—	- -	- -	- -	⊕	- -	—	- -
Stromatoporoids			⊕	⊕	⊕	- -	- -	- -	⊕	—	⊕	⊕	⊕	⊕	—	
Chaetetids			⊕	⊕	⊕	⊕	⊕	⊕	⊕	⊕	⊕	⊕	⊕	⊕	⊕	
Tabulate corals			⊕	⊕	⊕	⊕	⊕	⊕								
Corals			⊕	⊕	⊕	⊕	⊕	⊕	⊕	⊕	⊕	⊕	⊕	⊕	⊕	⊕
Bryozoa		⊕	⊕	⊕	⊕	⊕	⊕	⊕	⊕	⊕	⊕	⊕	⊕	⊕	⊕	⊕
Brachiopoda		/	/	/	/	/	/	/	/	/	—	—	—	—	—	—
Gastropoda			—	—	/	—	—	/	/	—	—	/	/	—	/	/
Pelecypoda		—	—	—	—	/	/	/	/	/	/	/	/	/	/	/
Filaments									×	×	×	×				
"Nautilids"			/	/	—	—	—	—	—	—	—	—	—	—	—	—
Ammonites					—	—	—	—	/	/	/	/	—	—		
Belemnites						—	—	—	—	/	/	/	/	/	—	
Tentaculites			—	×	×											
Serpulids			—	—	—	—	—	—	/	—	/	/	—	—	/	×
Trilobites		×	×	/	—	—	—	—								
Crustacean coprolites								/	/	/	/	/	/			
Ostracods		—	—	/	/	—	—	—	/	—	—	—	—	—	/	/
Cystoids, Blastoids			/	—	—	—	—	—								
Sessile crinoids			×	×	×	×	/	/	/	×	/	/	—	—	—	—
Planktonic crinoids									/	—	—	/	—	—	—	—
Echinids			—	/	/	/	/	/	/	/	/	/	/	/	/	/
Ophiurids				—	—	—	—	—	/	/	—	/	—	—	—	—
Holothurians						—	—	—	—	/	—	—	—	—	—	—
Blue-green algal "mats"	×	×	×	×	×	×	×	×	×	×	×	×	×	×	×	×
Porostromates		×	×	×	×	×	/	/	/	/	/	×	×	/		
Green algae:																
Codiaceae		/	/	/	/	/	/	/	/	/	/	×	×	×	×	×
Dasycladaceae		/	/	/	/	/	/	/	/	/	/	/	/	/	/	/
Characeae					/	/	—	—	/	—	—	/	/	/	/	/
Rhodophyceae:																
Solenoporaceae		/	/	×	×	/	/	/	×	/	/	×	×	/	/	
Gymnocodiaceae								/	/	- -	- -	- -	/			
Squamariaceae													/	/		
Corallinaceae												—	—	/	/	/
"Ungdarellaceae"						—	×	/								
"Ancestral Corallinaceae"						—	/	/								
"Phylloid algae"							/	×								
Calcispheres					/	/	- -	- -	- -	- -	- -	- -	/	×		
Globochaete									/	×	/	/				
Spores/pollen		—	—	—	—	—	/	—	—	—	—	+	—	—	—	—
Nannoconids												/	/			

5.2 Key to Identification

Keys to the identification of skeletal grains, which are usually cut quite randomly during thin-section preparation, have been designed by Horowitz and Potter (1971), Füchtbauer (1974: 212), and Majewske (1969, especially for pelecypod shells). Milliman (1974) has published a Key for identifying Recent carbonate particles in transmitted and direct light.

The following Key to Identification, was originally developed to serve the practical needs of the microfacies courses in the Paleontology Department of Erlangen University. It was found to be quite necessary, especially for beginners, to reduce the confusing mass of possible thin-section figures to a few basic forms, thus simplifying the first introduction to the problems involved. The sizes listed in the Key represent only the most common values, and hence must be regarded as just rough approximations.

The Key to Identification considers only two-dimensional "thin-section figures" and reduces them to the following major shapes and structures:

A. Circular and elliptical . . . . Key Nr. 1
B. U and V shaped 12
C. Curved 16
D. Net-like 23
E. Layered 31
F. Chains of segments 34

The letters or numbers in parentheses should aid in retracing your original steps in the identification process.

A. Circular and Elliptical

1 a Scattered figures 2
b Interconnected figures 11
(1 a) 2 a Several openings of more or less the same size:
a_1 Subrounded micritic particles, diameter about 0.15–0.50 mm; regular pattern of points which are actually cross-sections of canals, sometimes filled with sparry calcite (Plate 18/13) . . . *Crustacean coprolites*
a_2 Large opening with an outer ring of smaller openings (Plate 28/7) . . . . *Characeae*
b One central opening 3
c No central opening . . . . recrystallized *sponge spicules* (rhaxes) which cannot be more closely identified, *radiolarians,* planispiral *foraminifera* or *calcispheres,* or . . 10
(2 b) 3 a Diameter < 0.10 mm:
Calcite rings without microstructures, together with U and V shaped figures . . . . *oblique sections of calpionellids*
b Diameter generally < 0.50 mm 4
c Diameter often > 0.50 mm 5
d Diameter usually several mm 6
e Diameter > 1 cm, up to several cm 7

(3 b) 4 a Microcrystalline or sparry circular figures; original inner boundary often very indistinct, outer boundary sometimes obscured by jagged edges (Plate 53/2) . . . . *calcitized radiolarians*

b Often circular sections with unclear periphery, wall may consist of various kinds of layers; sometimes with fine radial pores (Plate 20/6–7) . . . . *calcisphaeres*

(3 c) 5 a Large central opening:
Thin calcitic wall elongated to regular pointed spines; sometimes spiral structure and outer ring of smaller openings; diameter about 3 mm (Plate 28/7) . . . . *Characeae-Oogonia and gyrogonites*

b Transparent, often yellowish; sometimes spiky outline around a circular figure; maximum diameter 1 mm . . *foraminifera, e.g., Nodosaria*

c Very complex geometrical pattern of radial and tangential elements around a central opening, sometimes ornamented, sometimes finely perforated single crystal (Plate 27/1, 2, 4) . . . . . *Echinoid spine*

d Central opening with walls of varying thickness, which are often transversed with pore canals which often appear dark. The pore canals may be of uniform width, or may taper either to the outside or the inside boundaries (Plate 29/6). . . *Dasycladaceae* or *Gymnocodiaceae*

e Large central opening, dentate outline (outside ornamentation); diameter up to about 1 mm (Plate 26/3) . . . . . . *tentaculites*

f Sharply defined circular and elliptical figures with very small central opening, calcite or SiO_2; usually together with rod-shaped figures; sometimes with a pointed end; without any internal structure (Plate 50/3) . . . . *sponge spicules*

g Calcitic circular and elliptical figures with a narrow central opening; wall with lamellar microstructure . . *brachiopod spines* (e.g. *productids*)

h Calcitic phosphate, often dark-colored circular figures; diameter mm-sized; sometimes irregularly elongated to an elliptical form; often together with "shepherd's crooks" – see 21; (Plate 26/7) . . . *trilobites*

(3 d) 6 a Large, calcitic figures of encrusted shells; microstructure divided into two parts: outer concentric lamellar structure, clear inner layer; figures usually in clusters (Plate 26/5) . . . . *serpulids* (see 22 c)

b Large calcitic figures, microstructure divided into three parts: differentiated prismatic middle layer and clear inner and outer layers; figures often overlap each other like the tubes of a telescope . . . *scaphopods*

(3 c) 7 a Central opening very small or absent 8

b Central opening large 9

(7 a) 8 a Small central lumen with various shapes (circular, pentagonal, etc.), surrounded by very fine meshwork; often yellowish single crystal (Plate 53/1) . . . . *crinoid plates*

b Large circular figure (diameter 3–10 mm) with very small (or absent) central opening, surrounded by alternating layers of light- and dark-colored rings and/or large radiating prisms of calcite (Plate 26/4) . . . . *belemnite rostrum*

(7 b) 9 a Large central opening with radiating elements which usually do not extend to the center; variously differentiated walls which may or may

not be very well defined; sometimes a dark dot in the center (axial structure). Microstructures (of septa) like bundles of fibers ("herring bone structure") or with dark lines down the middle (Plate 10) . . . . *solitary corals* (*Rugosa* and *Scleractinia*)

b Large central opening; thick shell wall usually recrystallized; longitudinal sections show cylindrical figures containing partitions with a single pore in the center . . . . *Nautiloidea*

c Large circular figure indented on one side, with an element somewhat thicker towards the center. Shell usually recrystallized or cross-lamellar (Plate 25/1) . . . . *gastropods* (*oblique section of columella*)

d Circular figure with poorly defined wall exhibiting vermiform structures (Plate 23/4) . . . . *calcareous sponges*

(2c) 10a Subrounded particles; very fine meshwork, dark brown to black in transmitted light; often differentiated into a thick peripheral part and a gauze-like central part (spongy material), phosphatic . . . . *remains of vertebrates*

b Subrounded micritic particles without any distinct internal structure; diameter generally < 1 mm . . . . *peloids*

(1b) 11a Generally a few circular figures or segments of circles linearly arranged or in spherical aggregates. Chambers round or angular, often with spine-like ornamentation, total diameter 0.5–1 mm, sometimes larger (Plate 20/2) . . . . *planktonic foraminifera*

b Circular segments with matching protuberances, lying contiguous to each other as mirror images – corresponds to two calcite plates which extinct at different positions . . . . *ophiuroid vertebra*

c Circular segments and kidney-shaped figures often growing linearly along a rod-shaped element; sometimes in groups; diameter about 0.05 mm. Wall not differentiated (Plate 16/1) . . . . *Globochaete*

d Symmetrical linear figure. Round chamber in the middle with concentric chamber segments becoming larger towards the ends

d_1 Microscopic sizes (Plate 18/12): . . . . *foraminifera* (*planispiral* and *trochospiral*)

d_2 Macroscopic sizes (Plate 26/2) . . . *juvenile ammonites* and *gastropods*

e Unsymmetrically arranged chambers:

e_1 Microscopic sizes (Plate 15/5) . . . . *encrusted foraminifera*

e_2 Macroscopic sizes (Plate 26/5) . . . . *serpulids*

B. U and V Shaped

12a Length generally < 0.20 mm:
U and V shaped figures sometimes with flange or constriction near the open end; sometimes with thorn-shaped elongation. Wall calcitic or with dark-colored (organic) lamella (Plate 22) . . . *calpionellids/tintinnids*

b Length generally > 0.50 mm 13

(12b) 13a V-shaped figures 14

b Broad U-(and V-)shaped figures 15

(13 a) 14 a Narrow V figures, sometimes serrated outer edge (ornamentation), sometimes also inner ornamentation, sometimes "segmented". Length in mm and cm range (Plate 26/3; 36/2) . . . . . . *tentaculites*

b V figures stacked inside each other, edges extended (Upper Jurassic/Cretaceous) (Plate 30/6) . . . . *"Bankia" striata*

c Broad V shaped figures with symmetrical bulbous thickenings at the base or along the sides, single crystals, often smaller than 14 a (Plate 27/6) . . . . *pelagic crinoids*

d Long, relatively narrow V shaped figures with narrow V shaped inner cavity; wall of inorganic (calcitic) and organic (dark-colored) lamellae; diameter up to cm size (Plate 26/4) . . . . *belemnites*

(13 b) 15 a Large central opening, distinct wall with pores which are uniformly thick or taper towards the inside or the outside (oblique longitudinal section); in tangential section circular pores may also be visible; sometimes clearly segmented (Plate 29/6–8) . . . . *dasycladaceans*

b Serial associations of barrel-shaped or ovoid segments, generally thin walls with numerous very fine pores which radiate obliquely outwards. Inner space wide, usually filled with sediment or sparite; rarely primary longitudinal structures (Plate 29/5) . . . *gymnocodiaceans*

C. Curved

16 a Generally solid 17

b Generally hollow 22

(16 a) 17 a Slightly curved 18

b Very curved, recurved, and coiled 19

(17 a) 18 a Very thin (about 0.03 mm); relatively long; often deposited parallel to each other and in great abundance. Lamellar and fine prismatic microstructure (Plates 15/6; 16/1; 25/5) . . . . . . *"filaments"*

b Thicker, sometimes corrugated; often various microstructures within a shell (lamellar, prismatic, crossed-prismatic) (Plates 51/1; 52/1) . . . . *pelecypod valves*

c Thick jagged convex edge . . . . *Aptychus*

d Curved shells; lamellar microstructure; calcium phosphate (brownish, very low birefringence); shell thickness to 0.20 mm . . . . *inarticulate brachiopods*

e Thick, often very curved and bent elements with (and without) dark-colored vertical and oblique pores (diameter about 0.03 mm), and conspicuous fine lamellar microstructures (Plates 12/3; 25/2, 3, 4) . . . . *articulate brachiopods*

f Calcitic; generally slightly curved elements with peripheral inner structures (pores in the cortex zone) and generally strong recrystallization; coarse sparitic central zone (Plates 28/6; 42/2) . . . . *"phylloid" algae (Squamariaceae?)*

g Half-moon elements, single crystals . . . . *arm plates of crinoids* or *dorsal and lateral shields of ophiuroids*

(17 b) 19 a Spiral 20
b Bent at one end 21
(19 a) 20 a Unpartitioned spiral, center often thickened. Crossed-prismatic microstructure usually destroyed (Plate 25/1) . . . *gastropod cross-section*
b Partitioned spiral; mm-sized; wall and surface elements variously differentiated (Plate 18/5; 21/4) . . . . *foraminifera*
c Partitioned spiral; cm-sized; partitions with an opening; walls usually recrystallized . . . . *juvenile ammonites*
(19 b) 21 a Figure bent at one end like a shepherd's crook, prismatic microstructure, generally distinct dark-colored edges (Plate 26/7) . . . . *trilobites* (*thorax segments*)
b Thin valves or shells, edges recurved; size 0.5 to 1 mm; fine prismatic microstructure (Plate 26/1) . . . . *ostracods*
(16 b) 22 a Very small (diameter 0.10 to 0.05 mm), short and curved tubes usually occurring in clusters; often embedded in micritic layers; sometimes linked in chains which may be intertwined (Plate 28/1) . . . . *porostromate algae* (*Cyanophyceae,* e.g., *Girvanella*)
b Slightly curved, longer, sometimes simple bifurcated tubes with microgranular wall; diameter usually about 0.10 mm; often occur in clusters; frequently together with reef builders (Late Triassic) (Plate 18/8) . . . . *Microtubus*
c Well-defined curved or coiled calcitic tubes; often cemented; microstructure divided into 2 parts (inside: parallel lamination; outside: cone-in-cone lamination); mm-sized, tube diameter about 1 mm (Plate 26/5) . . . . *serpulids*

D. Net-like

23 a Regular well-defined meshwork 24
b Irregular meshwork of various shapes 30
(23 a) 24 a Meshwork rectangular, square or polygonal 25
b Cystose meshwork 29
c Meshwork extremely regular, very fine; single crystals! (Plate 27/5) . . . . *echinoderms*
(24 a) 25 a Network structures 26
b Open meshwork of thin rod-shaped elements interlocked at distinct angles (SiO_2 or calcite) (Plate 50/1) . . . . *sponges*
c Bifurcated "tubes" whose angles are all the same within a "colony" (Plates 28/3; 52/4) *porostromate filamentous algae* (*Type "Cayeuxia"*)
d Colonies of tubes more or less parallel to each other 27
e Shells composed of prisms lying parallel to each other 28
(25 a) 26 a Very fine net structure of generally rectangular cells, which appear black in transmitted light. Differentiation into areas with larger and smaller meshes; sparry-filled circular figures embedded in the net (sporangia). Cell diameter about 5–15 microns (Plates 28/8; 29/1) . . . . *Corallinaceae*

b Cm-sized colonies of long thin vertical threads; horizontal elements, if present, may be irregularly or regularly spaced (zonation). No other differentiation. Mesh diameters between 10 and 30 microns (Plate 29/3) . . . . *Solenoporacean red algae*

c Fragments with a fine net structure of various-sized cells in the center and cyclic alternations of cell sizes on the periphery. Cell walls sometimes quite thickened; sometimes with carbonized organic substance, sometimes mineralized by infiltration of SiO_2, FeS_2, $CaCO_3$ . . . . *remains of wood*

d Fragments with networks of sometimes fine, porous, very distinct "horizontal elements" and connective elements that are either closely or widely spaced; usually microgranular fine structure; often serial associations of chambers (Plate 19) . . . *Fragments of Larger foraminifera*

e Larger colonies with net structures of horizontal, sometimes undulating elements (laminae) and vertical pillars and spines of various lengths; extremely differentiated microstructures (Plates 24/1; 41/5) . . . . *stromatoporoids*

(25 d) 27 a Colonies or fragments of colonies of uniform or various-sized tubes containing flat and curved transverse plates. Colony shape often ramose (with tubes leading obliquely outwards). Tube diameter generally < 0.5 mm. Wall structure lamellar, rarely fibrous (Plate 24/7, 9) . . . . *bryozoans*

b Reticulated colonies of uniformly large, angular meshes (in tangential sections sometimes with radial elements or septa) or vertical tubes with transverse plates, sometimes wall pores. Fibrous microstructure sometimes consisting of groups of radiating fibrous bundles. Tube diameter usually > 0.5 mm, often several mm (Plate 24/3) . . . . *Colonial corals* and *chaetetids*

(25 e) 28 a Shells with "cellular-prismatic" structure, very regular rectangular and radial pattern. Prism size approximately 0.05–0.1 mm (Plate 25/7) . . . . *rudist lamellibranch shells*

b Shells with "honeycomb structure" in tangential sections, diameter of the prisms to 0.10 mm (Plate 25/6) . . . *pelecypods* (*Inocerams*)

(24 b) 29 a Colonies of stacked hemispherical cysts; sometimes with spine-like columnal elements . . . . *stromatoporoids*

b Cystose structure containing short vertical tubes with horizontal elements . . . . . *bryozoans* (tube diameter generally < 0.50 mm) or *tabulate corals* (tube diameter generally > 0.5 mm, often several mm)

c Vertically arranged cysts, outer periphery of the total structure very distinct (Plate 23/2, 3) . . . .
peripheral and longitudinal sections of corals and sphinctozoan sponges

(23 b) 30 a Wormlike elements in a micritic matrix which is usually dark-colored; often whole structure may appear as a circular figure (Plate 23/6) . . . . *calcareous sponges*

b Flat fragments or conical figures with irregular network of dark-colored elements; sometimes distinct chambers (Plate 30/7–8) . . . . *reticulate foraminifera* (e.g., *lituolids*)

c Sessile colonies; irregular network structure consisting chiefly of vertical and horizontal elements; often star-shaped channel systems and tabulate tubes of various widths . . . (Mesozoic) *hydrozoans*

d Fragments of calcium phosphate, brownish or black-colored (low birefringence); meshes of various sizes . . . . . *remains of bones*

E. Layered

31 a Encrusted, planar distribution on the substrate 32

b Non-encrusted particles 33

(31 a) 32 a Irregular, micritic laminated fabric with spar-filled cavities parallel to the bedding plane; mm sizes (Plate 13/2) . . . . . *"algal mats"*

b Structures made up of micrite and sparite curved in various shapes; sometimes with terrigenous accessory elements (e.g., quartz) and overgrowth of sessile benthonic organisms. Basic geometric form, comparable to a hemisphere; cm- to m-sized (see Fig. 50) . . *stromatolites* (May be confused with "caliche" layers, see 2.3.2.2!)

(31 b) 33 a Irregular nodules and spherical particles. Outline often lobular; irregular layers not parallel to each other, sometimes overlapping (usually micrite, algae, foraminifera), surrounding biogenic or inorganic "nuclei". mm- to cm-sized (Plate 42/1) . . . . *oncoids* (May be confused with "caliche oncoids", see Table 4.)

b Phosphatic

b_1 Regularly outlined (often rhombic) particles with distinct very concentric lamellar structure; usually dark in transmitted light . . . . *fish scales*

b_2 Toothlike, often irregular outline, dark or light gray color; distinct lamellar structure in various orientations. Total size between 1 and 3 mm . . . . *conodonts*

F. Chains of Segments

34 a Uniserial 35

b Biserial or triserial 36

(34 a) 35 a Microscopic 37

b Macroscopic 38

(34 b) 36 Small chambers becoming successively larger . . . . *foraminifera*

(35 a) 37 a Small chambers becoming successively larger; uniserial, (Plate 42/3) . . . . *foraminifera*

b Basin-shaped segments stacked within each other, with central pore (Late Triassic; Plate 18/7; 48/1) . . . . *Cheilosporites*

(35 b) 38 a Straight chambered shells, cm-sized; transverse elements with pores (sipho) . . . . *"orthoceratids"*

b Ladder-like structure with slightly curved transverse elements . . . . *tangential sections of ammonites*

c Arched, overlapping segments, sometimes with perforated roofs; sometimes vesicular structures in the segments (Plate 23/1) . . . . *segmented calcisponges* (*Sphinctozoa*)

5.3 Selected Groups of Organisms

5.3.1 Foraminifera (Plates 10; 15/4, 5; 17; 18/1–6, 9–11; 19; 20/1–5; 21; 30/5, 7–9; 38/2; 41/1; 42/3, 6–8)

Systematic position: Protozoa, Sarcodina. Foraminifera are distinguished from other Sarcodina by their mineralized shells.

Morphology: Foraminifera are differentiated according to wall structure, chamber shape and chamber arrangement, apertural character and modification, pores and ornamentation. Sizes of the tests between 0.05 mm and about 2 cm.

The original *wall structure* is of primary importance for the classification of the major groups. Several microstructural types are distinguished:

- Agglutinated wall structure: particles (mineral grains, tests of microorganisms) are cemented onto an organic layer; in addition locular protoplasm, calcite, silica or ferruginous material may bind the arenaceous material. In cross-sections the wall of simple agglutinated forms consists of a simple layer composed of particles grading from finer on the inside to coarser on the outside (e.g., *Gaudryina*). In more complex agglutinated genera an outer microgranular wall may cover a complex inner layer. The inner wall appears spongy; portions of the inner wall may be folded inward into the chambers (labyrinthic or alveolar structures; e.g., *Cyclammina*). Important agglutinated foraminifera, common in thin-sections of limestones, are many Ammodiscacea (e.g., Astrorhizidae, Saccaminidae, and Ammodiscidae) and Lituolacea [e.g., Hormosinidae; Lituolidae, especially important in Mesozoic shallow-water limestones are Cyclammininae and Spirocyclininae; Textulariidae; Ataxophragmiidae and Pavonitinidae (important index fossils in zonations based on coenozones, see Sect. 5.4); Dicyclinidae, and Orbitolinidae (especially important for the biozonation of Cretaceous shelf carbonates, see Plate 19)].
- Microgranular wall structure: characterized by micrite grains cemented by a calcareous cement. In thin-sections the wall frequently appears dark. When the crystals are regularly aligned, a pseudofibrous structure can be seen. Advanced forms show a differentiation of the wall into 2 or more layers (e.g., Endothyracea and Fusulinacea). Endothyracea (e.g., Nodosinellidae and Palaeotextulariidae) and Fusulinacea are of special interest in connection with the study of Carboniferous and Permian shelf carbonates (see Plate 17).
- Hyaline calcareous wall structure; characterized by minute perforations in the wall; the walls may be composed of Low-Mg calcite, High-Mg calcite or arago-

nite. The crystals may be arranged with their c-axis normal to the test surface (radial hyaline microstructure) or oriented randomly (granular hyaline). In cross-sections of hyaline walls organic layers or calcareous laminations (lamellae, e.g., Rotaliidae) may be visible. In thin-sections hyaline walls appear light-colored in transmitted light, and dark in direct light. Foraminifera with hyaline tests (Rotaliina) include many groups, usually seen in thin-sections of subtidal or pelagic limestones. Examples are the Orbitoidacea (especially important in Tertiary shallow-water carbonates; e.g., Discocyclinidae and Lepidocyclinidae; and in Cretaceous limestones, e.g., Orbitoididae), the planktonic Globigerinacea (e.g., Globotruncanidae, Upper Cretaceous; Globigerinidae, Maastrichtian to Recent; and Globorotaliidae, since Paleocene), and the benthonic Rotaliacea (which include very important groups as the Nummulitidae and the Miogypsinidae; index-fossils in Tertiary shallow-water limestones, see Plate 21).

- Porcelaneous calcareous wall structure: The shiny, smooth appearance of the test is caused by randomly oriented micrite grains or by crystallites which are organized in a brickwork-like pattern. In thin-sections porcelaneous tests appear white in direct light, and dark in transmitted light. Under crossed nicols the calcite crystallites appear as tiny flecks. The wall consists of High-Mg calcite, and generally has two layers (organic inner layer and calcitic wall). This type is represented by the Miliolina, which include many groups determinable only in thin-sections (e.g., Soritidae and Alveolinidae, important in Cretaceous and Tertiary carbonates).

Plate 17. Foraminifera from the Late Paleozoic

1. Smaller Foraminifera: These foraminifera, which can occur abundantly in thin-sections, are very useful for biostratigraphic zonation in the Upper Devonian and Lower Carboniferous. Pictured is *Endothyranopsis* sp. Lower Carboniferous (Dinantian): Basin of Dinant, Belgium

2. Smaller Foraminifera: Waulsortian limestone (Viséan) with numerous sections of endothyrid Smaller foraminifera and a few calcispheres (center of the picture). Lower Carboniferous (Waulsortian): Büsbach near Aachen, Germany

3. Fusulinids: *Fusulinella subsphaerica* Toriyama. Axial section. Carboniferous (Westfalian): Shisidedai, Akiyoshi plateau, Japan

4. *Pseudoschwagerina (Zellia) heritschi* Kahler and Kahler. Upper Pseudoschwagerina limestone, Lower Permian: Zottachkopf, Carnic Alps, Austria

5. *Parafusulina kaerimizensis* (Ozawa). Lower Permian (Akiyoshi limestone): Kaerimizui, Akyoshi plateau, Japan

6. *Neoschwagerina craticulifera* (Schwager): Middle Permian: Kaerimizui, Akyoshi plateau, Japan

7. *Verbeekina verbeeki* (Geinitz). Upper Middle Permian. Sisopho, area near Battambang, Cambodia

8. *Yabeina multiseptata gigantea* (Gubler). Upper Middle Permian. Sisopho, area near Battambang, Cambodia

Scale for all figures 1 mm

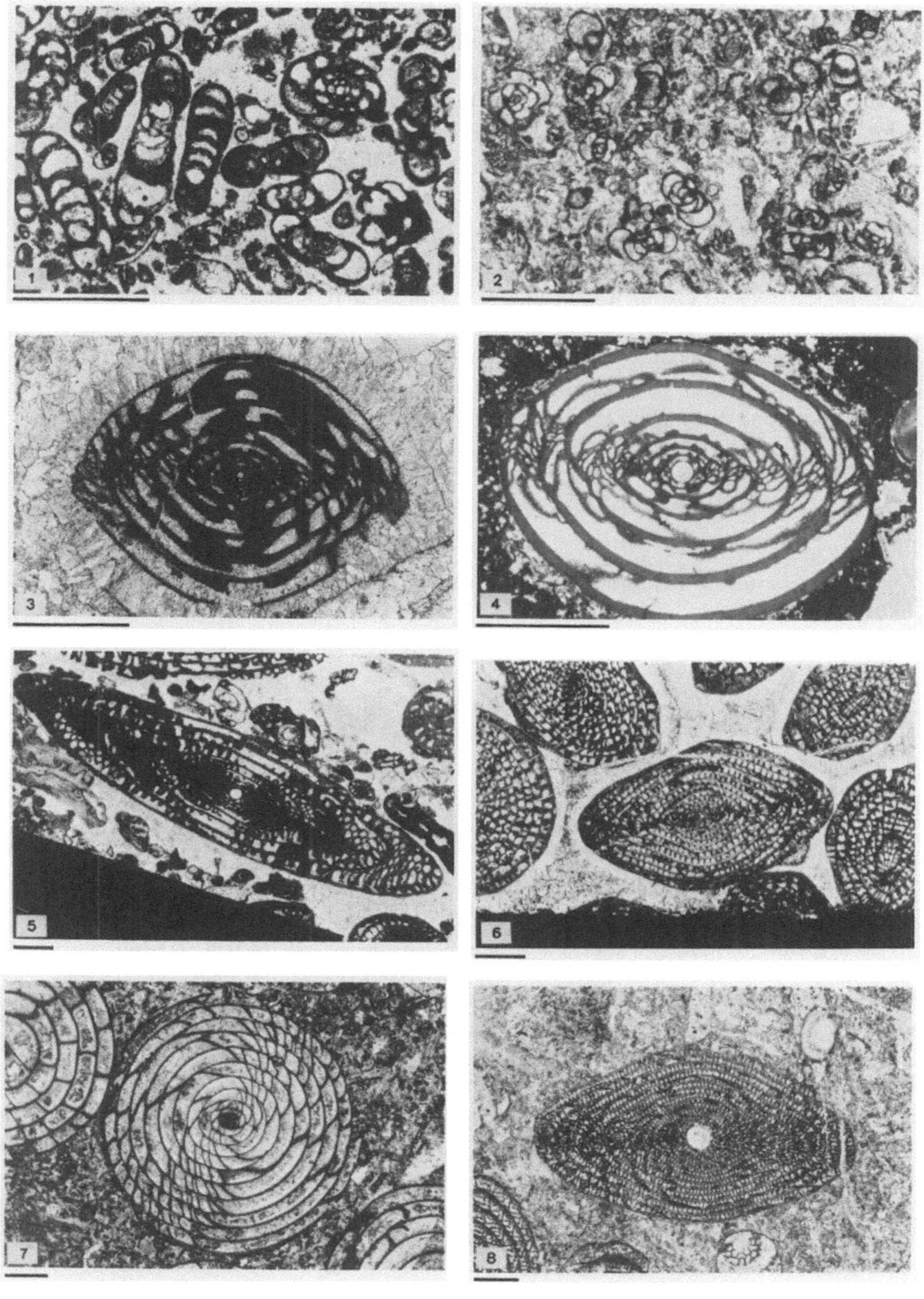
1
2
3
4
5
6
7
8

Foraminifera in thin-sections: The determination of foraminifera in thin-sections of limestones can be made difficult by randomly oriented sections, and obliteration of the original wall structure by micritization and recrystallization. Triserial tests may appear as uniserial or biserial sections; diagenetic alterations may result in the formation of a "miliolid" wall structure (see Piller and Hohenegger, 1974). Some of these problems may be diminished by considering the following possibilities:

- Comparison of free tests and sections using a coordinate system: Cummings (1955, 1958) has published line drawings showing possible sections of Late Paleozoic uniserial foraminifera (Paleotextulariidae). The description can be facilitated by the use of a coordinate system similar to that used in crystallographic studies.
- Comparison of hypothetical figures with random figures actually found in thin-sections: Examples are given in Neumann (1967), and for Mesozoic Larger foraminifera in Hottinger (1967, 1971).

Plate 18. Facies Diagnostic Organisms – Microfossils in Reef Limestones of the Alpine Late Triassic

1. *Involutina communis* (Kristan). Dachstein reef limestone (Norian): Gosaukamm, Upper Austria. Scale is 250 μm

2. *Sigmoilina* sp. Upper Rhaetian reef limestone: Adnet, Salzburg, Austria. Facies environment: central reef. Scale is 250 μm

3. *Ophthalmidium leischneri* (Kristan-Tollmann). Upper Rhaetian reef limestone: Rötelwand near Hallein, Salzburg, Austria. Facies environment: central reef. Scale is 250 μm

4. *Austrocolomia canaliculata* (Kristan). Upper Rhaetian reef limestone: Adnet, Salzburg, Austria. Facies environment: central reef. Scale is 250 μm

5. *Triasina hantkeni* Majzon. Dachstein limestone (Norian): Wopfing, Lower Austria. Facies environment: back reef. Scale is 250 μm

6. *Galeanella panticae* Brönnimann, Cadet, Ricout and Zaninetti. Upper Rhaetian reef limestone: Feichtenstein, Salzburg, Austria. Facies environment: central reef. Scale is 250 μm

7. *Cheilosporites tirolensis* Wähner, a microproblematicum constructed of basin-shaped segments nested in each other (probably a sphinctozoan). Upper Rhaetian reef limestone: Gruber reef near Hintersee, Salzburg. Scale is 5 mm

8. *Microtubus communis* E. Flügel, a microproblematicum which occurs abundantly in the reef core facies of Late Triassic reefs (worm tubes, foraminifera or calcareous algae?). Upper Rhaetian reef limestone: Feichtenstein, Salzburg, Austria. Scale is 1 mm

9. *Diplotremina* sp. Dachstein limestone (Rhaetian): Sauwand near Gußwerk, Styria, Austria. Facies environment: back reef. Scale is 1 mm

10. *Alpinophragmium perforatum* E. Flügel. Upper Rhaetian reef limestone: Adnet, Salzburg, Austria. Central reef facies. Scale is 250 μm

11. *Ophthalmidium triadicum* (Kristan). Upper Rhaetian reef limestone: Adnet, Salzburg, Austria. Central reef facies. Scale is 250 μm

12. *Glomospirella? pokornyi* (Salaj). Dachstein limestone (Norian): Leanyvar, Gerece mountains, Hungary. Back reef facies. Scale is 250 μm

13. Anomuran coprolites: *Parafavreina thoronetensis* Brönnimann, Caron and Zaninetti. Kössen beds (Rhaetian): Feichtenstein near Hintersee, Salzburg, Austria. Scale is 250 μm

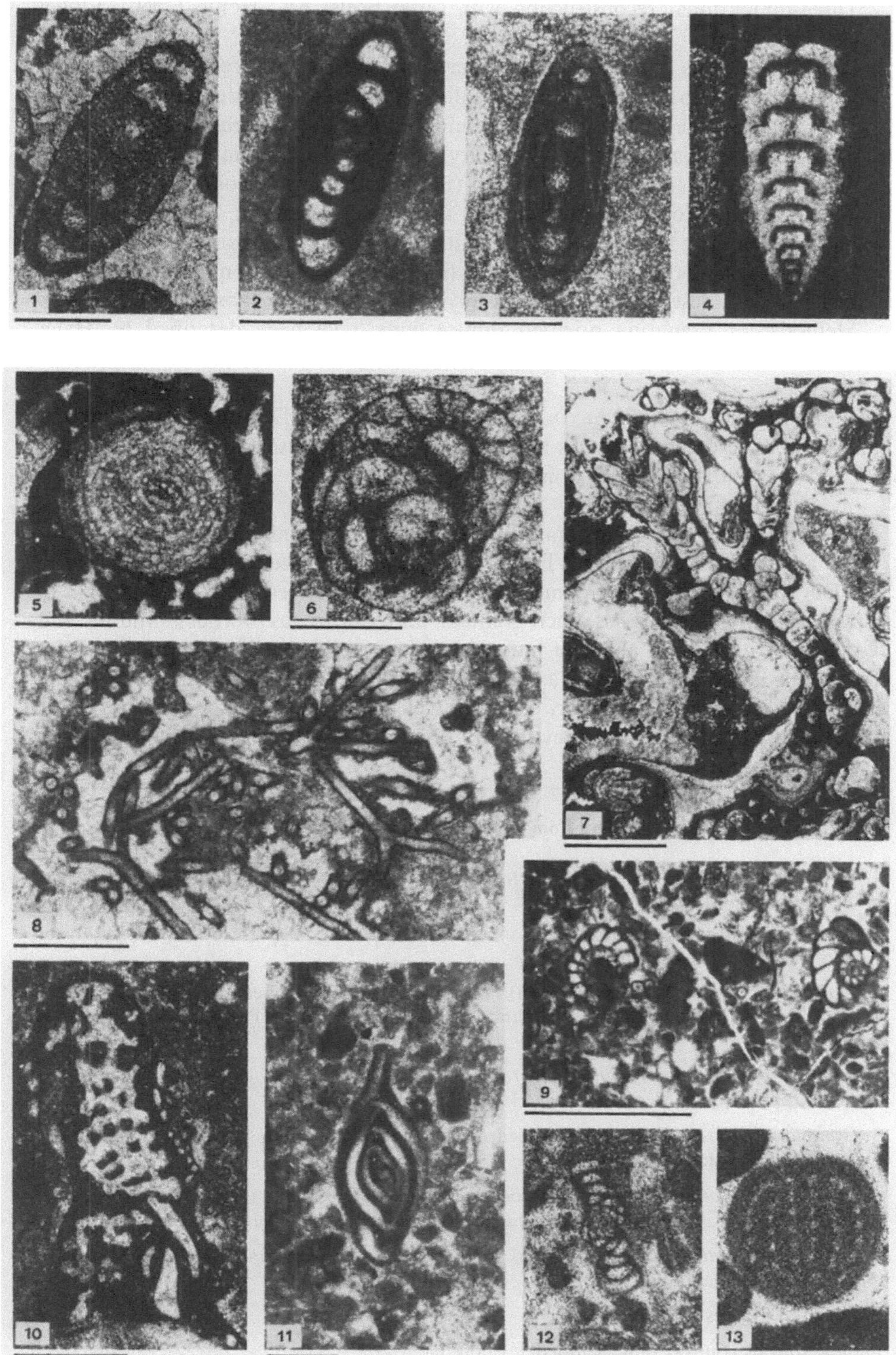
1
2
3
4
5
6
7
8
9
10
11
12
13

- Three-dimensional reconstruction with serial peels: e.g., Honjo (1960), Flügel (1967).
- Oriented sections (sse Plate 17; 19): Several standard cuts are necessary for the study of Larger foraminifera with complicated internal structures (e.g., larger lituolids in Mesozoic limestones; orbitolinids in Cretaceous carbonates; Carboniferous and Permian fusulinids; Tertiary orbitoids and nummulitids). Common names for these cuts are axial section (through the axis of coiling, or along the axis of growth; corresponds to vertical or transverse section), equatorial section (perpendicular to the axis of growth or in the plane of coiling or equatorial layer; known also as horizontal or median section), tangential section (tangential to the surface of the test), and oblique (in any plane other than those defined above). Methods for preparing oriented thin-sections have been described by Hagn (1953), Arnold (1965), and Sachs (1965).

Importance of foraminifera in microfacies analysis: Foraminifera found in thin-sections of carbonate rocks have proved to be most useful for biostratigraphic subdivisions and for paleoenvironmental analysis.

Paleozoic: Lower Paleozoic foraminiferal faunas are dominated by single-chambered and agglutinated genera. The first major radiation seems to have occurred in the Devonian with the development of multi-chambered tests and the evolution of new wall types (calcareous microgranular and calcareous fibrous). Middle and Upper Devonian shelf and reef limestones, deposited in shallow-marine environments may yield characteristic assemblages composed of Parathuramminacea and Endothyracea (e.g., Chuvashov, 1965; Flügel and Hötzl, 1971; Neumann et al., 1975; Kettenbrink and Toomey, 1975; Toomey, 1972). These foraminifera are

Plate 19. Orbitolinid Foraminifera from the Cretaceous

1. *Orbitolinopsis capuensis* (de Castro). Axial section. Lower Barremian: Djebel Debar (Constantinois, northeastern Algeria)

2. *Neoiraqia convexa* Danilova. Macrosphere. Axial section, Uppermost Albian to Lower Cenomanian: Caserta, Campania, Italy

3. *Palaeodictyoconus arabicus* (Henson). Horizontal section. Upper Barremian to Lower Aptian: Grands Banks, Eastern Canadian Shelf (Deep Sea Drilling Project, leg 43, hole 384, core 21)

4. *Orbitolina (Orbitolina) duranddelgai* Schroeder. Macrospheric embryonic stage, axial section. Upper Albian: Sierra de Aulet, Huesca Province, Spain

5. *Orbitolina (Orbitolina) duranddelgai* Schroeder. Tangential section. A form. Upper Albian: Sierra de Aulet, Huesca Province, Spain

6. *Orbitolina (Mesorbitolina) aperta* Schroeder. Macrosphere; oblique section of the embryonic stage. Lowermost Cenomanian: Montagne de Tauch, Département Aude, France

7. *Palorbitolina lenticularis* (Blumenbach). Macrospheric embryonic stage, axial section. Lower Aptian: Bari, Puglia, Italy

Scale for all Figures 0.5 mm

Photographs by courtesy of R. Schroeder, Frankfurt

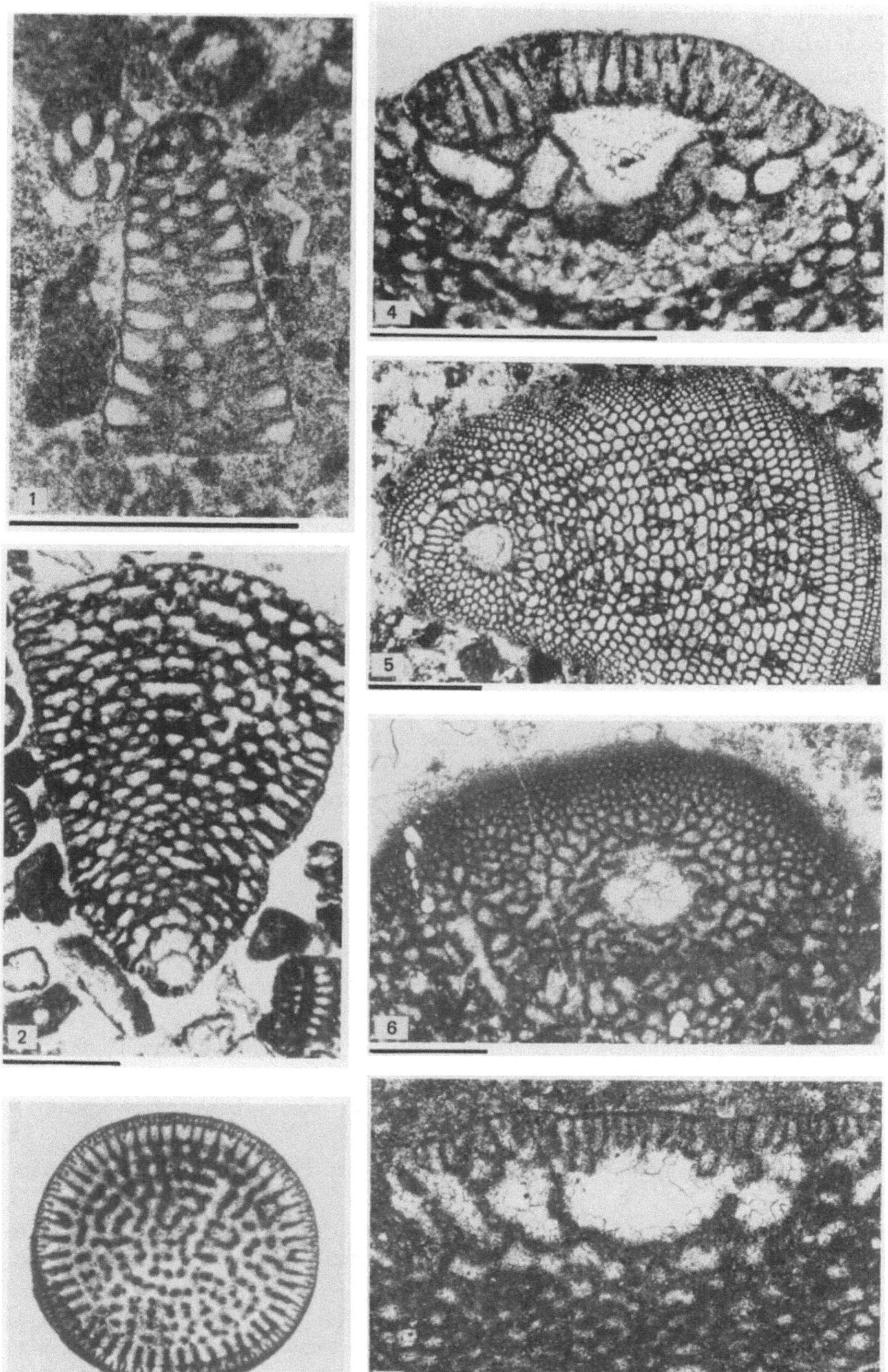
1
2
3
4
5
6
7

useful for the zonation of the Frasnian and the Famennian, as well as for the differentiation of paleoenvironments (e.g., reef and inter-reef facies, see Plate 41). In addition to bona fide foraminifers, many debatable taxa have been described from thin-sections which may be of algal affinity (see Sect. 5.3.13).

The Carboniferous was a period of radiation among the new microgranular families from the Devonian. The endothyrids built tests of increasing complexity and size, giving rise during the Viséan to the family Fusulinidae, the most important group of all Paleozoic foraminifera. During the Carboniferous foraminifera became more significant members of invertebrate communities in shallow-water and deeper-water environments. Fourteen biostratigraphic zones based on agglutinated and microgranular Smaller Foraminifera can be distinguished in the Carboniferous (Mamet and Skipp, 1970) which allow world-wide correlations (e.g., Conil and Lys, 1964, 1968; Conil et al., 1976; Mamet, 1970), especially of the Dinantian.

In the Upper Paleozoic there are about ten fusulinid zones based on groupings of genera. More than 5,000 species are known (catalogued by Kahler and Kahler, 1966/1967), which can be used for the designation of faunal provinces in the Permian (e.g., Kahler, 1974; Ross, 1967, 1970). Abundant fusulinid limestones occur in the late Pennsylvanian and Permian around the world, well-known examples are the Permian of Texas, central and eastern Asia, and southern Europe (see Plates 17 and 42).

Mesozoic: Triassic foraminifera (Plate 18) are widely distributed in shallow-water and basinal carbonates, especially in the Tethyan region (Alpine-Mediterranean area). Foraminifera found in thin-sections can be used for biostratigraphic zonations (e.g., Kristan-Tollmann, 1964; Zaninetti, 1976), but also in facies studies. The facies control of Triassic foraminifera has been documented by Hohenegger and Lobitzer (1971), Hohenegger (1974), Hohenegger and Piller (1975), Piller (1978),

→

Plate 20. Planktonic Foraminifera and Calcispheres

1. *Globigerina subhelvetica* Haeusler. Upper Jurassic (Oxfordian): Apennines, Italy

2. *Globotruncana* sp. and *Pseudotextularia* sp. (top right). Upper Cretaceous (Maastrichtian): E Delphi, Greece

3. *Globotruncana stuartiformis* Dalbiez and *Globotruncana* ex gr. *G. arca* (Cushman), top right. Upper Cretaceous (Lower Maastrichtian): E Delphi, Central Greece

4. *Globotruncana,* cf. *G. tricarinata* (Querreau). Upper Cretaceous (Lower Maastrichtian): E Delphi, Central Greece

5. *Orbulina universa* and other species. Tertiary (Miocene): Pietraroia, Southern Apennines, Italy

6. Calcisphaerulids: chiefly Stomiosphaeres, sporadic *Cadosina* (left, with black wall). Echinoid spine in the middle of the picture. Upper Cretaceous (Pläner, Turonian): Salzgitter, Germany

7. Calcisphaerulids: primarily *Stomiosphaera sphaerica* (Kaufmann), sporadic *Stomiosphaera* cf. *S. conoidea* Bonet. Upper Cretaceous (Seewen limestone, Lower Turonian): Seewen, Switzerland

Scale for all Figures 0.5 mm

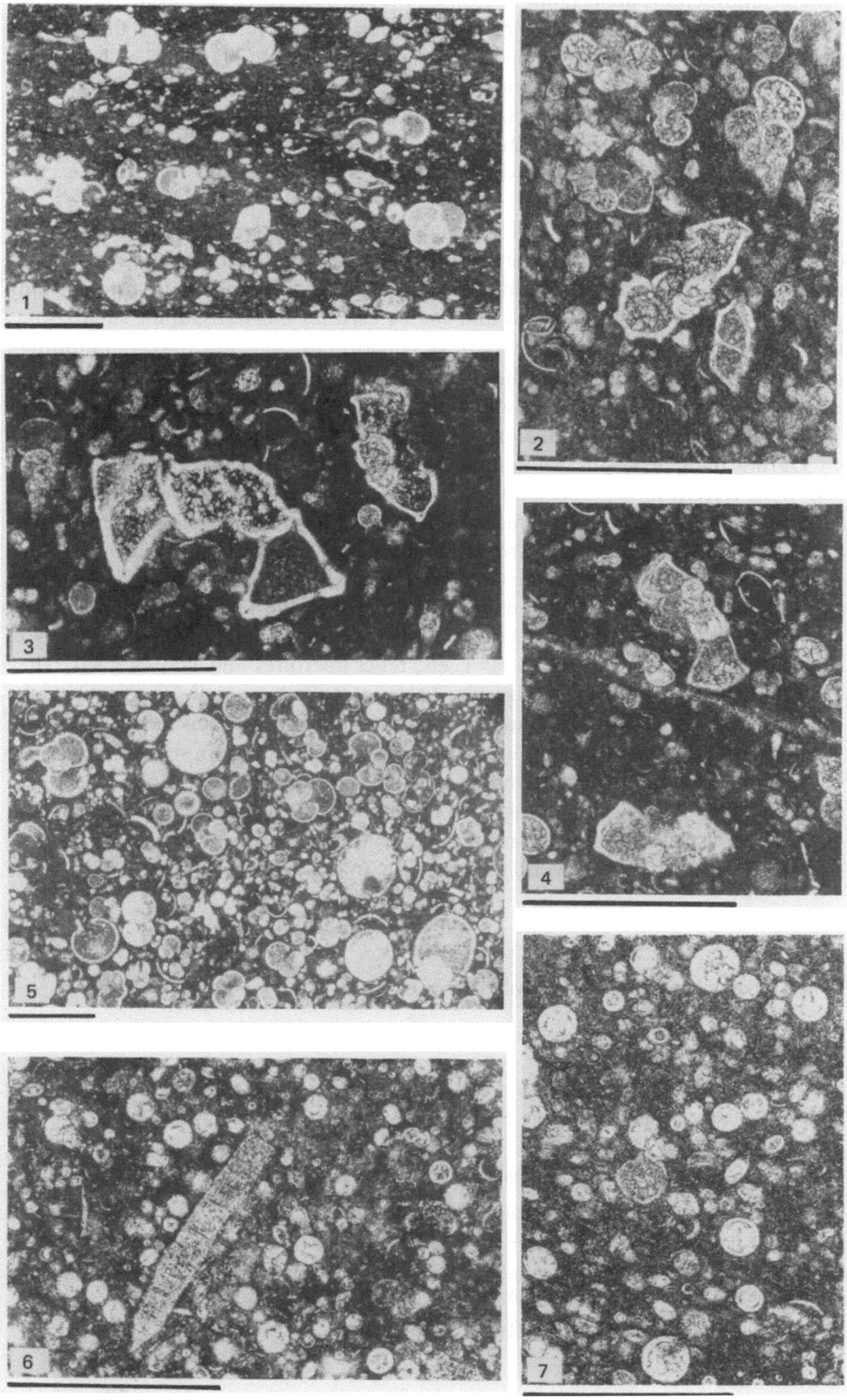
1
2
3
4
5
6
7

Schäfer and Senowbari-Daryan (1978), and Dullo (1979). Even considering the often poor preservation and strong recrystallization of some foraminifera, there is now sufficient evidence for recognition of various facies units on platforms and in basins by distinct foraminiferal associations. It is possible to differentiate various units of the shelf environment and special ecological niches within the central reef area by distributional patterns of foraminifera. All these studies are based on a quantitative survey in which all individuals are counted within a defined thin-section area (9 cm^2).

A major step in foraminiferal evolution was the adaption to a planktonic mode of life during the Jurassic. In the epicontinental seas both calcareous and agglutinated benthic foraminifera flourished (especially lituolids and nodosariids). Many groups are important index-fossils and can be used for biostratigraphic subdivisions in the context with the differentiation of coenozones (see Sect. 5.4). Foraminifera and calcareous algae are often the only fossils which permit the dating of Jurassic and Cretaceous platform carbonates (see Cita, 1965; Crescenti et al., 1969; Gusić, 1969; Hottinger, 1971; Sartoni and Crescenti, 1962).

During the Cretaceous, foraminifera occupied the vast variety of niches known also from Recent forms. Lagoonal, back-reef, and shallow neritic deposits primarily contain agglutinated foraminifera, particularly cuneolinids, dicyclinids, orbitolinids and valvulinids. Orbitolinids are of major importance for the biostratigraphic subdivision of Lower and Middle Cretaceous shelf deposits (see Schröder, 1979). On carbonate platforms porcelaneous foraminifera (e.g., the family Alveolinidae) are accompanied by larger orbitoid genera. Planktonic foraminifera exhibit a variety of new morphological criteria, such as keels, relict apertures and double keels. They are extremely useful for biostratigraphy because of their generally short ranges, the widespread distribution, their abundance in marine sediments, and the occurrence of planktonics in sediments of different water depths. Biostratigraphic zonations are based primarily on the evolution of the globotruncanids (see Bolli, 1966), see Plate 20. The stratigraphic ranges of planktonic foraminifera have been well established, especially since the recovery of continuous sequences of Mesozoic and Cenozoic strata from the world oceans (see Initial Reports, DSDP). Planktonic

Plate 21. Foraminifera from the Early Tertiary

1. Packstone with *Nummulites* sp. Tripolitza Limestone, Middle Eocene. Central Crete, Greece. Scale 2 mm

2. Packstone with *Assilina exponens* (Sowerby). Adelholzen Beds, Lutetian. Gergartsreiter Graben southeast Siegsdorf, Bavaria. Scale 2 mm

3. Packstone with *Discocyclina* sp. Adelholzen Beds, Middle Eocene. Road Rohrdorf to Langweid near Neubeuern am Inn, Bavaria. Scale 2 mm

4. *Fasciolites* sp. (= *Alveolina* auctorum). Zwischenschichten, Upper Cuisian. St. Pankraz near Haunsberg north of Salzburg, Austria. Scale 2 mm

5. Biomicrite with planktonic foraminifera (globigerinids). Uppermost Paleocene. Southeastern slope of Monte Altissimo di Nago, Monte Baldo, Northern Italy. Scale 1 mm

Photographs by F. Höck; courtesy of H. Hagn, Munich

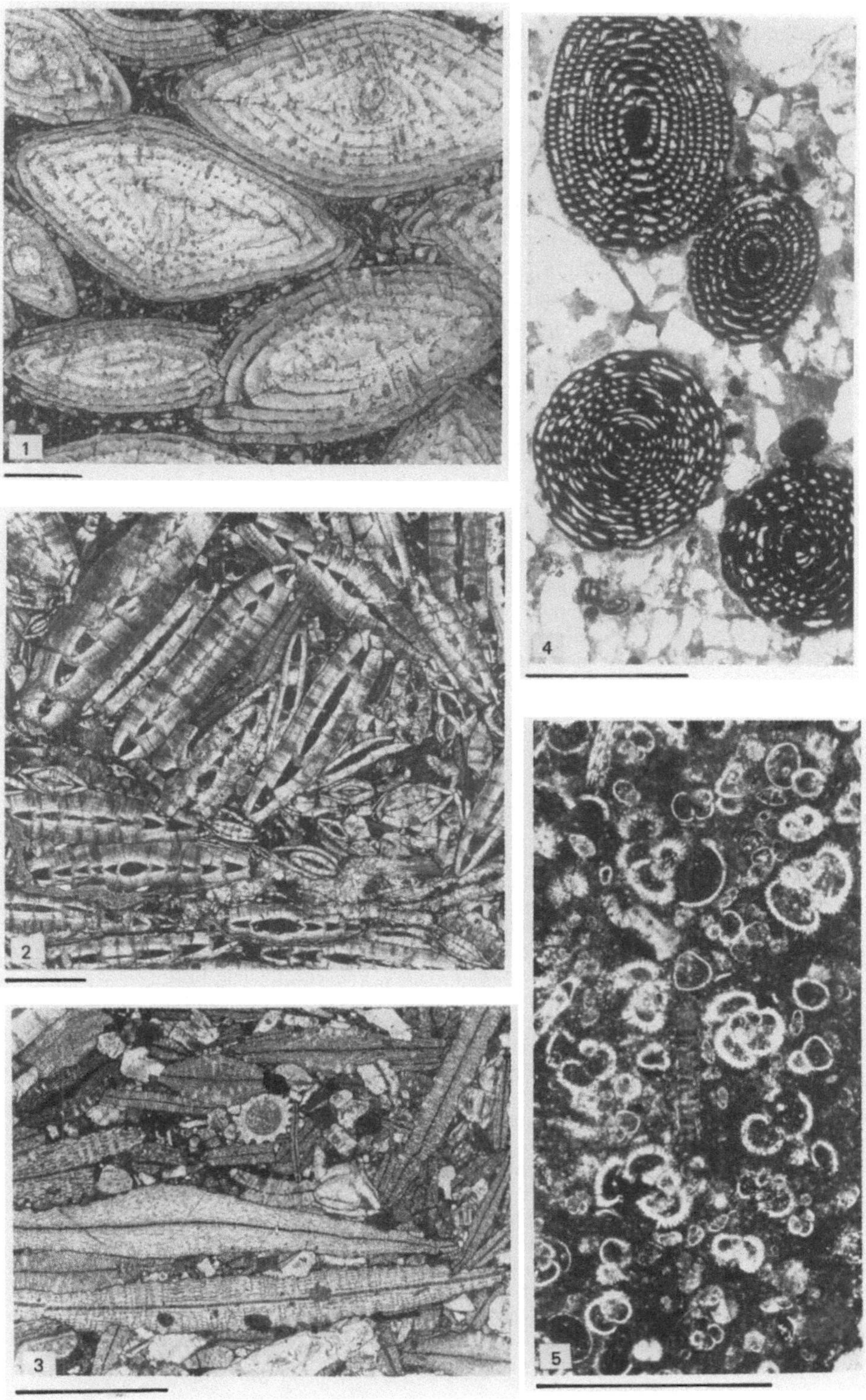
1
2
3
4
5

foraminifera also constitute a valuable means for interpreting Ancient climatic and oceanographic conditions. Alternating distributional patterns as a result of changing climatic conditions can be used to reconstruct paleoclimate maps. Oxygen isotope studies of Cretaceous and Cenozoic planktonic foraminifera tests permit the determination of paleotemperatures. Paleodepths may be estimated from solution patterns of foraminiferal assemblages (Berger, 1971).

Cenozoic: The Tertiary is characterized by the continuation of Mesozoic families. Modern benthic and planktonic foraminifera seem to have evolved during the Miocene. Forty-six biostratigraphic zones, based on planktonic foraminifera, have been distinguished in the Tertiary (Bergren, 1972; Blow, 1969; Bolli, 1966). Larger benthic foraminifera (e.g., nummulitids, miogypsinids, discocyclinids, and lepidocyclinids; see Plate 21) are important index-fossils for the subdivision of the Paleocene to Miocene interval; see Papp and Turnovsky, 1970.

References: Foraminifera

Arnold, Z. M. (1965): An improved instrument for sectioning microfossils. In: Kummel, B., Raup, D. (eds.): Handbook of paleontological techniques, 232–237. San Francisco–London: Freeman

Bé, A. W. H. (1977): An ecological, zoogeographic and taxonomic review of Recent planktonic foraminifera. In: Ramsay, A. T. S. (ed.): Oceanic Micropaleontology **1,** 1–88, London: Academic Press

Berger, W. H. (1971): Sedimentation of planktonic foraminifera. Marine Geol. **11,** 325–358, Amsterdam

Bergren, W. A. (1972): A Cenozoic time-scale – some implications for regional geology and paleobiogeography. Lethaia **5,** 195–215, Oslo

Bignot, G., Neumann, M. (1962): La structure des tests des foraminiferes. Analyse bibliographique. Rev. Micropal. **4**/4, 237–248, Paris

Bless, M. J. M., et al. (1976): Dinantian rocks in the subsurface North of the Brabant and Ardenno-Rhenish massifs in Belgium, the Netherlands and the Federal Republic of Germany. Meded. Rijks. Geol. Dienst. N. S. **27**/3, 81–195, 20 Pls., 3 supps., Roermond

Blow, W. H. (1969): Late Middle Eocene to Recent Planctonic Foraminiferal Biostratigraphy. Proc. 1st Intern. Conf. Plankt. Mikrofoss., Geneva, 199–222, Leiden: Brill

Boersma, A. (1978): Foraminifera. In: Haq, B. U., Boersma, A. (eds.): Introduction to marine micropaleontology, 19–77, New York: Elsevier

Bolli, H. (1966): Zonation of Cretaceous to Pliocene marine sediments based on planktonic Foraminifera. Boll. Inform. Ass. Venezolana Geol. Min. Petrol. **9**/1, 3–32, Caracas

Chuvashov, B. I. (1965): see 5.3.12.7

Cita, M. B. (1965): see 5.5

Conil, R., Lys, M. (1964): see 5.3.13

Conil, R., Lys, M. (1968): Utilisation stratigraphique des foraminifères du Dinantien. Ann. Soc. Géol. Belgique **91,** 491–558, Bruxelles

Conil, R., Papproth, E. (1968): Mit Foraminiferen gegliederte Profile aus dem nordwestdeutschen Kohlenkalk und Kulm. Decheniana **119,** 1/2, 51–94, 6 Pls., Bonn

Crescenti, U., Crostella, A., Donzelli, G., Raffi, G. (1969): Stratigrafia della serie calcarea dal Lias al Miocene nella regione Marchigiano-Abruzzense. Parte II. Litostratigrafia, Biostratigrafia, Paleogeografia. Mem. Soc. Geol. Italiana **8,** 343–420, 3 Plates, 64 Figs., Pisa

Cummings, R. H. (1958): The faunal analysis and stratigraphic application of Upper Palaeozoic smaller foraminifera. Micropal. **4**/1, 1–24, Pls. 1, 6 Figs., New York

Dilley, F. C. (1973): Larger foraminifera and seas through time. Spec. Paper Paleont. **12,** 155–168, London

Douglas, R. G. (1973): Benthonic foraminiferal biostratigraphy in the Central North Pacific, Leg. 17, Deep Sea Drilling Project. Initial Rep. Deep Sea Drilling Project **17,** 607–672, Washington

Dullo, W.-Chr. (1980): Paläontologie, Fazies und Geochemie der Dachstein-Kalke (Ober-Trias) im südwestlichen Gesäuse, Steiermark, Österreich. Facies **2,** 55–122, Plates 9–13, 10 Figs., Erlangen

Flügel, E. (1967): Eine neue Foraminifere aus den Riff-Kalken der nordalpinen Ober-Trias: *Alpinophragmium perforatum* n.g., n.sp. Senck. leth. **48**/5, 381–402, 2 Pls., Frankfurt/M.

Flügel, E. (1970): Palökologische Interpretation des Zottachkopf-Profiles mit Hilfe von Kleinforaminiferen (oberer Pseudoschwagerinen-Kalk, unteres Perm. Karnische Alpen). Carinthia II, Sonderheft **28,** 61–96, 4 Pls., 3 Figs., 2 Tab., Klagenfurt

Flügel, E., Hötzl, H. (1971): see 5.3.12.7

Gusić, I, (1969): Some new and inadequately known Jurassic foraminifers from Central Croatia. Geol. Vjesnik **22,** 89–97, Zagreb

Hagn, H. (1953): Ein neues Verfahren zur Anfertigung orientierter Dünnschliffe kleiner paläontologischer Objekte. Paläont. Zeitschrift **27,** 26–31, 1 Fig., Stuttgart

Hedley, R. H., Adams, C. G. (1974): Foraminifera. 276 p. London: Academic Press

Hohenegger, J. (1974): Über einfache Gruppierungsmethoden von Fossil-Vergesellschaftungen am Beispiel obertriadischer Foraminiferen. Neues Jahrbuch Geol. Paläont. Abhandlungen **146,** 263–297, Stuttgart

Hohenegger, J., Lobitzer, H. (1971): Die Foraminiferen-Verteilung in einem obertriadischen Karbonatplattform-Becken-Komplex der östlichen Kalkalpen. Verh. geol. Bundesanst. **1971**/3, 458–485, 3 Pls., 4 Figs., Wien

Hohenegger, J., Piller, W. (1975): Diagenetische Veränderungen bei obertriadischen Involutinidae (Foraminifera). N. Jb. Geol. Paläont. Mh. **1975**/1, 26–39, 3 Figs., Stuttgart

Hohenegger, J., Piller, W. (1975): Wandstrukturen und Großgliederung der Foraminiferen. Sitzungsberichte Österr. Akademie Wissenschaften, math.-naturwiss. Klasse, Abteilung 1, v. **184,** 67–96, Wien

Hohenegger, J., Piller, W. (1975 b): Ökologische und systematische Stellung der Foraminiferen im gebankten Dachsteinkalk (Obertrias) des Nordöstlichen Toten Gebirges (Oberösterreich). Paleo-3 **18,** 241–276, Amsterdam

Honjo, S. (1960): A study of some primitive Neoschwagerina by a new serial section technique. J. Fac. Sci. Hokkaido University, ser. 4, **10**/3, 457–470

Hottinger, L. (1960): Über paleocaene und eocaene Alveolinen. Eclogae geol. Helvetiae **53**/1, 265–283, 21 Pls., 4 Tab., Basel

Hottinger, L. (1967): Foraminifères imperforés du Mésozoique marocain. Notes et Mém. Serv. géol. Maroc **209,** 1295, Rabat

Hottinger, L. (1971): Larger Foraminifera of the Mediterranean Jurassic and their stratigraphic use. Ann. Inst. Geol. Publ. Hungar. **54**/2, 497–504, 1 Plate, Budapest

Kahler, F. (1974): Fusuliniden aus Tien-schan und Tibet. Mit Gedanken zur Geschichte der Fusuliniden-Meere im Perm. Rep. Sci. Exp. North-Western Prov. China Sven Hedin, Sino-Swedish Exped. Publ. **52,** 148 p., 2 Figs., 2 Pls., many Tab., Stockholm

Kahler, F., Kahler, G. (1968): Fossilium Catalogus, I, Animalia, **111–114,** Fusulinida (Foraminiferida), 974 p., s'Gravenhage: Junk

Kettenbrink, E., Toomey, D. (1975): Distribution and paleoecological implications of calcareous foraminifera in the Devonian Cedar Valley formation of Iowa. J. Foram. Res. **5**/3, 176–187, 2 Pls., 2 Figs.

Kristan-Tollmann, E. (1964): Zur Charakteristik triadischer Mikrofaunen. Paläont. Zeitschrift **38,** 66–73, Plates 6–7, 3 Figs., Stuttgart

Luterbacher, H. P. (1972): Foraminifera from the Lower Cretaceous and Upper Jurassic of the northwestern Atlantic. Initial Rep. Deep Sea Drilling Project **11,** 561–576, Washington

Mamet, B. L. (1970) Carbonate Microfacies of the Windsor Group (Carboniferous), Nova Scotia and New Brunswick. Geol. Surv. Canada, Dept. Energy Min. Res. Paper **70**/21, 121 pp., 19 Pls., 15 Figs., Ottawa

Mamet, B., Skipp, B. (1970): Lower Carboniferous foraminifera: preliminary zonation and stratigraphic implications for the Mississippian of North America. Proc. Intern. Congr. Strat. Geol. Carboniferous, Sheffield, **3**, 1129–1146, Sheffield
Neumann, M. (1967): Manuel de micropaleontologie des foraminifères, Systematique Stratigraphie. 297 pp., 182 Figs., 60 Pls., Paris: Gauthier-Villars
Neumann, M., Pozaryska, K., Vachard, D. (1975): Remarques sur les microfacies du Dévonien de Lubin (Pologne). Rev. Micropaléont. **18**, 38–52, Paris
Papp, A., Turnovsky, K. (1970): see 5.5
Piller, W. (1978): Involutinacea (Foraminifera) der Trias und des Lias. Beiträge Paläont. Österreich **5**, 1–164, Wien
Pokorny, V. (1958): Grundzüge der zoologischen Mikropaläontologie, Vol. 1, 582 pp., 549 Figs., Berlin: Deutscher Verlag der Wissenschaften
Postuma, J. (1971): Manual of planktonic foraminifera. 420 p. Amsterdam: Elsevier
Ross, Ch. A. (1967): Development of fusulinid (Foraminiferida) faunal realms. J. Paleont. **41**/6, 1341–1354, Tulsa
Ross, Ch. A. (1970): Concepts in Late Paleozoic correlations. Geol. Soc. Amer. Spec. Paper **124**, 7–36, 10 Figs., Washington
Sachs, K. N., Jr. (1965): Preparing paired thin sections from single Larger foraminifera. In: Kummel, B., Raup, D.: Handbook of paleontological techniques, 238–240, San Francisco – London: Freeman
Sartoni, S., Crescenti, U. (1962): see 5.5
Schäfer, P., Senowbari-Daryan, B. (1978): Die Häufigkeitsverteilung der Foraminiferen in drei obertriadischen Riffkomplexen der Nördlichen Kalkalpen (Salzburg/Österreich). Verh. geol. Bundesanst. **1978**, 73–96, Wien
Schröder, R. (1979): Les Orbitolines de l'Aptien: définitions, origine et évolution. Geobios, Mém. spéc. **3**, 289–299, 1 Pl., 1 Fig., Lyon
Toomey, D. F. (1972): Distribution and paleoecology of Upper Devonian (Frasnian) algae and foraminifers from selected areas in western Canada and the northern United States. 24th Internat. Geol. Congr., sect. 7, Paleontology, 621–630, Montreal
Toomey, D. F., Mamet, B. L. (1979): Devonian protozoa. Spec. Pap. Paleont. **23**, 189–192, London: Paleont. Ass.
Towe, K. M., Cifelli, R. (1967): Wall ultrastructure in the calcareous foraminifera: crystallographic aspects and a model for calcification. J. Paleont. **41**, 742–762, Tulsa
Zaninetti, L. (1976): Les Foraminifères du Trias, Essai de synthèse et correlation entre les domaines mesogéens européen et asiatique. Riv. Ital. Paleont. Strat. **82**/1, 1–158, Pls. 1–24, Milano

5.3.2 **Radiolaria** (Plate 53/2)

Systematic position: Protozoa, a subclass of the class Actinopoda. Geologists are concerned almost exclusively with one major radiolarian group, the Polycystina, which are characterized by skeletons of amorphous silica (opal). The Acantharia with skeletons composed of strontium sulfate are no longer included among radiolarians.

Morphology: The radiolarian skeleton is encased in the soft cytoplasm. The bulk of the test usually lies in the ectoplasm, with pseudopodia extending outward to encase any protruding spines. The siliceous skeleton of the Polycystina is typically constructed of a network of bars (elongate elements connected at both ends to other elements) and spines (elongated elements attached at one end only). The majority of species have a complex skeleton with a definite enclosing wall (latticed, spongy, or perforate plate wall). The two major groups of the Polycystina are characterized by different skeletal shapes: Spheres are the most common shape among the Spumellaria; the skeleton often consists of two or more nested spheres, which are connected by radial bars. Nasselaria are characterized by axial symmetry of the basic skeletal element which may represent spines and branches or latticed cup-like chambers.

Radiolaria in thin-sections: For exact determinations specimens must be etched from the sediment. Calcitized radiolarians may be differentiated into Spumellaria and Nasselaria according to the globular or vase-like appearance of the tests; better results can be obtained from differential etching with dilute acids (Riedel and Sanfilippo, 1974). Calcitized radiolarians may be confused with calcisphaeres (see Sect. 5.3.13) or with cross-sections of sponge spicula, but can be recognized often by a jagged margin along the sparite-filled test; size of the tests up to 1 mm, usually 0.10–0.20 mm.

Paleoecology: Radiolarians are planktonic, exclusively marine organisms. They live in open-marine environments and are found at all depths of Recent oceans. The bulk of species and individuals occur in the upper few hundred meters with a rapid decline in abundance in deeper waters. Radiolarians play an important role in the silica-cycle in the oceans; the accumulation of radiolarian-rich sediments (radiolarian ooze with 20–30% biogenous silica; e.g., in equatorial Pacific ocean bottoms) depends on complex interrelated processes (productivity in the water, transfer to the ocean bottom, preservation in sediments and dilution). Ancient radiolarian cherts may be interpreted as deposits in moderately deep to very deep basins. In cases where radiolarian cherts overlie pelagic limestones, bathymetric interpretations in terms of the fluctuation of the carbonate compensation depth are appropriate (Bosellini and Winterer, 1975; Garrison, 1974).

Radiolarians are found in pelagic limestones (e.g., Upper Triassic basinal Hallstatt facies of the Northern Alps; Jurassic of the Western Mediterranean, see Kuhry et al., 1976; Cretaceous Majolica limestones of the Southern Alps), but also in shallow-water limestones.

Biostratigraphy: Radiolarians have been recorded from the Cambrian onwards. They are now recognized as an important group for long-range age correlations, especially in the Cenozoic and Mesozoic (see Kling, 1978).

References: Radiolaria

Bosellini, A., Winterer, E. L. (1975): Pelagic limestone and radiolarite of the Tethyan Mesozoic: a genetic model. Geology **3**/5, 279–282, 2 Figs., Boulder

Calvert, S. E. (1974): Deposition and diagenesis of silica in marine sediments. Spec. Publ. intern. Ass. Sediment. **1**, 273–299, 7 Figs., Oxford

Casey, R. E. (1971): Radiolarians as indicators of past and present water-masses. Micropal. Oceans, 331–341, 3 Pls., 2 Figs., Cambridge

Colom, G. (1955): Jurassic-Cretaceous pelagic sediments of the Western Mediterranean and the Atlantic area. Micropal. **1**/2, 109–124, Pls. 1–5, 4 Figs., New York

Foreman, H. P., Riedel, W. R. (1972): Catalogue of polycystine Radiolaria. Series 1 (1834–1900), Vol. 1 (Meyer, 1834 – Bury, 1862), Parts 1 and 2. Spec. Publ. Amer. Mus. Natural Hist., New York
Garrison, R. E. (1974): Radiolarian cherts, pelagic limestones, and igneous rocks in eugeosynclinal assemblages. Spec. Publ. Intern. Ass. Sediment. **1,** 367–399, 6 Figs., Oxford
Garrison, R. E., Fischer, A. G. (1969): Deep-water limestones and radiolarites of the Alpine Jurassic. In: Friedman, G. M. (ed.): Depositional environments in carbonate rocks. Soc. Econ. Paleont. Min. Spec. Publ. **14,** 20–56, Tulsa
Grunau, H. R. (1965): Radiolarian cherts and associated rocks in space and time. Eclogae geol. Helv. **58,** 157–208, Basel
Kling, S. A. (1978): Radiolaria. In: Haq, B. U., Boersma, A. (eds.): Introduction to marine micropaleontology, 203–244, New York: Elsevier
Kuhry, B., de Clercq, S. W. G., Decker, L. (1976): Indications of current actions in Late Jurassic Limestones. Radiolarian Limestones, Saccocoma Limestones and associated rocks from the Subbetic SE Spain. Sed. Geol. **15,** 235–258, 3 Pls., 2 Figs., Amsterdam
Riedel, W. R., Sanfilippo, A. (1978): Stratigraphy and evolution of tropical Cenozoic radiolarians. In: Riedel, W. R., Sait, T. (eds.): Marine Plankton and Sediments, Proc. Third Planktonic Conf. Kiel, 1974, Micropaleont. **23**/1, 61–96, 10 Plates, New York
Steinmann, G. (1925): Gibt es fossile Tiefseeablagerungen von erdgeschichtlicher Bedeutung? Geol. Rundschau **16,** 435–468, Berlin

5.3.3 Calpionellids (Plate 22)

Systematic position: Protozoa, compared with Ciliata. Calpionellids were thought to be fossil tintinnids because of the overall resemblance of their tests to the lorica of some tintinnids. An interpretation of these as fossil tintinnids is not possible because of the lack of mineralized tests in Recent tintinnids and also in ciliates as a whole (Remane, 1978). Debatable fossil tintinnids with organic tests are extremely rare (e.g., Eicher, 1965; Chennaux, 1967).

Morphology: Calcitic bell-shaped test (lorica) with axial-symmetry. The oral opening is large and surrounded by an oral collar of variable form. Longitudinal sections U-shaped or V-shaped. Total length of the test 40–150 μm.

Microstructure: Wall consisting of spirally arranged calcite prisms normal to the surface of the lorica (Aubry et al., 1976). Agglutinated forms may occur.

Determination: Calpionellids can be differentiated in thin-sections only. The "geometric variation" (Remane, 1964) must be considered as well as the possibility of tectonically deformed tests (Remane, 1964b). A statistical treatment of the frequency distribution of the various species seen in the thin-sections is recommended.

Major groups: Two families are distinguished (Remane, 1978). The Colomiellidae (with the genus *Colomiella* Bonet), known from the Upper Aptian to the Albian, are characterized by very short, aborally rounded lorica and a long cylindrical collar separated by a suture from the lorica. Lorica and collar are hyaline and calcitic. Some transverse sections show a black cross at a 45° angle, which is not seen in the Calpionellidae. The Calpionellidae (Middle Tithonian to Middle Valanginian) are defined by calcitic lorica with axial symmetry, with a wide oral opening which is surrounded by a differentiated collar. Most Calpionellidae have a purely hyaline wall composed of radially fibrous calcite; in the earliest representatives (*Chitinoidella*) the microgranular wall appears brownish opaque in transmitted light, and white in reflected light. *Praetintinnopsella* Borza (Middle Tithonian) has an outer microgranular and an inner hyaline layer. Stratigraphically important genera of the Calpionellidae are *Chitinoidella* Doben (Middle Tithonian to lowermost Upper Tithonian), *Tintinnopsella* Mureganu and Filipescu (Upper Tithonian to Middle Valanginian), *Crassicollaria* Remane (Upper Tithonian to Lower Berriasian), *Calpionella* Lorenz (Upper Tithonian to uppermost Berriasian), *Calpionellopsis* Colom (Middle to Upper Berriasian, locally until Lower Valanginian), and *Calpionellites* Colom (Lower Valanginian).

STAGES			STANDARD CALPIONELLID ZONES	
LOWER CRETACEOUS	VALANGINIAN	UPPER		
		LOWER	Calpionellites	E
	BERRIASIAN		Calpionellopsis	D
			Calpionella	C
				B
UPPER JURASSIC	TITHONIAN	UPPER	Crassicollaria	A
		MIDDLE	Chitinoidella	CHIT.
			?	?

Fig. 46. Standard Calpionellid Zones. *Right column:* calpionellid zones of the Western Alps. After Remane (1978)

Paleoecology: Calpionellids were marine planktonic organisms. They occur typically in very fine-grained limestones which are rich in nannoplankton (coccoliths, nannoconids) and microplankton (radiolarians) but poor in benthic fossils. Calpionellid limestones are a characteristic facies of the uppermost Jurassic and Lower Cretaceous pelagic facies of the Tethys (e.g., Alpine and Mediterranean regions), often characterizing deep-marine, basinal depositional environments.

Biostratigraphy: The present knowledge of calpionellid biostratigraphy is shown in Fig. 46. Calpionellids can be used in world-wide correlations (see Allemann et al., 1971).

Pre-Jurassic "tintinnids": Sections with an overall resemblance to calpionellids or tintinnids have been described from Paleozoic and Triassic limestones (e.g., Hermes, 1966 (Ludlowian); Teichert, 1965, Cuvillier and Sacal, 1963, and Murray and Taylor, 1965 (Devonian); Cuvillier and Barreyre, 1964 (Viséan), and Borza, 1971 (Upper Triassic)). These are large forms (more than 1 mm long) which can not be compared in size, morphology and habitat with Jurassic/Cretaceous calpionellids.

The "aberrant tintinnids" (Radoičić, 1969), widely distributed in Uppermost Jurassic and Lowermost Cretaceous shallow-water carbonates, must be assigned to dasycladacean algae (see Sect. 5.3.13).

References: Calpionellids

Allemann, F., Catalano, R., Fares, F., Remane, J. (1971): Standard calpionellid zonation (Upper Tithonian-Valanginian) of the Western Mediterranean Province. Proc. II Planktonic Conference, Roma 1970, 1337–1340, 1 Tab., Rome
Standard zonations (important)

Aubry, M. P., Bignot, G., Bismuth, H., Remane, J. (1975): Premiers résultats de l'observation au M. E. B. de la Lorica des Calpionelles et de quelques microfossiles qui leur sont associés. Rev. Micropol. **18**/3, 127–133, Pls. 1–2, Paris

Bonet, F. (1965): Zonification microfaunistica de las calizas cretacias del este de Mexico. Bol. Asoc, mexicana de geologos petroleros **8**, 7/8, VI 3–102, 30 Pls., Mexico
Zonation of calpionellids in the Lower Cretaceous of Mexico. Two groups are distinguished within the tintinnids: 1) Calpionelloidea (calcitic test), 2) Tintinnidea (organic test)

Plate 22. Calpionellids from the Tithonian and Lower Cretaceous

1. *Calpionella alpina* Lorenz. Large specimen, characteristic for the time interval A 2/A 3. Upper Tithonian (subzone A 2): Serres-Montclus, France

2. *Crassicollaria intermedia* (Durand-Delga). Upper Tithonian (subzone A 2): Serres-Montclus, France

3. *Crassicollaria brevis* Remane. Upper Tithonian (subzone A 3): Serres-Montclus, France

4. *Crassicollaria parvula* Remane. Berriasian (zone B): Chevallon, France

5. *Calpionella alpina* Lorenz. Characteristic form from the lower part of zone B. Lowermost Berriasian (zone B): Saint-Pontcrasse (Isere), France

6. *Calpionella elliptica* Cadisch. Berriasian (zone C): Berrias, France

7. *Calpionellopsis simplex* (Colom). Berriasian: Mallorca, Spain

8. *Calpionellopsis simplex* (Colom). Berriasian (subzone D 1): Berrias, France

9. *Tintinnopsella carpathica* (Murgeanu and Filipescu). Berriasian (zone C): Serres-Montclus, France

10. *Tintinnopsella carpathica* (Murgeanu and Filipescu). Berriasian (zone C): Berrias, France

11. *Calpionellopsis oblonga* (Cadisch). Berriasian (subzone D 2): Berrias, France

12. *Lorenziella plicata* Remane. Berriasien (upper part of subzone D 2): Berrias, France

13. *Remaniella cadischiana* (Colom). Small specimen. Berriasian (subzone D 3): Berrias, France

14. *Tintinnopsella longa* (Colom). Lowermost Valanginian (D 3): Berrias, France

15. *Lorenziella hungarica* Knauer. Berriasian (D 3): Berrias, France

16. *Chitinoidella boneti* Doben. Zone with *Chitinoidella;* Tithonian pebble within sediments of Berriasian age: Pontaux (Drome), France

Scale for all Figures 20 μm

Photographs by courtesy of J. Remane, Neuchâtel

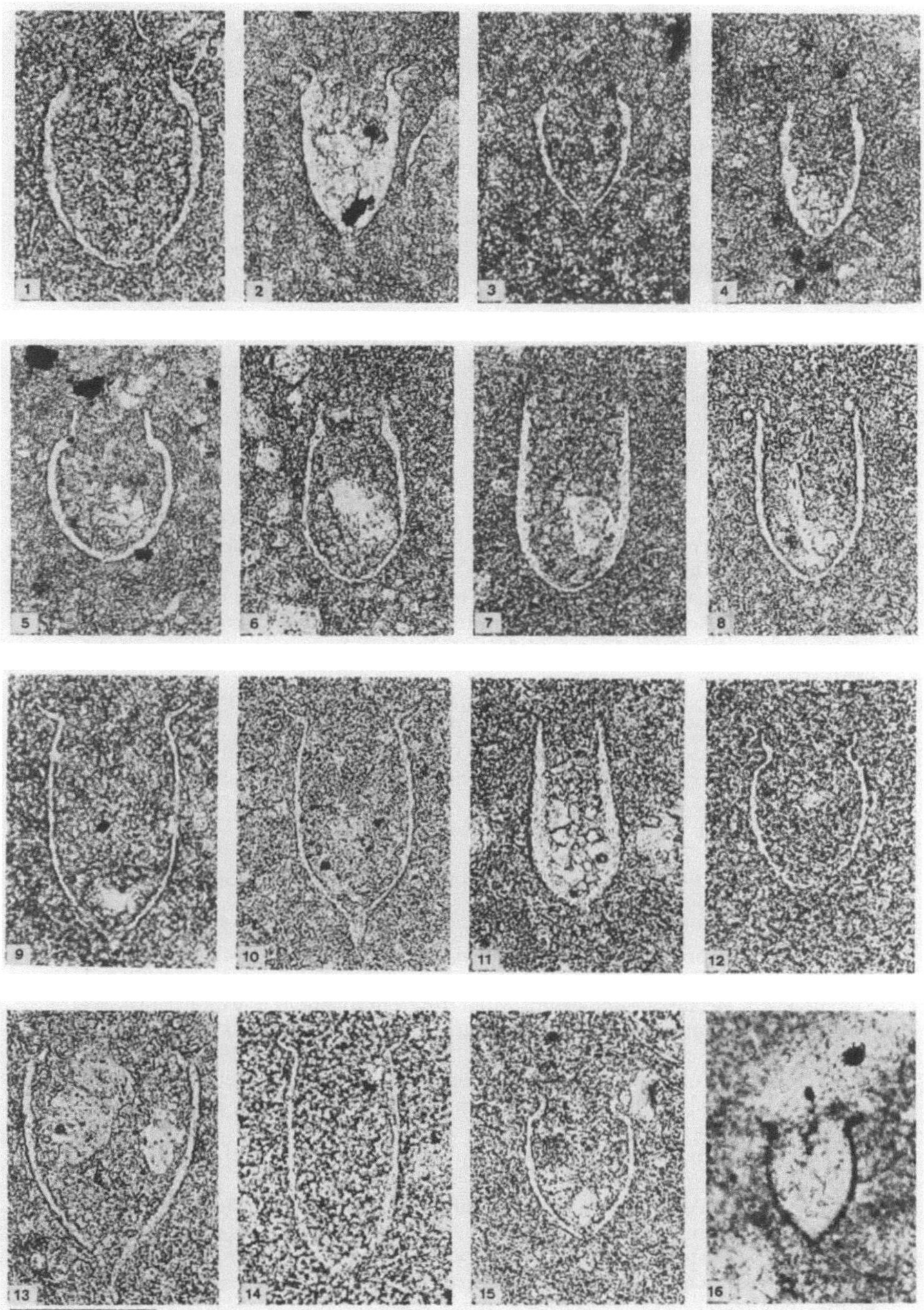
1
2
3
4
5
6
7
8
9
10
11
12
13
14
15
16

Borza, K. (1969): see 5.5
Borza, K. (1971): *Praecalpionellopsis gemeriensis* n. gen., n. sp., aus der oberen Trias der Westkarpaten. Geol. Zbornik, Geol. Carpathica **22**/1, 131–135, 6 Figs., Bratislava
Hallstatt limestones, Norian limestones
Colom, G. (1948): Fossil tintinnids: loricated infusoria of the order of the Oligotricha. J. Paleont. **22**, 233–266, Menasha
A classic monograph of the Calpionellidae; ecology and systematics are discussed in great detail.
Colom, G. (1965): Essais sur la biologie, la distribution géographique et stratigraphique des Tintinnoidiens fossiles. Eclogae geol. Helvetiae **58**/1, 319–334, 3 Pls., 2 Figs., 4 Tab., Basel
Very essential work; many references
Cuvillier, J., Barreyre, M. (1964): Presence des Tintinniens dans le Viseen des Asturies. Rev. Micropaléont. **7**/1, 80–81, Paris
Upper Visean of Asturias, Spain; circular figures with diameters of 30–40 μm; U-shaped figures 30–40 μm long, maximum width of 60 μm; wall 10 μm thick, structureless. Compared to stomiospheres
Cuvillier, J., Sacal, V. (1963): Présence de Tintinnoidiens dans le Devonien supérieur du Sahara septentrional. Rev. Micropal. **6**/2, 73–75, 1 Pl., 1 Fig., Paris
Famennian of the Sahara; goniatite limestones, *Vautrinella lapparenti* n. gen. n. sp., 200–250 μm long
Doben, K. (1963): Über Calpionelliden an der Jura/Kreide-Grenze. Mitt. Bayer. Staatssammlg. Paläont. hist. Geol. **3**, 35–50, Pls. 5–6, 1 Tab., München
Upper Tithonian up to and including the Middle Berriasian (the profile studied is thus more extensive than the profile described by Remane in Southern France). Six zonations distinguished. Description of tintinnids with organic walls: *Chitinoidella* n. gen., with *C. boneti* n. sp. from the Upper Middle Tithonian. Important work for stratigraphic interpretations!
Eichler, D. L. (1965): Cretaceous tintinnids from the western interior of the United States. Micropaleontographica **11**/4, 449–456, Pl. 1, 3 Figs., New York
Preserved body fossils of tintinnids from the shales of a marginal marine facies in the Albian and Cenomanian. The new species are referred to Recent tintinnid genera. *Dicloepella* n. gen.
Hermes, J. J. (1966): Tintinnids from the Silurian of the Betic Cordilleras, Span. Rev. Micropaleont. **8**/4, 211–214, 1 Pl., Paris
Ludlowian, red *Orthoceras* limestones. Round to oval figures, sometimes with collars; 0.19 mm long, 0.09 mm wide
Holzer, H. L. (1968): Stratigraphie und Lithologie der Jura-Kreide-Folge im nördlichsten Pechgraben-Steinbruch (OÖ), Mitt. naturwiss. Ver. Steiermark **98**, 47–57, 4 Figs., 1 supp., Graz
Zonation with the aid of calpionellids; good survey of zonations in various areas
Loeblich, A. R., Tappan, H. (1968): Annotated index to genera, subgenera and supra-generic taxa of the ciliate order Tintinnida. J. Protozool. **15**/1, 185–192
Lorenz, T. (1902): Geologische Studien im Grenzgebiet zwischen Helvetischer und Ostalpiner Fazies. 2. Der südliche Rhätikon. Ber. naturforsch. Ges. Freiburg i. Breisgau **12**, 35–93, Freiburg
Calpionella n. gen. with *C. alpina* n. sp. within the matrix of the Falknis Breccia (Tithonian). Description of the first calpionellids!
Radoičić, R. (1969): Aberantna Grana Fosilnih Tintinnina (Podred Tintinnina). La Branche Aberrantes des Tintinnines Fossiles (Sous-Ordre Tintinninina). Palaeontologia Jugoslavica **9**, 5–71, 8 Pls., Zagreb
Remane, J. (1962): Zur Calpionellen-Systematik. N. Jb. Geol. Paläont. Mh. **1962**/1, 8–24, 46 Figs., Stuttgart
Remane, J. (1964a): Untersuchungen zur Systematik und Stratigraphie der Calpionellen in den Jura-Kreide-Grenzschichten des Vocontischen Troges. Palaeontographica A **123**, 1/3, 1–57, Pls. 1–6, 1 Supp., 18 Figs., Stuttgart
Most important work about calpionellid stratigraphy and identification criteria (geometric range of variation). Zonation: Zone A, *Crassicollaria* Remane (Early Upper Tithonian);

Zone B, *Calpionella* Lorenz (Late Upper Tithonian to Lower Berriasian); Zone C, *Tintinnopsella* Colom (Upper Berriasian).
Remane, J. (1964b): Über zwei interessante Deformationstypen bei Calpionellen-Gehäusen (Protozoa, Tintinnina?), N. Jb. Geol. Paläont. Mh. **1964**/11, 664–672, 4 Figs., Stuttgart
Remane, J. (1965): Neubearbeitung der Gattung *Calpionellopsis* Col., 1948 (Protozoa, Tintinnina?). N. Jb. Geol. Paläont. Abh. **122**/1, 27–49, Pls. 5–6, 7 Figs., Stuttgart
New definition of *Calpionellopsis*. This genus characterizes Zone D (around the Middle Berriasian).
Remane, J. (1969): Les possibilités actuelles pour une utilisation stratigraphiques des Calpionelles (Protozoa incertae sedis, Ciliata?). Proc. First Inter. Cont. Planktonic Microfoss., Geneva 1967, **2**, 559–573, 7 Figs., Leiden: Brill
Remane, J. (1974): Les Calpionelles, Cours de IIIe cycle en science de la terre, Paléontologie, Université de Genève, partie II, 58 pp., 23 Figs., Genf
Survey of the morphology, systematics, and stratigraphic distribution of calpionellids.
Remane, J. (1978): Calpionellids. In: Haq, B. U., and Boersma, A. (eds.): Introduction to marine micropaleontology, 161–170, New York: Elsevier
Important review of the present knowledge of calpionellids.
Tappan, H., Loeblich, A. R. Jr. (1968): Lorica composition of modern and fossil Tintinnida (Ciliata, Protozoa), systematics. Geologic distribution and some new Tertiary taxa. J. Paleont. **42**/6, 1378–1394, Pls. 165–174, 1 Fig., Tulsa
Reevaluation of arguments in favor of including calpionellids in the order Tintinnida, with a discussion of the biology of living tintinnids.

5.3.4 Sponges (Plates 23; 43/1–3; 50/1)

Sponges are the most primitively organized multicellular organisms. In most sponges spezialized cells secrete a mineral skeleton which may consist of calcareous spicules; siliceous spicules; siliceous spicules and organic matter (spongin); or spongin fibers only.

The subdivision of the phylum Porifera into siliceous and calcareous sponges is anything but correct (see Hartman et al., 1980), but at least easy and feasible. "Siliceous sponges" include the Desmospongia and the Hexactinellida, the "Calcareous sponges" are represented by the Calcarea and the Sclerospongia.

Siliceous Sponges

Morphology: The shape of siliceous sponges is highly variable (e.g., cylindrical, saucer-shaped, irregular crusts); the size varies between a few centimeters and more than 50 cm. Major systematic groups are distinguished by the morphology, number of axes or rays present, and by the size (megascleres and microscleres). Megascleres have a size of 1 mm or larger; they occur isolated or are fused together to form compound skeletons. Microscleres with a size of about 0.1 mm are never fused together and rarely preserved as fossils. Classifications of siliceous sponge spicules are found in Laubenfels (1955), for Upper Jurassic spicules in Reif (1967). In cross-sections spicules often appear circular. Non-calcitized siliceous spicules may show a central canal.

Fossilization: Biostratinomical processes cause the decay of organic matter and the disintegration of skeletal components, followed by an autochthonous, but more often allochthonous embedding of the spicules in the sediment. Diagenetic

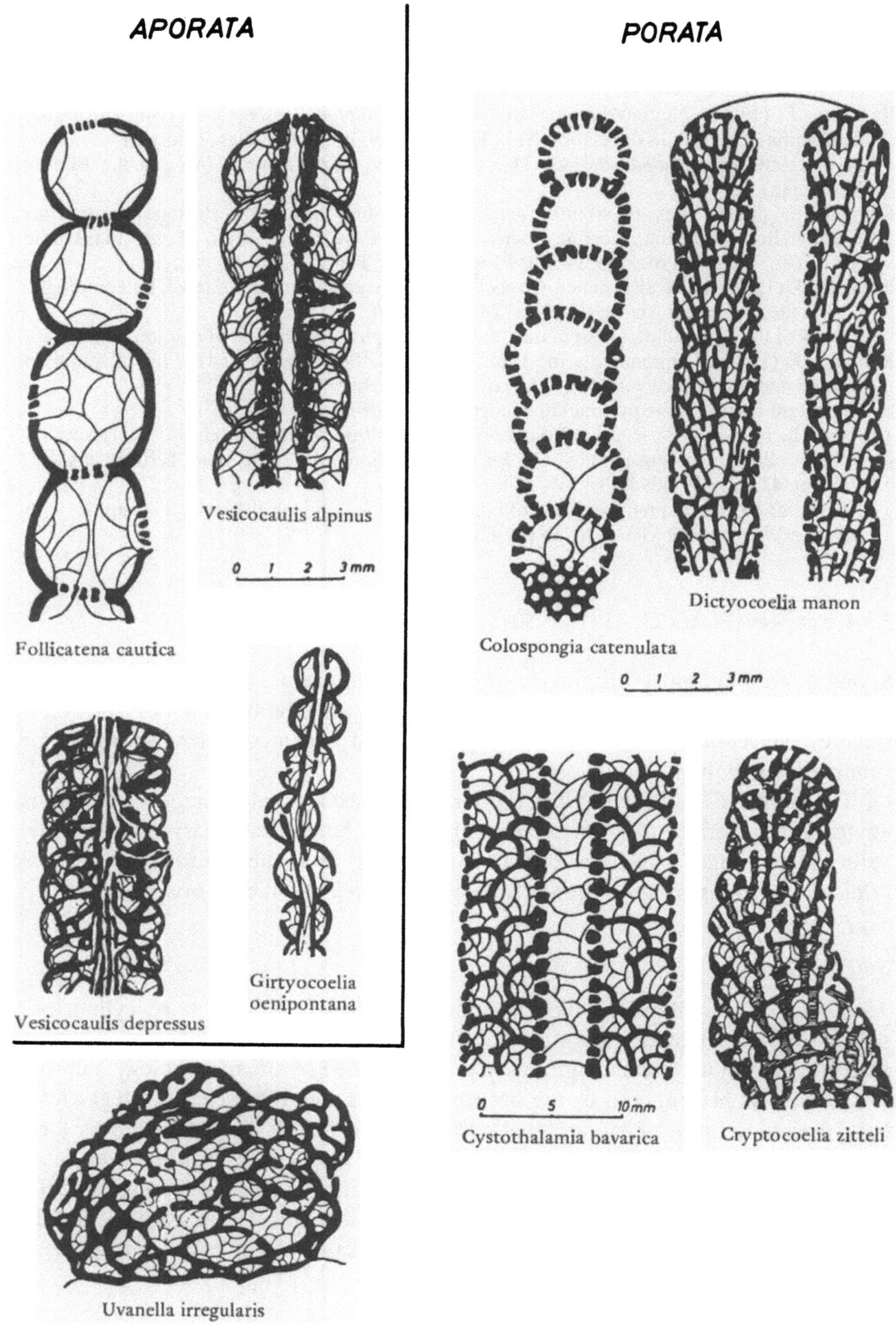

Fig. 47. Common segmented calcisponges (sphinctozoa) from the reef facies of Wetterstein limestones (Ladinian, Middle Triassic) in the Northern Alps. After Ott (1967)

alterations of the skeletal material may be connected with a transformation of opaline into crystalline silica, solution of the spicules and replacement by calcite, and the formation of carbonate concretions caused by the decay of organic matter, which accelerates carbonate precipitation around siliceous sponges (see below).

Stratigraphic distribution: Both Desmospongia and Hexactinellida have been known since the early Cambrian. An informative review of the development of siliceous sponges through time is given by Wiedenmayer (1980). During the Paleozoic, siliceous sponges flourished in quiet-water environments (sometimes contributing to the formation of mud mounds) and in shallow agitated environments (shelly facies). Triassic siliceous sponges seem to have concentrated on lower reef-slopes and at the margins of inter-reef basins. Liassic siliceous sponges are known from soft bottom and deeper water environments formed after the disintegration of vast carbonate platforms. Much more abundant and diverse are the siliceous sponges of the Upper Jurassic, which contribute to the development of mud mounds ("Sponge reefs"). Pelagic limestones (e.g., Aptychus limestone of the Northern Alps) are often rich in siliceous, partly pyritized spicules. During the late Cretaceous a major turnover took place in the desmosponges, demonstrated by the highly diverse sponge faunas of the Chalk facies in Northwestern Europe. The adaption of siliceous sponges to deeper-water environments was continued during the Tertiary.

Calcareous Sponges (Plates 23; 43/1–3)

Many authors consider the calcareous sponges to be a polyphyletic group, represented by the Calcinea and Calcaronea (with some fossil records from the Mesozoic), Heteractinida (an important Paleozoic group), and the Pharetronida (known since the Carboniferous). The pharetronid calcisponges include the non-segmented Inozoa and the segmented Sphinctozoa (which are regarded as a group which has a problematic systematic position in zoological classifications).

Morphology: Most calcareous sponges lack well-defined growth forms, except for the sphinctozoans. The external shape of the inozoans may be cylindrical, branched, spherical or irregular; many calcisponges encrust other organisms. Sphinctozoans are generally segmented; the chambers are arranged serially (with or without a central cavity) or in irregular clusters. Many Permian and Triassic porate sphinctozoans are characterized by filling tissues within the chambers (see Fig. 47). Fossil calcisponges seem to have developed a non-spicular skeleton (see Wendt, 1978).

Stratigraphic distribution and palecology: Inozoan calcisponges appear in the uppermost Lower Permian and reach an initial maximum of diversity in the lowermost Upper Triassic. During the Middle and Upper Jurassic high diversity is again reached. The last inozoans with rigid non-spicular skeletons are known from the uppermost Cretaceous. Sphinctozoan calcisponges appear in the Upper Carboniferous; they are important framework builders in Permian and Triassic reefs (e.g., Permian Reef Complex, Texas; Middle Triassic and Cordevolian Wetterstein Limestone reefs of the Northern Alps, see E. Ott, 1967; Upper Triassic Dachstein Limestone reefs of the Northern Alps, see E. Flügel, 1981).

Sponges and Sediment

Sponges may contribute to sedimentation by mass accumulations of siliceous spicules (especially in slope and basinal environments), by trapping and baffling suspended sediment (siliceous sponges), by stimulating biochemical precipitation of carbonate, and by construction of rigid reef frameworks (calcisponges). A terminology for sponge rocks was proposed by Geyer (1962).

Siliceous sponge *spicules* (Plate 50/3) are common in Lower Jurassic pelagic, cherty limestones of the Alpine-Mediterranean region (Wiedenmayer, 1980), which have been deposited in deeper-water basins and on swells. Topographic highs seem to have favored the growth of spezialized desmosponges; the accumulation of the spicules may be caused by redeposition. Stagnant basins are suitable environments for the dissolution of spiculites, which in turn may be responsible for the early diagenetic formation of cherts in limestones (see Correns, 1950; Newell et al., 1953).

Trapping and baffling of suspended sediments contributed to the formation of the Franconian-Swabian "sponge-algal reefs" in the Upper Jurassic (see Gwinner, 1976; Meyer, 1975; Wagenplast, 1972), but the sediment binding capacity of blue-green algae seems to have been more important in the growth of these buildups (Flügel and Steiger, 1981).

The term *tuberoid* (Fritz, 1958) refers to rounded, nodular, lobular, or spherical dark calcareous lumps of various sizes (about 0.2 to 5 mm) which show diverse internal microstructures (micritic; peloidal; fragments of sponge skeletons). These particles are believed to have been formed from the decay and synsedimentary calcification of sponge fragments and of algal crusts (Plate 50/3, p. 573).

Calcisponges are the main *framebuilders* in the more protected parts of Upper Triassic reefs (Schäfer and Senowbari-Daryan, 1981; Flügel, 1981), i.e., the central reef area, upper slope environments, or in slightly deeper water (mud-mound

Plate 23. Calcisponges

1. Longitudinal section of *Cystothalamia* sp., showing vesicular chambers and perforated walls as well as vesiculae. Sedimentary cavities are lined with several generations of fibrous, submarine cement; interior of cavities filled with vadose silt. Middle Permian: Pietra di Salomone, Sosio valley, Sicily

2. *Cystothalamia bavarica* Ott. Carnian Reef Limestone (Pantokrator Limestone): Hydra island, Greece

3. *Vesicocaulis multisiphonatus* Kovacs. Carnian Reef Limestone (Pantokrator Limestone): Hydra island, Greece

4. *Peronidella austriaca* Flügel, a non-segmented calcisponge. Zlambach Beds, Upper Triassic: Rohrmoos, Gosaukamm, Austria

5. *Colospongia* sp., a segmented (sphinctozoan) calcisponge; intergrown with tabulozoans (1) and spongiostromate algal crusts. Bottom right (2) *Disjectopora* sp. (hydrozoan). Upper Rhaetian Reef Limestone: Adnet near Hallein, Salzburg, Austria

6. Cross-section of thickly encrusted calcisponges. Trogkofel Limestone, Lower Permian: Forni Avoltri, Carnia, Italy

Scale for all Figures 5 mm

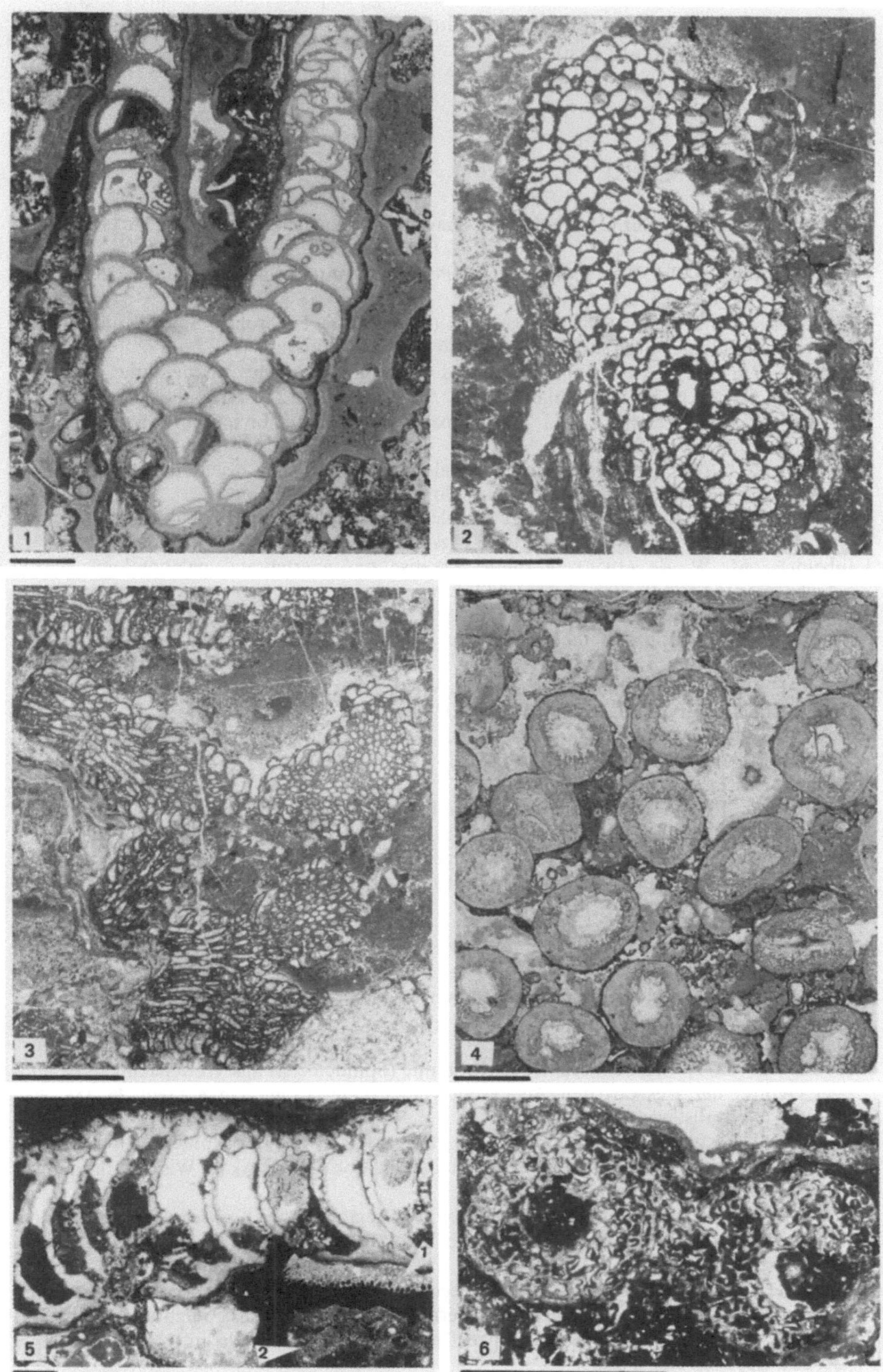
1
2
3
4
5
1
2
6

stage). Other examples for the framebuilding importance of calcisponges are Permian reefs in southern Tunisia and the Cassian patch reefs (Carnian) in South Tyrol (see Hartman et al., 1980).

References: Sponges

a) Surveys, Sedimentology

Correns, C. W. (1950): Zur Geochemie der Diagenese. Geochem. Cosmochim. Acta **1,** 49–54, Oxford

Fry, W. G. (ed.) (1976): The Biology of the Porifera. Symp. Zool. Soc. London **25,** 490 pp., London: Academic Press

Geyer, O. F. (1962): Über Schwammgesteine (Spongiolith, Tuberolith, Spiculit und Gaizit). Hermann-Aldinger-Festschrift, 51–59, Pls. 6–8, 1 Fig., Stuttgart

Gramann, F. (1962): Schwamm-Rhaxen und Schwamm-Gesteine (Spongiolithe, Spiculite) aus dem Oxford N. W. Deutschlands. Geol. Jb. **80,** 213–220, 2 Pls., Hannover

Hartman, W. D., Wendt, J. W., Wiedenmayer, F. (1980): Living and fossil sponges. Sedimenta **8,** 274 pp., Miami

Jones, W. C. (1970): The composition, development, form and orientation of calcareous sponge spicules. Symposia, Zool. Soc. London **25,** 91–123, London

Lane, N. G. (1981): A nearshore sponge spicule mat from the Pennsylvanian of West-Central Indiana. J. Sed. Petrol. **51**/1, 197–202, 3 Figs., Tulsa

Laubenfels, M. W. de (1955): Porifera. Treatise on invertebrate, paleontology, part E, Archaeocyatha and Porifera, E 21–E 112, Lawrence

Newell, N. D., Fischer, A. G., Whiteman, A. J., Hickox, J. E., Bradley, J. S. (1953): The Permian reef complex of the Guadalupe Mountains Region, Texas and New Mexico. 236 p., San Francisco: Freeman

Reif, W. F. (1967): Schwammspicula aus dem Weissen Jura Zeta von Nattheim (Schwäbische Alb). Palaeontographica, ser. A, **127,** 85–102, Stuttgart

Rigby, J. K. (1971): Sponges and Reef and Related Facies through Time. Symp. North American Paleont. Conv. **1969,** part I. 1374–1388

Wendt, J. (1978): see 5.3.5

Wiedenmayer, F. (1978): Modern sponge bioherms of the Great Bahama Bank and their likely Ancient analogues. Coll. Int. C.N.R.S., **291,** Biologie des Spongiaires, 289–296, Paris

Wiedenmayer, F. (1980): Spiculites and sponges in the Lower Jurassic of the western Tethys. In: Hartman, W. D. et al.: Living and fossil sponges. Sedimenta **8,** 135–145, Miami

b) Alpine Triassic

Cuif, J. P. (1973): Histologie de quelques Sphinctozoaires (Porifères) triasiques. Geobios **6**/2, 115–125, Lyon

Flügel, E. (1981): see 10.3.9

Jablonsky, E. (1973): Segmentierte Kalkschwämme (Sphinctozoa) aus Wettersteinkalken einiger Gebirge der Westkarpaten. Acta Geol. Geogr. Univ. Comenianae, Geol. **26,** 189–202, Pls. 67–68, Bratislava

Ott, E. (1967): Segmentierte Kalkschwämme (Sphinctozoa) aus der alpinen Mitteltrias und ihre Bedeutung als Riffbildner im Wettersteinkalk. Abh. Bayer. Akad. Wiss. math.-naturw. Kl. N.F. **131,** 96 pp., 10 Pls., München

Ott, E. (1974): *Phragmocoelia* n.g. (Sphinctozoa), ein segmentierter Kalkschwamm mit neuem Füllgewebetyp aus der alpinen Trias. N. Jb. Geol. Paläont. Mh. **1974**/12, 712–723, 4 Figs., Stuttgart

Schäfer, P., Senowbari-Daryan, B. (1981): Facies development and paleoecological zonation of four Upper Triassic patch-reefs, Northern Calcareous Alps near Salzburg, Austria. In: Toomey, D. F. (ed.): European fossil reef models. Soc. Econ. Paleont. Min. Spec. Paper **30,** 241–259, 10 Figs., Tulsa

Zankl, H. (1969): Der Hohe Göll. Aufbau und Lebensbild eines Dachsteinkalk-Riffes in der Obertrias der nördlichen Kalkalpen. Abh. Senckenberg. Naturforsch. Ges. **519,** 114 pp., 15 Pls., 74 Figs., Frankfurt a. M.

c) Southern German "sponge-reefs", Upper Jurassic

Behr, K., Behr, H.-J. (1976): Cyanophyten aus oberjurassischen Algen-Schwamm-Riffen. Lethaia **9,** 283–292, 8 Figs., Oslo
Draganescu, A. (1976): Constructional to corpuscular sponge algal, algal and coralalgal facies in the Upper Jurassic Carbonate formation of Central Dobrogea (the Casimcea Formation). In: Patrulius, D. et al.: Carbonate Rocks and Evaporites. Guidebook. Intern. Coll. Carb. Rocks/Evapor. Romania, Guidebook Ser. **15,** 13–43, 2 Pls., 8 Figs., Bucaresti
Flügel, E., Steiger, T. (1981): An Upper Jurassic sponge-algal buildup from the Northern Frankenalb, West Germany. In: Toomey, D. F. (ed.): European fossil reef models. Soc. Econ. Paleont. Min. Spec. Paper **30,** 371–397, 23 Figs., Tulsa
Fritz, G. K. (1958): Schwammstotzen, Tuberolithe und Schuttbreccien im Weißen Jura der Schwäbischen Alb. Arb. Geol. Paläont. Inst. TH Stuttgart **13,** 1–119, 24 Figs., 5 Pls., Stuttgart
Gwinner, M. P. (1976): Origin of the Upper Jurassic limestones of the Swabian Alb (Southwestern Germany). Contrib. Sed. **5,** 1–75, Stuttgart
Hiller, K. (1964): Über die Bank- und Schwammfazies des Weißen Jura der Schwäbischen Alb (Württemberg). Arb. Geol. Paläont. Inst. TH Stuttgart **40,** 1–190, 38 Figs., 26 Pls., Stuttgart
Meyer, R. K. (1975): Mikrofazielle Untersuchungen in Schwamm-Biohermen und -Biostromen des Malm Epsilon (Ober-Kimmeridge) und des obersten Malm-Delta der Frankenalb. Geol. Bl. NO-Bayern **25**/4, 149–177, 14 Figs., Erlangen
Meyer, R. K. (1977): Mikrofazies im Übergangsbereich von der Schwammfazies zur Korallen-Spongiomorphiden-Fazies im Malm (Kimmeridge-Tithon) von Regensburg bis Kelheim. Geol. Jb. A **37,** 33–69, 5 Pls., 1 Fig., 2 Tab., Hannover
Nitzopoulos, G. (1973): Faunistisch-ökologische, stratigraphische und sedimentologische Untersuchungen am Schwammstotzen vom Spielberg am Hahnenkamm (Ob. Oxfordien, Südliche Frankenalb). Stuttgarter Beiträge Naturkunde, B, **16,** 1–143, Stuttgart
Roll, A. (1934): Form, Bau und Entstehung der Schwammstotzen im süddeutschen Malm. Z. Paläont. **16,** 197–246, Berlin
Wagenplast, P. (1972): Ökologische Untersuchungen der Fauna aus der Bank- und Schwammfazies des Weißen Jura der schwäbischen Alb. Arb. Geol. Paläont. Inst. TU Stuttgart **67,** 1–99, 10 Figs., 18 Pls., Stuttgart

5.3.5 Stromatoporoids, Tabulozoans, Hydrozoans (Plates 24/1–5; 41/5; 43/7)

Several groups of extinct organisms, commonly regarded as coelenterates until now, are now believed to bear close relationships to calcareous sponges. These are the Stromatoporoidea, formerly attributed to the hydrozoans, and the Tabulata, which were interpreted as corals.

Stromatoporoidea (Plates 24/1, 2; 41/5; 52/3)

Morphology: The massive, laminar or dendroid skeleton (coenosteum) ranges from a few centimeters to several meters in diameter. It is constructed of vertical elements (pillars) and transversal elements (laminae or cyst plates) which form a regular or irregular meshwork (size of the interlaminar spaces 0.10–0.20 mm). Globular and laminar stromatoporoids are commonly marked by growth bands a few millimeters thick (latilaminae). Many stromatoporoids are pierced by vertical and transverse canals which are connected with the open framework of the skeleton and open onto the surface. Here they ramify into stellate grooves (astrorhizae) which may be centered on elevations (mamelons). Astrorhizal patterns are also visible in the interior of the coenosteum spread out over the latilaminae. Thin horizontal or upwardly convex plates (dissepiments, tabulae) separate major parts of the coenosteum and astrorhizal canals, showing that the living part of the animal usually covered the skeleton only as a thin veneer (see Stearn, 1975). The underside of stromatoporoids may be covered by a solid layer with growth lines (epitheca).

Microstructure: Many microstructure types have been described; most of these appear to be a result of diagenetic modifications of trabecular and spherulitic aragonite skeletons in the process of inversion to calcite. Only irregular, clinogonal, spherulitic and orthogonal arrangements of crystals are considered to be original (see Wendt, 1979; Stearn, 1966).

Systematic position: Stromatoporoids have been attributed to various groups of invertebrates (especially to Hydrozoa and Porifera), but also to algae (Kazmierczak, 1981). The comparison with encrusting calcareous sponges has received support

---

Plate 24. Stromatoporoids (Figs. 1–2), Chaetetids (Fig. 3), Hydrozoans (Figs. 4–5), and Bryozoans (Figs. 6–10)

1. Stromatoporoids: *Clathrocoilona spissa* (Lecompte), overgrowing rugose corals (*Acanthophyllum* sp.). Schwelm Limestone, Middle Devonian: quarry east of Letmathe, Sauerland, Germany. Scale 5 mm

2. Stromatoporoids: *Amphipora rudis* Lecompte. Cross-sections and longitudinal sections of cylindrical coenostea, showing the characteristic reticulation. Schwelm Limestone, Givetian, Middle Devonian: Steltenberg quarry near Hohenlimburg, Sauerland, Germany. Scale 5 mm

3. Chaetetids: *Ptychochaetetes globosus* Koechlin. Plassen Limestone, Upper Jurassic: Krahstein near Bad Mitterndorf, Styria, Austria. Scale 5 mm

4. Hydrozoans: Longitudinal section through a branched colony of *Ellipsactinia caprense* Steinmann. On the right a section through *Actinostromaria* sp. Plassen Limestone, Upper Jurassic: Röthelstein, Styria, Austria. Scale 5 mm

5. Hydrozoans: *Spongiomorpha ramosa* Frech, cross-sections. Upper Rhaetian Reef Limestone: Steinplatte near Waidring, Tyrol. Scale 5 mm

6. Bryozoans: Cystoporate bryozoan – *Fistulipora* sp., encrusting on ostracod. Tarvis Breccia, Lower Permian: Tarvisio, Italy. Scale 2 mm

7. Bryozoans: Cryptostomid bryozoan – *Streblotrypa* sp. Tarvis Breccia, Lower Permian: Tarvisio, Italy. Scale 2 mm

8. Bryozoans: Fragments of fenestrate cryptostomids. Trogkofel Limestone, Lower Permian: Kühweg Alm, Carnic Alps, Austria. Scale 2 mm

9. Bryozoans: Cryptostomid bryozoan – *Ascopora* sp. Carboniferous: Yasin Valley, Northeastern Pakistan. Scale 2 mm

10. Bryozoans: Trepostomid bryozoan – *Pseudobatostomella* sp. Tarvis Breccia, Lower Permian: Tarvisio, Italy. Scale 2 mm

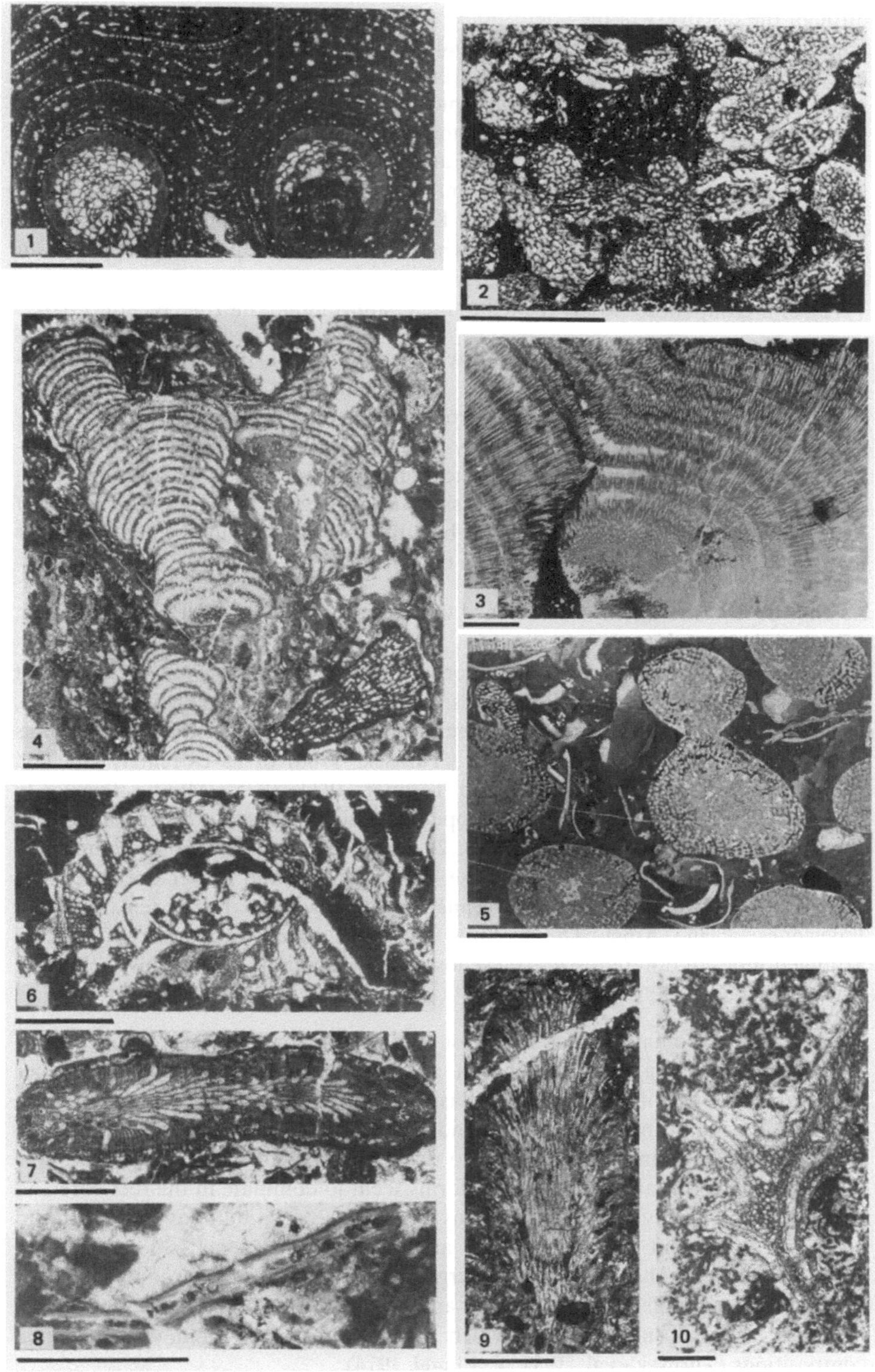
1
2
3
4
5
6
7
8
9
10

through the discovery of the sclerosponges (Hartman and Goreau, 1970), that have remarkable similarities to stromatoporoids (e.g., growth form, construction of the non-spicular skeleton, microstructure types, astrorhizae). The skeleton of Sclerospongiae is characterized by siliceous spicules enclosed in soft tissue or in a non-spicular aragonitic or Mg-calcitic skeleton. Fossil forms attributed to the Sclerospongiae on the basis of growth forms and/or the occurrence of spicule pseudomorphs are known from the Triassic and the Cretaceous (e.g., Cuif, 1974; Dieci et al., 1977; Kazmierczak, 1974, 1979; Kazmierczak and Hillmer, 1974; Wendt, 1974). The question is still open as to whether the Stromatoporoidea represent an order of the Sclerospongiae, or an independent subphylum of the Porifera (Stearn, 1972), or a group more closely related to the Pharetronidae (Inozoa; Wendt, 1980).

Classification: The classification of the stromatoporoids is based on the gross structure of the pillars, laminae, cysts, and amalgamate tissue, and also on the microstructure of these elements. There is no generally accepted classification, although the classification proposed by Stearn (1980) may be useful for the differentiation of the genera. This classification considers only Paleozoic stromatoporoids. Mesozoic stromatoporoids are similar with regard to the gross structure of the skeletal elements, but there seem to be differences in the consistently trabecular microstructure and in the differentiation of astrorhizae.

Stratigraphic distribution: Cambrian?; Ordovician to Lowermost Carboniferous; Middle Triassic to Upper Cretaceous.

Several groups related to stromatoporoids (or to hydrozoans?) may be abundant in thin-sections of Mesozoic limestones: 1) *Spongiomorphidae* (Plate 24/5; 48/1), originally assigned to the hexacorals, then to the Hydrozoa, and today regarded as a distinct group of the Cnidaria (e.g., Turnšek, 1968) or as extremely calcified sponges (Ischyrospongia; Termier and Termier, 1974, 1975). Middle Triassic to Lower Cretaceous, particularly in Upper Triassic reefs and Upper Jurassic to Lower Cretaceous platform carbonates; 2) *Ellipsactinidae* (Plate 24/4), common in shelf and shelf-margin carbonates of the Upper Jurassic and Lower Cretaceous; 3) *Heterastriidae,* confined to Norian basinal limestones of the Tethys (e.g., Hallstatt limestones of the Alps). The globular or ovoid coenostea consist of radially arranged tubes of different sizes; these fossils may represent hydrozoans (belonging to the Milleporoidea; Cuif, 1971).

Paleoecology: Stromatoporoids are most common in shallow-water environments (especially reefs; e.g., in the Silurian and Devonian), and in slightly deeper lagoonal areas or on upper slopes. Basic growth forms of stromatoporoids are ecologically significant (see Kershaw and Riding, 1978; Kobluk, 1978; Lecompte, 1970; Tsien, 1977) and may be used to distinguish paleobathymetric zones. Low dome-shaped forms and thick laminar forms seem to dominate in the reef-core and shallow fore-reef environments of many Devonian reefs; bulbous and dendroid types are prominent in the back-reef and bank environments [e.g., *Amphipora* (Plate 24/2), which characterize low-energy, possibly restricted environments].

In Mesozoic reefs stromatoporoids and related forms are found especially in reef and lagoonal environments, which may be recognized from various faunal compositions (Turnšek et al., 1981; E. Flügel, 1974).

Tabulata or Tabulozoa (Plate 24/3)

Tabulata are generally regarded as corals. The discovery of Recent chaetetids (Hartman and Goreau, 1975), belonging to the sclerosponges, and of spicule-pseudomorphs in Carboniferous, Triassic and Cretaceous chaetetids (Dieci et al., 1977; Kazmierczak, 1979, 1980) has stimulated discussion on the systematic position of the fossil Tabulata, especially of chaetetids and favositids (see Sokolov, 1962; H. W. Flügel, 1976). With the present knowledge available, generalization about the systematic position of the Tabulata should be avoided (see Oliver, 1979). Only chaetetids (first known from the Ordovician) and similar forms from Mesozoic shallow-water limestones seem to be related to sponges. O. Kühn (1943) proposed the name *Tabulozoa* for those "colonial" organisms, characterized by densely packed calcareous tubes with tabulae. Tabulozoans are common in Triassic and Jurassic reef limestones.

References: Stromatoporoids, Tabulozoans, Hydrozoans

Abbott, B. M. (1973): Terminology of stromatoporoid shapes. J. Paleont. **47**/4, 805–806, 1 Tab., Lawrence

Cuif, J.-P. (1971): Structure et position systematique de genre Heterastridium Reuss, 1865 (Hydrozoaire). Geobios **4**/1, 69–79, Pls. 3–5, Lyon

Cuif, J.-P. (1974): Role des sclérosponges dans la faune récifale du Trias des Dolomites (Italie du Nord). Geobios, **7**/2, 139–153, Plates 29–31, 5 Figs., Lyon

Dieci, G., Russo, A., Russo, F. (1974): Revisione del genre Leispongia d'Orbigny (Sclerospongia triassica). Boll. Soc. Paleont. Ital. **13,** 135–146, Plates 41–53, Modena

Dieci, G., Russo, A., Russo, F., Marchi, M. S. (1977): Occurrence of spicules in Triassic chaetetids and ceratoporellids. Boll. Soc. Paleont. Ital. **16,** 229–238, Modena

Embry, A. F., Klovan, J. E. (1972): Absolute water depth limits of late Devonian Paleocological zones. Geol. Rdsch. **61**/2, 672–686, 10 Figs., Stuttgart

Fenninger, A., Flajs, G. (1974): Zur Mikrostruktur rezenter und fossiler Hydrozoen. Biomin. Forsch. Ber. **7,** 69–99, Pls. 1–10, 10 Figs., Stuttgart – New York

Flügel, E. (1974): Fazies-Interpretation der Cladocoropsis-Kalke (Malm) auf Karaburun, W-Anatolien. Archiv Lagerstättenforschung Ostalpen, Sonderband **2,** 79–94, 4 Pls., Leoben

Flügel, E. (1975): Fossile Hydrozoen – Kenntnisstand und Probleme. Paläont. Z. **49**/4, 369–406, 13 Figs., 3 Tab., Stuttgart

Flügel, E., Flügel-Kahler, E. (1968): Stromatoporoidea (Hydrozoa palaeozoica). Fossilium Catalogus **115–116,** 681 pp., 's-Gravenhage: Junk

Flügel, H. W. (1976): Ein Spongienmodell für die Favositidae. Lethaia **9,** 405–419, Oslo

Galloway, J. J. (1957): Structure and classification of Stromatoporoidea. Bull. Amer. Paleont. **37,** 345–480, Pls. 31–37, Ithaca

Hartman, W. D. (1969): New genera and species of coralline sponges (Porifera) from Jamaica. Postilla **137,** 39 p., New Haven

Hartman, W. D. (1979): A new sclerosponge from the Bahamas and its relationship to Mesozoic stromatoporoids. Coll. internat. C.N.R.S. **291,** 467–474

Hartman, W. D., Goreau, T. F. (1972): *Ceratoporella* (Porifera: Sclerospongiae) and the chaetetid "corals". Transact Connecticut Acad. Arts Sci. **44,** 131–148, 26 Figs., New Haven

Hartman, W. D., Goreau, T. F. (1975): A Pacific tabulate sponge, living representative of a new order of sclerosponges. Postilla **167,** 21 p., New Haven

Hartman, W. D., Goreau, T. F. (1976): A new ceratoporellid sponge (Porifera: Sclerospongiae) from the Pacific. In: Harrison, F. W., Cowden, R. R. (eds.): Aspects of Sponge Biology, 329–347, New York: Academic Press

Hartman, W. D., Wendt, J. W., Wiedenmayer, F. (1980): Living and fossil sponges. Sedimenta **8,** 274 p., Miami

Kazmierczak, J. (1974): Lower Cretaceous sclerosponge from the Slovakian Tatra Mountains, Paleont. **17,** 341–347, Pls. 45–46, London

Kazmierczak, J. (1976): Cyanophycean nature of stromatoporoids. Nature (London) **264,** No. 5581, 49–51, 2 Figs., London

Kazmierczak, J. (1979): Sclerosponge nature of chaetetids evidenced by spiculated Chaetetopsis favrei (Deninger, 1906) from the Barremian of Crimea. Neues Jahrbuch Geol. Paläont. Monatshefte **1979,** 97–108, 4 Figs., Stuttgart

Kazmierczak, J. (1981): Evidences for Cyanophyte origin of Stromatoporoids. In: Monty, Cl. (ed.): Phanerozoic Stromatolites. Case Studies, 230–241, 3 Pls., Berlin – Heidelberg – New York: Springer

Kazmierczak, J., Hillmer, G. (1974): Sclerosponge nature of the lower Hauterivian "bryozoan" Neuropora pustulosa (Roemer, 1839) from Western Germany. Acta Paleont. Polonica **19,** 443–453, Warsaw

Kershaw, S., Riding, R. (1978): Parameterization of stromatoporoid shape. Lethaia **11,** 233–242, Oslo

Kobluk, D. R. (1978): Reef stromatoporoid morphologies as dynamic populations: application of field data to a model and the reconstruction of an Upper Devonian reef. Bull. Canadian Petrol. Geol. **26,** 218–236, Calgary

Kühn, O. (1943): Zur Kenntnis des Rhät von Vorarlberg. Mitt. alpenländ. geol. Ver. Wien **33,** 111–157, 2 Pls., 6 Figs., Wien

Lecompte, M. (1951/1952): Les Stromatoporoides du Dévonien moyen et supérieur du Bassin de Dinant. Inst. Roy. Sci. Natur. Belgique mém. **116–117,** 1–215, 216–369, Pls. 1–35 and 36–70, Bruxelles

Lecompte, M. (1970): Die Riffe im Devon der Ardennen und ihre Bildungsbedingungen. Geologica Palaeontologica **4,** 25–71, Marburg

Milan, A. (1969): Faziesverhältnisse und Hydrozoenfauna des Malm im Küstenland des Nördlichen Velebit und Velika Kapela. Geol. Vjesnik **22,** 11/16, 135–217, 8 Figs., 2 Tab., 1 geol. map, 22 Pls., Zagreb

Mori, K. (1968; 1972): Stromatoporoids from the Silurian of Gotland. Stockholm Contrib. Geol. **19,** 1–100, **22,** 1–112, Stockholm

Mori, K. (1976): A new Recent Sclerosponge from Ngargol, Palau Islands and its fossil relatives. Sci. Rep. Tohoku Univ. Sendai, Second Ser., Geol. **46**/1, 1–9, 6 Pls., Sendai

Oliver, W. A., Jr. (1979): Sponges, they are not. Paleobiol. **5**/2, 188–190

Scherer, M., Wendt, J. (1978): Diagenese oberpermischer Kalkschwämme aus Patch-Reefs des Djebel Tebaga (S-Tunesien). Neues Jahrbuch Geol. Paläont. Abh. **157,** 196–202, Stuttgart

Sokolov, B. S. (1962): Group Chaetetida and Subclass Tabulata. In: Sokolov, B. S. (ed.): Fundamentals of Paleontology **5,** 169–176, 192–266, Acad. Sci. USSR (English ed. 1971, Israel Progr. Sci. Transl., Jerusalem)

Stearn, C. W. (1966): The microstructure of stromatoporoids. Paleont. **9,** 74–124, London

Stearn, C. W. (1972): The relationship of the stromatoporoids to the sclerosponges. Lethaia **5,** 369–388, 9 Figs., Oslo

Stearn, C. W. (1975): The stromatoporoid animal. Lethaia **8,** 89–100, Oslo

Stearn, C. W. (1980): Classification of the Paleozoic Stromatoporoids. J. Paleont. **54**/5, 881–902, 5 Figs., Tulsa

Termier, G., Termier, H. (1975): Spongiomorphides et Spongiaires hypercalcifiés. Comptes Rendus Acad. Sci. Paris **280,** Ser. D, 1963–1965, 2 Figs., Paris

Termier, H., Termier, G. (1974): Stromatopores, Sclérosponges et Pharétrones: les Ischyrospongia. Ann. Mines Geol. **26,** 285–297, Paris

Tsien, H. H. (1977): Morphology and development of Devonian reefs and reef complexes in Belgium. Int. Coral Reef Symp. Miami, Proc. **2,** 191–200, Miami

Turnšek, D. (1968): Some hydrozoans and corals from Jurassic and Cretaceous stratas of Southwestern Jugoslavia. Slovenska Akad. Znanosti Umetnosti, Cl. IV, **11**/9, 351–376, 9 Pls., Ljubljana

Turnšek, D., Buser, St., Ogorelec, B. (1981): An Upper Jurassic reef complex from Slovenia, Yugoslavia. In: Toomey, D. F. (ed.): European Fossil Reef Models. Soc. Econ. Paleont. Min. Spec. Paper **30,** 361–369, 6 Figs., Tulsa
Veizer, J., Wendt, J. (1976): Mineralogy and chemical composition of Recent and fossil skeletons of calcareous sponges. Neues Jahrbuch Geol. Paläont. Monatshefte **1976**/9, 558–573, 8 Figs., Stuttgart
Wendt, J. (1975): Aragonitische Stromatoporen aus der alpinen Obertrias. N. Jb. geol. paläont. Abh. **150**/1, 111–125, 8 Figs., 1 Tab., Stuttgart
Wendt, J. (1976): Der Skelettbau mesozoischer und rezenter Kalkschwämme. Zentralblatt Geol. Paläont., Teil II, **1976,** 321–325, Stuttgart
Wendt, J. (1978): Development of skeletal formation, microstructures, and mineralogy of rigid calcareous sponges from the Late Palaeozoic to Recent. Coll. internat. C.N.R.S., **291,** Sponge Biology, 449–457, 2 Figs., Paris

5.3.6 Corals (Plates 10; 37/1; 43/4–6, 8; 45/8)

Morphology: Solitary and colonial corals, exoskeleton with outer wall, septa, dissepiments, partitions, and axial structures.

Microstructures: In Recent hexacorals (= Scleractinia, recorded from the Middle Triassic onwards) the exoskeleton is made of aragonite; sclerodermites radiating from calcification centers secrete a spherulitic primary structure. In thin-sections of Ancient *hexacorals* and *rugose corals* (= tetracorals, Ordovician to Permian, Plate 41/5) the septa often exhibit a herring-bone pattern in transmitted light. A survey of interpretations of the microstructures in Rugosa can be found in H. W. Flügel (1975). Longitudinal and oblique sections of corals often exhibit a "net-like structure" of septa with lateral spines (= ornamentation) and dissepiments. Corals may be confused with hydrozoans (particularly spongiomorphids), calcareous sponges, and sometimes prism structures of pelecypods (e.g., rudists)!

In contrast to the Scleractinia, Rugosa, and Tabulata (Ordovician – Permian, or perhaps Eocene, Plate 41/5), the skeletons of Octocorallia (since the Cretaceous) develop as scattered sclerites (spicules) embedded in organic tissue. Sclerites are straight, slightly curved, crossed, rarely smooth elements; usually with various kinds of spines. Size: for the Stolonifera (e.g., *Tubifera*) diameter 0.1 to 1.0 mm, for Alcyonaria and Gorgonia length 0.01–0.10 mm; very rare in thin-sections.

Thin-sections: Generally an exact identification of corals can be made only in oriented sections (transverse and longitudinal sections). Often in microfacies thin-sections no more than a classification according to major systematic groups is possible – due to rapid diagenetic alterations (see Pingitore, 1976) as well as to recrystallization of the reef limestone (e.g., "Thamnoporids" in Devonian limestones; "Thecosmilia" in Mesozoic reef limestones).

Distribution: Hermatypic corals are the most important elements in bioherms and biostromes. Rugose and tabulate corals form the Middle and Lowermost Upper Devonian "reefs" in the Eifel and Sauerland in Western Germany; hexacorals, e.g., the Upper Rhaetian reefs in the Northern Limestone Alps or the Tithonian reefs in the Southern Franconian Alb. From the Mesozoic ahermatypic corals start to appear (see Teichert, 1958): Recent branched deeper-water corals (e.g., *Lophelia*) form 10 to 20 m wide patches of living colonies (1 m high in depths of 200–400 m; Rockall Bank, Northeast Atlantic; Scoffin et al., 1980). These corals form an annular zone around the bank margin. Since the bioerosion of the coral framework is more rapid than the rate of burial of the upright colonies, the corals will be preserved as irregular bands of broken corals, embedded within pelagic muds.

References: Corals

Flügel, H. W. (1975): Skelettentwicklung, Ontogenie and Funktionsmorphologie rugoser Korallen. Paläont. Z. **49**/4, 407–431, 10 Figs., Stuttgart

Oliver, W. A., Jr. (1979): Corals. In: Fairbridge, R. W., Jablonski, D. (eds.): The Encyclopaedia of Palaeontology, 226–235, 12 Figs., Stroudsburg: Dowden, Hurchinson and Ross

Pingitore, N. E. (1976): Vadose and phreatic diagenesis: processes, products and their recognition in corals. J. Sed. Petrol. **46**/4, 985–1006, 10 Figs., Tulsa

Scoffin, T. P., Alexandersson, E. T., Bowes, G. E., Clockie, J. J., Farrow, G. F., Milliman, J. D. (1980): Recent, temperate, sub-photic carbonate sedimentation, Rockall Bank, Northeast Atlantic. J. Sed. Petrol. **50**/2, 331–356, 19 Figs., Tulsa

Teichert, C. (1958): Cold- and deep-water coral banks. Bull. Amer. Ass. Petrol. Geol. **42**/5, 1064–1082, 3 Figs., Tulsa

5.3.7 **Bryozoans** (Plates 24/6–10; 45/5, 7)

Fragments of bryozoans are most abundant in shelf limestones and reef carbonates as well as in calcareous shales and shelly marls. Paleozoic bryozoans are generally associated with articulate brachiopods, echinoderms, and solitary corals. Post-Palaeozoic bryozoans are found together with mollusks, foraminifera, sponges and octocorals.

Morphology: Colonies may appear bush-like, fungiform or encrusting, forming mats on hard substrates (shells, hardground surfaces, rocks). The colonies (zoaria) vary in size from a few mm to massive forms as much as 1 m across. The zoaria are composed of zooecia which represent the exoskeletons secreted by individual zooids. These zooecia may reflect a distinct polymorphism of the zooids; this can be seen by tubes of different size and of different structure. Walls are generally thin; most bryozoans have skeletons composed of calcium carbonate (calcite, except for the Cheilostomata which may have a secondary thickening of aragonite). The zooecia are seldom larger than 0.50 mm in diameter.

Major groups: Three classes are recognized (Stenolaemata and Gymnolaemata – almost exclusively marine; Phylactolaemata, exclusively freshwater). The Stenolaemata include the Cyclostomata (Ordovician to Holocene), Trepostomata (Ordovician to Permian, perhaps also Triassic), Cystoporata (Ordovician to Permian), and Cryptostomata (Ordovician to Permian). The Gymnolaemata comprise the Ctenostomata (Ordovician to Holocene), preserved only as delicate stolon-like traces within shells) and the Cheilostomata (Upper Jurassic to Holocene).

Cyclostomata consist of bundles or expansions of distally open calcareous tubes, which may be unpartitioned or have dividing calcareous diaphragms; the zooecia are connected by pores. Zoarial form is extremely varied: encrusting, flat, unilaminar or arborescent, cylindrical or ribbon-like, branched or massive. Cyclostomata are chiefly post-Palaeozoic, they reached their acme during the Jurassic and the Lower Cretaceous.

Trepostomata usually have more massive, heavily calcified zoaria. In branched zoaria the zooecia tubes budded off in the axial zone where they are thin-walled and subparallel to the branch axis. In the surrounding exozone, the zooecia are thicker walled and bend outward to meet the surface of the zoarium almost at right angles. Diaphragms are more frequent in the exozone. Between larger zooecia (autozooecia), smaller modified zooecia occur. These may be narrower tubes crowded with transverse partitions (mesopores), or small acanthopores which project above the zoarial surface as short spines. Trepostomata are especially common in Ordovician and Silurian shallow-water deposits.

Cystoporata are characterized by many cystopores (with abundant cysts) between the autopores. The zooecia are interconnected by pores. The group was established to include bryozoans formerly placed in the Trepostomata, e.g., the Fistuliporidae.

The zoaria of the *Cryptostomata* are cylindrical or ribbon-shaped, or, commonly form reticulate expansions with zooecia opening on one side of the zoarium only (fenestellid bryozoans). The zooecial tubes are short and thin-walled nearer the origin of growth and have thicker walls near the zoarial surface; mural pores are missing. Acanthopores may be common. The Cryptostomata were most numerous in Carboniferous and Permian deposits, sometimes forming reef structures by trapping sediment or by building organic frameworks (e.g., mud mounds of the Lower Carboniferous Waulsortian facies, Central Europe; Upper Permian Zechstein reefs, Germany and Poland).

The *Cheilostomata* usually form unilaminar encrustations, uni- or bilaminar foliose expansions, or branched stems. The zooecia are short and box-like. Ovicells are common in addition to normal zooecia; other modified zooecia are the avicularia and vibraculae. The zooecial aperture on the frontal side is closed by a hinged operculum. Cheilostomata are most important in Cretaceous rocks, e.g., in the Chalk.

Bryozoans in thin-sections: Trepostomid bryozoans may be confused with tabulate corals, hydrozoans or chaetetids, and with red algae, but the diameter of corallites is generally larger (5–10 mm), and that of red algae considerably smaller. Bryozoans may be distinguished by the presence of polymorphic tubes of various widths and by the arrangement of the tubes in the axial zone and near the surface of the zoaria. Trepostomata and Cryptostomata often show a characteristic microstructure in thin-sections (lamellar or fibrous).

Paleoecology: Recent bryozoans occur mainly in sediments of the continental shelves and in coral reefs (Cuffey, 1972). The distribution of bryozoans is controlled by the nature of the substrate (most common on stable bottoms, e.g., hard bottoms or water plants), water turbulence (encrusting and massive colonies in turbulent waters, delicately built erect forms in calm waters), salinity (most species are fairly stenohaline), and water temperatures. Fossil bryozoans occur mostly as fragments within calcarenites or together with other sessile organisms and algae in micritic limestones, contributing to the development of banks and mounds (e.g., Danian of Scandinavia; Cheetham, 1971).

References: Bryozoans

Bassler, R. S. (1953): Treatise on invertebrate paleontology, part G, Bryozoa. 253 p., Lawrence: Univ. Kansas Press
Boardman, R. S. (1960): Trepostomatous Bryozoa of the Hamilton Group of New York State. U.S. Geol. Surv. Prof. Paper **340,** 87 p., Washington
Boardman, R. S., Cheetham, A. H. (1969): Skeletal growth, intracolony variation, and evolution in Bryozoa. J. Paleont. **43,** 205–233, Pls. 27–30, 8 Figs., Tulsa
Brood, K. (1972): Cyclostomatous Bryozoa from the Upper Cretaceous and Danian in Scandinavia. Stockholm Contrib. Geol. **26,** 1–464, Stockholm
Brood, K. (1978): Bryozoa. In: Haq, B. U., Boersma, A. (eds.): Introduction to marine micropaleontology, 189–201, 29 Figs., New York: Elsevier
Cheetham, A. H. (1971): Functional morphology and biofacies distribution of cheilostome Bryozoa in the Danian Stage (Paleocene) of southern Scandinavia. Smithsonian Contrib. Paleobiol. **6,** 1–85, Washington
Cuffey, R. J. (1972): The roles of bryozoans in modern coral reefs. Geol. Rdsch. **61**/2, 542–550, 1 Fig., Stuttgart
Schopf, T. J. M. (1969): Paleoecology of ectoprocts (bryozoans). J. Paleont. **43,** 234–244, Tulsa
Wass, R. E., Conolly, J. R., MacIntyre, R. J. (1970): Bryozoan carbonate sand continuous along southern Australia. Marine Geol. **9,** 63–73, Amsterdam
Taverner-Smith, R., Williams, F. R. S. (1971): The secretion and structure of the skeleton of living and fossil Bryozoa. Phil. Trans. Roy. Soc. London **264,** No. 859, 97–159, London

5.3.8 Shells of Brachiopods (Plates 12/3; 25/2–4; 41/5), Mollusks (Plates 12/2; 16/2; 25/1, 5–7; 26/2, 4; 47/1; 51/1, 4, 7; 52/1), and Serpulids (Plates 26/5; 51/2)

The structures and microstructures of shells in thin-section are the primary means of *differentiating bioclasts according to specific groups of organisms.*

The following criteria can be used for differentiation (Majewske, 1969; Horowitz and Potter, 1971; see 5.5):

a) *Original shell mineralogy* (aragonitic shells are generally more strongly inverted and recrystallized than primary calcitic shells). Description of various stages of dissolution can be found in Catalov (1971).

b) *Shell-layering,* to separate higher systematic units – useful for brachiopods and serpulids as well as pelecypods.

c) *Microstructures,* e.g., for brachiopods and serpulids; also arthropods (in this group microstructures occur which are rarely or not at all present in mollusks).

d) *Pores in the shells* (position of the pores, example: brachiopods; size of the pores, example: trilobites and ostracods, see Table 27).

Bivalve shell microstructures and shell mineralogy are determined primarily through biological (i.e., evolutionary) processes. Definable trends from aragonitic nacreous ancestors to aragonitic porcelaneous and calcitic foliated descendants have been described (e.g., Waller, 1978). Mineralogy and microstructures are controlled by physiological factors and environmental influences. Most recent investigations of shell microstructures are based on SEM studies; a comparison of criteria as seen in thin-sections with the classification system developed for mollusk ultrastructures is rather difficult. Overviews of the "microstructures" as given by Bøggild (1930), Majewske (1969) or Horowitz and Potter (1971), therefore, are in some respects outdated. The recognition of the original microstructure is further complicated by diagenetic alterations related to dissolution in submarine and meteoric environments, and to anaerobic acid production by the mollusks themselves.

Table 26 gives a survey of the critera for differentiation in thin-sections of brachiopods, mollusks, gastropods, cephalopods, and serpulids. Table 27 lists the characteristics of trilobites and ostracod shells.

Microgastropods, which may be common in shelf carbonates and basinal limestones, have been described in thin-sections of Lower Triassic carbonates (Zaninetti et al., 1972).

Brachiopod spines are common in many Paleozoic pelletoidal and bioclastic wackestones and grainstones which were deposited in shallow lagoonal and inner platform environments (Vachard and Tellez-Giron, 1978). Spines from reticulariacean brachiopods (indented, hollow, tube-like or vase-like elements with very thin simple or double lamellar walls; cross-section diameter of about 100 μm, often with a "central bridge") have been described as *Magnella* (see Neumann et al., 1975). Productid spines are characterized by circular or oval cross-sections, exhibiting sharp boundaries (appear dark in transmitted light) which surround densely lamellar zones and a rather small hollow area.

References: Shells

Bachmann, G. H. (1973): Die karbonatischen Bestandteile des oberen Muschelkalkes (Mittlere Trias) in Südwestdeutschland und ihre Diagenese. Arb. Inst. Geol. Paläont. Univ. Stuttgart, N.F. **68,** 1–99, 67 Figs., Stuttgart

Bathurst, R. G. C. (1964): The replacement of aragonite by calcite in the molluscan shell wall. In: Imbrie, J., Newell, N. (eds.): Approaches to Paleoecology, 357–376, New York: Wiley

Boggild, O. G. (1930): The shell structure of the molluscs. Kgl. Danske Vitenskab. Selskab. Mat. Fys. Medd. **9,** 231–325, Copenhagen

Broglio Loriga, C. (1976): Aspetti paleobiologici e paleogeografici della facies a "Lithiotis" (Giurese inf.). Riv. Ital. Paleont. Strat. **82**/4, 651–706, Pls. 85–86, 8 Figs., Milano

Brönnimann, P., Zaninetti, L. (1972): On the occurrence of the Serpulid *Spirorbis* Dandin, 1800 (Annelida, Polychatia, Sedentaria) in thin sections of Triassic rocks of Europe and Iran. Riv. Ital. Paleont. Strat. **78**/1, 67–84, Pls. 10–12, Milano

Catalov, G. (1971): Certain diagenetic processes in the progress of lithification of the Triassic limy sediments in the Teteven Anticlinorium. Bull. Geol. Inst. Ser. Strat. Lithol. **20,** 157–176, 3 Pls., 2 Figs., Sofia

Carter, J. G. (1980): Environmental and biological controls of bivalve shell mineralogy and microstructure. In: Rhoads, D. C., Lutz, R. A. (eds.): Skeletal growth of aquatic organisms, 69–113, 6 Figs., New York – London: Plenum Press

Carter, J. G. (1980b): Guide to bivalve shell microstructures. In: Rhoads, D. C., Lutz, R. A. (eds.): Skeletal growth of aquatic organisms, 645–673, 50 Figs., New York – London: Plenum Press

Crenshaw, M. A. (1980): Mechanisms of shell formation and dissolution. In: Rhoads, D. C., Lutz, R. A. (eds.): Skeletal growth of aquatic organisms, 115–132, 2 Figs., New York – London: Plenum Press

Table 26. Shell structures of brachiopods, mollusks and serpulids

	Brachiopods since the Cambrian, only marine Pl. 25/2, 3; 41/5	Pelecypods since the Ordovician Pl. 25/5, 6, 7; 37/1; 40/5; 51/1, 7; 52/1	Gastropods since the Cambrian Pl. 25/1; 47/1; 51/4	Nautiloids since the Ordovician	Ammonoids Lower Devonian to Cretaceous Pl. 16/2; 26/2	Belemnites Lower Carboniferous to Eocene Pl. 26/4	Serpulids since the Ordovician Pl. 26/5; 51/2
Hard Parts	2 unequal valves; internal elements (brachidium; spicules)	2 valves, hinged along one edge; appearance in thin-section dependent on shape, orientation and ornamentation	Coiled shells, sometimes operculum	Straight or coiled shells, chambered	Coiled shells, chambered	Straight, 3 elements	Curved or spirally coiled unpartitioned tubes; prefers hard substrate or aquatic plants; sometimes form bioherms
Size	Usually 10 cm max., rarely larger	Usually 1–15 cm, max. 2 m	mm- to cm-sized	Usually > 10 cm	Usually cm-sized	cm-sized	1 mm in diameter, 0.2–10 cm long
Shell Mineralogy	Inarticulata: Chitinophosphatic with minor amounts of $CaCO_3$; Articulata: calcitic	Periostracum consists of organic conchiolin, below it a calcitic ostracum and an aragonitic hypostracum	Outer layer conchiolin (= periostracum), below it: Prosobranchia-calcitic outer layer, aragonitic inner layer; Opisthobranchia and Pulmonata – both layers aragonitic; operculum of conchiolin	External shell consists of calcite or calcite + aragonite or just aragonite; cameral deposits of calcite or aragonite	External shell usually aragonite, aptychi of calcite; anaptychi of conchiolin	Rostrum of radially oriented calcite prisms; proostracum of conchiolin and phragmocone aragonitic or organic	Calcite, also aragonite
Layering of the Shell	Usually 2 layers: a) very thin finely granular *outer layer* (usually not preserved); b) thicker *inner layer* made of several microstructural units	2 or 3 calcareous layers. Anisomyaria: outer layer of prismatic or foliated calcite, nacreous inner layer of aragonite (in oysters only thin sheets of variously oriented calcite); Homomyria: aragonitic, complex crossed-lamellar structure in inner and outer layers	1–5 calcareous layers (usually 2–3) with different mineralogy and microstructure	Up to 3 layers; prismatic outer and inner layer; nacreous middle layer	3 layers, similar to nautiloids	absent	2 layers: cone in cone outer layer, laminated inner layer

Pores in the shell	a) Punctate (or endopunctate) = pores in the inner layer, e.g., *Terebratula;* b) Impunctate = no pores, e.g., *Rhynchonella;* c) Pseudopunctate = distinct rods in the innermost shell layer	absent	absent	absent	absent	absent	absent
Microstructures	Inner layer consists of either an outer fibrous or an outer lamellar structure, and an inner prismatic structure	Most commonly *foliated* (typical for Monomyaria, e.g. Ostreacea, Pectinicea, Limacea) and *prismatic;* homogeneous; granular; crossed-lamellar (less common)	a) *Calcitic* layers: normal prismatic; foliated (rare, *Bellerophon*); crossed-lamellar (rare); irregularly granular (rare) b) *Aragonitic* layers: crossed-lamellar (very common); prismatic (rare); homogeneous; nacreous (7 families)	External shell often nacreous; internal partitions usually homogeneous or prismatic	Similar to nautiloids	Rostrum: prisms normal to wall	Longitudinal section: outer layer with parabolic structure, inner layer thinner laminated structure
Classification according to Microstructure	3 groups of Articulata: 1. porous primary layer of calcite crystallites, fibrous secondary layer; 2. porous primary, fibrous secondary layer, prismatic tertiary layer; 3. cross-bladed fabric	1. nacreous-prismatic (prismatic outer layer, nacreous inner layer); 2. foliated (1 or more layers); 3. crossed-lamellar (crossed-lamellar outer layer, complex-lamellar inner layer)	–	–	–	–	–
Diagnostic Criteria	Punctata and Impunctata – microstructure of inner layer shows rectangular or rhombic prisms; Pseudopunctata holes (punctae) within lamellar structures	Higher systematic units (sometimes Families) may be distinguished according to shape of fragments, shell layering, and microstructure – see Key to Identification in Majewske (1969)!	Several layers with different microstructural units; often leached shells	Often intensely recrystallized or leached shells because of original aragonitic composition		Rostrum: alternation of calcitic and ferruginous organic layers	Differ from vermetid gastropods because of outer microstructure (usually radial-prismatic in snails)

Cuvillier, J. (1969): Calcaires à "Filaments". Proc. First Int. Conf. Planctonic Microfossils **2,** 123–125, 2 Pls., Leiden: Brill

Neugebauer, J. (1978): Zur Erhaltung von Inoceramen in der Schreibkreide. Neues Jahrbuch Geol. Paläont. Monatshefte **1978**/6, 353–370, 5 Figs., Stuttgart

Neumann, M., Pozaryska, K., Vachard, D. (1975): see 5.3.1

Strausz, L. (1959): Über Gastropoden-Gehäuseschnitte. Acta Geol. **6**/1–2, 209–229, 3 Pls., 22 Figs., Budapest

Vachard, D., Tellez-Giron, C. (1978): Espines de Brachiopodes reticulacriacea dans les microfacies du Paléozoique supérieur. Rev. Inst. Mexicano Petrol. **10**/2, 16–30, 3 Pls., Mexico

Waller, T. R. (1978): Morphology, morphoclines and a new classification of the Pteriomorphia (Mollusca: Bivalvia). Phil. Trans. Roy. Soc. London, Ser. B, **284,** 345–365, London

Wilbur, K. M. (1964): Shell structure and mineralization in molluscs. Calcification in biological systems. Amer. Ass. Adv. Sci. Publ. **64,** 15–40

Williams, A. (1968): Evolution of the shell structure of articulate brachiopods. Paleont. Assoc. Spec. Papers Paleont. **2,** 55 pp., London

Zaninetti, L., Brönnimann, P., Bozorgnia, F., Huber, H. (1972): Étude lithologique et micropaléontologique de la formation d'Elika dans la coupe d'Aruh, Alborz central, Iran septentrional. Archiv Sci. **25**/2, 215–249, 4 Pls., 11 Figs., Geneva

Plate 25. Shells of Brachiopods and Mollusks

1. Gastropods and pelecypod valves. The snails (*Holopella* sp.) have shells made of an organically impregnated lamellar outer layer and a thicker prismatic inner layer. In some parts of the microsparitic matrix a coarse dolomite mosaic is recognizable. Early Triassic (Werfen limestone, Scythian): Kühweger Törl, Naßfeld area, Carnic Alps, Austria. Scale is 1 mm

2. Brachiopod with median septum. The shell shows the very fine pore patterns characteristic of punctate Articulata. Carbonate cements have grown on the outside surface of the shell as well as within the cavity enclosed by the valves; the "sparry roof" indicates the up direction during sedimentation. The inside of the shell acted as a local quiet water area (indicated by the similar orientation of the tiny shells). Lias fissure filling: Maurach, Tyrol, Austria. Scale is 2 mm

3. Micritic brachiopod limestone. The brachiopods are filled with micrite at the bottom and sparite at the top. Lias fissure filling: Maurach, Tyrol, Austria. Scale is 2 mm

4. Punctate brachiopod shell, tangential section. Lias fissure filling: Maurach, Tyrol, Austria. Scale is 2 mm

5. "Filaments" (probably prodissoconchs of planktonic pelecypods) and radiolarians. Middle Triassic (Reifling limestone): Zauchenbach valley, Salzkammergut, Styria, Austria. Scale is 2 mm

6. Pelecypod shells. Prism structures in the shells derived from inocerams (top left), section parallel to shell surface. The shell has been bored by pelecypods (still preserved in the boring on the right) and filled with peloids. Serpulids can be seen at the bottom of the larger boring on the left. Upper Jurassic (Kimmeridgian): Rothengst quarry, Osterwald, Lower Saxony, Germany. Scale is 2 mm

7. Pelecypod shells. Wall fragments of rudists, with characteristic prismatic microstructure, made of rectangular or rhombic meshes. Cretaceous reworked particles found in the Lower Tertiary (Eocene): Hallthurm near Reichenhall, Bavaria, Germany. Scale is 2 mm

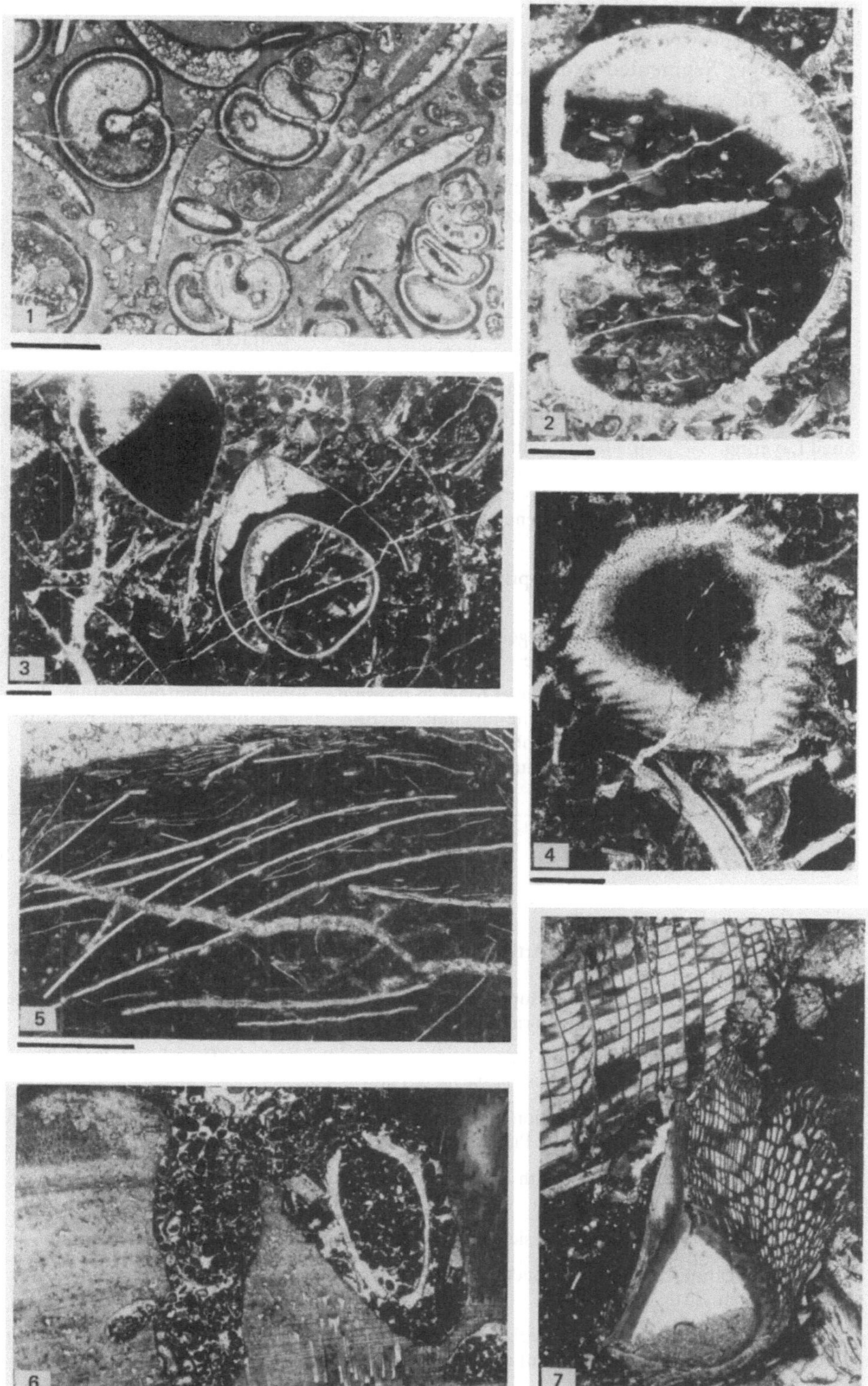
1
2
3
4
5
6
7

5.3.9 Trilobites (Plates 12/1, 3; 15/3; 26/6, 7; 36/2; Table 27)

Morphology: Segmented exoskeleton, dorsal carapace consisting of a cephalon, a thorax, and a pygidium. The segmented carapace, which is generally less than 1 mm thick, consists of chitin which is strongly impregnated with calcite and calcium phosphate. The microstructure of the shell cuticle, which generally consists of two or three layers, is characterized by a fine prismatic structure and many fine pores perpendicular to the shell surface (particularly in the Ordovician).

Table 27. Important characteristics of trilobites and ostracods in thin-sections

	Trilobites	Ostracods
Shell Mineralogy	Chitin, heavily impregnated with calcite and calcium phosphate	Chitin, chitin + calcite or Mg-calcite, calcium phosphate
Shell Layering	2 or 3 layers: thin outer layer consisting of vertically oriented prisms; thicker inner layer of lamellae. Sometimes a homogeneous or lamellar middle layer	Usually only 1 or a few layers, maximum of 9 layers
Microstructure	Homogeneous-prismatic or fine-lamellar	Homogeneous-prismatic or fine-prismatic
Pores in Shells	Smaller and larger canals normal to the shell surface	Very tiny
Diagnostic Criteria	Organic outer layer (dark in transmitted light) and fine-prismatic microstructure; "shepherd's crook" figures; "spongy" structure as a result of canals	Shell outline, duplicature, thin homogeneous or fine prismatic valves

Plate 26. Mollusks, Serpulids and Arthropods ►

1. Ostracod shell with distinct duplicature and homogeneous prismatic shell structure (crossed nicols). Lower Permian: Forni Avoltri, Carnia, Italy. Scale is 0.5 mm

2. Baby ammonites and very fine filaments in micritic matrix. Late Triassic (Hallstatt limestone, Norian): Hallstatt, Upper Austria. Scale is 3 mm

3. Tentaculites. Longitudinal section and cross-sections of *Nowakia* sp. Middle Devonian (Daleje beds, Eifelian): Hlubocepy, Prague basin, Czechoslovakia. Scale is 1 mm

4. Belemnite rostrum, cracked, with distinct growth layers. Middle Jurassic: Würgau well, Franconia, Germany. Scale is 1 mm

5. Serpulid limestone. Middle Jurassic: Würgau well, Franconia, Germany. Scale is 1 mm

6. Trilobite carapace. Marginal section. Ordovician (*macrourus* limestone, Caradocian): Öland, Sweden. Scale is 3 mm

7. Trilobite carapaces in characteristic "shepherd's crook" figures. Ordovician (*limbata* limestone, Arenigian): Köppingsklint quarry, Öland, Sweden. Scale is 1 mm

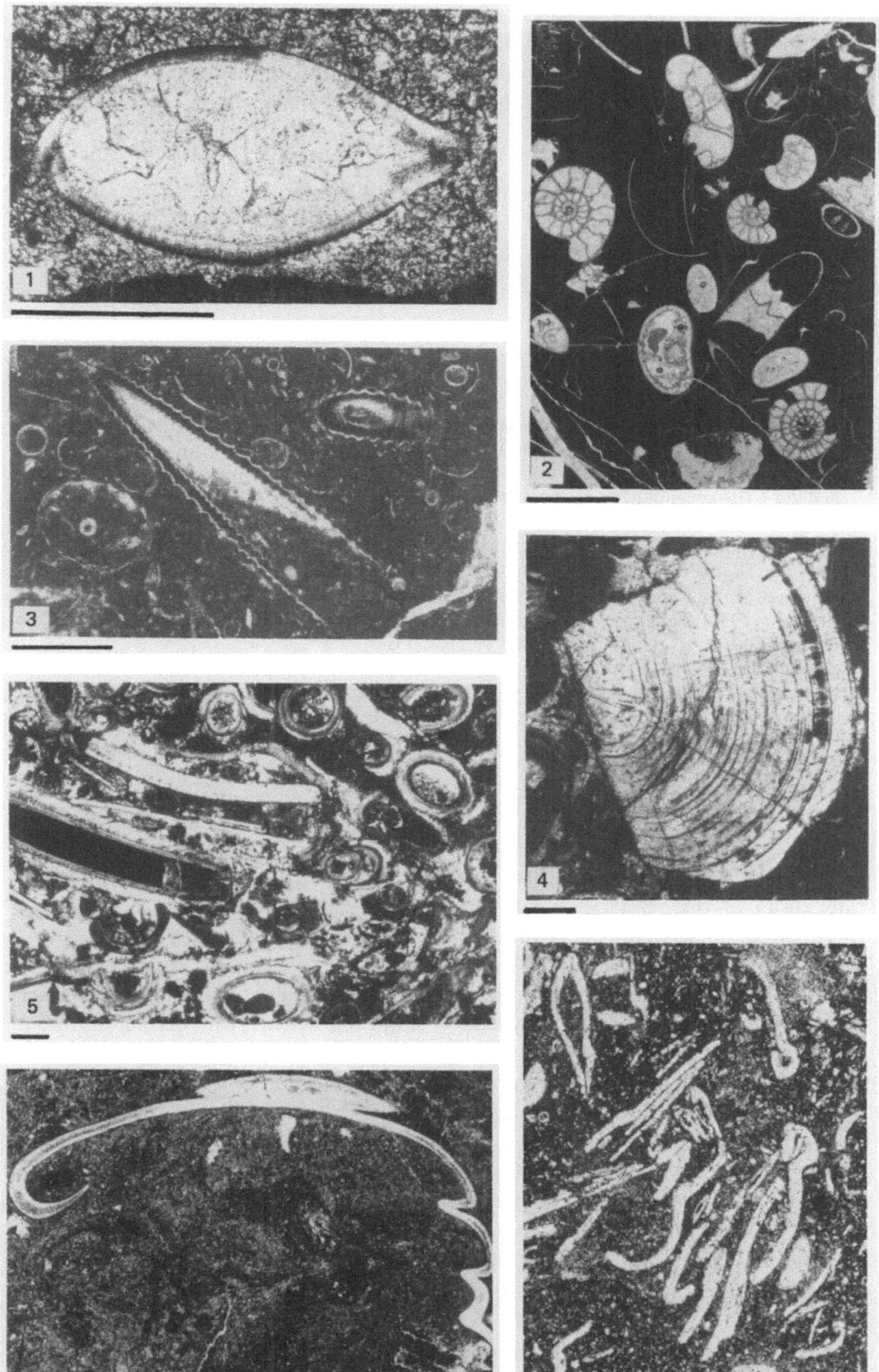
1
2
3
4
5
6
7

Tangential sections of the very recurved segments resemble "shepherd's crooks" (Plate 26/7). Cross sections of trilobite spines can be differentiated from brachiopod spines by the homogeneous prismatic structure of trilobites and by the layering of brachiopod spines (thin light-colored or clear outer layer, thicker lamellar inner layer).

Identification Problems: May be confused with brachiopods, mollusks, ostracods.

Distribution: from the Cambrian to the Permian. More abundant in Cambrian or Ordovician shelf carbonates.

References: Trilobites

Dalingwater, J. E. (1973): Trilobite cuticle microstructure and composition. Paleont. **16,** 827–839, Pls. 107–109, Tab. 1, London
Dudich, E. (1931): Systematische und biologische Untersuchungen über die Kalkeinlagerungen des Crustaceenpanzers im polarisierten Licht. Zoologica **30**/80, 1–154
Harrington, H. J. (1959): General description of Trilobita. In: Moore, R. C.: Treatise on invertebrate paleontology, part O, Arthropoda **1,** O 38–O 117, Figs. 27–85, Lawrence
Sohn, I. G., Kornicker, L. S. (1978): Contribution of Crustacea to the carbonate budget. Abstracts, Int. Congr. Sed., p. 627, Jerusalem
Størmer, L. (1930): Scandinavian Trunucleidae. Skrift. utgitt Norske Videnskaps, Akad. Oslo, I, Mat.-Naturv. Kl. **4,** 111 pp., 14 Pls., 47 Figs., Oslo
Teigler, D. J., Towe, K. M. (1974/75): Microstructure and composition of the trilobite exoskeleton. Fossils and Strata, Spec. Vol. Evolution and morphology of the Trilobita, Trilobitoidea, and Merostomata, Oslo

5.3.10 Ostracods (Plates 24/6; 26/1; 44/1, 3, 5)

Morphology: Two thin valves with inwardly recurved shell margins (duplicature; Plate 26/1) at one end; shell size often 1 mm (to 60 mm). Surface smooth or ornamented. Shells made of calcium phosphate or chitin or chitin + calcite.

Microstructure: Mineralogical composition and layering of the shells appears to vary over geological time and may exhibit up to nine layers. Microstructure fine or homogeneously prismatic. An axial cross appears in polarized light. Internal pores are generally not visible in thin-sections.

Determinations of different ostracods are generally not possible in microfacies thin-sections. An exception is the Upper Devonian and Lowermost Carboniferous pelagic group Entomozoacea; these forms may be recognized by ornamented, spinose valves (see Buchroithner et al., 1980).

Identification Problems: May be confused with thin-shelled pelecypods (*Posydonomya,* "filaments") and hyaline foraminifera tests.

Distribution: since the Cambrian, aquatic environments, marine water, brackish and freshwater.

References: Ostracods

Bate, R. H., East, B. A. (1972): The structure of the ostracod carapace. Lethaia **5,** 177–194, 11 Figs., Oslo

Buchroithner, M., Flügel, E., Flügel, H. W., Stattegger, K. (1980): Mikrofazies, Fossilien und Herkunft der Kalk-Gerölle im Karbon-"Flysch" der Betischen Kordilleren, Spanien. Facies **2,** 1–54, Pls., 1–8, 2 Figs., Erlangen

Langer, W. (1973): Zur Ultrastruktur, Mikromorphologie und Taphonomie des Ostracoda-Carapax. Palaeontographica, A, **144,** 1–54, Stuttgart

Levinson, S. A. (1961): Identification of fossil ostracods in thin-section. In: Moore, R. C. (ed.): Treatise on invertebrate paleontology, Part Q, Arthropoda **3,** Q70–Q73, Fig. 31, Lawrence

Pokorny, V. (1978): Ostracodes. In: Haq, B. U., Boersma, A. (eds.): Introduction to marine micropaleontology, 109–149, 61 Figs., New York: Elsevier

5.3.11 **Echinoderms** (Plates 12/1; 15/1, 7; 27; 32/1; 53/1)

The skeletal construction of the exclusively marine echinoderms is characterized by numerous individual, geometrically well-defined, calcitic elements, which are frequently disarticulated (Example: crinoid columnals). Individual elements may be < 1 mm to a few cm.

Diagnostically important features of echinoderms *in thin-sections* are:

a) *Optical character of the individual elements:* Each individual element (plates, columnals, etc.) acts optically as a single crystal of calcite. Consequently, each element will uniformly extinct under crossed nicols in a polarizing microscope. This uniform optical reaction of the skeleton elements is caused by the parallel arrangement of the c-axes of many of the crystals which constitute the skeletal elements.

b) *Meshwork structure* (Plates 27; 32/1). The plates and spines of many echinoderms exhibit an open meshwork structure in the center and a somewhat denser meshwork towards the margin; up to 60% of the hard parts are interspersed with organic tissue. Because of the large pore spaces, in the course of diagenesis rapid formation of calcite crystals with scalenohedric planes of cleavage occurs; they sparkle in fractured limestones in direct light.

c) *Gray or yellowish color in transmitted light* (Plate 27/7). In transmitted light crinoid and echinoid fragments often exhibit a uniform gray color or a honey yellow color.

d) *Characteristic outlines:* Many echinoderm elements are immediately recognizable because of their characteristic outlines (e.g., pentagonal); this is especially true for echinoid and crinoid remains.

e) *Typical microstructures* (Plate 27/1–3): Cross-sections of echinoid and ophiuroid spines are recognizable because of their sharply differentiated internal radial patterns.

f) *Diagenesis of echinoid fragments:* Calcite grows as syntaxial rim cement in optical continuum with the single crystals and within the partially hollow single crystals (stereom) of the echinoderms (see Lucia, 1962; Evamy and Shearman, 1965; Neugebauer and Ruhrmann, 1978; Neugebauer, 1979); see p. 75 and Plate 27/1.

Mineralogy of the echinoderm skeletons: High-Mg-calcite. The Mg content is dependent on temperature, although differences within the various skeletal parts occur (Raup, 1966). Sr, Mn, Fe, and Na content in Recent echinoids also seem to be temperature dependent (Harris and Pilkey, 1966).

Major Subdivisions: The Echinodermata include the subphyla Crinozoa (Ordovician to Holocene), Blastozoa (Lower Cambrian to Permian), Homalozoa (Middle Cambrian to Upper Devonian), Asterozoa (Lower Ordovician to Holocene), and Echinozoa (Lower Cambrian to Holocene).

The *Crinozoa* are mainly stemmed echinoderms with cup-like theca; they bear feeding structures (arms). The mouth is normally directed upwards. The class *Crinoidea* includes four orders, of which the first three are confined to the Paleozoic: Inadunata, Flexibilia, Camerata and Articulata. The inadunate, flexible, and camerate crinoids occur mainly in the Middle and Late Paleozoic (especially Mississippian), the articulates first appear in the Lower Triassic.

The *Blastozoa* were stemmed echinoderms with bud-like theca bearing feeding structures (brachioles). This subphylum includes the *Blastoidea* (Silurian to Permian), *Parablastoidea* (Middle Ordovician), *Cystoidea* (Cambrian to Devonian), and *Eocrinoidea* (Lower Cambrian to Upper Ordovician).

The *Homalozoa* are represented by wholly extinct, asymmetrical forms, formerly called "carpoids". Four classes are included in this subphylum: *Homostelea* (Middle Cambrian), *Homoistelea* (Upper Cambrian to Lower Devonian), *Stylophora* (Middle Cambrian to Upper-Devonian; sometimes regarded as a subphylum of the Chordata (Calcichordata; Jefferies, 1967), and *Ctenocystoidea* (Middle Cambrian).

Plate 27. Echinoderms

1. Echinoid spine. Cross-section. Type *Cidaris* (see Lucas et al., 1976). The white, somewhat oblate figures are cross-sections of sponge rhaxes. Upper Jurassic (Tithonian): Arnegg, Baden-Württemberg, Germany, Scale is 1 mm

2. Echinoid spine with micrite crusts. Type *Echinus.* Lower Permian: Forni Avoltri, Carnia, Italy. Scale is 1 mm

3. Crinoid elements with very fine meshwork, partially bored. Very small broken off bryozoan branches with angular and oval openings (far right). Lower Permian (Trogkofel limestone): Devils gorge, Karawanken, Slovenia, Yugoslavia. Scale is 1 mm

4. Echinoid spine. Cross-section. Type *Diadema.* Late Triassic (Dachstein limestone, Norian): Gosaukamm range, Upper Austria. Scale is 1 mm

5. Planktonic roveacrinids. Oblique section of arm plates, with bone-shaped outlines of *Osteocrinus rectus* (Frizell and Exline). Late Triassic (Hallstatt limestone, Carnian): Feuerkogel near Aussee, Salzkammergut, Styria, Austria. Scale is 1 mm

6. Planktonic crinoids. Various elements (antler-like secundibrachial plates, etc.) of *Saccocoma* Agassiz in micrite. Upper Jurassic (Kimmeridgian): Central Apennines, Italy. Scale is 1 mm

7. Echinoid spine. Longitudinal section with the very fine meshwork characteristic of echinoderms and the typical gray color in transmitted light. Lower Permian: Lika, Croatia, Yugoslavia. Scale is 1 mm

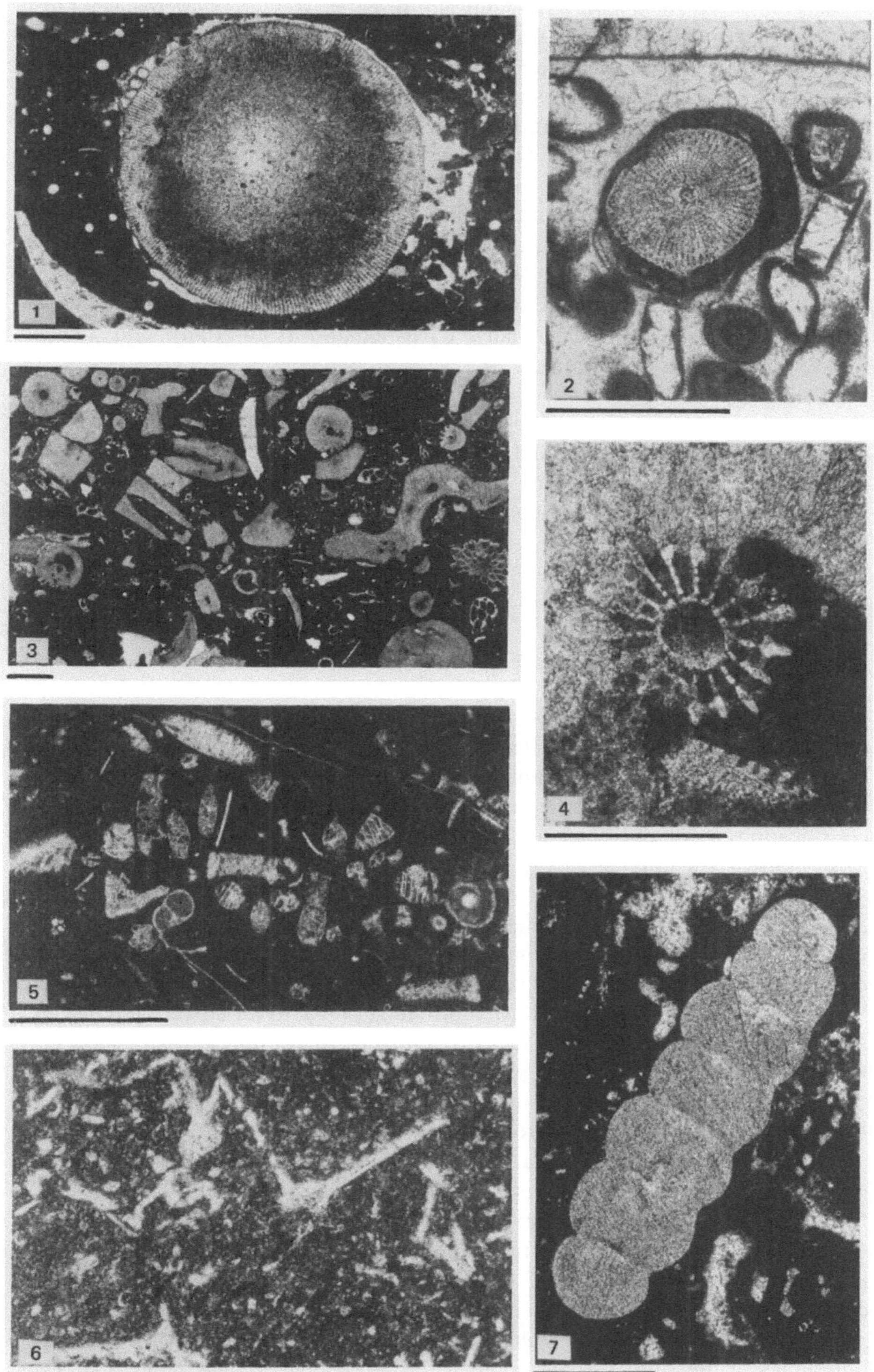
1
2
3
4
5
6
7

The star-shaped *Asterozoa* belong to three classes: *Somasteroidea* (Lower Ordovician to Holocene), *Ophiurioidea* (Lower Ordovician to Holocene), and *Asteroidea* (Lower Ordovician to Holocene).

The subphylum *Echinozoa* contains all the basically pentamerous, mostly free-living (eleutherozoic) echinoderms that are not star-shaped but generally globular. This subphylum comprises the *Echinoidea* (sea urchins; Ordovician to Holocene) including the more primitive, so-called regular echinoids (globular in shape, mouth at the center of the under surface, anus at the apex) and the more advanced irregular echinoids (with a secondary bilateral symmetry, mostly burrowers), *Holothuroidea* (sea cucumbers; Middle Cambrian to Holocene), *Edrioasterioidea* (Lower Cambrian to Carboniferous), *Helicoplacoidea* (Lower Cambrian), *Ophiocistioidea* (Ordovician to Devonian), *Cyclocystoidea* (Cambrian), and *Camptostromatoidea* (Lower Cambrian).

Echinoderms in thin-sections: Crinoid and echinid fragments are most common in shelf limestones; holothurian and ophiuroid elements are less abundant in thin-sections, but may be important microfossils found in acid residue samples of shallow-water and basinal limestones.

1. Crinoids (Plates 27/3, 5, 6; 32/1)

Morphology: Root structures, stem, calyx divided into dorsal cup and tegmen, arms.

Classification of dissociated elements (after Moore, 1939; Moore and Jeffords, 1968; Tab. 28): Dissociated crinoid plates and other crinoid elements are stratigraphically useful (e.g., Rasmussen, 1961: Biol. Skrifter Dansk Vitensk. Selskap *12;* subdivision of the Cretaceous by using columnals of isocrinids). Up to now more exact classifications have not been undertaken in thin-sections (with the exception of Mišik, 1964; see 4.3.2).

Distribution: from the Ordovician to the Recent. In the Paleozoic and to some extent in the Mesozoic, predominantly sessile benthic organisms occur in shallow marine environments. Important components of reef and shelf fauna. The genesis of "crinoid limestones" was studied by Ruhrmann (1971 a, b) for far-reef and near-reef sedimentation; the paleoenvironment of Jurassic crinoid limestones in the Tethys was investigated by Jenkyns (1971).

Of particular significance are the *"planktonic crinoids"* known from the Mesozoic; two groups are of interest in thin-sections of limestones:

a) *Triassic Roveacrinids* (Plate 27/5). Kristan-Tollmann (1970) and Mostler (1972) have pointed out the rock-building and biostratigraphic significance of free-swimming crinoids, which are recognizable in limestone thin-sections of basinal sediments of the Alpine Mediterranean Triassic. Several species of *Osteocrinus* are important in the lower part of the Upper Triassic (Carnian); usually just the centrodorsal plate is preserved (central plate of the dorsal cup in stemless forms).

b) *Jurassic free-swimming crinoids* (*"Saccocoma"*), Plate 27/6. In Upper Jurassic limestones of the Alpine Mediterranean region elements of the genus *Saccocoma*

Table 28. Classification of crinoid elements

Skeletal elements	Function	Outline	Pattern
Columnalia (1–10 mm in diameter)	Stem plates	Circular, oval	Axial canal (lumen) with different shapes, peripheral radial patterns (articulating surfaces)
Apicalia	Basal and infrabasal plates at the base of the cup	Pentagonal or circular, often not yet dissociated	If circular, central canal
Facetalia (Radialia)	Plates for attaching the lowest arm segments	Curved, rounded lunate shape	–
Pinnata (Brachialia)	Plates of free branchlets	Sometimes U-shaped	Ligament bases with ridges, central canal
Polygonalia	Miscellaneous group for all plates whose function is unclear	Circular, oval, etc.	Indistinct; no articulating surfaces, no lumen

Agassiz occur, sometimes in rock forming abundance (Upper Jurassic to Upper Cretaceous). Lombard (1937) first described these forms in thin-sections as "formes découpeés". Brönnimann (1955) recognized that they belonged to the echinoderms and created the genus *Lombardia;* this should be abolished according to Verniory (1954, 1956), because this name refers to Facetalia and Brachialia of *Saccocoma* as well as to ophiuroid fragments (Groiss, 1964).

Distribution: Often the *Saccocoma* facies is found below a Calpionellid facies and above limestones with globochaetes (planktonic algae), e.g., Farinacci and Sirna (1959; Sicily and Umbria, Italy), Lefeld and Radwánski (1960; Tatra, Poland). Mass accumulations of *Saccocoma* appear to be limited to the time span between the Kimmeridgian and the Middle Tithonian (Borza, 1969).

2. *Echinoids* (Plate 27/1, 2, 4, 7)

Morphology: Numerous calcareous plates make up a test, generally having five ambulacral areas with pores and usually five interambulacral areas without pores. The plates are covered with various kinds of spines (for protection and locomotion).

Thin-section: Generally cross-sections and longitudinal sections of *spines* or *plates* are easily recognizable in limestones:

a) *Ambulacral and interambulacral plates:* Similar to crinoid plates but distinguished by the absence of a radial pattern; considerable structural variation ("cells" of various shapes).

b) *Spines:* Generally consist of a shaft, neck, and base. *Grouped* in cross-section (actually easier to see in rarer longitudinal sections) according to the wall of the spine (consisting of a cortex, perforate septa, and regularly spaced cross members) and the axial zone (hollow canal or a complex of interconnected vertical tubes or spongy structures). According to Hesse (1900) sea urchin spines exhibit differences specific to the genus; see also Lucas et al. (1976: 194). Sea urchin spines can be found in shallow water carbonates as well as in deep water carbonates. Known from the Ordovician to the Recent.

Identification Problems: May be confused with dasycladaceans, small corals and bone tissue of vertebrates.

3. Holothuroids

Morphology: Internal skeleton made of sclerites embedded in the soft body wall. The individual sclerites of the skeleton consist of calcite crystals, which are constructed in a characteristic reticulate pattern with long fenestrae or holes piercing the bulk. Other sclerites have the shape of anchors, wheels, crosses, etc.

Thin-section: Almost all holothurian sclerites described up to now refer to acid insoluble material (e.g., Alpine Triassic, Mostler, 1972). In thin-sections holothurian remains are usually overlooked for the following reasons:

a) tiny size (smaller than 1 mm, usually 0.25 to 0.50 mm).

b) in two-dimensional sections only a few groups can be classified. Holothurian sclerites are known from thin-sections of reef limestones (Dachstein reef limestones) as well as from basinal carbonates (Hallstatt limestones) in the Alpine Triassic.

4. Ophiuroids

Morphology: Flattened body composed of a central disc with a ventral mouth (with jaw-like projections on the margin); the central body disc is surrounded by slender arms which propel it by snake-like movements. These arms consist of a series of articulating vertebral ossicles; each vertebrae consists of two halves which are mirror images of each other. Vertebrae are protected by lateral shields, dorsal shields, and ventral shields; each shield bears a row of small scale-like spines.

Thin-section: Glazek and Radwânski (1968) observed symmetrically connected calcite plates which extinct differently under crossed nicols in thin-sections from the Polish Rhaetian, and identified them as vertebrae. The lateral shields are curved in thin-sections. Ophiuroid spines in cross-sections are difficult to distinguish from echinid spines (Means of differentiation in Mostler, 1971).

References: Echinoderms

a) Echinoderms

Caster, K. E., Eaton, T. H. J. R. (1956): Microstructure of the plates in the carpoid Echinoderm Paranacystis. J. Paleont. **30**, 611–614, Tulsa

Evamy, B. D., Shearman, D. J. (1965): The development of overgrowth from Echinoderm fragments. Sedimentology **5**, 211–233, Amsterdam
Harris, R. S., Pilkey, O. H. (1966): Temperature and salinity control of the concentration of skeletal Na, Mn, and Fe in *Dendraster excentricus*. Pacific Sci. **20**, 235–238
Jeffries, R. P. S. (1967): Some fossil chordates with echinoderm affinities. Symp. Zool. Soc. London **20**, 163–208, London
Lucas, G. (1953): Étude, au microscope polisarant, des hydrospires des Blastoides. In: Piveteau, J. (ed.): Traité de Paléontologie **3**, 635–637, Paris: Masson
Lucas, G., et al. (1976): see 5.5
Lucia, J. (1962): Diagenesis of a crinoidal sediment. J. Sed. Petrol. **32**, 848–865, Tulsa
Neugebauer, J. (1979): Drei Probleme der Echinodermendiagenese: Innere Zementation, Mikroporenbildung und der Übergang von Magnesiumcalcit zu Calcit. Geol. Rundschau, **68**/3, 856–875, 25 Figs., Stuttgart
Neugebauer, J., Ruhrmann, G. (1978): Experimentelle Karbonatzementation: Syntaxialer Calcit auf Echinodermen. Neues Jahrbuch Geol. Paläont. Monatshefte **1978**/9, 545–555, 3 Figs., Stuttgart
Nissen, H. V. (1963): Röntgengefügeanalyse in Kalzit von Echinodermenskeletten. N. Jb. Paläont. Abh. **117**, 230–234, Stuttgart
Raup, D. N. (1959): Crystallography of Echinoid calcite. J. Geol. **67**/6, 661–674, Chicago
Raup, D. N. (1966): The endoskeleton: In: Boolootiani, R. A. (ed.): The Physiology of Echinodermata, 379–395, New York: Wiley
Rioult, M. (1961): Les vistiges microscopiques d'Echinodermes dans les sédiments jurassiques de Normandie. Bull. Soc. Linnéenne Normandie, sér. 9, v. **10**, 32–36, Caen

b) *Crinoids*

Borza, K. (1969): Die Mikrofazies und Mikrofossilien des Oberjuras und der Unterkreide der Klippenzone der Westkarpaten. 301 pp., 88 Pls., Bratislava: Slowak. Akad. Wiss.
Brönnimann, P. (1955): Microfossils incertae sedis from the Upper Jurassic and Lower Cretaceous of Cuba. Micropal. **1**/1, 28–51, Pls. 1–2, 10 Figs., 1 Tab., New York
Farinacci, A., Sirna, G. (1959): Livelli a *Saccocoma* nel Malm dell'Umbria e della Sicilia. Boll. Soc. Geol. Ital. **79**/3, 1–23, Pls. 1–6, 2 Figs., Rom
Jenkyns, H. C. (1971): Speculations on the genesis of crinoidal limestones in the Tethyan Jurassic. Geol. Rdsch. **60**, 471–488, Stuttgart
Kristan-Tollmann, E. (1970): Die Osteocrinusfazies, ein Leithorizont von Schwebcrinoiden im Oberladin-Unterkarn der Tethys. Erdöl und Kohle **23**, 781–789, 13 Figs., 1 Pl., Hamburg
Kristan-Tollmann, E. (1977): Zur Gattungsunterscheidung und Rekonstruktion der triadischen Schwebcrinoiden. Paläont. Z. **51**, 185–198, 8 Figs., Stuttgart
Lefeld, J., Radwánski, A. (1960): Les Crinoides planctoniques *Saccocoma* Agassiz dans le Malm et le Neocomien Haut Tatrique des Tatras polonaises. Acta Geol. Pol. **10**, 593–614, Pls 38–41, 1 Fig., Warsaw
Lombard, A. (1937): Microfossiles d'attribution incertaine du Jurassique supérieur alpin. Eclogae geol. Helv. **30**/2, 320–331, Basel
Lombard, A. (1945): Attribution de microfossiles du Jurassique supérieur alpin à des Chlorophycées (Proto- et Pleurococcalées). Eclogae geol. Helv. **38**, 163–173, Basel
Moore, R. C. (1939): The use of fragmentary crinoidal remains in stratigraphic paleontology. Bull. Sci. Lab. Denison Univ. **33**, 165–250, 4 Pls., 14 Figs., Denison
Moore, R. C., Jeffords, R. M. (1968): Classification and nomenclature of fossil crinoids based on studies of dissociated parts of their columns. Univ. Kansas Paleont., Inst. Echinodermata, Art. **9**, Ser. No. 46, 1–86, Pls. 1–28, 6 Figs., Lawrence
Mostler, H. (1972): Die stratigraphische Bedeutung von Crinoiden-, Echiniden- und Ophiuren-Skelettelementen in triassischen Karbonatgesteinen. Mitt. Ges. Geol. Bergbaustud. **21**, 711–728, 3 Pls., 3 Figs., Innsbruck
Roux, M. (1970): Introduction à l'étude des microstructures des tiges de Crinoides. Geobios **3**/3, 79–98, Lyon

Ruhrmann, G. (1971a): Riff-ferne Sedimentation unterdevonischer Krinoidenkalke im Kantabrischen Gebirge (Spanien). N. Jb. Geol. Paläont. Mh. **1971**/4, 231–248, 12 Figs., Stuttgart
Ruhrmann, G. (1971b): Riff-nahe Sedimentation paläozoischer Krinoiden-Fragmente. N. Jb. Geol. Paläont. Abh. **138**/1, 56–100, 22 Figs., 3 Tab., Stuttgart
Verniory, R. (1954): *Eotrix alpina* Lombard, Algue ou Crinoide? Arch. Sci. **7,** Genf
Verniory, R. (1956): La création du genre *Lombardia* Brönnimann est elle justificée? Arch. Sci. **9,** Genf

c) *Echinids*

Hesse, E. (1900): Die Mikrostruktur der fossilen Echinoideenstacheln und deren systematische Bedeutung. N. Jb. Min. Geol. Paläont. Beil.-Bd. **13,** 185–264, 12 Pls., Stuttgart
Lucas, G., et al. (1976): see 5.5
Manze, U., Richter, D. K. (1979): Die Veränderung des C^{13}/C^{12}-Verhältnisses in Seeigelcoronen bei der Umwandlung von Mg-Calcit unter meteorisch-vadosen Bedingungen. Neues Jahrbuch Geol. Paläont. Abhandlungen **158**/3, 334–345, 3 Figs., Stuttgart
Richter, D. K. (1974): Zur subaerischen Diagenese von Echinidenskeletten und das relative Alter pleistozäner Karbonatterrassen bei Korinth (Griechenland). Neues Jahrbuch Geologie Paläont. Abhandlungen **146,** 51–77, Stuttgart

d) *Holothuroids*

Frizzel, D. L., Exline, H. (1955): Monograph of fossil Holothurian sclerites. Bull. Univ. Missouri School Mines Metallurgy **89,** 203 p., Rolla
Mostler, H. (1972): Holothuriensklerite in der alpinen Trias und ihre stratigraphische Bedeutung. Mitt. Ges. Geol. Bergbaustud. **21,** 729–744, Innsbruck
Rioult, M. (1965): Sclérites d'Holothuries tertiaires (Eocène du bassin de Paris et Miocène du bassin d'Aquitaine). Rev. Micropaléont. **8**/3, 65–74, Paris

e) *Ophiuroids*

Glazek, J., Radwánski, A. (1968): Determination of brittle stars in thin sections. Bull. Acad. Polon. Sci., Ser. Sci. Geol. geogr. **16,** Warsaw
Groiss, J. Th. (1964): Echinodermenreste in Schlämmproben aus dem Weißen Jura der Franken-Alb. Geol. Bl. NO-Bayern **14**/2, 43–53, 3 Figs., 1 Tab., Erlangen
Mostler, H. (1971): Ophiurenskelettelemente (äußere Skelettanhänge) aus der alpinen Trias. Geol. Paläont. Mitt. Innsbruck **1**/9, 1–35, 5 Pls., 27 Figs., Innsbruck

5.3.12 Calcareous Algae (Plates 28; 29; 30/1–4, 6; 44/1–3; 47/2, 5, 6; 52/4)

About 6% of the Recent algae species display thalli which are capable of calcification. Although "calcareous algae" in the true sense of the word are only benthonic and planktonic algae with a biochemically precipitated $CaCO_3$ skeleton, often some algae are designated "calcareous algae", because they contribute to carbonate precipitation and the formation of laminated fabrics (algal mats and stromatolites) by interacting in physical and biological processes. These "non-skeletal" algae are known as early as the Precambrian (approximately 2.4 billion years ago); the true calcareous algae have been known only since the Cambrian.

Table 29. Classification of benthonic calcareous algae. After Wray (1977)

Group (Division)	Class	Order	Family	Stratigraphic Range	Recent Environment	Recent Mineralogy
Cyanophyta (blue-green algae)	Cyanophyceae			since Precambrian	Non-marine or marginal marine	Primarily calcite
Rhodophyta (red algae)	Rhodophyceae	Cryptonemiales	Solenoporaceae	Cambrian-Neogene	–	–
			Gymnocodiaceae	Permian-Paleocene	–	–
			Squamariaceae	Carboniferous-Permian, since Cretaceous	Marine	Aragonite
			Corallinaceae	since Jurassic	Marine	Mg-calcite
Chlorophyta (green algae)	Chlorophyceae	Siphonales	Codiaceae	since Cambrian	Marine	Aragonite
		Dasycladales	Dasycladaceae	since Cambrian	Marine	Aragonite
	Charophyceae	Charales	Characeae etc.	since Silurian	Non-marine	Calcite

5.3.12.1 Classification

Algae with thalli (plant bodies) capable of calcification are found in several major groups. In addition to benthonic species there are planktonic groups as well, such as the coccolithoporids (Chrysophyta) from the Mesozoic to the Recent, globochaetes (Triassic to Recent, perhaps even as early as the Paleozoic, see p. 350). Table 29 shows the classification of benthonic calcareous algae. The Cyanophyceae are regarded as bacteria ("Cyanobacteria", Stanier and Cohen-Bazire, 1977) by several authors (see W. E. Krumbein, 1979).

The small number of *algal spores* described up to now in limestone sections have not been taken into consideration in the subsequent discussion of calcareous algae:

a) Lower Permian of the Southern Alps: circular and kidney-shaped figures with a thin calcitic wall and a diameter between 60 and 160 μm, mean value 150 μm; interpreted as dasycladacean spores (E. Flügel, 1966). These spores are of rock-building abundance in lagoonal limestones.

b) Upper Jurassic of the Southern Franconian Alb and the Northern Alps: oval figures with thin laminated calcitic walls and an internal space divided by up to 4 septa, diameter between 130×170 μm and 208×415 μm; interpreted as rhodophycean spores (Fay and Dirsche, 1975).

Plate 28. Calcareous Algae

1. Porostromate Cyanophyceae. *Girvanella ducii* Wethered. Middle Devonian (Schwelm limestone, Givetian): Letmathe, Sauerland, Germany. Scale is 0.5 mm

2. Porostromate Cyanophyceae. *Rothpletzella devonica* (Maslov) together with encrusted foraminifera(?). Lower Devonian (Currajong limestone): NW Canberra, Australia. Scale is 1 mm

3. Cyanophyceae. Type *Ortonella* or *Cayeuxia*. Bifurcated filamentous tubes with acute angles. Upper Jurassic (Lower Tithonian): Kapfelberg quarry near Kelheim, Southern Franconian Alb, Germany. Scale is 1 mm

4. Codiaceae. Fragments of *Halimeda* sp. Well-developed central medulla and peripheral cortex area. *Halimeda* in Recent lagoons is an important producer of carbonate silt and carbonate mud. Recent: Kenya, East Africa. Scale is 1 mm

5. Dasycladaceans. Cross-sections and longitudinal sections of *Heteroporella zankli* (Ott). Late Triassic: Cozzo di Lupo near Palermo, Sicily. Scale is 1 mm

6. Phylloid algae. The curved plates encrusted by foraminifera (black) exhibit a narrow external zone of pores and a broad recrystallized area in the center. *Eugonophyllum johnsoni* Konishi and Wray. Phylloid algae contribute greatly to the formation of "mud mounds" in the North American Pennsylvanian and in the Lower Permian of the Southern Alps (see Sect. 10). Lower Permian (Lower Rattendorf formation): Hüttenkofel, Carnic Alps, Austria. Scale is 1 mm

7. Characeae: Cross-sections of characean stems. Recent Characeae are restricted to non-marine environments. Ostracods (left) often co-occur with these algae. Uppermost Jurassic (Münden beds, Purbeckian): Lower Saxony, Germany. Scale is 1 mm

8. Corallinaceae. *Lithothamnium roveretoi* Airoldi. In the center of the picture is a balanid section (truncated cone). Upper Oligocene: Pohlkotte near Osnabrück, Germany. Scale is 100 μm

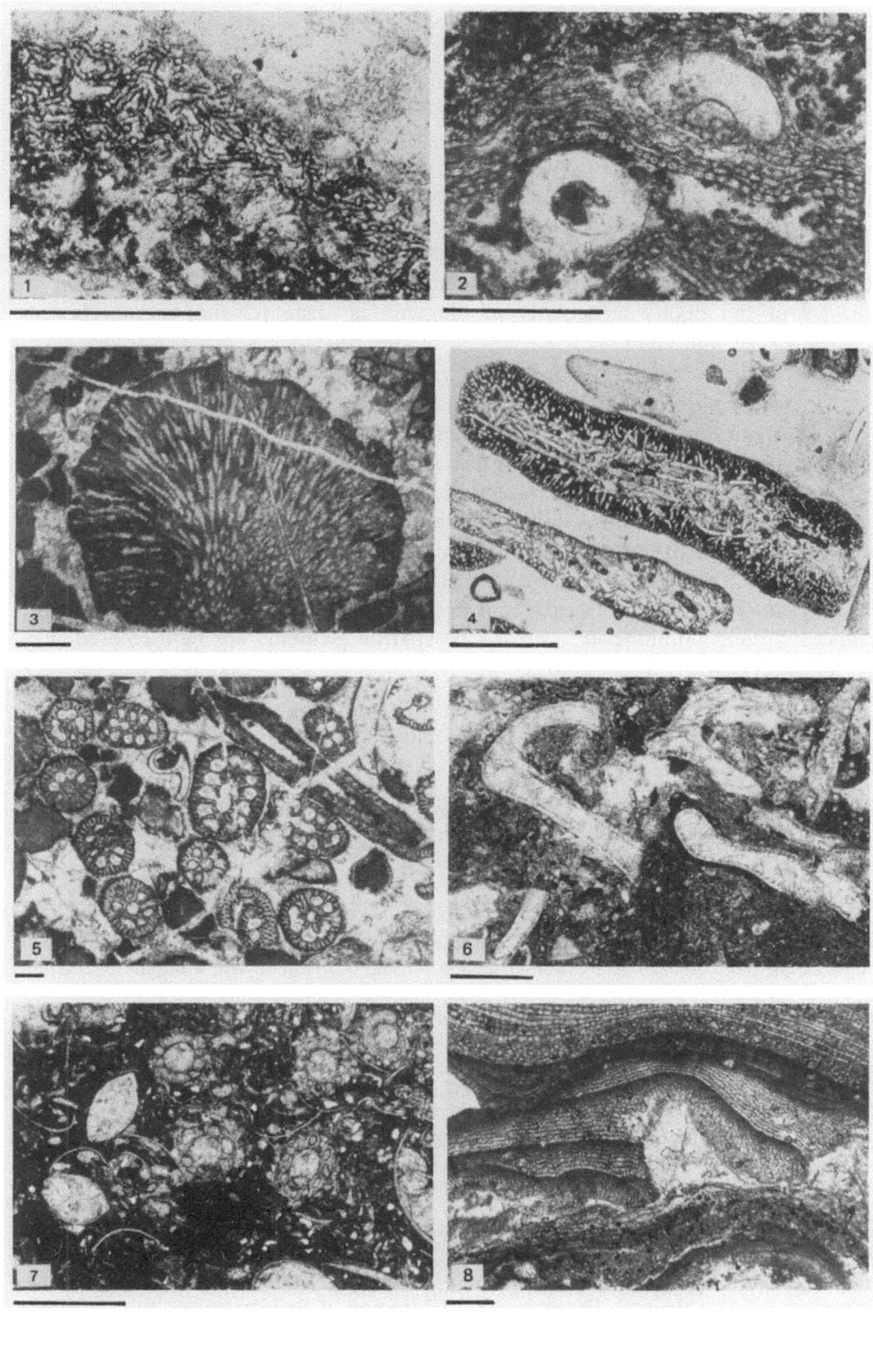
1
2
3
4
5
6
7
8

5.3.12.2 Mode of Calcification and Morphology

$CaCO_3$ is found within the algal cells (in single-celled algae, e.g., in coccolithoporids and in the dasycladacean genus *Acetabularia*), associated with the cell wall (e.g., in red algae), and outside the cell within the mucilaginous sheath (e.g., in dasycladaceans) or as an encrustation on the thallus (in many blue-green algae). Nine structural types can be differentiated according to the mineralogical composition, the ultrastructure, and the pattern of calcification (extracellular or within the cell wall) (Flajs, 1977). However, it is very difficult to separate extracellular $CaCO_3$ precipitated in the mucilaginous sheath from $CaCO_3$ encrusted on the surface of the thalli, particularly in filamentous algae (Cyanophyceae) (see E. Flügel, 1977c, and Plate 2/2).

The *morphology* of calcareous algae can be described by:

1. Growth forms and geometrical shape (encrusted, erect – e.g., dasycladaceans; micron-sized filaments, e.g., in blue-green algae).
2. Internal structures in thin-sections (e.g., cell layer in Corallinaceae, segmentation of the thallus in Codiaceae, thallus shape and types of pores in dasycladaceans, reproductive organs).
3. Microstructure and ultrastructure (rarely well-preserved in fossils). In thin-sections microgranular elements appear dark in transmitted light (e.g., in Corallinaceae, often "minimicrite", see 4.1.2); sparitic elements are light-colored (e.g., dasycladaceans).

To *identify genus and species,* the following terms may be useful (see also Fig. 48 and Table 30):

Articulated – thalli subdivided into segments, as in Corallinaceae (Corallinoideae)
Aspondyle – branches irregularly arranged, not whorled, (e.g., dasycladaceans)

Fig. 48. Morphology and criteria for identification of a few important groups of calcareous algae. *Blue-green algae: 1* Longitudinal-section through a filament with a bundle of trichoms lying in the center and coated by the mucilage sheath. *2* The branching pattern of the "tubes" of Ancient filamentous Cyanophaceae, which are abundant in marginal marine environments with restricted ecological conditions, is important for taxonomic differentiation. *Red algae: 3* Articulated Corallinaceae; *A,* characteristic growth form. *B,* Section with hypothallus and perithallus. *4* The arrangement of the cells (*C* and *D* with many layers, *E* and *F* included in one layer) and the position of the conceptacle as well as the structure of the conceptacle roofs are all significant in identification. *5* (*G* arranged serially; *H* many pores; *I* one pore). *H* and *P,* hypothallus and perithallus. *Green algae: 6* In dasycladaceans the shape and arrangement of the branches or the "pores" in the wall and their pattern of branching have substantial taxonomic significance. *7* Cross-section *A* and longitudinal section *B* of an articulated dasycladacean genus (*Cymopolia*). The magnification allows the primary branches (Pb), the secondary branches (Sb) and the sporangia lying within the minor branches to be recognized. Charophyceae carry calcified reproductive organs (*8*) in the bifurcation of their branches. *9* A few Mesozoic Gyrogonites (*A,* Characeae; *B,* Sycidiaceae; *C,* Trochiliscaceae; *D,* Clavatoraceae). (*10*) Erect, highly branched Codiaceae are very important for the production of bioclastic carbonate. The systematic position of the "phylloid algae", which are restricted to the Late Paleozoic, is still under discussion. Simplified after Wray (1977)

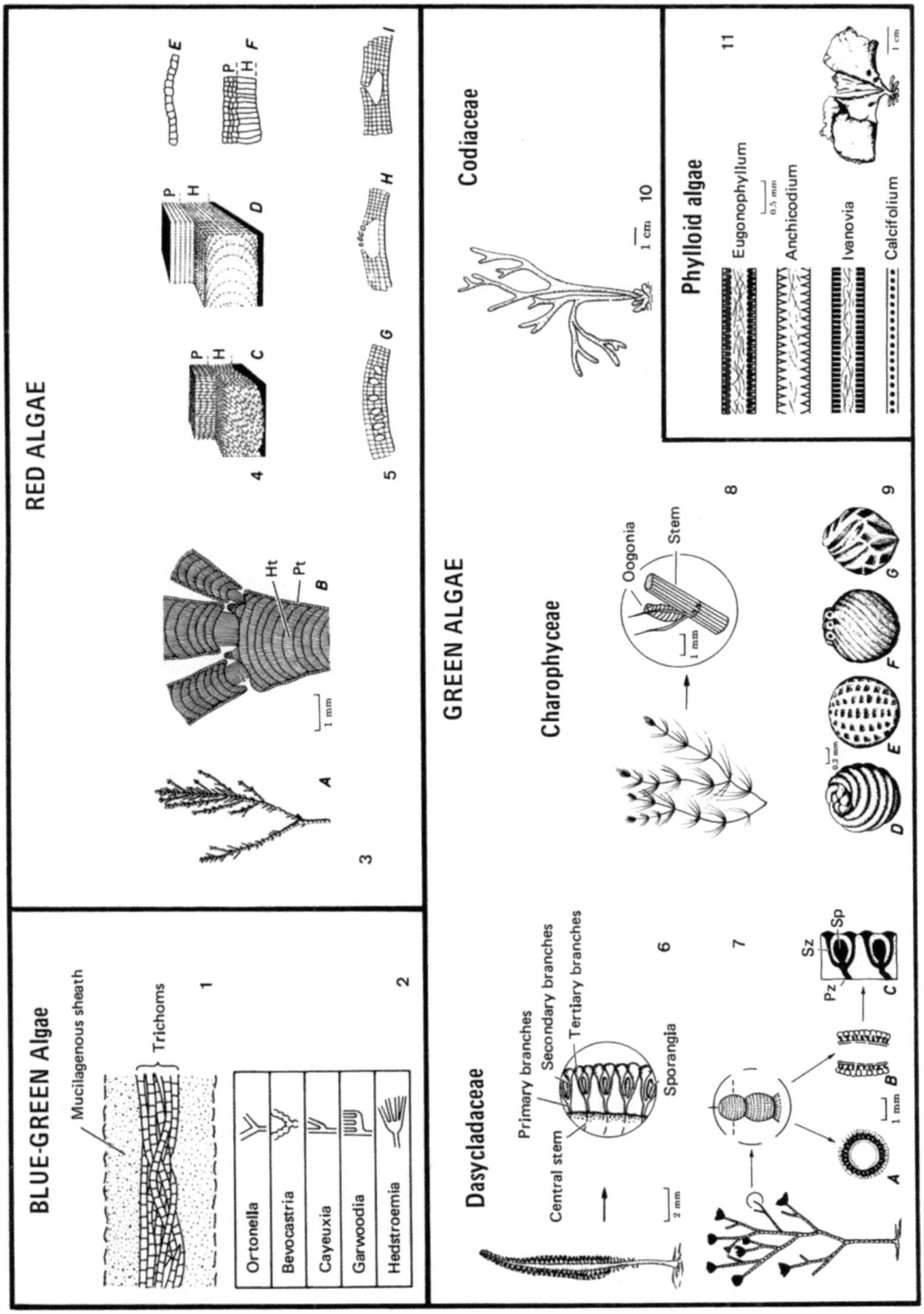
BLUE-GREEN Algae
Mucilagenous sheath
Trichoms
1
Ortonella
Bevocastria
Cayeuxia
Garwoodia
Hedstroemia
2
RED ALGAE
A
1 mm
Ht
Pt
B
3
P
H
C
D
E
F
4
G
H
I
5
GREEN ALGAE
Dasycladaceae
Central stem
Primary branches
Secondary branches
Tertiary branches
Sporangia
2 mm
6
1 mm
A
B
Pz
Sz
Sp
C
7
Charophyceae
Oogonia
Stem
1 mm
8
0.2 mm
D
E
F
G
9
Codiaceae
1 cm
10
Phylloid algae
Eugonophyllum
0.5 mm
Anchicodium
Ivanovia
Calcifolium
1 cm
11

Table 30. Important characteristics for the identification of genera and species of Ancient calcareous algae (dasycladaceans after Bassoulet et al., 1977; Corallinaceae after Poignant, 1979). See also Fig. 48

Major Group	Group	Diagnostic Criteria
Cyanophyta	Calcified filamentous algae Pl. 28/1–3; 41/2; 44/1; 47/5; 52/4	1. Arrangement of the filaments (encrusted or erect); 2. Branching of the filaments themselves or the calcareous tubes around the filaments; 3. Diameter of the tubes; 4. Shape of the tubes
Chlorophyta	Codiaceae P. 28/4; 41/3	1. Growth form (segmented, unsegmented); 2. Size, shape, and arrangement of the filaments in the central medulla and peripheral cortex; 3. Reproductive organs
	Dasycladaceae Pl. 5/5; 28/5; 29/6–8; 30/1–3, 6; 31/4; Pl. 33; Pl. 34; 42/9; 47/2; 48/2; 52/2	1. Arrangement of the primary branches; 2. Shape of the thallus (segmented, etc.); shape and arrangement of the secondary branches, that is "pores" and reproductive organs = generic criteria; 3. Dimensions, degree of calcification, and inner boundary of the central stem
Rhodophyta	Solenoporaceae Pl. 10; 29/3; Pl. 33; 37/1	1. Presence or absence of filamentous partitions (horizontal elements); 2. Regular or irregular spacing of the partitions; 3. Shape of the "cells" in cross-section; 4. Shape of the filaments (string of pearls, etc.); 5. Dimensions
	Gymnocodiaceae Pl. 29/5	1. Size of the thallus segments; 2. Density and size of the pores in the wall
	Corallinaceae Pl. 3/4, 6; 4/6; 28/8; 29/1, 2; 38/2; 40/3, 4; 45/5, 6	1. External morphology of the thallus; 2. Type and position of the reproductive organs (conceptacle); 3. Shape of the cells and arrangement in the central hypothallus; 4. Shape of the cells and arrangement in the peripheral perithallus; 5. Heterocysts; 6. Dimensions

Axis – the stem-like part that supports the branches (e.g. dasycladaceans) or the central part in Codiaceae.

Central stem – the central element of dasycladaceans which bears the primary branches.

Conceptacle – cavity in the Corallinaceae skeleton in which reproductive organs, e.g., sporangia, are developed (see Fig. 48).

Cortex – cells or tissues of the thalli located peripherally or on the margin around a central axis, particularly in Codiaceae (e.g., *Halimeda,* see Fig. 48).

Euspondyle – primary branches regularly arranged in whorls (like the spokes of a wagon wheel), as in some dasycladaceans.

Filaments – parallel rows of cells, branched or unbranched, with or without a mucilaginous sheath.

Gyrogonites – small, spiral, calcified parts of the oogonia of Characeae (see Fig. 48).

Hypothallus – basal or central part of the thalli of red algae with cells which are arranged parallel to the substrate and which are larger in comparison to the peripheral areas (see Fig. 48).
Medulla – the central region of an internally differentiated thallus, e.g., in Codiaceae (see Fig. 48).
Metaspondyle – the primary branches are arranged into clusters of regular whorls, as in some dasycladaceans.
Oogonium – single-celled female reproductive organs of Charophyceae.
Perithallus – cell layer over or around the hypothallus, generally consisting of very tiny cells oriented essentially perpendicular to the substrate.
Phylloid – platy or leaf-like growth forms, "phylloid algae".
Primary branches – branches which shoot off from the central stem, as in dasycladaceans.
Secondary branches – branches and groups of branches which shoot off from primary branches, as in some dasycladaceans.
Sporangia – spore producing structure, spore case.
Spores – specialized mobile or immobile cells which are released from the parent plant and are capable of giving rise to a new adult individual without a union with other cells.
Thallus – the entire plant body or the entire algal colony.
Whorls – a circle of branches in the same plane around a stem, usually equally spaced, as in dasycladaceans.

5.3.12.3 Ecological Factors

The distribution of Recent algae is determined by:

a) Physical factors such as the intensity and quality of light, water temperature, water movement, condition of the substrate, geometrical shape of the niche

b) Chemical factors, such as the chemism of the water, salinity, dissolved gases, and pH values.

c) Biological factors, such as competing organisms and the abundance of herbivorous organisms

Table 31 lists the essential ecological factors for Recent calcareous algae. Attention should be paid to the possibility of a change in the environment over geological time (e.g., blue-green algae; Richter et al., 1979).

The intensity of the light necessary for photosynthesis decreases with increasing water depth, whereby the zone of light penetration is dependent upon the geographical latitude. About 50% of the light penetrating clear water is absorbed in the uppermost 5–10 m. Despite this, algae occur in greater depths of water if they have special pigments which can utilize the minimal intensity of light in deeper oceans. Because the wave lengths of light are selectively absorbed at various depths of water, Recent red algae which have adapted to blue and green sectors can also be found in deeper water depths (down to about 400 m), whereas the calcareous green algae which assimilate most strongly in the red sectors are preferentially bound to the shallow seas (down to about 100 m).

Table 31. Ecological factors of Recent calcareous algae

	Salinity	Water temperature	Water Depth	Water Energy	Substrate	Biotope
Cyanophyceae	Freshwater, brackish water, hypersaline, not normal marine	Warm and cold water	Shallowest tidal zones to the greatest depths	Generally low	Mud and sand bottoms (sediment binders); borers in hard grounds	Terrestrial, lacustrine, tidal zone
Chlorophyceae: Codiaceae	Normal marine	Tropical (app. 25° C), few subtropical species	Lower tidal zone, usually < 10 m; to 100 m; range of species dependent on depth	Low, generally quiet waters	Mud and sand bottoms, very few species on hard bottoms	Protected lagoons and deeper fore-reef slope
Dasycladaceae	Usually normal marine, a few species tolerate brackish water and hypersaline environments	Tropical to subtropical; a few species in warm temperate water. Important factor!	Below the tidal zone to about 30 m; max. about 90 m; usually 3–5 m	Low, generally quiet waters (below wave base or in protected lagoons)	Mud and sand bottoms, a few species on hard bottoms (rocks; reef flats)	Protected lagoons, protected reef flats
Charophyceae	Fresh and brackish water, hypersaline	Warm water and cooler water	To about 10 m, usually very shallow	Standing or sluggish water	Mud and sand bottoms	Shores of freshwater lakes; in near-coastal brackish water
Rhodophyceae: Squamariaceae	Normal marine	Warm water (tropics, subtropics)	Lower tidal area down to a few meters below mean sea level, max. about 90 m	Low	Usually hard bottoms	Reef areas, open marine
Corallinaceae	Usually normal marine; only a few species tolerate restricted salinity in coastal areas	Warm and cold water (tropics to polar region); stenothermic species and genera	Tidal zone to 250 m; usually < 25 m; range of species dependent on depth	Varies: a few taxa in high-energy environments, others in low-energy environments	Usually hard grounds (reefs, underwater cliffs), but also soft bottoms or unstable substrates	Open marine platforms and bays with banks or reefs (outer reef margin); shelf slopes (some rhodolites)

The use of *Ancient algae as indicators of depth* must, therefore, be based on the following assumption (see also Riding, 1974):

1. The algae must be referred to a major taxonomic group (that is, not just "algae", but rather dasycladaceans, melobesians, etc.).
2. Allochthonous deposition (transport by currents, e.g., Elliott, 1975) must be excluded.
3. If possible, one should attempt to apply the bathymetrical zoning patterns of Recent genera and species to Ancient algae as well. This appears possible for Tertiary and perhaps for Mesozoic calcareous algae if studies are very carefully conducted.

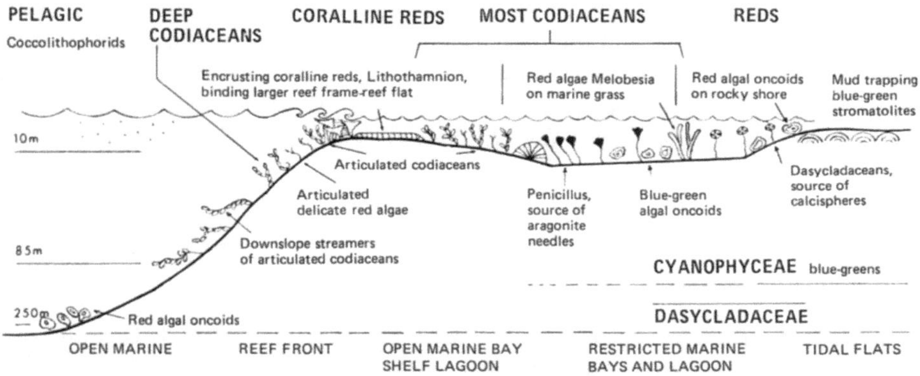

Fig. 49. Ecology of calcareous marine algae; depositional environments along an idealized profile of a carbonate shelf margin. After Wilson (1975). This diagram shows only the main areas of spatial distribution for some algal groups. Recent dasycladacean algae, for instance, are also found in various restricted environments (lagoons, protected reef flats, mangrove zones) as well as in near-coastal hard bottoms, where they grow on rocks (see Valet, 1979)

Recent benthonic calcareous algae usually display their greatest diversity in shallow waters near the coast (a few meters to tens of meters). Filamentous and coccoid blue-green algae prefer tidal zones; green algae are frequently found in depths down to about 20–30 m and red algae predominantly in tidal or shallow subtidal areas. Of course, species of encrusting Corallinaceae as well as Codiaceae occur which have adapted to water of more than 50 m in depth (e.g., *Lithothamnium*-rhodolites; *Halimeda* in the deeper fore-reef area of the Caribbean reefs). The horizontal distribution pattern of many algae is a result of the dependence on depth (examples in Milliman, 1977). Ancient shelf and reef areas can be subdivided according to facies-controlled algal genera or associations (Wray, 1979; Flügel, 1979). Examples of Devonian reef complexes are found in Wray (1972), Lower Permian shelf and shelf margin areas in E. Flügel (1977 b), Middle Triassic reef complexes in E. Ott (1972), Late Mesozoic shelf zones in Conrad (1977), and Tertiary platform carbonates in Elliott (1968).

On the basis of the very generalized distribution pattern of Recent calcareous algae (Fig. 49), algae are regarded as facies indicators of warm water regions relatively near the coast and shallow waters. Deviations are, however, the norm!

5.3.12.4 Algae and Carbonate Sedimentation

The interrelationships between benthonic algae and sediment formation can be summarized as follows:

A. Sediment formation

1. *Sedimentation of algal bioclasts:* The remains of dasycladaceans, Codiaceae (*Halimeda*) and articulated Corallinaceae comprise about 20%–60% of the arenitic grain fraction in Recent lagoonal sediments. The "maerl", that is the carbonate sand and gravel sediments of the Mediterranean and Atlantic, consists of up to 75% Corallinaceae. This sediment acts as a substrate for a diverse epiflora consisting of green and red algae (Cabioch, 1974). Similar algal bioclast accumulations are also known in Late Paleozoic and Mesozoic lagoonal sediments (e.g., *Mizzia* limestones of the Middle Permian; *Diplopora* limestones of the Alpine Middle Triassic).

2. *Sedimentation from micron-sized organic detritus:* After the disintegration of their thalli, Codiaceae (*Penicillus, Halimeda*) and other green algae release

Plate 29. Calcareous Algae

1. Corallinacean red algae. Nodular thalli of *Mesophyllum* sp. with relatively large conceptacles, overgrown by *Archaeolithophyllum* sp. (characterized by sporangia arranged in rows). Leitha Limestone, Miocene: Wildon, Styria, Austria. Scale 1 mm

2. Rhodolites. Oncoids are designated as "rhodolites" when they are formed by an encrustation of Corallinaceae (together with sessile foraminifera, bryozoans, serpulids, etc.). In the Recent such oncoids are preferentially bound to shallow unstable substrates, which also occur sometimes in greater water depths (down to about 250 m!). Pictured are dark Corallinaceae layers with clearly defined rows of conceptacles (*Archaeolithothamnium* sp.) and light-colored rows of cells from sessile foraminifera (*Miniacina multiformis* Scheibner). Miocene: Pietraroia, Southern Apennines, Italy. Scale is 10 mm

3. Solenoporacean red algae. *Solenopora cassiana* Flügel. Cassian Beds, Early Upper Triassic: Seeland Alm near Cortina d'Ampezzo, Southern Alps. Scale 1 mm

4. Ancestral Corallinaceae. *Ungdarella uralica* (Maslov) serve as nucleus of a bio-oncoid. Lower Permian: Forni Avoltri, Carnic Alps, Italy. Scale is 1 mm

5. Gymnocodiaceae. This group, which is referred to red algae, is distinguished from dasycladaceans by densely packed pores oblique to the outer surface of the wall and to central structures. *Gymnocodium bellerophontis* (Rothpletz). Upper Permian: Kapiskaya, Western Anatolia, Turkey. Scale is 1 mm

6. Dasycladaceans. Oblique-sections of *Oligoporella pilosa* Pia. The species of the generic group *Physoporella-Oligoporella* with biserial whorls characterize the Hydaspian, Pelsonian and the Lower Illyrian of the Anisian stage of the Lower Alpine Middle Triassic (see Sect. 5.3.12.5). Middle Triassic (Pelsonian, Anisian): Silicka Brezova', Slovensky kras, Carpathian Mountains, Czechoslovakia. Scale is 1 mm

7. Dasycladaceans. *Griphoporella curvata* Gümbel. Late Triassic (Dachstein limestone) Haindlmauergößß, Gesäuse, Styria, Austria. Scale is 1 mm

8. Dasycladaceans. *Teutloporella socialis* Praturlon. Upper Jurassic: Monte Carlo, Abruzzi, Italy. (Type locality!). Scale is 1 mm

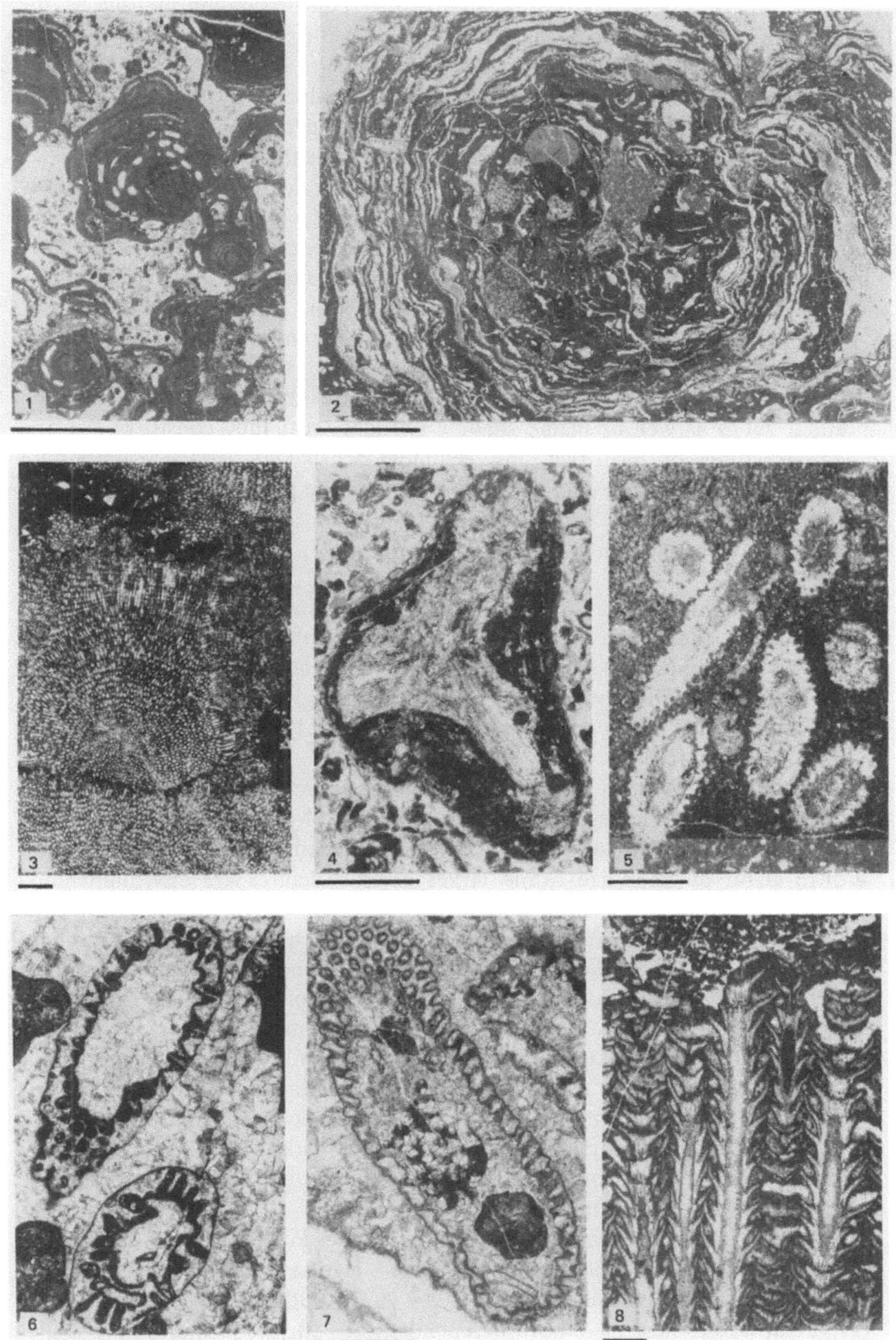
1
2
3
4
5
6
7
8

aragonitic fibers a few microns in size, which contribute enormously to the formation of calcareous mud in shallow water – and after transport – also in deeper environments (see Fig. 30). The annual production lies between 25 g/m^2 in reefs and 90 g/m^2 in lagoons. Similar quantities are also produced by other algae, such as epiphytic red algae of the genus *Melobesia* (40 g/m^2/year). Ancient algae comparable to Recent Codiaceae have been known since the Upper Triassic.

3. *Formation of biogenic frameworks* by encrusting calcareous algae, particularly by Corallinaceae. Recent examples are provided by *Lithothamnium* ridges in the Pacific coral reefs and the hard grounds in the Mediterranean known as the "biocenoses coralligenes". Such Ancient frameworks are known in the Late Mesozoic and the Tertiary (e.g., Buchbinder, 1977; see Plate 28/8).

4. *Biochemically induced carbonate precipitation* around blue-green algae which can build mats and bind sediments. Biosedimentary structures arise which are called *stromatolites* (Fig. 50). These structures are formed in intertidal and supratidal zones as well as in the shallow subtidal area; they consist of flat and domed structures of algal-rich and sediment-rich layers, and are important indicators of the depth and energy of the water and the proximity to the coast (see Reference list 5.5, and especially Walter, 1976, and Purser, 1980). The algal oncoids also belong to the stromatolites (see 4.1.3.4).

The stromatolite groups can be differentiated quantitatively by means of their morphologic attributes like the shapes of the lamina profiles, cross-sections (plan views), and the longitudinal sections (silhouettes), see Hofmann (1977).

Some widely accepted views concerning the ecology of stromatolites must be revised in the light of recent studies (see Awramik et al., 1978):

- Stromatolite morphology is not entirely and exclusively the product of environmental factors like the degree of water turbulence. Morphogenesis of small-scale macrostructures, domes, and columns may be the consequence of the biological growth habit of the algae, while larger macrostructures are environmentally controlled.
- Carbonate stromatolites are not restricted exclusively to intertidal and supratidal zones but occur in shallow marine subtidal environments (Bermuda); shallow hypersaline to periodically exposed marine environments (Shark Bay and Hamelia Pool, Western Australia); lagoonal marine environments (Laguna Mormona, Mexico); freshwater marshes (Andros Island, Bahamas); meromictic lakes (Green Lake, New York); saline and hypersaline lakes (Great Salt Lake, Utah; Solar Lake, Gulf of Aqaba); streams (Little Connestoga Creek, Pennsylvania); and thermal springs (Yellowstone National Park). Fossil stromatolites are believed to have been formed in normal marine subtidal environments (e.g., Upper Permian Zechstein algal stromatolite reefs of the Harz Mountains, Germany; Paul, 1980, and of Poland, Peryt and Piatkowski, 1977); shallow hypersaline to periodically exposed regions; lagoonal marine environments; perhaps also freshwater marshes, and lakes. A few authors conjecture that some cryptalgal fabrics may have been independent of bathymetry and light (Monty, 1977; Massari, 1980); this may be true for stromatolitic structures occurring in red nodular limestones (see Sect. 4.2.3.6).

- Stromatolites are not built solely by blue-green algae. Eucaryotic algae may participate in algal mat formations, particularly in subtidal or permanently submerged environments. Bacteria are commonly viewed as decomposers of accreted algal mat organics, but they also contribute to the formation of the laminae at the surface of some biomats.

Different terms are in use to designate fossil algal crusts without definite structures. Pia (1927) introduced the name "Spongiostromata" to characterize algal crusts without distinct organic structures but with special growth forms. These fossils are attributed to an extracellular precipitation of carbonate. According to Pia the Spongiostromata include the stromatolites as well as the oncolites. Wolf (1965a) proposed the name "Pseudostromata" for the products of these crustose algal growths; the products are irregular masses or patches of dense, grumous, cel-

Continued p. 340

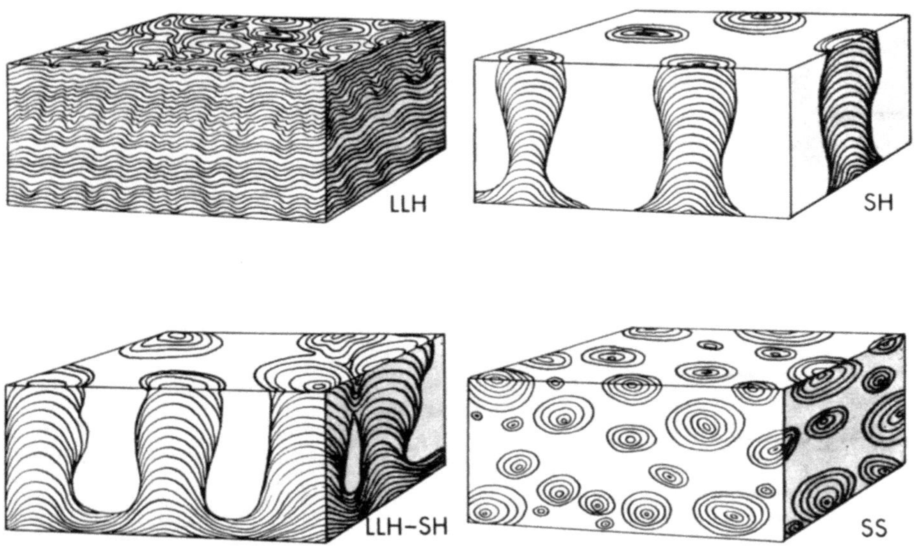

Fig. 50. Growth forms of stromatolites. The shape of many stromatolites can be described by basic geometric forms (Logan et al., 1964), whose arrangement, however, is not controlled entirely by environmental conditions, such as the intensity of the water energy. Biological factors (varied interactions of different species of algae with sediment deposition and carbonate precipitation) are very significant (see Walter, 1976). The following symbols may be used in the field as well as in the laboratory to describe thin-sections and polished sections: *LLH* – laterally linked hemispheroids with lamellae whose domes are either closely packed or spaced somewhat apart (subtypes LLH-C and LLH-S). *SH* – stacked hemispheroids form columns which are separated by non-laminated sediment. The domes have either a constant diameter or various widths (subtypes SH-C and SH-V). *SS* – spheroidal structures around a nucleus (oncoids). Subtypes SS-C (normal oncoid, concentric structure), SS-R (laminae randomly overlapped) and SS-I (inverted; laminae facing each other as concentric hemispheres. The symbols can be linearly combined to indicate mixed geometric forms (e.g., *LLH-SH*) or expressed as fractions to indicate microstructures, whereby: numerator = macrostructures classified in the field and in hand specimens; denominator = microstructural forms seen in hand specimens and thin-sections. After Anstey and Chase (1974)

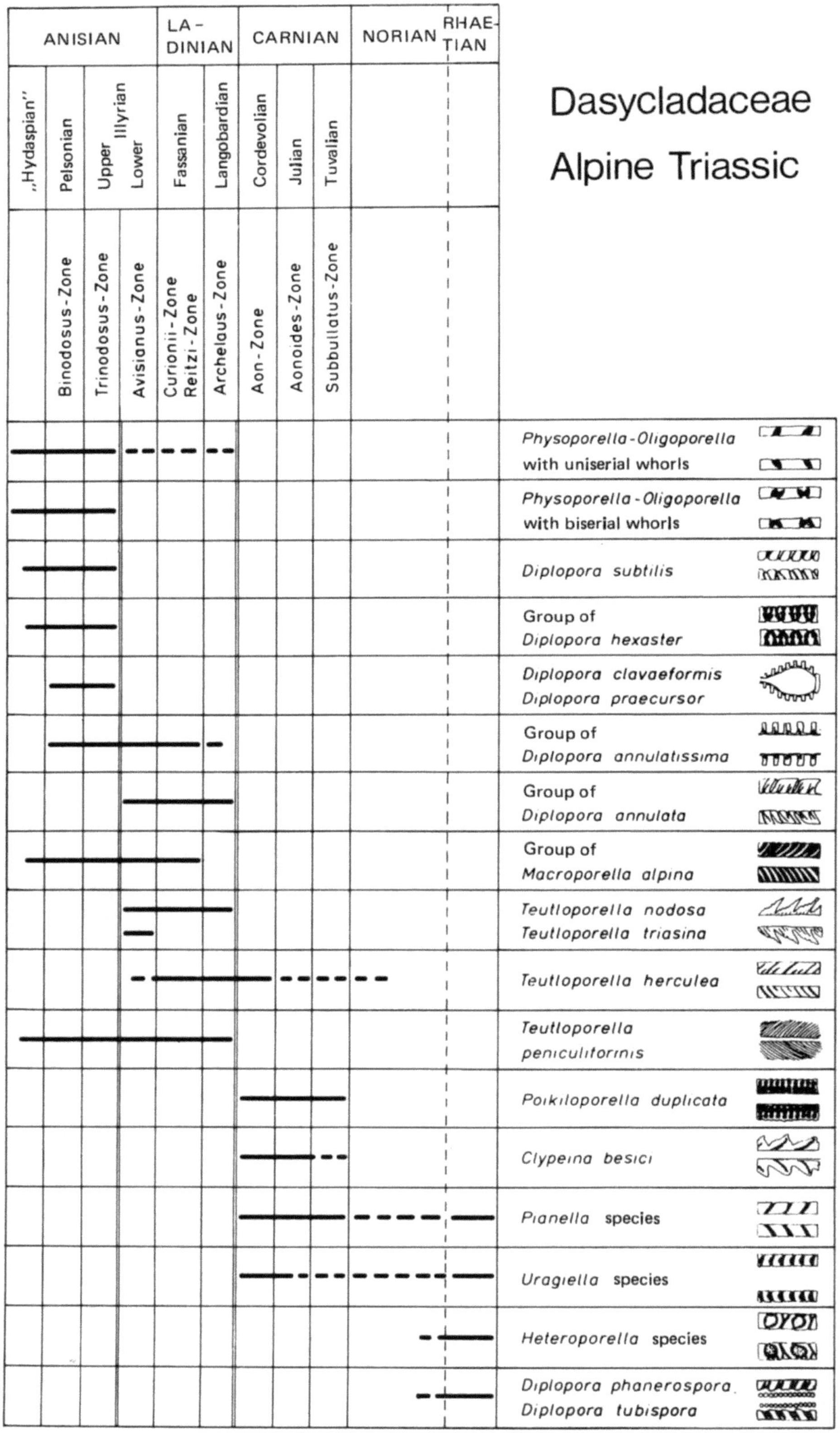
Dasycladaceae
Alpine Triassic
ANISIAN
LA-DINIAN
CARNIAN
NORIAN
RHAE-TIAN
„Hydaspian"
Pelsonian
Upper Illyrian
Lower
Fassanian
Langobardian
Cordevolian
Julian
Tuvalian
Binodosus-Zone
Trinodosus-Zone
Avisianus-Zone
Curionii-Zone
Reitzi-Zone
Archelaus-Zone
Aon-Zone
Aonoides-Zone
Subbullatus-Zone
Physoporella-Oligoporella with uniserial whorls
Physoporella-Oligoporella with biserial whorls
Diplopora subtilis
Group of Diplopora hexaster
Diplopora clavaeformis
Diplopora praecursor
Group of Diplopora annulatissima
Group of Diplopora annulata
Group of Macroporella alpina
Teutloporella nodosa
Teutloporella triasina
Teutloporella herculea
Teutloporella peniculiformis
Poikiloporella duplicata
Clypeina besici
Pianella species
Uragiella species
Heteroporella species
Diplopora phanerospora
Diplopora tubispora

Table 32. Significance of individual groups of calcareous algae in the geologic record. Characeae have been deliberately left out. The systematic position of many fossil genera attributed to the Cyanophyta is still a matter of debate (e.g., *Epiphyton* – referred to the Rhodophyta?; *Girvanella* – algae incertae sedis). The boldface vertical lines indicate the stratigraphic range. X = sometimes more abundant; (X) = only of local significance; B = aids in formation of bioherms in quiet water environments (mud mounds); R = reef framework builder; BR = back-reef, generally protected environments; S = shelf region

	Cyanophyta					Chlorophyta		Rhodophyta							
	"Girvanella"	*"Renalcis"*	*"Epiphyton"*	*"Ortonella"*, *"Cayeuxia"*	Other types	Codiaceae	Dasycladaceae	Solenoporaceae	Gymnocodiaceae	Squamariaceae	Encrusting Corallinaceae	Segmented Corallinaceae	*"Ungdarella"*	Ancestral Corallinaceae	Phylloid Algae
Recent				X	X	S	S			R	R	S			
Tertiary						S	S				R	S			
Cretaceous				X		S	S	X	S	R	R	S			
Jurassic				X	B	X	SBR	X				X			
Triassic	X			BR		(X)	SBR	RBR							
Permian						X	SBR		S				X	R	R
Carboniferous				S		(X)	X	X					S	S R	B S
Devonian	R	R				(X)	(X)	BR							
Silurian	R			X		X	X	R							
Ordovician	B	B	B			X	X	X							
Cambrian	B	B	B			X	X	X							
Precambrian															

◄

Fig. 51. Zonation of the Middle and Late Triassic with dasycladaceans in the Alpine-Mediterranean region. After E. Ott (1972)

lular, spongy, tubular, pelletoid and granuloid algal micrite with some admixtures of detrital particles.

The preservation potential of cryptalgal structures such as algal laminites, stromatolites, oncolites or thrombolites in the rock record may be lessened by eroding tides and waves, by bioturbation, or by grazing metazoans. Bioturbation and a destruction of the algal mats by grazers may be suppressed if metazoans are restricted by overly rigorous environmental conditions, such as hypersalinity or desiccation. The most important factor for the preservation of cryptalgal fabrics is an early and rapid cementation and lithification which in Recent settings occurs in hypersaline, brackish, and freshwater environments as well as in intertidal zones with normal marine conditions (Pratt, 1979).

B. Sediment lithification

5. *Carbonate cement formation:* Alexandersson (1974, 1977) has described the possibilities of submarine cementation in Corallinaceae frameworks induced by bacterial decay of the organic matter or as a result of algal assimilation.

C. Sediment destruction

6. Consolidated sediments and calcareous shells are loosened by *boring algae* (usually Cyanophyceae and Chlorophyceae) and are sometimes totally destroyed. Algal micrite ("algal dust", Wood, 1941) or micritic envelopes and grains result (see 4.1.3.7). Because fungi and deep water algae may also act as borers, a paleobathymetric interpretation of boring algae is difficult. The specific identification of endoliths on the basis of fossilized microborings in Ancient rocks is difficult because of morphological alternations caused by complicated dissolution and precipitation processes during early diagenesis (Harris et al., 1979).

5.3.12.5 Biostratigraphic Significance (Fig. 51)

Of all the calcareous algae groups, the dasycladaceans, despite considerable difficulties involved in the identification and systematics (see Herak et al., 1977; Bassoulet et al., 1977), have major biostratigraphic significance. This is particularly true for the Late Paleozoic (e.g., Mississippian, Mamet and Roux, 1975) and for the Mesozoic shelf carbonates of the Tethys. One example of this is the dasycladacean zonation established by J. Pia and modified by E. Ott (1972) for the Alpine Middle and Upper Triassic (Fig. 51); other examples are the "microfacies zones", defined primarily according to the occurrence of calcareous algae, in the Jurassic of the Mediterranean region (see Sect. 5.4).

5.3.12.6 Key to Identification of the Major Groups of Calcareous Algae

A General shape ill-defined, flat layers and crusts with distinct or indistinct mm-thick layers . . . . *sediment binding algae* (see Tab. 22)

B General shape well-defined 1

(B) 1 a Distinct microstructure in thin-sections 2
b Indistinct or no microstructure in thin-sections 8
c Shape spherical or egg-shaped, sometimes with external ornamentation of furrows and ridges arranged in spirals . . . . *Characeae* (Devonian-Recent) (Plate 28/7)

(1 a) 2 a Microstructure of distinct "reticulated cells" 3
b Microstructure of many small tubes 4
c Microstructure with one central tube 6

(2 a) 3 Cells rectangular in longitudinal sections
a Thallus differentiation in the hypothallus (center – larger cells) and perithallus (periphery – smaller cells); sporangia . . . . *Corallinaceae* (from the Jurassic)
a_1 Thallus solid, encrusted, unsegmented . . . . *Melobesoideae* (Jurassic-Recent) (Plates 28/8; 45/5, 6)
a_2 Thalli dendroid, segmented . . . . *Corallinoideae* (Lower Cretaceous-Recent)
b No thallus differentiation, no sporangia, transverse elements may be reduced . . . *Solenoporaceae* (Middle Cambrian-Neogene) (Plate 29/3)

(2 b) 4 a Tubes more or less homogeneous, short or long 5
b Tubes of various sizes, arranged like a string of pearls . . . . *Rothpletzella,* Cyanophyceae (Silurian-Triassic) (Plates 28/2; 41/2)
c Tubes large in the center of the thallus, becoming smaller towards the periphery . . . . *Codiaceae*
c_1 Thallus usually segmented, oval outline of segments . . . . Type *Halimeda* (Triassic-Recent) (Plate 28/4)
c_2 Thallus plate-like, wavy to crinkled; internal differentiation in marginal pore zone (cortex) and broad, usually recrystallized central zone (medulla) . . . . Type *"phylloid algae"* (Squamariaceae?), Carboniferous-Permian (Plates 28/6)

(4 a) 5 a Short curved tubes without cross-partitions, often fused into aggregates; often nodular thalli . . . . *Girvanella* (Cyanophyceae) (Cambrian to Cretaceous) (Plate 28/1)
b Long closely packed tubes, forking into certain specific angles which are important for the identification of genera; thallus nodular or crust-like . . . *porostromate Cyanophyceae* (Cambrian-Recent) (Plate 28/3; 47/5)

(2 c) 6 a Wall interspersed with pores, sometimes has whorled branches . . . 7
b Wall not perforated, but sometimes has a ring of smaller circles around a larger central circle; always ornamented . . . . oogonia of *Characeae* (Plate 28/7)

(6 a) 7 a Cylindrical or spherical thalli; hollow calcareous segments with wall interspersed with well-separated pores . . . . *Dasycladaceae* (Cambrian-Recent) (Plate 29/6–8; 30/1, 2)
b Large, generally segmented thalli; hollow calcareous segments with thin walls (interspersed with *densely packed* pores oblique to the surface of the segments) and a large central cavity originally containing loosely arranged cell filaments, however, often recrystallized beyond recognition . . . . *Gymnocodiaceae* (Permian-Paleocene) (Plate 29/5)

(1 b) 8 a Diverse growth forms; more or less parallel, micritic laminated fabric with open-space structures concordant to the bedding plane . . . . some *algal mats* ("Loferite", see Table 20)

b Geometric growth forms with pronounced shape 9

(8 b) 9 Often large "colonies" (cm- to m-sized), more or less regularly layered structures . . . . *Spongiostromata* (Cyanophyceae/bacteria community, from the Precambrian)

a "Colonies" attached to substrate, *dome-shaped* growth forms, more or less distinct laminated structures, no distinct microstructure . . . . *Stromatolites* (Fig. 50)

b "Colonies" free, not attached to substrate, surrounding nucleus, generally irregular laminated structures . . . *Oncolites* (oncoids) (Plate 39/2)

5.3.12.7 Supplementary Reading: Calcareous Algae

Surveys of the present state of knowledge of Ancient algae are found in:

Flügel, E. (ed.) (1977 a): Fossil Algae. Recent results and developments, 375 pp., Berlin – Heidelberg – New York: Springer
37 articles on blue-green algae and stromatolites; Development and classification of dasycladaceans; Red algae; Problems of affinity in Paleozoic algae; Biometric analyses; Ultrastructure of calcareous algae; Algae and sedimentation.

Ginsburg, R., Rezak, R., Wray, J. L. (1972): Geology of Calcareous Algae (Notes for a Short Course). Sedimenta **1,** 78 pp., Comp. Sed. Lab. Marine Geol. Geophys., Univ. Miami
Brief description of the morphology, calcification, reproduction, classification, and ecology of Recent calcareous algae as well as the significance of Ancient calcareous algae as reef and stromatolite builders.

Monty, Cl. (ed.) (1981): Phanerozoic Stromatolites. Case Studies. 249 p., 10 Pls., 121 Figs., Berlin – Heidelberg – New York: Springer
Results from the 2nd International Symposium on Fossil Algae (Paris, 1979). The papers included in this important book describe case histories of Ancient and Recent stromatolites.

Poignant, A. F., Deloffre, R. (eds.) (1979): 2nd International Symposium on Fossil Algae. Bull. Cent. Rech. Explor.-Prod. Elf-Aquitaine **3,** 385–885, Pau
An important collection of 51 papers presented at the 2nd International Symposium on Fossil Algae (Paris, 1979). The papers deal with systematics, paleoecology and phylogeny of calcareous algae with the exception of stromatolites (see Monty, 1981).

Walter, M. R. (ed.) (1976): Stromatolites. Dev. Sed. **20,** 790 pp., Amsterdam – Oxford – New York: Elsevier
Important standard work on Recent and Ancient stromatolites. Methodology and systematics; Abiogenic stromatolite-like structures; Biology; Fabrics and microstructure; Morphogenesis; Biostratigraphy; Environmental interpretation – Recent models; Geological examples; Stromatolites as facies indicators; Mineralization of stromatolites; Geophysical interferences from stromatolite lamination.

Wray, J. L. (1977): Calcareous Algae. Dev. Paleont. Strat. **4,** 185 pp., 170 Figs., Amsterdam – Oxford – New York: Elsevier
Good survey of morphology, classification, geological distribution, and ecology of calcified Cyanophyceae, Rhodophyceae, and Chlorophyceae. Other sections discuss algal mat sediments, environmental factors, sediment producing algae, and examples of algal communities of various ages.

Bibliographies

Johnson, J. H. (1943): Geologic importance of calcareous algae with annotated bibliography. Quart. Colorado School of Mines **38,** Golden

Johnson, J. H. (1957): Bibliography of fossil algae, 1942–1955. Quart. Colorado School of Mines **52,** Golden

Johnson, J. H. (1967): Bibliography of fossil algae, algal limestones, and the geological work of algae, 1956–1965. Quart. Colorado School of Mines **62**/4, 148 pp., Golden

Surveys according to geological time periods (J. H. Johnson)

Cambrian (Quart. Colorado School of Mines, **61**/1, 1966), Ordovician (**56**/2, 1961), Silurian (**54**/1, 1959), Devonian (**53**/2, 1959), Mississippian (**51**/4, 1956), Pennsylvanian and Permian (**58**/3, 1963), Jurassic (**59**/2, 1964), Lower Cretaceous (Prof. Contrib. Colorado School of Mines **6,** 1969).

Larger Surveys

Fott, B. (1971): Algenkunde, 2. Aufl., 581 pp., 303 Figs., Stuttgart: Fischer
Basic overview of taxonomy, systematics and ecology of Recent algae. Good illustrations, numerous references.

Johnson, J. H. (1961): Limestone-Building Algae and Algal Limestones. 297 pp., 139 Pls., Golden (Colorado School of Mines)

Pia, J. (1927): Thallophyta. In: Hirmer, M. (ed.): Handbuch der Paläobotanik **1,** 31–136, München – Berlin: Oldenbourg

Round, F. E. (1975): Biologie der Algen, 2. Aufl., 342 pp., 12 Pls., 77 Figs., Stuttgart: Thieme
Compact description of the morphology, reproduction, cytology, genetics, ecological physiology, economic significance, and taxonomy of Recent algae. Of interest for paleontological studies are: $CaCO_3$ and SiO_2 deposits by algae.

Identification

1. Early Paleozoic

Chuvashov, B. I. (1965): Foraminifery i vodorosli iz verkhnedevonskikh otloshenii zapadnogo sklona srednego i yushnogo Urala. Akad. nauk. SSSR, Uralskii, Trudy inst. geol. **74,** 3–153, Pls. 1–28, Sverdlovsk
Algae from the Frasnian and Famennian.

Copper, P. (1976): The Cyanophyte *Wetheredella* in Ordovician reefs and off-reef sediments. Lethaia **9,** 273–281, 3 Figs., Oslo

Elliott, G. F. (1972): Lower Palaeozoic green algae from Southern Scotland, and their evolutionary significance. Bull. Brit. Mus. Nat. Hist., Geol. **22**/4, 357–376, 10 Pls., London

Emberger, J. (1978): Les Algues (Chlorophyceae, Prasinophyceae, Rhodophyceae) du Devonien. Essai d'un inventaire bibliographique, géographique, stratigraphique. Bull. Inst. Geol. Bassin d'Aquitaine, Num. Spec., 94 p., Talence

Gnilovskaja, M. B. (1972): Izvestkovye vodorosli srednego i pozdnego ordovika vostochnogo Kazachstana. Akad. nauk. SSSR. Inst. geol. geochron. dokembriya, 195 pp., 15 Pls., Leningrad: Nauka
Description of Ordovician green and red algae, new genera and species.

Guilbault, J. P., Mamet, B. L. (1976): Codiacées (Algues) ordoviciennes des Basses-Terres du Saint-Laurant. Can. J. Earth Sci. **13**/5, 636–660, 7 Pls., 7 Figs., Ottawa

Johnson, J. H. (1964): Lower Devonian algae and encrusting foraminifera from New South Wales. J. Paleont. **38**/1, 98–108, Pls. 25–29, 1 Fig., Tulsa
Codiaceae, blue-green algae; description of a few wide-spread species in the Lower Devonian.

Kazmierczak, J. (1975): Colonial Volvocales (Chlorophyta) from the Upper Devonian of Poland and their paleoenvironmental significance. Acta Paleont. Polonica **20**/1, 73–85, Pls. 17–20, Warsaw
Spherical organic remains in lagoonal Devonian shallow-water carbonates, which were once termed "calcispheres", are interpreted as colonies of green algae (Volvocaceae). The biotope is probably characterized by a low salinity. Important for the interpretation of Devonian and Lower Carboniferous limestones with calcispheres!
Korde, K. B. (1973): Vodorosli kembriya. Akad. nauk SSSR, Trudy paleont. inst. **139,** 349 pp., 59 Pls., Moscow: Nauka
Basic work on the morphology, systematics, distribution, ecology, and evolution of Cambrian algae.
Machielse, S. (1972): Devonian algae and their contribution to the western Canadian sedimentary basin. Bull. Can. Petrol. Geol. **20**/2, 187–237, 19 Pls., 8 Figs., Calgary
Interesting description of sediment binding and sediment producing significance of Devonian algae. Certain algae groups and genera are connected with certain sedimentary environments. Poor quality illustrations.
Maslov, V. P. (1956): Iskopaemye izvestkovye vodorosli SSSR. Trudy Inst. Geol. Akad. Nauk SSSR **160,** 361 pp., 86 Pls., 136 Figs., 9 Tab., Moscow (Russian; French translation).
Basic work about Ancient algae from all systems, especially about Paleozoic forms.
Pantić, S. (1973): New species of Devonian algae from Klek, Southeastern Bosnia. Ann. Geol. Péninsule Balkanique **38,** 259–276, 3 Pls., 1 Fig., Beograd
Poncet, J. (1976): Hypothèse relative à la morphologie du thalle de *Renalcis* (algue calcaire-Paléozoique), et affinité possible avec les Rivulariacées actuelles. Geobios **9**/3, 345–351, 5 Figs., Lyon
Shuysky, V. P. (1973): Izvestkovye rifoobrazuyushchie vodorosli nizhnego devona urala. Akad. Nauk SSSR, Uralskii nauchnyi tsentr. institut geologii i geochemii, 155 pp., 34 Pls., Moscow: Nauka
Description of Lower Devonian calcareous algae from reef biotopes in the Ural Mountains.
Wray, J. L. (1967): Upper Devonian calcareous algae from the Canning Basin, Western Australia. Colorado School of Mines, Prof. Contrib. **3,** 76 pp., Golden
Wray, J. L. (1971): Algae in reefs through time. Proc. North American Paleont. Convention, Sept. 1969, Part **J,** 1358–1373, 20 Figs.
Wray, J. L., Playford, P. E. (1970): Some occurrences of Devonian reef-building algae in Alberta. Can. Petrol. Geol. Bull. **18,** 544–555, Calgary

2. *Late Paleozoic*

Emberger, J. (1976): Les Algues (Chlorophyceae, Prasinophyceae, Rhodophyceae) du Carbonifère et du Permien. Essai d'un inventaire bibliographique, géographique, stratigraphique. Bull. Inst. géol. Bassin d'Aquitaine, Num. Spec., 168 pp., 35 Tab., Talance
Güvenç, T. (1974): Repartition stratigraphique des Algues Calcaires du Carbonifère et du Permian de la Tethys. Sept. Congr. Intern. Strat. Géol. Carbonifères, Krefeld 1971, C.R. **3,** 143–159, 6 Figs., Krefeld
Güvenç, T. (1975): Discussion des Genres de Porostromata Pia 1927 et leurs representants dans la Carbonifère et le Permien de Turquie. Congr. Earth Sci., 50th Anniv. Turkish Republic, Papers, 241–256, 2 Pls., 1 Fig., 3 Tab., Ankara
Homann, W. (1972): Unter- und tief-mittelpermische Kalkalgen aus den Rattendorfer Schichten, dem Trogkofel-Kalk und dem Treßdorfer Kalk der Karnischen Alpen (Österreich). Senck. leth. **53,** 3/4, 135–313, 10 Pls., 10 Figs., Frankfurt/M.
Mamet, B., Roux, A. (1975): Dasycladacées devoniennes et carboniferes de la Tethys occidentale. Rev. Espan. Micropal. **7**/2, 245–295, 13 Pls., 5 Figs., Madrid

3. *Triassic*

Bystricky, J. (1964): Slovensky kras. Stratigrafia a Dasycladaceae mezozoika Slovenskeho krasu. Ustred. ustav. geol., 104 pp., 38 Pls., Bratislava (Slowakian with German summary)
Essential work on identification; should be supplemented in a few particulars by the articles of E. Ott.

Flügel, E. (1975): Kalkalgen aus Riffkomplexen der alpin-mediterranen Obertrias. Verh. Geol. Bundesanst. **1974**, 2/3, 297–346, 4 Pls., 11 Figs., Wien
Description of red and green algae from Norian and Rhaetian reef limestones in the Alps, the Aegean, and in Hungary.

Ott, E. (1967): Dasycladaceen (Kalkalgen) aus der nordalpinen Obertrias. Mitt. Bayer. Staatsslg. Paläont. histol. Geol. **7**, 205–226, Pls. 12–13, München

Ott, E. (1972): Die Kalkalgen-Chronologie der alpinen Mitteltrias in Angleichung an die Ammoniten-Chronologie. N. Jb. Geol. Paläont. Abh. **141**/1, 81–115, 2 Figs., 1 Tab., Stuttgart

Pia, J. (1912): Neue Studien über die triadischen Siphoneae verticillatae. Beitr. Paläont. Geol. Österr.-Ung. u. d. Orients **25**, 25–81, Pls. 2–8, Wien

Pia, J. (1920): Die Siphoneae verticillatae vom Karbon bis zur Kreide. Abh. zool. botan. Ges. Wien **11**/2, 1–263, 8 Pls., Wien

4. *Jurassic and Cretaceous*

Emberger, J., Jaffrezo, M. (1975): Compléments à un essai d'inventaire des algues Dasycladacées du Jurassique et du Crétacé inférieur. Bull. Inst. Geol. Bassin d'Aquitaine **18**, 59–132, Talance
Bibliography of all the dasycladaceans described from the Jurassic to the Lower Cretaceous

Bassoullet, J. P., Bernier, P., Conrad, M. A., Deloffre, R., Jaffrezo, M. (1978): Les Algues dasycladales du Jurassique et du Crétacé. Geobios, Mém. spec. **2**, 330 p., 40 Pls., Lyon
Catalogue of all dasycladacean species described from Jurassic and Cretaceous deposits. The catalogue includes the original and a revised description, complete synonym lists, references on the stratigraphic distribution, and an English Abstract which summarizes the state of present knowledge. The book is most important for the identification of dasycladacean algae.

5. *Tertiary*

Adey, W. H., MacIntyre, I. G. (1973): Crustose coralline algae: A re-evaluation in the geological sciences. Geol. Soc. Amer. Bull. **84**, 883–904, 31 Figs., New York

Buchbinder, B. (1975): Lithogenesis of Miocene reef limestones in Israel with particular reference to the significance of the red algae. State of Israel, Min. Comm. Industry, Geol. Surv. Oil Res. Dev. Rep. OD/**3**/75, 170 pp., 4 Pls., Jerusalem

Johnson, J. H., Adey, W. H. (1965): Studies of *Lithophyllum* and related algal genera. Quart. Colorado School of Mines **60**/2, 105 pp., 12 Pls., Golden

Maslov, V. P. (1962): Iskopaemye bagryanye vodorosli SSSR i ikh svyaz a fatsiyami. Akad. Nauk SSSR, Trudy geol. in-ta **53**, 222 pp., 36 Pls., 128 Figs., Moscow (russ.)

Mastrorilli, V. I. (1967): Nuovo contributo allo studio delle Corallinacee dell'Oligocene Ligure-Piemontese: i reperti della Tavetta Ponzone. Atti Ist. Geol. Univ. Genova, 1967 **5**/2, 153–406, 42 Pls., Genua

6. *Algae and Sediment*

There is a great amount of information on this subject, but the book on stromatolites by Walter (1976) deserves special attention; the following articles have been selectively mentioned:

Aitken, J. D. (1967): Classification and environmental significance of cryptalgal limestones and dolomites, with illustrations from the Cambrian and Ordovician of Southwestern Alberta. J. Sed. Petrol. **37**/4, 1163–1178, 22 Figs., Tulsa

Awramik, S. M., Gebelein, C. D., Cloud, P. (1978): Biogeologic relationship of Ancient stromatolites and modern analogs. In: Krumbein, W. E. (ed.): Environmental Biogeochemistry and Geomicrobiology, v. **1**, The Aquatic Environment, 165–178, 1 Fig., Ann Arbor: Ann Arbor Science

Bathurst, R. G. C. (1967): Subtidal gelatinous mat, sand stabilizer and food, Great Bahama Bank. J. Geol. **75**/6, 763–783, Chicago

Cabioch, J. (1974): Un fond de maerl de l'Archipel de Madèra et son peuplement végétal Soc. Phycol. France Bull. **19,** 74–82, 2 Pls., Brest

Gebelein, C. D. (1969): Distribution, morphology, and accretion rate of recent subtidal algal Stromatolites, Bermuda. J. Sed. Petrol. **39**/1, 49–69, 19 Figs., Tulsa

Geldsetzer, H. H. J. (1978): Subtidal Stromatolites: Examples from the Proterozoic and Paleozoic of Canada. Abstract, Tenth Intern. Congr. Sed., 243, Jerusalem

Hofmann, H. J. (1975): Stratiform Precambrian Stromatolites, Belcher Islands, Canada: Relations between silicified microfossils and microstructure. Amer. J. Sci. **275,** 1121–1132, 4 Pls., Washington

Hofmann, H. J. (1977): On Aphebian Stromatolites and Riphean Stromatolite Stratigraphy. Precambrian Res. **5,** 175–205, 20 Figs., Amsterdam

Kerkmann, K. (1966): Über „Oolithe" und „Stromatolithe" und die Beteiligung von Algen an der Kalksteinbildung. Wiss. Z. Hochschule f. Archit. u. Bauwesen, Weimar **13**/3, 293–302, 19 Figs., Weimar

Logan, B. W., Rezak, R., Ginsburg, R. N. (1964): Classification and environmental significance of algal stromatolites. J. Geol. **72**/1, 68–83, 5 Figs., Chicago

Massari, F. (1980): Cryptalgal fabrics in the Rosso Ammonitico sequences of Venetian Alps. Symp. Rosso Ammonitico, Roma, Abstracts, 4 pp., Rome

Monty, C. (1977): Evolving concepts on the nature and the ecological significance of stromatolites. In: Flügel, E. (ed.): Fossil Algae, 15–35, Berlin – Heidelberg – New York: Springer

Paul, J. (1980): Upper Permian algal stromatolite reefs, Harz Mountains (F. R. Germany). Contrib. Sed. **9,** 253–268, 14 Figs., Stuttgart

Peryt, T. M., Piatkowski, T. (1977): Stromatolites from the Zechstein Limestone (Upper Permian) of Poland. In: Flügel, E. (ed.): Fossil Algae, 124–135, Berlin – Heidelberg – New York: Springer

Richter, D. K., Herforth, A., Ott, E. (1979): Pleistozäne, brackische Blaugrünalgenriffe mit Rivularia haematites auf der Perachorahalbinsel bei Korinth (Griechenland). Neues Jahrbuch Geol. Paläont. Abhandlungen **159**/1, 14–40, 9 Figs., Stuttgart

Schröder, J. H. (1972): Calcified filaments of an endolithic alga in recent Bermuda reefs. N. Jb. Geol. Paläont. Mh. **1972**/1, 16–33, 18 Figs., Stuttgart

Walther, J. (1885): Die gesteinsbildenden Kalkalgen des Golfes von Neapel und die Entstehung strukturloser Kalke. Z. dtsch. geol. Ges. **37,** 229–357, Berlin

Wolf, K. H. (1962): The importance of calcareous algae in limestone genesis and sedimentation. N. Jb. Geol. Mh. **1962**/5, 245–261, 1 Tab., Stuttgart

Wray, J. L. (1971): Algae in reef through time. Proc. N. Amer. Conv., Part **J,** 20 Figs., 1358–1373

Wray, J. (1979): Paleoenvironmental constructions using benthic calcareous algae. Bull. Cent. Rech. Explor.-Prod. Elf-Aquitaine **3**/2, 873–879, 6 Figs., Pau

7. Other references cited in Section 5.3.12

Alexandersson, T. (1974): Carbonate cementation in coralline algal nodules in the Skagerrak, North Sea: biochemical precipitation in undersaturated waters. J. Sed. Petrol. **42,** 441–460, Tulsa

Alexandersson, T. (1977): Carbonate cementation in recent coralline algal constructions. In: Flügel, E. (ed.): Fossil Algae, 261–269, 2 Figs., Berlin – Heidelberg – New York: Springer

Bassoullet, J. P., Bernier, P., Deloffre, R., Genot, P., Jaffrezo, M., Poignant, A. F., Segonzac, G. (1977): Classification criteria of fossil Dasycladales. In: Flügel, E. (ed.): Fossil Algae, 154–166, Berlin – Heidelberg – New York: Springer

Böhm, L., Fütterer, D. (1978): Skeletal ultrastructure of the capitular filaments of some Codiaceae. Abstracts, Tenth Int. Congr. Sed., 76–77, Jerusalem

Buchbinder, B. (1977): see 5.3.12.7/5

Conrad, M. A. (1977): The Lower Cretaceous calcareous algae in the area surrounding Geneva (Switzerland): Biostratigraphy and depositional environments. In: Flügel, E. (ed.): Fossil Algae, 295–300, 2 Figs., 3 Tab., Berlin – Heidelberg – New York: Springer

Elliott, G. F. (1968): Permian to Paleocene calcareous algae (Dasycladaceae) of the Middle East. Brit. Mus. (Nat. Hist.) Bull. Geol. Suppl. **4,** 111 pp., 24 Pls., London

Elliott, G. F. (1975): Transported algae as indicators of different marine habitats in the English Middle Jurassic. Paleont. **18**/2, 351–366, Pls. 48–50, London

Fay, M., Dirsche, V. (1975): First evidence of red algal spores in Upper Jurassic carbonate sediments of southern Germany and the Austro-Bavarian Alps. N. Jb. Geol. Paläont. Mh. **1975**/10, 586–592, 9 Figs., 1 Tab., Stuttgart

Flajs, G. (1977): Die Ultrastrukturen des Kalkalgenskelettes. Palaeontogr. (B) **160**/4–6, 69–128, 17 Pls., 16 Figs., 1 Tab., Stuttgart

Flügel, E. (1966): Algen aus dem Perm der Karnischen Alpen. Carinthia II, Sonderheft **25,** 3–76, 11 Pls., 15 Figs., 12 Tab., Klagenfurt

Flügel, E. (1977a): Environmental models for upper Paleozoic benthic calcareous algal communities. In: Flügel, E. (ed.): Fossil Algae, 314–343, 4 Pls., 3 Figs., 8 Tab., Berlin – Heidelberg – New York: Springer

Flügel, E. (1977b): Verkalkungsmuster porostromater Algen aus dem Malm der Südlichen Frankenalb. Geol. Bl. NO-Bayern **27**/3, Erlangen

Flügel, E. (1979): Paleoecology and microfacies of Permian, Triassic and Jurassic algal communities of platform and reef carbonates from the Alps. Bull. Centre Rech. Explor.-Prod. Elf-Aquitaine **3**/2, 569–587, 3 Pls., 5 Figs., Pau

Ginsburg, R. N., Schroeder, J. (1973): Growth and submarine fossilization of algal cup reefs, Bermuda. Sedimentology **20,** 575–614, Amsterdam

Herak, M., Kochansky-Devidé, V., Gusić, I. (1977): The development of the Dasyclad algae through the ages. In: Flügel, E. (ed.): Fossil Algae, 143–153, 2 Figs., Berlin – Heidelberg – New York: Springer

Krumbein, W. E. (ed.) (1979): Cyanobakterien – Bakterien oder Algen? 1. Oldenburger Symp. Cyanobakterien, 1977, 130 pp., Univ. Oldenburg/Germany

Mazzullo, S. J., Friedman, G. M. (1977): Competitive algal colonization of peritidal flats in a schizohaline environment: The Lower Ordovician of New York. J. Sed. Petrol. **47**/1, 398–410, 11 Figs., Tulsa

Poignant, A.-F. (1979): Détermination générique des Corallinacées mésozoiques et cénozoiques. Bull. Centre Rech. Explor.-Prof. Elf-Aquitaine **3**/2, 757–765, 2 Pls., Pau

Riding, R. (1974): *Girvanella* and other algae as depth indicators. Lethaia **8,** 173–179, Oslo

Stanier, R. Y., Cohen-Bazire, G. (1977): Phototrophic procaryotes: The cyanobacteria. Ann. Rev. Microbiol. **31,** 225–274

Valet, G. (1979): Approche paléoécologique du monde des Dasycladales a patir de l'écologie des forms actuelles. Bull. Centre Rech. Explor.-Prof. Elf-Aquitaine **3**/2, 859–866, Pau

Wood, A. (1941): "Algal dust" and the finer-grained varieties of Carboniferous limestones. Geol. Mag. **78**/3, 192–200, London

Wray, J. (1972): Environmental distribution of calcareous algae in Upper Devonian reef complexes. Geol. Rdsch. **61,** 578–584, Stuttgart

5.3.13 Microproblematica

In many thin-sections of Paleozoic and Mesozoic limestones microfossils can be seen which are difficult to assign to a major systematic group. These include:

a) *Paleozoic:* *Calcisphaeres*
Umbellina
b) *Mesozoic:* *Calcisphaeres*
"Bankia" striata (Carozzi)
Globochaete
Filaments
Microtubus communis Flügel *and other Triassic microproblematica*
Tubiphytes Maslov

A. Paleozoic Calcisphaeres (Plates 17/2; 37/2)

Hollow spheres with walls exhibiting various microstructural types which may be partially caused by diagenetic effects. Diameters smaller than 0.10–0.50 mm.

First described by Williamson (1879) from the Lower Carboniferous of England and from the Devonian of North America, Paleozoic calcispheres today are known throughout the world, particularly in the Upper Devonian of Belgium (Lombard and Monteyne, 1952), Canada (Konishi, 1958), U.S.A. (Stanton, 1967), and Mexico (Brunner, 1975). Calcispheres are widely distributed in Devonian shelf carbonates of the USSR (Chuvashov, 1965; Reitlinger, 1957) and in Tournaisian and Viséan limestones of the Waulsortian facies in Western Europe (Dervillé 1950, 1960; Conil and Lys, 1964).

Systematic position: Calcisphaeres have been regarded as algae (Gyrogonites of Characeae; dasycladacean spores, e.g., Rupp, 1967; Marszalek, 1975) green algal colonies (e.g., *Eovolvox* Kazmierczak, 1975) and as foraminifera (e.g., Conil and Lys, 1964). Many Russian authors have described single-chambered "foraminifera" (e.g., *Cribrosphaeroides, Archaesphaera, Proninella*) which are better included within algal calcisphaeres (see Toomey and Mamet, 1979). These and similar forms are common in restricted lagoonal environments of Devonian reef-complexes; see Flügel and Hötzl, 1972.

B. Umbellinaceans

The umbellinaceans include small flasklike or globular microfossils of Devonian and Mississippian age which are widely distributed in Europe, Asia, Australia and North America, but whose relationships are obscure. They have been attributed to the foraminifera or to the algae (charophytes); many authors consider the umbellinaceans to be either primitive charophyte gyrogonia, or perhaps utricles. The diameter of the tests is about 400–450 microns.

The group is restricted to the littoral facies, probably to areas of decreased salinity; umbellinaceans are rare in strata containing foraminifers.

C. *Mesozoic Calcisphaeres* (Plates 8/5; 20/6, 7)

Hollow spheres with or without aperture(s) and differentiated walls. Diameter in thin-sections between about 20 and 150 microns. Described in 1865 by Kaufmann from the Turonian Seewen Limestone of Switzerland ("Oligostegina"), the abundant spheres were named *Pithonella* by Lorenz (1902). Several genera have been proposed according to differences in the shape of the calcispheres and in the wall structures (e.g., *Fibrosphaera* Colom, 1935; *Stomiosphaera* Wanner, 1940; *Cadosina,* Wanner, 1940; *Calcisphaerula,* Bonet, 1956). These genera are abundant in Mesozoic pelagic limestones. Three families (Cadosinidae, Stomiosphaeridae, Calcisphaerulidae) were distinguished using the structures seen in thin-sections (Borza, 1972).

SEM studies indicate that many "species" and "genera" separated by thin-section criteria represent diagenetic forms only, and that the Cretaceous calcisphaeres exhibit an extremely high morphological variability (Keupp, 1981). It is doubtful that the existence of "double-walled" or "single-walled" microstructures, or of "pores" is of any use for systematic classifications. With the present state of knowledge calcisphaeres described by thin-section investigations and calcisphaeres studied by SEM can not be compared.

Systematic position: Mesozoic calcisphaeres have been regarded as foraminifera or as algae. Many single-layered spheres (type *Pithonella* Lorenz) represent calcareous dinoflagellate cysts (Keupp, 1978; 1981).

Paleoecology: Upper Jurassic and Cretaceous calcisphaeres are abundant in pelagic carbonates, in association with coccoliths, as well as in shelf carbonates (e.g., Adams et al., 1967; Locker, 1967; Voigt and Häntzschel, 1964; Borza, 1964, 1972). They have been used by some authors as bathymetric indicators for shelf-margin to slope environments. Rock-forming occurrences are known from Liassic pelagic limestones (e.g., *Schizoporella*), from the Upper Jurassic, and from various stages of the Cretaceous (e.g., Pläner of Westphalia, Germany).

D. *"Bankia" striata (Carozzi)* (Plate 30/6)

Calcitic, V-shaped sections nested in each other with an extended outer edge.

Described in 1927 by Favre and Richard in the Purbeck of the Jura Mountains as "organisme C" (interpreted as pteropods). In 1954 Carozzi proposed the genus *Vaginella* together with the species *V. striata* (interpreted as pteropods). Radoičić described them as "aberrant tintinnids" in the Valangian of Yugoslavia in 1959, and later in more detail in 1966. In 1963 Farinacci interpreted the species *"striata"* as pallets of teredinid boring clams and recognized the stratigraphic value of this species. Barthel (1960) and Gollestaneh (1975) classified the species under dasycladaceans. The latter author called the new genus *Carozziella,* and Bernier (1975) called it *Campbelliella.*

Distribution: Abundant in shallow water limestones of the Upper Jurassic and the Lower Cretaceous of the Alpine Mediterranean region.

Significance: Index fossil for the "cenozone with *Bankia striata"*; (Upper Jurassic – Upper Jurassic/Cretaceous transition).

E. Globochaete (Plate 16/1)

In 1937 Lombard described scattered spherical and deeply indented kidney-shaped organic remains, about 0.05 mm in size, from Upper Jurassic limestones of the Swiss Jura and the Swiss Alps (*Globochaete alpina* after Lombard, 1945); they may also be serially arranged (*Eothrix alpina* after Lombard, 1945). According to Lombard these were planktonic single-celled green algae. These forms are known today from the Anisian to the Cretaceous, possibly to the Eocene, with a maximum in the Lias and in the Upper Jurassic. Globochaetes occur particularly in micritic deeper water limestones, with free-swimming crinoids and calpionellids (in the Upper Jurassic), and with filaments, radiolaria, echinoderms, and thin-shelled mollusks and ostracods, but also with reef organisms, in the Triassic and in the Lias.

F. "Filaments" (Plates 15/6; 16/1; 25/5)

Occur in mass accumulations; about 0.10 mm thick, curved calcitic elements up to about 5 mm long. Usually in micritic limestones („Microlumachelle limestones").

At first regarded as filaments of Cyanophyceae (Cuvillier, 1952, 1959) and named *Paleotrix* or *Axothrix* (Ferasin, 1956; Nagy, 1971), today their interpretation as remains of thin-shelled pelecypods is commonly accepted (Peyre, 1959; Kuhry, 1975). Their very large number may possibly be interpreted as prodissoconches of planktonic drifting larvae. According to Kuhry (1975) the stratigraphic range of the typical thin-shelled filaments coincides with that of the pelecypod *Bositra buchi* (Toarcien to Oxfordian). Stratigraphically older filaments may possibly be traced back to *Halobia,* etc.; some Upper Jurassic and Cretaceous filaments may be atypical sections of *Saccocoma* or ostracods.

Significance: The filament facies, which is particularly wide-spread in the Middle Jurassic, is often overlain by a *Saccocoma* facies (see Kuhry, 1975) or by radiolarian siliceous limestones. On the whole, filaments are known from the Lower Middle Triassic (Anisian) to the Cretaceous, but they are most abundant in the basin facies of the Alpine Triassic (e.g., Reifling limestones, Hallstatt limestones), in the red limestones of the Alpine Lias, in the Middle Jurassic, and in the lower and middle part of the Upper Jurassic. Sedimentation occurred in subtidal to bathyal environments.

G. Microtubus communis E. Flügel (Plate 18/8) *and other Microproblematica in the Alpine Triassic*

Very often tubiform remains of organisms (about 0.10 mm in diameter) occur in the central reef facies of Upper Triassic reef limestones in association with calcareous sponges, corals, and other frame builders; these tubiform elements may be interpreted as worm tubes, foraminifera or blue-green algae (E. Flügel, 1964; Senowbari-Daryan, 1980). They are characteristic facies fossils.

Other microproblematica in the Alpine Triassic which are sometimes useful as facies or index fossils have been described by E. Flügel (1972) and Borza (1975). These are, among others, hollow tubes (*Microtubus* as well as *Lamellitubus* Ott and *Aeolisaccus* Elliott); globular elements (calcisphaeres, *Ladinosphaera* Oberhauser); irregular globular aggregates (*Globochaete, Baccanella* Pantić); crusts with vertically perforated layers (problematicum A in Ohlen, 1959); cell associations (*Thaumatoporella* Raineri – Plate 30/4, *Bacinella* Radoičić, *Pycnoporidium? eomesozoicum* Flügel); serial associations of basin-shaped segments nested within each other (*Cheilosporites* Wähner, Plate 18/7; 48/1). Mesozoic tunicate spicula have been described by Mišik and Borza (1978).

H. Tubiphytes Maslov (Plates 42/5; 49/1, 2; 50/3)

These encrusted micritic skeletal grains with subrounded to lobular outlines possess a very fine web-like structure and one or more sparry filled tubes. Milky white in direct light, dark in transmitted light.

Distribution: Lower Carboniferous to Jurassic. Very abundant in Permian reefs (e.g., Trogkofel limestone of the Carnic Alps, Austria) and in Middle Triassic Wetterstein limestone reefs of the Alps. Similar forms also occur in biostromal limestones of the Franconian Alb (Upper Jurassic, Germany; Flügel, 1981).

Interpretation: Described as a blue-green alga; sometimes regarded as a red alga or as a calcareous sponge, but it is absolutely out of the question to call it a hydrozoan.

References: Microproblematica

Calcispheres: Paleozoic

Birina, L. M. (1948): Novye vidy izvestkovykh vodoroslei i foraminifer pogranichnykh sloev devona i karbona. Sovetskaya geol. **28,** 154–159, 2 Pls., Moscow

Brunner, P. (1975): Estudio estratigrafico del Devonico en el area de El Bisani, Carborca, Sonora. Rev. Inst. Mexicano Petroleo **7**/1, 16–45, 6 Pls., Mexico

Chuvashov, B. I. (1965): Foraminiferi i vodorosli verkhnedevonskikh otlozhenii zapadnogo sklona srednego i iushnogo Urala. Akad. Nauk SSSR Uralski filial. Trudy inst. geol. **74,** 153 pp., 29 Pls., 3 Figs., Sverdlovsk

Conil, R., Lys, M. (1964): Materieux pour l'étude micropaléontologique du Dinantien de la Belgique et de la France (Avesnois). Parties 1–2. Algues et Foraminifères. Mem. Inst. Géol. Univ. Louvain **23,** 1–296, Pls. 1–42, Louvain

Dervillé, H. (1950): Contribution à l'étude des Calcisphères du Calcaire de Bachant. Ann. Soc. géol. Nord **70,** 273–285, Pl. 14, Lille

Dervillé, R. P. H. (1960): Contribution à l'étude des calcisphères végétales. Bull. Serv. Carte géol., Als. Lorr. **13**/4, 169–177, 2 Pls., Strasbourg

Flügel, E., Hötzl, H. (1971): Foraminiferen, Calcisphaeren und Kalkalgen aus dem Schwelmer Kalk (Givet) von Letmathe im Sauerland. N. Jb. Geol. Paläont. Abh. **137,** 358–395, 6 Pls., Stuttgart

Kazmierczak, J. (1975): Colonial Volvocales (Chlorophyta) from the Upper Devonian of Poland and their paleogeographical significance. Acta Palaeont. Polonica **29**/1, 73–85, Pls. 17–20, Warsaw

Konishi, K. (1958): Devonian calcareous algae from Alberta, Canada. Quart. Colorado School of Mines **53**/2, 85–109, 4 Pls., Golden

Lombard, A., Monteyne, R. (1952): Calcisphères dans le Frasnien de Bois de Villers (Namur). Bull. Soc. Belg. Géol. Paléont. Hydrol. **61,** 13–25, Bruxelles
Marszalek, D. S. (1975): Calcisphere ultrastructure and skeletal aragonite from the alga *Acetabularia antillana.* J. Sed. Petrol. **45,** 266–271, Tulsa
Reitlinger, E. A. (1957): Sfery devonskikh otlozhenii russkoi platformy. Doklady Akad. nauk SSSR **115**/4, 774–776, 1 Pl., Moscow
Rupp, A. W. (1967): Origin, structure and environmental significance of recent and fossil calcispheres (abs.). Geol. Soc. Amer. Spec. Paper **101,** 186
Stanton, R. H. Jr. (1967): Radiosphaerid calcispheres in North America and remarks on calcisphere classification. Micropaleont. **13**/4, 465–472, Pl. 1, 1 Tab., New York

Umbellinaceans

Bykova, E. V., Polenova, E. N. (1955): Foraminiferi, radiolarii i ostrakody devona Volgo-Uralskoi oblast., VNIGRI, Trudy, n.s. **87,** 141 pp., 24 Pls., Moscow
Mamet, B. (1970): Sur les Umbellaceae. Can. J. Earth Sci. **7**/4, 1164–1171, 1 Pl., 1 Fig., Ottawa
Peck, R. E. (1974): On the systematic position of the umbellids. J. Paleont. **48,** 409–414, 1 Fig., Tulsa

Calcisphaeres: Mesozoic

Adams, T. D., Khalilli, M., Khosrovi Said, A. (1967): Stratigraphic significance of some oligosteginid assemblages from Lurestan Province, northwest Iran. Micropal. **13**/1, 55–67, Pl. 1, 4 Figs., New York
Bein, A., Reiss, Z. (1976): Cretaceous *Pithonella* from Israel. Micropal. **22**/1, 83–91, 3 Pls., New York
Bolli, H. (1974): Jurassic and Cretaceous Calcisphaerulidae from DSDP, leg 27, Eastern Indian Ocean. In: Veevers, J. J., et al. (eds.): Initial Rep. Deep-Sea Drilling Project **27,** 843–907, 24 Pls., 5 Figs., Washington
Bolli, H. M. (1978 a): Cretaceous and Paleogene Calcisphaerulidae from DSDP Leg 40, Southeastern Atlantic. Initial Rep. Deep Sea Drilling Project **40,** 819–837, 7 Pls., Washington
Bolli, H. M. (1978 b): Upper Jurassic Calcisphaerulidae from DSDP Leg 44, Hole 391 C, Blake-Bahama Basin, Western North Atlantic. Initial Rep. Deep Sea Drilling Project **44,** 911–919, 3 Pls., Washington
Bolli, H. M. (1980): Calcisphaerulidae and Calpionellidae from the Upper Jurassic and Lower Cretaceous of Deep Sea Drilling Project Hole 416 A, Moroccan Basin. Initial Rep. Deep Sea Drilling Project **50,** 525–543, Pls 1–7, Washington
Bonet, F., Trejo, M. (1956): Nuevos datos sobre la familia Calcisphaerulidae (Protozoa). Ann. Esc. Nac. Ciencias Biol. **9**/1, 43–48, Mexico
Borza, K. (1964): Die Gattung *Stomiosphaera* Wanner, 1940, in den Westkarpaten. Geol. Sbornik **15**/2, 189–195, Pls. 1–2, Bratislava
Borza, K. (1972): Neue Arten der Gattungen *Cadosina* Wanner, *Pithonella* Lorenz und *Palinosphaera* Reinsch aus der oberen Kreide. Geol. Zborn. Slov. Akad. Vied. **23**/1, 139–150, Bratislava
Fütterer, D. (1976): Kalkige Dinoflagellaten ("Calciodinelloideae") und die systematische Stellung der Thoracosphaeroideae. Neues Jahrbuch Geologie Paläont. Abhandlungen **151**/2, 119–141, 24 Figs., Stuttgart
Jafar, S. (1979): Taxonomy, stratigraphy and affinities of calcareous nannoplankton genus *Thoracosphaera* Kamptner. Proc. IV. Internat. Palynol. Conf., Lucknow, v. **2,** 1–21, Pls. 1–6, Lucknow
Kaufmann, F. J. (1865): Die Polythalamien des Seewerkalkes, des Gault und des Schrattenkalkes. In: Heer, O.: Die Urwelt der Schweiz, 194–203, Figs. 104–115, Zürich: Schulthess
Keupp, H. (1977): see 4.1.7.1

Keupp, H. (1978): Calcisphaeren des Untertithons der südlichen Frankenalb und die systematische Stellung von *Pithonella* Lorenz 1901. N. Jb. Geol. Paläont. Mh. **1978,** 87–98, 14 Figs., Stuttgart
Keupp, H. (1981): Die kalkigen Dinoflagellaten-Zysten der borealen Unter-Kreide (Unter-Hauterivium bis Unter-Albium). Facies **5,** 1–190, 53 Plates, Erlangen
Locker, S. (1967): Die Sphaeren der Oberkreide und die sogenannten Orbulinaritfazies. Geologie **16,** 850–859, 2 Pls., 2 Figs., Berlin
Masters, B. A., Scott, R. W. (1978): Microstructure, affinities and systematics of Cretaceous calcispheres. Micropaleont. **24**/2, 210–211, Pls. 1–2, New York
Orlsson, R. K., Youssefnia, I. (1979): Cretaceous Calcisphaerulidae from New Jersey. J. Paleont. **53**/5, 1085–1093, 2 Pls., 2 Figs., Tulsa
Pflaumann, U., Krashennikov, V. A. (1978): Cretaceous Calcisphaerulids from DSDP Leg 42, Eastern North Atlantic. Initial Report Deep Sea Drilling Project **41,** 817–839, Pls. 1–8, Washington
Villain, J.-M. (1975): "Calcisphaerulidae" (incertae sedis) du Crétacé supérieur du Limbourg (Pays-bas), et d'autres regions. Palaeontograph., A **149,** 193–242, 9 Pls., Stuttgart
Villain, J.-M. (1977): Les Calcispherulidae: Architectures, Calcification de la paroi et Phylogenese. Palaeontographica, A. **159,** 139–177, Pls. 42–47, 11 Figs., Stuttgart
Vogler, J. (1941): Ober-Jura und Kreide von Misol (Niederländisch-Ostindien). Palaeontographica, Suppl. **4**/4, 243–293, Pls. 19–24, Stuttgart
Voigt, E., Häntzschel, W. (1964): Gradierte Schichtung in der Oberkreide Westfalens. Fortschr. Geol. Rheinland Westfalen **7,** 495–548, 18 Pls., Krefeld
Wall, D., Dale, B. (1968): Quaternary calcareous dinoflagellates (Calciodinellideae) and their natural affinities. J. Paleont. **42**/6, 1395–1408, Pl. 172, 3 Figs., Tulsa
Wanner, J. (1940): Gesteinsbildende Foraminiferen aus dem Malm und der Unterkreide des östlichen Ostindischen Archipels. Paläont. Z. **22,** 75–99, Pls. 1–2, 37 Figs., Berlin

"Bankia" striata

Barthel, K. W. (1969): Die obertithonische regressive Flachwasser-Phase der Neuburger Bankkalke in Bayern. Bayer. Akad. Wiss., math.-naturwiss. Kl., Abh., N.F. **142,** 174 pp., 14 Pls., 33 Figs., München
Bernier, P. (1974): *Campbelliella striata* (Carozzi): Algue dasycladacée? Une nouvelle interprétation de l'"Organisme C" Favre et Richard, 1927. Geobios **7**/2, 155–175, Pls. 32–34, Lyon (cum lit!).
Farinacci, A. (1963): L'"organismo C" Favre 1927 appartiene alle Teredinidae? Geol. Romana **2,** 151–178, 5 Pls., 6 Figs., Rome
Favre, J., Richard, A. (1927): Etude du Jurassique supérieur de Pierre-Chatel et de la Cluse de la Balme (Jura méridional). Mém. Soc. Paléont. Suisse **46,** 31–39, 3 Pls., 14 Figs., Genf
Gollesstaneh, A. (1965): A micropaleontological study of the Upper Jurassic and Lower Cretaceous of Southern Iran. Thesis Doct. Phil. Univ. London, 629 pp., 240 Pls., 23 Figs., London
Radoičić, R. (1969): Aberantne grana fossilnih Tintinnina (Podred Tintinnina). Paleont. Yugoslavica **9,** 71 pp., 31 Figs., 8 Pls., Zagreb

Globochaete

Lombard, A. (1937): see 5.3.11
Radwánski, A. (1968): Studium petrograficzne i seymentologiczne retyku wierchowegi Tatr. Zakld naukgeol. Polskiej. Akad. nauk. Studie Geol. Polonica **25,** 146 pp., 54 Pls., Warsaw
Schäfer, P., Senowbari-Daryan, B. (1980): Globochaeten-Zoosporen aus obertriadischen Riffkalken südlich von Salzburg (Nördliche Kalkalpen). Verhandlungen geol. Bundesanstalt Wien **1980,** 97–103, 1 Pl., 1 Fig., Wien
Tellez-Giron, B., Trejo, M. (1974): Nuevos datos sobre *Globochaete alpina*. Rev. Inst. Mex. Petrol. **6**/1, 98–103, 2 Pls., Mexico

Zawidzka, K. (1972): *Globochaete alpina* Lombard in the Lower Muschelkalk of Lower Silesia. Acta Geol. Polon. **22,** 467–472, 2 Pls., 2 Figs., Warsaw

Filaments

Cuvillier, J. (1952): see p. 10
Cuvillier, J. (1969): Calcaires à filamentes. Proc. 1st Intern. Conf. Planctonic microfoss. Geneva **1976**/2, 123–125, Leiden: Brill
Ferasin, F. (1956): Sulla presenza di supposte alghe filamentose nel giurese del Veneto. Riv. Ital. Paleont. Strat. **62,** 198–204, Pls. 17–18, 1 Fig., Milano
Kuhry, B. (1975): Observations on filaments from the Subbetic of SE Spain. Rev. Espan. Micropaleont. **7**/2, 291–343, Madrid
Mišik, M., Borza, K. (1978): *Gemeridella, Didemnoides, Didemum* und Körperchen ähnlicher Gestalt aus dem Mesozoikum der Westkarpaten. Geol. Zborn. Geol. Carpathica **29**/2, 307–326, 5 Pls. 3 Figs., Bratislava
Nagy, I. (1971): Le probleme de *Paleotrix.* Le rôle des algues filiformes dans le Jurassique supérieur de la Montagne de Mecsek. Magyar all. Földt, Inteze, Evi Jelentese **1971,** 299–326, Budapest
Peyre, Y. (1959): Études sur l'organisme du Jurassique présentant l'aspect du filaments. Rev. Micropaléont. **2**/2, 80–87, 2 Pls., 2 Figs., Paris

Microtubus communis, Triassic Microproblematica, Tubiphytes

Borza, K. (1975): Mikroproblematika aus der oberen Trias der Westkarpaten. Geol. Zbornik, Geol. Carpathica **26**/2, 199–236, 8 Pls., Bratislava
Flügel, E. (1964): Mikroproblematika aus den rhätischen Riffkalken der Nordalpen. Paläont. Z. **38,** 74–87, Pls. 8–9, 1 Fig., Stuttgart
Flügel, E. (1972): Mikroproblematika in Dünnschliffen von Trias-Kalken. Mitt. Ges. Geol. Bergbaustud. **21,** 957–988, 2 Figs., 2 Tab., Innsbruck
Flügel, E. (1981): „Tubiphyten" aus dem fränkischen Malm. Geol. Blätter NO-Bayern **31,** 126–142, 10 Figs., Erlangen
Senowbari-Daryan, B. (1980): See 10.3.9

5.4 Microfacies Zones (Plate 30)

Thin-section studies of Mesozoic carbonates in the Mediterranean, in the Carpathians, and in the Alps have led in recent years to the idea of "cenozones" (biostratigraphically related assemblage zones) in Jurassic and Cretaceous carbonate sequences. The basis for these are the empirically determined ranges of individual species of organisms; however, consideration has also been given to the peak zones of the individual species or associations (e.g., Colom, 1955) and "mass abundances" dependent on ecological factors. Thus, ecological factors have been introduced into stratigraphy.

Zonations of shallow water facies are based particularly on calcareous algae (dasycladaceans) and foraminifera as well as on a few hydrozoans and microproblematica. Because the organisms often exhibit a strong facies control, stratigraphic zonation can often only be applied to similar facies types. Pelagic limestones are zoned with the aid of planktonic organisms (calpionellids, free-swimming crinoids, calcisphaeres, globochaetes, foraminifera).

Table 33. Cenozones in the Jurassic and in the Lower Cretaceous of the Southern Apennines, Italy

		Sartoni and Crescenti (1962)	Farinacci and Radoičić (1964)		de Castro (1962)	Crescenti (1969)
Cretaceous	Upper	*Cuneolina pavonia parva* and *Dicyclina schlumbergeri*	Rudist biostromes			*Cuneolina pavonia parva* and *Dicyclina schlumbergeri*
	Middle	*Cuneolina pavonia parva*	*Cuneolina pavonia parva* and *Nummoloculina heimi*			*Cuneolina pavonia parva*
	Lower	*Cuneolina camposaurii*	Orbitolinids + *Salpingoporella dinarica*			*Cuneolina camposaurii*
			Salpingoporella annulata			
Jurassic	Upper Jurassic	*Salpingoporella appeninica*				*Salpingoporella appeninica*
		Clypeina jurassica and *"Vaginella" striata*	*Clypeina jurassica* and *"Bankia" striata*		*"Vaginella" striata*	*Clypeina jurassica* and *"Vaginella" striata*
					Clypeina jurassica	
		Kurnubia palastiniensis	*Cladocoropsis mirabilis*		(complex with chert layers)	*Kurnubia palastiniensis*
		Macroporella sellii			*Cladocoropsis mirabilis*	
	Middle Jurassic	*Pfenderina salernitana*	Trochaminids, Verneulinids and Textularians	Pfenderinids	*Pfenderina neocomiensis*	*Pfenderina salernitana*
		Thaumatoporella parvovesiculifera		*Selliporella donzelli*	Complex with oolitic levels	*Thaumatoporella parvovesiculifera*
				Trochaminids, Verneulids and Textularians		
			Atypical area			
	Lias	*Palaeodasycladus mediterraneus*	*Palaeodasycladus mediterraneus*		*Palaeodasycladus mediterraneus*	*Palaeodasycladus mediterraneus*
		Orbitopsella praecursor				
		Atypical zone				

Table 34. Microfacies zones in various Jurassic sedimentary environments of the Central Apennines, Italy (after Crescenti, 1971)

<table>
<tr><th rowspan="2"></th><th colspan="4">Depositional Environments</th></tr>
<tr><th>Shelf-Lagoon</th><th colspan="2">Shelf-Margin</th><th>Basin (Pelagic Facies)</th></tr>
<tr><td rowspan="4">Upper Jurassic</td><td>Salpingoporella apenninica</td><td colspan="2">Tubiphytes morronensis</td><td>Calpionella alpina</td></tr>
<tr><td rowspan="2">Clypeina jurassica and “Vaginella” striata</td><td rowspan="5">Protopeneroplis striata</td><td rowspan="2">Pianella pygmaea</td><td>Saccocoma</td></tr>
<tr><td rowspan="5">Globochaete alpina</td></tr>
<tr><td>Kurnubia palastiniensis</td><td>Labyrinthina mirabilis</td></tr>
<tr><td rowspan="2">Middle Jurassic</td><td>Pfenderina salernitana</td><td rowspan="2"></td></tr>
<tr><td>Thaumatoporella parvovesiculifera</td></tr>
<tr><td>Lias</td><td>Palaeodasycladus mediterraneus</td><td colspan="2">“Dictyoconus” cayeuxi</td></tr>
</table>

Plate 30. Index Species for Microfacies Zonation in the Alpine-Mediterranean Jurassic ►

1. *Pianella pygmaea* (Gümbel). Upper Jurassic: Krahstein, Salzkammergut, Styria, Austria. Abundant in the Late Upper Jurassic

2. *Salpingoporella annulata* (Carozzi): Krahstein, Salzkammergut, Styria, Austria

3. *Clypeina jurassica* Favre. Upper Jurassic (Plassen limestone): Röthelstein, Salzkammergut, Styria, Austria

4. *Thaumatoporella parvovesiculifera* (Raineri). Upper Jurassic (Plassen limestone): Krahstein, Salzkammergut, Styria, Austria. This problematic species, referred by some authors to red algae, is known in the interval between the Anisian and the Upper Cretaceous. Relatively abundant in the Lias and Upper Jurassic

5. *Protopeneroplis striata* Weynschenk. Upper Jurassic (Plassen limestone): Krahstein, Salzkammergut, Styria, Austria. Abundant in the Middle Jurassic and the Early Upper Jurassic

6. *"Bankia" striata* (Carozzi), enclosed in a micritic oncoid. This species, whose systematic position is controversial (see Sect. 5.3.13), occurs preferentially in the Upper Jurassic and in the Lowermost Cretaceous. Late Upper Jurassic: Central Apennines, Italy

7. *Labyrinthina mirabilis* Weynschenk. Upper Jurassic (Plassen limestone): Krahstein, Salzkammergut, Styria, Austria

8. *Kurnubia palastinensis* Henson. Early Upper Jurassic: Central Apennines, Italy

9. *Orbitopsella praecursor* (Gümbel). Middle Lias: Central Apennines, Italy

Scale for all Figures 0.5 mm

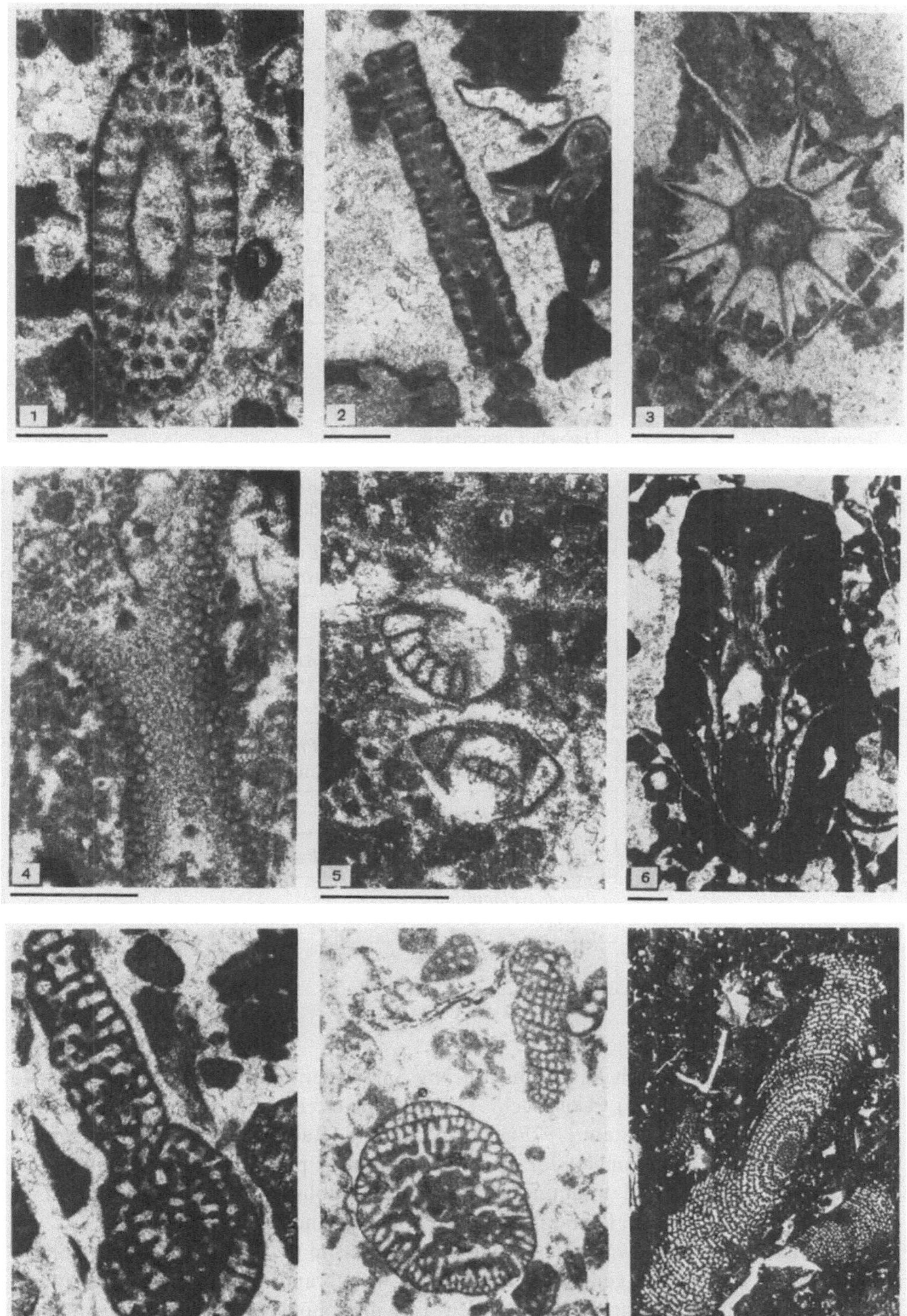

Table 35. Biozonation in the Jurassic and Lower Cretaceous of the Mesogean area, based on foraminifera and algae (after Jaffrezo, 1980)

Stages	Biozones	Subzones
Gargasian Bedoulian	*Mesorbitolina texana*	*Palorbitolina*
	Palorbitolina lenticularis	*Praeorbitolina*
Barremian		Orbitolinids
	Orbitolinids and dasyclads	Orbitolinids abundant
Hauterivian		Algae abundant
Valanginian	*Eclusia moutyi*	
	Valdanchella miliani	
Berriasian	Dasyclad association	
Portlandian	*Clypeina jurassica*	*Anchispirocyclina lusitanica*
Kimmeridgian		*Everticyclammina virguliana*
	Kurnubia palastiniensis	*Alveosepta jaccardi*
Oxfordian		
		Pfenderina salernitana and *Praecurnubia crusei*
Callovian		
	Pfenderina salernitana and *Praekurnubia crusei*	
Bathonian		*Pseudocyclammina maynci*
Bajocian	*Pseudocyclammina maynci*	
Aalenian		

Fundamental research has been done by Sartoni and Crescenti (1961), Crescenti (1966, 1969, 1971), Colacicchi (1967), Colacicchi and Praturlon (1965), Radoičić (1966), Dragastan (1975), Farinacci and Radoičić (1964) and Borza (1969). Important zonations of Jurassic and Cretaceous platform carbonates based on calcareous algae and foraminifera have been published by Chiocchini et al. (1979; Lazio, Apennines), Peybernes (1979; Pyrenées), and Jaffrezo (1980).

Tables 33 and 34 show a few microfacies zonations in the Mediterranean region. Because the facies belts generally change their stratigraphic position in time, the microfacies zones can hardly be regarded as synchronous units. A few common index species are listed in Plate 30. A comprehensive review of all the microfacies biozonations developed for Jurassic and Lower Cretaceous carbonate sequences is found in Jaffrezo (1980), together with a new zonation for the Mesogean area, based on benthic foraminifera and dasyclad algae (see Table 35).

References: Microfacies Zones

Borza, K. (1969): see 5.5

Canerot, J. (1979): Les algues et leur environment dans le Malm et le Crétacé inférieur des Chaines ibérique et Catalane (Espagne). Bull. Centre Rech. Explor.-Prod. Elf-Aquitaine, **3**/2, 505–518, 4 Figs., Pau

Castro, P. de (1962): Il Giura-Lias dei Monti Lattari e dei rilievi ad ovest della valle dell' Irno e della Piana di Montoro. Boll. Soc. Naturalisti Napoli **71,** 73 pp., 19 Pls., 5 Figs., Napoli

Chiocchini, M., Mancinelli, A. (1977): Microbiostratigrafia del Mesozoico in facies di piattaforma carbonatica dei Monti Aurunci (Lazio centre-meridionale). Studi Geol. Camerti **3,** 109–152, 48 Pls., 7 Figs., Perugia

Chiocchini, M., Mancinelli, A., Molinari-Paganelli, V., Tilia-Zuccari, A. (1979): Répartition stratigraphique des algues dasycladales et codiacées dans les successions mésozoiques de plate-forme carbonatée du Lazio centre-méridional (Italie). Bull. Centre Rech. Explor.-Prod. Elf-Aquitaine **3**/2, 525–535, 2 Pls., 1 Fig., Pau

Colacicchi, R. (1967): Geologia della Marsica orientale. Geol. Romana **6,** 189–316, 72 Figs., 2 Pls., Rome

Colacicchi, R., Praturlon, A. (1965): Stratigraphical and palaeogeographical investigations on the Mesozoic shelf-edge facies in Eastern Marsica (Central Apennines, Italy). Geol. Romana **4,** 89–118, 11 Figs., 3 Pls., Rome

Colom, G. (1955): Jurassic-Cretaceous pelagic sediments of the western Mediterranean zone and Atlantic area. Micropaleont. **1,** 109–119, New York

Crescenti, U. (1966): Osservazioni sulla stratigrafia dell'Apennino meridionale alla luce delle recenti ricerche micropaleontologiche. Boll. Soc. Geol. Ital. **85,** 541–579, 3 Figs., Rome

Crescenti, U. (1969): Biostratigrafia delle facies mesozoiche dell'Appennino centrale: Correlazioni. Geol. Romana **8,** 15–40, 22 Figs., 1 Pl., Rome

Crescenti, U. (1971): Biostratigraphic correlations in the Jurassic facies of central Italy by means of the microfossils. Ann. Inst. Geol. Publ. Hungarici **54**/2, 209–213, 2 Figs., Budapest

Dragastan, O. (1975): see 5.5

Dragastan, O. (1981): Mesozoic Dasycladaceae from Romania: Distribution and Biostratigraphic Importance. Facies **4,** 165–196, Pl. 9, 8 Figs., Erlangen

Farinacci, A., Radoičić, R. (1964): Correlazione fra serie gioresi e cretacee dell'Apennino Centrale e delle Dinaridi esterne. Rendiconti **7**/2, Ser. 2, 269–300, Pls., Rome

Jaffrezo, M. (1980): Les formations carbonatées des Corbières (France) du Dogger a l'Aptien: Micropaléontologie stratigraphique, biozonation, paléoecologie. Extension des resultats a la Mésogée. Thèse Doctorat d'Etat, Univ. Pierre et Marie Curie, Paris VI, 613 p., many plates, Paris

Peybernes, B. (1979): Les Algues jurassique et du Crétacé inférieur des Pyrenées franco-espagnoles. Intéret biostratigraphique et paléoecologique. Bull. Centre Rech. Explor.-Prod. Elf-Aquitaine **3**/2, 733–741, 3 Figs., Pau

Radoičić, R. (1966): see 5.5

Sartoni, S., Crescenti, U. (1962): see 5.5

5.5 Annotated Bibliography: Microfacies Monographs

Listed below are the publications with numerous plates of thin-section photographs, which can be used for identifying "thin-section fossils" or interpreting facies.

Abbreviations: C = Cenozoic, M = Mesozoic, P = Paleozoic, TF = important articles on "thin-section fossils".

(M) Bachmann, G. H. (1973): Die karbonatischen Bestandteile des Oberen Muschelkalkes. Arb. Inst. Geol. Paläont. Univ. Stuttgart, N.F. **68,** 1–99, 67 Figs.
Very exact study, based on over 1000 thin-sections and 500 peels, of the formation, genesis, and diagenesis of Upper Muschelkalk (Middle Triassic). Good photographs of thin-sections. The diagenesis of shells is given special treatment.

(M) Bissell, H. J. (1970): Petrology and petrography of Lower Triassic marine carbonates of Southern Nevada (U.S.A.). Intern. Sed. Petrograph. Ser. **14,** 27 pp., 82 Pls., 13 Figs., Leiden: Brill
Shelf, transitional, and basinal environments in the Lower Triassic of Nevada. Good photographs of thin-sections.

(M) Bonet, F. (1952): La Facies Urgoniana del Cretacio Medio de la Region de Tampico. Bol. Asoc. Mex. Geol. Petrol. **4**/5–6, 153–261, 50 Figs., Mexico

(M) Bouroullec, J., Deloffre, R. (1976): Rélations facies-environment au Crétacé moyen en Aquitaine occidentale. Bull. Centre Rech. Pau, SNPA **10**/2, 535–583, 34 Pls., 5 Figs., 2 Tab., Pau
Vraconian, Cenomanien, various environments (supratidal zone, lagoons, deeper marine). Good photographs of thin-sections.

(M) Borza, K. (1969): Die Mikrofazies und Mikrofossilien des Oberjuras und der Unterkreide der Klippenzone der Westkarpaten. 301 pp., 88 Pls., Bratislava (Slowak. Akad. Wiss.)
Very exact paleontological description of the pelagic microfossils occurring in the Upper Jurassic and Lower Cretaceous limestones of the inner Klippen Belt of the Western Carpathians, as seen in thin-sections. The main focus of this excellent book is the study and stratigraphic evaluation of spheres (*Cadosina, Stomiosphaera, Calcisphaerula, Pithonella*) and tintinnids. Good photomicrographs.

(P, M, C) Bozorgnia, F., Banafti, S. (1964): Microfacies and micro-organisms of the Paleozoic through Tertiary sediments of some parts of Iran. 22 pp., 158 Pls., Teheran: Iranian National Oil Co.
Cambrian to the Late Tertiary (3 plates Early Paleozoic, 37 Late Paleozoic, 3 Triassic, 22 Jurassic, 41 Cretaceous, 52 Tertiary). Short stratigraphic survey; 1 geological map of Iran. Photographs generally good. Fossil identifications not always up to standard (atypical fusulinid sections; algae usually not very accurately identified; Jurassic good).

(M) Carozzi, A. V., Bourroullec, J., Deloffre, R., Rumeau, J. L. (1972): Microfacies du Jurassique d'Aquitaine. Bull. Centre Rech. Pau, CMFA Vol. Spec. **1,** 594 pp., 200 Pls., Pau
Detailed microfacies analyses combined with chemical and sedimentological studies of Jurassic carbonates (Lias to Upper Jurassic) in Aquitaine. The classification of the cores, which was carried out very precisely with the aid of thin-sections, quantitative and qualitative analyses of insoluble residues and trace elements as well as petrophysical data have made possible a reconstruction, based on four theoretical models, of the depositional environment and the paleogeography. Photographs of fossils (primarily only foraminifera and calcareous algae) are very good but not described in any detail (compare the articles by Bouroullec and Deloffre, 1969 and 1970, Bull. Centre Rech. Pau *SNPA,* vol. 3 or 4).

(C) Carozzi, A. V., Reyes, M. V., Ocampo, V. P. (1976): Microfacies and Microfossils of the Miocene Reef Carbonates of the Philippines. Spec. Publ. **1,** Philippine Oil Dev. Co., 79 pp., 20 Pls., Manila

(P) Carozzi, A. V., Textoris, D. A. (1967): Paleozoic carbonate microfacies of the Eastern stable interior (U.S.A.). Intern. Sed. Petrograph. Ser. **11,** 41 pp., 100 Pls., 13 Figs., Leiden: Brill

Ordovician to the Pennsylvanian, geological stratigraphic survey according to systems, each with paleogeographic interpretations; photographs generally good. Fossil identifications missing or too general (only groups). "Environmental interpretation" also much too general.

(TF) Cayeux, L. (1935): Les roches sédimentaires de France. Roches carbonatées. 463 pp., 26 Pls., Paris: Masson
Classic work on the sedimentological significance of fossils in carbonates; good illustrations.

(M, C) Christodoulou, G., Tsaila-Monopolis, St. (1975): Eastern Hellenic zone microfacies. National Inst. Geol. Mining Res. **17**/1, Geol. Geophys. Red., 63 pp., 75 Pls., 6 Figs., Athen (Greek with English Plate legends)
Good survey of thin-section fossils and microfacies (also methods) from the Skythian to the Paleocene.

(M, C) Cita, M. B. (1965): Jurassic, Cretaceous, and Tertiary microfacies from the Southern Alps (Northern Italy). Intern. Sed. Petrograph. Ser. **8,** 99 pp., 117 Pls., 17 Figs., Leiden: Brill
Lias to Late Tertiary. Book divided up equally into descriptions of the Jurassic to the Cretaceous and the Tertiary; short survey of the stratigraphy, paleogeography, and tectonics of the Southern Alps. pp. 59–63: "Meaning of Microfacies and its Application to Stratigraphy" – worthwhile reading! Photographs of varying quality; fossil identifications good to very good.

(M, C) Cuvillier, J., Sacal, V. (1951): Correlations stratigraphiques par microfacies en Aquitaine occidental. Intern. Sed. Petrogr. Ser. **2,** 16 pp., 90 Pls., Leiden: Brill
This classic work has gone through three editions in the meantime (now English). Lias to the Tertiary; good photographs; generally good fossil identifications.

(M) Derin, B., Reiss, Z. (1966): Jurassic microfacies of Israel. Israel Inst. Petrol., Spec. Publ. 43 pp., 320 Figs., 1 supp., Tel Aviv
Lias to the Upper Jurassic (post-Oxfordian). pp. 5–8: "Significance of Microfacies" – worthwhile reading! Survey of the Jurassic in Israel. Correlation tables; standard sedimentological classification (after Folk). Photographs of varying quality; fossil identifications generally good. Important article for the Jurassic.

(C) Doebl, F., Mowahed-Awal, H., Rothe, P., Sonne, V., Tobien, H., Weiler, H., Weiler, W. (1972): Ein „Aquitain"-Profil von Mainz-Weisenau (Tertiär, Mainzer Becken). Mikrofaunistische, sedimentpetrographische und geochemische Untersuchungen zu seiner Gliederung. Geol. Jb. A **5,** 141 pp., 17 Pls., 4 Figs., 13 Tab., Hannover
Important work on a Late Tertiary limestone-marl sequence.

(M) Draganescu, A. (1976): Lower Cretaceous carbonate and carbonate-evaporite-sedimentation in the East-Wallachian Sector of the Moesian Platform (Eastern Romanian Plain). An. Inst. Geol. Geofiz. **48,** 5–56, 38 Pls., 3 Figs., Bucharest
Upper Tithonian/Lower Berriasian to the Valanginian. Micritic limestones of the inner shelf. Carbonate-evaporite sequence in tidal zones. Good photographs of thin-sections, particularly of *Favreina* limestones and pelsparites.

(M) Dragastan, O. (1975): Upper Jurassic and Lower Cretaceous microfacies from the Bicaz Valley Basin (East Carpathian). Mem. Inst. Geol. Geophys. **21,** 87 pp., 95 Pls., Bucharest
Kimmeridgian to Lower Aptian. Locally useful zonation with the aid of "microfacies zones". Description of foraminifera, calcareous algae, calpionellids, and other microfossils in thin-sections. Good photographs of thin-sections. Important work for the microfacies study of Upper Jurassic and Lower Cretaceous sequences.

(M) Fabricius, F. H. (1966): Beckensedimentation und Riffbildung an der Wende Trias/Jura in den Bayerisch-Tiroler Kalkalpen. Intern. Sed. Petrogr. Ser. **9,** 143 pp., 27 Pls., 24 Figs., 7 Tab., Leiden: Brill
Upper Triassic (Norian) to the Lower Lias. Diverges very positively from the point of content and description from other volumes in the series. Excellent survey of paleogeography. Photographs good; fossil identifications good as well.

(P) Falke, H. (ed.) (1972): Rotliegend. Essays on European Lower Permian. Intern. Sed. Petrogr. Ser. **15,** 299 pp., 66 Figs., Leiden: Brill
Survey of the facies and stratigraphic development of the Lower Permian sedimentation in Central and Western Europe.

(M) Fenninger, A., Holzer, H. L. (1972): Fazies und Paläogeographie des ober-ostalpinen Malm. Mitt. Geol. Ges. Wien **63,** Jg. 1970, 52–141, 15 Figs., 2 supps., 19 Pls., 1 Tab., Wien
Summary of the studies of the Upper Jurassic facies of the Upper Austro-Alpine unit, with a detailed investigation of microfacies. Deals with all the essential sequences developing in shallow water (e.g., Plassen limestones), on sills (e.g., Steinmühl limestone), and in basins (e.g., Oberalm beds) as well as sediments from the transitional environments between these depositional areas.

(M) Ford, N., Houbolt, J. J. (1963): The microfacies of the Cretaceous of Western Venezuela. Intern. Sed. Petrograph. Ser. **6,** 52 pp., 55 Pls., Leiden: Brill
Mainly Upper Cretaceous; planktonic foraminifera. Photographs generally good; fossil identifications of varying quality.

(M) Gazdzicki, A. (1974): Rhaetian microfacies, stratigraphy and facial development in the Tatra Mts. Acta Geol. Polonica **24**/1, 17–96, Pls. 1–52, Warsaw
Microfacies and biostratigraphic studies of the carbonate Krizna unit which can be assigned with the aid of foraminifera to the Lower Rhaetian (*pokornyi* and *friedli* zone) and the Upper Rhaetian (*hantkeni* zone). Numerous good photographs of thin-sections; exact descriptions especially of foraminifera classified in the thin-sections. An important work for microfacies studies in the Alpine Rhaetian!

(P) Glintzboeckel, C., Rabate, J. (1964): Microfaune et microfacies du Permo-Carbonifer du Sud Tunisien. Intern. Sed. Petrograph. Ser. **7,** 45 pp., 108 Pls., Leiden: Brill
Three cores from the Middle to Upper Permian and Namurian to Upper Permian. Stratigraphic correlation tables. Photos generally good; fossil identifications of fusulinids only go to genus; algae are usually missing.

(M) Grunau, H. R. (1959): Mikrofazies und Schichtung ausgewählter jungmesozoischer radiolarit-führender Sedimentserien in den Zentralalpen. Intern. Sed. Petrogr. Ser. **4,** 179 pp., 95 Figs., Leiden: Brill
Excellent, stimulating work on the sedimentological and stratigraphic problems of the Southern Alpine Late Mesozoic. Important for thin-section fossils of nannoconids and calpionellids.

(M, C) Hagn, H. (1955): Fazies und Mikrofauna der Gesteine der Bayerischen Alpen. Intern. Sed. Petrogr. Ser. **1,** 174 pp., 71 Pls., 8 Figs., Leiden: Brill
Triassic to the Miocene; the main emphasis is on the Cretaceous and the Early Tertiary. Good photographs; good fossil identifications. The book is due to be published again in a revised edition.

(P, M, C) Hanazawa, F. (1961): Facies and micro-organisms of the paleozoic, mesozoic and cenozoic sediments of Japan and her adjacent Islands. Intern. Sed. Petrogr., Ser. **5,** 117 pp., 148 Pls., Leiden: Brill
Silurian to the Holocene. Good survey of the stratigraphy and tectonics of the Japanese islands. Good photographs; fossil identifications good, usually go to species. Important for designation, especially for Early Paleozoic microfossils.

(M, C) Henson, F. R. S. (1950): Cretaceous and Tertiary reef formation and associated sediments in Middle East. Amer. Ass. Petrol. Geol. Bull. **34,** 215–238, 1 Tab., 14 Figs., Tulsa
Differentiation of the reef-complex in Cretaceous-Tertiary carbonates with the aid of microfacies types. Classic work from a methodological standpoint.

(TF) Horowitz, A. St., Potter, P. E. (1971): Introductory petrography of fossils. 302 pp., 100 Pls., 28 Figs., Berlin – Heidelberg – New York: Springer
Short introduction to carbonate petrography; surveys of the fossil groups found in thin-sections of limestones. First monograph on the subject; long reference list (with annotations). Photographs of thin-sections generally good to very good (also available as slides).

(P) Khvorova, I. A. (1958): Atlas karbonatikh porod srednego i verkhnego karbona Russkoi platformy. Trudy Akad. nauk. SSSR, Geol. Inst., 165 pp., 67 Pls., Moscow
Middle and Upper Carboniferous. The main emphasis is on sedimentological data (dolomitization textures!). Fossils are only generally identified (usually only to groups), except for the fusulinids.

(P, C, M) Lehmann, E. P., et al. (1957): Microfacies of Libya. Tripoli, Petrol. Explor. Soc. Libya, 80 pp., 37 Pls., 1 Fig., Tripoli
Cambrian to Miocene. Identifications to genera and species.

(TF) Lucas, G., Cros, P., Lang, J. (1976): Les roches sédimentaires. 2. Etude microscopique des roches meubles et consolidées. 503 pp., 400 Figs., 40 Tab., Paris: Doin
Short description of carbonate petrography. Very extensive and richly illustrated survey of the fossil groups appearing in thin-sections as well as of the diagenesis of carbonates. Very instructive photographs of thin-sections, but printing quality leaves much to be desired.

(TF) Majewske, O. P. (1969): Recognition of invertebrate fossil fragments in rocks and thin-sections. Intern. Sed. Petrogr. Ser. **13,** 101 pp., 106 Pls., Leiden: Brill
Description and good illustrations of the morphology and microstructures of fossils and Recent invertebrates in thin-sections. Excellent schematic drawings of the morphology of main groups. The main emphasis of the book is on the description (essentially based upon the monograph by Bøggild, 1930) and classification of molluskan shells (Key to Identification). Foraminifera are not discussed.

(P) Mamet, B. (1975): An atlas of microfacies in Carboniferous carbonates of the Canadian Cordillera. Bull. Canada Geol. Surv. **255,** 131 p., 95 Pls., 3 Figs., Ottawa
An important work dealing with microfacies and microzonation (based on Smaller Foraminifera and algae) of Carboniferous shelf carbonates.

(TF) Maslov, V. P. (1937): Atlas karbonatnykh prod. **1.** Inst. Sci. Rech. Min. appl., 54 pp., 55 Pls., Moscow – Leningrad, (Russian; summary and plate legends in French).
Paleozoic and Mesozoic. Good photographs; useful for an introductory survey of fossils in thin-sections.

(C) Massieux, M. (1973): Micropaléontologie stratigraphique de l'Eocene des Corbières septentrionales (Aude). Cahiers de Paleontologie, Centre nationale de recherche scientifique. 139 pp., 29 Pls., 17 Figs., Paris
A thin-section study of the Eocene carbonates south of the Montagne Noire, with special consideration of foraminifera, especially Nummulitidae and algae.

, M, C) Mehrnush, M., Partoazar, H. (1977): Selected microfauna of Iran. Geol. Min. Survey Iran Report **33,** 396 p., 117 Pls., Tehran
Deals with the stratigraphic distribution of Late Devonian to Miocene microfossils determined in thin-sections. Good photographs.

, M, C) Microfacies italiane (dal carbonifero al miocene medio) (1959). 35 pp., 145 Pls., Milano: AGIP MINERARIA, San Donato-Milanese
Upper Carboniferous to Miocene. Excellent photographs of thin-sections; fossils unfortunately often wrongly or insufficiently identified. (The identifications are being revised in individual articles, usually in Riv. Ital. Paleont. Strat.).

(M, C) Mišik, M. (1966): Microfacies of the Mesozoic and Tertiary limestones of the West Carpathians. 269 pp., 110 Pls., 3 Figs., Bratislava: Slov. Akad. Vied (English and Slowakian)
Scythian to the Late Tertiary; main emphasis on the Late Mesozoic and Tertiary. Geological stratigraphic survey of the Western Carpathians. Photographs vary in quality. The descriptions of the pictures are very good, because exact fossil identifications are correlated with sedimentological data, followed by statements about the depositional environment and related references.

(M) Mostler, H. (ed.) (1972): Symposium Innsbruck, 20.–23. März 1972, Mikrofazies und Mikropaläontologie der Alpinen Trias und deren Nachbargebiete. Mitt. Ges. Geol. Bergbaustud. Österreich **21**/1–2, 1–1016, many plates and figs., Innsbruck
Results of a symposium on microfacies and micropaleontological work in the Alpine Triassic. Numerous important detailed studies and surveys!

(P) Ota, M. (1968): The Akiyoshi limestone group: A geosynclinal organic reef complex. Bull. Akiyoshi-dai Sci. Mus. **5,** 44 pp., 31 Pls., 15 Figs., Akiyoshi
Lower Carboniferous to the Upper Permian. Reefs growing on submarine vulcanoes. Many good photomicrographs.

(M) Pantić-Prodanović, S. (1975): Les microfacies triassiques des Dinarides. Le Montenegro, la Bosnie occidentale et l'Herzegovine et Serbie occidentale. Soc. Sci. Arts Montenegro, Monograph. **4,** Cl. Sci. Nat. 4, 257 pp., 100 Pls., Titograd
Important compilation of thin-section fossils which can be stratigraphically evaluated, especially foraminifera, calcareous algae, and microproblematica. 12 Pls.: Skythian, 54 Pls.: Middle Triassic, 34 Pls.: Late Triassic.

(TF) Papp, A., Turnovsky, K. (1970): Anleitung zur biostratigraphischen Auswertung von Gesteinsschliffen (Microfacies Austriaca). Jb. Geol. Bundesanst., Sonderbd. **16,** 50 pp., 88 Pls., 11 Figs., 3 Tab., Wien
General survey of a few fossils which occur commonly in thin-sections. Paleozoic (Plates 13–21: Silurian chitinozoans and hystrichosphaerids, Late Paleozoic fusulinids and Smaller foraminifera); Mesozoic (Plates 22–62); Tertiary (Plates 63–88). Classification of intervals in the Cretaceous and in the Early Tertiary by planktonic and benthonic Larger foraminifera (Tables 2 and 3!). In some cases much too generalized; today somewhat out of date (especially Triassic and Jurassic). Thin-section photographs of varying quality.

(M) Perconig, E. (1968): Recognition of the Triassic and Jurassic sediments of Spain. Intern. Sed. Petrogr. Ser. **10,** 63 pp., 123 Pls., 11 Figs., Leiden: Brill

(C) Popescu, B. (1976): Sedimentology of Priabonian carbonate rocks, Jibou area, NW Transsylvanian basin. An. Inst. Geol. Geofiz. **48,** 117–140, 11 Pls., 8 Figs., Bucharest
Upper Eocene; coralline algae, foraminifera, mollusks, echinoids, bryozoans, corals. Platform, shelf lagoon, no detailed fossil identifications.

(M, C) Radoičić, R. (1960): Microfacies du Cretacé et du Paléogene des Dinarides externes de Yougoslavie. Inst. Rech. Geol. R. P. Crna Gora, Ser. A **4**/1, 171 pp., 67 Pls., Titograd
Valanginian to Eocene. Good photographs; good fossil identifications (rather general in the Eocene, more exact in the Lower and Middle Cretaceous). Important work on identification.

(M) Radoičić, R. (1966): Microfacies du Jurassique des Dinarides externes de la Yougoslavie. Geologija **9,** 377 pp., 165 Pls., 11 Figs., 2 supps., Ljubljana
Lias to Upper Jurassic/Lower Cretaceous; summarizing description of profiles for various Jurassic occurrences in Western Yugoslavia and Montenegro. Distribution tables for foraminifera and algae. Good photographs; fossil identifications good, often based on previous detailed studies. Very important work for thin-section fossils in the Jurassic.

(M) Radwânski, A. (1968): Petrographical and sedimentological studies of the High-Tatric Rhaetic in the Tatra Mountains. Studia Geol. Polonica, Zaklad. Nauk Geol. **25,** 146 pp., 54 Pls., Warsaw (Polish and English summaries)
Detailed description of the clastic Lower Rhaetian Tormanowa beds and the carbonate Rhaetian sequences, based on thin-sections. Good photographs of thin-sections with comments in English.

(M, C) Rey, M., Nouet, G. (1958): Microfacies de la région prérifaine et de la Moyenne Moulouya. Intern. Sed. Petrogr. Ser. **3,** 27 pp., 97 Pls., Leiden: Brill
Triassic to Miocene. Good photographs; fossil identifications generally rather short. Main emphasis on the Lias/Middle Jurassic, Cenomanian/Senonian and Oligocene/Miocene.

(C) Rusu, A., Draganescu, A. (1976): Facies-zoned carbonate sedimentology at the time of the Hoia Limestone (Upper Tongrian) in NW Transsylvania (Romania) An. Inst. Geol. Geofiz. **48,** 141–169, 20 Pls., 2 Figs., Bucharest
Middle Oligocene (Upper Tongrian). Arenaceous corallinacean facies = agitated shallow water, muddy miliolid-corallinacean facies = shelf margin and shelf slope with coral bioherm, muddy globigerinid-bryozoan facies = marine basin facies. Species identification of corals and coralline algae. Thin-section photographs of varying quality.

(P) Sacal, V., Cuvillier, J. (1964): Microfacies di Paléozoique Saharien. Com-franc. Petroles Notes et Mém. **6,** 30 pp., 100 Pls., Paris.
Cambrian to Carboniferous. Excellent (glossy!) photographs of thin-sections; fossil identifications only partially sufficient, usually only identifications of groups. Quite useful for the first study of Early Paleozoic (particularly Devonian) limestones in thin-sections.

(P, M, C) Sampo, M. (1969): Microfacies and microfossils of the Zagros area, Southwestern Iran (from Permian to Miocene). Intern. Sed. Petrogr. Ser. **12,** 102 pp., 105 Pls., Leiden: Brill
Exact microfacies information based on profile studies of carbonate sequences in the Southwest (Triassic to Pleistocene) and Northeast (Pre-Permian to Pleistocene) Zagros range. Differentiation of "faunal zones" according to benthonic and pelagic foraminifera as well as algae, particularly from the Jurassic to the Miocene. Good fossil identifications; good detailed photographs of foraminifera in thin-sections.

(M, C) Samuel, O., Borza, K., Köhler, E. (1972): Microfauna and lithostratigraphy of the Paleogene and adjacent Cretaceous of the Middle Vah Valley (West Carpathian). 246 pp., 180 Pls., Bratislava (Geol. ust. Dionyza Stura)
Study of Early Tertiary and Cretaceous sequences containing stratigraphically older lithoclasts in conglomerates of the Vah valley NE Bratislava, Czechoslovakia, with special emphasis on the organisms identifiable in thin-sections (foraminifera, calcispheres, algae). Excellent photographs of thin-sections.

(M) Sartoni, S., Crescenti, U. (1962): Recherche biostratigrafiche nel Mesozoico dell'Appennino meridionale. Giorn. Geol. (2), **29,** 161–304, Pls. 11–52, Bologna
Lias to Paleocene. Zonation of the carbonate sequences in ten cenozones (fossil assemblage zones) and three subzones (which in the meantime have been found again all over the Mediterranean region). Good paleontological identifications and exact descriptions. Today a "classic" (should be supplemented by Crescenti, 1966, Boll. Soc. Geol. Ital. **85**).

(M) Schulze, K.-H. (1975): Mikrofazielle, geochemische und technologische Eigenschaften von Gesteinen der Oberen Heersumer Schichten des Korallenoolith (Mittleres bis Oberes Oxfordium, NW-Deutschland) zwischen Weser und Leine. Geol. Jb. D **11,** 3–102, 6 Figs., 6 Tab., 20 Pls., Hannover
Semi-quantitative evaluation of over 1700 thin-sections. Comparison with geochemical and technological data. The technological properties of rocks, particularly weathering criteria, are correlated with 21 microfacies types.

(P) Tellez-Giron, C. (1970): Microfacies y microfosiles paleozóicos del area de ciudad Victoria. Tamps, N. E. de Mexico. Inst. Mexicana Petroleo, Ser. Monografía **1,** 127, 57 Pls., 1 Fig., 1 Tab., Mexico
Ordovician, Silurian, Carboniferous, and Permian. Generally only identification of fossil groups; identification to genus for Silurian bryozoans, Carboniferous and Permian Smaller foraminifera and fusulinids. Illustrations of various kinds of calcispheres. Microfossils often overmagnified; photographs sometimes blurred; no fossil descriptions.

(M, C) Tsaila-Monopolis, St. (1977): Micropaleontological and stratigraphical study of the Tripolitza (Gavrovo) zone in the Peloponnesus. Inst. Geol. Mining Res., Geol. Geophys. Res. **20**/1, 106, 52 Pls., Athens
Foraminifera and calcareous algae, found in thin-sections of Upper Triassic to Eocene limestones, are shown in photographs; the stratigraphic distribution within the profiles is indicated.

(P) Zamarreño, I. (1972): Las litofacies carbonatadas del Cambrico de la zona cantabrica (NW España) y su distribución paleogeografía. Fac. Ciencias, Univ. Oviedo, 102 pp., 17 Pls., Oviedo
Analysis based on exact profiling, paleontological and sedimentological data as well as evaluation of numerous thin-sections, of Lower and Middle Cambrian sequences. Good photomicrographs.

6 Classifications of Carbonate Rocks

Carbonate rocks can be subdivided according to:

a) chemical and mineralogical composition, e.g., Chilingar (1960), Pettijohn (1957), or Füchtbauer (1959),
b) fabric features – "groundmass" (= matrix and/or cement) and particles,
c) special physical parameters, e.g., porosity, Choquette and Pray (1970); see 9.2.1.

Almost all the systems of classifications used today are based upon the criteria characterizable either in thin-sections or hand specimens, such as matrix, cement, and particles. Hence, the following discussion of these classifications will emphasize fabric features.

6.1 Principles of Classification

Table 36 lists the advantages and disadvantages of some of the modern systems of classification. Dissimilarities in the systems are due less to the different basic concepts than to the consideration of various kinds and quantities of particles types (some authors refer "lumps" to intraclasts; others consider them to be an independent group). Dissimilarities are also due to the various percentages used as boundary values, to a divergent emphasis on size, sorting, and rounding of transported or transportable particles, and to somewhat specialized problems which call for modified systems of classification (e.g., coquina limestones, Schäfer, 1970).

6.2 Systems of Classification

Important suggestions for systems of classification are found in *Classification of Carbonate Rocks,* a collection of symposium papers edited by Ham in 1962 (Mem. Amer. Ass. Petrol. Geol. 1). The most commonly used classifications are those of Folk (1959, 1962), Dunham (1962), Leighton and Pendexter (1962) and Bissell and Chilingar (1967, modified by Füchtbauer, 1974). A discussion of all the suggestions published up to 1967 is found in Baisert (1968).

Table 36. Comparison of the usefulness of a few modern systems of limestone classification

	Application	Preliminary Knowledge	Advantages	Disadvantages
Dunham (1962)	Field, hand specimen, thin-section	Knowledge of texture types	Rapid "spot" identification in the field	–
Folk (1959, 1962)	Thin-section	Knowledge of particle types, matrix	Sensible subdivision of textural spectra	Sometimes too time consuming, some subdivisions too artificial
Füchtbauer (1970)	Field, hand specimen, thin-section	Knowledge of particle types, matrix, cement	Rapid classification	Mixture of descriptive and genetic criteria
Leighton and Pendexter (1962)	Field, hand specimen	Knowledge of particle types, matrix	Rapid rock designation	Mixture of descriptive and genetic criteria; percentage boundary values sometimes too rigorous
Monty (1963)	Thin-section	Detailed knowledge of particle types	Well-devised scheme	Nomenclature systems and terms too complicated; concept similar to Folk
Plumley et al. (1962)	Hand specimen, thin-section	Knowledge of texture types, amounts of insoluble residues	Rapid classification of "genetic" types	A few basic concepts somewhat questionable
Todd (1966)	Thin-section	Knowledge of particles, matrix, cement, texture, insoluble residues, chemical composition	Attempt at a "comprehensive" classification	Too involved for facies analyses

6.2.1 Folk Classification (1959, 1962) (Table 37, Figs. 52, 53)

The system of classification suggested by Folk is based upon the fact that, in principle, carbonate rocks are comparable to sandstones and shales, in regard to sedimentation.

I. Major constituents of limestones are:

A. *Allochems* (carbonate grains or particles):

1) *Intraclasts* – synsedimentary resediments, e.g., mud pebbles, grapestones, bahamites.
2) *Pellets* – subrounded, spherical to elliptical grains of micrite, generally devoid of any internal structure. Sizes range between 0.03 mm and 0.15 mm, most commonly 0.04–0.08 mm. These are probably fecal pellets in most cases.

3) *Ooids.*
4) *Fossils and skeletal grains.*
5) The Folk classification must be modified to include *oncoids* as terminologically independent particles. The occurrence of more than 10% of oncoids should be marked in the name of the rock type (e.g., oncomicrite or oncosparite); see Table 37.

B. *Matrix* (micrite)

C. *Sparite*

Continued p. 371

Plate 31. Classifications of Limestones

A = Classification after Folk, 1959 or 1962; B = Classification after Dunham, 1962; C = Energy Index Classification after Plumley et al., 1962; D = Classification after Füchtbauer, 1974

1. A – pelsparite or sorted pelsparite with echinoderm fragments; B – grainstone or pelletoidal grainstone; C – IV/2; D – cemented pellet-limestone with echinoderms. Lower Cretaceous: mouth of the Rhone, Geneva, Switzerland. Scale is 1 mm

2. A – biomicrite or sparse biomicrite; B – bioclastic wackestone; C – I/3; D – limestone rich in skeletal grains or filaments. Late Triassic (Kössen beds): Gaissau near Hallein, Salzburg, Austria. Scale is 10 mm

3. A – oosparite or poorly sorted oosparite with skeletal grains; B – oolith lime grainstone; C – IV/2; D – oolitic limestone. Lower Permian: Forni Avoltri, Carnia, Italy. Scale is 1 mm

4. A – dasycladacean algal biomicrite; B – dasycladacean lime wackestone; C – II/1; D – fossiliferous limestone rich in matrix. Skeletal grains include Dasycladaceae (*Diplopora adnetensis* Flügel) as well as echinoderm fragments. This is possibly a "textural inversion" in the sense of Folk (1962), because the dasycladaceans probably did not live in the low energy sedimentary environment indicated by the fine-grained micrite. Late Triassic (Upper Rhaetian limestone): Gruber reef, Feichtenstein, Salzburg, Austria. Scale is 1 mm

5. A – extraclast-bearing sparite with echinoderms; B – lithiclast lime grain/rudstone; C – V/2; D – detrital fossiliferous limestone. The very well-rounded micritic particles should be regarded as extraclasts. Evidence of this are: high degree of rounding, truncation of the particles in the lithoclasts on the lithoclast boundaries, different cements in and between the pebbles. Lower Carboniferous: Kücu Tepe near Balya, Anatolia, Turkey. Scale is 1 mm

6. A – intrasparite; B – lithoclastic grainstone; C – IV/2; D – Sparry lump limestone. Particles are typical aggregate grains. Dachstein Limestone, Upper Triassic: Dachstein, Styria, Austria. Scale 1 mm

7. A – biolithite; B – agal-foraminiferal boundstone or framestone; C – V/3; D – reef-limestone. The biogenic framework is composed of the agglutinated foraminifera *Haddonia heissigi* Hagn (top center), various Corallinaceae (gray and black crusts), corals (bottom and top left, partly bored), and calcareous sponges (top left). At the bottom right *Discocyclina* sp. Tertiary (Upper Eocene): quarry in the Northeast flank of the Eisenrichterstein peak near Hallthurn, SE Bad Reichenhall, Germany. Scale is 1 mm

8. A – biopelsparite; B – bioclastic lime grainstone; C – IV/2; D – cemented pellet-limestone with foraminifera. The sections of *Galeanella tollmanni* (Kristan) are quite conspicuous among the foraminifera. This species is restricted to small cavities in the framework of Upper Triassic reefs. Late Triassic (Dachstein reef limestone): Sauwand near Gußwerk, Styria. Austria. Scale is 1 mm

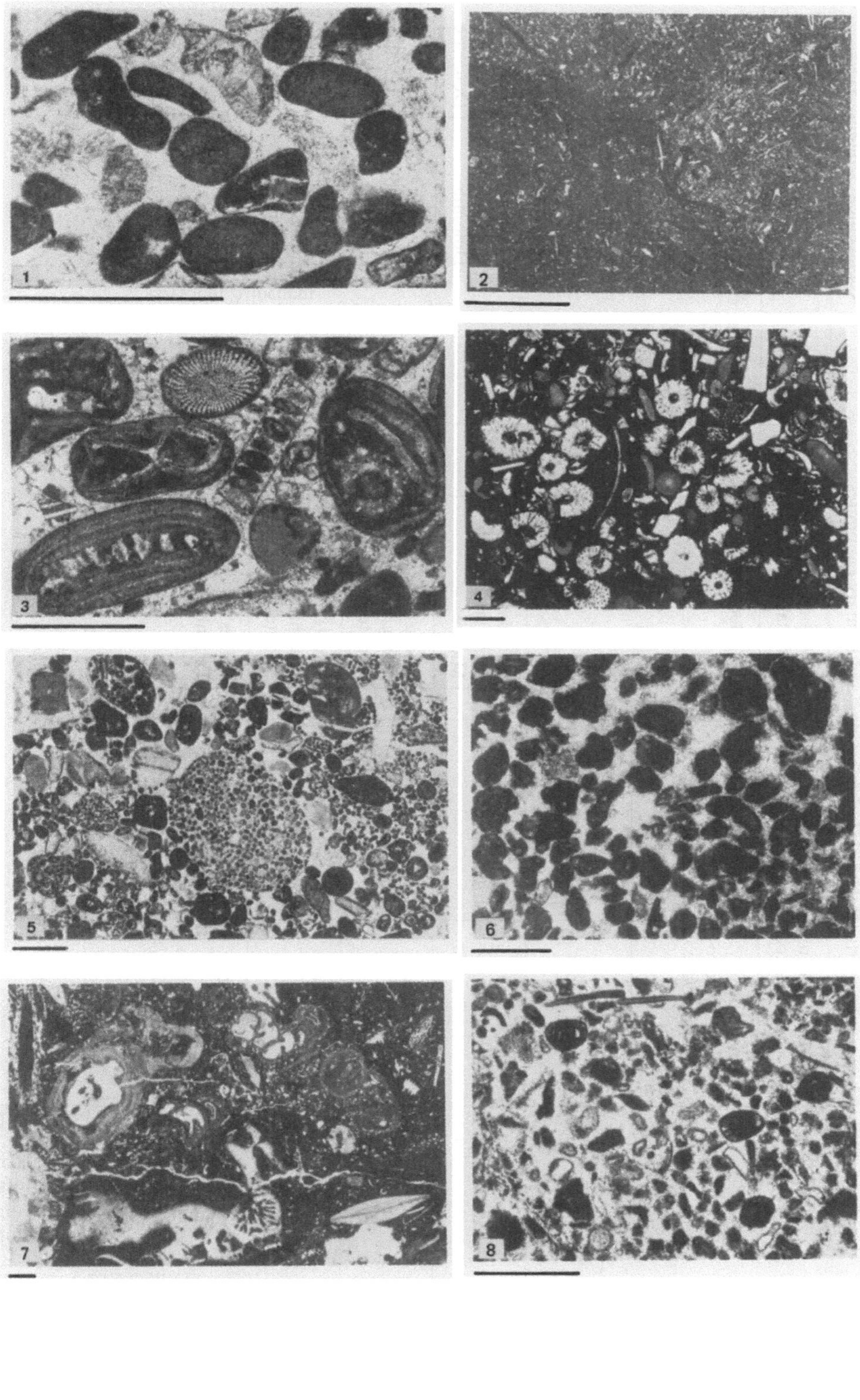
1
2
3
4
5
6
7
8

Table 37. Classification of Carbonate Rocks after Folk (1959, 1962), slightly modified

	Limestones, Partly Dolomitized Limestones, and Primary Dolomites					Replacement Dolomites (V)	
	> 10% Allochems Allochemical Rocks (I and II)		< 10% Allochems Microcrystalline Rocks (III)				
	Sparry Calcite Cement > Microcrystalline Ooze Matrix	Microcrystalline Ooze Matrix > Sparry Calcite Cement	1–10% Allochems	< 1% Allochems	Undisturbed Bioherm Rocks (IV)	Allochem Ghosts	No Allochem Ghosts
	Sparry Allochemical Rocks (I)	Microcrystalline Allochemical Rocks (II)					
Volumetric Allochem Composition: > 25% Intraclasts	Intrasparrudite Intrasparite	Intramicrudite Intramicrite	Most Abundant Allochems are: Intraclasts: Intraclast-bearing Micrite	Micrite; if disturbed, Dismicrite; if primary dolomite, Dolomicrite	Biolithite	Evident Allochem: Finely Crystalline Intraclastic Dolomite	Medium Crystalline Dolomite
< 25% Intraclasts: > 25% Oolites	Oosparrudite Oosparite	Oomicrudite Oomicrite	Oolites: Oolite-bearing Micrite			Coarsely Crystalline Oolitic Dolomite	Finely Crystalline Dolomite
< 25% Oolites: Volume Ratio of Fossils to Pellets > 3 : 1	Biosparrudite Biosparite	Biomicrudite Biomicrite	Fossils: Fossiliferous Micrite			Aphanocrystalline Biogenic Dolomite	
3 : 1–1 : 3	Biopelsparite	Biopelmicrite	Pellets: Pelletiferous Micrite			Very Finely Crystalline Pellet Dolomite	etc.
< 1 : 3	Pelsparite	Pelmicrite					
Volume Ratio of Fossils to Oncoids 1 : 2	Oncosparite Oncosparrudite	Oncomicrite Oncomicrudite	Oncoids: Oncoid-bearing Micrite				

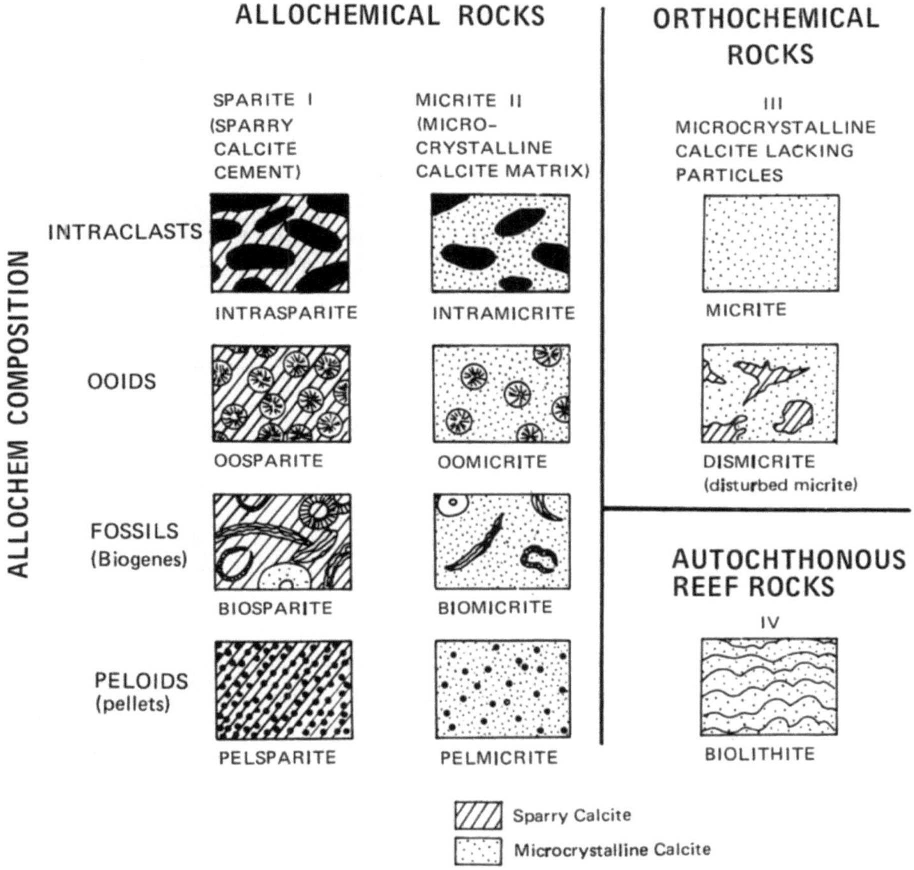

Fig. 52. Folk classification (Folk, 1959). Basically carbonate rocks rich in particles (allochemical rocks) and those poor in or devoid of particles (orthochemical rocks) as well as autochthonous reef rocks are distinguished. The names are formed by combining terms for matrix (micrite), cement (sparite), and particles (intraclasts, etc.)

II. The rock names obtained from combination with micrite (see Fig. 52), sparite, and particles may be *modified* by adding the following:

A. *Grain sizes,* terrigenous admixtures, etc. (e.g., "medium biosparrudite" = limestone with bioclasts 4–16 mm in size; "sandy glauconite foraminifera biosparite" = limestone with more than 10% of foraminifera, more than 10% of siliceous admixtures with glauconite, and carbonate cement);

B. *Field properties* (e.g., "black, hard, massive micrite", "white, hard cross-bedded oosparite").

III. Subdivision according to textural spectra (see Fig. 53): The three main limestone groups can be subdivided after Folk (1962) into *8 groups* with textural spectra which reflect the various degrees of water energy during deposition:

1. *Pure micrite or dismicrite with less than 1% of particles,* e.g., Solnhofen lithographic limestone
2. *Micrite or dismicrite with 1%–10% of particles,* e.g. fossiliferous micrite or pellet-bearing micrite
3. *Biomicrite, intramicrite, oomicrite and pelmicrite with 10%–15% of particles.* The particles are scattered throughout the matrix (sparse biomicrites, etc.) and are loosely packed. See Plate 31/2.
4. *Biomicrite, pelmicrite, etc. with more than 50% particles.* The particles appear "densely packed" (packed biomicrites, etc.). See Plate 31/4.
5. *Poorly washed biosparite* with about subequal micrite and sparite; arises because of weak or changing current intensity.
6. *Unsorted biosparite,* etc. All or almost all of the micrite is washed out and the poorly sorted particles are cemented with sparite. More than two thirds sparite. See Plates 31/3.
7. *Sorted biosparite,* etc. The particles are well-sorted but only slightly abraded and rounded. See Plate 31/1, 6.
8. *Rounded biosparite.* The majority of the particles is well-rounded. Generally takes place in the surf zone.

These groups may reflect the various carbonate sediments in basinal, shallow shelf, and coastal environments. However, it should be remembered that primary textural inversions are common (e.g., transport of carbonate grains from a reef flat into the carbonate mud lagoons behind it; in this way a sediment corresponding to Group 3 or Group 4 arises, in which the carbonate grains are well-sorted). Oomicrites are formed when carbonate ooids are washed out of a shallow water environment and later deposited into a somewhat deeper protected mud

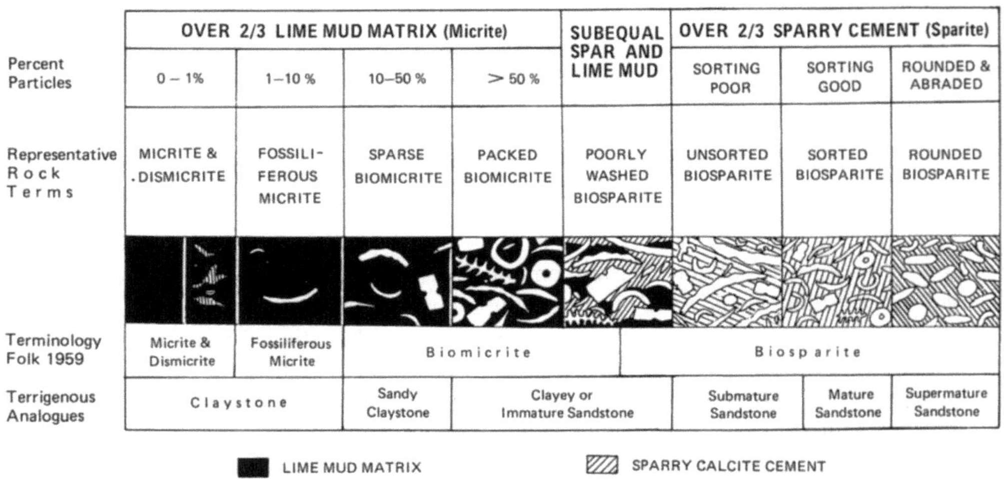

Fig. 53. Textural spectrum of carbonate rocks. In general, "low-energy" sediments on the left, with successively "higher-energy" sediments leading to the right, analogous to a Recent environmental sequence representing the change from a basin up onto a shallow shelf and then to a coastal area. After Folk (1962)

environment. Even bioturbation can lead to a mixture of textural types! Plate 31/4 gives an example of textural inversion.

A statistical comparison of the Folk system of classification with the distribution of sediment on the Andros platform in the Bahama Banks, done by Imbrie and Purdy (1962), shows that Groups 5–8 are easily recognizable with factor analyses; however, the subdivision of the micritic carbonates into oomicrite, intramicrite, biomicrite, biopelmicrite, and pelmicrite is too finely spun. For instance, in the Bahama Banks only "skeletal mud facies" and a "pellet mud facies" could be differentiated.

6.2.2 Dunham Classification (1962). Expanded by Embry and Klovan (1972) (Table 38)

This subdivision is based on the particle fabric and on the kind of particle binding during sedimentation. In the former a distinction is made between mud-supported and grain-supported fabrics. "Mud support" corresponds to carbonate mud with floating particles which generally do not touch each other (similar to pebbles in a mud puddle) (see Plate 36/2; 46/3). "Grain support" exists then when the particles support each other (similar to grains in a pile of sand). (Plate 32).

The particle contacts of a grain support fabric are not always reflected in thin-sections because of the two dimensional perspective. Criteria for "grain supports" are concave-convex particle contacts and a packing index < 2, generally around 1.2 (see Sect. 4.2.3.1), and often sparitic cements.

Dunham uses names that combine the names of fabric types with the names of the grain types (e.g., ostracod lime mudstone; ostracod-lithoclast lime wackestone; Crinoid lime packstone; Coated-grain lime packstone; Oolitic lime grainstone). It should be stressed that many limestones can be described more meaningfully by a combination of two names, e.g., bioclastic grain/packstone; a limestone with a predominantly grainstone fabric, but with somewhat densely packed grains.

The terms floatstone and rudstone, introduced by Embry and Klovan in connection with the investigation of reef limestones, should be used only for allochthonous limestones in which the significant size of the detrital particles has been caused by erosion and redeposition (e.g., large lithoclasts eroded at the shelf-margin and transported into basins, or fore-reef detritus). There is little point in speaking of "oncolitic rudstones" because the size of algal oncoids is mainly controlled by biological factors and not caused by redeposition. The same holds true for bioclastic limestones with particularly large whole shells (e.g., double-valved pelecypods).

Differentiation of autochthonous carbonates (boundstones) is based on the various interactions between sessile organisms and sediments; these interactions are characterized by baffling, binding, and framework building processes. Bafflestones originate in local quiet water environments between intergrown branches of corals or between sponges ("*Thecosmilia*" in the micrite of Rhaetian Kössen beds; calcisponges in the reef-core facies of Late Triassic Dachstein limestones, see Plate 43/9). Other organisms, especially algae, but also sessile foraminifera or lamellar stromatoporoids contribute to the formation of bindstones (see Table 23).

Table 38. Classification of carbonate rocks according to depositional texture. After Dunham (1962) and Embry and Klovan (1972)

Original components not bound together during deposition				Original components were bound together during deposition			Original components not bound together during deposition	
Generally *smaller* grains (arenite and silt size)							More than 10 percent *larger* grains (rudite size)	
Contains mud (micrite matrix)		Lacks mud (sparite matrix)		Organisms act as sediment bafflers (e.g., dendroid corals)	Organisms act as sediment binders (e.g., algal mats)	Organisms act as frame-builders (e.g., intergrown reef corals)	Contains mud (micrite matrix)	Lacks mud (sparite matrix)
Less than 10 percent grains	More than 10 percent grains							
Mud-supported		Grain-supported		Boundstone			Matrix-supported	Grain-supported
Mudstone	Wackestone	Packstone	Grainstone	Bafflestone	Bindstone	Framestone	Floatstone	Rudstone
Plate 45/2	Plate 6/5, 12/1, 15/6, 26/2, 27/3, 31/4, 36/1, 46/3, 51/4, 53/1	Plate 9/2, 11/3, 20/5, 21/2, 21/3, 32/1, 36/2, 37/1, 41/5, 44/5, 50/3, 51/6, 52/2, 53/2	Plate 9/1, 17/1, 17/2, 28/5, 31/1, 31/6, 32/2, 37/2, 38/1, 42/1, 47/1, 47/4	Plate 24/5, 43/5, 43/9, 46/1, 48/3	Plate 13/1, 13/2, 14/1, 14/3, 29/2, 39/1, 45/3, 47/5, 49/1	Plate 23/4, 31/7, 43/1, 43/3, 45/5, 48/2, 50/1, 52/3	Plate 39/2, 51/7	Plate 48/1, 53/3

Indications of algal binding are the existence of pseudostromata, skeletal intergrowth frameworks (e.g., by red algae), constructed voids, laminations contrary to gravity, algal mat laminations, packstone fabric composed of algal-derived particles (algal peloids, etc.) and in situ growth of algae.

Typical framestones are formed by closely intergrown primary and secondary reef builders. Examples of the subdivision of the boundstone groups especially important for studies of reef-complexes are found in Embry and Klovan (1972, Devonian of Canada).

6.2.3 Classifications of Leighton and Pendexter (1962), Bissell and Chilingar (1967), and Füchtbauer (1974) (Plate 31)

These classifications are similar with respect to the characteristics used in differentiating rock-types. Differences are represented by the boundaries used for the percentages of grains to micritic material.

Leighton and Pendexter suggested that the grain/micrite ratio and the percentage of grains were of value for a textural classification of limestones. The grain/micrite ratio can be calculated as the sum of the percentages of grains divided by the percentage of micrite. The authors distinguish the following particle types: detrital grains (including rock fragments and intraclasts), skeletal grains, pellets, lumps, and coated grains (including ooids, pisoids and algal- or foraminiferal-encrusted grains). Boundaries used to distinguish the main textural limestone types are 10%, 50%, and 90% grains. The name of the limestone type is formed in the following manner:

Oolitic Limestone: Limestone with ooids, and a grain/micrite ratio of 9:1; oolitic-micritic limestone: grain/micrite ratio of about 1:1; micritic-oolitic limestone: grain/micrite ratio 1:9. Other types are called detrital limestones, skeletal limestones, pellet limestones, lump limestones, or algal limestones (if organic framebuilders are of importance). Oolitic limestones correspond to "densely packed oosparites" (Folk) or "oolitic grainstone" (Dunham); oolitic-micritic limestones are "oomicrites" or "oolitic wacke/packstones"; micritic-oolitic limestones can also be called "ooid-bearing micrites" or "ooid-bearing mudstones".

Bissell and Chilingar (1967) modified the classification by introducing additional boundaries for the percentages of grains: Six groups are distinguished according to the existence of less than 10% grains (micrite limestone), 10–25% grains (oolitic-micritic limestone), 25–50% (oolitic-micritic limestone), 50–75% (micritic-oolitic limestone), 75–90% (micritic-oolitic limestone), more than 90% grains (oolitic limestones).

The same boundaries were used by Füchtbauer (1970, 1974) who calls a microcrystalline limestone with no ooids or less than 10% ooids "calcilutite" or "micrite" or "micritic limestone with some ooids"; 10–25% grains: "ooid-bearing limestone"; 25–50% ooids: "limestone rich in ooids"; 50–75% grains: "oolitic limestone rich in matrix"; 75–90% grains: "matrix-bearing oolitic limestone"; more than 90% ooids: "oolitic limestone". Instead of matrix the term cement can be used for sparitic limestones. The existence of several grain types may be indicated by a modification of the rock name: A limestone containing 20% ooids and 15% peloids may be described as "rich in ooids and peloids", or "oolitic, peloid-bearing limestone" (if the particles are regarded separately).

6.2.4 Energy Index Classification (Plumley et al., 1962) (Plate 31)

This classification (Table 39) takes into account the fact that different water agitation generates different structures and textural types. Crucial genetic factors are the wave base (water depth below which the movement of the water caused by surface waves does not move the sediment – dependent on wave amplitude, storms, and the bottom topography of the sea). The energy level is a result of the kinetic energy on the sea floor due to wave or current action at the depositional interface and a few feet above; this can be implied from the fabric of the transported particles as well as from the amount of micrite or sparite. It is fundamentally important to recognize the fact that the *particles have been transported. Criteria* for this may be:

a) Fragments of partially indurated sediment (intraclasts) or stratigraphically older rocks (extraclasts) ranging in size from silt to boulders; angular or rounded.
b) Rounded fragments of fossils that were not originally round.
c) A poorly sorted matrix, e.g., silt grains in a fine-grained matrix (can also be caused by bioturbation!).
d) A mixture of carbonate and siliceous particles of the same size e.g., bioclasts and quartz sand.
e) A mixture of diverse ecologically incompatible organism assemblages, (e.g., fusulinids in algal biomicrites, dasycladaceans in crinoid biosparites).

Plate 32. Limestone Fabrics: Grain-Support in Packstones and Grainstones

1. Echinoderm packstone with abundant echinoderm fragments and some foraminifers. Grain-support fabric is indicated by the dense packing of bioclasts and by the free-standing burrow, which would have collapsed in a mud-supported fabric. Packstones are grain-supported micritic limestones which show properties of sediments deposited in quiet water (micrite) and also properties of sediments deposited in agitated water (grain-support; co-occurrence of sparite and micrite). Overly close packing of particles suggests that grain-support of some packstones was aquired during compaction. Some packstones clearly are of diagenetic origin (e.g., stylobreccias, see Sect. 3.7). Others are of sedimentary origin, as indicated by interstices filled with mud (recording an early or late infiltering of previously deposited mud-free sediment), or by a patchy distribution of micrite in predominantly sparry packstones (due to burrowers which mix originally interbedded sand and mud together). Red nodular Adnet Limestone, Liassic: Adnet near Hallein, Salzburg, Austria

2. Oolitic grainstone with redeposited quiet-water ooids. Grain-support is indicated by the presence of sparry cement and by close packing of the ooids. A quiet-water depositional environment of the ooids can be inferred from the radial microstructure (see Sect. 4.1.3.5), the irregular shape of the ooids, and from the occurrence of broken and abraded grains (indicating a redeposition within a high-energy environment). Lower Cretaceous platform carbonates: Southern Apennines, Italy

Scale for both Figures 2 mm

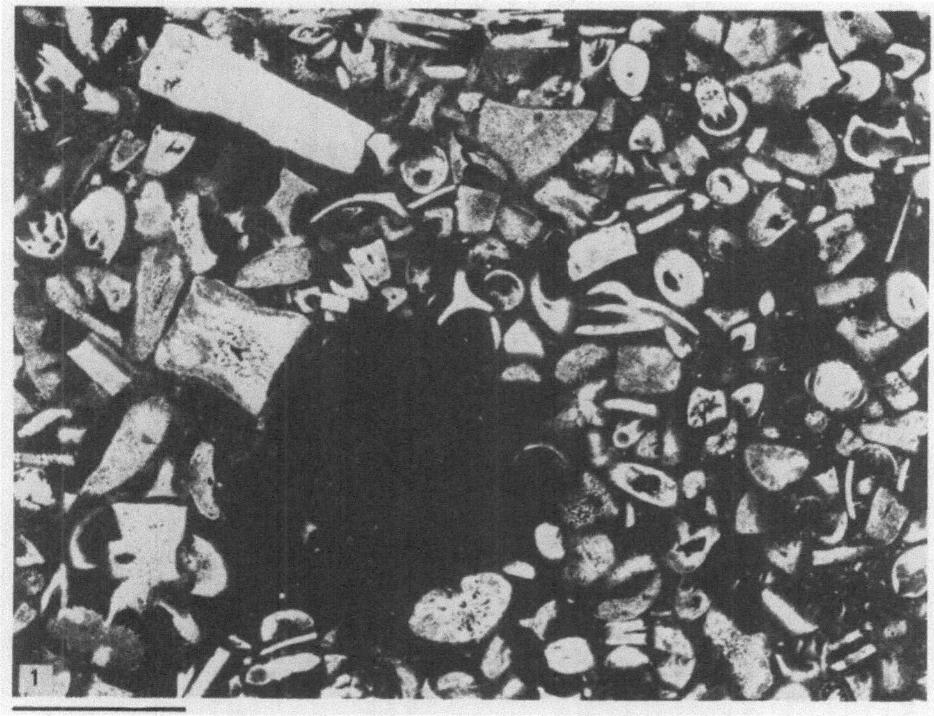

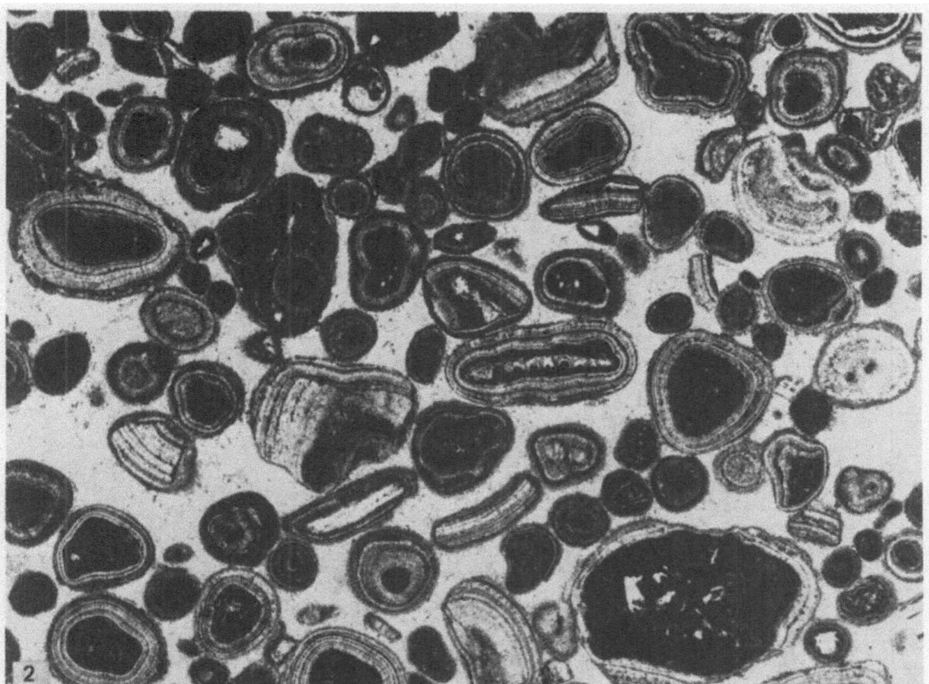

Table 39. Energy Index Classification of Limestones (after Plumley et al., 1962)

Limestone Type According to Energy Index	Limestone Sub-Types	Mineralogy	Texture			Fossil Abundance and Complexity	Characteristic Fossils Fossil Associations Fossil Preservations
			Size	Sorting	Roundness		
Quiet I Deposition in quiet water	I_1	Calcite Clay (15 to 50%) Detrital quartz (< 5%)	Microcrystalline carbonate (< 0.06 mm) or any size fossil fragments in a microcrystalline carbonate matrix (matrix < 50%)	Matrix: good; Fossils: poor	Original fossil shapes; angular fragments if broken	Barren to moderately fossiliferous Simple assemblages	Crinoids; echinoids; bryozoans (fragile branching types); solitary corals; ostracodes; thin-shelled brachiopods, pelecypods, and gastropods; Foraminifera; sponge spicules; tubular, encrusting, and sediment-binding algae; fecal pellets of bottom scavengers. Common fossil associations are crinoid-bryozoa assemblages, bivalve shell assemblages, Foraminifera assemblages (predominantly planktonic). Many fossils are whole and unbroken and are not mechanically abraded. Any fragmentation of fossil material probably is due to disarticulation upon death, to predatory (boring, opening, and breaking) activity and scavenger activity, or to solution.
	I_2	Calcite (predominant) Clay (< 15%) Detrital quartz (< 5%)					
	I_3		Any size fossil fragments in microcrystalline matrix (matrix < 50%)	Matrix: good: Fossils: moderate to good		Moderately to abundantly fossiliferous Simple assemblages (coquinoid limestone)	
Intermittently agitated II Deposition alternately in agitated water and in quiet water	II_1	Calcite (predominant) Clay (< 25%) Detrital quartz (< 50%)	Microcrystalline matrix (> 50%), Micrograined to medium-grained clastic carbonate and terrigenous material	Matrix: good; Clastic material: poor to good	Clastic carbonate material subangular to rounded. Roundness of terrigenous clastics is principally a function of size. Oolites may be present	Barren to moderately fossiliferous. Moderately simple assemblages	Characteristic fossils and fossil associations are similar to Type I limestones. Fossil materials are more fragmental than those in Type I limestones and also may be more or less rounded by wave action. Scattered fragments of fossils from rougher water environments may be present.
	II_2		Microcrystalline matrix (> 50%), Coarse to very coarse-grained clastic carbonate and terrigenous material				
	II_3		Interbedded microcrystalline carbonate and any size clastic. Microscale rhythmic bedding	Sorting good within individual lamina		Barren to moderately fossiliferous. Moderately complex assemblages	

Slightly agitated III Deposition in slightly agitated water	III_1	Calcite (predominant) Detrital quartz (up to 50%)	Micrograined clastic carbonate (< 0.06 mm) predominates	Matrix: good; Clastic material: moderate to good	Clastic material subrounded to well rounded. Fine-grained oolites may be present	Barren to sparsely fossiliferous Simple assemblages	Echinoderm, bryozoan, and bivalve shell debris; Foraminifera; encrusting algae. Common fossil associations are Foraminifera abraded bivalve shell fragment assemblages. Fossil materials comminuted from larger fossil structures are well abraded by wave and current action.
	III_2		Very fine-grained clastic carbonate (0.06 to 0.125 mm) predominates			Barren to moderately fossiliferous Simple assemblages	
	III_3		Fine-grained clastic carbonate (0.125 to 0.25 mm) predominates	Matrix: poor; Clastic material: moderate to good		Barren to abundantly fossiliferous Simple to moderately complex assemblages	
Moderately agitated IV Deposition in moderately agitated water	IV_1	Calcite (predominant) Detrital quartz (up to 50%)	Medium-grained clastic carbonate (0.25 to 0.5 mm) predominates	Matrix: poor; Clastic material: moderate to good	Clastic material subrounded to well rounded. Oolites may be present	Moderately to abundantly fossiliferous Simple to moderately complex assemblages	Crinoids, echinoids, bryozoans, brachiopod and pelecypod shell fragments, colonial coral fragments, stromatoporoid fragments (Silurian and Devonian predominantly); tubular algal fragments, colonial algal fragments (rare), encrusting algae. Common fossil associations are similar to associations of Types I, II, and III, or they are mixtures of these associations. Fossil materials are generally broken and abraded.
	IV_2		Coarse-grained clastic carbonate (0.5 to 1.0 mm) predominates				
	IV_3		Very coarse-grained clastic carbonate (1.0 to 2.0 mm) predominates			Moderately to abundantly fossiliferous Moderately complex to complex assemblages	
Strongly agitated V Deposition and growth in strongly agitated water	V_1	Calcite (predominant) Clay (<5%) Detrital quartz (<25%)	Gravel-size clastic carbonate (rock fragments and fossil material >2.0 mm) predominates	Matrix: poor; Clastic material: poor to moderate	Clastic material subrounded to well rounded. Pisolites may be present	Sparsely to moderately fossiliferous Complex assemblages	Crinoids; echinoids; encrusting bryozoans; thick-shelled brachiopods, pelecypods, and gastropods; colonial coral fragments; stromatoporoid fragments (Silurian and Devonian predominantly); colonial algal fragments; rudistid fragments (Cretaceous predominantly). Fossil associations are similar to Type IV associations. Fossil materials are generally broken and abraded.
	V_2		Gravel-size conglomeratic or brecciated carbonate (>2.0 mm) Tectonic breccias excluded	Matrix: poor; Clastic material: poor	Clastic material angular to well rounded	Barren to sparsely fossiliferous Complex assemblages	
	V_3	Calcite	Not applicable	Not applicable	Not applicable	Abundantly fossiliferous Simple assemblages (fossil colonial growth in place)	Colonial corals, stromatoporoids, colonial algae (principally the Rhodophyta or red algae and some genera of the Cyanophyta or blue-green algae).

f) Ooids (careful! – may originate in agitated waters, but also in quiet waters, see 4.1.3.5).
g) Wave-resistant colonial organisms in situ (evaluation of growth forms, e.g., in corals and stromatoporoids as well as Corallinaceae).
h) Characteristic sedimentary structures such as small-scale cross-bedding, imbrication structures.

This classification generally facilitates a rapid description of the possible water movement at the time of sedimentation. It is difficult to differentiate the proposed subgroups, because the non-carbonate constituents must be known quantitatively – which is not possible in thin-sections. One should be wary of overrating the energy-index classification, for the following reasons:

a) The particle sizes are brought into direct dependence on the water movement. These sizes, however, are controlled not only by the water energy but also by the supply available (e.g., size-dependent disintegration of skeletal grains, see Fig. 30); the sizes are also influenced by other factors in the environment (kind and intensity of the cement formation in grapestones; growth rate of algae in oncoids).

b) The abundance of detrital quartz is not necessarily dependent upon water energy. Due to eolian transport, carbonate is produced in quiet water environments with up to 50% of quartz – in Plumley et al. a figure of only 5% is mentioned.

c) As in all classifications, the occurrence of microcrystalline matrix does not prove *a priori* that quiet-water conditions existed, because sediment binding organisms can fix the finest sediment grains even in agitated water. This is especially true of algae.

d) The energy ranges given for the individual groups of organisms are too limited. Corals and calcareous algae also occur, for example, in Energy Types I and II. Thus, the fossil assemblages can only be marginally employed in the determination of the energy index.

e) The conjectured energy ranges must be checked if possible against observations in hand specimens and outcrops, because current marks, cross-bedding, etc. are often not clearly identifiable in thin-sections.

The advantage of the classification is that the vertical and horizontal facies changes can be quickly and simply described. As long as the different energy types are correlated with various water depths (careful!), simplified sedimentation models are produced (examples in Plumley et al., 1962, and Skupin, 1973 – trochite limestone, German: "Muschelkalk"). A critical evaluation of this classification has been undertaken by Bolliger and Burri (1970) in conjunction with work on the Jurassic limestones in Central Switzerland.

Examples for the Energy Indices (EI I–V) are shown in the following Plates

I: Plates 31/2; 36/1; 45/1
II: Plates 12/1; 31/4; 32/1; 46/3; 50/3; 51/1
III: Plates 35; 51/4, 6
IV: Plates 11/3; 21/2; 32/2; 38/1
V: Plates 31/7; 37/1; 48/1; 53/3

Consideration of the changes in the minimum and maximum energy indices (EI $\log_{min}$, EI $\log_{max}$) in profiles makes the recognition and description of the cyclic sedimentation pattern much easier (Catalov, 1972).

6.3 Discussion and Examples (Plate 31)

In the following *checklist for classification,* a compilation has been made of the *criteria* in thin-sections which are important for nomenclature:

A. Allochthonous limestones (e.g., "detrital limestones")
1. Which groundmass types occur? – micrite, microsparite, sparite (orthosparite or pseudosparite)?
2. Which of the following particles dominates? Which particles are second in abundance? – fossils or bioclasts, peloids, aggregate grains, oncoids, ooids, cortoids, intraclasts, extraclasts, terrigenous particles?
3. What is the grain/micrite ratio, i.e. grain/sparite ratio? The boundary values 1:9, 1:1, and 9:1 are often used. The ratio 9:1 is hardly ever achieved in nature. Hence, other boundary values such as 3:7, 1:1, and 7:3 may be more meaningful (see K. A. Schäfer, 1970).
4. Are the particles supported by mud or by other particles? What is the packing index?
5. In which range of sizes do the most abundant particles occur? Siltite, arenite, or rudite?
6. Are those particles, which can be rounded, actually round, or angular?

B. Autochthonous carbonates (e.g., "reef limestones")
1. Do the sessile organisms form relatively flat crusts stacked tightly together or biogenic encrustations (bindstone)?
2. Do the sessile organisms form vertical structures with sediment-filled interstitial spaces (bafflestone)?
3. Do the sessile organisms build "framework structures", where the individual colonies are bound together by encrustations of organisms (framestone)?
4. Which groups of organisms are dominant?
5. Are conspicuous communities of organisms present?

6.4 References: Classification of Carbonate Rocks

Baisert, H. (1967): Vergleich der Kalkklassifikationen. Z. angew. Geol. **12,** Berlin
Bissell, H. J., Chilingar, G. V. (1967): Classification of sedimentary carbonate rocks. In: Chilingar, G. V., Bissell, H. J. Fairbridge, R. W. (eds.): Carbonate Rocks. Dev. Sed. **9 A,** 87–168, 16 Pls., Amsterdam: Elsevier
Bramkamp, R. A., Powers, R. W. (1958): Classification of Arabian carbonate rocks. Geol. Soc. Amer. Bull. **69,** 1305–1318, New York
Bolliger, W., Burri, P. (1970): see 4.1.7.2
Catalov, G. (1971): Structural-genetic classification of the limestones. Bull. Geol. Inst. Sev. Strat. Lithol. **20,** 133–156, Pls. 1–6, 1 Tab., Sofia (Bulgarian, English summary)
Catalov, G. A. (1972): An attempt at energy index (EI) analysis of the Upper Anisian, Ladinian and Carnian carbonate rocks in the Teteven Anticlinorium (Bulgaria). Sed. Geol. **8,** 159–175, 4 Pls., 5 Figs., Amsterdam
Chilingar, G. V. (1960): Notes on classification of carbonate rocks on basis of chemical composition. J. Sed. Petrol. **30**/1, 157–158, 3 Tab., Tulsa
Choquette, P. W., Pray, L. C. (1970): Geologic nomenclature and classification of porosity in

sedimentary carbonates. Bull. Amer. Ass. Petrol., Geol. **54**/2, 107–250, 13 Figs., Littleton, Col. 80121
Dunham, R. J. (1962): Classification of carbonate rocks according to depositional texture. Mem. Amer. Ass. Petrol. Geol. **1,** 108–121, 7 Pls., Tulsa
Embry, A. F., Klovan, E. J. (1972): Absolute water depths limits of Late Devonian paleoecological zones. Geol. Rdsch. **61**/2, 10 Figs., Stuttgart
Feray, D. E., Heuer, E., Hewatt, W. G. (1962): Biological, genetic and utilitarian aspects of limestone classification. Mem. Amer. Ass. Petrol. Geol. **1,** 20–32, 3 Figs., Tulsa
Folk, R. L. (1959): Practical petrographical classification of limestones. Amer. Ass. Petrol. Geol. Bull. **43**/1, 1–38, 41 Figs., Tulsa
Folk, R. L. (1962): Spectral subdivision of limestone types. Amer. Ass. Petrol., Geol. Mem. **1,** 62–84, 1 Pl., 7 Figs., Tulsa
Füchtbauer, H. (1959): Zur Nomenklatur der Sedimentgesteine. Erdöl u. Kohle **12,** 605–613, 7 Figs., Hamburg
Füchtbauer, H. (1974): see 1.4.1
Füchtbauer, H., Müller, G. (1970): Sedimente und Sedimentgesteine. In: Sediment-Petrologie **3,** 378 pp., 133 Figs., 55 Tab., Stuttgart: Schweizerbart
Grabau, A. W. (1904): On the classification of sedimentary rocks. Amer. Geol. **33,** 228–247
Imbrie, J., Purdy, E. G. (1962): Classification of modern Bahamian carbonate sediments. Mem. Amer. Ass. Petrol. Geol. **1,** 253–272, 13 Figs., Tulsa
Leighton, M. W., Pendexter, C. (1962): Carbonate rock types. Mem. Amer. Ass. Petrol. Geol. **1,** 33–61, 9 Pls., 2 Figs., Tulsa
Mamet, B. (1961): Reflexions sur la classification des calcaires. Bull. Soc. belge Géol. Paléont. Hydrol. **70,** 48–64, 7 Figs., Bruxelles
Mišik, M. (1972): Lithologische und fazielle Analyse der mittleren Trias der Kerngebirge der Westkarpaten. Acta Geol. Geogr. Univ. Comenianae, Geol. **22,** 5–154, Pls. 1–54, 9 Figs., Bratislava
Monty, Cl. (1963): Bases d'une nomenclature des roches calcaires marines. Ann. Soc. Géol. Belgique **86,** 1962/63, Mem. 2 B 87–122, 1 Pl., 1 Tab., Bruxelles
Müller-Jungbluth, W. V., Toschek, P. H. (1969): Karbonatsedimentologische Arbeitsgrundlagen. 2. Aufl., Veröff. Univ. Innsbruck **8,** Alpenkundl. Studien 4, 32 pp., 3 Pls., Innsbruck
Pettijohn, J. (1957): see 1.4.1
Pirlet, H. (1965): Classification des Calcaires de la Série des Areno-Cryptites. Ann. Soc. Géol. Belgique **88,** 1964/65, Bull. 7–10, B 395–B 410, 2 Figs., 1 Tab., Bruxelles
Plumley, W. J., Risley, G. A., Graves, R. W., Kaley, M. E. (1962): Energy index for limestone interpretation and classification. Mem. Amer. Ass. Petrol., Geol. **1,** 85–107, 4 Pls., 5 Figs., Tulsa
Purser, B. H. (1972): Subdivision et interprétation des sequences carbonateés. Mem. Bur. Rech. Geol. Mines **77,** 679–698, Paris
Robinson, R. B. (1966): Classification of reservoir rocks by surface texture. Bull. Amer. Ass. Petrol. **50,** 3/1, 547–559, Tulsa
Rutte, E. (1954): Eine Klassifikation der karbonatischen Süßwassergesteine mit Beispielen aus Südwestdeutschland. N. Jb. Geol. Paläont. Abh. **100,** 208–246, Stuttgart
Sander, B. (1936): Beiträge zur Kenntnis der Anlagerungsgefüge (rhythmische Kalke und Dolomite aus der Trias). Tschermaks Min. Petrogr., Mitt. **48,** 1/2, 3/4, 27–139, Leipzig
Sander, N. J. (1967): Classification of carbonate rocks of marine origin. Bull. Amer. Ass. Petrol. Geol. **52,** 3/1, 325–336, Tulsa
Schäfer, K. A. (1970): Zur Klassifikation der bioklastischen Karbonatgesteine des Unteren Hauptmuschelkalkes in Baden-Württemberg. N. Jb. Geol. Paläont. Mh. **1970**/2, 102–115, 5 Figs., 2 Tab., Stuttgart
Skupin, K. (1973): see 4.3.2
Todd, T. W. (1966): Petrogenetic classification of carbonate rocks. J. Sed. Petrol. **36**/2, 317–340, 3 Figs., 10 Tab., Tulsa
Wolf, K. H. (1961): An introduction to the classification of limestones. N. Jb. Geol. Paläontol. Mh. **1961**/5, 236–250, 2 Figs., 1 Tab., Stuttgart
Zuffa, G. G. (1980): Hybrid arenites: Their composition and classification. J. Sed. Petrol. **50**/1, 21–29, 3 Figs., Tulsa

7 Microfacies Types

If thin-sections of limestones have not been analyzed with just paleontological or petrographic determinations in mind, basic "types" may be defined in order to compare samples and achieve reproducible results.

7.1 Prerequisites

An essential part of facies analyses is the differentiation of microfacies (=MF) types which can be interpreted genetically. The criteria for defining these types should preferably be those features whose existence and frequency are due to relatively few

Table 40. Microfacies variations in three shelf limestone beds; Lower Permian of the Carnic Alps, Austria (compare Fig. 54)

Bed	Lithology	Bed Thickness	Lateral Sampling Interval	Variation (after point-counting analysis)
D 64	Biomicrite or biosparite with fusulinids, phylloid algae and Smaller foraminifera	50 cm	Alternately 30 cm and 3 m (30 samples total)	Deviation percentages of skeletal grains: a) 30 cm interval/15 sample pairs – 6 pairs 10% deviation, 8 pairs 10–20%, 1 pair 50%; b) 3 m interval/15 pairs – 12 pairs 20%, 1 pair 20–30%, 2 pairs 50%. Result: The intervals for representative samples must be less than 3 m; mean percentage of the bed is *not* significant for microfacies. If the samples were classified after Folk, 4 types would be distinguished!
B 274	Biosparite with echinoderms, bryozoans, and foraminifera	40 cm	4 m (15 samples)	Very slight differences in the quantitative composition of the individual samples; mean percentage of the bed is also representative for the microfacies; 2 types after Folk
B 300	Biomicrite with echinoderms and Smaller foraminifera	45 cm	20 cm (10 samples)	Differences chiefly in the matrix/particle ratio; mean percentage of the bed *not* significant; 2 types after Folk

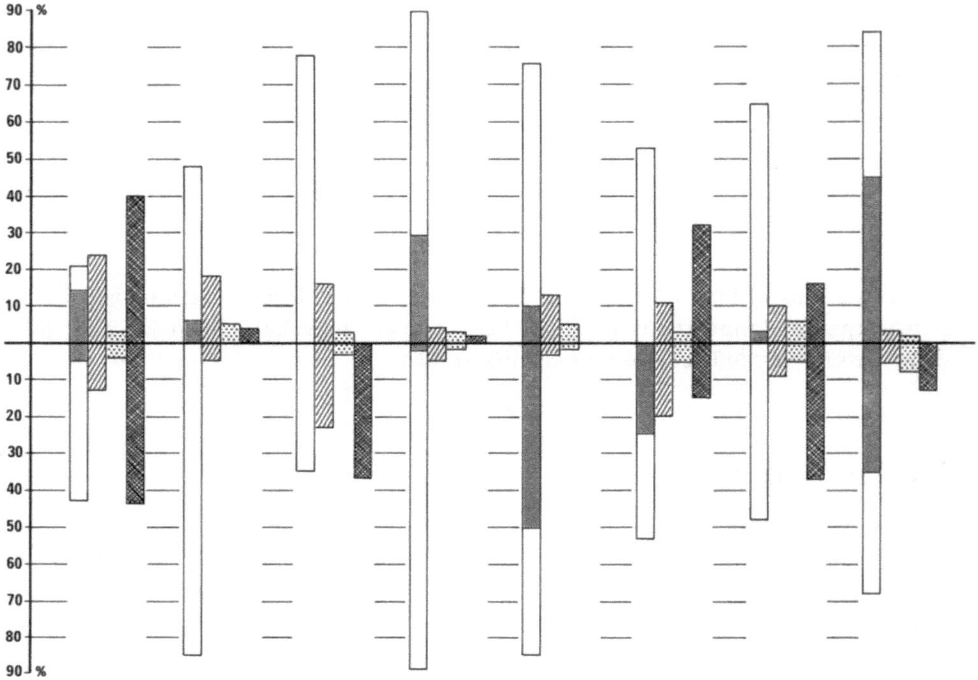

Fig. 54. Microfacies variation in a limestone bed. Each of the samples was taken at a horizontal interval of 30 cm (graphed above and below the middle line of the figure) and a horizontal interval of 3 m between the pairs of samples; the modal composition was described by point-counting analysis. Compare the results with Table 40 (bed 64 D). Legend: *white* – sparite; *stippling* – micrite; *diagonals* – fusulinids; *dots* – Smaller foraminifera; *cross-hatching* – algae, especially phylloid algae. This example comes from the Lower Permian Rattendorf formation (Lower Pseudoschwagerina limestones) of the Carnic Alps, Austria. These limestones represent sediments of the inner shelf environment

environmental factors or which are linked with certain sedimentary environments ("Micromechanisms of Sedimentation", Carozzi, 1958). Essentially these characteristics comprise:

- matrix, particles, cement, size and shape of particles, kinds and growth forms of organisms, associations of organisms, particle fabrics

The frequency of the particles is often referred to as an essential criterion for differentiating MF types. Types defined in this manner coincide in some cases with groups found in classifications of limestones. However, as long as no other paleontological indices are available, these carbonate types often cannot be used for genetic interpretations.

Categories of thin-section samples are linked with the question of *"microfacies variation":*

Included in this is the scattering of characteristics used for microfacies classification around a reference standard within a sampling unit (e.g., in a limestone bed). Basically it must be assumed that the composition of a limestone bed, and hence the MF types, changes laterally. This problem is exemplified by the Leavenworth limestone unit which stretches over 540 km (Pennsylvanian, Kansas, Toomey, 1966); here a quantitative survey of the thin-section characteristics leads to three facies types, which in turn must be assigned to various depositional environments. Another example from the Lower Permian of the Carnic Alps, Austria (Table 40, Fig. 54) shows that the quantitative and qualitative differences which can be determined in limestone beds vary: For instance, in samples of uniformly composed and well-sorted calcarenites in Bed 274, the deviation in percentile compositions between samples is so small that the calculated arithmetic mean of their total can be considered as representative of the composition of the whole bed. Conversely, the standard deviations of the samples taken from beds D 64 and B 300 are greater than 15%; thus, the calculated arithmetic mean of the whole bed can not be used to define a microfacies type. A classification of the samples solely after Folk is very unlikely to lead to groups which can be interpreted in terms of depositional environments, etc. Additional help in such cases can be obtained from statistical methods, for example, the sequence test, Eltgen (1969, 1970); see Sect. 7.3.1.

Microfacies variation is present especially in tide-dominated carbonate sediments (see Smosna and Warshauer, 1979); this makes it difficult to apply standard carbonate classification systems to these limestones and necessitates the use of numerical methods. A primarily patchy distribution of the biota may be another cause for striking microfacies variations, as shown by discriminant function analysis of the lateral and vertical composition of shallow-marine Ordovician limestones (Harris and Martin, 1979).

7.2 Qualitative Categorization

Most MF types are differentiated by qualitative criteria such as the presence or absence, or relative frequency of selected features. Generally, the "primary" features expressed by depositional fabrics are preferred and are usually documented by representative thin-section photographs. Such classifications are highly subjective, but the weighting of the criteria, which precedes the classification, appears to be very advantageous indeed for genetic interpretations.

Basically it can be assumed that carbonate environments actually have a rather limited number of MF types: This can be seen in a comparison of the number of facies types in Recent carbonate environments with the number of MF types in Ancient limestones of various ages. In the shallow marine environments of the Bahama Banks, no more than 6 facies types can be distinguished; in the Persian Gulf, no more than 12. The number of facies types distinguished in deeper marine environments is generally even lower. If we base the comparison of the Ancient MF types on the data of Wilson (1975), supplemented by our own observations, we obtain the following numbers:

Devonian, Central Europe: basinal facies 2 MF types; platform with open circulation 2; platform slope 3; platform margin 3; near-coastal inner platform 3.

Permian reef-complex, Texas: basin 1, deeper shelf margin 1, platform slope 1, reef area 3, platform margin 4, protected lagoon 2.

Upper Triassic reefs of the Northern Alps: reef core 2, reef slope 1, back-reef area with open circulation 2, shoals in high-energy environment 3, protected shelf lagoon 5, protected shoals in the lagoon 3.

Upper Jurassic platform and basinal facies in the Northern Alps: platform with open circulation 4, near-coastal inner platform in high-energy environment 3, seamounts 3, basinal facies 3.

Cretaceous of Mexico: basin and marginal basin 1, slope 2, reef 1, shelf margin with open circulation 1.

Miocene reef facies in the Philippines (Carozzi et al., 1976): tidal flat 1, coastal zone 1, lagoon 6, back-reef 2, reef 1, fore-reef 1, lower fore-reef slope 2, open marine area 1, deeper area with carbonate turbidites 1.

Various authors, such as Klovan (1964), Müller-Jungbluth and Toschek (1969) and E. Flügel (1972) have developed work sheets for MF analyses. The application of such work sheets depends upon the problems involved and upon the number of criteria described in the thin-sections. In practice, though, most people prefer to make up their own work sheets.

To facilitate the categorization process for individual research, the checklist in Section 7.4 is strongly recommended; MF types which indicate depositional environments and the mechanisms of deposition can then be more easily defined.

7.3 Quantitative Categorization

As long as the frequency of certain particles, that is, the modal composition is considered to be important facies criteria, and sufficient significant data are available, statistical comparisons of the similarity of MF samples may be carried out. Many methods are used, among them cluster and factor analyses, whereby either quantitative data (e.g., Imbrie and Purdy, 1962) or semi-quantitative data may be used (presence-absence data, e.g. Bonham-Carter, 1965, 1967 a; Buchbinder, 1975; Hohenegger, 1974).

7.3.1 Statistical Methods

The purpose of using statistical methods for facies analyses is to reduce the large numbers of criteria observed and thus make the data to be interpreted more manageable.

The basis for such analyses is generally a large number of thin-sections, data on the frequency of microfacies features, and the desire to replace subjective classifications with "objective" ones. The frequency data can be based either on point-counting methods, or on the presence or absence of certain criteria, or on the relative abundance of the criteria (rare, common, abundant = semi-quantitative data).

These data are the basis for analyses of the similarity or dissimilarity of samples. By using various correlation coefficients or the ordination method (see Ekdale et al., 1976), either the total body of samples (Q-mode) or n variables (R-mode) can be compared with each other. Cluster and factor analyses are used for these comparisons.

To describe random samples and quantitatively defined relations between variables, the correlation coefficient (usually the product-moment correlation or the Pearson-Bravaissche correlation), the distance coefficient (e.g., Harbaugh, 1964), and the cosine-theta coefficient (e.g., Imbrie, 1963) are available. Each of the three coefficients can be used in Q-mode as well as in R-mode analyses. Comparative calculations show that the selection of the similarity measure is not a major problem, because all three coefficients lead to similar classifications of variables and random samples (see E. Flügel and Hötzl, 1976; Dörhöfer, 1974). The correlation coefficient, however, is somewhat more advantageous for factor analysis.

If semi-quantitative data are used, the cluster analysis is based on binary coefficients (see Cheetham and Hazel, 1969), such as the Jaccard coefficient (see Bonham-Carter, 1965; Buchbinder, 1975; Ekdale et al., 1976; Jaquet, 1973 a).

The results of cluster analyses are plotted on dendrograms in which all the statistically possible variables or random samples are shown in groups.

Cluster analyses are an efficient way of displaying complex relationships among many individual variables. Factor analysis gives a survey of the causal factors hidden behind a large number of co-related variables. In other words, from a large number of observed criteria hypothetical models can be deduced which reproduce the possible relations between the variables. Under certain circumstances these "factors" can be interpreted as real causal relations, such as environmental conditions. Such interpretations, however, should be critically tested using facies analyses and, if possible, supported by paleoecological investigations. The advantage of factor analysis in facies analyses lies in the fact that the "possibilities" to be interpreted are reduced in quantity and the factors are weighted for each individual criterion.

Cluster and factor analyses produce classifications which in ideal situations contain "reaction groups" as basic units. Purdy (1963) understands these as groups defined by certain particle associations and similar reactions to changes in the environment (see Keskin, 1966).

In addition to these methods, there is an increasing tendency to apply time-trend analyses to MF studies in order to make changes in stratigraphic profiles (Fox and Brown, 1965; Rao and Carozzi, 1971) clearer.

Simulation methods seem to be of particular importance in that they generate the further statistical development of a series of genetic models, which in turn make the possible sedimentary processes more obvious (Harbaugh and Bonham-Carter, 1970). Harbaugh (1966 a) described the advantages of this modeling process, illustrating it with a marine shallow water facies influenced by various stages of algal growth; Lumsden (1975) demonstrated the possibilities of calcarenite formation.

The EDP method suggested by Charollais and Davaud (1974), in which potentially all the theoretically possible microfacies data are described, requires a uniform codification system. The computer program designed by these authors starts with a total of 772 or 876 parameters which serve to identify and describe the MF sample. Here structural and mineralogical data as well as various carbonate particles and skeletal grains are taken into consideration. The program EXTRACT permits the classification of samples after Dunham (1962) and Folk (1959). Other programs help to follow the variation in profiles, to carry out cluster analyses (up to 1000 samples, each with 200 variables), and to graph similarity groups in dendrograms, or to calculate them in the form of binary coefficients.

With the EDP method large numbers of samples can be rapidly processed and ordered. But objectification of MF classification cannot be achieved solely by exhausting multivariate analyses as long as the resultant groups are still dependent to a large degree on the kind of method used. McCammon (1968) vividly demonstrated how different facies groups and facies maps can be, when different multivariate methods of classifications are used.

The following list of references contains the most important and stimulating articles on methods.

7.3.2 Annotated Bibliography: Multivariate Facies Analysis

M = methodologically important articles, C = quantitative facies analyses of carbonates

(C) Behrens, E. W. (1965): Environment reconstruction for a part of the Glen Rose Limestone, Central Texas: Sedimentology **4,** 65–111, 10 Pls., 11 Figs., Amsterdam
Lower Cretaceous. 15 m thick limestone sequences. 199 samples from 24 localities as well as about 300 randomly chosen samples. Description of MF criteria by point-counting as well as mineral constituents and non-carbonate residues. Calculation of "reaction groups" for the samples (R-mode, product-moment correlation) led to a differentiation of 4 facies and a total of 6 subfacies units occurring in intertidal and subtidal zones, and in hypersaline areas.

(M) Bonham-Carter, G. F. (1965): A numerical method of classification using qualitative and semi-quantitative data, as applied to the facies analysis of limestones. Bull. Canad. Petrol. Geol. **13**/4, 482–502, 3 Figs., 9 Tab., Calgary
The samples are described according to the presence or absence or the relative abundance of the criteria. Hence, not measurable properties (e.g. rock color) can also be considered. The reciprocal relation of all the sample pairs (Q-mode, association coefficient after Jaccard) is described by cluster analysis and plotted on a dendrogram. A comparison of the results of semi-quantitative methods with those of point counting data and correlation analyses (Example Bahama Bank, Purdy) led to differences in facies classifications for only about 12% of the samples.

(C) Bonham-Carter, G. F. (1967 a): An example of the analysis of semi-quantitative petrographic data. Proc. Seventh World Petrol. Congr., Mexico City, 567–583
Pennsylvanian, Canada. Semi-quantitative categorization according to the methods described in 1965.

(M) Bonham-Carter, G. F (1967 b): Fortran IV program for Q-mode cluster analysis of non-quantitative petrographic data using IBM 7098/7094 computers. State Geol. Surv. Kansas, Computer Contrib. **17,** 28 pp., Lawrence

(M) Bouroullec, J., Deloffre, R. (1974): Essai d'utilisation statistique des microfacies dans le Crétacé moyen d'Aquitaine occidentale (France). Bull. Centre Rech. Pau-SNPA **8**/2, 469–509, 5 Pls., 8Figs., 6 Tab., Pau

Albian and Cenomanian. Material from 23 oil wells. Thin-sections of samples taken every 2 m were quantitatively evaluated and the frequency of the MF criteria in all the samples of the lithostratigraphic units were compared with the frequency data of other units. The evolution of the sedimentary environment expressed by these changes in frequency can be interpreted as deltaic sedimentation or platform sedimentation.

(M, C) Buchbinder, B. (1975): Lithogenesis of Miocene reef limestones in Israel with particular reference to the significance of the Red Algae. Geol. Surv. Israel, Oil Res. Div., Rep. OD/3/75, 173 pp., 6+6 Pls., Jerusalem
Upper Miocene. Investigation of 249 biocalcarenite samples according to the following scheme: visual estimation of the particle frequency in the thin-section; frequency histograms and calculation of means as well as the standard deviation; subdivision into three frequency classes and processing of the material according to presence/absence; classification of the samples by association analyses and cluster analyses (Q-mode, Jaccard coefficient); check of the relative importance of the groups in each classification (Chi-square test); comparison of the classifications obtained from the various methods: Association analysis permitted detailed groupings and led to more uniform groups than cluster analysis.

(M) Burr, E. J. (1968): Cluster sorting with mixed character types. I. Standardization of character values. Aust. Computer J **1**, 97–99

(M) Burr, E. J. (1970): Cluster sorting with mixed character types. II. Fusion strategies. Aust. Computer J. **2**, 98–103

Carozzi, A. V. (1958): Micro-mechanisms of sedimentation in the epicontinental environment. J. Sed. Petrol. **28**/2, 133–150, 13 Figs., Tulsa

Carozzi, A. V., Reyes, M. V., Ocampo, V. P. (1976): Microfacies and microfossils of the Miocene reef carbonates of the Philippines. Spec. Publ. **1**, Philippine Oil Dev. Co., 79 pp., 20 Pls., Manila

(M, C) Carss, B. W. Carozzi, A. V. (1965): Petrology of Upper Devonian pelletoidal limestones, Arrow Canyon Range, Clark County, Nevada. Sedimentology **4**, 197–224, 2 Pls., 12 Figs., 3 Tab., Amsterdam
Lower Late Devonian. About 360 m thick limestone sequences; sample interval 60 cm. Thin-section study (clasticity indices and frequency indices for all detrital particles; frequency indices of the skeletal grains). On the basis of the arithmetic means of these indices, 10 MF types were distinguished, which could be assigned to different water depths. To check the accuracy with which the samples were assigned to a certain MF type and to determine the reciprocal relations between the MF types, correlation analyses were used.

(M) Charollais, J., Davaud, F. (1974): Microfaciès et ordinateur. 224 pp., 188 Figs., Paris: Editions Technique
The authors suggest standardized methods for the codification and quantification of microfacies criteria in sedimentary rocks so that computer analyses can be carried out. The book is divided into three sections: sample designation and microfacies data; codification; programs for evaluating data (e.g. classification of new samples, plots of MF criteria). The programs are written in Fortran IV.

(M) Cheetham, A. H., Hazel, J. E. (1969): Binary (Presence-Absence) similarity coefficients. J. Palaeont. **43**/5, 1130–1136, Tulsa
Critical survey of the coefficients used in association analyses, by which the similarities or differences in samples can be described.

(M) Davis, J. C. (1973): Statistics and data analysis in geology. 550 pp., New York: Wiley

(M) Demirmen, F. (1969): Multivariate procedures and Fortran-IV program for evaluation and improvement of classifications. Kansas State Geol. Surv., Computer Contrib. **31**, 1–51, Lawrence

(C) Dörhöfer, S. (1974): Mikrofazielle Untersuchungen der Kalksteine des Mündener Mergels von Thüste. Unveröff. Diplomarb. TU Hannover, 40 pp., 9 Pls., 12 Figs., 1 encl., Hannover (typescript)
Upper Jurassic, Westphalia. Differentiation of microfacies types with factor and cluster analyses (use of distance and correlation coefficients produces similar results). Comparison of groupings based on sedimentary and diagenetic criteria with a method of analysis that

disregards diagenetic criteria produces a clearer separation into facies units, hence, a clearer interpretation.

Dunham, R. J. (1962): see 6.4

(M, C) Ekdale, A. A., Ekdale, S. E., Wilson, J. L. (1976): Numerical analysis of carbonate microfacies in the Cupido Limestone (Neocomian-Aptian), Coahuila, Mexico, J. Sed. Petrol. **46**/2, 362–368, 7 Figs., Tulsa
Lower Cretaceous. Samples evaluated on the basis of 16 thin-section criteria were subjected to a cluster analysis, where only the presence or absence of criteria were noted. The Q-mode leads to a differentiation of 15 groups (= microfacies). A comparison of sample similarity based on the ordination method indicated the possible original relationships between the MF types.

(C) Ellenor, D. W. (1975): Sedimentology of the Middle Devonian Timor Limestone, Northeastern New South Wales, Australia. Sed. Geol. **13,** 125–152, 3 Pls., 12 Figs., 1 Tab., Amsterdam
Eifelian/Givetian. 14 profiles; sample interval 1.5 m; 697 thin-sections, 13 MF criteria per thin-section; point-counting analyses. Differentiation of 5 facies types and several subtypes by cluster analysis.

(M) Eltgen, H. (1969): Statistische Methoden in der Mikro-Biofazies. Sedimentology **11,** 259–274, 8 Figs., Amsterdam

(M) Eltgen, H. (1970): Statistische Methoden in der Mikro-Biofazies-Analyse. Clausthaler Tekton. H. **10,** 363–374, Clausthal
Upper Jurassic and Cretaceous, Westphalia. Check of the variation in the microfacies by comparison of the mean values of the organogenic and inorganic constituents (circle graphs indicate only the general abundance of the particles; coordinate systems indicate spatial developments); correlation analyses (results unsatisfactory; reasons: the distribution of MF criteria does not correspond to a normal distribution); discriminance analysis, linked with variance analysis and the Chi-squares test as well as sequence analyses allow the degree of homogeneity of the limestone beds to be determined. The necessary number of samples can be determined by the sequence tests. The variation of the particles within the individual beds is not any greater than between the beds.

(M) Everitt, B. (1974): Cluster Analysis. London: Heinemann

(C) Fenninger, A. (1970): Faktorenanalyse nordalpiner Malmkalke. Verh. geol. Bundesanst. **1970**/4, 618–636, 7 Figs., 6 Tab., Wien
Upper Jurassic carbonates from shallow water, sills, and basins. 365 samples, 15 criteria; point counting method; cluster analysis (R-mode) and factor analysis accentuate variable interrelationships and thus permit a better rating of the criteria as facies indicators.

(M) Flügel, E. (1972): Mikrofazielle Untersuchungen in der alpinen Trias. Methoden und Probleme. Mitt. Ges. Geol. Bergbaustud. Österreich **21**/1, 7–64, 9 Pls., 1 supp., 1 Fig., 1 Tab., Innsbruck

(M) Flügel, E., Hötzl, H. (1976): Palökologische und statistische Untersuchungen in mitteldevonischen Schelf-Kalken (Schwelmer Kalk, Givet; Rheinisches Schiefergebirge). Bayer. Akad. Wiss., Math-naturwiss. Kl., Abh., N.F. **156,** 70 pp., 29 Figs. 17 Tab., München
Givetian, Sauerland, Germany. Comparative studies of statistical methods in regard to their suitability for facies analyses. The following methods were compared: point-counting and line methods; correlation coefficients, cosine-theta coefficients, and distance coefficients used in cluster analyses (no significant differences). No advantages between oblique factor analysis and rotated orthogonal factor analysis.

Folk, R. L. (1959): see 6.4

(M) Forgotson, J. M. Jr., Iglehart, C. F. (1967): Current uses of computers by exploration geologists. Bull. Amer. Ass. Petrol. Geol. **51**/7, 1202–1224, 27 Figs., 3 Tab., Tulsa

(M) Fox, W. T., Brown, J. A. (1965): The use of time-trend analysis for environmental interpretation of limestones. J. Geol. **73**/3, 510–518, 4 Figs., Chicago
Upper Ordovician, Indiana/USA, 28.5 m thick limestone-marl sequences; 105 limestone samples; point-counting analysis. Traces changes in frequency of the MF criteria in the vertical profile by using time-trend curves with various equations.

(C) Griffith, L. S., Pitcher, M. G., Rice, G. W. (1969): Quantitative environmental analysis of a Lower Cretaceous reef complex. Soc. Econ. Paleont. Min. Spec. Paper **14,** 120–138, 34 Figs., Tulsa

A comparison of the Recent sedimentary conditions off the Florida coast with the Lower Cretaceous El Abra reef in Mexico and the carbonates of the Lower Cretaceous Edward formation in Texas, which is known only from boreholes. The results of factor analyses, which took into consideration all the data available, and factor analyses, which only considered the more important faunal elements, were in almost complete agreement.

(M) Griffiths, J. C. (1967): Scientific Method in Analysis of Sediments. 508 pp., McGraw Hill Int. Ser. Earth Planetary Sci., New York

(M) Harbaugh, J. W. (1964): Balgol programs for calculation of distance coefficients and correlation coefficients using an IBM 7090 computer. Spec. Distrib. Publ., State Geol. Surv. Kansas **9,** 33 pp., Lawrence

(M, C) Harbaugh, J. W. (1966 a): Mathematical simulation of marine sedimentation with IBM 7090/7094 computers. Computer Contrib. **1,** State Geol. Surv. Kansas, 52 pp., 22 Figs., Lawrence

(C) Harbaugh, J. W. (1966 b): Computer simulation of marine-organism community environments. Bull. Amer. Ass. Petrol. Geol. **50**/3, 616–617, Tulsa
Mathematical simulation of inorganic processes and processes controlled by organisms in a shallow marine sedimentary environment. A comparison with the facies features observed in Pennsylvanian algal limestones allowed the model to be checked.

(M) Harbaugh, J. W., Bonham-Carter, G. (1970): Computer Simulation in Geology. 575 pp., New York: Wiley

(C, M) Harbaugh, J., Demirmen, F. (1964): Application of factor analysis to petrologic variations of Americus limestone (Lower Permian), Kansas and Oklahoma. State Geol. Surv. Kansas, Spec, Distrib. Publ. **15,** 41 pp., 15 Figs., Lawrence
30 cm to 1.5 m thick unit followed over 450 km; 97 samples from 27 localities; point-counting analyses. Check of the variation occurring within and between the samples by factor analyses (Q-mode and R-mode) and trend-surface analyses.

(M) Harbaugh, J., Merriam, D. F. (1968): Computer Applications in Stratigraphic Analysis. 282 pp., New York: Wiley

(C) Harris, F. W., Martin, W. D. (1979): Benthic community development in limestone beds of the Waynesville (Upper Dillsboro) Formation (Cincinnatian Series, Upper Ordovician) of southeastern Indiana. J. Sed. Petrol. **49**/4, 1295–1306, 12 Figs., Tulsa
Vertical and lateral changes in composition within 13 individual limestone beds are investigated using certain variables and discriminant analysis. Differences can be recognized for the tops, bottoms and centers of the beds indicating changes in community successions.

(M) Hohenegger, J. (1974): Über einfache Gruppierungsmethoden von Fossilvergesellschaftungen am Beispiel obertriadischer Foraminiferen. N. Jb. Geol. Paläont. Abh. **146**/3, 263–297, 11 Figs., Stuttgart

(M) Hriskevich, M. E (1970): Middle Devonian reef production, Rainbow Area, Alberta, Canada. Bull. Amer. Ass. Petrol. Geol. **54**/12, 2260–2281, 27 Figs., Tulsa
Mathematical simulation as an aid to prospecting

(C, M) Imbrie, J., Purdy, E. G. (1962): Classification of modern Bahamian carbonate sediments. Mem. Amer. Ass. Petrol. Geol. **1,** 253–272, 13 Figs., Tulsa
Factor analysis of the samples from the Bahama Banks, which were studied and classified by Purdy (1963), allow a classification based on the "equivalency" of all criteria and the Folk classification to be compared.

(M, C) Jaquet, J.-M. (1973 a): Définition de microfaciès par classification automatique et comparison des descripteurs. Arch. Sci. Geneve **26**/2, 137–171, 2 Pls., 7 Figs., 7 Tab., Geneva
Upper Berriasian, Jura mountains/France. 700 samples; 41 non-numerical criteria. Calculation of the similarity coefficients on the basis of presence or absence: dendrograms after Sokal and Sneath; 7 MF types were differentiated. The determination of the variations between the variables of a sample and between the samples themselves permit the differentiation of sensible facies units.

(M, C) Jaquet, J.-M. (1973 b): Analyse statistique de la variation verticale des descripteurs dans la formation de Vions (Berriasien supérieur, Jura méridional). Arch. Sci. Geneve **26**/1, 246–283, 3 Pls., 8 Figs., 4 Tab., Genf

Study of the vertical variations of the carbonate, quartz, and clay content, the mean quartz grain diameter etc., with trend and auto-correlation analyses.
(M) Jöreskog, K. G., Klovan, J. E., Reyment, C. (1976): Geological factor analysis. 178 p., New York: Elsevier
(C) Keskin, C. (1966): Microfacies study of the Pinarhisar Reef Complex, Istanbul Univ. Fen Fak. Mecmuasi, Ser. B **31**/3–4, 109–146, 12 Pls., 6 Figs., 6 Tab., Istanbul
Eocene, Thrace/Turkey. 1000 samples; point-counting analyses. Calculation of "reaction groups" for 730 samples, factor analyses. Five facies types were determined (bryozoan-algae F, coral-hydrozoan F, coastal F, mud F with echinoid fragments, foraminifera facies).
(C) Klovan, J. E. (1964): Facies analysis of the Redwater Reef Complex, Alberta, Canada. Bull. Canadian Petrol. Geol. **12**/1, 1–100, 9 Pls., Calgary
(M) Klovan, J. E., Billings, G. K. (1967): Classification of geological samples by discriminant-function analysis. Bull. Canadian Petrol. Geol. **15**/3, 313–330, Calgary
(M) Klovan, J. E., Imbrie, J. (1971): An Algorithm and Fortran-IV program for large-scale Q-mode factor analysis and calculation of factor scores. J. Math. Geol. **3**/1, 61–77, New York
(M) Krumbein, W. C. (1968): Statistical models in sedimentology. Sedimentology **10**/1, 7–23, 4 Figs., 4 Tab., Amsterdam
(M) Krumbein, W., Graybill, F. (1965): An Introduction to Statistical Models in Geology. 475 pp., New York: McGraw Hill
(M) Lumsden, D. N. (1975): Computer simulation experiments with clastic limestones. J. Geol. **83**/3, 375–386, 8 Figs., Chicago
In order to explain the development of detrital limestones, the possible relationships between the water energy at the bottom of the sea and five main constituents of limestones (micrite, sparite, skeletal grains, cortoids, ooids) were used for mathematical simulations.
(C) Lumsden, D. N. (1975): Markov chain analysis of computer-simulated limestone sequences. Bull. Geol. Soc. Amer. **86**, 1067–1072, 7 Figs., New York
No relationships could be found between statistical characteristics and the paleogeographic position of simulated limestone sequences.
(M) Marsal, D. (1967): Statistische Methoden für Erdwissenschaftler. 152 pp., 40 Figs., Stuttgart: Schweizerbart
(M) McCammon, R. B. (1968): Multiple component analysis and its application in classification of environments. Bull. Amer. Ass. Petrol. Geol. **52**/11, 2178–2196, 10 Figs., 13 Tab., Tulsa
Comparison of various methods useful for classifying facies samples [association analyses, similarity analyses based on factor analyses (Q-mode) and multiple component analysis], as illustrated by the Bahama Banks, the method mentioned last being the preferred one.
(M) McCammon, R. B. (ed.) (1975): Concepts in Geostatistics. 168 pp., Berlin – Heidelberg – New York: Springer
(M) McCammon, R. B., Wenninger, G. (1970): The Dendrograph. Computer Contrib. State Geol. Surv. Kansas **48**, 28 pp., Lawrence
(M) Melguen, M. (1973): Correspondence Analysis for Recognition of Facies in Homogeneous Sediments off an Iranian River Mouth. In: Purser, B. H. (ed.): The Persian Gulf, 99–113, 11 Figs., Berlin – Heidelberg – New York: Springer
(M) Melguen, M. (1974): Facies analysis by "correspondence analysis": Numerous advantages of this new statistical technique. Marine Geol. **17**, 165–182, 10 Figs., Amsterdam
Holocene drill cores from the Persian Gulf. Areas of "homogeneous" sedimentation can be differentiated into facies using correspondence analysis (modified factor analysis). Advantages: rapid synthesis of vast numbers of data, correlation of microfossils with environmental factors, determination of significant facies parameters, objective classification of sedimentary particles according to their genesis and sedimentary history.
(M) Mello, J. F., Buzas, M. A. (1968): An application of cluster analysis as a method of determining biofacies. J. Paleont. **42**/3, 747–758, 6 Figs., Menasha
44 Recent foraminifera species from 99 samples in three traverses off the Texas coast are combined into sensible ecological groups using the Jaccard coefficients in cluster analyses.
(M) Merriam, D. F. (ed.) (1976): Quantitative techniques for the analysis of sediments. 186 p., Elmsford: Pergamon Press
(M) Miller, R. L., Kahn, J. St. (1962): Statistical Analysis in the Geological Sciences. New York – London: Wiley
Müller-Jungbluth, W. V., Toschek, P. H. (1969): see 6.4

(C) Nowak, F. J., Carozzi, A. V. (1973): Microfacies of the Upper Bird Spring Group (Pennsylvanian-Permian) Arrow Canyon Range, Clark County, Nevada. Arch. Sci. **25**/3, 343–382, 2 Pls., 2 Figs., Geneva
200 m thick limestone sequences; 907 samples; sample interval 22 cm. Thin-sections: frequency and clasticity indices of crinoid fragments, ooids, lithoclasts, and terrigenous quartz as well as frequency indices of bryozoans and brachiopods, agglutinated foraminifera and peloids. Computer printout of data. Comparison of the number of MF types produced by "continuous" sample taking, and sample intervals of about 20 cm produce statistically significant differences.

(M) Ojakangas, D. R. (1970): Fortran IV program for simulating geological development of sedimentary basins. Computer Contrib. **49,** State Geol. Surv. Kansas, 51 pp., 15 Figs., Lawrence

(C) Osborne, R. H. (1967): The American Upper Ordovician standard. VIII. R-mode factor analysis of Cincinnatian limestone. J. Sed. Petrol. **37**/2, 649–657, 5 Tab., Tulsa

(C) Osborne, R. H. (1969): The American Upper Ordovician standard. XI. Multivariate classification of typical Cincinnatian calcarenites. J. Sed. Petrol. **39**/2, 769–776, 3 Figs., Tulsa

(C) Osborne, R. H. (1973): The American Upper Ordovician standard. XVII. Areal variations of limestone frequencies in the Kope and Fairview Formations, Hamilton County, Ohio. J. Sed. Petrol. **43**/1, 137–146, 4 Figs., Tulsa

(M) Parks, J. M. (1967): Cluster analysis applied to multivariate geologic problems. J. Geol. **74,** 703–715, 5 Figs., Chicago

(C, M) Purdy, E. G. (1963): Recent calcium carbonate facies of the Great Bahama Bank. 1. Petrography and reaction groups. J. Geol. **71**/3, 334–355, 5 Pls., 5 Figs., Chicago
218 Recent sediment samples impregnated and studied in thin-sections. Point-counting analyses. Correlation analyses led to the determination of "reaction groups", defined as groups of sedimentary constituents which react similarly to environmental changes.

(C) Radke, B. M. (1976): Hierarchical classification and vector ordination in the distribution of limestones in the Fairfield Group, Canning Basin, Western Australia. J. Austral. Geol. Geophys. **1**/1, 63–76, 10 Figs., Sidney
Differentiation of Upper Devonian and Lower Carboniferous limestones by multivariate analysis of thin-section data and rare elements.

(C, M) Rao, C. P., Carozzi, A. V. (1971): Application of computer techniques to the petrographic study of oolithic environments. Ste. Genevieve Limestone (Mississippian), southern Illinois and eastern Missouri. Arch. Sci. **24**/1, 17–55, 1 Pl., 18 Figs., Geneva
Mississippian. 800 thin-sections. Frequency and clasticity indices. Differentiation of 8 MF types as well as autochthonous and allochthonous ooids. Trend analyses indicate the relations between the clasticity and frequency indices of ooids and non-ooids.

(C) Rao, C. P., Mann, C. J. (1972): Quantitative environmental analysis of Ste. Genevieve Limestone (Mississippian), Southern Illinois and Eastern Missouri. J. Geol. **80,** 464–480, 7 Figs., Chicago
Discriminance analysis shows that the clasticity indices are better facies criteria than the clasticity indices + frequency indices. The results of the discriminance analysis, the relations between the 8 MF types and the consideration of the depositional structures permit a differentiation of 21 "energy groups", which can be listed in 5 major groups.

(C) Rao, C. P., Mann, C. J., Carozzi, A. V. (1973): Factor analysis testing microfacies and interpretated environmental factors: Ste. Genevieve Limestone (Mississippian), Illinois and Missouri. J. Geol. **81**/1, 65–80, 7 Figs., Chicago
Factor analyses (Q-mode) show that just four factors explain about 87% of the data appearing in microfacies types and in the conjectured energy types.

(C) Rincon, A., Lopez de Azcona, C., Alvarez, C. (1977): Estudio geomatematico de las rocas carbonatadas de la provincia de Segovia. Estudios geol. **33,** 581–587, 7 Figs., Madrid
Investigation of the distribution of rare elements in limestones by factor analysis.

(M) Sarntheim, M., Walger, E. (1973): Classification of modern marl sediments in the Persian Gulf by factor analysis. In: Purser, B. H. (ed.): The Persian Gulf, 81–98, 10 Figs., Berlin – Heidelberg – New York: Springer
170 bottom samples. Study of 38 features each by factor analysis, using the Q and R modes, leading to a definition of mappable facies units.

(M) Schwarzacher, W. (1978): Quantitative theoretische Methoden in der stratigraphisch-sedimentologischen Analyse. Geol. Rundschau **67**/3, 809–822, 4 Figs., Stuttgart
(M) Smith, F. G. (1966): Geological data processing using Fortran IV. 284 pp., New York: Harper and Row
(C) Smosna, R., Warshauer, S. M. (1978): The evolution of a carbonate shelf, Silurian McKenzie formation, West Virginia: a cluster analytic approach. J. Sed. Petrol. **48**/1, 127–142, 9 Figs., Tulsa
(C) Smosna, R., Warshauer, S. M. (1979): A scheme for multivariate analysis in carbonate petrology with an example from the Silurian Tonoloway limestone. J. Sed. Petrol. **49**/1, 257–272, 5 Figs., Tulsa
Two methods of exploratory data analysis – a numerical classification system applied in conjunction with an ordination technique – were used to facilitate environmental interpretation.
(C) Smosna, R., Warshauer, S. M. (1981): Rank exposure index on a Silurian tidal flat. Sed. **28**, 723–731, 5 Figs., Oxford
(M) Sokal, R. R., Michener, C. D. (1958): A statistical method for evaluating systematic relationship. Univ. Kansas Sci. Bull. **38**, 1409–1438, Lawrence
(M) Sokal, R. R., Sneath, P. (1963): Principles of Numerical Taxonomy. 359 pp., San Francisco: Freeman
(C) Strasser, A. (1979): Betlis-Kalk und Diphoyides-Kalk (± Valanginian) in der Zentral- und Ost-Schweiz. Mitt. Geol. Inst. ETH Univ. Zürich, A. F. **225**, 209 pp., 4 Pls., 30 Figs., 8 Tables, Zürich
Factor analysis and cluster analysis of more than 800 thin-sections based on 22 microfacies criteria of terrigenous influenced platform and deeper shelf-slope limestones. 12 microfacies types are distinguished.
(C) Toomey, D. F. (1966): Application of factor analysis to a facies study of the Leavenworth Limestone (Pennsylvanian – Virgilian) of Kansas and Environs. Spec. Distrib. Publ. Kansas Geol. Surv. **27**, 28 pp., 13 Figs., 6 Tab., Lawrence
Determination of reaction groups on the basis of point-counting analyses and foraminifera counting; three facies types were identified. Because the observed criteria were equally weighted, the facies types recognized seemed meaningful. The classification obtained on the basis of foraminifera frequency does not coincide with the facies classification.
Wilson, J. L. (1975): see 1.4.1
(M) Yükler, A., Cornford, C., Welte, D. (1978): One-dimensional model to simulate geologic, hydrodynamic and thermodynamic development of a sedimentary basin. Geol. Rundschau **67**/3, 960–979, 14 Figs., Stuttgart

7.4 Checklist for Microfacies Studies

The most important questions which may arise during a microfacies analysis have been compiled in the following list:

1. Relative proportions of the main constituents

a) Matrix (micrite and microsparite)
b) Sparite (pore-filling cement, neomorphic sparite)
c) Carbonate particles (Table 13)
d) Non-carbonate particles (quartz, clay, cherts, etc.)
e) Authigenic minerals
f) Dolomite
g) Other constituents

2. *Matrix*

a) Homogeneous or inhomogeneous?
b) Micrite, in a restricted sense (micrite I, grains less than 4 microns) or micrite II, i.e. microsparite (grain sizes up to 20 microns)?
c) In inhomogeneous micrites look for calcisiltite (very fine detritus), "clotted fabric" (e.g., recrystallized peloids), bioturbation, lamination
d) Micrite together with sparite (about subequal; primary feature – due to incomplete winnowing of very fine grains of sediment, or secondary feature – due to recrystallization, etc.)?

3. *Skeletal grains*

a) Whole skeletal grains (biomorpha) or fragments (bioclasts)?
b) Systematic designation according to groups, genera and species
c) Monomictic or polymictic assemblages?
d) Ecological classifications (sessile or vagile benthonic organisms, plankton, nekton; various growth forms)
e) Classification of algal groups
f) The function of organisms as sediment binders, sediment bafflers, framework builders or essential rock building constituents
g) Biogenic encrustation of bioclasts and other particles?
h) Micrite rims around bioclasts?
i) Intragranular pores filled with micrite or sparite?
j) Are aragonitic hard parts preferentially dissolved, or are they missing altogether?
k) Size, sorting, and rounding of bioclasts
l) Frequency of all the skeletal grains and the individual groups of organisms.

4. *Peloids*

a) Shape and rounding
b) Frequency
c) Size and "sorting"
d) Occurrence of peloids (scattered; when in beds, in lenses or as laminated fabrics; sometimes nested)? – indication of pseudopeloids or fecal pellets
e) Color and grain size of the micrite in the peloids and the matrix (same or different) – indication of pseudopeloids or fecal pellets
f) Occurrence together with endobiontic organisms or with bioturbated structures? – indication of fecal pellets
g) Occurrence together with algae or aggregate grains? – indication of algal peloids
h) Shelly bioclasts micritized or recrystallized? – indication of pelletoids
i) Do the peloids exhibit regular structures (stripes, grid-like patterns of dots)? – crustacean fecal pellets

5. Aggregate grains

a) Grapestones, algal lumps, or lumps?
b) Which particles are agglutinated?
c) Are the particles in the aggregate grains truncated on the edges?
d) Do the aggregate grains occur in association with algae or algal mats?
e) Is the type of cement in the aggregate grains the same as that between the aggregate grains?
f) Size
g) Percentage of the aggregate grains in the modal composition or in the total sum of particles?
h) Do the aggregate grains occur in sparry or micritic limestones?

6. Oncoids

a) Algal oncoids (porostromate oncoids or rhodolites)?
b) Foraminiferal-algal oncoids, i.e., "zoophytic" bio-oncoids or micritic oncoids?
c) Shape of the oncoids
d) Do the oncoids exhibit a more or less regular structure or are discontinuities in growth visible?
e) Do truncated or very differently oriented laminae occur?
f) Are the oncoid laminae planar or crinkled?
g) Size of the oncoids
h) Are the oncoids deformed?
i) Do the nuclei of the oncoids correspond to the material available in the sedimentary environment?
j) Are the oncoids embedded in a fine-grained matrix or in sparite?
k) Do barrel-shaped oncoids have orientation patterns?
l) What is the frequency in comparison with other particles?

7. Ooids

a) Superficial ooids or normal ooids?
b) Frequency, size, and sorting
c) Do tangential structures, radial structures, or micritic rings occur?
d) Are the rings regularly formed or are there indications of algal borings and algal coatings?

Plate 33. Exercise for Determining Microfacies Criteria

This must be done in four steps: (1) Describe the microfacies criteria by answering the questions in the checklist (p. 394); (2) Classify the limestone according to Folk and to Dunham; (3) Determine the Standard Microfacies Type according to Wilson; (4) Diagnose the facies and classify it using a facies model: a) possible water depth, b) water energy, c) salinity, d) possible environments in the Irwin model and the Wilson model. The answers as well as the data for the paleontological identification are found in Section 12, p. 581. Scale is 10 mm

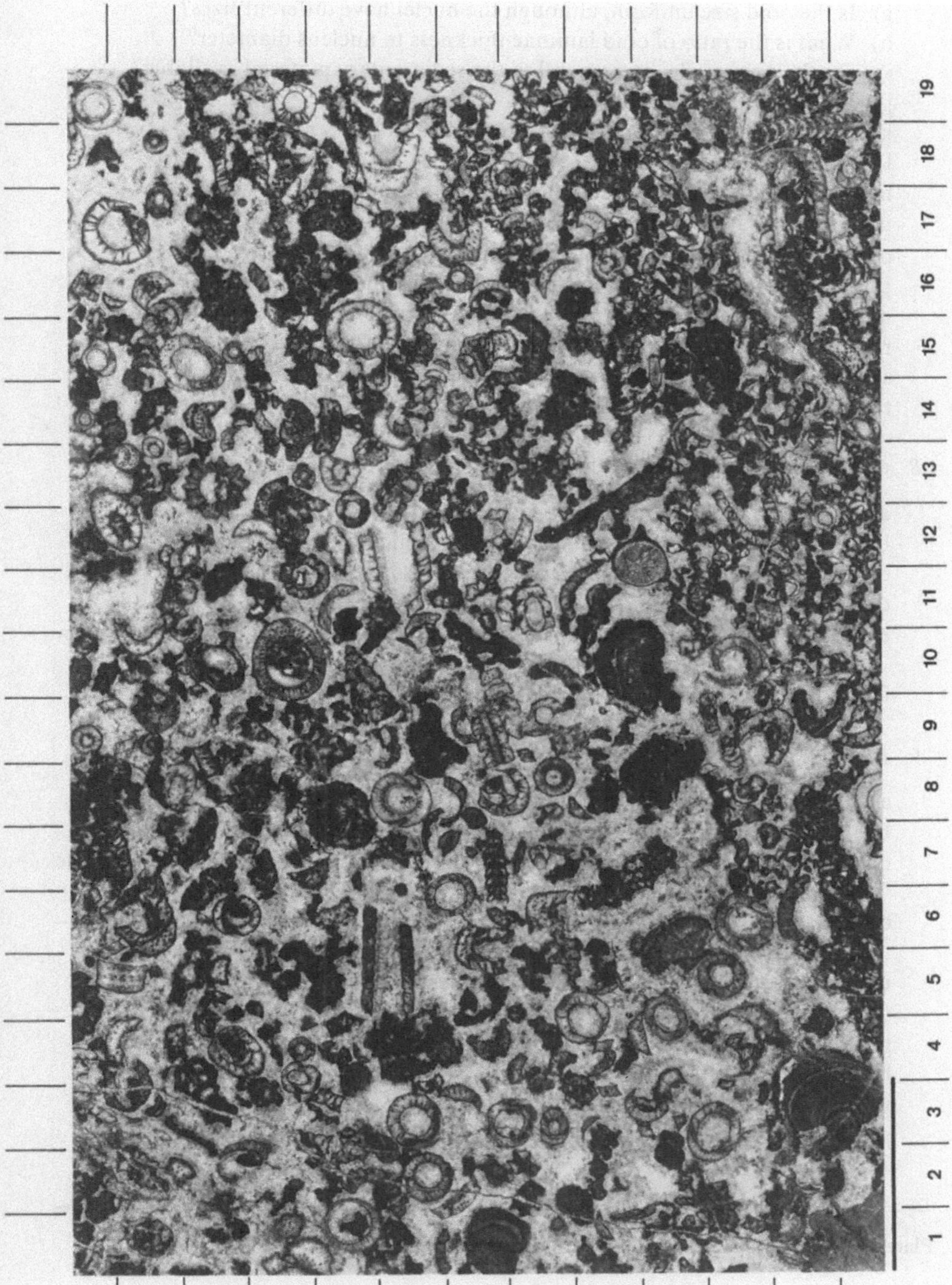

L
K
J
I
H
G
F
E
D
C
B
A
1
2
3
4
5
6
7
8
9
10
11
12
13
14
15
16
17
18
19

e) How thick are the ooid laminae?
f) How big are the nuclei of the ooids?
g) Is the ooid size uniform, although the nuclei have different sizes?
h) What is the ratio of ooid laminae thickness to nucleus diameter?
i) Do the ooid nuclei correspond to the sedimentary material available?
j) Do compound ooids occur?
k) Are several generations of ooids recognizable?
l) Do broken ooids occur?
m) Are ooids which are already present used as nuclei for new ooids?
n) Are there deformed, snouted ooids?
o) Do leached ooids occur? (half-moon ooids)
p) Do the ooids have an asymmetrical shape? – indication of quiet water ooids
q) Are ooids found together with carbonate crusts? – indication of pisoids
r) Do ooids occur in sparry or micritic limestones?
s) Do ooids occur alone, or are they mixed with other particles?
t) Is the packing index high or low?

8. Cortoids (Coated Grains)

a) Which particles exhibit micrite envelopes?
b) Do all or only a part of the particles have micrite envelopes?
c) Can "micritization" be proved (are borings recognizable)?
d) Are the cortoids found together with algae?
e) Is the diversity of the organisms which occur together with the cortoids low or high?
f) Is there any indication of the cortoids having been transported?

9. Lithoclasts

a) Can intraclasts and extraclasts be differentiated?
b) Are the particles in the clasts truncated along the edges of the lithoclasts?
c) Do the particles in the lithoclasts differ from the other particles in the sample?
d) Do the lithoclasts have a color different from the other particles? – indication of extraclasts ("black pebbles")
e) Do the lithoclasts exhibit compaction features?
f) Are the cements in the lithoclasts and in the other particles similar?
g) Shape and size
h) Are the clasts flat?
i) Are the clasts rounded?
j) Do the clasts exhibit parallel orientation or cross-bedding?
k) Frequency

Plate 34. Exercise for Determining Microfacies Criteria

1) Describe the MF features in the structurally different areas of the thin-section using the checklist; 2) Classify the limestone according to Folk and to Dunham; 3) Interpret the facies. The answers as well as the paleontological data are found in Section 12, p. 584. Scale is 10 mm

S
R
Q
P
O
N
M
L
K
J
I
H
G
F
E
D
C
B
A
1
2
3
4
5
6
7
8
9
10
11
12
13

10. Terrigenous particles

a) Which minerals or rock fragments occur?
b) Are the mineral grains rounded?
c) Frequency
d) Are the minerals concentrated in certain zones?
e) Are the terrigenous particles enclosed by other particles?

11. Fabric

a) Do the particles exhibit "grain-support" or "mud-support"?
b) What is the packing index or the kind of particle contact?
c) Do the particles exhibit a preferred orientation pattern?
d) Is the fabric homogenized by bioturbation?
e) Is a current-dependent imbricate structure recognizable?
f) Do geopetal structures exist?
g) Do the particles exhibit normal or inverse grain grading?
h) Is there an indication of textural inversion (high-energy particles in a micritic matrix)?

12. Laminated fabrics and open-space structures, discontinuity planes

a) Is lamination faintly or clearly marked?
b) Are there laminites or rhythmites?
c) Do laminated or laminoid fabrics occur?
d) Which type do the laminoid fabrics correspond to?
e) Do the fenestral fabrics correspond to LF-A or LF-B fabrics?
f) Are there thrombolitic fabrics?
g) Do the voids correspond to birdseyes or Stromatactis?
h) Is the internal sediment micrite, laminated micrite, or silt?
i) Does the internal sediment exhibit cross-bedding or slump structures?
j) What kind of carbonate cement is in the top part of the small voids?
k) Are there discontinuity planes with Fe-Mn enrichments?
l) Are there limonite crusts with "stromatolitic" structures ("cauliflower structures")?
m) Do the discontinuity planes exhibit encrustations of sessile organisms?
n) Do bores occur?
o) Are there subsolution fragments?
p) Are there solution structures?

Plate 35. Exercise for Determining Microfacies Criteria

1) Describe the MF facies using the checklist in 7.4, p. 394; 2) Describe a) the conditions of deposition, b) the diagenesis, c) the possible paleogeographic situation. The answers can be found in Section 12, p. 586. Scale is 2 mm

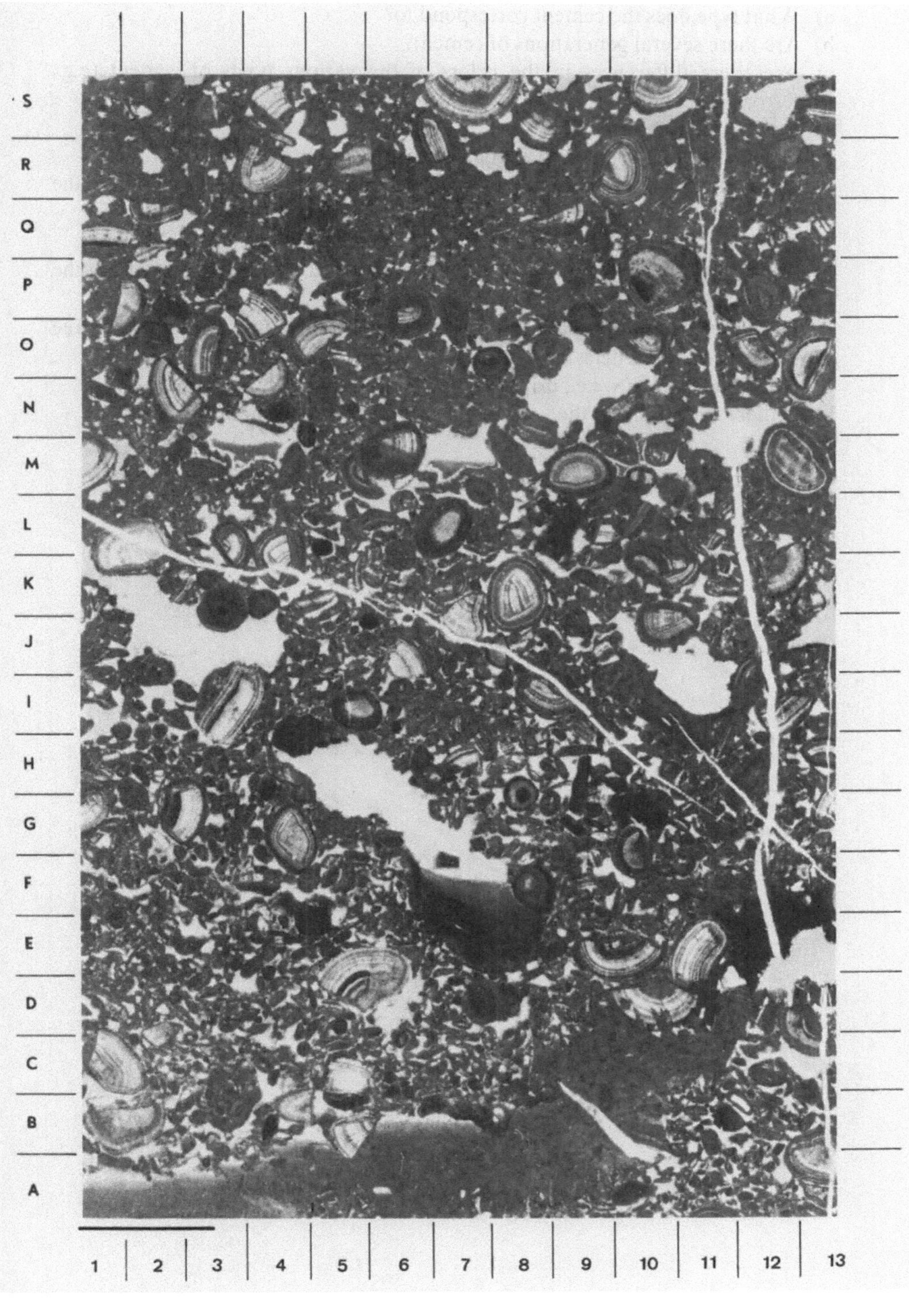
S
R
Q
P
O
N
M
L
K
J
I
H
G
F
E
D
C
B
A
1
2
3
4
5
6
7
8
9
10
11
12
13

13. Carbonate cement (sparite) – see Sect. 3.3.1

a) What type does the cement correspond to?
b) Are there several generations of cement?
c) Are there differences in the colors of the various types of cement (e.g., ferroan cement B)?
d) Is there micritic cement?
e) Does rim cement surround echinoderms?
f) Does the cement between the particles correspond to the cement in the particles?
g) What kind of contact is there between the cements and the particles?
h) Are the crystal sizes uniform or do the crystals increase in size away from the pore walls?
i) Does the edge of the sparry mosaic correspond to what was originally a free surface on the particles?
j) Do vadose cements, i.e., dripstone or meniscus cements, occur?
k) Is there vadose silt in the voids which are lined with cements?

8 Standard Microfacies Types (Plates 36–39)

With the aid of sedimentological as well as paleoecological data, the microfacies types of limestone of various ages can be combined into major types which reflect the depositional and ecological conditions in a certain sedimentary environment. This concept, which was developed on Upper Triassic reef carbonates in the Alpine Mediterranean region (E. Flügel, 1972), was expanded by Wilson (1975) and made more specific for numerous Paleozoic and Mesozoic examples. These Standard Microfacies Types (SMF Types) should, however, not be overrated, although they often prove useful as a basis for discussing depositional environments.

8.1 Principles

The first requirement for recognizing Standard Microfacies Types, as designated by Wilson (1975: 63), is the description of the microfacies criteria using the Dunham classification system (see p. 373). The 24 SMF Types are attributed to the carbonate types differentiated by Dunham in Table 41. This Table may be used as an identification key for the Standard Microfacies Types, but it should be taken into account that the limestone types defined by depositional fabrics provide only a rough guideline for the determination of the SMF types.

8.2 SMF Types (Table 42; Fig. 55)

Some SMF types seem to be too generalized because no differentiations are made with regard to the various possible origins of carbonate particles: SMF Type 15 is characterized by oolitic grainstones which should indicate a high-energy shallow-water environment. This may be true for oosparites with well-developed, tangentially structured ooids, but not for oosparites composed of radially structured ooids. These ooids may have been formed in quiet-water environments (see Sect. 4.1.3.5) and later redeposited in high-energetic areas. SMF Type 15 also does not correspond to oomicrites which may be the result of textural inversion or of an accumulation of quiet-water ooids. SMF Type 7 includes framestones, bindstones and bafflestones, but bindstones are also represented by the SMF Types 19, 20, and 21. In addition, diagenetic bindstones, formed by an interaction of binding effects

Continued p. 410

Table 41. Standard Microfacies Types 1–24 and depositional limestone types

<table>
<tr><th>Mudstone or
Bindstone (c, d, e)</th><th>Wackestone
Floatstone</th><th>Packstone</th><th>Grainstone
Rudstone</th><th>Boundstone</th></tr>
<tr><td rowspan="3">a) with spicula 1
b) with pelagic microfossils 3
c) laminoid fenestral fabrics 19
d) algal stromatolites 20
e) spongiostrome fabric 21
f) no fossils 23</td><td>a) with bioclasts 9
b) with oncoids 22
c) with sessile organisms 8</td><td>a) with bioclasts and extraclasts 4</td><td rowspan="2">a) with cortoids 11
b) with oncoids 13
c) with ooids 15
d) with peloids 16
e) with aggregate grains 17
f) with abundant dasycladaceans or abundant foraminifera 18
g) rudstone with reef debris 6
h) rudstone or floatstone with clasts consisting of micrite or calcisiltite without fossils 24</td><td rowspan="3">Sessile organisms in situ form organic structures (framestones or bafflestones) 7</td></tr>
<tr><td colspan="2">d) cortoids and worn bioclasts from high-energy environments in micrite 10</td></tr>
<tr><td></td><td colspan="2">b) microbioclastic calcisiltite, very small bioclasts and peloids 2
c) bioclasts, internal sediments abundant, sometimes developed as floatstone 5
d) bioclast accumulations, e.g., pelecypod coquinas, or echinoderms 12</td></tr>
</table>

Plate 36. Standard Microfacies Types ►

1. SMF Type 3. Pelagic wackestone with abundant planktonic foraminifera (*Globotruncana, Globigerina*). This SMF type characterizes basinal and lower slope environments. Scaglia rossa, Maastrichtian: Lake Garda region, Northern Italy

2. SMF Type 9. Styliolinid packstone with abundant sections of tentaculites (*Nowakia, Styliolina*) as well as some echinoderm fragments, ostracods and trilobite shells. This SMF type is common in shallow neritic water with open circulation (Standard Facies Belt 2, open sea shelf) or in shelf lagoons (Standard Facies Belt 7). Another location of the depositional environment seems to be submarine swells within the subtidal or bathyal shelf (see p. 499). Givetian, Middle Devonian: Menorca, Spain

Scale for both Figures 2 mm

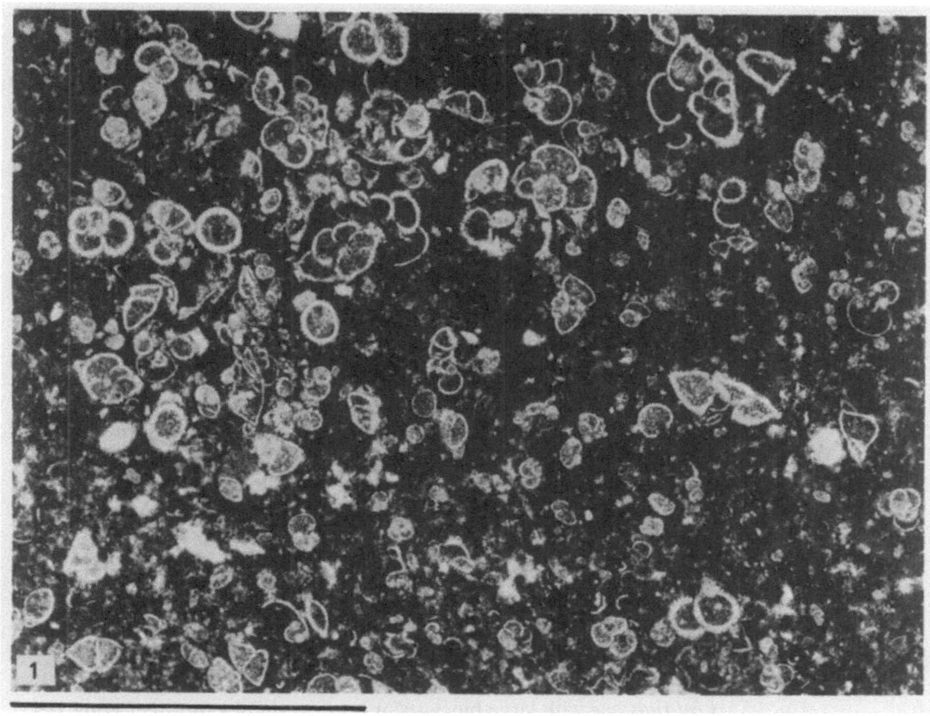
1

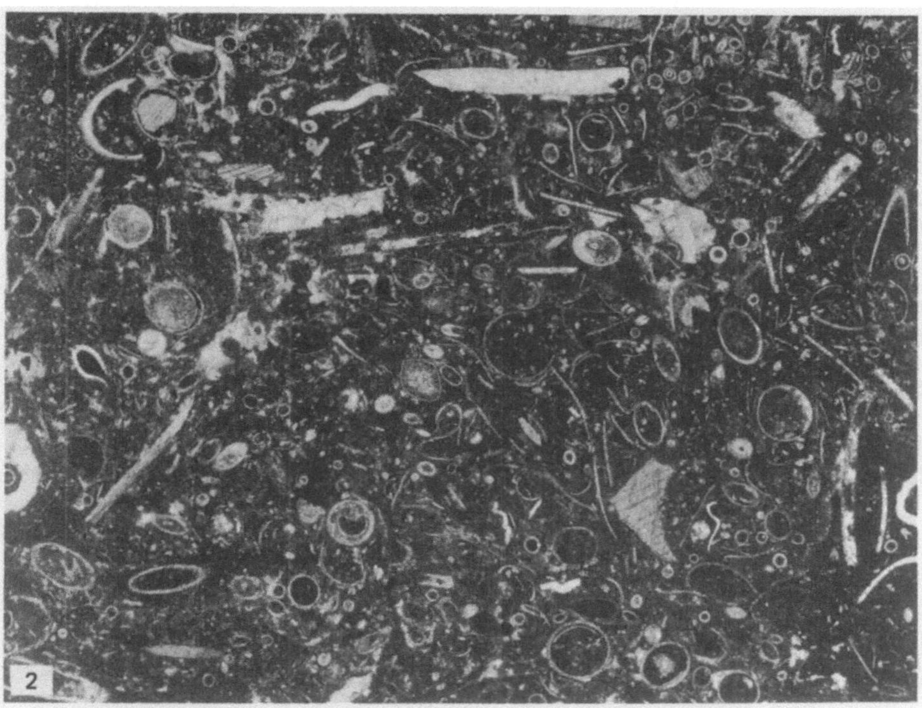
2

Table 42. Standard MF types. After Wilson (1975) and E. Flügel (1972)

SMF type	Features	Facies Zones (comp. Fig. 55)
1	*Spiculite.* Dark clayey mudstone or wackestone rich in organic substance, or siliceous spiculitic calcisiltite. Spicules usually oriented, generally siliceous monaxons, commonly replaced by calcite	FZ 1 basin, deep water environment with slow sedimentation
2 (Pl. 11/1)	*Microbioclastic calcisiltite.* Small bioclasts and peloids in very finegrained grainstone or packstone; mm ripple cross-bedding common	FZ 1; FZ 2 – open sea shelf near the lower slope; FZ 3 – deeper shelf margin
3 (Pl. 15/6, 16/1, 20/5, 26/2, 36/1, 38/2)	*Pelagic mudstone and wackestone.* Micritic matrix containing scattered pelagic microfossils (e.g. radiolarians or globigerinids) or megafauna (e.g. graptolites or thin-shelled pelecypod fragments)	FZ 1 and 3
4 (Pl. 21/2, 21/3, 31/5, 52/1, 2)	*Microbreccia* or *bioclastic-lithoclastic packstone.* Worn grains, often graded. Polymict or monomict in origin. Also quartz, cherts, and carbonate detritus.	FZ 3; FZ 4 – fore-slope talus. These sediments include the "allodapic limestones" after Meischner (1965)
5 (Pl. 41/5)	*Grainstone-packstone* or *floatstone* with bioclasts derived from reef dwellers and reef builders. Geopetal filling and umbrella effects from infiltered finer sediment	FZ 4 – typical reef flank facies
6 (Pl. 48/1)	*Reef rudstone* with large bioclasts or broken colonies of framework builders; no matrix material	FZ 4 – fore-reef slope, debris from the reef; commonly in high-energy zone
7 (Pl. 23/1, 23/4, 24/5, 31/7, 38/2, 42/2, 43/3, 48/2, 52/3)	*Boundstone.* Sessile organisms in situ. Subtypes framestone, bindstone, or bafflestone.	FZ 5 – organic reef, often found on platform margins
8 (Pl. 24/2)	*Wackestone with whole organisms* which are rooted in micrite. Only a few bioclasts. Well-preserved infauna and epifauna.	FZ 2; FZ 7 – shelf lagoon with open circulation; quiet water below normal wave base
9 (Pl. 6/3, 6/4, 12/1, 36/2, 46/3, 51/2)	*Bioclastic wackestone* or bioclastic micrite. Fragments of diverse organisms which have been texturally homogenized through bioturbation. Bioclasts may be micritized.	FZ 2; FZ 7 – shallow waters with open circulation at or just below wave base
10 (Pl. 37/1)	*Packstone-wackestone* with *coated* and worn *bioclasts*	FZ 2; FZ 7 – textural inversion; dominant particles are from high-energy environment on shoals and have moved down local slopes to be deposited in quiet water
11 (Pl. 25/1, 47/1, 11/3)	*Grainstones* with *coated bioclasts* in sparry *cement*	FZ 6 – winnowed platform edge sands; areas with constant wave action, at or above wave base

Table 42. (continued)

SMF type	Features	Facies Zones
12 (Pl. 21/1, 26/5, 32/1, 51/6)	*Coquina, bioclastic packstone, grainstone or rudstone* with *concentrations of organisms,* whereby certain types of organisms dominate (e.g., dasyclads, shells, or crinoids)	FZ 6 – commonly on slopes and shelf edges
13 (Pl. 9/8, 42/1)	*Oncoid biosparite grainstone*	FZ 6 – moderately high-energy area, very shallow water
14	*Lags.* Coated and worn particles, in places mixed with ooids and peloids which are blackened and iron-stained; with phosphate; also allochthonous lithoclasts; usually thin beds	FZ 6 – slowed accumulation of coarse material in zone of winnowing
15 (Pl. 9/1, 47/4)	*Oolites* of well-sorted, well-formed ooids with tangential microstructures, commonly from 0.5 to 1.5 mm in diameter; fabric usually overpacked; always cross-bedded	FZ 6 – high-energy environment on oolite shoals, beaches, and tidal bars
16 (Pl. 31/1, 37/2)	*Grainstone* with *pellets.* Probably fecal pellets, in places admixed with concentrated ostracod tests or foraminifera	FZ 7; FZ 8 – very warm shallow water with only moderate water circulation
17 (Pl. 31/6, 38/1)	*Grapestone, pelsparite* or *grainstone* with aggregate grains (grapestones and lumps), isolated and agglutinated peloids, some coated particles	FZ 7; FZ 8 – shelf with restricted water circulation and tidal flats
18 (Pl. 17/1, 33)	*Foraminiferal or dasycladacean grainstones* with concentrations of their skeletal grains	FZ 7; FZ 8 – in tidal bars and channels of lagoons
19 (Pl. 13/1, 14/1, 14/3)	*Loferite, laminated* mudstone-wackestone, grading occasionally into pelsparite with fenestral fabrics. Often ostracod-peloid assemblage, sporadic foraminifera, gastropods and algae	FZ 8 – very restricted bays and ponds
20 (Pl. 13/2)	*Algal stromatolite mudstone*	FZ 9 – commonest in the intertidal zone
21 (Pl. 39/1)	*Spongiostrome mudstone.* Tufted algal fabric in fine-grained micrite lime mud sediment	FZ 8 – in tidal ponds
22 (Pl. 39/2)	*Micrite with large oncoids,* wackestone or floatstone	FZ 8 – quiet water environments, shallow water, back-reef; often on the edges of ponds or channels
23 (Pl. 45/1)	Unlaminated, *homogeneous unfossiliferous pure micrite,* sometimes crystals of evaporitic minerals	FZ 8; FZ 9 – in hypersaline tidal ponds
24	*Rudstone or floatstone with coarse* lithoclasts and bioclasts. Clasts usually consist of unfossiliferous micrite or calcisiltite; sometimes imbricate texture and cross-bedding; matrix sparse	FZ 8 – formed as a lag deposit in tidal channels ("intraformational breccia")

	1	2	3	4	5	6	7	8	9	
	Basin (Fondotherm)	Open Sea Shelf (Deep Undathem)	Deep Shelf Margin or Basin Margin (Clinothem)	Foreslope (Clinothem)	Buildups or Platform Margin	Winnowed Platform Edge Sands	Open Platforms (Shelf Lagoon) (Shallow Undathem)	Restricted Platforms	Platform Evaporites (Sabkha)	Land

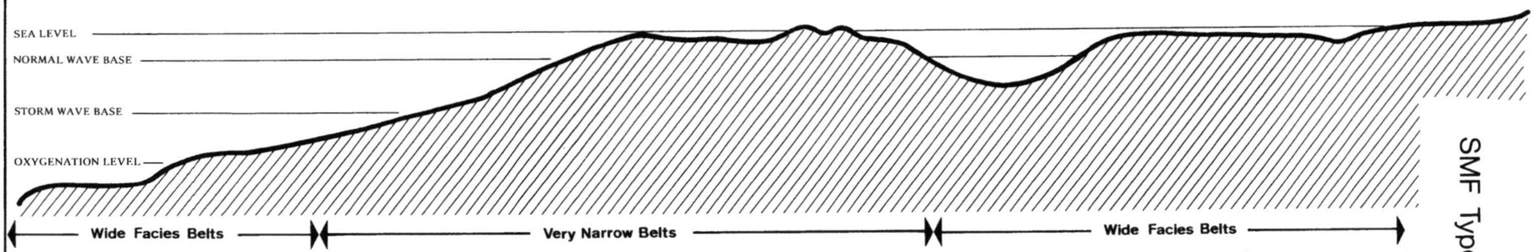

	1	2	3	4	5	6	7	8	9
SMF Types	1, 2, 3	2, 8, 9, 10	2, 3, 4	4, 5, 6	7, 11, 12	11, 12, 13, 14, 15	8, 9, 10, 16, 17, 18	16, 17, 18, 19, 20, 21, 22, 24	20, 23
Facies Belts	Deep marine as are a) leptogeosynclines with calcareous turbidites, b) miogeosynclines	Deep marine neritic sedimentation	Sedimentation of carbonate detritus and pelagic material	Instable sedimentation of debris	Organic reefs	Winnowed carbonate sands on shoals, coastal areas or tidal bars, eolian dune sands on islands	Bays and open lagoons behind the outer platform edge	Cut-off lagoons and coastal ponds with restricted circulation	Gypsum and anhydrite formation by evaporation of intermittently flooded supratidal areas, arid climate
Lithology	Dark shales and siltstones, thin-bedded limestones (mudstones and calcisiltites): very even mm lamination: rhythmic bedding	Very fossiliferous limestone (bioclastic wackestone and whole fossils wackestones, calcisiltite) interbedded with marls, thin to medium, wavy to nodular beds	Fine-grained limestone (dark mudstone, some calcisiltite), cherty in some cases, some shaly and silty intercalations, often massive units, lenses of graded sediments, some large blocks	Sedimentary breccias and resedimented limestones, some shale, siltstone, etc. Slump structures, exotic blocks, some bioherms	Massive limestone and dolomites. No terrigenous sediments	Calcarenite (well-sorted grainstones) with ooids, bioclasts, etc., some dolomite, quartz sand admixtures. Cross-bedding	Various limestone types (grainstone to mudstone), some thin-bedded terrigenous clastics	Dolomite and dolomitic limestones. Mudstones with peloids, algal mats, lithoclastic wackestones. Carbonate and terrigenous clastics in well-segregated beds. Fenestral fabrics, mm lamination	Irregularly laminated dolomite and anhydrite, may grade to red beds, caliche, terrigenous clastics sometimes important
Organisms	Exclusively nekton and plankton preserved in local abundance on bedding planes sometimes condensed fauna	Very diverse shelly fauna preserving both infauna and epifauna, bioturbation common	Redeposited shallow marine bioclasts together with autochthonous benthonic and planktonic organisms, some ghost fauna	Rebedded colonies from Belt 5. Often very fossiliferous.	Frame-building organisms with growth forms determined by water energy	Concentrations of worn and abraded shells of benthonic animals living in Belts 4 and 5. Few autochthonous elements (gastropods, foraminifera, dasyclads)	Very diverse fauna, stenohaline elements rare. Sediment stabilization by abundant marine grasses, occasional patch reefs, abundant endobionts and bioturbation	Very limited fauna, mainly gastropods, algae, certain foraminifera (e.g., miliolids), and ostracods	Only stromatolitic algae, almost no fauna

Fig. 55. Facies scheme after Wilson (1975). Sequence of the Standard Microfacies Types

Table 43. Standard Microfacies Types (SMF) and Standard Facies Zones (FZ)

FZ →	1	2	3	4	5	6	7	8	9
SMF Type ↓	Basin	Open-Sea Shelf	Deep Shelf Margin	Fore-slope	Organic Buildup	Winnowed Edge Platform Sands	Shelf Lagoon, Open Circulation	Shelf and Tidal Flats, Restricted Circulation	Sabkhas with Evaporites, Salinas
1	X								
2	X	X	X						
3	X		X						
4			X	X					
5				X					
6				X					
7					X				
8		X					X		
9		X					X		
10		X					X		
11					X	X			
12					X	X			
13						X			
14						X			
15						X			
16							X	X	
17							X	X	
18							X	X	
19								X	
20									X
21								X	
22								X	
23								X	X
24								X	

of organisms and of a rapid synsedimentary cementation are not included in the SMF Type 7. This type must be differentiated by paleontological data. Other generalizations concern those types which are defined by the abundance of particular fossil groups (e.g., SMF Types 12 and 18). The origin of foraminiferal grainstones is often due to autochthonous accumulations subsequent to particle size differentiation.

8.3 SMF Types and Facies Belts

J. L. Wilson assigns the 24 SMF Types to 9 "Standard Facies Belts" (Figs. 55 and 67), whereby individual SMF Types may appear in several of the facies belts (Table 43).

Difficulties may arise when using SMF Types to differentiate facies belts 2 (open sea shelf) and 7 (open shelf lagoon), and sometimes when separating belt 3 (deep shelf margin) and belt 1 (basin).

8.4 Discussion and Examples (Plates 36–39)

1. Plate 37/1 (limestone from the Zlambach beds (see p. 532), Late Triassic; Fischerwiese near Alt-Aussee, Styria, Austria):

Matrix micrite. Densely packed, broken, not rounded bioclasts (primarily corals, Solenoporaceae, pelecypods, and a few echinoderms, serpulids, and bryozoans). Dunham Classification: packstone. Integration into the SMF scheme: The fragmentation of skeletal grains which are bound to the well-lighted high-energy environments, and the deposition of skeletal grains in fine-grained argillaceous micrite indicate a "textural inversion"; namely, that particles from a shallow water and high-energy environment are deposited in areas with quiet water sedimenta-

Plate 37. Standard Microfacies Types

1. SMF Type 10. Packstone with fragmented densely packed bioclasts in micritic matrix. Chiefly remains of sessile benthonic organisms: corals –1, bryozoans –2, hydrozoans (spongiomorphids –3), serpulids –4, pelecypod shells –5, calcareous algae (Solenoporaceae –6). Flora and fauna indicate an originally high-energy biotope and later allochthonous deposition of these particles in a protected low-energy environment. Late Triassic (Zlambach limestones): Fischerwiese near Lupitsch, Salzkammergut, Styria, Austria. Scale is 10 mm

2. SMF Type 16. Peloidal grainstone or pelsparite with frequent peloids and some calcispheres. Peloids correspond to fecal pellets and to small reworked mud particles (pseudopeloids, see Sect. 4.1.3.2). Peloids may have been trapped by algal crusts (which are indicated by patchy distributed open-space structures, arrow). This sediment is common on tidal flats (Standard Facies Belt 8), but is found also in open-marine platform limestones (Standard Facies Belt 7). The example figured comes from restricted shelf-lagoon carbonates. Lower Carboniferous: Belgium. Scale 2 mm

1
2
3
4
5
6

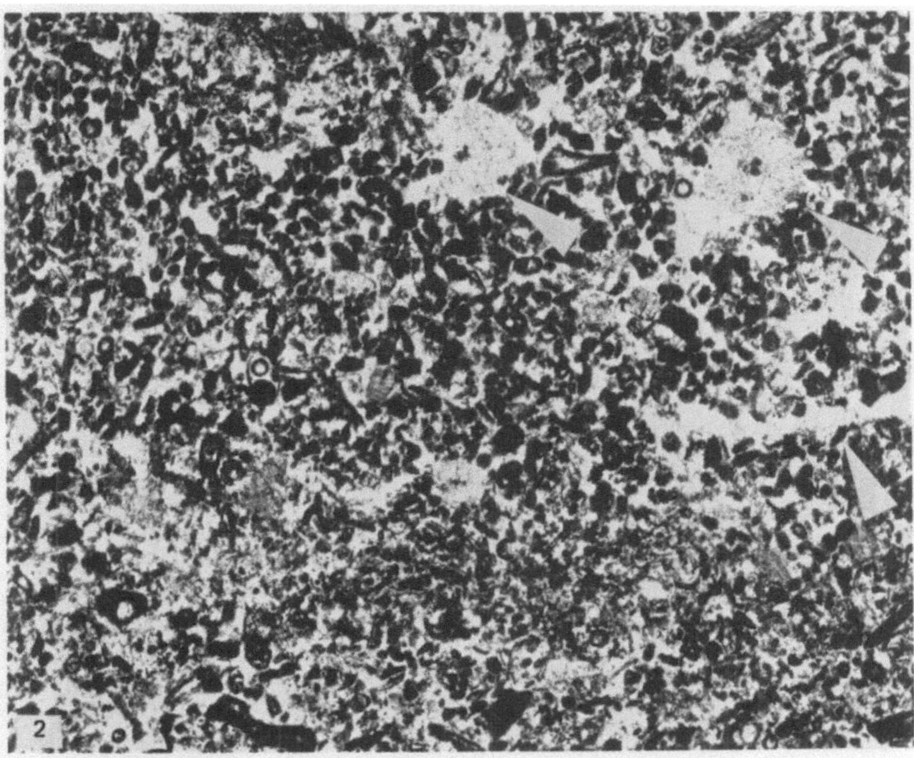
2

tion. Thus, we assign the sample to SMF Type 10. Sediments corresponding to this type occur, after Wilson, in open marine shelf environments as well as in shelf lagoons with open water circulation. The paleogeographic situation (see Bolz, 1974) indicates that the micritic calcarenites correspond to the reef detritus deposited in low-energy environments between local patch reefs.

2. Plate 53/1 (red Adnet nodular limestone, Lias alpha 3/Lias gamma; Kiefer quarry, NE Adnet near Hallein, Salzburg, Austria):

Matrix micrite. The particles consist of echinoderms (crinoids, echinoids), foraminifera, a few brachiopod shells, and juvenile gastropods. The irregular patchy distribution of the organic remains is very conspicuous. Dunham Classification: wackestone. SMF Type: The irregular distribution of the organisms indicates bioturbation; the fossil content argues for a highly diverse bottom fauna. The sample should be assigned to SMF Type 9 after Wilson, because the benthos is indigenous. Thus, it would be possible to classify the sample in the marine shelf or in the environment of a shelf lagoon with open water circulation. Because of the paleogeographic situation (see Hudson and Jenkyns, 1969) as well as the distinct solution structures which appear in thin-sections (residual clay seams; the solution boundaries truncate the skeletal grains), sedimentation in a deep neritic, normal marine (echinoderms!) and well-oxygenated environment is quite probable. How much the "conglomerate" fabric of the limestone has been affected by solution at the bottom of the ocean (subsolution) or by pressure solution on the particles of submarine slump sediments is not clear.

Plates 36–39 can be used to practice determining SMF Types.

The SMF Types can be used as a first orientation in the early stages of interpretation, but on no account should they be regarded as a "Key to Identification" in facies interpretations.

Plate 38. Standard Microfacies Types

1. SMF Type 17. Grainstone with aggregate grains as well as a few foraminifera and shells. The composition of the sediment, the high carbonate content of the limestone (non-carbonate insoluble residues $<0.4\%$ by weight) and the scarcity of fossils, i.e., the low diversity, allow a comparison with the "grapestone" sedimentation on the Great Bahama Banks (see Sect. 4.1.3.3). Late Triassic (Dachstein limestone, Norian): Western Gosaukamm, Upper Austria

2. SMF Type 7 (corallinacean boundstone) and SMF Type 3 (pelagic wackestone with abundant globigerinid foraminifera). Borings within the corallinacean thalli (*Mesophyllum* sp.) are sometimes filled with planktonic foraminifera indicating a synchronous mixing of two SMF types which normally would indicate completely different depositional environments. Miocene: Pietraroia, Southern Apennines, Italy

Scale for both Figures 2 mm

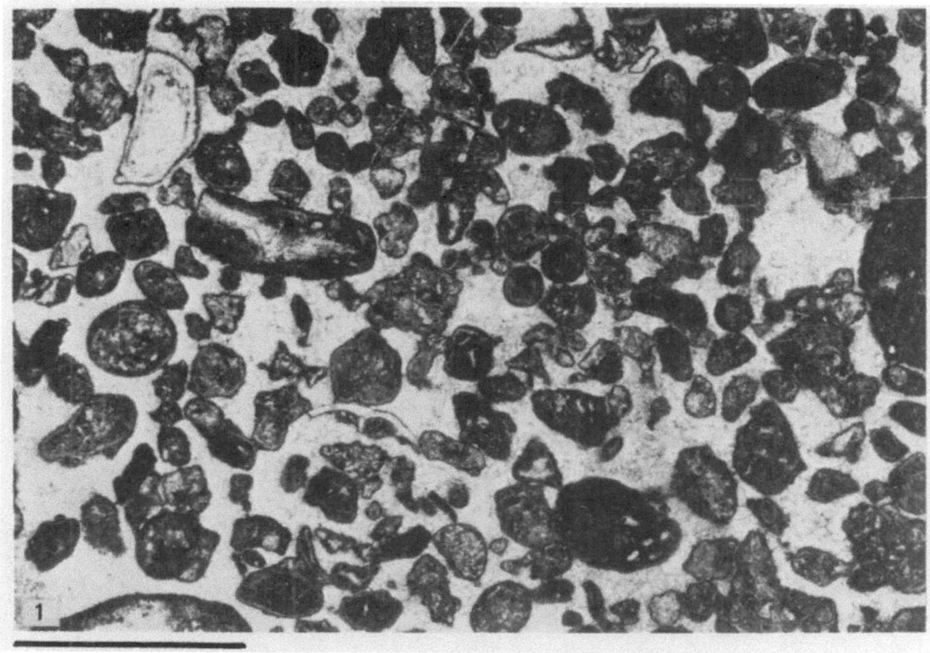
1

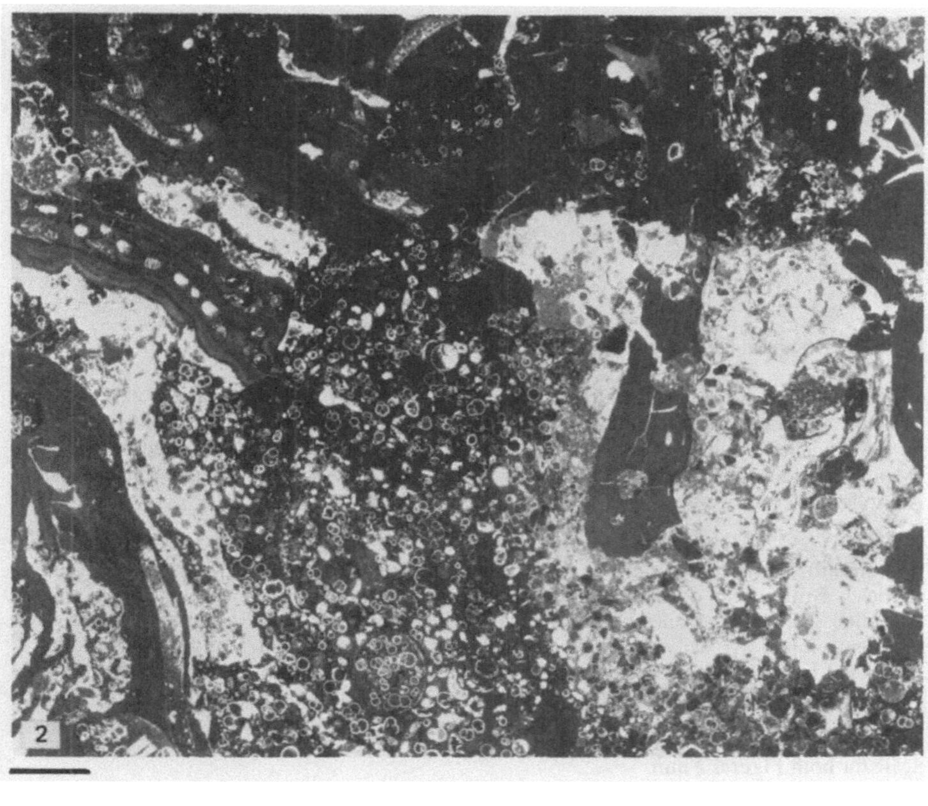
2

8.5 References: Standard Microfacies Types

Bolz, H. (1974): Die Zlambach-Schichten (alpine Obertrias) unter besonderer Berücksichtigung der Ostrakoden. 2.: Zur Stratigraphie und Fazies der Zlambach-Schichten. Senck. leth. **55,** 325–361, 4 Pls., 2 Figs., 3 Tab., Frankfurt

Dunham, R. J. (1962): see 6.4

Flügel, E. (1972): Mikrofazielle Untersuchungen in der Alpinen Trias. Methoden und Probleme. Mitt. Ges. Geol. Bergbaustud. **21,** 9–64, 9 Pls., Innsbruck

Hudson, J. D., Jenkyns, H. C. (1969): Conglomerates in the Adnet Limestones of Adnet (Austria) and the origin of the "Scheck". N. Jb. Geol. Paläont. Mh. **1969**/9, 552–558, 2 Figs., Stuttgart

Wilson, J. L. (1975): see 1.4.1

Plate 39. Standard Microfacies Types

1. SMF Type 21. Spongiostrome bindstone, characterized by algal laminations exhibiting an irregular anastomosing fabric between algal peloid grains. Existence of algal tufts is indicated by the branched outline of some open spaces (arrows). This SMF type is formed on tidal flats (Standard Facies Belt 8). Nohutalan Beds, Upper Triassic: Karaburun peninsula, Western Anatolia, Turkey

2. SMF Type 22. Micrite with large oncoids. The fabric would correspond to floatstones, but it seems to be more meaningful to restrict this term to detrital carbonates with ruditic particles. Grain size of oncoids is controlled by biological and environmental factors, grain size of ruditic detrital limestones by mechanical effects. SMF type 22 characterizes a quiet-water sediment, found typically on edges of ponds or channels within Standard Facies Belt 8. Restricted conditions may occur in the intertidal area but also in marine or brackish coastal environments. The sample is from brackish limestones. Lusitanian, Upper Jurassic: Fatima, Portugal

Scale for both Figures 2 mm

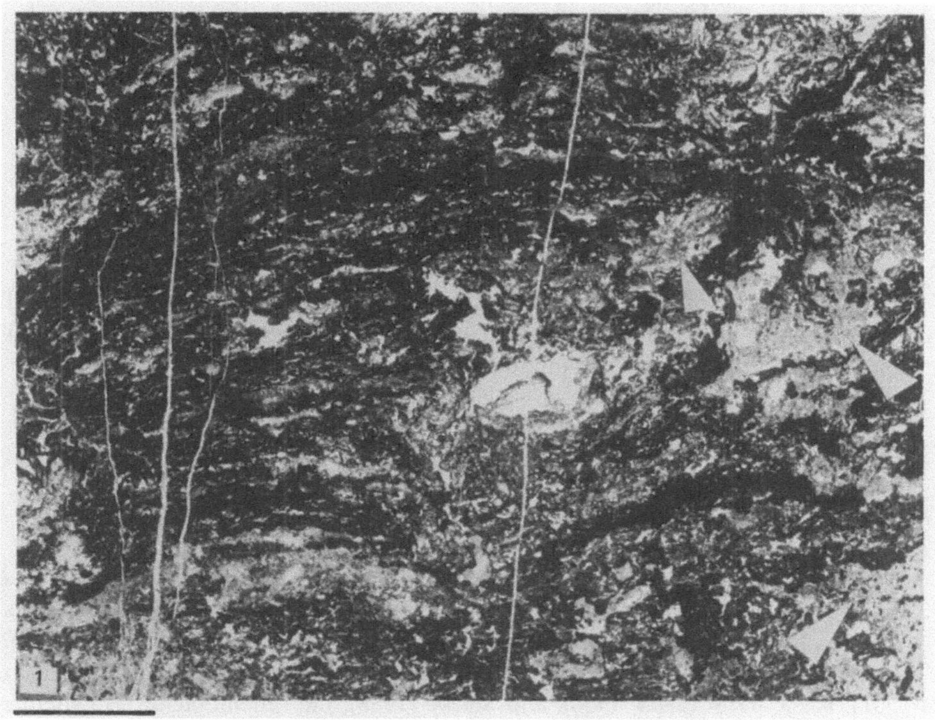
1

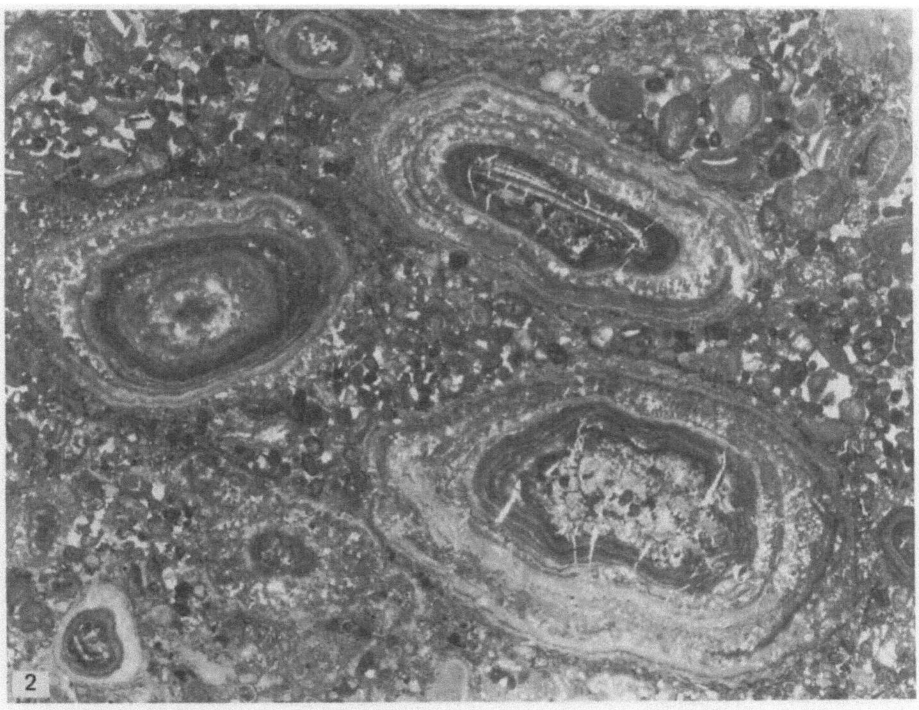
2

9 Complementary Methods

The interpretation of microfacies categories can be supplemented by geochemical studies and by description of physical criteria. Sections 9.1, 9.2 and 9.3 should hopefully stimulate more interest in these methods.

9.1 Relationship Between Microfacies and Geochemical Criteria

Carbonate sediments and rocks consist of two groups of elements: a) elements bound to a carbonate phase (Ca, Mg, Sr, Mn, sometimes Ba, etc.) and b) elements occurring in minerals of acid insoluble non-carbonate residues (e.g., Si, Al, Fe, B). As shown in the comprehensive papers of Degens (1968), Ernst (1970), Ingerson (1961), Krejci-Graf (1964, 1966), and Starke (1968), the "chemofacies" is valuable as an indicator of the formation and diagenesis of the sediments. Admittedly, though, with the present state of knowledge it seems absolutely necessary to critically re-examine the usefulness of geochemical methods for facies studies (see discussion in Wedepohl, 1970).

Up to now the amount and mineralogical composition of the insoluble residues as well as the distribution and abundance of trace elements have been relied upon for facies interpretations.

9.1.1 Insoluble Residues of Carbonate Rocks

The abundance of terrigenous components depends on the influx from the continents and on the dilution by biogenous sediments as well as other sediments. Influx is increased during regressions and periods of orogenesis, uplift, and high denudation rates in the hinterland; it is decreased during transgressions.

The "absolute" or the "relative" amount (see Seibold, 1952) as well as the composition of the residues can provide meaningful information about the process of sedimentation. The methods of clay mineral analysis are described by G. Müller (1964) and Millott (1970).

9.1.1.1 Amount of Clastic Material

Quantitative descriptions of the insoluble residue content do the following:

a) *Provide a check on the energy indices derived from MF types* (see 6.2.4): Theoretically, orthosparitic limestones should contain fewer residues than low-energy micritic limestones. This is in fact the case for limestones such as the "Muschelkalk" of the Odenwald (low values in coquina limestones of the *Spiriferina* bed; higher values in the micritic limestones, with values fluctuating sharply within the individual beds; Backhaus and E. Flügel, 1971). However, in other cases (Lower Permian shelf limestones of the Carnic Alps; reef core of the Northern Alpine Late Triassic), no correlation between the amount of matrix and the amount of residue can be determined. This may be due to the similar hydrodynamic features of quartz grains and carbonate particles. As shown by Bausch (1980), sparry limestones can contain up to an average of 1% insoluble residues, in contrast to biomicrites and pelmicrites (1–3%), and micrites (4–5%).

b) *Provide a check on the course of deposition:* Checking the carbonate content and the amounts of residues in profiles can sometimes reveal a *"cyclic sedimentation"* which had not been seen in the MF types (Examples in Reed and Burchett, 1964; E. Flügel et al., 1971). Changes in the *rate of sedimentation* become quite obvious in diagrams which take into account bed thickness and absolute clay and carbonate content (Seibold, 1952; H. W. Flügel, 1968). *Interruptions of sedimentation* which are linked with solution, bottom formation, and new deposition are often characterized by an abrupt change from bed to bed in the amount of insoluble residues, by different clay minerals, and by an increase in the coarse fraction (Hayes, 1963). Of course, the sometimes considerable range of variation within the bed must also be taken into consideration.

c) *Indicate a near-coastal tidal area:* In comparison to the values of normal marine limestones, markedly higher residue contents are frequently found in stromatolitic micrites and pelmicrites of carbonate mud flats (Example: Norian Hauptdolomit of the Lechtal Alps, Germany; Fruth and Scherreiks, 1975; Cambrian of Morocco, Buggisch et al., 1979). Solution in the supratidal zone may lead to a concentration of silicate particles as well as of trace elements (Fe, Mn, Zn) in lower-energy tidal zones.

d) Aid in *stratigraphic correlations:* Profiles, especially of basinal carbonates which are rather difficult to subdivide into microfacies, can be compared with the aid of residue percentages or grain-fraction quotient curves and can even, to some extent, be correlated. Examples from the Alpine Upper Jurassic have been described by Bausch and Poll (1976).

9.1.1.2 Mineralogical Composition of Residues

The residues of many marine limestones consist of silicates. Among the most common are: illite, mixed-layer minerals, kaolinite, chlorite, muscovite, as well as quartz and albite (see Robbins and Keller, 1952; Weaver, 1963; Millicott, 1970). An interpretation of the qualitative differences must depend on whether the total residue content can be explained as terrigenous detritus, which seems to be the case for a major portion of the clay minerals contained in the limestones, or whether transformation or neoformation of minerals is involved (see Engelhardt, 1961; Millott, 1970).

Kaolinite is formed in non-marine environments with tropical climates. A diagenetic origin can be neglected for kaolinites found in limestones, because of the protective effect of the carbonate matrix against the formation of an acid milieu

(Bausch, 1980). The most abundant marine clay mineral is illite, which may be the object of multiple redeposition (indicated by very high radiometric ages of illites from Recent sediments). An interpretation of illite in limestones is difficult, because a diagenetic origin cannot be excluded. Mixed-layer minerals are formed in connection with the transformation of clay minerals. Many smectites found in limestones may be the result of the transformation of illites to smectites in soils and of subsequent transport to marine basins. Chlorites are formed during low metamorphosis, but also in salinar sediments; in Recent oceans chlorites are found predominantly in Arctic regions.

The views of various authors are divided on the value of clay minerals as facies indicators:

a) *Indication of climatic conditions in erosion areas and of paleogeographic configurations:* Often clay mineral provinces characterized by a dominance of kaolinite are considered to indicate a near-coastal shallow sea sedimentation (e.g., Becher, 1965; Griffin, 1962; Bausch, 1971); deeper offshore depositional areas, on the other hand, often exhibit associations of illites and chlorites (Rateev et al., 1969). Other patterns of distribution, though, also occur, as shown in Persoz and Remane (1976).

b) *Paleosalinity:* Degens (1968) is rather cautious about this, whereas Kübler (1963) attempts to find a relation between the clay mineral distribution and the salinity in various parts of the ocean. Discussion of this topic can be found in Krumm (1965) and Griffin and Ingram (1955).

c) *Illite crystallinity:* This index is used in numerous articles as a measure of the depth of subsidence (Kubler, 1968; Dunoyer de Segonzac, 1970). Persoz and Remane (1976) point out the special problems involved in the evaluation of illites in carbonate rocks. An interesting example of the use of illite crystallinity in clay minerals found in the insoluble residues of Carboniferous limestones is described by Gill et al. (1977).

d) *Large-scale "clay mineral provinces"* seem to reflect variations in the climate of the erosion area, in the source material and in sedimentation rates. An interesting example of clay mineral provinces based on insoluble residues of Upper Jurassic limestones is found in Bausch (1980).

9.1.2 Trace Elements

Trace elements occurring in concentrations between 10^{-2} and 10^{-3} wt.%, or less, are either bound to the carbonate phase (and are then very dependent on the kind of biomineralization and the change during diagenesis) or to the non-carbonate phases (dependent on the process of deposition, on the grain size and mineralogy as well as the organic substances). According to Wedepohl (1970), the majority of the trace elements known in carbonate rocks are bound to the detrital silicate-oxidic fraction of the limestones. Mg, Sr, and Mn are linked in a specific way with the carbonate phase, depending on the similarity of the crystal chemistries of main components of more abundant carbonate minerals (Sr and Ca, Mg and Ca, Mn and Mg) and on the relatively high concentrations in ocean and pore waters. It is understandable that a major portion of the facies analyses based on trace element distribution are concerned with these elements (see surveys in Friedman, 1969; Schroll, 1967, 1976).

Opinions are divided as to the diagnostic information provided by trace elements for the genetic interpretation of carbonate rocks. To some extent this is due to the fact that up to now very few studies have been done on the correlations between limestone types (MF types) and geochemical parameters (e.g., Cerny, 1976, 1978) and many conjectured "correlations" are not statistically supported. Another reason may be the lack of information about the amount and the mineralogical composition of the insoluble residues. Trace element analysis has been used in the following situations:

a) *Subdivision of shallow sea environments:* The object is to separate reef sediments from non-reef sediments, to recognize various parts of reef-complexes, and to differentiate and better interpret them:

Reef and non-reef sediments or *various parts of reef-complexes:* The studies carried out in Recent reef environments (Table 44) show that the abundance of trace elements, especially Sr, is different in lagoons, in the central reef area, and in the fore-reef; this is partly due to the different distributions of aragonitic bioclasts. The results of Till (1971) lead in another direction (trace element distribution is due to the distribution of clay minerals and organic substances).

The differences determined in Ancient reefs do not exhibit this same distribution pattern, although often different Sr values are observed in lagoonal, reef and basinal sediments (see Table 44). The variations are due to (a) differences during the incorporation of Sr into the carbonate minerals of the shells of organisms [dependent upon the Sr/Ca ratio in sea water, the carbonate mineralogy (aragonite has more Sr than calcite), water temperature, salinity, vital effects of organisms] and (b) depletion of the Sr content during diagenesis in closed or open systems with or without inflowing fresh water (see H. W. Flügel and Wedepohl, 1967; Kinsman, 1969; Veizer et al., 1971; Veizer, 1977).

b) *Differentiation of shallow-water and deep-water limestones* (Table 45): Most papers demonstrate that the Mn content and the varying amounts of aluminosilicates or trace elements contained in them such as Co, Ni, Ca, Zn, Mo, and Pb, can be useful in differentiating shallow-water and deep-water limestones. Low and high Sr values in shallow-water and deep-water limestones appear to be linked with sparry and micritic limestone types. These varying values are due to differences in the chemism of the interstitial water and in the rate of diagenetic solution transfer. Just as for Sr, the factors responsible for the Mn distribution are still under discussion: One of the problems is whether or not Mn is bound to the carbonate phase or the silicate phase of limestones (see Lütke, 1976; Cerny, 1976). A relatively high Mn content can be caused by reduced clay sedimentation, submarine vulcanism, enrichment during periods of subaerial exposure. Dolomitization does not appear to influence the Mn content as substantially as it does the Sr content. Within platform carbonates the higher Mn content may be expected in lagoonal environments where the precipitation of Mg-calcite predominates over aragonite; coexisting calcite can incorporate more manganese than aragonite (see Annovi et al., 1980).

Different facies types of Permian and Mesozoic shelf and reef carbonates can also be differentiated by the Na content of the carbonates (Fritz and Katz, 1972; Land and Hoops, 1973; Kitano, Okumura and Idogaki, 1975; Ogorelec and Rothe, 1979). Similar to Recent carbonates, the sodium content of Ancient limestones

Table 44. Facies subdivisions, using trace elements, in Recent and Ancient shallow water carbonates, i.e. reef carbonates

Author	Age Locality	Analyzed Trace Elements	Results
Stehli and Hower (1960)	Recent: Florida	Sr	Sr content highest in central reef; decreases towards the back-reef and fore-reef
Siegel (1961)	Recent: Florida	Sr	Similar to Stehli and Howe
Millimann (1967)	Recent: Hogsty Atoll, Bahamas	Sr	Sr content highest in the lagoon; increases slightly from the back-reef over the reef to the fore-reef
Hoskin (1968)	Recent: Alacran Reef, Mexico	Mg, Sr	Maximum Mg value near red algal sediments; highest Sr content near Sr-rich corals and codiacean algae
Billings and Ragland (1968)	Recent: Belize, British-Honduras	Sr, Ca, Mg, Fe, Mn, Zn, Na, K, Li, Rb	Different trace element distribution in reef and lagoon
Till (1971)	Recent: Belize, Ancient: Devonian/Alberta, Canada		Discussion of data by Chester and by Cameron: more differences in occurrence of clay minerals and organic substances than in trace elements
Davies (1972)	Upper Jurassic (Corallian): Yorkshire, England and Switzerland	Sr, Ba, Si, Al, Ti, Co, Mn, Fe, Ni	Coral reef limestones low in trace elements, but much Sr (> 2000 ppm); underlying oolites have little Sr
H. W. Flügel and Wedepohl (1967)	Upper Jurassic: Plassen limestone of Northern Alps	Sr	Constantly low Sr values in shallow water facies (100 – 200 ppm) due to diagenesis
Sternberg et al. (1959)	Late Triassic (Upper Rhaetian): Steinplatte reef, Tyrol	Sr	Sr content increases from lagoon over reef to fore-reef
E. Flügel and E. Flügel-Kahler (1963)	Late Triassic (Rhaetian): Sauwand reef, Styria, Austria	Sr	Sr increases slowly from back-reef and then markedly on the fore-reef
Kollmann and Kurat (1970)	Late Triassic (Dachstein limestone, Norian): Gosaukamm, Upper Austria	Sr	No significant differences in various parts of the reef (because patch reefs instead of reef-complexes were investigated)
Campbell and Lerbekmo (1965)	Mississippian: Alberta, Canada	Sr	Increased Sr content near the reefs
Weaver (1968)	Upper Devonian: Alberta, Canada	Sr, Ba	Sr distribution bound to reef talus; Ba distribution to non-carbonate minerals

Table 44. (continued)

Author	Age Locality	Analyzed Trace Elements	Results
Chester (1965)	Upper Devonian: Sturgeon Lake reef, Alberta, Canada	Ni, Co, Cr, V, Ba, Sr, Ni/V, Cr/Ni, Ba/Sr, Pb/Ga	Non-reef facies characterized by Ni, Co, Cr, Ba, Sr and ratios of Ni/Co, Cr/N, Pb/Ga, and Cu/Co
Veevers (1969)	Upper Devonian and Lower Carboniferous: Bonaparte Gulf basin, NW Australia	Ca, Mg, Fe, P, Sr, Ba, Mn, Pb, Zn	Trace elements are bound to certain limestone types; Sr content low in coastal lagoon, in reef, and in fore-reef; high in lagoons far from the coast

seems to be controlled by various salinities. High values (80–240 ppm, mean 150 ppm) have been found in restricted lagoonal environments with evaporites, whereas low values (30–90 ppm) are distributed in the limestones of the open-marine shelf and in coral reef limestones. Early diagenetic dolomites with stromatolites and flat pebble conglomerates yield up to 440 ppm Na; late diagenetic dolomites are poor in sodium (< 100 ppm).

c) *Indication of the kind of sedimentation:* As demonstrated in the papers by Matthews (1966), Hoskin (1968), Friedman (1968), the constituents of Recent lime mud and lime sands can be defined by their origin or source area, with the aid of Sr and Mg contents (high-Sr and low-Sr aragonite). Veizer and Demovic (1974) and Veizer et al. (1971) have tried to use the Sr values together with other criteria (insoluble residues, grain fabric) to differentiate physico-chemically or biochemically formed carbonate sediments from microbioclastic carbonates (automicrite vs allomicrite, see 4.1.2).

The primary mineralogical composition of skeletal grains may be reflected by trace elements (Brand and Veizer, 1980). Analysis of the diagenetic textures of Paleozoic limestones and of Sr and Mn content of brachiopods, crinoids and rugose corals suggests that diagenetic equilibrium of carbonate components increases with meteoric water, which is also indicated by the textural alteration from micrite to microspar to minor amounts of pseudospar and/or by precipitation of interparticle sparite cement. This is paralleled by the decrease of strontium and an increase of manganese; the relatively large difference in strontium concentrations of bioclasts consisting originally of aragonite and high-Mg calcite is partially preserved in their respective diagenetic low-Mg calcites. Low-Mg calcite particles exhibit strontium concentrations which are only slightly altered by diagenesis.

d) *Support of interpretations of MF types with geochemical data:* The attempt is more often made now to check information about the depositional environment and biotopes, which have been derived from the fossil content and microfacies, with geochemical criteria. Examples can be found in Carozzi et al. (1972) – various Mesozoic shelf and basinal environments, Carozzi (1971) – internal subdivision of Middle Devonian back-reef carbonates, and MacQuown and Bloxam (1972) – Lower Carboniferous shallow-water shelf with different degrees of terrigenous

Table 45. Geochemical criteria of marine shallow water and deeper water carbonates

Author	Age Locality	Results	
		Shallow Marine	Deeper Marine
Veizer (1967)	Upper Jurassic-Urgonian: High Tatra, Czechoslovakia	High Mg and Sr content	Increased values of Al, Si, Fe, Mn; occurrence of Ti, Ni, Co
Bencini and Turi (1974)	Rhaetian to Neocomian: Lima valley, N. Apennines, Italy	Low Mn content	High Mn content
H. W. Flügel and Wedepohl (1967)	Upper Jurassic: Northern Alps	Relatively low Sr values (Plassen limestone)	Relatively high Sr content (Oberalm limestone
Bausch (1965)	Upper Jurassic: Southern Germany	Relatively low Sr values (sponge limestone)	Relatively high Sr content (bedded limestone with ammonites)
Bonnefous and Kubler (1968)	Lias to Tithonian: Tunisia	Low Mo, Pb, Zn, Cu contents in platform carbonates	Increased values of Mo, Pb, Zn, Cu in deeper water carbonates
Buggisch (1972)	Early Upper Devonian (Kellwasser limestone): Rhenish Massif, Germany	Low Mn content	High Mn content
Franke (1975)	Late Upper Devonian (cephalopod limestones and platform carbonates): Brilon, Sauerland, Germany	Low Mn content	Very high Mn content
Lütke (1976)	Givetian and Upper Devonian (Flinz): Harz mountains, Germany	Low Mn content in shallow water carbonates (mean 510 ppm), in carbonates formed on submarine rises (cephalopod limestones: 1200 ppm)	High Mn content in Flinz limestones (mean 2100 ppm)
Tucker (1973)	Upper Devonian (*Clymenia* limestone): Rhenish Massif, Germany	Low Mn content	High Mn content in cephalopod limestones (Schwellen facies)

influx; Fenninger and Holzer (1978) – Devonian platform carbonates. The comparison of geochemical data (Mn, insoluble residues) and microfacies types recognized in Lower Permian platform and shelf-edge carbonates of the Southern Alps indicates that Mn content can also be used to differentiate carbonate shallow-water environments (W. Buggisch, 1980); see Fig. 56.

e) *Explanation of early diagenetic and late diagenetic solution processes:* Sodium is a possible index to the salinity of diagenetic solutions in carbonate sediments (see

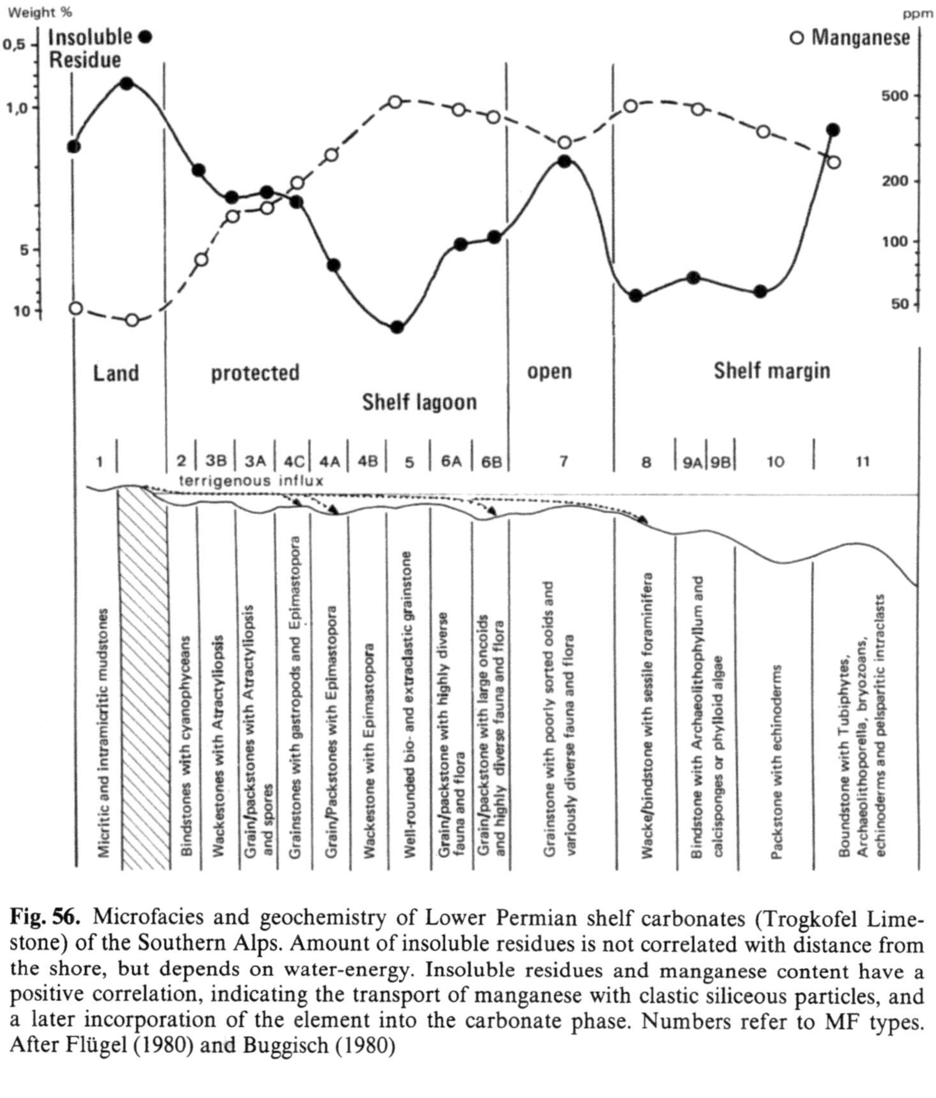

Fig. 56. Microfacies and geochemistry of Lower Permian shelf carbonates (Trogkofel Limestone) of the Southern Alps. Amount of insoluble residues is not correlated with distance from the shore, but depends on water-energy. Insoluble residues and manganese content have a positive correlation, indicating the transport of manganese with clastic siliceous particles, and a later incorporation of the element into the carbonate phase. Numbers refer to MF types. After Flügel (1980) and Buggisch (1980)

Land and Hoops, 1973; Scherer and Wendt, 1978; Mirsal, 1978). Another element often studied is Strontium. The Sr distribution in Ancient carbonates is decisively influenced by diagenesis, dolomitization, the clay content of the carbonates, and by the original salinity of the depositional environment (see Kranz, 1976):

Diagenesis: In an open-system model, if one assumes sea water to be a diagenizing solution with a Sr/Ca mole ratio of 0.9×10^{-2} and accepts the value of 0.055 for the Sr calcite-liquid partition coefficient, then the final Sr content in the carbonate sediment will be 430 ppm (Annovi et al., 1980). In the case of fresh water with a Sr/Ca mole ratio of 0.32×10^{-2} in the diagenetic solution, a Sr content of about 150 ppm should be expected in carbonates influenced by meteoric diagenesis

(Cioni et al., 1973). The depletion of Sr (Recent up to more than 10,000 ppm, in Pre-Tertiary limestones generally < 1000 ppm, mean 460 ppm – Wedepohl, 1979) which takes place during the recrystallization of primary metastable carbonate minerals (transformation aragonite – calcite) can, when results of experimental investigations are compared (Kinsman and Holland, 1969), reveal the processes involved in lithification and cement formation in open or closed systems of water circulation (freshwater, brackish water, marine water). Discussion in H. W. Flügel and Wedepohl (1967), Veizer and Demovic (1974), Folk and Land (1975), and Veizer (1977). Quantitative estimates of the changes of trace element concentrations in limestones during solution and recrystallization have been made by Morrow and Mayers (1978).

If Sr values in limestones are to be used as diagenetic tools, one must consider the primary differences in the Sr content of aragonite and low-calcite carbonate particles (about 1200 ppm, e.g., planktonic foraminifera, coccolithophorids, Recent pelagic calcareous muds).

The partitioning behavior of zinc and manganese can also serve as an important indicator of early and late diagenetic processes in limestones (Pingitore, 1978). Where there is no significant external source of these trace elements, later calcites should theoretically be enriched in Mn and Zn but depleted of Sr and Mg.

Salinity: Carbonates and sulfates in the Northern German Zechstein show that limestones within evaporitic sequences have relatively high Sr values. Kranz (1976) also demonstrated similar relationships in the lagoon facies of Middle Triassic Wetterstein limestones. High Sr values are, however, also found in lacustrine carbonates. Different salinity conditions can be inferred for Lower Cretaceous dasycladacean and charophycean limestones using Sr, Na, K, and Mg as indicators of normal salinity or hypersalinity, and Fe/Mn as indicators of continental influence (Jaffrezo and Renard, 1979).

The analysis of rare elements (Mg, Na, Sr, K) and of isotope ratios (O^{18} and C^{13}) makes it possible to recognize environmental factors such as salinity, redox potential and water temperature, which affect the formation of carbonate cements. Rhythmic variations of these parameters within successive generations of carbonate cements formed in open-void structures of limestones reflect changing salinity conditions within open, semi-closed, and closed systems (Mirsal, 1978; Gökdag, 1974).

Influence of clay content: According to Bausch (1965, 1968), the Sr content of limestones increases with an increased amount of clay. The reason for this he sees to be the adsorption on clay minerals of a portion of the Sr which has been set free during diagenesis. Finely dispersed clay can inhibit coalescive neomorphism and slow down the diagenetic depletion of Sr. A few authors, such as Veizer et al. (1974) and Knoblauch and Weiskirchner (1979) disagree with the theory that Sr is adsorbed on clay particles.

f) *Statements about ecologically efficient environmental factors:* Various elements, particularly boron, which is incorporated into clay minerals, are used to characterize paleosalinity (Ernst, 1966, 1970; Harder, 1970; Cody, 1971) although the information thus provided is rather controversial. Such an analysis of clay minerals from insoluble residues of carbonate rocks runs into difficulties because there is usually not enough unaltered material available. Other elements used as salinity indicators are Sr/Ca, Ca/Mg as well as Fe, Mn, and P; examples can be found in Aust (1969) and especially in the very interesting paper of Gittinger (1969), where paleontological and geochemical data are compared.

9.1.3 Stable Isotopes

The isotopic composition of a solid precipitated from solution is dependent upon the isotopic composition of the precipitating medium. In the $CaCO_3$–CO_2–H_2O system, the stable isotopes of oxygen and carbon, O^{16}, O^{18}, C^{12}, and C^{13}, are incorporated into the carbonates. Knowing the distribution of stable isotopes within a carbonate can provide valuable information concerning the nature of the medium from which the carbonate was precipitated. Carbon and oxygen stable isotopes of marine and non-marine carbonate sediments and calcareous shells have been studied over several decades (Graf, 1960; Hoefs, 1980; Hudson, 1977; Savin, 1977; Land, 1980). Numerous other studies deal with Holocene to Pleistocene carbonates of shallow-water environments and with the isotopic signals reflecting the reaction between marine sediment of varied mineralogy and meteoric waters of the vadose or shallow phreatic zone (e.g., Choquette, 1968; Gross and Tracey, 1966; Land, 1967, 1977; Shackleton and Matthews, 1977). Ancient limestones and the isotopic composition of carbonate cements have been studied by various authors, including Brennecke (1978), Davies and Krouse (1975), Dickson and Coleman (1980), Fritz (1971), Gökdag (1974), Hudson and Coleman (1978), Jux and Manze (1974, 1976, 1978), Keith and Weber (1964, a source of information), Meyers and James (1978), Moore and Druckman (1981), Mountjoy and Walls (1977), Murata et al. (1969), Rothe et al. (1974), Scholle and Arthur (1980), and Tan and Hudson (1974). A great deal of information can be found in papers dealing with petrography and diagenesis of deeper-marine carbonates (e.g., Anderson and Schneiderman, 1973; Berger, 1977, 1979; Coplen and Schlanger, 1973; Lawrence, 1973; Matter et al. 1975; McKenzie et al., 1978). Variations in the stable isotopes in individual skeletons of marine invertebrates (foraminifera, corals, echinids) have become increasingly important for our understanding of environmental fluctuations and paleoecological data (e.g., Wefer and Berger, 1980; Wefer et al., 1981).

Assuming that a carbonate is precipitated in equilibrium with the ambient environment, the stable isotope composition of the carbonate will be dependent upon

- Isotopic composition of the seawater and salinity (variations in the isotopic composition of the surface seawater exist due to net evaporation or freshwater contamination; Craig and Gordon, 1965). Variations in O^{18} and C^{13} in open marine environments are small; freshwater exhibits considerably lower O^{18} and C^{13} contents than do marine waters
- Mineral phase of the carbonate
- Temperature of the water (the temperature dependence of the O^{18}/O^{16} ratio is used to measure paleotemperatures; see Bowen, 1966; Epstein et al., 1953, Spaeth et al., 1971)

Non-equilibrium deposition may occur in organisms which use metabolic CO_2 (enriched in the lighter isotopes) for carbonate formation, or which preferentially incorporate individual isotopes. The effect of metabolic fractioning is especially critical in the deposition of C^{13}; changes in the δC^{13} of the oceanic bicarbonate reservoir with time might have produced secular trends in the isotope composition of carbonates (Veizer and Hoefs 1976). Errors in the interpretation of isotope signals of carbonates may be also introduced by vital effects upon carbonate precipitation (Friedman, 1968; Vinot-Bertoville and Duplessy, 1973; Wefer et al., 1981).

Implicit in the interpretation of the isotope ratios of limestones and calcareous fossils is the assumption that the overall isotopic composition of the oceanic water has remained essentially constant through time (Epstein and Mayeda, 1953; Lowenstam, 1961; Keith and Weber, 1964), and that the isotopic variation seen in Ancient carbonates is a function of isotope fractionation due to temperature changes, local salinity changes, and diagenetic alternations. Significant short-term carbon isotope fluctuations have been found in Cretaceous pelagic limestones (Scholle and Arthur, 1980), possibly connected with changes in the amount of organic carbon preserved in marine sediments.

Later alterations of the stable isotope composition of carbonates may be caused by

- Recrystallization of the sediment in contact with isotopically lighter freshwater fluids. An increase of lighter isotopes (O^{16}, C^{12}) is caused by the inversion of aragonite, the transformation of high-Mg calcite to low-Mg calcite, and by the cementation of open voids within a freshwater environment (Land, 1970)
- Re-equilibration of the carbonate-water isotopic system at higher temperatures by geothermal heat flow may lead to lighter isotopic values of oxygen and carbon (Anderson, 1969)

Isotope studies of pelagic sequences from several DSDP sites have shown that there is often a depletion in O^{18} or negative shift in the δO^{18} value of the bulk carbonate sediment with increasing age (or burial); corresponding changes in the δC^{13} ratio with depth (age) were not found (except by contribution of carbon from a secondary source). A similar situation has also been described for shallow-marine calcareous fossils (Dickson and Coleman, 1980). These isotopic depletions may represent either a gradual readjustment during early diagenetic stages (Matter et al., 1975; McKenzie et al., 1978) or after burial (Lawrence, 1973), or they may record temperature influences (Andersen and Schneiderman, 1973), or world-wide changes in the paleoenvironment (Coplen and Schlanger, 1973).

During diagenesis oxygen isotope values may show considerably larger variations than carbon isotope values. This is caused by the mobilization of oxygen during the transformation of aragonite to calcite, by the influence of meteoric waters, and also perhaps by weathering effects on limestones. The ratio of oxygen in pore water to oxygen in the rock is high in carbonate oozes which are initially extremely porous; this is the inverse of the situation for carbon. A significant temperature fractionation can play an important role during burial-diagenetic cementation (Friedman and O'Neil, 1977).

The oxygen isotope composition might have been shifted towards a higher O^{18} content since the Cambrian (Keith and Weber, 1964). This is also indicated by paleowatertemperatures obtained from oxygen isotope measurements of Paleozoic fossils (Jux and Manze 1976). The temperatures are generally too high when compared with Mesozoic and Cenozoic fossils (see also Brand and Veizer, 1981).

The δC^{13} values of marine limestones and bioclasts which have not undergone near-surface freshwater diagenesis are similar to those of modern carbonate sediments and shells (Hudson, 1977; Dickson and Coleman, 1980). The influence of temperature fluctuations on carbon isotope variations can be discounted, because the temperature fractionation effect is minor for carbon (Emrich et al., 1970). Changes in porewater chemistry and freshwater diagenesis cause large alterations in the isotopic composition; marine carbonates generally become lighter in carbon isotopic composition. Limestones without this isotopic depletion might have been cemented mainly during the compactional phase of burial diagenesis (Hudson, 1977).

Lighter δC^{13} values may also be produced by the oxidation of organic matter after burial in the sediment (e.g., Murata et al. 1969, 1972; Irwin et al., 1977), by bioturbation in slowly deposited pelagic carbonates (Scholle, 1980), and by the deposition of carbonate sediments under regions of variable upwelling (Berger et al., 1977, 1979).

Use of Stable Isotope Studies of Carbonates in Sedimentology and Paleoecology: Oxygen isotope investigations have dealt primarily with the determination of paleotemperatures and paleosalinities of Ancient water masses; carbon isotope data have largely been used in studies of diagenesis and paleosalinity. Stable isotope ratios in carbonate minerals yield the following information:

1. Mode of origin of carbonate minerals and carbonate particles of Recent sediments: Organic or inorganic formation of the aragonite needles in the bottom sediments of the Bahamas (Lowenstam and Epstein, 1957); importance of algae in ooid formation (Fabricius, 1977), and in micritization and submarine cementation (Lloyd, 1971).

2. Temperature of seawater (paleowatertemperatures): Interpretation of oxygen isotope ratios as indicators of paleowatertemperatures must take into account the strong influence of the vital effect of organisms, which is connected with the physiological fractionation during the deposition of skeletal carbonate; the possibility of the secretion of the skeleton during only part of a yearly temperature cycle, and differences as well in life habitats of mobile organisms. Other problems are the possible shift of the oxygen isotope composition towards O^{18} since the Cambrian and alterations connected with the recrystallization of shells (Spaeth et al., 1971).

3. Salinity of water: Assuming only small temperature effects, salinity trends derived from carbon and oxygen isotopic composition are fairly consistent with paleontological data. Examples are known from Devonian, Permian, Tertiary, and Holocene fossils and carbonate sediments (Jux and Manze, 1974; Rössler et al., 1968; Rothe et al., 1974; Lloyd, 1964).

4. Depth of deposition: Dissolution effects on stable isotope compositions of Holocene and Pleistocene planktonic and benthonic foraminifera are reflected by increasing δO^{18} values with increasing water depth. These alterations can be used to describe changes in paleo-water depth and paleoclimate (Berger and Killingley, 1977).

5. Diagenetic environments: The C^{13}/C^{12} and O^{18}/O^{16} ratios of limestones appear to be useful for determining their diagenetic history when the geologic relationships are relatively simple, and when the original isotope compositions of the carbonates and waters are known. Changes in the isotope ratios of diagenetically altered limestones are the result of the precipitation of carbonate as cement, veins and casts in the limestones, as well as alterations of the individual constituent grains. Most alterations occur in the freshwater zone. Different diagenetic environments (early submarine, early subaerial, and late cementation) should have left their isotopic imprints.

Isotope studies are most important for the explanation of cementation processes. The original isotopic composition may be preserved in calcitic cements in spite of many later diagenetic and even tectonic events (Hudson and Coleman, 1978; Gökdag, 1974). The widespread opinion that the oxygen-isotopic record of early diagenetic events in limestones was "washed out" by later meteoric-derived waters seems to be erroneous.

Oxygen and carbon isotope values are characteristically more negative in meteoric cements than those found in marine carbonates. This fact has been used successfully in the investigation of Ancient carbonate cements (Mirsal and Zankl, 1979; Gökdag, 1974, see Sect. 3.3.3). As shown by Gökdag, no isotopic differences seem to exist between micritic limestones and recrystallized limestones; alteration of stable isotope composition seems to take place only in connection with replacement processes and the formation of new minerals.

6. Dolomitization: Studies based on isotope chemistry are widely used in order to separate late diagenetic dolomites from early diagenetic or "primary" dolomites (see Bausch and Hoefs, 1972; Milliman, 1974). However, at the moment there is considerable critical discussion about the interpretation of the isotope composition of dolomites because of the lack of enough parallel petrographic studies and facies analysis of dolomite rocks (see Land, 1980).

9.1.4 References: Isotopes

Anderson, T. F. (1969): Self-diffusion of carbon and oxygen in calcite by isotopic exchange with carbon dioxide. J. Geophys. Res. **74,** 3918–3932

Anderson, T. F., Schneidermann, N. (1973): Stable isotope relationships in pelagic limestones from the Central Caribbean, Leg 15, Deep-Sea Drilling Project. Initial Rep. Deep Sea Drilling Project **15,** 795–803, Washington

Bausch, W., Hoefs, J. (1972): Die Isotopenzusammensetzung von Dolomiten und Kalken aus dem süddeutschen Malm. Contrib. Min. petrol. **37,** 121–130, Heidelberg

Berger, W. H. (1977): Carbon dioxide excursions and the deep sea record: aspects of the problem. In: Anderson, N. R., Malahoff, A. (eds.): The fate of fossil fuel CO_2 in the ocean. 505–542, New York: Plenum Press

Berger, W. (1979 a): Stable isotopes in foraminifera. In: Lipps, J. H., Berger, W., Buzas, M. A., Douglas, C. A. (eds.): Foraminiferal ecology and paleoecology. Soc. Econ. Paleont. Min. Short Course **6,** 156–198, Tulsa

Berger, W. H. (1979 b): Impact of deep-sea drilling on paleo-oceanography. In: Talwani, M., Hay, W., Ryan, W. B. F. (eds.): Deep drilling results in the Atlantic ocean: continental margins and paleo-environment. Maurice Ewing Ser. **3,** Amer. Geophys. Union, 297–314, 12 Figs., Chelsea, Michigan

Berger, W. H., Killingley, J. S. (1977): Glacial-Holocene Transition in deep-sea carbonates: selective dissolution and the stable isotope signal. Science **197,** 563–566, Washington

Bowen, R. (1966): Paleotemperature analysis. 256 pp., Amsterdam: Elsevier

Brand, U., Veizer, J. (1981): Chemical diagenesis of a multicomponent carbonate system – 2: Stable Isotopes. J. Sed. Petrol. **51**/3, 987–997, 4 Figs., Tulsa

Brennecke, J. C. (1978): A comparison of the stable oxygen and carbon isotope composition of Early Cretaceous and Late Jurassic carbonates from DSDP Sites 105 and 367. Initial Rep. Deep Sea Drilling Project **41,** 937–955, Washington

Choquette, P. W. (1968): Marine diagenesis of shallow-marine lime-mud sediments: insights from δO^{18} and δC^{13} data. Science **161,** 1130–1132, Washington

Coplen, T. B., Schlanger, S. O. (1973): Oxygen and carbon isotope studies of carbonate sediments from Site 167, Magellan Rise, Leg 17. Initial Rep. Deep Sea Drilling Project **17,** 505–509, Washington

Craig, H. (1953): The geochemistry of stable carbon isotopes. Geochim. Cosmochim. Acta **3,** 53–92, Oxford

Craig, H. (1965): The measurement of oxygen isotope paleotemperature. In: Tongiorgi, E. (ed.): Stable isotopes in oceanographic studies and paleotemperature. Pisa: Consiglio Naz. Ricc., Lab. Geol. Nucleare

Craig, H., Gordon, L. I. (1965): Isotopic oceanography – deuterium and oxygen 18 variations in the ocean and the marine atmosphere. In: Tongiorgi, E. (ed.): Stable isotopes in oceanographic studies and paleotemperatures, Spoleto, 1965. Consiglio Naz. Ricerche, Laboratorio Geol. Nucleare, 9–130, Pisa

Davies, G. R. Krouse, H. R. (1975): Carbon and oxygen isotopic composition of late Paleozoic calcitic cements; Canadian arctic archipelago – preliminary results and interpretation. Paper Geol. Surv. Canada **75,** 1 B, 215–220, Ottawa

Degens, E. T., Epstein, S. (1964): Oxygen and carbon isotope ratios in coexisting calcites and dolomites from recent and ancient sediments. Geochim. Cosmochim. Acta **28,** 23–44, Oxford

Deuser, W. G., Degens, E. T. (1969): O^{18}/O^{16} and C^{13}/C^{12} ratios of fossils from the Hot Brine deep area of the central Red Sea. In: Degens, E. T., Ross, D. A. (eds.): Hot Brines and Recent Heavy Metal Deposits in the Red Sea, 336–347, Heidelberg: Springer

Dickson, J. A. D., Coleman, M. L. (1980): see 9.1.5

Dodd, J. R., Stanton, R. J. (1975): Paleosalinities within a Pliocene Bay, Kettleman Hills, California. Geol. Soc. Amer. Bull. **86,** 51–64, Boulder

Duplessy, J. C. (1978): Isotope studies. In: Gribbin, J. (ed.): Climatic Change. 46–67, London: Cambridge Univ. Press

Emiliani, C., Hudson, H. J., Shinn, E. A., George, R. J. (1978): Oxygen and carbon isotopic growth record in a reef coral from the Florida Keys and a deep-sea coral from Blake Plateau. Science **202,** 627–629, Washington

Emrich, K., Ehhalt, D. H., Vogel, J. C. (1970): Carbon isotope fractionation during the precipitation of calcium carbonate. Earth Planet. Sci. Lett. **8,** 363–371

Epstein, S. (1959): The variations of the O^{18}/O^{16} ratio in nature and some implications. In: Abelson, P. H. (ed.): Researches in geochemistry. 217–240, New York: Wiley

Epstein, S., Buchsbaum, R., Lowenstam, H., Urey, H. C. (1953): Revised carbonate-water isotopic temperature scale. Bull. Geol. Soc. Amer. **64,** 1315–1325, New York

Epstein, S., Graf, D. L., Degens, E. T. (1963): Oxygen isotope studies on the origin of dolomites. In: Isotopic and cosmic chemistry, 169–180, Amsterdam: North Holland Publ.

Epstein, S., Mayeda, T. (1953): Variation of O^{18} content of waters from natural sources. Geochim. Cosmochim. Acta **4,** 213–224, Oxford

Erez, J. (1978): Vital effect on stable-isotope composition seen in foraminifera and coral skeletons. Nature **273,** 199–202, London

Fairbanks, R. G., Dodge, R. E. (1979): Annual periodicity of the $^{18}O/^{16}O$ and $^{13}C/^{12}C$ ratios in the coral Montastraea annularis. Geochim. Cosmochim. Acta **43,** 1009–1020, Oxford

Faure, G. (1977): Principles of Isotope Geology. 464 pp., New York: Wiley

Friedman, G. M. (1964): see 3.9

Friedman, G. M. (1968): see 3.9

Fritz, P. (1971): Oxygen and carbon isotopic composition of carbonates from the Jura of S. Germany. Canad. J. Earth Sci. **4,** 1247–1267

Fritz, P., Smith, D. G. W. (1970): The isotopic concentration of secondary dolomites. Geochim. Cosmochim. Acta **34,** 1161–1163, Oxford

Gökdag, H. (1974): see 3.9
Goodney, D. E. (1977): Non-equilibrium fractionation of the stable isotopes of carbon and oxygen during precipitation of calcium carbonate by marine phytoplankton. PhD Thesis, Univ. Hawai, 146 p.
Graf, D. L. (1960): Geochemistry of Carbonate Sediments and Sedimentary Carbonate Rocks. Part IV-A, Isotopic Composition, Chemical Analysis. Illinois State Geol. Surv. Circ. **308,** 42 p., Urbana
Gross, M. G. (1964): Variations in the $^{18}O/^{16}O$ and $^{13}C/^{12}C$ ratios of diagenetically altered limestones in the Bermuda Islands. J. Geol. **72,** 170–194, Chicago
Gross, M. G., Tracey, J. I., Jr. (1966): Oxygen and carbon isotopic composition of limestones and dolomites, Bikini and Eniwetok atolls. Science **151,** 1082–1084, Washington
Hoefs, J. (1980): Stable isotope geochemistry. 140 pp., Amsterdam: Elsevier
Hudson, J. D. (1975): Carbon isotopes and limestone cements. Geol. **3,** 19–22, Boulder
Hudson, J. D. (1977): Stable isotopes and limestone lithification. Geol. Soc. London Quart. J. **133,** 637–660, London
Hudson, J. D., Coleman, M. L. (1978): Submarine cementation of the Scheck Limestone Conglomerate (Jurassic, Austria): isotopic evidence. N. Jb. Geol. Paläont. Mh. **1978**/9, 534–544, 2 Figs., Stuttgart
Irwin, H., Curtis, C., Coleman, M. (1977): Isotopic evidence for source of diagenetic carbonates formed during burial of organic-rich sediments. Nature **269,** 209–213, London
Jux, U., Manze, U. (1974): Milieu-Indikatoren im Devon des Bergischen Landes mittels Kohlenstoff-Isotopen. N. Jb. Geol. Paläont. Mh. **1974**/6, 353–373, 6 Figs., Stuttgart
Jux, U., Manze, U. (1976): Milieu-Indikatoren aus einem mitteldevonischen biostromalen Riff des Bergischen Landes mittels Kohlenstoff-Isotopen. Decheniana **129,** 245–262, Bonn
Jux, U., Manze, U. (1978): Milieu-Indikatoren aus einem mitteldevonischen biohermalen Riff des Bergischen Landes mittels C- und O-Isotopen. Decheniana **131,** 300–324, 7 Figs., Bonn
Keith, M. L., Anderson, G. M., Eichler, R. (1964): Carbon and oxygen isotopic composition of mollusk shells from marine and freshwater environments. Geochim. Cosmochim. Acta **28,** 1757–1786, Oxford
Keith, M. L., Parker, R. H. (1965): Local variation of ^{13}C and ^{18}O content of mollusk shells and the relatively minor temperature effect in marginal marine environments. Marine Geol. **3,** 115–129, Amsterdam
Keith, M. L., Weber, J. N. (1964): Isotopic composition and environmental classification of selected limestones and fossils. Geochim. Cosmochim. Acta **28,** 1787–1816, Oxford
Keith, M. L., Weber, J. N. (1965): Systematic relationships between carbon and oxygen isotopes in carbonates deposited by modern corals and algae. Science **150,** 498–501, Washington
Kroopnick, P. M., Margolis, S. V., Wong, C. S. (1977): $\delta^{13}C$ variations in marine carbonate sediments as indicators of the CO_2 balance between the atmosphere and the oceans. In: Anderson, N. R., Malahoff, A. (eds.): The fate of fossil fuel CO_2 in the oceans, 295–321, New York: Plenum Press
Land, L. S. (1970): Phreatic versus vadose meteoric diagenesis of limestones: Evidence from a fossil water table. Sed. **14,** 175–185, Oxford
Land, L. S. (1980): The isotopic and trace element geochemistry of dolomite: the state of the art. Soc. Econ. Paleont. Min. Spec. Publ. **28,** 87–110, Tulsa
Land, L. S., Lang, J. C., Smith, B. N. (1975): Extension rate: a primary control on the isotopic composition of West Indian (Jamaican) scleractinian reef coral skeletons. Marine Biol. **33,** 221–233
Lawrence, J. R. (1973): Interstitial water studies, Leg 15. Stable oxygen and carbon isotope variations in water, carbonates, and silicates from the Venezuela Basin (Site 149) and the Aves Rise (Site 148). Initial Report Deep Sea Drilling Project **20,** 891–899, Washington
Lloyd, R. M. (1964): Variations in the oxygen and carbon isotope ratios of Florida Bay mollusks and their environmental significance. J. Geol. **72**/1, 84–111, 19 Figs., Chicago
Lloyd, R. M. (1971): Some observations of recent sediment alteration ("micritization") and the possible role of algae in submarine cementation. In: Bricker, O. P. (ed.): Carbonate Cements. John Jopkins Univ. Stud. Geol. **19,** 72–79, Figs. 37–42, Baltimore

Lowenstam, H. A. (1961): Mineralogy, O^{18}/O^{16} ratio, and strontium and magnesium contents of Recent and fossil brachiopods and their bearing on the history of the oceans. J. Geol. **69**/3, 241–260, 6 Figs., Chicago

Lowenstam, H. A., Epstein, S. (1957): On the origin of sedimentary aragonite needles of the Great Bahama Bank. J. Geol. **65**, 364–375, Chicago

Manze, U., Richter, D. K. (1979): Die Veränderung des C^{13}/C^{12}-Verhältnisses in Seeigelcoronen bei der Umwandlung von Mg-Calcit in Calcit unter meteorisch-vadosen Bedingungen. N. Jb. Geol. Paläont. Abh. **158**/3, 334–345, 3 Figs., Stuttgart

Matter, A., Douglas, R. G., Perch-Nielsen, K. (1975): see 3.9

McKenzie, J., Bernoulli, D., Garrison, R. E. (1978): Lithification of pelagic-hemipelagic sediments at DSDP Site 372: oxygen isotope alteration with diagenesis. Initial Rep. Deep Sea Drilling Project **42**/1, 473–478, 4 Figs., Washington

Meyers, W. J., James, A. T. (1978): Stable isotopes of cherts and carbonate cements in the Lake Valley formation (Mississippian), Sacramento Mountains, New Mexico. Sed. **25**, 105–124, Oxford

Milliken, K. L., Land, L. S., Loucks, R. G. (1981): History of burial diagenesis determined from isotopic geochemistry, Frio Formation, Brazoria County, Texas. Amer. Ass. Petrol. Geol. Bull. **65**, 1397–1413, 14 Figs., Tulsa

Milliman, J. D. (1974): see 1.4.1

Milliman, J. D., Müller, J. (1977): Characteristics and genesis of shallow-water and deep-sea limestones. In: Anderson, N. R., Malahoff, A. (eds.): the fate of fossil fuel CO_2 in the oceans, 655–672, New York: Plenum Publishing Corporation

Mirsal, J. A., Zankl, H. (1979): Petrography and geochemistry of carbonate void-filling cements in fossil reefs. Geol. Rundschau **68**/3, 920–951, Stuttgart

Mountjoy, E. W., Walls, R. A. (1977): see 10.3.9

Murata, K. J., Friedman, I., Madsen, B. M. (1969): Isotopic composition of diagenetic carbonates in marine Miocene formations of California and Oregon. U.S. Geol. Surv. Prof. Paper **614**-B, 1–24, Washington

Moore, C. H., Druckman, Y. (1981): Burial diagenesis and porosity evolution, Upper Jurassic Smackover, Arkansas and Louisiana. Bull. Amer. Ass. Petrol. Geol. **65**/4, 597–628, 13 Figs., Tulsa

Park, K., Epstein, S. (1960): Carbon isotope fractionation during photosynthesis. Geochim. Cosmochim. Acta **21**, 110–126, Oxford

Rankama, K. (1954): Isotope geology. 535 p., London: Pergamon Press

Robinson, M., Clayton, R. N. (1969): Carbon-14 fractionation between aragonite and calcite. Geochim. Cosmochim. Acta **33**, 997–1002, Oxford

Roessler, H. J., Pilot, J., Harzer, D., Krüger, M. (1968): Isotopenchemische Untersuchungen (O, S, C) an Salinar- und Sapropelsedimenten Mitteleuropas. 23rd Int. Geol. Congr., **6**, 89–100, Prag

Rothe, P., Hoefs, J., Sonne, V. (1974): The isotope composition of Tertiary carbonates from the Mainz Basin: an example of isotopic fractionation in "closed basins". Sed. **21**, 373–395, 6 Figs., Oxford

Savin, S. M. (1977): The history of the earth's surface temperature during the last 100 million years. Ann. Rev. Earth Planetary Sci. **15**, 319–355

Scholle, P. A. (1974): see 3.9

Scholle, P. A., Arthur, M. A. (1980): Carbon isotope fluctuations in Cretaceous pelagic limestones: potential stratigraphic and petroleum exploration tool. Bull. Amer. Ass. Petrol. Geol. **64**/1, 67–87, 12 Figs., Tulsa

Shackleton, N. J., Kennett, J. P. (1975): Paleotemperature history of the Cenozoic and the initiation of Antarctic glaciation: oxygen and carbon isotope analysis in DSDP Sites 277, 279, and 281. Initial Rep. Deep Sea Drilling Project **29**, 743–755, Washington

Shackleton, N. J., Matthews, R. (1977): Oxygen isotope stratigraphy of Late Pleistocene coral terraces in Barbados. Nature **268**, No. 5621, 618–620, 1 Fig., London

Shackleton, N. J., Vincent, E. (1978): Oxygen and carbon isotope studies in Recent foraminifera from the southwest Indian Ocean. Micropaleont. **3**, 1–13, New York

Sharma, T., Clayton, R. N. (1965): Measurements of O^{18}/O^{16} ratios of total oxygen of carbonates. Geochim. Cosmochim. Acta **29**, 1347–1353, Oxford

Spaeth, C., Hoefs, J., Vetter, U. (1971): Some aspects of isotope composition of belemnites and related paleotemperatures. Geol. Soc. Amer. Bull. **82,** 3139–3150, New York

Tan, F. C., Hudson, J. D. (1974): Isotopic studies on the paleoecology and diagenesis of the Great Estuarine Series (Jurassic) of Scotland. Scottish J. Geol. **10,** 91–128, Edinburgh

Tappan, H. (1968): Primary production, isotopes, extinctions and the atmosphere. Paleo-3, **4,** 187–210, Amsterdam

Tarutani, T., Clayton, R. N., Mayeda, T. K. (1969): The effect of polymorphism and magnesium substitution on oxygen isotope formation between calcium carbonate and water. Geochim. Cosmochim. Acta **33,** 987–996, Oxford

Veizer, J., Hoefs, J. (1976): The nature of O^{18}/O^{16} and C^{13}/C^{12} secular trends in sedimentary carbonate rocks. Geochim. Cosmochim. Acta **40,** 1387–1395, Oxford

Vergnaud-Grazzini, C. (1979): Non-equilibrium isotopic composition of shells of planktonic foraminifera in the Mediterranean Sea. Paleo-3, 263–276, Amsterdam

Vinot-Bertouille, A. C., Duplessy, J. C. (1973): Individual isotopic fractionation of carbon and oxygen in benthic foraminifera. Earth Planetary Sci. Lett. **18,** 247–252

Vogel, J. C. (1959): Über den Isotopengehalt des Kohlenstoffs in Süßwasser-Kalkablagerungen. Geochim. Cosmochim. Acta **16,** 236–242, Oxford

Weber, J. N., Woodhead, P. M. J. (1970): Carbon and oxygen isotope fractionation in the skeletal carbonate of reef-building corals. Chem. Geol. **6,** 93–117, Amsterdam

Wefer, G., Berger, W. H. (1980): Stable isotopes in benthic foraminifera: seasonal variation in large tropical species. Science **209,** 803–805, Washington

Wefer, G., Killingley, J. S., Lutze, G. F. (1981): Stable isotopes in Recent larger foraminifera. Paleo-3, **33,** 253–270, 3 Figs., Amsterdam

9.1.5 References: Geochemistry of Carbonate Rocks

Al-Hashimi, W. S. (1976): Significance of Strontium distribution in some carbonate rocks in the Carboniferous of Northumberland, England. J. Sed. Petrol. **46**/2, 339–376, 4 Figs., Tulsa

Amiel, A. J., Friedman, G. M., Miller, D. S. (1973): Distribution and nature of incorporation of trace elements in modern aragonitic corals. Sedimentology **20,** 47–64, Oxford

Atrup, P., Gauthier, J. (1967): Etude géochemique d'une séquence des "terres noires". Application au problème de l'écologie de Spongiaires constructeurs. Bull. Soc. Géol. France (7), **8,** 1966, 405–412, 2 Figs., Paris

Aust, H. (1969): Lithologie, Geochemie und Paläontologie des Grenzbereiches Muschelkalk-Keuper in Franken. Arb. Naturwiss. Ver. Würzburg **10,** 3–155, 33 Figs., 4 Pls., 3 Tab., Würzburg

Backhaus, E., Flügel, E. (1971): Fazielle und geochemische Untersuchungen am unteren Muschelkalk des Michelstädter Grabens (Odenwald). Notizbl. Hess. Landesamt Bodenforsch. **99,** 200–213, Pl. 15, 5 Tab., Wiesbaden

Bausch, W. M. (1965): Strontiumgehalte in süddeutschen Malmkalken. Geol. Rdsch. **55**/1, 86–96, 9 Figs., Stuttgart

Bausch, W. (1968): Outlines of distribution of strontium in marine limestones. In: Müller, G., Friedman, G. M. (eds.): Recent Developments in Carbonate Sedimentology in Central Europe, 106–115, 6 Figs., Berlin–Heidelberg–New York: Springer

Bausch, W. (1971): Tonmineralprovinzen im europäischen Malm. Ann. Inst. Geol. Publ. Hungaricae **54**/2, 333–334, Budapest

Bausch, W. M. (1980): Tonmineralprovinzen in Malmkalken. Erlanger Forschungen, B, Naturwiss. Medizin, **8,** 78 pp., 33 Figs., Erlangen

Bausch, W. M., Poll, K. G. (1976): Nichtkarbonatische Rückstände von Malmkalken der Ostalpen. Geol. Rdsch. **65**/2, 579–592, 10 Figs., Stuttgart

Becher, A. (1965): Eine Tonmineralfolge vom Beckenrand zum Beckeninneren im Buntsandstein Nordost-Bayerns. Beitr. Min. Petrogr. **11,** 586–613, 17 Figs., Heidelberg

Behrens, M., Krumsiek, K., Meyer, D. E., Schäfer, A., Siehl, A., Stets, J., Thein, J., Wurster, P. (1978): Sedimentationsabläufe im Atlas Golf (Kreide-Küstenbecken, Marokko). Geol. Rundschau **67**/2, 424–453, 7 Figs., Stuttgart

Bencini, A., Turi, A. (1974): Mn distribution in the Mesozoic carbonate rocks from Lima Valley, northern Apennines. J. Sed. Petrol. **44**/3, 774–782, 2 Figs., Tulsa

Billings, G. K., Ragland, P. C. (1969): Geochemistry and mineralogy of the recent reefs and lagoonal sediments South of Belize (British Honduras). Chem. Geol. **3**, 135–153, 9 Figs., 4 Tab., Amsterdam

Bolliger, W., Burri, P. (1967): Versuch einer Zeitkorrelation zwischen Plattformcarbonaten und tiefmarinen Sedimenten mit Hilfe von Quarz-Feldspat-Schüttungen (mittlerer Malm des Schweizer Jura). Eclogae geol. Helvet. **60**/2, 491–507, 4 Figs., Basel

Bonnefous, J., Jubler, B. (1968): Etudes stratigraphiques et géochimiques d'échantillons de terrains jurassiques de Tunisie (région nord-est). Bull. Centre Rech. Pau. SNPA **2**/2, 283–319, 6 Figs., 7 Pls., 1 Tab., Pau

Bordine, B. W. (1965): Paleoecologic implications of strontium, calcium and magnesium in Jurassic rocks near Thistle, Utah. Brigham Young Univ. Geol. Studies **12**, 91–120, 13 Figs.

Bouroullec, J., Deloffre, R., Rao, C. P. (1978): Caractérisation statistique des environment et de leur polarité en plate-forme carbonatée application au réservoir albo-aptien de Lacq. Bull. Cent. Rech. Explor.-Prod. Elf-Aquitaine **2**/2, 219–336, 11 Figs., Pau

Bowen, R. (1966): Paleotemperature analysis. 265 p., Amsterdam: Elsevier

Bruder, H. J. (1968): Ökologische, geochemische und sedimentologische Untersuchungen im Lias zeta (Oberes Toarcium) Schwabens, mit Berücksichtigung des obersten Lias epsilon. Arb. Geol. Paläont. Inst. Univ. Stuttgart, N. F. **56**, 165 pp., 21 Pls., 6 Figs., 29 Tabs., Stuttgart

Buggisch, W. (1972): Zur Geologie und Geochemie der Kellwasserkalke und ihrer begleitenden Sedimente (Unteres Oberdevon). Abh. hess. Landesamt Bodenforschg. **62**, 68 pp., 13 Pls., 19 Figs., 6 Tab., Wiesbaden

Buggisch, W. (1980): Die Geochemie der Kalke in den Trogkofel-Schichten der Karnischen Alpen. Carinthia II, Sonderheft **36**, 101–111, 5 Figs., Klagenfurt

Buggisch, W., Marzella, Cl., Hügel, Ph. (1979): Die fazielle und paläogeographische Entwicklung der infrakambrischen bis ordovizischen Sedimente im mittleren Antiatlas um Agdz (S-Marokko). Geol. Rundschau **68**/1, 195–224, 21 Figs., Stuttgart

Butler, G. P. (1972): Strontium geochemistry of modern and ancient calcium-sulphate-sediments. In: Purser, B. H. (Ed.): The Persian Gulf, 423–462, Berlin–Heidelberg–New York: Springer

Campbell, F. A., Ethier, V. G. (1972): Mineralogy and trace-element distribution, B and D Pools, Rainbow Lake Area, Alberta. Bull. Canadian Petrol. Geol. **20**/2, 286–300, 87 Figs., Calgary

Campbell, F. A., Lerbekmo, J. F. (1965): Chemical composition of Mississippian carbonate rocks from Throton Creek, Alberta. Bull. Canadian Petrol. Geol. **13**, 229–237, Calgary

Carozzi, A. V. (1971): Geochemical data on back-reef carbonates; Traverse group (Givetian) of the northern part of the Southern Peninsula of Michigan. Bull. Centre. Rech. Pau. SNPA **5**/2, 213–222, Pau

Carozzi, A. V., et al. (1972): see 5.5

Casati, P., Bertozzi, P., Cita, M. B., Longinelli, A., Damiani, V. (1976): Stratigraphy and paleoenvironment of the Messinian "Columbacci" formation in the Periadriatic trough. A pilot study. Mem. Soc. Geol. Ital. **16**, 173–195, 13 Figs., Rome

Cerny, I. (1976): Vergleich des Anis beiderseits der Periadriatischen Naht im Raum von Eisenkappel (Kärnten). Diss. Univ. Graz

Cerny, I. (1978): Geochemie "anisischer" Sedimentgesteine in den Nordkarawanken. Carinthia II, v. **88**, 55–70, 8 Figs., 10 Tables, Klagenfurt

Chester, A. (1965): Geochemical criteria for differentiating reef from non-reef facies in carbonate rocks. Bull. Amer. Ass. Petrol. Geol. **49**, 258–276, Tulsa

Cioni, R., Innocenti, F., Mazzuoli, R., Radicati di Brozole, F. (1973): Sr-distribution and carbonate mineralogy in the "Calcare Masiccio" formation of non-metamorphic Tuscan series, North of Arno River. Boll. Soc. Geol. **92**, 363–389, 13 Figs., 2 Tab., Rome

Cody, R. D. (1971): Adsorption and the reliability of trace elements as environment indicators for shales. J. Sed. Petrol. **41**/2, 461–471, 2 Figs., Tulsa

Couch, E. L. (1971): Calculation of paleosalinities from boron and clay mineral data. Bull. Amer. Ass. Petrol. Geol. **55**/10, 1829–1837, 5 Figs., Tulsa

Curtis, C. D. (1964): Studies on the use of B as a paleoenvironmental indicator. Geochim. Acta **28,** 1125–1135, 1 Tab., Oxford

Davies, P. J. (1972): Trace elements distribution in reef and subreef rocks of Jurassic age in Britain and Switzerland. J. Sed. Petrol. **42**/1, 183–194, 3 Figs., Tulsa

Degens, E. T. (1968): Geochemie der Sedimente. 282 pp., 75 Figs., 20 Tab., Stuttgart: Enke

Delfaud, J., Gauthier, J. (1968): Contribution à la connaissance de l'environment de déposits carbonates de plate-forme. Caractères géochimiques et sédimentologiques du Jurassique Nord-Aquitain. Bull. Centre. Rech. Pau. SNPA **2**/2, 347–383, 2 Pls., 8 Tab., 7 Figs., Pau

Demovic, R. (1974): Geochemie der mittel- und obertriassischen Kalke des slowakischen Karstes, der Krizna- und der Choc-Decke. Geol. Zbornik Slov. Akad. Vied. **25**/2, 335–354, Bratislava

Dickson, J. A. D., Coleman, M. L. (1980): Changes in carbon and oxygen isotope composition during limestone genesis. Sed. **27,** 107–118, 5 Figs., Oxford

Engelhardt, W. v. (1961): Neuere Ergebnisse der Tonmineralforschung. Geol. Rdsch. **51**/2, 457–477, 11 Tab., Stuttgart

Ernst, G. (1963): Stratigraphische und gesteinschemische Untersuchungen im Santon und Campan von Lägersdorf (SW-Holstein). Mittl. Geol. Staatsinst. Hamburg **32,** 71–127, 8 Figs., Pls. 13–14, Hamburg

Ernst, W. (1966): Stratigraphisch-fazielle Identifizierung von Sedimenten auf chemisch-geologischem Wege. Geol. Rdsch. **55**/1, 21–29, 3 Figs., Stuttgart

Ernst, W. (1970): Geochemical facies analysis. Methods Geochem. Geophys. **11,** 152 pp., Amsterdam–London–New York: Elsevier

Faure, G. (1977): Principles of isotope geology. 464 pp., New York: Wiley

Faure, G., Assereto, R., Tremba, E. L. (1978): Strontium isotope composition of marine carbonates of Middle Triassic to Early Jurassic age, Lombardic Alps, Italy. Sed. **25,** 523–543, 4 Figs., Oxford

Fenninger, A., Holzer, H.-L. (1978): Die Genese der Dolomitsandstein-Folge des Grazer Paläozoikums. Mitt. österr. geol. Gesellschaft **69,** 109–162, 10 Pls, 11 Figs., Wien

Flügel, E., Flügel-Kahler, E. (1963): Mikrofazielle und geochemische Gliederung eines obertriadischen Riffes der nördlichen Kalkalpen (Sauwand bei Gußwerk, Steiermark, Österreich). Mitt. Mus. Bergbau, Geol. Technik **24,** Landesmus. Joanneum Jg. 1962, 129 pp., 10 Pls., 11 Figs., 19 Tab., Graz

Flügel, E., Homann, W., Tietz, G.-F. (1971): Litho- und Biofazies eines Detailprofils in den Oberen Pseudoschwagerinen-Schichten (Unter-Perm) der Karnischen Alpen. Verh. geol. Bundesanst. Wien **1971**/1, 10–42, 6 Figs., 4 Tab., Wien

Flügel, H. W. (1968): Some notes on the insoluble residues in limestones. In: Müller, G., Friedman, G. (eds.): Recent Developments in Carbonate Sedimentology in Central Europe, 46–54, 2 Figs., Berlin–Heidelberg–New York: Springer

Flügel, H. W., Wedepohl, K. H. (1967): Die Verteilung des Strontiums in oberjurassischen Karbonatgesteinen der Nördlichen Kalkalpen. Ein Beitrag zur Diagenese von Karbonatgesteinen. Contr. Mineral. Petrol. **14,** 229–249, 6 Figs., 7 Pls., Heidelberg

Fornaseri, M., Grandi, L. (1963): Contenuto in stronzio di serie calcares italiane. Giorn. geol (2), **31,** 171–198, Bologna

Franke, W. (1975): Sedimentchemische und palökologische Aspekte stabiler Schwellen. Ber. SFB **48**/A, 31–50, 6 Figs., Göttingen

Frey, M., Niggli, E. (1971): Illit-Kristallinität, Mineralfazien und Inkohlungsgrad. Schweizer min.-petrograph. Mitteilungen **58,** 229–234, Bern

Friedman, G. M. (1968): Geology and geochemistry of reefs, carbonate sediments and waters, Gulf of Aqaba (Elat), Red Sea. J. Sed. Petrol. **38**/3, 895–919, 32 Figs., Tulsa

Friedman, G. M. (1969): Trace elements as possible environmental indicators in carbonate sediments. Spec. Publ. Soc. Econ. Paleont. Min. **14,** 193–198, 3 Figs., Tulsa

Fritz, P., Katz, A. (1972): The sodium distribution of dolomite crystals. Chem. Geol. **10,** 237–244, Amsterdam

Fruth, I., Scherreicks, R. (1975): Facies and geochemical correlations in the Upper Hauptdolomit (Norian) of the Eastern Lechtaler Alps. Sediment. Geol. **13,** 27–45, Amsterdam

Galle, O. K. (1969): Chemical analysis of some standard carbonate rocks. Chem. Geol. **5**, 143–146, Amsterdam

Gasser, U. (1974): Zur Struktur und Geochemie der stratiformen Sulfidlagerstätte Meggen (Mitteldevon, Rheinisches Schiefergebirge). Geol. Rundschau **63**/1, 52–73, Stuttgart

Geyh, M. A. (1971): Die Anwendung der ^{14}C-Methode. Clausthaler Tekton. Hefte **11**, 118 p., 12 Figs., Clausthal

Gill, W. D., Khalaf, F. I., Massoud, M. S. (1977): Clay minerals as an index of the degree of metamorphism of the carbonate and terrigenous rocks in the South Wales coalfield. Sed. **24**, 675–691, 6 Figs., Oxford

Gittinger, K. (1969): Geochemische Faziesuntersuchungen im oberen Hauptmuschelkalk und unteren Keuper Luxemburgs. Publ. Serv. Geol. Luxemburg **19**, 109 pp., 41 Figs., Luxemburg

Graf, D. L. (1960): Geochemistry of carbonate sediments and sedimentary carbonate rocks. Part. III. Minor elements distribution; Part IV, Bibliography (55 pp.!). Div. Illinois State Geol. Surv. Circ. **301** and **309**, Urbana

Griffin, G. M. (1962): Regional clay-mineral facies-products of weathering intensity and current distribution in the Northeastern Gulf of Mexico. Geol. Soc. Amer. Bull. **73**, 737–768, 1 Pl., 19 Figs., New York

Griffin, G. M., Ingram, R. L. (1955): Clay minerals of the Neuse River Estuary. J. Sed. Petrol **25**/3, 194–200, Tulsa

Grotjohann, H. (1978): Die Korrelation von Geochemie und Faziesdifferenzierung im erzhöffigen Muschelkalk (Mitteltrias) der Ostalpen. Berliner Geowiss. Abh., A, **3**, 1–70, 5 Pls., 34 Figs., 10 Tab., Berlin

Guillevin, Y., Renard, M. (1978): Apports de la géochimie des éléments en traces (Sr, Mg, Na) à l'étude des carbonates rècifaux dano-montien du bassin de Paris (Vigny-Montainville). Essai de reconstitution du paléo-environment recifal. Bull. Inf. Géol. Bassin Paris **15**/2, 53–63, Paris

Harder, H. (1970): Boron content of sediments as a tool in facies analysis. Sed. Geol. **4**/2, 153–175, 2 Figs., 9 Tab., Amsterdam

Hayes, J. B. (1963): Clay minerals of Mississippian strata of southeast Iowa. Clay and Clay Minerals **10**, 413–425, 7 Figs., Oxford: Pergamon Press

Hoskin, Ch. M. (1968): Magnesium and strontium in mud fraction of recent carbonate sediment, Alacran Reef, Mexico. Bull. Amer. Ass. Petrol. Geol. **52**/11, 2170–2177, 5 Figs., Tulsa

Hudson, J. D. (1977): Stable isotopes and limestone lithification. Geol. Soc. London Quart. J. **133**, 637–660, London

Imreh, J., Imreh, G., Mihalka, S. (1965): Zur Verbreitung des Strontiums in tertiären Kalken. N. Jb. Min. Abh. **103**/3, 286–292

Ingerson, E. (1961): Problems of the geochemistry of sedimentary carbonate rocks. Geochim. Cosmochim. Acta **26**, 815–847, New York

Jaffrezo, M., Renard, M. (1979): Eléments en traces de calcaires a Dasycladales et Charophytes. Bull. Centre Rech. Explor.-Prod. Elf-Aquitaine **3**/2, 639–649, 4 Figs., 2 Tables, Pau

Jux, U., Manze, U. (1978): Milieu-Indikationen aus einem mitteldevonischen biohermalen Riff des Bergischen Landes mittels C- und O-Isotopen. Decheniana **131**, 300–324, 9 Figs., Bonn

Jux, U., Manze, U. (1979): Glazialeustatisch gesteuerte Sedimentationsabläufe auf dem kaledonischen Schelf (Mittelschweden) an der Wende Ordovicium-Silur, Neues Jahrbuch Geol. Paläont. Monatshefte **1979**/3, 155–180, 2 Pls, 7 Figs., 8 Tab., Stuttgart

Katz, A., Sass, E., Starinsky, A., Holland, H. D. (1972): Strontium behaviour in the aragonite-calcite transformation. An experimental study at 40–98 °C. Geochim. Cosmochim. Acta **36**, 481–496, Oxford

Kinsman, D. J. J. (1969): Interpretation of Sr^{2+} concentration in carbonate minerals and rocks. J. Sed. Petrol. **39**/2, 486–508, 4 Figs., Tulsa

Kinsman, D. H., Holland, H. D. (1969): The co-precipitation of cations with $CaCO_3$. – IV. The co-precipitation of Sr^{2+} with aragonite between 16 ° and 96 °. Geochim. Cosmochim. Acta **33**, 1–17, 1 Fig., Oxford

Kitano, Y., Okumura, M., Idogaki, M. (1975): Incorporation of sodium, chloride, and sulfate with calcium carbonate. Geochem. J. **9,** 75–84
Knoblauch, G., Weiskirchner, W. (1979): Ein Beitrag zur Frage der Sedimentation im Malm der Schwäbischen Alb. Jahresberichte Mitteilungen oberrhein. geol. Ver. **61,** 321–329, 3 Figs., Stuttgart
Knox, G. J. (1977): Caliche profile formation, Saldanha Bay (South Africa). Sed. **24,** 657–674, 9 Figs., Oxford
Köhler, M. (1973): Mikrofazielle, geochemische und paläogeographische Untersuchungen des Plattenkalkes und der Kössener Schichten der mittleren Gailtaler Alpen (Kärnten). Veröff. Univ. Innsbruck **86,** Festschrift Heissel, 129–180, Innsbruck
Kollmann, H., Kurat, G. (1970): Zur Strontiumverteilung im Riff und den anschließenden Gebieten des Gosaukammes. Ann. Naturhist. Mus. Wien **74,** 409–415, 3 Figs., Wien
Kolmer, H. (1975): Spurenanalytische Untersuchungen an Karbonatgesteinen des Grazer Paläozoikums. Mitt. naturwiss. Verein Steiermark **105,** 53–69, 3 Figs., 4 Tab., Graz
Kranz, J. R. (1973): Die Strontium-Verteilung in den Arlbergschichten (Oberes Ladin) des Klostertales (Vorarlberg/Nördliche Kalkalpen). Neues Jahrbuch Geol. Paläont. Monatshefte **1973,** 170–187, Stuttgart
Kranz, J. R. (1976): Strontium – ein Fazies-Diagenese-Indikator im oberen Wettersteinkalk (Mittel-Trias) der Ostalpen. Geol. Rdsch. **65**/2, 593–615, 9 Figs., 4 Tab., Stuttgart
Krejci-Graf, K. (1964): Geochemical diagnosis of facies. Proc. Yorkshire Geol. Soc. **34,** 469–521, Leeds
Krejci-Graf, K. (1966): Geochemische Faziesdiagnostik. Freiberger Forschungshefte C **224,** Geol., 80 pp., 4 Tab., Leipzig
Krejci-Graf, K. (1968): Zur Chemie von Gesteinen mit Korallen und Organismen ähnlicher Umweltbedingungen. Chemie der Erde **27,** 121–142, 9 Tab., Jena
Krejci-Graf, K., Klein, K., Kreher, A., Rosswurm, H., Wenzel, G. (1965): Versuche zur geochemischen Fazies-Diagnostik. Chemie der Erde **24**/2, 115–146, Jena
Krumm, H. (1965): Mineralbestand und Genese fränkischer Keuper- und Lias-Tone. Beitr. Min. Petrographie **11,** 91–137, 24 Figs., 3 Tab., Heidelberg
Krumbein, W. C., Garrels, R. M. (1952): Origin and classification of chemical sediments in terms of pH and oxidation-reduction potentials. J. Geol. **60**/1, 1–33, Figs. 1–8
Kubanek, F., Parekh, P. (1976): A study of the trace element distribution in an interlaminated limestone-dolostone. Geol. Jahrbuch, D, **20,** 23–39, 1 Pl., 8 Figs., 9 Tab., Hannover
Kübler, B. (1963): Untersuchungen über die Tonfraktion in der Sahara. Fortschr. Geol. Rheinl. Westf. **10,** 319–324, Krefeld
Kübler, B. (1968): Evaluation quantitative du métamorphisme par la cristallinité de l'illites. Etat des progres réalisés ces dernières années. Bull. Centr. Rech. Pau SNPA **2,** 385–397, Pau
Kühn, W. (1963): Geochemische Untersuchungen bei der Kalkerkundung südwestlich Elbingerode (Harz). Z. angewandte Geol. **9**/9, 458–461, Berlin
Land, L. S., Hoops, G. K. (1973): Sodium in carbonate sediments and rocks: a possible index to the salinity of diagenetic solutions. J. Sed. Petrol. **43,** 614–617, Tulsa
Lerman, A. (1966): Boron in clays and estimation of paleosalinities. Sedimentology **6,** 267–286, 6 Figs., 3 Tab., Amsterdam
Long, D. T., Angino, E. E. (1977): Chemical specification of Cd, Cu, Pb and Zn in mixed freshwater, seawater and brines solutions. Geochem. Cosmochim. Acta **41,** Oxford
Ludwig, G. (1962): Eine Methode zum Nachweis hydrodynamisch-paläogeographisch bedingter Metallverteilung (texturell-metallfazielle Integralmethode). Erzmetall **15,** 201–206, Stuttgart
Lütke, F. (1976): Sedimentologische und geochemische Untersuchungen zur Genese der Flinzfazies im Harz (Givet und Oberdevon). Z. dtsch. Geol. Ges. **127,** 499–508, 4 Figs., Hannover
MacQuown, W. C., Jr., Bloxam, T. W. (1972): Depositional history of Carboniferous (Middle Visean) limestones from Bristol and parts of South Wales. Bull. Amer. Ass. Petrol., Geol. **56**/12, 2392–2414, 11 Figs., 7 Tab., Tulsa
Marchig, V. (1974): Zur Geochemie rezenter Sedimente des Indischen Ozeans. „Meteor" Forschungserg. C.

Marshall, J. D., Ashton, M. (1980): Isotopic and trace element evidence for submarine lithification of hardgrounds in the Jurassic of eastern England. Sed. **27,** 271–289, 15 Figs., Oxford

Matthews, R. K. (1966): Genesis of recent lime mud in southern British Honduras. J. Sed. Petrol. **36,** 428–454, Tulsa

Meyers, W. J., James, A. T. (1978): Stable isotopes of cherts and carbonate cements in the Lake Valley Formation (Mississippian), Sacramento Mountains, New Mexico. Sed. **25,** 105–124, Oxford

Milliman, J. D. (1967): Carbonate sedimentation on Hogsty Reef. A Bahamian Atoll. J. Sed. Petrol. **37**/2, 658–676, 19 Figs., Tulsa

Millot, G. (1970): Geology of Clay Minerals. Weathering, Sedimentology, Geochemistry. 425 pp., 85 Figs., New York–Heidelberg–Berlin: Springer

Morrow, D. W., Mayers, I. R. (1978): Simulation of limestone diagenesis – a model based on strontium depletion. Canadian J. Earth Sci. **15,** 376–396, Ottawa

Müller, G. (1964): see 1.4.1

Müller, G. (1967): Strontium distribution in recent Indian Ocean sediments off the Eastern Coast of Somalia. J. Sed. Petrol. **37**/3, 957–960, 2 Figs., 1 Tab., Tulsa

Müller, G. (1969): Sedimentary phosphate method for estimating paleosalinities: Limited applicability. Science **163,** No. 3869, 812–813, 1 Tab., Washington

Ogorelec, B., Rothe, P. (1979): Diagenetische Entwicklung und faziesabhängige Na-Verteilung in Karbonatgesteinen Sloweniens. Geol. Rundschau **68**/3, 965–978, 2 Pls., 3 Figs., Stuttgart

Parkeh, P. P., Müller, P., Dulski, P., Bausch, W. (1977): Distribution of trace elements between carbonate and non-carbonate phases of limestone. Earth Planet. Sci. Letter **34,** 39–50

Pascal, A. (1979): Utilisation des éléments traces dans la caractérisation des paléomilieux sédimentaires urgoniens basco-cantabriques (Espagne). Geobios, Mem. spec. **3,** 331–345, 4 Figs., Lyon

Persoz, F., Remane, J. (1976): Mineralogie et géochemie des formations à la limite Jurassique-Crétacé dans le Jura et le Basin vocontien. Eclogae geol. Helvet. **69**/1, 1–38, 25 Figs., 2 Tab., Basel

Pingitore, N. E. (1978): The behaviour of Zn^{2+} and Mn^{2+} during carbonate diagenesis: theory and applications. J. Sed. Petrol. **48**/3, 799–814, 5 Figs., Tulsa

Porthault, B. (1979): Profil geochimique de la plate-forme urgonienne au bassin vocontien (Sud-est de la France). Geobios, Mem. spec. **3,** 347–359, 2 Tab., 5 Figs., Lyon

Rateev, M. A., Gorbunova, Z. N., Lisitzyn, A. P., Nosov, G. L. (1969): The distribution of clay minerals in the oceans. Sedimentology **13,** 1/2, 21–43, 6 Figs., Amsterdam

Reed, E. C., Burchett, R. R. (1964): Stratigraphic sequences in the Pennsylvanian of Nebraska and their relationships to cyclic sedimentation. Bull. Kansas Geol. Surv. **169**/2, 441–447, 3 Figs., Lawrence

Renard, M. (1972): Interpretation des teneurs en strontium des carbonates du Lutetien supérieur à Saint-Vaast-les-Mélo (Oise): mise en évidence de la valeur de cet élément comme indicateur des conditions de diagenese et de sédimentation des carbonates. Bull. Inf. Ass. Geol. Bassin Paris **34,** 19–29, Paris

Robbins, C., Keller, W. D. (1952): Clay and other non-carbonate minerals in some limestones. J. Sed. Petrol. **22**/3, 146–152, 2 Figs., Tulsa

Rye, D. M., Sommer, M. A. (1980): Reconstructing paleotemperature and paleosalinity regimes with oxygen isotopes. In: Rhoads, D. C., Lutz, R. A. (eds.): Skeletal growth of aquatic organisms, 169–202, 13 Figs., New York–London: Plenum Press

Savin, S. M. (1977): The history of the earth's surface temperature during the last 100 million years. Ann. Rev. Earth Planetary Sci. **15,** 319–355

Scheriau-Niedermayr, E. (1973): Geochemische Untersuchungen über die Verteilung von Zink und Blei in den mesozoischen Gesteinen des Drauzuges und der Karawanken (Kärnten, Österreich). Carinthia II, **163**/83, 147–168, 4 Figs., 4 Tab., Klagenfurt

Schneidermann, N. (1970): Genesis of some Cretaceous carbonates in Israel. Israel J. Earth-Sci. **19,** 97–115, Plates 1–13, 5 Figs., Jerusalem

Scholle, P. A. (1977): Chalk diagenesis and its relation to petroleum exploration: Oil from chalks, a modern miracle? Amer. Ass. Petrol. Geol. Bull. **61**/7, 982–1009, 13 Figs., Tulsa

Scholle, P. A., Arthur, M. A. (1980): Carbon isotope fluctuations in Cretaceous pelagic limestones: Potential stratigraphic and petroleum exploration tool. Amer. Ass. Petrol. Geol. Bull. **64**/1, 67–87, 12 Figs., 1 Tab., Tulsa

Schroeder, J. H. (1969): Experimental dissolution of calcium, magnesium, and strontium from recent biogenic carbonates: a model of diagenesis. J. Sed. Petrol. **39**/3, 1057–1073, 11 Figs., Tulsa

Schroeder, J. H., Miller, D. S., Friedman, G. M. (1970): Uranium distributions in recent skeletal carbonates. J. Sed. Petrol. **40**/2, 672–681, 12 Figs., Tulsa

Schroeder, J. H., Siegel, F. R. (1969): Experimental dissolution of calcium, magnesium and strontium from Holocene biogenic carbonate: A model of diagenesis. Bull. Amer. Ass. Petrol. Geol. **53**/3, 740, Tulsa

Schroll, E. (1967): Über den Wert geochemischer Analysen bei stratigraphischen und lithologischen Untersuchungen von Sedimentgesteinen am Beispiel ausgewählter Profile der postalpinen Trias. Geol. Sbornik **18**/2, 315–330, 9 Figs., Bratislava

Schroll, E. (1976): Analytische Geochemie, 2, Grundlagen und Anwendung. 374 pp., 126 Figs., 70 Tab., Stuttgart: Enke

Schüttle, M. (1978): Geochemische Untersuchungen an rezenten Sedimenten des Kontinentalrandes von NW-Afrika zwischen den Kanarischen und Kapverdischen Inseln unter besonderer Berücksichtigung der Gehalte von Schwermetallen. Giessener Geol. Schriften **17**, 188 pp., Giessen

Schweizer, V. (1971): Zum Kalk-, Eisen- und Mangangehalt einiger Profile aus dem Lias gamma (Carixium, Untere Schwarzjura-Mergel) Süddeutschlands. N. Jb. Geol. Paläont. Abh. **138**/1, 119–133, 7 Figs., 2 Tab., Stuttgart

Seibold, E. (1952): Chemische Untersuchungen zur Bankung im unteren Malm Schwabens. N. Jb. Geol. Paläont. Abh. **95**/3, 337–370, 11 Figs., 2 Tab., Stuttgart

Siegel, F. R. (1961): Variations of Sr/Ca ratios and Mg contents in recent carbonate sediments of the northern Florida Keys area. J. Sed. Petrol. **31**/3, Tulsa

Siesser, W. G. (1977): Chemical composition of calcareous nannofossils. South African J. Sci. **73**/9, 283–285, 3 Figs.

Starke, R. (1968): Mineralogisch-geochemische Methoden der Faziesdiagnostik. Ber. dtsch. Ges. geol. Wiss. B.: Min.- u. Lagerstättenforsch. **13**, 219–226, Berlin

Stehli, F. G., Hower, J. (1961): Mineralogy and early diagenesis of carbonate sediments. J. Sed. Petrol. **31**/3, 358–371, 11 Figs., Tulsa

Sternberg, E. T., Fischer, A. G., Holland, H. D. (1959): Strontium contents of calcite from the Steinplatte Reef Complex, Austria. Bull. Geol. Soc. Amer. **70**, 12/2, p. 1681 (abstract), New York

Till, R. (1971): Are there geochemical criteria for differentiating reef and nonreef carbonate? Bull. Amer. Ass. Petrol. Geol. **55**/3, 523–530, 6 Tab., Tulsa

Truckenbrodt, W. (1976): Tonminerale im Muschelkalk Nordost-Bayerns. Thesis Univ. Erlangen-Nürnberg, 82 p., Erlangen

Turekian, K. K. (1964): The marine geochemistry of strontium. Geochim. cosmochim. Acta **28**/9, 1479–1496, Oxford

Turekian, K. K., Wedepohl, K. H. (1961): Distribution of the elements in some major units of the Earth's crust. Bull. Geol. Soc. Amer. **72**/2, 175–192, 1 Tab., New York

Tucker, M. E. (1973): Sedimentology and diagenesis of Devonian pelagic limestones (Cephalopodenkalk) and associated sediments of the Rhenohercyne Geosyncline, West Germany. N. Jb. Geol. Paläont. Abh. **142**, 320–350, 24 Figs., 1 Tab., Stuttgart

Veevers, J. J. (1969): Associations of fossils, grain-types, and chemical constituents in the Upper Devonian and Lower Carboniferous limestones of the Bonaparte Gulf Basin, Northwest-Australia. J. Sed. Petrol. **39**/3, 1118–1131, 4 Figs., Tulsa

Veizer, J. (1967): Employing of semiquantitative spectral analysis for stratigraphic and lithofacial investigation in limestones from the Mantle Series of the High Tatra Mts. Geol. Sbornik **18**/1, 79–85, 7 Figs., 1–2 Pls., Bratislava

Veizer, J. (1977): Diagenesis of pre-Quaternary carbonates as indicated by tracer studies. J. Sed. Petrol. **47**/2, 565–581. 13 Figs., Tulsa

Veizer, J. (1977): Geochemistry of lithographic limestones and dark marls from the Jurassic of southern Germany. Neues Jahrbuch Geol. Paläont. Abhandlungen **153**/1, 129–140, 6 Figs., 1 Tab., Stuttgart

Veizer, J., Demovič, R. (1973): Environmental and climatic controlled fractionation of elements in the Mesozoic carbonate sequences of the western Carpathians. J. Sed. Petrol. **43**/1, 258–271, 10 Figs., Tulsa

Veizer, J., Demovič, R. (1974): Strontium as a tool for facies analysis. J. Sed. Petrol. **44,** 93–115, 9 Figs., Tulsa

Veizer, J., Demovič, R., Turan, J. (1971): Possible use of strontium in sedimentary carbonate rocks as a palaeoenvironmental indicator. Sed. Geol. **5,** 5–22, 9 Figs., Amsterdam

Veizer, J., Hoefs, J. (1976): The nature of $^{18}O/^{16}O$ and $^{13}C/^{12}C$ secular trends in sedimentary carbonate rocks. Geochim. Cosmochim. Acta **40,** 1387–1395, Oxford

Wagner, G. H., Konig, R. H., Smith, D. A., Stele, K. F., Zachry, D. L. (1979): Geochemistry of Carboniferous limestone units in northwest Arkansas. Chem. Geol. **24,** 293–313, 8 Figs., 6 Tab., Amsterdam

Walker, C. T. (1968): Evaluation of Boron as a palaeosalinity indicator and its application of offshore prospects. Bull. Amer. Ass. Petrol. Geol. **52**/5, 751–766, 9 Figs., Tulsa

Walls, R. A., Ragland, P. C., Crisp, E. L. (1977): Experimental and natural early diagenetic mobility of Sr and Mg in biogenic carbonates. Geochim. Cosmochim. Acta **41**/12, 1731–1737, 6 Figs., Oxford

Wazny, H. (1978): Trace elements as indices of conditions of sedimentation of carbonate deposits. Przeglad Geol. **3/299,** 176–180, 3 Figs., Warsaw

Weaver, Ch. E. (1958): Geological interpretation of argillaceous sediments. Bull. Amer. Ass. Petrol. Geol. **42,** 245–271, Tulsa

Weaver, Ch. E. (1960): Possible uses of clay minerals in search for oil. Bull. Amer. Ass. Petrol. Geol. **44,** 1505–1518, Tulsa

Weaver, Ch. E. (1963): Discussion on the origin of clay minerals in sedimentary rocks. Clays and Clay Minerals **10,** 159–173, 4 Figs., Oxford–London–New York: Pergamon Press

Weaver, Ch. E. (1968): Geochemical study of a reef complex. Amer. Ass. Petrol. Geol. Bull. **52**/11, 2153–2169, 15 Figs., Tulsa

Weaver, C. E., Pollard, L. D. (1973): The chemistry of clay minerals. Dev. Sed. **15,** 213 p., 32 Figs., Amsterdam–London–New York: Elsevier

Weber, J. N. (1973): Incorporation of strontium into reef coral skeletal carbonate. Geochim. Cosmochim. Acta **37**/9, 2173–2190, Oxford

Wedepohl, K. H. (1969): Primäre und diagenetische Strontiumgehalte von Karbonatgesteinen. Ber. dtsch. Ges. geol. Wiss. B. Min. Lagerstättenforsch. **14**/1, 17–23, 2 Figs., 2 Tab., Berlin

Wedepohl, K. H. (1970): Geochemische Daten von sedimentären Karbonaten und Karbonatgesteinen in ihrem faziellen und petrogenetischen Aussagewert. Verh. geol. Bundesanst. **1970**/4, 692–705, 4 Figs., Wien

Wedepohl, K. H. (1979): Geochemische Aspekte der Diagenese von marinen Ton- und Karbonatsedimenten. Geol. Rundschau **68**/3, 833–847, 3 Figs., Stuttgart

Wigley, P. (1973): The distribution of strontium in limestones on Barbuda, West-Indies. Sedimentology **20,** 295–304, Amsterdam

Wirth, K. (1974): Spurenelementgehalte in Quellwässern und ihre Beziehung zum durchflossenen Gestein. Geol. Mitteilungen **12,** 367–388, Aachen

9.2 Relationship Between Microfacies and Physical Criteria

Apart from the well-known correlations between the porosity and permeability of carbonate rocks and the depositional facies there also seem to be relationships between particular microfacies types and technological properties which are important for the economic use of limestones.

9.2.1 Facies-Controlled Porosity in Carbonate Rocks

Facies-controlled porosity is of fundamental significance not only for oil exploration but also for problems connected with paleohydrogeological questions (see Flötgen and Milde, 1974). Subsurface location of stratigraphic-type hydrocarbon accumulations must rely on detailed facies analysis in conjunction with geophysical and geological methods. Examples of carbonate stratigraphic traps and associated depositional and diagenetic facies have been described by many authors (see "Selected Bibliography", in Mazullo, 1980; and Elf-Aquitaine, 1977). A variety of depositional environments is important in the formation of stratigraphic traps, including peritidal dolomites, bioclastic shelf limestones, oolite bars, organic buildups, and gravity flows. Shelf-edge reefs may represent important reservoir rocks due to their high primary porosity and their capacity for dolomitization and freshwater diagenesis, which may produce secondary porosity (e.g., Pennsylvanian

---

Plate 40. Primary Porosity in Limestones

1. Growth-framework porosity (intraskeletal porosity) within a coral boundstone. Porosity may have been enlarged by dolomitization of the corals. Miocene: Gulf of Suez region, Egypt. Scale 0.5 mm. ×20

2. Detail of Fig. 1. Dolomitized septa of the coral; interspaces between septa are partly closed by large calcite crystals of meteoric origin. Scale 0.5 mm. ×85

3. Ultrastructure of corallinacean algae. The thallus is composed of tiny open cells arranged in regular rows. Miocene reef limestone: Gulf of Suez region, Egypt. Scale 20 μm. ×180

4. Detail of Fig. 3. Intraskeletal porosity is caused by interconnected cells. Cell walls consist of rounded micrite crystals; intercrystalline porosity may be of some importance for porosity development. Permeability may be low because of the uniform and very small size of the pores. Scale 20 μm. ×4000

5. Ultrastructure of a rudist shell. Porosity is moderate and restricted to intercrystalline porosity within the prismatic layers of the shell, but may be enlarged by selective solution of crystals. Cretaceous rudist packstone: Arabia. Scale 1 μm. ×10 000

6. Intraparticle and interparticle porosity in a chalk rich in coccolithophorids. Under normal circumstances a typical nannofossil chalk ooze may have 70% porosity at the sediment-water interface, but this porosity becomes generally reduced towards areas of greater burial. Deviations from the general diagenetic pattern (see Scholle, 1977) are responsible for highly porous chalks which may act as reservoir rocks of hydrocarbons. Soft Chalk, Maastrichtian: Løgstør, Denmark. Scale 10 μm. ×3000

All Figures are SEM photographs

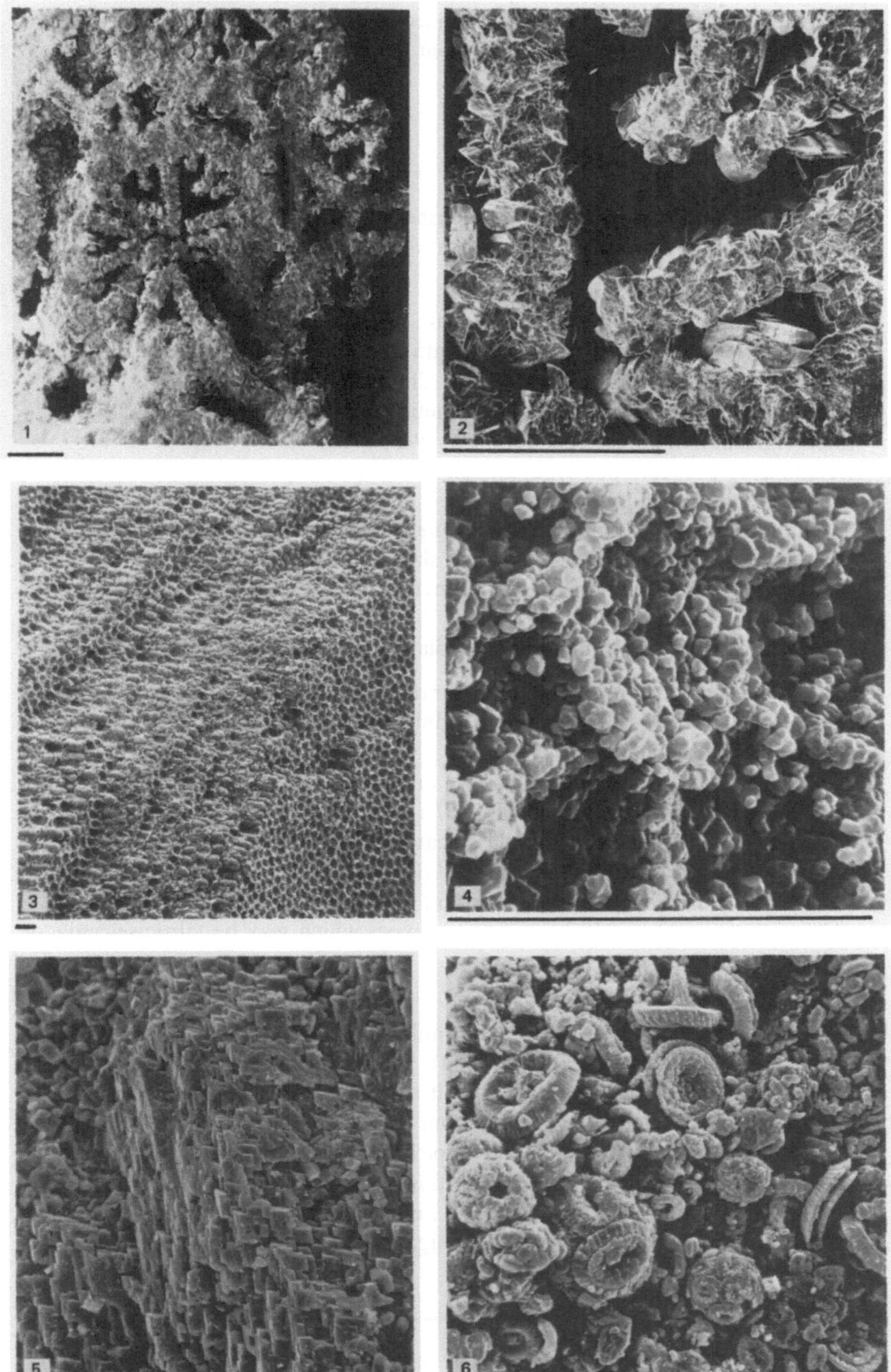
1
2
3
4
5
6

and Permian phylloid algal complexes of the Paradox and Permian basins; Cretaceous of the Gulf Coast region). Primary as well as secondary porosity can be related to the original fabric of the sediment (e.g., Zelten Field, Tertiary of Libya; Bebout and Pendexter, 1975).

9.2.1.1 Basic Porosity Types (Plate 40)

Many classifications have been proposed especially oriented to types and origin of porosity in carbonate rocks. These classifications are based on:

- Geometry of pores, described by
  - Size and orientation (Waldschmidt et al., 1956)
  - Relations between pore types and origin of pores (Levorsen, 1956; Harbaugh, 1967; Robinson, 1966; Klement, 1971; Choquette and Pray, 1970)
  - Relations between porosity and diagenetic effects (Baillie and Vecsy, 1972)
- Interconnection of pores (Theodorovich, 1943; 1960)
- Relationship between rock-types and porosity types, expressed by
  - Size of pores (Archie, 1952)
  - Relations between lithology and diagenetic alterations (Powers, 1962)
  - Relations between lithology and grain sizes of the matrix (Thomas, 1962)
  - Relations between lithology, granulometry, and diagenetic alterations (Sander, 1967)
  - Relations between granulometry and diagenetic alterations (Ball, 1968)

A general survey of these sometimes rather similar classifications is found in volume 2 of "Essai de caracterisation sédimentologique des dépots carbonatés", published by Elf-Aquitaine (1977).

Of all the various classifications the one proposed by Choquette and Pray (1970) has proved to be most useful (Fig. 57):

The four main elements of this classification include the "basic porosity types", and three sets of porosity modifiers: time of porosity origin, process and direction or stage of porosity modification; pore size (and shape); and pore abundance.

The authors recognize 15 basic porosity types which may be fabric selective or not fabric selective:

A. Fabric selective:

1. Interparticle porosity (or intergranular porosity): Porosity between particles (see Plates 38/1; 40/6; 44/4). Interparticle porosity is generally of depositional (primary) origin, but it also can form by selective solution of finer textured matrix or micrite from between larger particles. Interparticle porosity is the dominant porosity type in most carbonate sediments.
2. Intraparticle porosity: Porosity within individual particles or grains, particularly within skeletal grains (e.g., internal chambers and other openings within single or colonial skeletal organisms, such as corals or rudists; see

Fig. 57. Classification of porosity in sedimentary carbonates after Choquette and Pray (1970)

BASIC POROSITY TYPES

FABRIC SELECTIVE

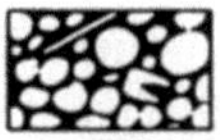
INTERPARTICLE BP

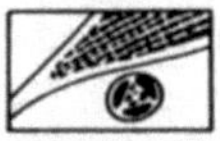
INTRAPARTICLE WP

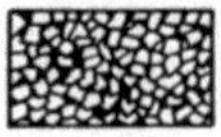
INTERCRYSTAL BC

MOLDIC MO

FENESTRAL FE

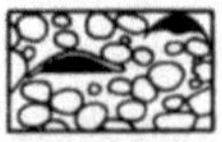
SHELTER SH

GROWTH-FRAMEWORK GF

NOT FABRIC SELECTIVE

FRACTURE FR

CHANNEL CH

VUG* VUG

CAVERN* CV

* Cavern applies to man-sized or larger pores of channel or vug shapes.

FABRIC SELECTIVE OR NOT

BRECCIA BR

BORING BO

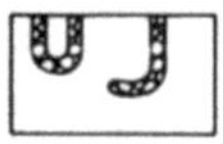
BURROW BU

SHRINKAGE SK

MODIFYING TERMS

GENETIC MODIFIERS

PROCESS		DIRECTION OR STAGE	
SOLUTION	s	ENLARGED	x
CEMENTATION	c	REDUCED	r
INTERNAL SEDIMENT	i	FILLED	f

TIME OF FORMATION

PRIMARY	P
pre-depositional	Pp
depositional	Pd
SECONDARY	S
eogenetic	Se
mesogenetic	Sm
telogenetic	St

Genetic modifiers are combined as follows:

PROCESS + DIRECTION + TIME

EXAMPLES:	
solution-enlarged	sx
cement-reduced primary	crP
sediment-filled eogenetic	ifSe

SIZE* MODIFIERS

CLASSES				mm⁺
				256
MEGAPORE mg	large	lmg		
				32
	small	smg		
				4
MESOPORE ms	large	lms		
				1/2
	small	sms		
				1/16
MICROPORE mc				

Use size prefixes with basic porosity types:

mesovug	msVUG
small mesomold	smsMO
microinterparticle	mcBP

* For regular-shaped pores smaller than cavern size.

⁺ Measures refer to average pore diameter of a single pore or the range in size of a pore assemblage. For tubular pores use average cross-section. For platy pores use width and note shape.

ABUNDANCE MODIFIERS

percent porosity (15 %)

or

ratio of porosity types (1:2)

or

ratio and percent (1:2) (15 %)

Plate 40/1, 2, 6). Intraparticle porosity may also be developed within peloids, ooids, grapestones, or intraclasts, sometimes caused by borings and solution.

3. Intercrystal porosity: between individual crystals either equal or approximately equal in size, as in many porous "sucrose" dolomites. (Plate 40/4, 5)
4. Moldic porosity: Porosity in pores formed by the selective removal, normally by solution, of a former individual constituent of the sediment or rock such as a shell or an ooid (biomoldic or oomoldic porosity). Especially common in limestones are molds of primary aragonitic constituents (e.g., gastropod shells). Molds in dolomite are usually formed by selective solution of either aragonite or calcite primary constituents, or by solution of anhydrite, gypsum, or halite. Dedolomitization may also result in the formation of molds. Other molds correspond to decomposed organic matter such as plant roots or stalks. "Stromatactis" (see Sect. 4.2.3.5) may represent another type of moldic porosity. (Plate 44/1)
5. Fenestral porosity: within the openings of fenestral fabrics (see Sect. 4.2.3.5); the openings within the mud-supported or grain-supported fabrics are larger than normal interparticle openings. Examples are intertidal algal mats or Ancient loferites. (Plate 35)
6. Shelter porosity: larger particles (e.g., shell fragments, phylloid algae) prevent the pore space they cover from being infilled by sediment. Shelter porosity is a special type of primary interparticle porosity, occurring in many oolitic or bioclastic grainstones as well as in shelly wackestones and packstones.
7. Growth-framework porosity (Plate 40/1): created by the in-place growth of a carbonate rock framework. This type applies to organic boundstones, but growth-framework porosity may also occur in travertine or in pisolitic caliche.

B. Not fabric selective:

8. Fracture porosity: formed by fracturing which may originate from collapse related to solution, slumping or tectonic deformation. Fracture porosity grades into breccia porosity with increasing dislocation.
9. Channel porosity: characterized by markedly elongate pores or irregular openings with a conspicuous elongation in one or two dimensions. Channel lengths are 10 times or more their cross-section diameter or width. Most channels originate by solution along fracture systems or by lateral coalescence of other types of pores.
10. Vug porosity (Plate 13/2): Vugs are pores that are somewhat equant, or not markedly elongate, large enough to be visible with the unaided eye (diameter greater than 1/16 mm), and generally not dependent on the fabric. Most vugs represent solution enlargement of fabric-selective pores.
11. Cavern porosity: characterized by large openings, or caverns. Cavern applies to man-sized or larger pores of channel or vug shapes. Much cavernous porosity is of solution origin.

C. Fabric selective or not:

12. Breccia porosity (Plate 53/3): interparticle porosity in a breccia.

13. Boring porosity: Openings created in relatively rigid rock or constituents by boring organisms. (Plate 12/2, 3)
14. Burrow porosity: created by organic burrows in relatively unconsolidated sediment.
15. Shrinkage porosity: produced by sediment shrinkage.

These basic porosity types may be modified by designations related to the genesis, the size and the abundance of pores (see Fig. 57). Any modifying term can be combined with the basic porosity type in the following sequence:

Genetic modifier + size modifier + basic porosity type + abundance, e.g., solution-enlarged, primary mesointraparticle porosity, 10%; or: cementation-reduced, eogenic, small mesointerparticle porosity, 20%.

Many carbonate rocks contain two or more basic types of porosity (compound pore systems), e.g., interparticle and intraparticle porosity, or moldic and intercrystalline porosity.

9.2.1.2 Examples

The primary sedimentary facies has a profound effect on the development of porosity and permeability of limestones and dolomites, because secondary porosity connected with dolomitization and freshwater leaching may be strongly controlled by the distribution of microfacies types and original porosity.

Most limestone oil reservoirs owe their porosity to solutions performed by meteoric waters, to dolomitization or to fracturing processes. Efficiency of these processes may be different for carbonate sediments deposited in different environments:

Marginal marine environments: Evaporitic carbonate shoreline sediments which accumulate predominantly as tidal flat deposits often exhibit low porosities in the microcrystalline dolomites of the supratidal and intertidal facies but higher porosity within dolomitic carbonates of the shallow subtidal facies. This may be explained by the precipitation of gypsum or anhydrite in arid and semiarid climates. The precipitation of evaporites in interstitial hypersaline waters within the sediment (resulting in the formation of nodules, replacement crystals, or pore-filling crystals) or from standing bodies of water (resulting in the formation of laminated sediments) produce a dolomitizing fluid which may give rise to an extensive reflux dolomitization. In supratidal and intertidal environments porosity is often plugged by secondary anhydrite cement (e.g., Ottman et al., 1976). Leaching of gypsum or anhydrite by freshwater may produce moldic, vuggy, and breccia porosity; dolomites may exhibit intercrystalline porosity (Lucia, 1972; Loucks and Anderson, 1980).

Shallow marine environments: Porosity in nonreefal carbonate shelf environments may be distinctly controlled by the original depositional fabric of the sediment. An interesting example has been described by Bebout and Pendexter (1975) from the highly porous shelf limestones of the Early Tertiary Zelten Member of Libya:

Five main types of porosity were recognized, including intragranular, intergranular, leached skeletal, leached micrite and dolomite vuggy porosity. Porosity and permeability are best developed in a Discocyclina-foraminiferal-calcarenite (inter- and intragranular porosity, modified by leaching), coralgal micrite (leached-skeletal and leached micrite porosity), foraminiferal calcarenite and micrite (leached-skeletal, leached-micrite, as well as inter- and intragranular porosity). Porosity is low in the argillaceous molluskan carbonates. The primary depositional fabric, especially the ratio of grains to mud, played an important role in later development of secondary or leached porosity: Porosity is best developed in sediments which show evidence of being grain-supported at an early stage. The calcarenite facies may have either carbonate mud or open spaces between the grains. Where carbonate mud is present, the original mud porosity (commonly more than 35%) is preserved because of the grain-supported nature of the sediment and the lack of compaction. Primary mud porosity provides avenues for later movement of ground water; thus, porosity is inversely proportional to the degree of post-depositional packing.

Oolitic shoals: Submarine topographic highs detached from the shoreline and composed of high-energy ooids may represent important reservoir rocks (see Asquith, 1979; Croft et al., 1980). Porosity generally corresponds to the intergranular and oomoldic types produced by freshwater leaching.

Reefs and organic banks: Reefs and banks are important petroleum reservoirs because of their topographic relief, their capacity for freshwater leaching and dolomitization, and because of the primary growth-framework porosity (which is generally less common than suggested because of rapid synsedimentary cementation). Porosity may be controlled predominantly by depositional facies (e.g., Devonian reefs in Canada, see Barss et al., 1970; Hemphill et al., 1970; Jardine et al., 1977; Jenik and Lerbekmo, 1968; Klovan, 1964), or by leaching and dolomitization processes which may also be dependent on the original depositional pattern. Examples are known from shelf-margin reefs (e.g., Edwards-Glen Rose Barrier Reef, Lower Cretaceous of Texas Gulf Coast (Bebout and Loucks, 1974; Bebout et al., 1977; Fisher and Rhodda, 1969; Schatzinger et al., 1980), and from carbonate mud buildups (e.g., phylloid algal reefs in the Pennsylvanian of the Four Corners area, Midwest U.S.A., Peterson, 1966; Pray and Wray, 1963; Gray, 1967). Porosity in these limestones resulted from leaching processes (forming calciruditic solution breccias) and from dolomitization which is controlled by the original porosity patterns (e.g., intercrystalline porosity of dolomites formed by dolomitization of predominantly micritic carbonates; thus, beds of initial low permeability may be changed to rocks with high permeability and porosity; see Longacre, 1980). Much of the so-called primary porosity in reef reservoirs may be evaporite solution porosity (Beales and Oldershaw, 1969; Petta, 1980), but evaporite solution often can only be inferred from the development of box-work textures, breccia-interstitial porosity textures, anhydrite replacement pseudomorphs, uniform base and irregular tops, laterally equivalent evaporites in subsurface, and penecontemporaneous dolomitization.

Many Paleozoic and Mesozoic reefs may be "cementation reefs" (Heckel, 1974) rather than ecologic reefs (see Schmidt et al., 1980; E. Flügel et al., 1981). These buildups are primarily products of pervasive cementation of mechanically deposited skeletal debris; porosity may be the result of eodiagenetic processes, i.e., replacement of carbonate cements by dolomite under evaporitic conditions.

Slope and basin carbonates: Basin margin carbonate debris fans may exhibit significant secondary porosity, such as vuggy porosity (caused by leaching of aragonitic fragments), intercrystalline porosity (by dolomitization), and fracture porosity (due to tectonics). An example is the Tamabra Limestone of the Poza Rica oilfield in Mexico (see Enos, 1974), or reservoirs within carbonate turbidites of the Scaglia formation in Italy (Duvernoy and Reulet, 1977).

Major quantities of oil are produced from chalks, having as much as 40% largely primary porosity at depths greater than 3 km (Scholle, 1977). These exceptional porosities may be explained by the lack of deep burial, by the exclusion of freshwater due to the retainment of marine waters, and by the reduction of effective compressive stresses by abnormally high pore fluid pressures.

9.2.2 Carbonate Rock Type and Log Response (Figs. 58; 59)

Methods used to determine carbonate lithologies from mechanical logs have been described by Asquith (1979). Several attempts have been made to establish relationships between log response and carbonate rock type:

Pickett (1977) uses cross-plots of different logs (interval transit time vs. neutron porosity; bulk density vs. neutron porosity; bulk density vs. interval transit time; deep Laterolog vs. neutron porosity) to identify log-response rock relationships. Petrographic analysis of reference cores or cuttings is necessary; log responses from these control wells are cross-plotted and areas outlined for different rock types. These clusters may be used to determine lithological variations and depositional environments for wells without cores or cuttings (see Figs. 58, 59).

Another possibility has been shown by Jodry (1972) who compared thin-section data of carbonate rocks with porosity, permeability, and capillary pressure data, and recognized distinct differences for the mercury capillary pressure curves for mud-supported, grain-supported, and leached rocks. Four main curve types are distinguished: nonreservoir rocks (micrites and sparse micrites; cemented calcarenites and calcirudites; leached and infilled micrites); poor reservoir rocks (biomicrites, dismicrites, and intramicrites with small vugs; biosparites, intrasparites, and calcarenites with voids filled with sparite); intermediate reservoir rocks (biomicrites, dismicrites, and intramicrites, composed of coarse calcilutite grains) biosparites, pelsparites, and calcarenites with sparite filled voids; coarse calcilutites with solution voids); and good reservoir rocks (pelsparites, pisosparites, and calcarenites with coarse particles; altered rocks, strongly leached, only secondary porosity).

9.2.3 Microfacies and Technological Properties of Carbonate Rocks

The use of limestones in highway construction requires parameters which characterize fracture behavior and frost resistance, such as toughness properties, compressive strength, apparent specific gravity, and the degree of saturation (summarized in: Forschungsgesellschaft für das Straßenwesen, 1963, 1970).

Relationships between microfacies types and technological properties of rocks: Schulze (1969, 1970, 1975) demonstrated that certain microfacies properties of

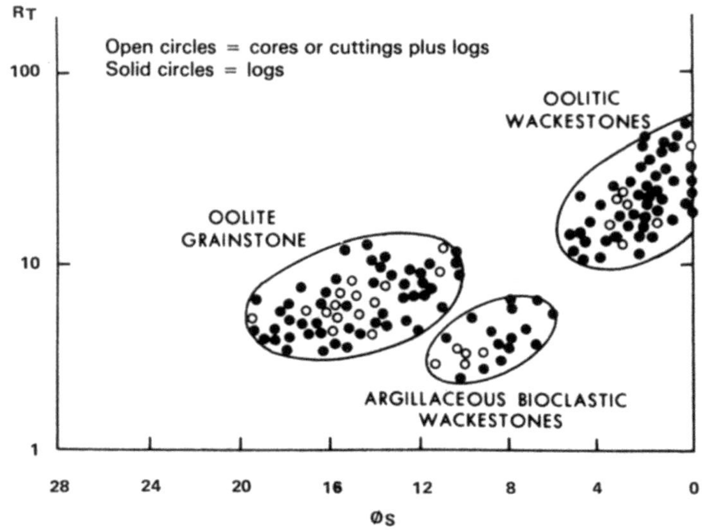

Fig. 58. Lithologic logging. Carbonate rock type identification based on differences in log response. Rt (deep induction log) versus s (sonic porosity, based on limestone matrix). Lower Permian Council Grove B – zone, Ochiltree County, Texas. After Asquith (1979)

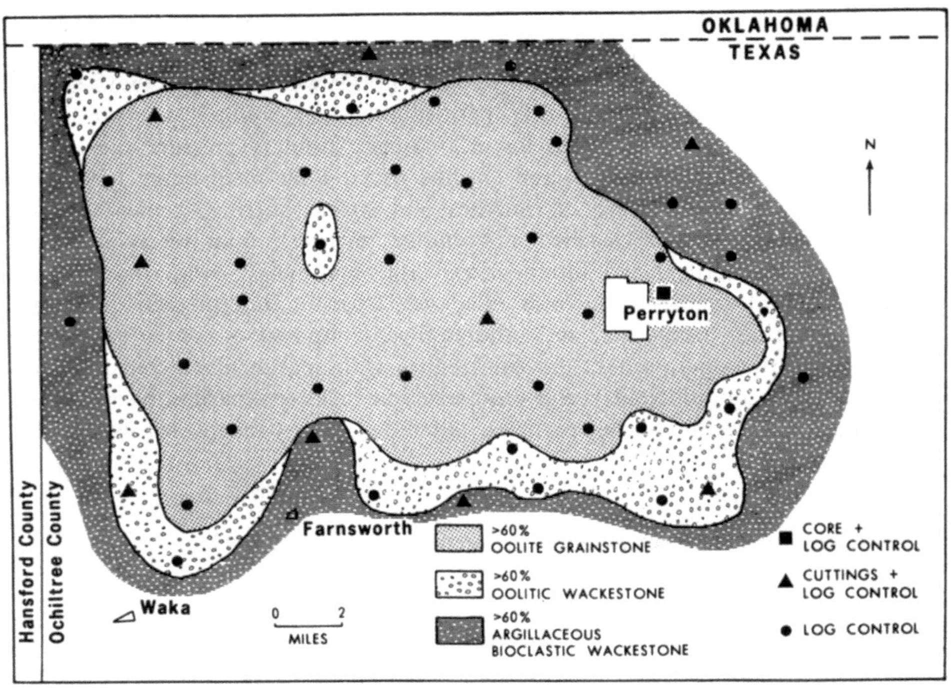

Fig. 59. Facies map of the Lower Permian Council Grove B – zone, Ochiltree County, Texas. Facies distribution based on facies clusters established from cross-plotted log data (compare Fig. 58). After Asquith (1979)

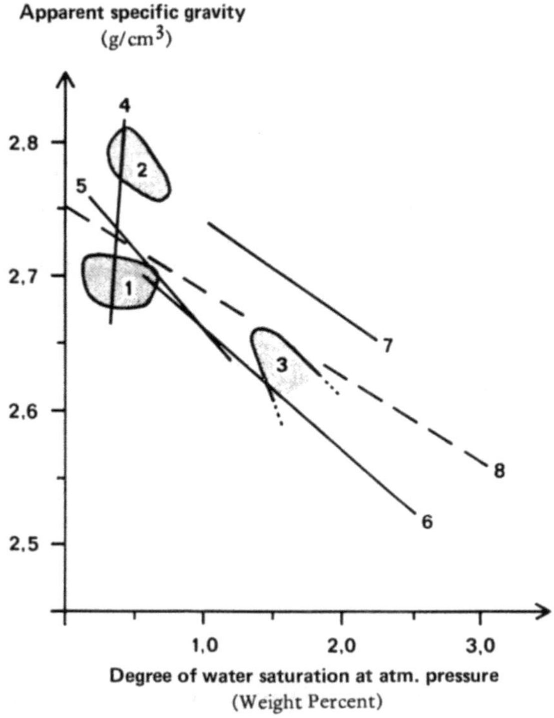

Fig. 60. Correlation between microfacies and technological properties of carbonate rocks. Regression lines of the apparent specific gravity and degree of water saturation of carbonate rocks from the Weser mountains, Germany. *1–7* Microfacies Types: *1* oolite; *2* calcitic dolomite; *3* porous calcitic dolomite; *4* bioclastic calcarenite; *5* micritic limestone, sometimes dolomitic; *6* dolomitic *humeralis* dolomite, sometimes dedolomitized; *7* quartz-bearing, argillaceous dolomite; *8* regression line for the values of all facies types. After Schulze (1970)

Jurassic limestones are responsible for different values in the apparent specific gravity and degree of saturation (Fig. 60) and that different behavior in the test for toughness also correspond to different MF types. Up to now very few comparative studies have been made which give a general picture of the relationships between MF types and the parameters mentioned above (e.g., Cox, 1978). One should not forget that a few properties, such as compression strength, can be strongly influenced by "non-sedimentary" factors like tectonic stress, and later diagenetic dolomitization or dedolomitization. These relationships should be pursued to a greater extent in regard to MF analyses. One such study has been done by Dörhöfer (1974), who characterized the calcarenites of the Upper Jurassic Münden marls (Lower Saxony, Germany), used for building materials, according to MF criteria. This has since led to directed quarrying of the rock.

9.3 Relationship Between Microfacies and Chemism of Carbonate Rocks

Although the dependence of various technological parameters on the chemico-mineralogical composition as well as the texture of carbonate rocks is known (see Siegel, 1967; Boynton, 1966), only in the last few years have directed studies been carried out. In the following paragraph an example is described, in order to encourage similar investigations.

Pure and very pure limestones and microfacies types: The $CaCO_3$ content or the amount of non-carbonate acid insoluble residues is extremely important for the use of limestones in the chemical industry. In accordance with the classification system used by E. Flügel and J. G. Haditsch (1977), the following groups of limestones can be distinguished: the purest limestone groups (>99% by weight $CaCO_3$), very pure (98%–99% by weight $CaCO_3$), and pure (95%–98% by weight $CaCO_3$).

Directed studies in the Late Triassic of the Gesäuse and the Grimming region (Styria, Austria) show that limestone types poor in residues are bound to certain genetically caused MF types. Although the loferite facies of Dachstein limestones, which is characterized by algal mats, and to some extent the calcareous algal-foraminifera detritus facies exhibit relatively high and widely varying residue contents, the reef-core facies, and particularly the aggregate grain facies, are characterized by low residue contents, that is, in the "pure" and "very pure" limestone categories. The fact that they are bound to an aggregate grain facies is then readily explained by the paleogeographic situation of these sediments (shallow marine backreef zone, possibly in tidal areas with restricted sedimentation and, thus, a reduced accumulation of terrigenous particles).

9.4 References: Physical and Technological Criteria of Limestones

Adams, J. E. (1953): Non-reef limestone reservoirs. Amer. Ass. Petrol. Geol. Bull. **37,** 2566–2569, Tulsa

Aoyagi, K. (1973): Petrophysical approach to the origin of porosity of carbonate rocks in Middle Carboniferous Windsor Group, Nova Scotia, Canada. Amer. Ass. Petrol. Geol. Bull. **57**/9, 1692–1702, 11 Figs., Tulsa

Archie, G. E. (1952): Classification of carbonate reservoir rocks and petrophysical considerations. Amer. Ass. Petrol. Geol. Bull. **36**/2, 278–298, Tulsa

Armstrong, A. K., Snavely, P. D., Jr., Addicott, W. O. (1980): Porosity evolution of Upper Miocene reefs, Almeria Province, Southern Spain. Amer. Ass. Petrol. Geol. Bull. **64**/2, 188–208, 17 Figs., Tulsa

Aschenbrenner, B. C., Chilingar, G. V. (1960): Theodorovich's method for determining permeability from pore-space characters of carbonate rocks. Amer. Ass. Petrol. Geol. Bull. **44,** 1421–1424, Tulsa

Asquith, G. B. (1979): Subsurface carbonate depositional models: a concise review. 121 p., Tulsa: Petroleum Publishing Company

Baillie, A. D., Vecsey, G. E. (1972): Diagenesis and porosity in carbonate rocks. 24th intern. Geol. Congr., Sect. 6, p. 171, Montreal

Ball, M. M. (1967): Carbonate sand bodies of Florida and the Bahamas. J. Sed. Petrol. **37,** 556–591, Tulsa

Barss, D. L., Copland, A. B., Ritchie, W. D. (1970): Geology of Middle Devonian reefs, Rainbow Area, Alberta, Canada. In: Halbouty, M. T. (ed.): Geology of Giant Petroleum Fields. Amer. Ass. Petrol. Geol. Mem. **14,** 18–49, Tulsa

Beales, F. W., Oldershaw, A. E. (1969): Evaporite-solution brecciation and Devonian carbonate reservoir porosity in Western Canada. Amer. Ass. Petrol. Geol. Bull. **53**/3, 503–512, 3 Figs., Tulsa

Bebout, D. G., Loucks, R. G. (1974): Stuart City Trend Lower Cretaceous, South Texas, a carbonate shelf-margin model for hydrocarbon exploration. Univ. Texas, Austin, Bur. Econ. Geol. Rept. Inv. **78,** 80 p., Austin

Bebout, D. G., Pendexter, C. (1975): Secondary carbonate porosity as related to early Tertiary depositional facies. Zelten Field, Libya, Amer. Ass. Petrol. Geol. Bull. **59**/4, 665–693, 25 Figs., Tulsa

Bebout, D. G., Schatzinger, R. A., Loucks, R. G. (1977): Porosity distribution in the Stuart City Trend, Lower Cretaceous, south Texas. In: Bebout, D. G., Loucks, R. G. (eds): Cretaceous carbonates of Texas and Mexico. Texas Bur. Econ. Geol., Rep. Invest. **89,** 97–126, Austin

Bebout, D. G., Schatzinger, R. A., Loucks, R. G. (1979): Porosity distribution in the Stuart City Trend, Lower Cretaceous, South Texas. In: Moore, C. H. (ed.): Geology of Carbonate Porosity, B-3–B-25, 20 Figs., Houston

Becher, J. W., Moore, C. H. (1979): The Walker Creek Field: a Smackover diagenetic trap. In: Moore, C. H. (ed.): Geology of Carbonate Porosity, A24–A26, 12 Figs., Houston

Beiersdorf, H. (1969): Druckspannungsindizien in Karbonatgesteinen Süd-Niedersachsens, Ost-Westfalens und Nord-Hessens. Geol. Mitt. **8,** 217–262, 28 Figs., Aachen

Bouroullec, J., Deloffre, R. (1977): Support sédimentologique a l'interprétation diagraphique. Example d'un paléorivage jurassique aquitain (France SW). In: Elf-Aquitaine: Essai de caractérisation sédimentologique des dépôts carbonatés. **2,** Eléments d'interprétation, 151–157, Figs. 24–30, Pau

Boynton, R. S. (1966): Chemistry and Technology of Lime and Limestone. 520 pp., New York–London–Sidney: Wiley

Chilingar, G. V., Mannon, R. W., Rieke, H. H. (1972): Introduction to carbonate reservoir rocks. In: Chilingar, G. V., Mannon, R. W., Riecke, H. H. (eds.): Oil and Gas Production from Carbonate Rocks, 1–34, New York: Elsevier

Chilingar, G. V., Mannon, R. W., Rieke, H. H. (eds.) (1972): Oil and gas production from carbonate rocks. 408 p., Amsterdam: Elsevier

Choquette, Ph., Pray, L. (1970): Geologic nomenclature and classification of porosity in sedimentary carbonates. Amer. Ass. Petrol. Geol. Bull. **54**/2, 207–250, 13 Figs., 3 Tab., Tulsa

Choquette, Ph., Traut, J. D. (1963): Pennsylvanian carbonate reservoirs, Ismay Field, Utah and Colorado. In: Bass, R. O., Sharps, S. L. (eds.): Shelf Carbonates of the Paradox Basin. Four Corners Geol. Soc. Symp. 4th Field Conf., 157–186, 5 Pls., Durango

Coogan, A. H., Bebout, D. G., Maggio, C. (1972): Depositional environments and geologic history of Golden Lane and Poza Rica trend, Mexico, an alternate view. Amer. Ass. Petrol. Geol. Bull. **56,** 1419–1447, Tulsa

Cox, F. C. (1978): An industrial classification of limestone for use in reserve and resource investigation. Abstracts, Tenth Int. Congr. Sed., v. **1,** 135–136, Jerusalem

Croft, W. S., Druckman, Y. C., Moore, C. H. (1980): Jurassic Smackover carbonate grainstone reservoir, North Haynesville Field, Claiborne Parish, Louisiana. In: Halley, R. B., Loucks, R. G. (eds.): Carbonate Reservoir Rocks, Notes Soc. Econ. Paleont. Min. Core Workshop **1,** 120–135, Denver

Cussey, R., Friedman, G. M. (1977): Patterns of porosity and cement in ooid reservoirs in Dogger (Middle Jurassic) of France. Amer. Ass. Petrol. Geol. Bull. **61**/4, 511–518, 14 Figs., Tulsa

Cussey, R., Grosdidier, E., Sulpice, L., Umbach, P. (1977): Un exemple de sédimentation carbonatée de plate-forme: Le sommet du Jurassique moyen du Bassin de Paris. In: Elf-Aquitaine: Essai de caractérisation sédimentologique des dépôts carbonatés, **2,** Eléments d'interprétation, 180–194, Figs. 45–51, Pau

Davies, G. R. (1979): Dolomite reservoir rocks: processes, controls, porosity development. In: Moore, C. (ed.): Geology of Carbonate Porosity, Amer. Ass. Petrol. Geol. Short Course, C 1–C 17, Houston

Donath, F. A., Carozzi, A. V., Fruth, L. S., Rich, D. W. (1980): Oomoldic porosity experimentally developed in Mississippian oolitic limestones. J. Sed. Petrol. **50**/4, 1249–1260, 7 Figs., Tulsa

Dörhöfer, S. (1974): see 7.3.2

Duvernory, B., Reulet, J. (1977): Réservoir de turbidite calcaire: La formation "Scaglia calcaire" (Bassins des Marches – Italie). In: Elf-Aquitaine: Essai de caractérisation

sédimentologique de dépôts carbonatés, **2,** Eléments d'interprétation, 225–231, Figs. 72–76, Pau
Elf-Aquitaine (1977): see 1.4.1
Elloy, R. (1977): Un exemple de roche carbonatée réservoir: Les récifs du Dévonien moyen du Nord de l'Alberta (Canada). In: Elf-Aquitaine: Essai de caractérisation sédimentologiques des dépôts carbonatés, **2,** Eléments d'interpretation, 158–169, Figs. 31–38, Pau
Engelhardt, W. v. (1960): Der Porenraum der Sedimente. Min. Petrographie in Einzeldarstellungen **2,** 207 pp., 83 Figs., Berlin: Springer
Enos, P. (1974): Tambara Limestone of the Poza Rica Trend, Cretaceous, Mexico. Soc. Econ. Paleont. Min. Spec. Publ. **25,** 273–314, 37 Figs., Tulsa
Enos, P., Sawatsky, L. H. (1981): Pore networks in Holocene carbonate sediments. J. Sed. Petrol. **51**/3, 961–985, 19 Figs., Tulsa
Evamy, B. D. (1967): Dedolomitization and the development of rhombohedral pores in limestones. J. Sed. Petrol. **37,** 1204–1215, Tulsa
Fisher, W. L., Rodda, P. U. (1969): Edwards Formation (Lower Cretaceous), Texas: Dolomitization in a carbonate platform system. Bull. Amer. Ass. Petrol. Geol. **53,** 55–72, Tulsa
Flötgen, F., Milde, G. (1974): Zur Rolle von Karbonatgesteinen in der Paläohydrogeologie. Z. angew. Geol. **20**/7, 313–318, 3 Figs., Berlin
Flügel, E. (1977): Untersuchungen über die Beziehungen zwischen mikrofaziellen und technologischen Merkmalen steirischer Dachstein-Kalke (Obertrias; Grimmingstock, Gesäuse). Mitt. Abt. Geol. Paläont. Landesmus. Joanneum, Graz, **38,** 47–58, 8 Figs., Graz
Flügel, E., Haditsch, J. G. (1977): Vorkommen hochreiner und reinster Kalke im Steirischen Salzkammergut. Arch. Lagerstättenforsch. Ostalpen **15,** 65–83, 6 Figs., Leoben
Flügel, E., Sadati, M., Wurm, D. (1981): Ecologic vs. stratigraphic reefs – Algal crusts and cement in Upper Triassic and Lower Permian Reefs. Abstracts Internat. Symp. Triassic Reefs, 26–30, Erlangen
Forschungsgemeinschaft für das Straßenwesen e.V. (ed.) (1963): Vorläufiges Merkblatt über Verwendung und Prüfung von Natursteinen für den Straßenbau. 14 pp., 2 Tab., Köln
Forschungsgemeinschaft für das Straßenwesen e.V. (ed.) (1970): Richtlinien für die Güteüberwachung von Straßenbaustoffen, Teil I: Straßenbausteine (RGS-I). 10 pp., 9 encls., Köln
Füchtbauer, H. (1964): Fazies, Porosität und Gasinhalt der Karbonatgesteine des norddeutschen Zechsteins. Z. dtsch. Geol. Ges. **114**/3, 484–531, 3 Pls., 10 Figs., 3 Tab., Hannover
Füchtbauer, H (1974): see 1.4.1
Gray, R. S. (1967): Cache Field – A Pennsylvanian algal reservoir in southwestern Colorado. Amer. Ass. Petrol. Geol. Bull. **51**/10, 1559–1978, 27 Figs., Tulsa
Halbouty, M. T. et al. (eds.) (1970): Geology of giant petroleum fields. Amer. Ass. Petrol. Geol. Mem. **14,** 575 pp., Tulsa
Harbaugh, J. W. (1967): Carbonate oil reservoir rocks. In: Chilingar, G. V., Bissell, H. J., Fairbridge, R. W. (eds.): Carbonate rocks, Part A, Origin, occurrence and classification. Dev. Sed. **9,** 349–398, Amsterdam: Elsevier
Harris, J. F. (1968): Carbonate rock characteristics and effect on oil accumulation in Mid-Continent area. Amer. Ass. Petrol. Geol. Bull. **52**/9, 1662–1669, 14 Figs., Tulsa
Heckel, P. H. (1974): Carbonate buildups in the geologic record: a review. Soc. Econ. Paleont. Min. Spec. Publ. **18,** 90–154, 9 Figs., Tulsa
Hemphill, C. R., Smith, R. I., Szabo, F. (1970): Geology of Beaverhill Lake Reefs, Swan Hills Area, Alberta. Amer. Ass. Petrol. Geol. Mem. **14,** 50–72, 27 Figs., Tulsa
Jacka, A. D., Brand, J. P. (1977): Biofacies and development and differential occlusion of porosity in a Lower Cretaceous (Edwards) reef. J. Sed. Petrol. **47,** 366–381, Tulsa
Jardine, D. E., Abdrew, D. P., Wishart, J. W., Young, J. W. (1977): Distribution and continuity of carbonate reservoirs. J. Petrol. Technology **10,** 873–885
Jenik, A. K., Lerbekmo, J. F. (1968): Facies and geometry of Swan Hills Reef Member of Beaverhill Lake Formation (Upper Devonian), Goose River Field, Alberta, Canada. Amer. Ass. Petrol. Geol. Bull. **52,** 21–56, Tulsa
Jodry, R. L. (1972): Pore geometry of carbonate rocks (basic geologic concepts). In: Chilingar, G. V., Mannon, R. W., Roecke, H. H. (eds.): Oil and Gas Production from Carbonate Rocks, 35–82, New York: Springer

Katuzi, M.-R. (1976): Untersuchung der erdölgeologischen Parameter von Kalksteinen der Tabur-Formation in der Provinz Fars/Iran. Thesis Freie Univ. Berlin, 116 p., Berlin

Klovan, J. E. (1964): see 7.3.2

Krebs, W. (1978): Aspekte einer potentiellen Kohlenwasserstoff-Führung in den devonischen Riffen Nordwestdeutschlands. Erdöl Erdgas Zeitschrift **94**/1, 15–25, 5 Figs., Hamburg

Levorsen, A. I. (1967): Geology of petroleum. 2nd edition. 724 pp., San Francisco: Freeman

Longacre, S. A. (1980): Dolomite reservoirs from Permian biomicrites. In: Halley, R. B., Loucks, R. G. (eds.): Carbonate Reservoir Rocks, Notes Soc. Econ. Paleont. Min. Workshop **1**, 105–117, Denver

Loucks, R. G. (1977): Porosity development and distribution in shoal-water carbonate complexes – subsurface Pearsall Formation (Lower Cretaceous), south Texas. Texas Univ. Bur. Econ. Geol. Report Inv. **89**, 97–126, Austin

Loucks, R. G., Anderson, J. H. (1980): Depositional facies and porosity development in Lower Ordovician Ellenburger Dolomite, Puckett Field, Pecos County, Texas. In: Halley, R. B., Loucks, R. G. (eds.): Carbonate reservoir rocks, Notes Soc. Econ. Paleont. Min. Workshop **1**, 1–31, Denver

Lucia, F. J. (1972): Recognition of evaporite-carbonate shoreline sedimentation. Soc. Econ. Paleont. Min. Spec. Publ. **16**, 160–191, 34 Figs., Tulsa

Lucia, F. J., Murray, R. C. (1967): Origin and distribution of porosity in crinoidal rocks. 7th World Petrol. Congr. Mexico City, Proc. **2**, 409–423, Mexico City

Mazzullo, S. J. (ed.) (1980): Stratigraphic traps in carbonate rocks. Amer. Ass. Petrol. Geol. Reprint Ser. **23**, 217 p., Tulsa

McQuillan, H. (1973): Small-scale fracture density in Asmari Formation of southwest Iran and its relation to bed thickness and structural setting. Amer. Ass. Petrol. Geol. Bull. **57**/12, 2367–2385, 21 Figs., Tulsa

Moore, C. A. (1964): Handbook of Subsurface Geology. New York: Harper

Moore, C. H. (1979): Porosity in carbonate rock sequences. In: Geology of carbonate porosity, Amer. Ass. Petrol. Continuing Education Course Notes **11**, A 1–A 124, Tulsa

Moore, C. H., Druckman, Y. (1981): Burial diagenesis and porosity evolution, Upper Jurassic Smackover, Arkansas and Louisiana. Amer. Ass. Petrol. Geol. Bull. **65**, 597–628, 13 Figs., Tulsa

Moore, C. H., Smitherman, J. M., Allen, S. H. (1979): Pore system evolution in Cretaceous carbonate beach sequence. In: Moore, C. H. (ed.): Geology of Carbonate Rocks, A 11–A 23, 6 Figs., Houston

Murray, R. C. (1960): Origin of porosity in carbonate rocks. J. Sed. Petrol. **30**, 59–84, Tulsa

Ottmann, R. D., Keyes, P. L., Ziegler, M. A. (1976): Jay Field, Florida – a Jurassic stratigraphic trap. Amer. Ass. Petrol. Geol. Mem. **24**, 276–292, 15 Figs., Tulsa

Peterson, J. A. (1966): Stratigraphic vs. structural controls on carbonate-mound hydrocarbon accumulation, Aneth area, Paradox basin. Amer. Ass. Petrol. Geol. Bull. **50**, 2068–2081, Tulsa

Petta, T. J. (1980): Silurian pinnacle reef diagenesis – Northern Michigan: Effects of evaporites on pore space distribution. In: Halley, R. B., Loucks, R. G. (eds.): Carbonate Reservoir Rocks, Notes Soc. Econ. Paleont. Min. Workshop **1**, 32–34, Denver

Pickett, G. R. (1977): Recognition of environments and carbonate rock types identification. Formation Evaluation Manual Unit II, Section Exploration Wells, 4–25, Oil and Gas Consultants Int. Inc.

Pirson, S. M. (1970): Geologic well log analysis. Houston: Gulf Publ.

Pittman, E. D. (1971): Microporosity in Carbonate Rocks. Amer. Ass. Petrol. Geol. Bull. **55**/10, 1873–1881, 9 Figs., Tulsa

Pittman, E. D. (1974): Porosity and permeability changes during diagenesis of Pleistocene corals, Barbados, West Indies. Bull. Geol. Soc. Amer. **85**, 1811–1820, New York

Powers, R. W. (1962): Arabian Upper Jurassic carbonate reservoir rocks. In: Ham, W. E. (ed.): Classification of carbonate rocks – a symposium. Amer. Ass. Petrol. Geol. Mem. **1**, 122–192, Tulsa

Pray, L. C., Wray, J. L. (1963): Porous algal facies (Pennsylvanian) Honacker Trail, San Juan Canyon, Utah. In: Bass, R. O. (ed.): Shelf carbonates of the Paradox basin, Four Corners Geol. Soc., 204–234, Durango

Purser, B. H. (1973): Relations entre les caractéristiques lithologiques et pétrophysiques des roches carbonatées. Publ. Inst. franc. Petrole, no. 21558, 243 pp., Paris

Reulet, J. (1977): Réservoir calcaire de plate-forme marine (formation Mishrif, Moyen Orient). In: Elf-Aquitaine: Essai de caractérisation sédimentologique des dépôts carbonatés, **2,** Eléments d'interpretation, 169–180, Figs. 39–44, Plate 17, Pau

Robinson, R. B. (1966): Classification of reservoir rocks by surface texture. Amer. Ass. Petrol. Geol. Bull. **50,** 547–559, Tulsa

Sander, N. J. (1967): Classification of carbonate rocks of marine origin. Amer. Ass. Petrol. Geol. Bull. **51,** 325–336, Tulsa

Schatzinger, R. A., Bebout, D. G., Loucks, R. G., Reid, A. M., III (1980): Gas reservoirs in Stuart City Trend along Lower Cretaceous shelf margin in South Texas. In: Halley, R. B., Loucks, R. G. (eds.): Carbonate reservoir Rocks, Soc. Econ. Paleont. Min. Notes Workshop **1,** 137–160, Denver

Schlumberger (1972): Log interpretation manual, **1,** principles. 113 p. New York: Schlumberger

Schmidt, V. (1965): Facies, diagenesis, and related reservoir properties in the Gigas Beds (Upper Jurassic), northwestern Germany. In: Pray, L. C., Murray, R. C. (eds.): Dolomitization and limestone diagenesis. Soc. Econ. Paleont. Min. Spec. Publ. **13,** 124–168, Tulsa

Schmidt, V., McDonald, D. A., McIlreath, I. A. (1980): Growth and Diagenesis of Middle Devonian Keg River cementation reefs, Rainbow Field, Alberta. In: Halley, R. B., Loucks, R. G. (eds.): Carbonate reservoir rocks, Soc. Econ. Paleont. Min. Notes Workshop **1,** 43–63, Denver

Scholle, P. A. (1977): Deposition, Diagenesis, and Hydrocarbon Potential of "Deeper-Water" Limestones. Amer. Ass. Petrol. Geol. Course Note Ser. **7,** 25 p., 2 Figs., Houston

Scholle, P. A. (1979): Porosity prediction in shallow versus deep water limestones – primary porosity preservation under burial conditions. Amer. Ass. Petrol. Geol. Course Note Ser. **11,** D 1–D 9, Houston

Schulze, K.-H. (1969): Porosität, Wasseraufnahme und Frostempfindlichkeit der Gesteine des Korallenoolith im Weser-Gebirge. Straßen- und Tiefbau **10,** 918–925, 9 Figs., Heidelberg

Schulze, K.-H. (1970): Untersuchung über die Abhängigkeit der Trockenrohdichte, Wasseraufnahme und Schotterschlagfestigkeit von petrographischen Eigenschaften der Gesteine des Korallenoolith im Weser-Gebirge. Naturstein-Industrie **6,** 2–10, 14 Figs., 2 Tab., Offenbach

Schulze, K.-H. (1975): see 5.5

Siegel, F. R. (1967): Properties and uses of the carbonates. In: Chilingar, G. V., Bissell, H. J., Fairbridge, R. W. (eds.): Carbonate Rocks. Dev. Sed. **9B,** 343–393, Amsterdam–London–New York: Elsevier

Stout, J. L. (1964): Pore geometry as related to carbonate stratigraphic traps. Amer. Ass. Petrol. Geol. Bull. **48,** 329–337, Tulsa

Terriere, T. R. (1976): Geology of Fairway field, East Texas. In: Braunstein, J. (ed.): North American oil and gas fields. Amer. Ass. Petrol. Geol. Mem. **24,** 157–176, Tulsa

Thomas, G. E. (1962): Grouping of carbonate rocks into textural and porosity units for mapping purposes. In: Ham, W. E. (ed.): Classification of carbonate rocks – a symposium. Amer. Ass. Petrol. Geol. Mem. **1,** 193–223, Tulsa

Thomas, G. E., Glaister, R. P. (1960): Facies and porosity relationships in some Mississippian carbonate cycles of western Canada basin. Amer. Ass. Petrol. Geol. Bull. **44**/5, 569–588, 5 Plates, 9 Figs., Tulsa

Tyler, A. N., Erwin, W. L. (1976): Sunoco-Felda Field, Hendry and Collier Counties, Florida. Amer. Ass. Petrol. Geol. Mem. **24,** 287–297, 12 Figs., Tulsa

Waldschmidt, W. A., Fitzgerald, P. E., Lunsford, C. L. (1956): Classification of porosity and fractures in reservoir rocks. Amer. Ass. Petrol. Geol. Bull. **40**/5, 953–974, Tulsa

Wardlaw, N. C. (1976): Pore geometry of carbonate rocks as revealed by pore casts and capillary pressure. Amer. Ass. Petrol. Geol. Bull. **60**/2, 245–257, 14 Figs., Tulsa

Wardlaw, N. C. (1979): Pore system in carbonate rocks and their influence in hydrocarbon recovery efficiency. Amer. Ass. Petrol. Geol. Course Note Ser. **11,** E 1–E 24, 12 Figs., Houston

Wardlaw, N. C., Taylor, R. P. (1976): Mercury capillary pressure curves and the interpretation of pore structure and capillary behaviour in reservoir rocks. Bull. Can. Petrol. Geol. **24**/2, 225–262, 31 Figs., Calgary

Wermund, E. G. (1975): Upper Pennsylvanian limestone banks, North Central Texas. Univ. Texas Bur. Econ. Geol. Circ. **75**/3, 34 p., Austin

10 Facies Diagnosis and Facies Models

Microfacies types document differences in the kind of and significance of sedimentological (affecting sediment) and biological (affecting organisms) environmental factors. Although during the genesis of facies units there is an interplay of several factors, some environmental factors – especially in marine areas – seem to be of major importance. Based on the ideas expressed by J. Walther (1893) and H. Schmidt (1956), and in keeping with the special conditions necessary for limestone formation, the following controls are active for:

1. Sedimentation:
   a) Organic activity
   b) Water movement or energy levels (energy produced by waves or currents in the boundary zone between water and the sedimentary substrate)
   c) Suspended terrigenous influx (influx from continents, sedimentation of non-carbonate particles and of carbonate extraclasts; see Sects. 4.1.3.8 and 4.1.4)
   d) Processes of cementation, particularly during early diagenesis (see Sect. 3), and during various stages of dissolution

2. Organisms (non-biological factors)
   a) Light and water depth
   b) Water energy and oxygen supply
   c) Salinity of the water
   d) Water temperature
   e) Composition and consistency of the substrate

In addition to the above factors, the distribution of organisms in the ocean is controlled by water density, hydrostatic pressure, the quantities of trace elements as well as CO_2 and other gases, and by biotic factors such as food supply, quantity of dissolved organic substances, and suspended matter as well as competition among the organisms themselves (see Friedrich, 1965; Seibold, 1974; Tait, 1971).

In Sect. 10.1 the range of possibilities will be demonstrated by which the above-mentioned environmental controls can be diagnosed on the basis of microfacies analyses, taking into account other criteria from the fields of paleontology, sedimentology, and geochemistry. A combination of the features, which indicate environmental factors, results in the facies criteria listed in Sect. 10.2, by which sedimentary environments can be recognized and subdivided. Sect. 10.3 deals with facies models by which certain recurring paleogeographic situations in the geologic record can be described and categorized.

10.1 Diagnosis of Environmental Factors

Because the role of organisms and the importance of diagenesis for the origin of different types of limestones has already been dealt with, the Sections below will discuss only the criteria for water movement, terrigenous influx, lighting and water depth, salinity and water temperature, and the composition of the substrate, i.e., the consistency of the sea floor. Following that, references will be made to features that explain an evident change in the conditions of sedimentation.

10.1.1 Water Energy and Oxygen Supply

All systems of carbonate rock classification are based on the fact that different hydrodynamic conditions lead to the formation of various carbonate structures and textures. This is particularly emphasized in the Energy-Index classification of Plumley et al. (1962); see Sect. 6.2.4. The probable theoretical relationships between water energy and limestone types as well as between water energy and the origin of certain particle categories are illustrated in Figs. 61 and 62:

– When judging particle sorting, major primary differences, which might be pro-matrix (clay- and silt-sized micrite and particles). Non-winnowed biomicrite (Stage 1, Fig. 61) in a normal situation, thus, indicates a low-energy depositional environment, although very weak currents may lead to a slight reworking of the calcareous mud. Incomplete winnowing can be interpreted as an indication of water energy which is stronger only locally (Stage 2). With more intense currents, winnowing may be more complete, but the grains remain poorly sorted, resulting in

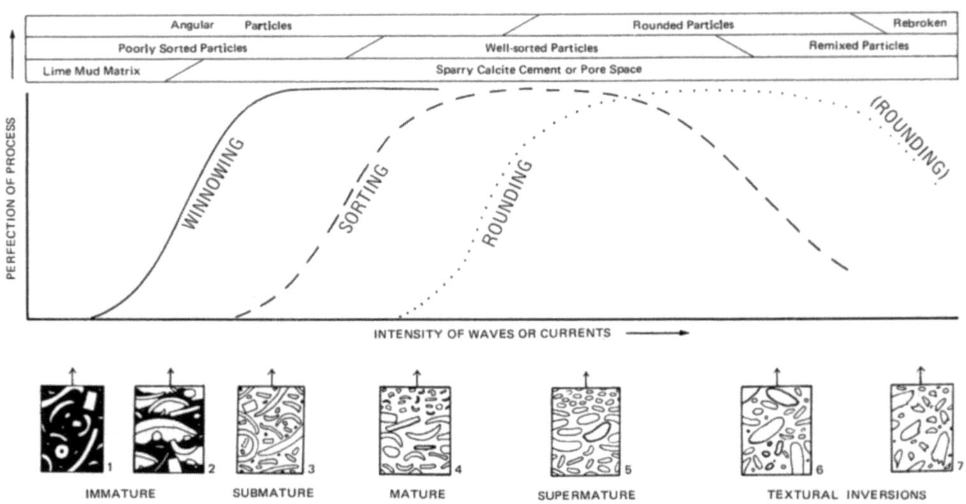

Fig. 61. Model of the relationships between the winnowing of fine-grained sediments and the sorting and rounding of particles in carbonate rocks. Explanation in the text. After Folk (1962)

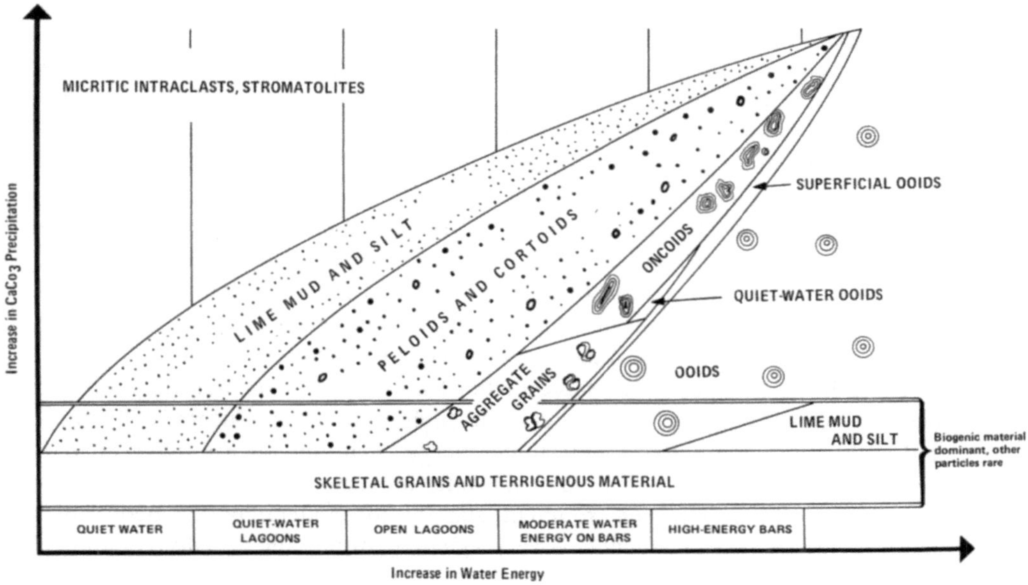

Fig. 62. Dependence of carbonate particles on the intensity of the water movement and on the supply of $CaCO_3$. Explanation in the text. Simplified after Kendall (1969)

"unsorted biosparite" (Stage 3). More intense wave action leads to the development of sorted biosparite (Stage 4). Extreme wave action, particularly in the surf zone, may produce increasingly rounded particles (Stage 5). Sporadically very strong wave action as well as organic activity (e.g., sediment burrowing) can also lead to a "textural inversion" (Folk's terminology) of particles produced in such a manner.

When Folk's diagram is used to interpret water energy, the following points should be noted:

- Winnowing of micrite can be greatly limited by sediment-binding organic films (chiefly algae and bacteria). This is particularly true for tidal sediments.
- When judging particle sorting, major primary differences which might be produced by the various supplies of bioclasts from various sized hardparts, must be taken into account.
- One should be cautious about making an all too generalized facies interpretation based on the rounding of carbonate particles, for the reasons listed in Sect. 4.2.2.
- Early cement formation in the intergranular pores of arentic carbonate sediments interrupts the processes of winnowing and sorting. The fabric of the limestone, under certain circumstances, would then indicate lower energy conditions than were actually present.

The scheme (Fig. 62) developed by Kendall (1969) shows the possible correlations between increasing water energy and increasing carbonate formation. As shown in a very simplified manner in the scheme, a change in the kind of particle is expressed by increasing water energy. Most peloids and cortoids in lagoonal environments occur together with relatively low water energies; aggregate grains form in areas of moving water; normal ooids originate in strongly agitated waters. The prerequisite for an interpretation of this scheme is the elimination of all modes of genesis of carbonate particles which are not controlled by water energy. Accordingly, the following particles must be excluded from consideration: fecal pellets and bahamites from the peloid group; diagenetically formed caliche-lumps from the aggregate grains; pisoids occurring in vadose environments from the ooids; autoclasts, pseudoclasts, and caliche intraclasts from the lithoclasts. The "quiet water" ooids and the superficial ooids occur in relatively confined environments; see Sect. 4.1.3.5. The formation of purely radial and radial-concentric or concentric ooids seems to be controlled by current velocities (Heller et al., 1980). Ooid texture and ooid sizes may be indicators of Ancient paleocurrent regimes.

Table 46 gives a survey of some of the features useful in indicating varying intensities of water energy. Here, deviating slightly from the concepts of Plumley et al. (1962) and Folk (1962), only three energy environments are differentiated: Environment A – low energy, little or no water movement near the bottom; Environment B – transitional environment slightly to moderately agitated water, changing locally; Environment C – very agitated, strong and usually continuous water movement. Based on Recent situations, the causes of water movement which come into question for Environment B are surface currents (wind-drift currents, effective in the open seas down to depths of about 100–200 m) and bottom currents. High-energy conditions are produced by tidal currents, shallow surface currents, and above all by wave turbulence caused by wind. The lower limit of wave action, known as the "wave base" or "critical wave base", which helps stir and distribute sediment on the ocean floor, lies for shallow protected parts of the sea between about 10 and 50 m; in an open-sea environment wave action has been proved to go as deep as about 200 m. The average lower limit of marine erosion is given by Bradley (1958) as about 10 m. Shepard (1963) emphasized that below about 15 m not much sediment transport usually takes place by wave action. Deviations from these values may be caused by the action of storm waves.

Areas of the ocean with Recent shelf carbonates are usually quite shallow; average water depths for subtidal regions are: Bahama Bank < 7 m; Florida < 10 m; Batabano Bay, Cuba 2–7 m; Abu Dhabi, Persian Gulf < 10 m. In these areas waves and tidal currents work together in the distribution of sediment.

The intensity of paleo-waves may be estimated by using the relationships between the amplitudes of waves, bottom current velocities, water depth and the mean size of detrital components, e.g., detrital quartz, as described by Weggel (1972). The possible strength of bottom currents may be hypothetically calculated by using the experimental relationships between the threshold of sediment movement under unidirectional flow and the mean size of particles, as found by Miller et al. (1977) and Miller and Komar (1977). According to these diagrams, bottom current velocities for detrital quartz grains with a mean size of 0.1 mm may be 30 cm/sec, for a detrital quartz grain with a mean size of 0.5 mm, about 50 cm/sec and for a hard calcareous peloid with a mean size of 0.1 mm, 9 cm/sec (see A. Strasser, 1979; Wanless et al., 1981).

Table 46 lists the microfacies indicators of various water energies at the time of sedimentation; these can be supplemented by paleontological criteria of which only the following will be mentioned: growth forms of reef-building benthonic organisms influenced by water energy (e.g., corals, stromatoporoids, coralline algae), types of feeding groups (Heckel, 1972), and special burial of mollusk and brachiopod shells (examples in Lecompte, 1962; Rhoads et al., 1972). High-energy environments may also be inferred from special adaptations of microfossils, e.g. sessile foraminifera and ostracods, or shell forms which facilitate a rolling transport without destruction. The marine biological aspects of water energy have been discussed by Riedl (1969).

We should also mention here the divisions into biofacies types undertaken by H. Schmidt (1935, 1958), which are based on the degree of oxygenation at the ocean bottom. The six types which are characterized by different oxygen levels are due primarily to various water energies. W. Schäfer (1962) expanded this system and differentiated five first order biofacies types according to structure, form, and spatial distribution of the particles of a facies originating in environments with agitated, slight, or even no water movement.

Anaerobic to dysaerobic and aerobic conditions may be inferred from the existence and frequency of bioturbation structures (see Sect. 4.2.3.4).

Storm sediments or tempestites (Table 47, Plate 11/2) are especially interesting because they illustrate sporadically occurring high-energy conditions. Characteristic features of this type of deposition are:

- Sedimentation of reworked material which is transported from a shallow sea environment (from depths of as much as 25 m!) by hurricanes in the coastal area.
- Deposition of various-sized sedimentary particles in near-coastal shallow water and carpets of laminated lime mud onto coasts flooded by storm waves.
- There are no facies relationships between sediments deposited in the supratidal zone and at some distance from the coasts.
- Storm sediments superimposed on various facies types exhibit sharp or erosional boundaries.
- Sediments contain bioclasts which may be derived from various ecological environments.
- In addition to mass accumulations of bioclasts, there are different kinds of intraclasts (micritic plasticlasts, laminated clasts, and in places, dolomite-crust intraclasts).
- Calcarenite deposited on the coast usually does not exhibit any preferred orientation of particles; current controlled imbricate structures and cross-bedding only occur sporadically.
- Reworked sedimentary particles from a normal marine environment.
- Mass concentrations of densely packed foraminifera or pelecypod shells washed in together can be better explained by storm floods than by tidal currents or waves produced by wind.
- Generally rather thin beds within fine-grained shallow-water low-energy sediments (micrites, marls).

The sediment building effect of hurricanes has been described by Ball et al. (1967), Hayes (1967) and Perkins and Enos (1968), Kumar and Sanders (1976) as well as by Ogg and Koslow (1978). Likely examples of Ancient carbonate storm

Table 46. Microfacies indications of various water energies; A = low-energy environment, B = transitional zone, C = high-energy environment

Features	Water Energy			Comments
	Environment A	Environment B	Environment C	
Micritic matrix	Common	Sometimes still present	"absent"	In high energy environments mud may be fixed by algal films
Particle : Micrite ratio	1 : 9	1 : 9 to 1 : 1	to 9 : 1	Exercise caution when considering textural inversion and "autochthonous" and "allochthonous" bioclasts
Peloids	Fecal pellets	Pseudopeloids (see 4.1.3.2) Fecal pellets		Fecal pellets are more rapidly cemented in low-energy environments and are thus better preserved there
Aggregate grains		Originate in areas with restricted water circulation		The absence of a micritic matrix in many limestones made of aggregate grains is probably due to non-sedimentation (see 4.1.3.3)
Oncoids		Sluggish currents changing in intensity		Great difficulties are involved in the comparison of Ancient and Recent oncoids, see 4.1.3.4
Ooids	Radial fibrous ooids	Radial fibrous ooids; micritic ooids; tangentially structured ooids, rarely superficial ooids	Normal and superficial ooids, tangential structures	Facies control of radial fibrous ooids is probable but not yet certain
Cortoids		Common	Present	Micrite envelopes caused by algal borings occur preferentially in relatively low-energy environments; the other modes of origin are unrelated to water energy
Lithoclasts		Plasticlasts common; Extraclasts present	Plasticlasts present, imbricate structure; extraclasts common	Exercise caution with textural inversion
Terrigenous Particles	Fine-grained material common	Fine-grained and coarse-grained material present	Fine-grained material absent, coarse-grained material abundant	Very different depending on erosional material, topographical relief and sedimentary conditions in the depositional area

Size and Sorting of the Particles	Poor sorting	Sorting poor to moderate	Sorting poor to very good	Not valid for particles whose primary sizes are given by biological factors (e.g., growth stages determined by age) or a special mode of origin (e.g., fecal pellets); useful for bioclasts
Shape of the Particles	Not rounded, low sphericity	Well-rounded to very well-rounded, increasing sphericity	Well-rounded to very well-rounded	Valid essentially only for bioclasts (see 4.2.2.1), textural inversion is frequent
Fabric	Mud support; particles usually not oriented	Mud support; particles somewhat oriented	Grain support, particles more often oriented (e.g., imbricate structures), sometimes cross-bedding	Very generalized
	Open-space structures present. Stromatactis abundant	Open-space structures present, birdseyes present	Open-space structures present, birdseyes abundant	Due to very different modes of origin, open-space structures are not good indicators of water energy

Table 47. Characteristics of carbonate storm deposits. Westphalia limestone (Pennsylvanian), Kansas. After Ball (1971)

Facies Criteria	Source Material	Storm Sediments Deposited in Shallow Coastal Waters	Storm Sediments Deposited in Supratidal Zone
Limestone Type	Wackestone with various kinds of invertebrates	Fusulinid packstone to grainstone	Carbonaceous ostracod mudstone
Bedding	Unknown	Usually absent	Irregularly laminated
Geometry		Lens-shaped to tabular	Lens-shaped
Thickness	About 3 m	Up to about 1.5 m	To about 0.45 m
Matrix	Micrite, clay	Micrite	Micrite, clay
Particles	Various skeletal grains, clay-sized terrigenous detritus; silt-sized quartz	Mass accumulations of fusulinids; brachiopods, pelecypods, gastropods, bryozoans, crinoids, echinids; mud pebbles	Ostracods, gastropods, plant litter, clay minerals, quartz grains
Fabric	Unknown	Particles usually not oriented, imbricate structures due to local currents	Particles not at all oriented
Cement	Absent	Sparite	Absent

sediments have been published from the Ordovician of Sweden (Bohlin, 1949; Lindström, 1979), the lower Devonian of Spain (Stel, 1975), the Pennsylvanian of Kansas (S. M. Ball, 1971), the Carboniferous of Morocco (Kelling and Mullin, 1975), the Alpine Middle Triassic (Germann, 1969), the German Muschelkalk (Middle Triassic; Aigner, 1979), and the Upper Jurassic of Morocco (Ager, 1974) as well as the Lower Cretaceous of Montana (storm-dominated coastline within an overall transition zone from a non-marine to a marine depositional setting; W. C. James, 1980).

The possible relationship between winnowing effects (based on wave length/water depth relations) and amount of grain-support in the fabrics produced is expressed by a model described by Specht and Brenner (1979). According to this model, grainstone fabrics result from wave energies that either transport bioclasts and mud-nodules, or cause a thorough flushing of mud-sized grains from near-surface sediments. Packstone and wackestone fabrics are formed in areas of lower wave energy. Storm waves whose magnitudes are larger than normal produce a sequence of undisturbed muds going to well-winnowed grainstone fabrics; this sequence may be misinterpreted as "turbidites" because of the graded particle-size distribution and muddy matrix.

10.1.2 Terrigenous Influx and Indications of Coastlines

It is very difficult to prove that there is a terrigenous control of marine carbonate sedimentation, solely by means of microfacies criteria. Exceptions are:

a) Special extraclasts, "black pebbles" – see 4.1.3.8,
b) Concentrations of unrounded extraclasts which are lithologically varied,
c) Increased quantity of medium- to coarse-grained quartz. Because water and wind transport must be reckoned with (example: Persian Gulf, Purser, 1973), different and sometimes very great transport distances must also be noted.

Better indications of coastal proximity are given by paleontological and sedimentological features:

a) Adapted benthic communities,
b) Ecological zonations by body fossils and trace fossils along shorelines (see Howard, 1972),
c) Accumulations of lithophages, epizoans, and encrusting organisms in the littoral zone (Radwánski, 1959, 1968, 1969),
d) Whole plants or much plant litter together with fossils,
e) Decreasing diversity towards the coastline (e.g., Stevens, 1968),
f) Sedimentary clay mineral provinces; dominance of kaolinite near the shore (see 9.1.1.2),
g) Distribution of stable carbon isotopes; indication of "near-shore – off-shore" (see Jux and Manze, 1974), see Sect. 9.1.3.

Basically it is easier to reconstruct the original position of the coastline by determining the supratidal and intertidal zones (see Sect. 10.2.2). However, it must be remembered that actually only the supratidal zone is a useful indicator, because large stretches of shore may have been covered by tidal sediments (see examples in Selley, 1970).

Terrigenous influx can also be recognized by the thorough investigation of *marls:* These sediments, composed of a friable mixture of silt- and clay-sized grains of carbonate and clay minerals, are known in marine as well as in lacustrine environments (Terlecky, 1974). The most wide-spread marine marls contain variable amounts of macro- and microfaunal elements, which often suggest their autochthonous origin. Many marls were deposited in calm marine environments of moderate depth, without intensive ocean current energy, or in association with near-coastal deltaic sediments (Sarntheim and Walger, 1973).

Another possible mode for the genesis of marls is the selective dissolution of carbonates during sedimentation (see H. Ernst, 1978). Many limestone/marl sequences with only thin marl layers may be caused by early diagenetic processes.

10.1.3 Light Penetration and Water Depth; Subaquatic and Subaerial Environments

Although water depth is not an ecological boundary and the sedimentological differences observed at various water depths can usually be explained by the level of water energy and the sedimentary process itself, a great deal of the scientific literature is concerned with the search for and determination of paleobathymetric criteria (see Hallam, 1967). When interpreting the depth of formation of carbonate sediments, "shallow-marine" and "deeper-marine" limestones are usually differentiated, with "shallow marine" commonly referring to the area bordered by the lower tidal zone and the edge of the continental shelf. Because the shelf edge in many regions now lies at about 200 m, and this depth coincides with the lower boundary of the euphotic or dysphotic zone, the range of the shallow marine zone just overlying the continental shelf can also be defined by biological factors. It must not be forgotten, however, that the lower boundary of marine plants dependent on light is often deeper (see Voigt, 1956).

Bearing in mind the present criteria on water depths, it seems sensible to distinguish the following areas:

1. Land or "subaerial" exposure (e.g., coastal region above the splash zone, emerged parts of reefs, carbonate cays)
2. Marginal marine areas without continuous inundation
   a) Supratidal zone (above the normal tidal zone – corresponds to the splash zone; intermittently moistened or flooded)
   a) Intertidal zone (tidal area between extreme high-water spring tides and extreme low-water spring tides; intermittently flooded and exposed).
3. Shallow marine environment (subtidal) below the tidal zone. Sometimes directly adjoining the supratidal area. The lower boundary limit is not clearly defined, often drawn around 200 m. It can be subdivided according to the significance of the phytal:
   a) Shallower subtidal zone, characterized by an abundance of algae and by invertebrates which are adapted to the phytal, such as foraminifera, sponges, hydroids, bryozoans, and polychaetes (Voigt, 1956). Epizoans and epiphytes are common. Important criteria for determining shallow subtidal environments are a high diversity of dasycladaceans and coralline algae as well as frequently observed associations of algae and encrusting organisms (foraminifera, bryozoans, serpulids). The lower boundary of this zone, after comparison with Recent zonations, can be drawn at about 30 m. Often, however, much shallower depths are observed.
   b) Deeper subtidal environment characterized by communities of organisms with various diversities. Calcareous algae still occur in addition to invertebrates; dasycladaceans, however, are absent.

   In Table 48 the lower and the upper subtidal zones can also be distinguished according to the occurrence of certain microfacies features. Aggregate grains, oncoids, and ooids as well as most cortoids (see Sect. 4.1.3.7) and fenestral fabrics appear to be restricted to the shallower subtidal region (and the intertidal zone). It is quite noticeable that all the particles and fabrics involved here are associated with algal activity. The upper photic zone and the lower photic and aphotic zones are differentiated by different directions of the penetrations of microborings (perpendicular vs. parallel to the substrate; Budd and Perkins, 1980).

Water depth in stagnant basins can be estimated from the different kinds of benthonic shelled epifauna and bioturbation within layered water zones charac-

Table 48. Depth range of Recent groups of organisms and microfacies criteria. x = allochthonous

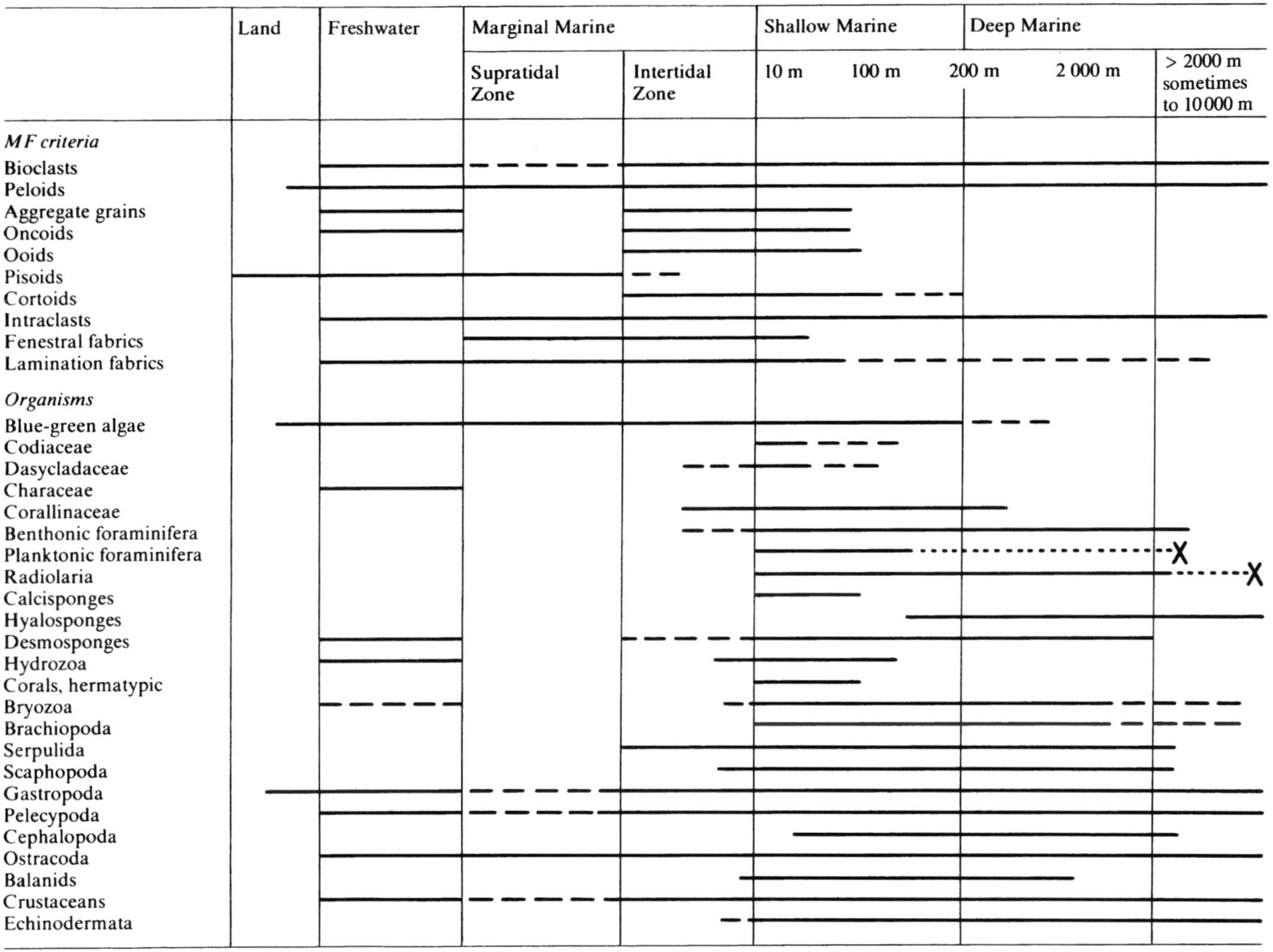

terized by different degrees of oxygenation (Rhoads and Morse, 1971). Within the surface zone (upper 50 m), conditions are nearly uniform, due to vertical mixing by waves and thermal overturn. The water throughout the surface layer is of nearly equal salinity and density, and atmospheric oxygen is continually mixed in at the surface. Calcareous epifauna and an intensive bioturbation of the sediment are characteristic for this zone, which corresponds to the shallow subtidal region. This aerobic zone is underlain by about a 100 m thick dysaerobic zone of rapid change in salinity and density (pycnocline) in which the water column is markedly stratified. The calcareous epifauna is missing within the pycnocline, whereas bioturbation still exists. The deepest water zone below the pycnocline is nearly uniform in salinity and density, and almost completely lacking in oxygen (anaerobic zone). This zone is characterized by the absence of calcareous epifauna as well as of bioturbation. Presence or absence of calcareous body fossils and bioturbation in response to oxygen content can be used, therefore, to determine paleodepth (Byers, 1977).

4. Deep marine environment (bathyal, abyssal, hadal) below the zone where light still penetrates. Extending down to water depths of over 10,000 m.

Because reliable, accurate paleobathymetric information is possible only in exceptional cases (fusulinid biotopes, Stevens, 1969; Devonian "deep-water" stromatolites, Playford et al., 1976), statements about the probable water depths are usually made using only general names for the bathymetric zones.

These zones are characterized by:

- Paleontological or ecological data
  a) Comparison of the depth ranges of Recent and Ancient organisms (usually involving many problems, see Hallam, 1967; Heckel, 1972).
  b) Interpretation of the different diversities (Information on Recent deep marine environments is still limited).
  c) Interpretation of trace fossils (see Seilacher, 1967) and trophic groups. These methods appear to be useful when the possible evolution of the ecosystems is included.
  d) Interpretation of special modes of life, such as the abundance of boring organisms indicating intertidal and shallow-marine environments.
  e) Interpretation (and overestimation) of algae and algal sedimentation. The difficulties involved here are described in Sect. 5.3.12.3
  f) Ratio of the absolute numbers of planktonic and benthonic foraminifera (see Grimsdale and Morkhoven, 1955; Lynts, 1971; Thiede, 1973; H. Ernst, 1978).

Paleobathymetrical estimations using paleontological data must take into account the taxonomic differences between Recent and pre-Tertiary organisms, changes of the fossilization potential, and an evolution of the diversity patterns over geological time.

- Sedimentological data
  a) *Depth of wave action.* In the research work of A. V. Carozzi and his students (e.g., Wanless et al., 1957) as well as in the grain-size diagrams developed by Passega (1964), attempts have been made to describe the lower depth limit at which wave action stirs the sediments at the bottom. Important for the

recognition of wave energy is, on the one hand, the quantity of fine-grained particles in calcarenites (example in Sarntheim, 1970, Persian Gulf) and, on the other hand, the diameter of the largest particle (clasticity index, see 4.2.1) as a measure of the strongest water energy dominating at the time of sedimentation. As Sarntheim has shown, the amount of fine-grained material decreases continuously with increasing water depth. In the Persian Gulf maximum wave bases of 50–70 m have been deduced from the grain-size distribution and the strong connection, in marl areas, between optimal echinoderm development and the maximum supply of nutrients near the wave base.

b) *Carbonate dissolution in the deep sea.* Deep sea drilling has provided new information about paleobathymetric evaluations (Berger and Winterer, 1974; Hesse and Butt, 1976). Because the $CaCO_3$ content of marine deep-sea sediments in areas with low rates of sedimentation and open circulation is controlled by hydrophysical, hydrochemical, and biological factors which are dependent on depth, the position of the CCD level yields important data about the depths of deposition and dissolution. Carbonate content and selective dissolution of carbonate minerals (example in W. Schlager, 1974, Mesozoic cephalopod shells) have proved to be valuable indicators. Recurring changes in the calcite compensation depth (Fischer and Arthur, 1977) may be used as partial evidence for rhythmic behavior in the oceans (with "polytaxic states" being characterized by a shoaling of the CCD relative to most "oligotaxic" periods – as in Pleistocene and Recent seas). Important contributions to the problem of carbonate dissolution in deep seas can be found in *Joides Initial Reports* (see 1.4.1), in Hsü and Jenkyns (1974), and in Burns (1974), and Hesse and Butt (1976).

c) *Geochemical criteria.* We refer here only to the depth dependent formation of various authigenic minerals such as glauconite and chamosite, phosphorite, and the manganese nodules in the deep seas. Surveys are found in Porrenga (1967), Bromley (1967), and Schopf (1980); the use of trace elements as paleobathymetric criteria is described in Nicholls (1967) and Veizer (1977). See also 9.1.2.

d) *Geological criteria.* Water depth of deposition may be estimated from the reconstruction of the primary submarine topography (e.g. Stevens, 1971). Paleobathymetrical reconstructions can also be made by seismic profiles which indicate the primary topography (e.g. Vail et al., 1977).

Depositional depth of Mesozoic and Cenozoic oceanic sediments can also be determined by the "backtracking method" (Berger, 1972; Berger and Winterer, 1974; Ramsay, 1977; Sclatter et al., 1977) which is based on the relationships between the age and the depth of the normal cooling oceanic crust. New crust formed at the crest of an oceanic ridge has a depth of 2400–2800 m (Veevers, 1977), after 20 m.y. 3800–4200 m, after 40 m.y. 4560–5100 m, after 80 m.y. 5100–5800 and after 100 m.y. 5200–6000 m. The sediments deposited upon the oceanic crusts are affected by the same subsidence as the crust. Carbonates are affected by the vertically fluctuating depth of the CCD, and also by the laterally changing zones with different plankton production rates (Lancelot, 1980).

In reference to the facies criteria compiled in the Tables in Sect. 10.2, the following microfacies and paleontological features can be used for the *Classification of subaerial or subaquatic environments:*

Land or exposed areas

a) Strongly to very strongly recrystallized or leached particles,
b) Microsparitic matrix (see Sect. 3.4) or blocky cement,
c) Zones with red or orange-red coloring (resulting from intense oxidation of Ancient bottoms permeated with rain water,
d) Black or dark-colored, angular lithoclasts on bedding planes or in the uppermost part of the bed (black pebbles, see Sect. 4.1.3.8),
e) Thin algal mats,
f) Surficial and internal reliefs ("microkarst") filled with partially dissolved and reworked calcareous clasts,
g) Thin, unfossiliferous limonite or hematite crusts upon the bedding planes of limestones, sometimes with shrinkage cracks and crinkly patterns,
h) Extremely irregular fabric with thin calcite veins which follow vertical and horizontal fractures,
i) Caliche crusts (see Sect. 2.3.2.2),
j) Pisoids (see Sect. 2.3.2.2 and 4.1.3.6) – no distinct criterion (Esteban and Pray, 1978), because pisoids may also be formed subaquatically,
k) Cm-sized fissures filled with non-carbonate sediment,
l) SiO_2 formation in the upper part of the bed (surficial weathering and redeposition of silica near the paleo-water table; see Rubin and Friedman, 1978),
m) Beach-rock (see Sect. 3.3)

Supratidal zone

a) Fossils rare or absent,
b) Assemblage of limestones, fine-grained dolomite and evaporites,
c) Cm-sized intraclasts common, sometimes rock-forming,
d) Sheet cracks concordant to the bedding,
e) "Zebra" fabrics (A. G. Fischer, 1964); parallel to and oblique to cracks in the bedding; filled with drusy cement,
f) Tepee structures (similar to inverted depressed Vs); cross-sections of very large desiccation polygons with prismatic cracks,
g) Irregular mm laminations, sometimes bituminous films,
h) Algal mats,
i) Stromatolites, particularly type LLH (see Sect. 5.3.12.4),
j) Birdseyes (see Sect. 4.2.3.5),
k) Fenestral fabrics (see Sect. 4.2.3.5),
l) Plant litter,
m) Caliche and caliche breccia,
n) Pisoids,
o) Half-moon ooids and pisoids,

p) Evaporite solution breccia,
q) Beach-rock,
r) Microborings (see J. Schneider, 1976).

Intertidal zone

a) Ecologically adapted organisms,
b) Algal mats and stromatolites, "spongiostromes", and LLH types,
c) Fenestral fabrics, "loferite",
d) Birdseyes,
e) Intraclasts common; some rounded, some imbricated,
f) Peloids common, usually in cm graded beds,
g) Aggregate grains,
h) Black pebbles (allochthonous),
i) Oncoids,
j) Plant litter,
k) Bioturbation common,
l) Vertical burrows dominant,
m) Ooids common,
n) Cortoids common,
o) Desiccation cracks, shrinkage cracks,
p) Assemblage of limestones and fine-grained dolomite,
q) Considerable mixing of sediment, rapidly changing MF types.

Subtidal zone

a) Often high diversity of organisms (benthonics abundant, planktonics rare),
b) Dasycladaceans and other calcareous algae common,
c) Oncoids,
d) Cortoids common,
e) Aggregate grains common,
f) Normal ooids common,
g) Indications of phytal common,
h) Organism abundance large to very large,
i) Horizontal burrows dominant,
j) Bioturbation common,
k) Flat- and cross-bedded calcarenites, often dominant calcilutites,
l) Assemblage of various MF types, indicating different water energies.

Deep marine

a) Algae and indications of algal activity missing,
b) LF fenestral fabrics absent,
c) Small-scale multicolored bedding,
d) Fe-Mn crusts and nodules,
e) Discontinuity planes common, often associated with condensed fauna,
f) Solution structures (subsolution fragments),

g) Planktonic organisms dominant, in contrast to benthonic organisms,
h) Small amounts of terrigenous quartz,
i) Often very small amounts of residual clay,
j) Hardground formations common,
k) Large proportion of micrite clasts or micritic resediments.
l) Glauconite and phosphorite occur in highest percentages of abundance,
m) Calcilutites dominate calcarenites.

The uncertainties associated with the above criteria will be discussed in Sect. 10.2.4 and Table 53.

In a very stimulating paper Liebau (1980) discusses the zonations of marine seas used by marine biologists, and the *possibilities of recognizing depth zones* by paleoecological criteria. Paleoecologically important boundaries are caused by light, water movement, and special thanatocoenoses. Several boundaries can be recognized within the euphotic and dysphotic zones:

Zostera boundary: lower distributional boundary of macrostenophotic sea grasses (*Zostera*); depth about 6–12 m.

Chlorocline: lower boundary of deeper living green algae (algal chlorocline; 60–90 m, maximum 110 m; sea grass chlorocline, deeper living marine Potamogetonaceae (phanerogams) like *Thalassia* and *Posidonia;* 60–80m, maximum 90–110m).

Rhodocline: lower boundary of multicellular algae (e.g. calcareous red algae); 150–200 m, maximum 250 m.

Lepidocline: lower boundary of deeper-water Larger foraminifera with algal symbiosis (*Cycloclypeus* and other Nummulitidae; down to 150 m or more?)

Ophthalmocline: "Growing blind" zone of benthonic ostracods (lower boundary for ostracods with normal eyes 500–600 m; upper boundary of blind ostracods about 800–600 m, maximum 400 m).

Pterygocline: upper distributional boundary of benthonic ostracods, which avoid agitated waters (in oceanic near-coastal areas 60±20 m; in land-locked seas 25±10 m; in protected seas 5–10 m).

Water agitation can produce several boundaries:
- high and low tide mark
- *mean wave base* (mean lower boundary of waves which can transport sand). Oceanic near-coastal environments 10–20 m; protected seas 4–6 m; lagoons 0.3–0.5 m.
- *storm wave base* (oceanic near-coastal environment; 40–80 m, maximum 200 m)

Some *thanatocoenoses* can be used as bathymetrical boundaries:
- coquina composed of small shells (corresponding to the storm wave base, the pterygocline, and the biotope boundary between worn and unworn micro-organisms)
- thanatocoenoses with 50% planktonic organisms.

Using these bathymetrical criteria, the following zones can be distinguished:
- Supralittoral (supratidal): between upper supratidal boundary and upper tide mark
- Mediolittoral (intertidal): between upper and lower tide marks
- Upper sublittoral (upper subtidal): between lower tide mark and mean wave base

- Middle sublittoral (middle subtidal): between mean wave base and storm wave base
- Lower sublittoral (lower subtidal): between storm wave base and upper boundary of the bathyal zone (with abundant planktonic organisms)

An "infralittoral zone" (corresponding to the deeper sublittoral zone as defined at the beginning of this Section) may be recognized between the low water mark and the green algal chlorocline or the hermacline; below this zone is the "circalittoral zone", lying between the green algal chlorocline and the rhodocline.

10.1.4 Salinity

As Table 49 shows, only stenohaline organisms can actually be used to indicate different salinities. The following parameters are important (see Hiltermann, 1966; Jordan, 1971):

- Comparison of salinity dependence in Recent and Ancient organisms
- Distributional patterns of dasycladacean algae in platform carbonates (Peybernes, 1979)
- Charophyte and evaporite-bearing units (e.g., the Purbeckian evaporites of Dorset (England) and the Tertiary evaporites of the Paris Basin), may be interpreted from seasonally varying salinity within Mediterranean-type saline lakes in semi-arid climatic zones (see Burne et al., 1980).
- Interpretation of the different numbers of species and genera, i.e., the different diversities (Remane curve) in different salinity zones
- Consideration of monomictic and polymictic associations
- Sizes of the organisms decrease from normal marine to lacustrine environments to brackish water
- Thinning of shells in lacustrine mollusks
- Specially adapted organisms
- Differences in the diversity of boring organisms (usually only microborings, algae, etc. in fresh water)

These paleontological features can be supplemented by geochemical criteria (see 9.1.2), among which the interpretation of boron content as well as carbon isotopes and oxygen isotopes is particularly significant.

Hypersalinity can be inferred from the following criteria:

a) Marine fauna of reduced diversity
b) Preservation of algal mat structures. Burrowers and grazers normally destroy algal mats in marine waters of normal salinity. Reinforced by the antipathetic relationship between algal mats and burrowing and grazing organisms, algal mats can, then, flourish under conditions of increased salinity (Cussey and Friedman, 1976; Friedman et al., 1973; Garrett, 1970; Mazzullo and Friedman, 1977).
c) Abundance of ooids and grapestones, which are more likely to occur in areas of higher salinities and warm waters

Table 49. MF criteria, distribution of organisms and salinity ranges

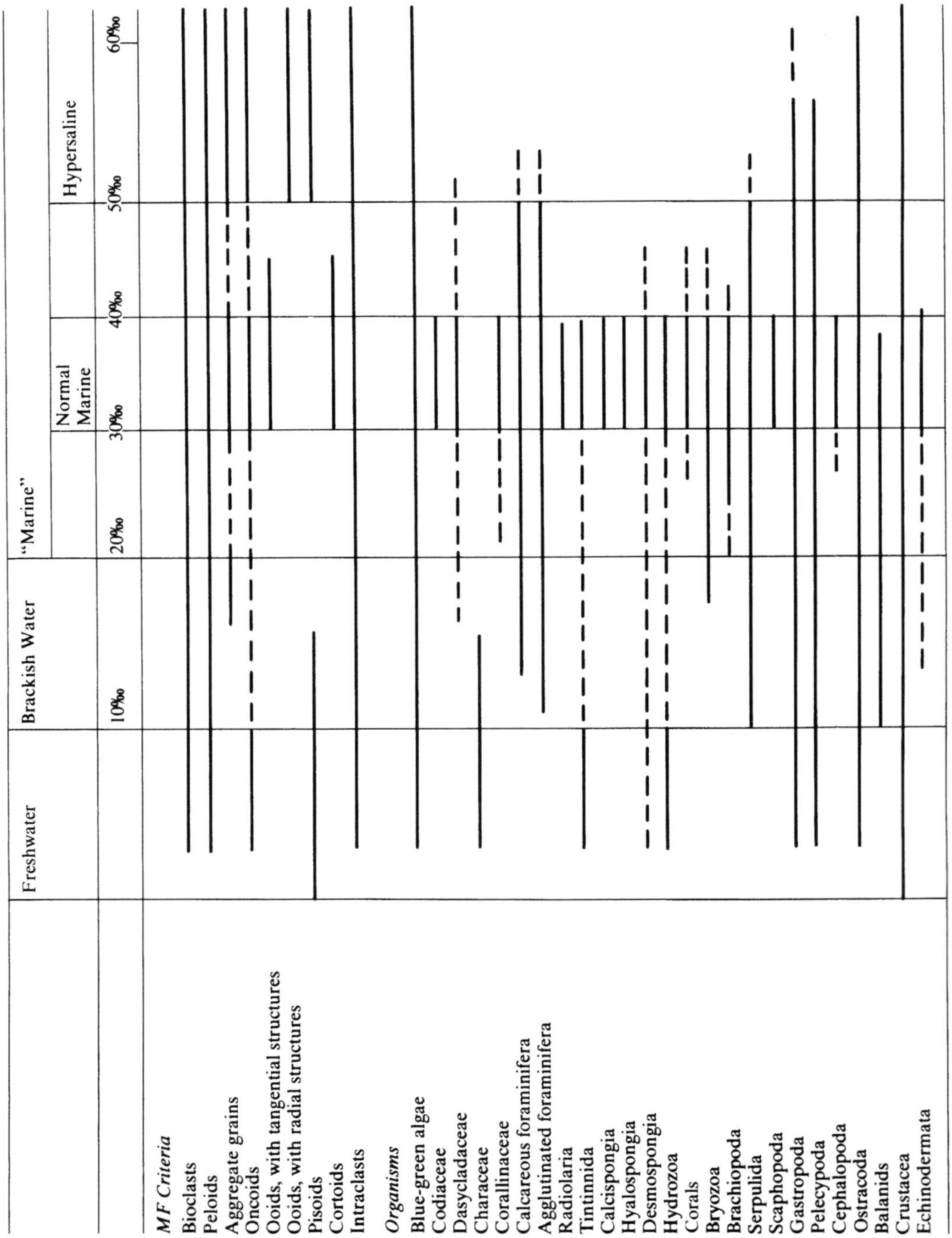

d) Synaeresis shrinkage cracks: Because of high evaporation rates, water salinities in shallow areas may rise quickly and exceed those of interstitial pore fluids of the substrate, producing synaeresis shrinkage cracks (Donovan and Foster, 1972) which are often subsequently filled with sedimentary material. The magnitude of polygon development parallels the declining salinity profile from hypersaline sabkha environments to normal marine environments (Baria, 1977).

Hypersaline environments may be subjected to episodic flushing by freshwater, resulting in environments with strongly fluctuating but never normal marine salinities ("schizohaline environments", Folk and Siedlecka, 1974).

Sediments deposited under *evaporitic conditions* are characterized by:

a) Silicified or calcitized nodules which have been formed by the replacement of sulfate nodules (nodular anhydrite of arid supratidal sabkha flats); see Friedman and Radke, 1979).
b) Microfolds within laminated dolostones, resulting from a dehydration of primary gypsum or the hydration of anhydrite (Maiklem, Debout and Glaister, 1969).
c) Half-moon ooids may owe their shape to a collapse of insoluble parts following the dissolution of envelopes composed of soluble salts, especially of anhydrite or gypsum (Carozzi, 1963; Folk and Pittman, 1971). Diagnostic criteria are relict laminae replaced by quartzine.
d) Finely laminated or thinly-bedded micron-sized dolostones, which may be interpreted as the result of penecontemporaneous dolomitization.
e) Relict evaporite minerals
f) Euhedral quartz, double-terminated subhedral quartz crystals (Grimm, 1962)
g) Quartzine (length-slow chalcedony) together with pseudomorph replicas of evaporites (Folk and Pittman, 1971; Folk and Siedlecka, 1974).
h) "Baroque" dolomite, characterized by curved crystal faces, undulose extinction and a generally clouded appearance (Folk, 1977), is formed as a diagenetic product in hypersaline carbonate deposits.
i) Authigenic K-feldspar. Weaver (1967) suggested a possible genetic relationship between the occurrence of authigenic K-feldspar and hypersaline environments.
j) Fluorite: In marginal sabkhas fluorite (Moore et al., 1972) and celestite are associated with early dolomitization of aragonitic sediments (Ziehr et al., 1980). During dolomitization fluorine is liberated from aragonite and precipitated as fluorite.
k) Magnesite: The diagenetic transformation of aragonite and calcite to magnesite seems to be possible only in a hypersaline evaporitic environment which occurs in marine or lacustrine settings (Niedermayr et al., 1979).
l) Dedolomitization, generally related to the weathering of surfaces, may be aided by the presence of sulfate ions which may come from evaporation of interstitial brines.
m) Collapse breccias may result after the dissolution of interbedded evaporites: They are characterized by stratiform breccia bodies with well-defined lower contacts and by in situ clasts in a fine matrix (Middleton, 1961).

10.1.5 Temperature

The interpretation of the paleontological material is also extremely important for the reconstruction of the paleoclimate (see Strauch, 1972). The following criteria have proved to be quite useful:

- Comparison of Recent and Ancient temperature dependent organisms with regard to close taxonomic relationships
- Shells of calcareous organisms are often thicker in warm water
- Sizes of organisms increase in tropical climates (very controversial)
- Rich ornamentation in tropical fauna, can usually be interpreted only when large numbers of fossils are present
- Shape adaptation
- Differences in the diversity or number of species
- Differences in population density
- Increase of epibiontic organisms in warm waters
- Differences in growth rhythms recognizable in molluskan shells
- Shell mineralogy

Of all the microfacies particles, only normal ooids and the cortoids are somewhat bound to warmer temperature ranges. This is only valid for bioclasts micritized by boring algae (see Sect. 4.1.3.7).

Refer to the general survey in Schwarzbach (1974) for other possibilities of describing paleotemperatures based on geochemical and sedimentological data.

10.1.6 Substrate

Fabricius (1968) described the significance of different carbonate mud and carbonate sand bottoms for colonizing by organisms and for the kind of lithification, as illustrated by Mesozoic shallow- and deep-water limestones. He distinguished:

1. Soft bottoms
   a) Unstable mud bottoms (sometimes hostile to life or with restricted environmental conditions, such as in marl limestones of the Rhaetian Kössen beds in the Alps)
   b) Stable mud bottoms (stabilization by organisms growing on the sediment, such as brachiopods, branched corals, siliceous sponges).
   c) Sand bottoms (calcarenite made up of bioclasts, e.g., near and in the backreef area of Upper Rhaetian reefs; ooid sand bottoms; sand bottoms made up of crinoid fragments, sometimes stabilized by rooted crinoids. Otherwise usually unstable; indications of currents in oolitic limestones)

2. Solid bottoms
   a) Biogenic hard bottoms
      i) Sand bottoms with organic encrustations (e.g., algae, sessile foraminifera, corals)
      ii) Reef frameworks with organic encrustations (organisms listed above as well as calcareous sponges, serpulids, bryozoans, etc.).

b) Solution bottoms formed by submarine carbonate solution during restricted or absent sedimentation. Often characterized by microreliefs.
c) Temporary hard bottom formed by alternating soft and hard bottoms. The mode of genesis can be explained by the subsolution model of Hollmann (1962):
   i) Sedimentation of carbonate mud during a very reduced rate of sedimentation
   ii) Period of non-deposition and lithification; change from soft to hard bottom
   iii) Intensive submarine solution; example: Adnet limestone, Lias; Pl. 15/2.

The following *parameters* reveal the *consistency* of carbonate mud bottoms or carbonate sand bottoms before or at the beginning of lithification and cementation:

a) Bioturbation and formation of pseudointraclasts as well as of colored micrite areas which may deviate in grain size from the matrix (indication of soft bottoms)
b) Autoclasts or protointraclasts (indication of soft bottoms)
c) Hard bottoms with sharp upper boundary and unclear lower boundary caused by bottom solution; boring by algae and various invertebrates, organic crusts, concentrations of epizoans (indication of hard bottoms)
d) Adaptation of the stromatolite shape to the contours of the substrate (indication of solid bottoms)
e) Calcarenite or grainstones with distinct current structures such as cross-bedding; absence of bioturbation, etc. (indication of unstable sand bottoms)

One should avoid putting too much emphasis on the presence or absence of bioturbation in sediments. Bioturbation may be absent because

- Very few or no benthonic organisms may live at the bottom of the ocean, because of stagnation, hypersalinity, or restricted conditions
- Flooding of the supratidal area occurs too rarely for a constant and heavy colonization of the mud bottoms by benthonic organisms
- Rapid and intermittent sedimentation of bioclastic particles or calcareous ooids prevents colonizing of unstable sand bottoms even in areas with low water energies.

Hard bottoms in deeper marine environments are generally formed during times of strongly reduced sedimentation. According to Fürsich (1979) different types of hard bottoms may be produced.

10.1.7 Continuous and Discontinuous Sedimentation

The alternation of microfacies in a profile illustrates more explicitly than lithological or even paleontological features the sometimes very rapid change in sedimentary conditions. This is very clearly seen in the studies carried out by Carozzi and his students, where samples were taken at close intervals (see Figs. 3 and 26). Of course, percentage differences in the abundance of particles or of certain organisms should not be overrated, as long as statistical analyses (e.g., time-trend analyses) have not been done. The following criteria in microfacies are more important for the recognition of a change in the sedimentary condition:

- Rapid transition from mud support to grain support (change of water energy)
- Occurrence of particles not seen in earlier adjacent samples, particularly aggregate grains (expressing a special depositional situation), ooids (indicating changed water energy or redeposition), cortoids (in great abundance), and extraclasts (change in intensity of erosion)
- Occurrence of skeletal grains displaying new ecological niches
- Conspicuous changes in organism diversity
- Change in the grain sizes and in the sorting and skewness of arenitic limestones
- Distinct changes in rock coloring
- Changes in the average bed thickness
- Changes in sedimentary structures.

10.1.8 References: Environmental Factors

Ager, D. V. (1974): Storm deposits in the Jurassic of the Moroccan High Atlas. Paleogeography, Paleoclimatology, Paleoecology **15,** 83–93, Amsterdam

Aigner, T. (1979): Schill-Tempestite im oberen Muschelkalk (Trias, SW-Deutschland). Neues Jahrbuch Geol. Paläont. Abhandlungen **157**/3, 326–343, 7 Figs., Stuttgart

Ball, M. M., Shinn, E. A., Stockman, K. W. (1967): The geologic effect of hurricane Donna in South Florida. J. Geol. **75**/5, 583–597, Chicago

Ball, St. M. (1971): The Westphalia Limestone of the Northern Midcontinent: a possible Ancient storm deposit. J. Sed. Petrol. **41**/1, 217–232, 11 Figs., Tulsa

Barbieri, G., Zanche, V., di Lallo, E., Miette, P., Sedea, R. (1977): Middle Triassic emersion phases in the Recoaro area (Vicentinian Alps, N.E. Italy). Neues Jahrbuch Geol. Paläont. Monatshefte **1977**/9, 523–531, Stuttgart

Bein, A., Weiler, Y. (1976): The Cretaceous Talme Yafe Formation: a contour current sedimentary prism of calcareous detritus at the continental margin of the Arabian Craton. Sed. **23,** 511–532, 15 Figs., Oxford

Berger, W., Winterer, E. L. (1974): Plate stratigraphy and the fluctuating carbonate line. Int. Ass. Sed. Pec. Publ. **1,** 11–48, Oxford

Bohlin, B. (1949): The Asaphus Limestone in northernmost Öland. Bull. Geol. Inst. Univ. Uppsala **33,** 529–571, Uppsala

Bradley, W. C. (1958): Submarine abrasion and wave-cut platforms. Bull. Geol. Soc. Amer. **69,** 967–974, New York

Bromley, R. G. (1967): Marine phosphorites as depth indicators. Marine Geol. **5,** 503–509, Amsterdam

Budd, D. A., Perkins, R. D. (1980): Bathymetric Zonation and Paleoecological Significance of Microborings in Puerto Rican Shelf and Slope Sediments. J. Sed. Petrol. **50**/3, 881–904, 9 Figs., Tulsa

Burne, R. V., Bauld, J., de Deckker, P. (1980): Saline Lake Charophytes and their Geological Significance. J. Sed. Petrol. **50**/1, 281–293, 6 Figs., Tulsa

Burns, D. A. (1974): see 10.2.5

Byers, Ch. W. (1977): Biofacies patterns in euxinic basins: a general model. Soc. Econ. Paleont. Min. Spec. Publ. **25,** 5–17, 8 Figs., Tulsa

Carozzi, A. V. (1963): Half-moon oolites. J. Sed. Petrol. **33,** 633–645, Tulsa

Donovan, R. N., Forster, R. J. (1972): Subaqueous shrinkage cracks from the Chaithness Flagstone Series (Middle Devonian) of Northeast Scotland. J. Sed. Petrol. **42,** 309–317, Tulsa

Ernst, H. (1978): Zu Bathymetrie und Sedimentstrukturen der Schreibkreide von Lägerndorf/Holstein (Conias-Santon): Eine quantitative Analyse der Foraminiferen-Faunen. Mitteilungen Geol. Paläont. Inst. Univ. Hamburg **48,** 53–78, Plate 8, 2 Figs., Hamburg

Fabricius, F. (1968): see 4.2.4.7

Fischer, A. G. (1964): see 4.2.4.5

Flügel, E. (1977): see 5.3.12.7

Fischer, A., Arthur, M. A. (1977): Secular Variations in the Pelagic Rocks. Spec. Publ. Soc. Econ. Paleont. Min. **25,** 19–50, 8 Figs., Tulsa

Folk, R. L. (1962): see 6.4

Folk, R. L. (1977): Peculiar forms of diagenetic carbonate from hypersaline and cave deposits, Ancient to Recent. West Texas Geol. Soc., Newsletter, p. 11

Folk, R. L., Pittman, J. S. (1971): Length-slow chalcedony: A new-testament for vanished evaporites. J. Sed. Petrol. **41,** 1045–1058, Tulsa

Folk, R. L., Siedlecka, A. (1974): The "Schizohaline" environment: its sedimentary and diagenetic fabrics as exemplified by Late Paleozoic rocks of Bear Island, Svalbard. Sed. Geol. **11,** 1–15, Amsterdam

Friedrich, H. (1965): see 1.4.1

Fürsich, F. T. (1979): Genesis, environments, and ecology of Jurassic hardgrounds. Neues Jahrbuch Geol. Paläont. Abhandlungen **158**/1, 1–63, Stuttgart

Garrett, P. (1970): Phanerozoic stromatolites: noncompetitive ecologic restriction by grazing and burrowing animals. Science **169,** 171–173, Washington

Germann, K. (1969): see 4.1.7.2

Grimm, V. W. D. (1962): Ausfällung von Kieselsäure in salinar beeinflußten Sedimenten. Zeitschrift deutsch. geol. Gesellschaft **114,** 590–619, Stuttgart

Grimsdale, T. F., Morkoven, F. P. C. M. van (1955): The Ratio between Pelagic and Benthonic Foraminifera as a Means of Estimating Depth of Deposition of Sedimentary Rocks. Proc. 4th World Petrol. Congr., Sect. 1/D, Paper **4,** 473–491, 9 Figs., Rome

Hallam, A. (ed.) (1967): see 1.4.1

Hayes, M. O. (1967): Hurricanes as geological agents: case studies of hurricanes Carla, 1961, and Cindy, 1963. Texas Univ. Bur. Econ. Geol., Rep. Inv. **61,** 56 pp.

Heckel, P. H. (1972): Recognition of ancient shallow marine environments. In: Rigby, J. K., Hamblin, W. K. (eds.): Recognition of Ancient Sedimentary Environments, 226–286, 15 Figs., Tulsa

Heller, P. L., Komar, P. D., Pevear, D. R. (1980): Transport Processes in Ooid Genesis. J. Sed. Petrol. **50**/3, 943–952, Tulsa

Hesse, R., Butt, A. (1976): Paleobathymetry of Cretaceous turbidite basins of the East Alps relative to the calcite compensation level. J. Geol. **34**/5, 505–533, 12 Figs., Chicago

Hiltermann, H. (1966): Klassifikation rezenter Brack- und Salinar-Wässer in ihrer Anwendung für fossile Bildungen. Z. Dtsch. Geol. Ges. **115**/2–3, 463–496, Pls. 11–12, 7 Figs., 2 Tab., Hannover

Hollmann, R. (1962): see 4.2.4.7

Howard, J. D. (1972): Trace fossils as criteria for recognizing shorelines in the stratigraphic record. Spec. Paper Soc. Econ. Paleont. Min. **16,** 215–225, 6 Figs., Tulsa

Hsü, K. H., Jenkyns, H. C. (eds) (1974): see 1.4.1

James, W. C. (1980): Limestone Channel Storm Complex (Lower Cretaceous), Elkhorn Mountains, Monatana. J. Sed. Petrol. **50**/2, 447–456, 8 Figs., Tulsa

Jordan, R. (1971): Zur Salinität des Meeres im höheren Ober-Jura Nordwest-Deutschlands. Z. Dtsch. Geol. Ges. **122,** 231–241, 2 Figs., Hannover

Jux, U., Manze, U. (1976): see 9.1.4

Kelling, G., Mullin, P. R. (1975): Graded limestones and limestone-quartzite couplets: possible storm-deposits from the Maroccan Carboniferous. Sed. Geol. **13,** 161–190, Amsterdam

Kendall, Ch. G. St. C. (1969): see 4.1.7.2

Kumar, N., Sanders, J. E. (1976): Characteristics of shoreface storm deposits: Modern and Ancient examples. J. Sed. Petrol. **46**/1, 145–162, 10 Figs., Tulsa

Land, L. S., Hoops, G. H. (1973): Sodium in carbonate sediments and rocks: a possible index to the salinity of diagenetic solutions. J. Sed. Petrol. **43,** 614–617, Tulsa

Lancelot, Y. (1980): Environments sédimentaires océaniques; développement de la paléo-océanographie. Mém. Soc. géol. France **10,** 351–362, Paris
Lecompte, M. (1962): Die Riffe im Devon der Ardennen und ihre Bildungsbedingungen. Geol. Palaeont. **4,** 25–71, 4 Pls., 20 Figs., 3 Tab., Marburg
Liebau, A. (1976): Entwurf einer paläobathymetrischen Flachmeer-Zonierung mit Hilfe von Ostracoden und anderen Mikrofossilien. Zentralbl. Geol. Paläont. Teil II, **1976,** 5/6, 442–447, 1 Fig., Stuttgart
Liebau, A. (1980): Paläobathymetrie und Ökofaktoren: Flachmeer-Zonierungen. Neues Jahrbuch Geol. Paläont. Abhandlungen **160,** 173–216, Stuttgart
Lindström, M. (1979): see 3.9
Lynts, G. W. (1971): Analysis of Planctonic Foraminiferal Fauna of Core 6275, Tongue of the Ocean, Bahamas. Micropaleont. **17**/2, 152–166, 4 Figs., New York
Maiklem, W. R., Debout, D. G., Glasiter, R. P. (1969): Classification of anhydrite – a practical approach. Bull. Canadian Petrol. Geol. **17,** 194–233, Calgary
Mazzullo, S. J., Friedman, G. M. (1977): Competitive algal colonization of peritidal flats in a schizohaline environment: The lower Ordovician of New York. J. Sed. Petrol. **47,** 398–410, Tulsa
Middleton, G. V. (1961): Evaporite solution breccias from the Mississippian of southwest Montana. J. Sed. Petrol. **31,** 189–195, Tulsa
Miller, M. C., McCave, I. N., Komar, P. D. (1977): Threshold of sediment motion under unidirectional currents. Sed. **24,** 507–527, Amsterdam
Moore, C. H., Smitherman, J. M., Allen, S. H. (1972): Pore system evolution in a carbonate beach sequence. Proc. 24th Int. Geol. Congr., Sect. 6, 124–136, Montreal
Nicholls, G. D. (1967): Trace elements in sediments: an assessment of their possible utility as depth indicators. Marine Geol. **5,** 539–555, Amsterdam
Niedermayr, G., Scheriau-Niedermayr, E., Beran, A. (1979): Diagenetisch gebildeter Magnesit und Dolomit in den Grödener Schichten des Dobratsch im Gailtal, Kärnten, Österreich. Geol. Rundschau **68**/3, 979–995, 9 Figs., Stuttgart
Ogg, J. G., Koscow, A. (1978): The impact of Typhoon Pamela (1976) on Guam's coral reefs and beaches. Pacific Science **32**/2, 105–118, 9 Figs.
Passega, R. (1964): see 4.2.1.2
Perkins, R. D., Enos, P. (1968): Hurricane Betsy in the Florida-Bahaman Area – Geologic effect and comparison with Hurricane Donna. J. Geol. **76**/6, 710–717, Chicago
Peybernes, S. (1979): see 5.4
Playford, P. E., Cockbain, A. E., Druce, E. C., Wray, J. L. (1976): Devonian stromatolites from the Canning Basin, Western Australia. In: Walther, M. (ed.): Stromatolites. Dev. Sed. **20,** 543–564, Amsterdam: Elsevier
Plumley, W. J., Risley, G. A., Graves, R. W., Kaley, M. E. (1962): see 6.4
Porrenga, H. D. (1967): see 10.2.5
Purser, B. H. (ed.) (1973): see 1.4.1
Radwánski, A. (1959): Littoral structures (cliff, clastic dikes, and veins, and borings of Potamilla) in the High-Tatric Lias. Acta Geol. Polonica **9,** 231–280, Pls. 20–24, 9 Figs., Warsaw
Radwánski, A. (1968): Tortonian cliff deposits at Zahorska Bystrica near Bratislawa (Southern Slovakia). Bull. Acad. Poln. Sci., Ser. Sci. geol. geogr. **16**/2, 97–102, 3 Figs., Bratislava
Radwánski, A. (1969): Lower Tortonian transgression onto the southern slopes of the Holy Cross Mountains. Acta Geol. Polonica **19**/1, 1–164, 42 Pls., 34 Figs., Warsaw
Ramsay, A. T. S. (1977): Sedimentological clues to paleo-oceanography. In: Ramsay, A. T. S. (ed.): Oceanographic Micropaleontology **2,** 1371–1453, London: Academic Press
Rhoads, D. C., Morse, J. W. (1971): Evolutionary and ecologic significance of oxygen-deficient marine basins. Lethaia **4,** 413–428, Oslo
Rhoads, D. C., Speden, I. G., Waage, K. M. (1972): Trophic group analysis of Upper Cretaceous (Maastrichtian) bivalve assemblages from South Dakota. Bull. Amer. Ass. Petrol. Geol. **56**/2, 1100–1113, 4 Figs., 3 Tab., Tulsa
Riedl, R. (1969): Marinbiologische Aspekte der Wasserbewegung. Marine Biol. **4,** 62–78, 20 Figs.
Sarntheim, M. (1970): see 4.2.1.2

Sarntheim, M., Walger, E. (1973): Classification of modern marl sediments in the Persian Gulf by factor analysis. In: Purser, B. H. (ed.): The Persian Gulf, 81–97, New York (Springer)
Schäfer, W. (1972): see 1.4.1
Schlager, W. (1974): Observation of cephalopod skeletons and carbonate dissolution on ancient Tethyan sea floors. Spec. Publ. intern. Ass. Sed. **1,** 49–70, 10 Figs., Oxford
Schmidt, H. (1935): Die bionomische Einteilung der fossilen Meeresböden. Fortschr. Geol. Paläont. **12**/38, Berlin
Schmidt, H. (1956): Zur Rangordnung der Faziesbegriffe. Mitt. Geol. Ges. Wien **49,** 333–345, 1 Fig., Wien
Schneider, J. (1976): see 2.4.1
Schopf, Th. J. M. (1980): Paleooceanography. 431 p., Harvard: University Press
Schwarzbach, M. (1974): see 1.4.1
Sclater, J. G., Hellinger, S., Tapscott, C. (1977): The paleobathymetry of the Atlantic Ocean from the Jurassic to present. J. Geol. **85**/5, 509–552, 21 Figs., Chicago
Seibold, E. (1974): see 1.4.1
Selley, R. C. (1970): see 1.4.1
Shepard, F. P. (1963): see 1.4.1
Sprecht, R. W., Brenner, R. L. (1979): Storm-wave genesis of bioclastic carbonates in Upper Jurassic epicontinental mudstones, East-Central Wyoming. J. Sed. Petrol. **49**/4, 1307–1322, 15 Figs., Tulsa
Stel, J. H. (1975): see 4.3.2
Stevens, C. H. (1969): Water depth control of fusulinid distribution. Lethaia **2**/2, 121–132, 13 Figs., Oslo
Stevens, C. H. (1971): Distribution and diversity of Pennsylvanian marine faunas relative to water depth and distance from shore. Lethaia **4,** 403–412, Oslo
Strauch, F. (1972): Zur Klimabindung mariner Organismen und ihre geologisch-paläontologische Bedeutung. N. Jb. Geol. Paläont. **140**/1, 82–127, 7 Figs., 9 Tab., Stuttgart
Tait, R. V. (1971): see 1.4.1
Terlecky, P. M. (1974): The origin of a Late Pleistocene and Holocene marl deposit. J. Sed. Petrol. **44,** 456–465, Tulsa
Thiede, J. (1973): Sedimentation Rates of Planctonic and Benthonic Foraminifera in Sediments from the Atlantic Continental Margin of Portugal and Morocco. Geol. Soc. Amer. Bull. **84,** 2749–2754, 3 Figs., Boulder
Vail, P. R., Mirchum, R. M., Todd, R. G., Widmier, J. M., Thompson III S., Sangrze, J. B., Hatllzlid, W. G. (1977): Seismic stratigraphy and global changes of sea level. Amer. Ass. Petrol. Geol. Mem. **26,** 49–212, Tulsa
Veevers, J. J. (1977): Paleobathymetry of the crest of spreading ridges related to the age of the ocean basins. Earth Planetary Sci. Letters **34,** 100–106
Veizer, J. (1977): see 9.1.5
Voigt, E. (1956): Der Nachweis des Phytals durch Epizoen als Kriterium der Tiefe vorzeitlicher Meere. Geol. Rdsch. **45**/1, 97–119, Pls. 1–4, 5 Figs., Stuttgart
Walther, J. (1893): see 10.2.5.4
Wanless, H. R., Burton, E. A., Dravis, J. (1981): Hydrodynamics of carbonate fecal pellets. J. Sed. Petrol. **51**/1, 27–36, 8 Figs., Tulsa
Wanless, H. R., Ziebell, W. G., Ziemba, E. A., Carozzi, A. V. (1957): Limestone texture as a key to interpreting depth of deposition. Congr. Geol. Intern. 20th Sess., Mexico, Sess. **5**/1, 65–82, 9 Figs., Mexico
Weaver, C. E. (1967): Potassium, illite and the ocean. Geochim. Cosmochim. Acta **31,** 2182–2196, Oxford
Weggel, J. R. (1972): An introduction to oceanic water motions and their relation to sediment transport. In: Swift, D. J. P., Duane, D. B., Pilkey, O. H. (eds.): Shelf Sediment Transport, 1–20, Stroudsburg: Dowden, Hurchinson, Ross
Ziehr, H., Matzke, K., Ott, G., Vouttsidis, V. (1980): Ein stratiformes Fluoritvorkommen im Zechsteindolomit bei Eschwege und Sontra in Hessen. Geol. Rundschau, **69**/2, 325–348, 11 Figs., 3 Tab., Stuttgart

10.2 Facies Criteria

When the environmental factors outlined in Sect. 10.1 are taken into consideration along with microfacies criteria and paleontological as well as sedimentological criteria, indications of the origin and diagenesis of limestones are given. It is obvious, though, that paleontological data which have received only cursory treatment here are, in many cases, extremely significant. Without an interpretation of the paleontological relationships, facies studies remain incomplete; valuable impulses for this aspect of facies analyses have been provided by the Tübingen Research Report (Seilacher, 1976).

10.2.1 Non-Marine Environments

"... lacustrine carbonate cannot be distinguished from marine carbonate on the basis of petrographic relationships" (Picard and High, 1972). This pessimistic attitude is quite understandable when MF criteria of fresh water limestones are compared with those of marine shallow-water limestones:

Micrite and sparite as well as all the other particle categories are found in both types of limestones. In addition, there are many similarities in the lamination of bedding fabrics and the fenestral fabrics.

The *identification of lacustrine carbonates* must, therefore, be supported primarily by paleontological evidence. To this can be added geochemical criteria (evaporites, authigenic minerals, clay minerals, trace elements such as boron; see Ernst, 1970). The invertebrates used to categorize lacustrine conditions are (very rare) freshwater sponges, gastropods, pelecypods, and various arthropods, especially ostracods. Lacustrine faunas are usually of low diversity.

Of the algae, only Characean algae prove to be useful indicators. However, it must be noted that in Devonian limestones, for example, Characean oogonia also occur together with marine fauna (washed in from the coast or indication of marginal marine conditions?). Other fresh-water algae which may be of value in the facies interpretation of lacustrine limestones belong to chlorophycean and cyanophycean genera, as well as to *Microcodium* Glück. In Ancient lacustrine limestones (except in stromatolite structures) green and blue-green algae are rarely preserved. Examples are known from the Lower Permian of the Southern Alps (E. Flügel and E. Flügel-Kahler, 1980), from Tertiary and Quaternary limestones in Afghanistan (Lang and Segonzac, 1979), and from the German Tertiary (Reiss, 1923).

Facies criteria which might possibly be used are carbonate cements (Riding, 1978, freshwater limestone in the Ries, Germany).

Other possibilities which may be developed appear to be the more precise investigation of blue-green algae and primitive green algae, which are very important for building lacustrine carbonates.

Considerable progress in the study of lacustrine carbonates is represented by the discussion of the Altenglan limestones (Lower Rotliegend Saarpfalz, Germany) by Stapf (1970). To characterize the microfacies spectrum of this "predominantly biochemically formed limestone gyttja in a eutrophic lake," Stapf introduced the following MF types:

- Detrital limestone made up of carbonate clasts, non-carbonate clasts and terrigenous minerals; matrix predominantly micritic
- Micritic limestone with relatively few carbonate and non-carbonate clasts as well as fine-bedding and lamination
- Dolosparite limestones with stratified concentrations of carbonate clasts and dolosparitic matrix
- Pellet limestone with detrital quartz, mica, and many peloids
- Oncoid limestones with numerous oncoids, peloids, and algal nodules. The oncoid nuclei are usually quartz, ore minerals, feldspars, and mica; the laminae consist of ankerite
- Limestones made up of mm-sized algal nodules, detrital quartz, and stromatolitic crusts
- Stromatolites with several basic geometric forms
- Dolostones (dolomite to ankerite)

These MF types involve chiefly micritic limestones rich in intraclasts, with a great deal of non-carbonate material brought in from land, and in situ algal carbonates, of which stromatolites and oncoids are especially important.

Another important investigation of Ancient lacustrine limestones was done by Donovan (1975), who stressed the importance of lateral facies differentiation at the margins of lake basins (stromatolites, massive limestones with birdseye structures, local limestone breccias, carbonate-rich laminites).

Some help in the interpretation of Ancient carbonates found in temperate lake deposits is given by the facies model developed by Murphy and Wilkinson (1980) for Littlefield Lake in Central Michigan; see Fig. 63: A complete sequence is composed of peat at the top and five carbonate units. Ostracod micrite is deposited in cold, relatively stagnant portions of the lake basin as a result of particle settling in quiet waters of the hypolimnion. This sediment is derived from resuspension of fine-grained material in shallow water by surface waves, and possibly from physico-chemical precipitation of fine-grained calcite in the epilimnion. In a thick sequence this facies should comprise the bulk of the carbonate sediments. Gastropod micrite is deposited on the lower portions of the bench slope in the area below the photic zone by processes similar to those which caused the deposition of ostracod micrites. Laminations are well-preserved in both of these facies. Sandy algal micrite is deposited on the upper portion of the bench slope, an area overgrown with thick stands of Chara which become thickly encrusted with calcite. Partial disintegration of these encrustations produces fine sediment, most of which is deposited in situ. Oncoidal gravel is formed on the flat bench platform in very shallow water. This sediment, composed of whole blue-green algal oncoids and fragments of calcareous peat, deposited on the landward edge of the bench platform, grades into both underlying oncoidal gravel and overlying peat which is deposited just above the level of the lake.

Better information appears to be available for *terrestrial carbonates:*

Calcareous sinter and *tufa* in thin-sections often exhibit algal filaments arranged in zones which are even recognizable in relatively thick travertines (see Plate 2/3). The key characteristics of travertine crusts are the dense laminar layers of vague internal color banding alternating with fenestral layers with a tubular, clotted or a

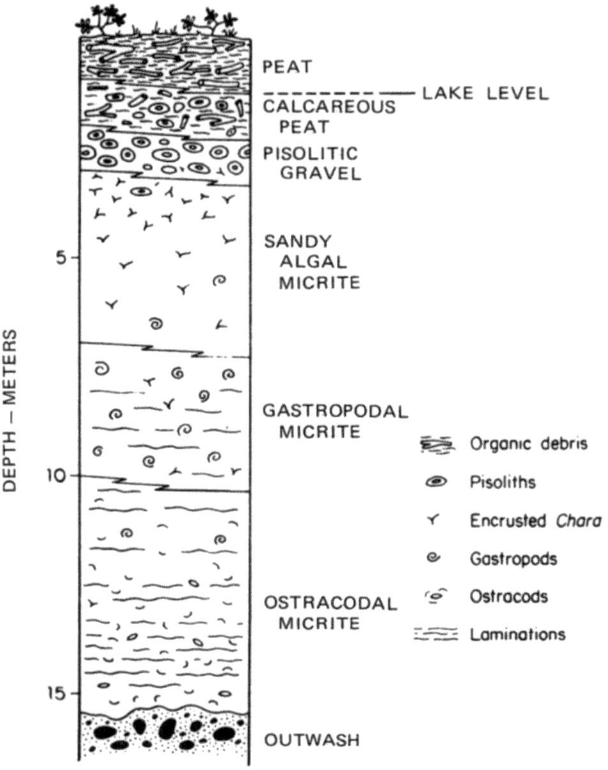

Fig. 63. Idealized stratigraphic column for lacustrine carbonate deposits in temperate regions (after Murphy and Wilkinson, 1980). The facies model includes six units: peat and five carbonate facies. Ostracodal micrite is deposited in the cold, relatively stagnant portions of the lake basin, gastropodal micrite on the lower portions of the bench slope in the area below the photic zone, and sandy algal micrite on the upper portion of the bench slope. Pisolitic gravel forms on the flat bench platform in very shallow water. Calcareous peat grades into both underlying pisolitic gravel and overlying peat which is deposited just above lake level. After Murphy and Wilkinson (1980)

vertical palisade fabric (Smoot, 1978), and with porous layers. *Caliche formations* are most easily recognizable, whereby the following criteria are important (see Sect. 2.3.2.2 and Plate 2/4–6):

a) Sparseness or absence of fossils
b) Usually micritic limestones
c) Commonly laminated crusts upon or surrounding fragments of caliche
d) Caliche-pisoids and caliche-peloids formed in situ
e) Partially brecciated lamellae structures filling in small depressions of a relief
f) Extremely fine, very regular lamellae in pisoids and "oncoids" themselves as well as in the cm-thick laminations

Many caliche crusts in thin-sections have a clotted texture with irregular spar-lined voids, spar-coated quartz grains, and disseminated quartz silt grains (Smoot, 1978).

Ancient pisolitic deposits may have been formed as caliche but also under a variety of other conditions ranging from freshwater streams and lakes to hypersaline sabkha environments. Features useful in distinguishing caliche from hypersaline pisolites are: regional geological setting, association with nodules and crust deposits, and the presence of calcite rhombs and/or bladed sparry calcite with triangular-shaped cross-sections (Chafetz and Butler, 1980). Hypersaline pisoids are characterized by a brickwork-like calcite texture and relict aragonite rays with square terminations.

The microfacies of Tertiary and Recent caliche crusts has been described by Rutte (1955), that of the Cornstone from the British New Red Sandstone by Steel (1974).

Paleospeleothemes can under certain circumstances be identified by special cement structures (see Sect. 2.3.2.3).

10.2.2 Marginal Marine and Shallow Marine Environments

Tables 48, 49, 50 summarize the criteria important for distinguishing supratidal and intertidal zones of marginal marine environments from shallow-marine environments. It then becomes quite apparent that supratidal and intertidal sedimentary conditions are relatively easily recognized in carbonate rocks. The carbonate tidal flat sequence is generally a limestone-evaporite sequence or a limestone-algal mat sequence with birdseyes, stromatolites, and intraclasts. Especially tidal sediments must be carefully studied because the time-lapse effect produced by the geologic record obscures the differences between diurnal and seasonal high tide/low tide fluctuations. Important papers on Recent and Ancient tidal sediments can be found in Ginsburg (1975), and Till (1978); a bibliography in Reineck (1973).

The interpretation of Ancient tidalites has been significantly advanced through the detailed and carefully illustrated description of the sedimentary structures, fabrics, and textures of the carbonate tidal flats of Northwest Andros Island, Bahamas (Hardie, 1977).

Interesting and informative articles, in regard to method, on Ancient marginal marine carbonates are those of Brown and Friedman (1969), Fischer (1964), Kahle and Floyd (1971), Laporte (1967), Matter (1967), Mazzulo and Friedman (1975), Müller-Jungbluth (1970), and Roehl (1967).

Wood and Wolfe (1969) as well as Lucia (1972) describe the evaporitic Sabkha development of supratidal areas. According to them, an ideal Sabkha cycle consists of three units:

Subtidal zone with sedimentation on shoals (algal boundstone and grainstone) and in lagoons (dolomite with birdseyes); intertidal zone (dolomite with algal mats); supratidal zone (nodular anhydrite).

One should exercise caution when interpreting the different ways in which evaporites may be formed (Friedman, 1978). Not all evaporites are formed in marine environments, such as sea-marginal sabkhas (characterized in arid climates by the extensive development of nodular anhydrite; e.g., Abu Dhabi, Butler, 1969; Kinsman, 1969), sea-marginal pools (e.g. Red Sea, Friedman and Sanders, 1978), or in semi-closed basins with hot dry climates (e.g., Red Sea, Mediterranean); some evaporites are deposited from the waters of inland desert basins (playas; e.g., Salt Flat Graben, West Texas; Friedman, 1966) or from an evaporitic concentration within the interstitial waters of sediment bodies located along the margins of the sea (continental sabkhas, e.g., Arava Valley between the Dead Sea and the Red Sea, Amiel and Friedman, 1971).

Excellent descriptions of Ancient sabkhas, mainly a carbonate-sulfate evaporite facies which merges laterally into red-beds, are given by West (1964, 1965; Lower

Table 50. Facies criteria for marginal marine and shallow marine environments

	Marginal Marine		Shallow Marine	
	Supratidal	Intertidal	Shallow Subtidal	Deep Subtidal
Limestones	Common	Common	Common	Common
Early diagenetic Dolomite	Present	Present	–	–
Evaporites	Present	Present	–	–
Micrite	Common	Present	Present	Present
Sparite	–	Rare	Present	Present
Vadose Silt	Common	Present	–	–
Particle : Micrite Ratio	1 : 9 to 1 : 1	1 : 9 to 1 : 1	1 : 9 to 9 : 1	1 : 9 to 9 : 1
Peloids	Rare to present	Common	Common	Present
Aggregate Grains	–	Present	Common	–
Oncoids	–	Present	Present	Present
Normal Ooids	Rare	Common	Common	–
Radial Fibrous Ooids	Present	Present	–	–
Vadose Pisoids	Common	Rare	–	–
Cortoids	–	Common	Common	Rare
Intraclasts: mud pebbles	Common	Present	Rare	Rare
Extraclasts: black pebbles	Present	Present	Rare	–
Terrigenous Particles	Common	Rare to present	Rare	Rare
Birdseyes	Common	Common	Rare	–
Fenestral Fabrics	Common	Common	–	–
Lamination	Common	Present	Rare	–
Desiccation Cracks	Common	Present	–	–
Collaps Breccia	Present	–	–	–
Cross-bedding	Rare	Present	Rare	Rare
Storm Deposits	Common	Common	Rare	–
Erosional Structures	Rare	Common	Present	Rare
Tepee Structures	Present	Rare	–	–
Bituminous Films	Present	Present	–	–
Algal Mats	Common	Common	Present	–
Stromatolites	–	Common	Present	–
Plant Litter	Present	Present	Rare	Rare
Organism Diversity	Very low	Low	High or low	Mostly high
Organism Abundance	Low or very high	High or low	High to very high	High
Endobiontic Organisms	–	Present	Common	Common
Bioturbation	–	Common	Present	Present
Vertical Burrows	–	Common	Present	–
Horizontal Burrows	–	–	Common	Present

Purbeck of Dorset, Southern England) and Schenk (1969, Middle Carboniferous, Maritime Provinces, Canada).

"Tepee-structures" or "pseudo-anticlines" have been recorded mainly from subaerial regions (peritidal tepee structures, caliche pseudo-anticlines) as buckled margins of saucer-like megapolygons. The megapolygons are formed by expansion of the cemented surface layer by as much as 15% as a result of several chemical processes in the vadose zone (Assereto and Kendall, 1977). Submarine tepees are much rarer (e.g., Persian Gulf; Shinn, 1969; Middle Jurassic of Normandy, Fürsich and Palmer, 1979).

10.2.3 Reef Complexes (Tables 51, 52)

Reef complexes growing well into turbulent water areas can often be *subdivided into three zones* (see Fig. 76):

1. *Fore-reef:* the side facing the wind, reef flank beds against the open sea. Abundant coarse m- to cm-sized reef debris; strong cement formation with characteristic open-space structures (see Krebs, 1969; J. Schroeder, 1972, 1973; Zankl, 1971); micritic or calcarenitic internal sediments. Specialized communities of organisms (Goreau and Land, 1974). Sediments, biota and diagenesis of the steep seaward margin of the Belize Reefs near Honduras provide most interesting possibilities for comparisons with Ancient fore-reef regions (James and Ginsburg, 1979).

2. *Reef flat* (central reef): the area of active growth of sessile framework building organisms (e.g., corals). A wealth of organisms with great diversity characterizes this area; concentrations of encrusting organisms and bioclastic micrite.

Modern and Ancient reef-flats may exhibit a geomorphic system of alternate spurs (ridges) and grooves (depressions) oriented in a parallel pattern. Spur and groove systems may be explained as the result of organic growth and erosion by which skeletal sand is redeposited between the reefal bodies (Sneh and Friedman, 1980). Grooves are believed to be the result of the backsurge of waves (Cloud, 1959; Shinn, 1963) or solution features of exposed reef-flats (Purdy, 1974).

3. *Back-reef or lagoon:* usually an extensive area protected by the central reef, with changing, sometimes restricted water circulation. The sediments are characterized by arenitic carbonate sands and carbonate mud; accumulations of peloids and bioclasts (calcareous algae, foraminifera, mollusks, etc) are common. Patch reefs may occur within the lagoon. Algal mats and stromatolites are found in shallow tidal zones at a greater distance from the reef.

Very often the reef complex is divided up into single reefs and in these inter-reef areas fine-grained sediments are deposited, sometimes admixed with remains of organisms derived from the reef.

In Table 51 a few facies criteria have been listed in very abbreviated form. It must be emphasized that these data can only be applied to reefs where a balance exists between biogenic framework building and destruction by biological and physical erosion.

Plate 41. Facies Diagnostic Criteria – Devonian ▶

1. Foraminifera can be used for zoning platform and reef carbonates. Examples known from the oil-bearing carbonates of Western Canada, from the Ural and from Australia (see 5.3.1). Pictured is *Parathurammina dagmarae* (Suleimanov). Middle Devonian (Schwelm limestone, Givetian): Maria quarry, Letmathe, Sauerland, Germany. Scale is 0.10 mm

2. Algae. Porostromate blue-green algae occur in Devonian reef complexes in reef cores and in the transition to the fore-reef (chiefly as sediment builders and secondarily as framework builders) as well as in oncoids in protected lagoons. Examples known from the Western Australian and Canadian Devonian as well as from the Middle and Upper Devonian of Belgium (Wray, 1977; Tsien and Dricot, 1977). Pictured is *Rothpletzella devonica* (Maslov) of the Middle Devonian (Eifelian): Mechra ben Abou, Moroccan Meseta. Scale is 0.10 mm

3. Algae. Green algae are relatively rare in Devonian carbonates. Exceptions are a few Codiaceae, such as *Lancicula alta* (Maslov), which appear to be limited in rock forming abundance to the Lower Devonian (Kuznetsk basin, USSR; Australia; Carnic Alps, Austria). Pictured are numerous saucer-shaped segments. Lower Devonian: E Kardamila, NE Chios island, Greece. Scale is 2 mm

4. Problematic algae or foraminifera are important facies indicators in reef and reef margin carbonates. Pictured is *Renalcis turbitus* Wray [Cyanophycean algae (Rivulariaceae), chlorophycean algae (Chroococcalean colonies) or foraminifera (Parathuramminaceae?)], Devonian: Iberg, Harz mountains, Germany. Scale is 1 mm

5. Stromatoporoids, tabulate corals and rugose corals are the most important framework builders of Devonian reef structures, whereby especially for the stromatoporoids, associations occur which are characteristic for certain parts of the reef (see 5.3.5). Pictured are stromatoporoids (*Actinostroma – 1, Clathrocoilona – 2, Stachyodes – 3*), tabulate corals (*Striatopora – 4*), rugose solitary corals – *5*, and brachiopods – *6*. Middle Devonian limestone: Brilon, Sauerland, Germany. Scale is 10 mm

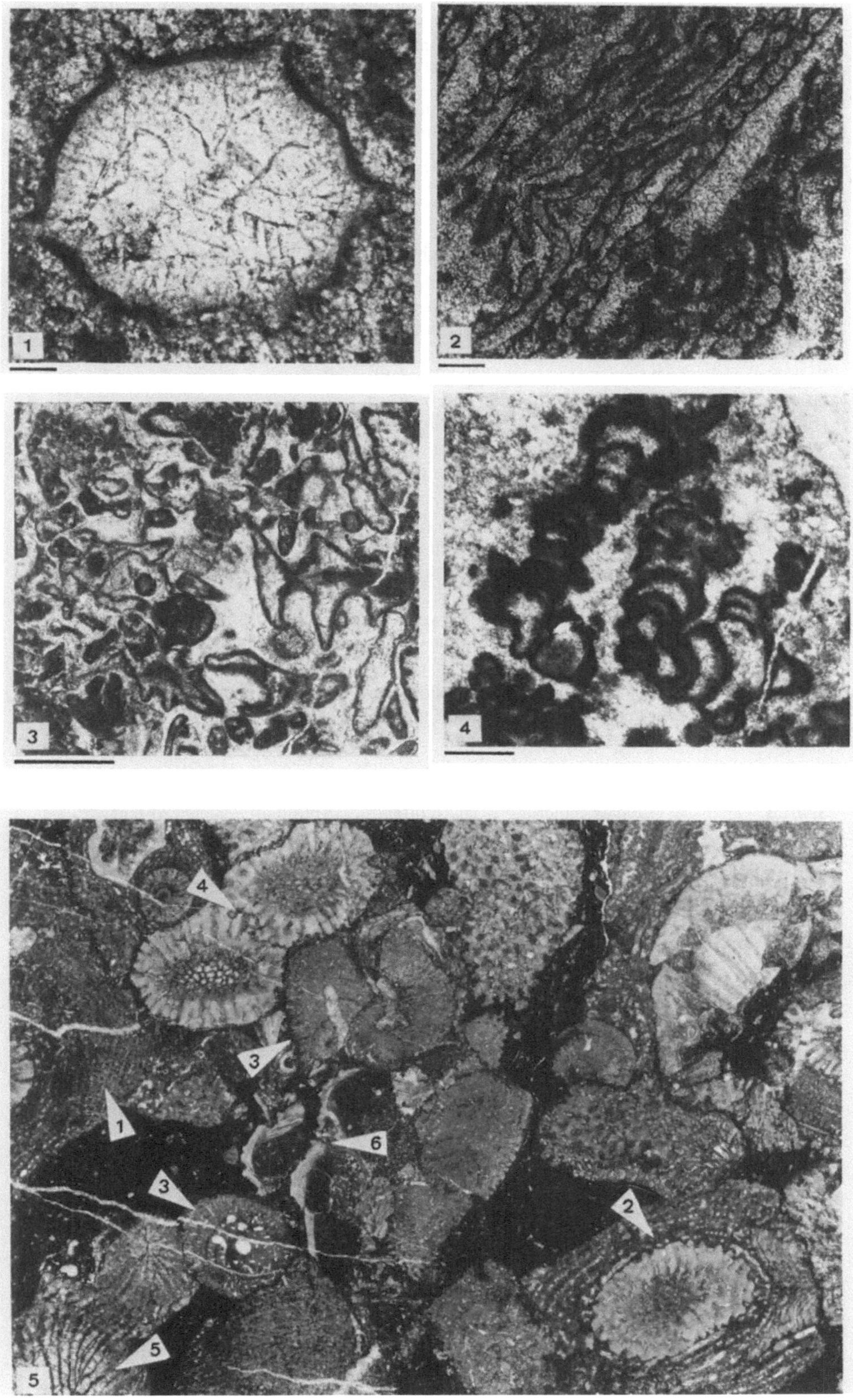
1
2
3
4
5
4
3
1
6
3
2
5

Table 51. Facies criteria for reef-complexes in high-energy environments

	Fore-reef	Reef Core	Back-reef Lagoon	Inter-reef
Organisms (Abundance)	Present	Abundant to very abundant	Present to abundant	Relatively rare
Ecological Types	Special associations, Ecological zoning	Primarily sessile benthic organisms, Ecological zoning	Special associations	Benthos, also nekton and plankton
Diversity	High	Very high	Low	Low
Preservation	Often redeposited skeletal grains from the reef core	Often in life position or parautochthonous	Often as detrital particles	Irregularly distributed
Biogenic encrustations around remains of organisms	Rare	Very abundant, on and between reef builders	Present	Generally absent
Boring organisms	Abundant	Very abundant	Rare	Generally absent
Organogenic Detritus	cm- to m-sized, often angular	mm- to cm-sized, between reef builders	mm- to cm-sized, often rounded, restricted to certain areas	Generally absent
Importance of the algae	Agglutination and framework building	Framework building	Sediment production and agglutination	–
Sediment (Limestone Types)	Floatstone, rudstone	Framestone, bafflestone, wackestone	Mudstone, wackestone, grainstone, bindstone	Mudstone, wackestone
Micrite	Present	Abundant	Present	Very abundant
Sparite	Often as cements in, sometimes large, intergranular pores	In small intergranular pores and in intragranular pores	Abundant	Absent
Peloids	Rare	Present	Often as fecal pellets in protected areas	Present
Aggregate Grains	Absent	Absent	Generally rare	Absent
Oncoids	Rare in deeper fore-reef area	Rare	Rare	Rare
Ooids	Absent	Absent	Rare	Absent
Cortoids	Absent	Present	Abundant	Absent
Resediments	Present	Abundant	Present	Rare
Particle Sorting	Poor	Generally poor	Moderate	Poor
Void Spaces, Birdseyes	Absent	Absent	Present	Absent
mm Lamination	Absent	Absent	Present	Absent
Bioturbation	Rare	Present	Present	Present
Sedimentary Structures	Indistinct bedding	No bedding or large-scale beds	Often bedding m- to cm-sized	Bedding common

Table 32. Processes and characteristics of facies in modern coral reef complexes (after Longman, 1981)

Facies	Process of Sedimentation	Types of Organisms Likely to be Preserved	Grain Size	Sorting	Amount of Framework	Typical Depth (m)	Dominant Rock Type
Lagoon	Low energy, much burrowing, sporadic currents and turbidity, possible terrigenous influx	Mollusks, echinids, miliolids, other foraminifera, ostracods	Mud mixed with coarse skeletal debris	Poor	0%	5 – 30	Wackestone
Back-reef-sand	Sporadic storms and currents across reef, saltation, gravity sliding	*Halimeda*, miliolids, minor red algae, sparse corals	Coarse	Moderate to good	0%	1 – 10	Grainstone
Reef-flat	Sporadic storms, good current circulation winnowing of mud	Corals, red and green algae, larger benthic foraminifera, massive corals	Coarse to very coarse	Moderate	0 – 10%	1 – 3	Grainstone with scattered corals
Reef-crest	High-wave energy, constant turbulence, good water circulation	Wave resistant corals and algae	Very coarse	Moderate to good	0 – 80%	0 – 2	Grainstone (minor Boundstone)
Reef-framework	Good water circulation high wave energy – sporadic at greater depths	Abundant corals, algae, mollusks, echinoderms, foraminifers	Framework and sand	Poor; mud in some cavities	20 – 80%	1 – 30	Boundstone
Reef-slope	Limited light, sporadic turbulence, gravity transport of reef debris	Soft corals, flattened coral plates, sponges	Mixed	Poor	5 – 40%	20 – 50	Packstone, Boundstone
Proximal talus	Sporadic turbulence, gravity transport, little light, unstable substrate	Few living organisms	Medium to coarse	Poor to good	0%	40 – 100	Grainstone, Packstone
Distal talus	Quiet water, no light, gravity sliding of sediments	Planktonic foraminifera	Fine	Moderate to good	0%	100 – 200	Packstone

Plate 42. Facies Diagnostic Criteria – Marine Permian ►

1. Cyanophyceae oncoids (*"Osagia"*) with Girvanella and sessile foraminifera. This microfacies in the Lower Permian of the Southern Alps is limited to protected shelf lagoons and turbulent environments on the margin of open shelf platforms (E. Flügel, 1977). Fusulinids and curved shell fragments are found loose with oncoids, but also serve as nuclei for oncoids. Limestone classification after Folk (modified): oncosparite; after Dunham oncoid-grainstone. SMF Type 13. Lower Permian (Grenzland formation): Rattendorf Sattel, Carnic Alps, Austria. Scale is 10 mm

2. Phylloid algae as sediment binding organisms contribute substantially to the formation of quiet water bioherms (mud mounds). The systematic position of these algae is uncertain (possibly Squamariaceae?); the high primary intraparticle porosity and secondary solution porosity are responsible for the importance of limestones with phylloid algae as oil reservoir rocks in the Southwest USA. In the Lower Permian of the Southern Alps these facies are preferentially found in the inner shelf region, sometimes also on the shelf margin. Pictured are phylloid algae encrusted with foraminifera and wall fragments of the problematic alga *Epimastopora.* Algal biomicrite after Folk, or algal lime wackestone after Dunham. SMF Type 7. Lower Permian: Forni Avoltri, Carnia, Italy. Scale is 5 mm

3. In addition to fusulinids and dasycladaceans Smaller foraminifera, particularly palaeotextularids, are important for the facies zoning of Late Paleozoic shelf carbonates. Pictured is *Climacammina* sp. Middle Permian: Eastern Iran. Scale is 1 mm

4. Algal associations with low diversity are abundant in protected as well as in open environments of the shelf region. Examples are fragments of *Epimastopora* Pia (irregular wavy plates with vertical pores – not dasycladaceans!). Lower Permian: Forni Avoltri, Carnia, Italy. Scale is 1 mm

5. *Tubiphytes obscurus* Maslov. In the Lower Permian of the Southern Alps the encrusting forms of this microproblematicum (see 5.3.13) form thick buildups corresponding to "stratigraphic reefs" in the shelf margin as well as on the boundary between the inner and outer shelf (see Sect. 11). Lower Permian: Matzenboden near Innichen, Sexten Dolomites, Southern Tyrol. Scale is 1 mm

6. *Bradyina* sp. Lower Permian (lower part of the Lower Pseudoschwagerina formation): Schulter, Carnic Alps, Austria. Scale is 1 mm

7. *Tuberitina* sp. These sessile foraminifera bind sediment grains. Lower Permian (Upper Pseudoschwagerina formation): Zottachkopf, Carnic Alps, Austria. Scale is 1 mm

8. *Tetrataxis* cf. *T. maxima* Schellwien. Lower Permian (Upper Pseudoschwagerina formation): Zottachkopf, Carnic Alps, Austria. Scale is 1 mm

9. Dasycladaceans. A dasycladacean association featuring a high species diversity characterizes the shallow outer shelf region with open circulation. This holds true for the Lower Permian of the Southern Alps as well as for the Middle Permian in the Dinaric Alps. Pictured are cross-sections of *Mizzia.* Middle Permian: Velebit, Croatia, Yugoslavia. Scale is 1 mm

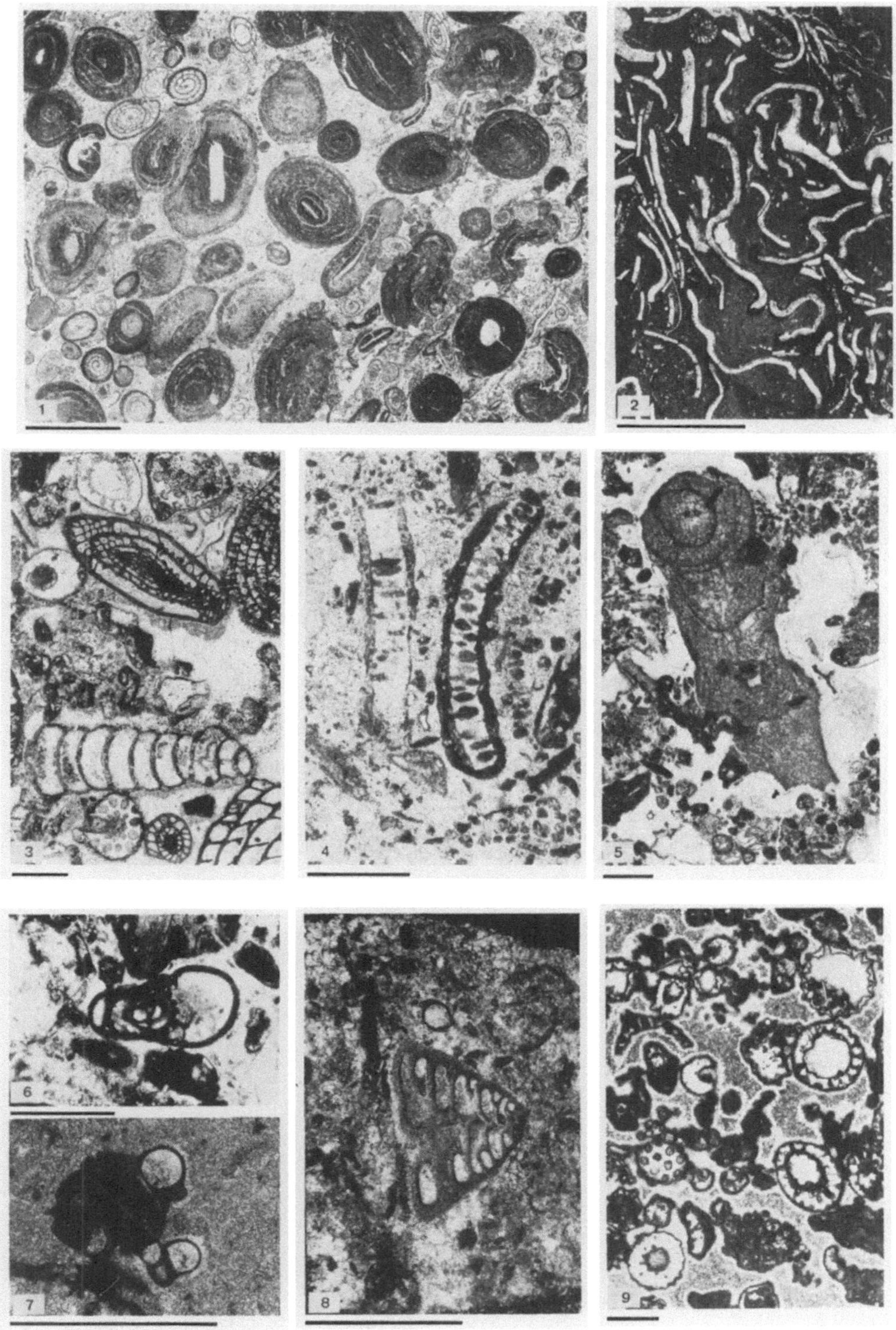

1
2
3
4
5
6
7
8
9

Plate 43. Facies Diagnostic Organisms in the Reef Facies of the Alpine Late Triassic ►

The distribution patterns of organisms in the Late Triassic reef complexes (Dachstein limestones and Upper Rhaetian limestones) of the Northern Alps aids in the differentiation of parts of ecological environments and the recognition of growth phases in the reefs. Plate 43 shows examples of more abundant reef building organisms. In Plate 18 examples of foraminifera and microproblematica are represented which are facies-controlled.

1. Segmented calcareous sponge (*Colospongia* sp.) with a biogenic overgrowth, characteristic of the reef core, consisting of bryozoans, sphinctozoan sponges and spongiostromate blue-green algal crusts. Upper Rhaetian reef limestone: Adnet, Salzburg, Austria

2. Sphinctozoan sponges (*Paradeningeria alpina* Senowbari-Daryan and Schäfer); micritic matrix. Upper Rhaetian reef limestone: Feichtenstein near Hintersee, Salzburg, Austria

3. *Follicatena irregularis* Senowbari-Daryan and Schäfer, one of the more abundant framework building sponges in the upper Rhaetian reefs. The irregularly formed chambers of the sponge also bind other reef organisms (bryozoans, hydrozoans). Biogenic overgrowth (sessile foraminifera) is common. Upper Rhaetian reef limestone: Gruber reef near Hintersee, Salzburg, Austria

4. Colonial corals (*Astrocoenia waltheri* Frech), cross-section. Echinoderm fragments (upper right). Upper Rhaetian reef limestone: Rötelwand near Hallein, Salzburg, Austria

5. *Thecosmilia clathrata* (Emmerich), a very abundant reef building colonial coral in the Norian and Rhaetian of the Northern Alps. Upper Rhaetian reef limestone: Feichtenstein near Hintersee, Salzburg, Austria

6. Colonial corals *Astraeomorpha confusa* (Winkler). Upper Rhaetian reef limestone: Gruber reef near Hintersee, Salzburg, Austria

7. Bryozoans belong to quantitatively important reef organisms in the Alpine Late Triassic, which, however, have been rather incompletely studied. The zoaria have been bored by pelecypods. Upper Rhaetian reef limestone: Feichtenstein near Hintersee, Salzburg, Austria

8. *Montlivaultia norica* Frech, a solitary coral common in limestone facies as well as in marl facies. Upper Rhaetian reef limestone: Gruber reef near Hintersee, Salzburg, Austria

9. Non-segmented calcareous sponges (*Peronidella* sp.). Upper Rhaetian reef limestone: Gruber reef near Hintersee, Salzburg, Austria

Scale for all Figures is 5 mm

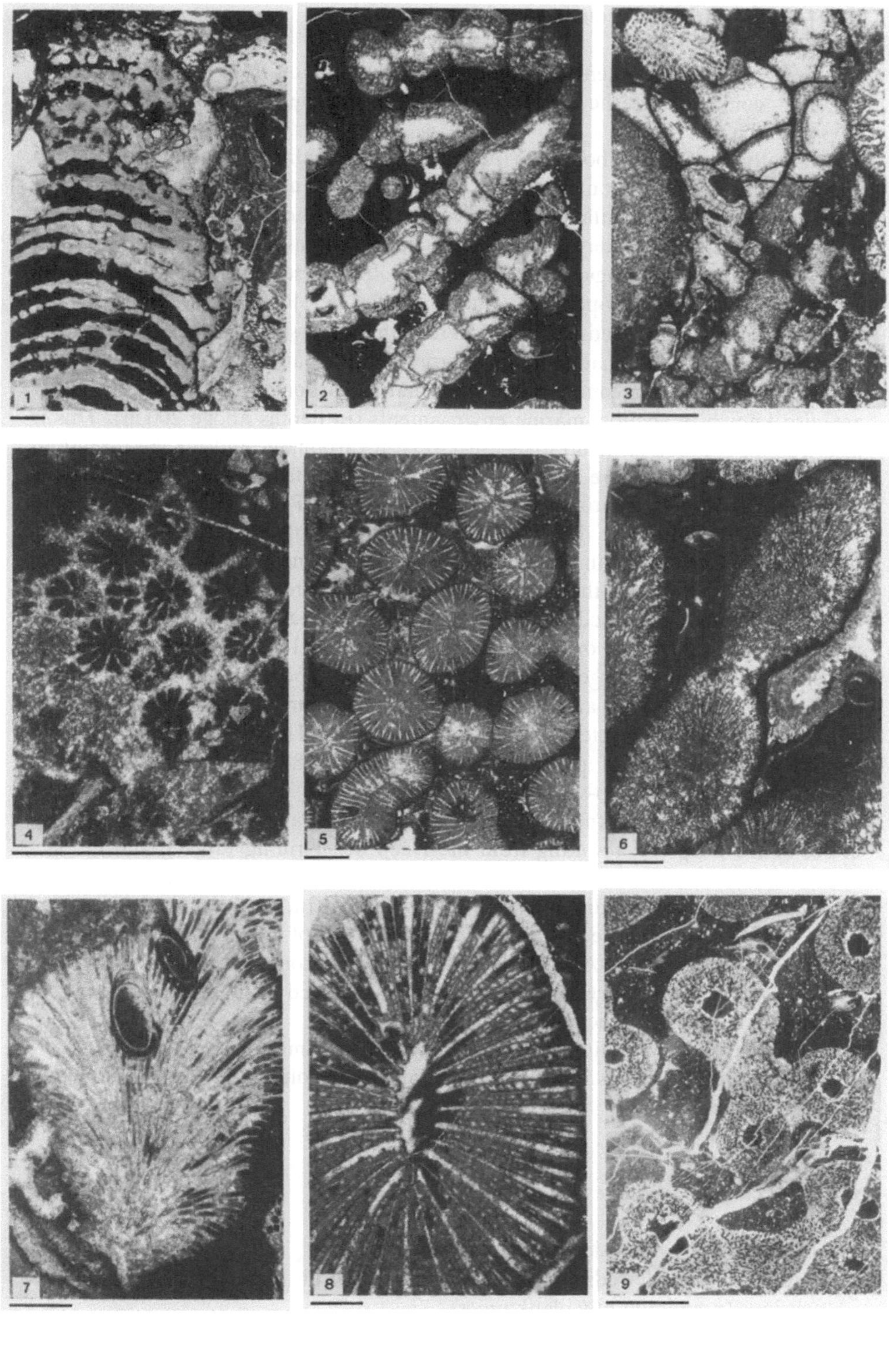

10.2.4 Deep-Marine Environments

"Deep marine" environments – without sharp delimitation from the lower subtidal regions – coincide with the deeper shelf margin, the shelf slope, and the bathyal, abyssal and hadal zones.

Basically, "pelagic carbonates" are defined by: a) sedimentation at moderate to great depths, b) deposition below the wave base and usually below the zone of water mixing, c) sedimentation of extremely fine-grained material from suspensions, and d) little or no terrigenous influx. In order to avoid semantic discussions, many authors use "pelagic" in a qualitative and descriptive sense, essentially synonymous with "of the open sea" (e.g., R. Riedl, 1963; Reading, 1978).

The facies criteria listed in Table 53 are being discussed at present in regard to their relevance to paleobathymetric interpretations. Plate 15 shows a few microfacies criteria occurring commonly in deep marine carbonates. Many facies criteria are also known in shallow-marine environments (in depths above 200 m). This is also true for solution structures (see Sect. 3.1) in carbonates and for the formation of hard bottoms, which are known in the deep sea and from shallow shelf areas as well as in subaerial environments (see Shinn, 1969; Rose, 1970; Goldring and Kazmierczak, 1964; Voigt, 1959; Bromley, 1975). The effects of carbonate dissolution at the CCD, ACD, and at the calcite lysoclines and the aragonite lysoclines may be better criteria for the paleobathymetrical interpretation of Ancient deeper-water limestones, provided that dissolution effects can be deduced from the preservation of aragonitic or calcitic shells (see Sect. 10.1.3). Applying the depth of Recent dissolution to the geological record is difficult, because of variations in the size of the Ancient oceans and greater fluctuations in climate (Berger and Winterer, 1974; Burns, 1974) connected with sea-floor spreading. The "backtracking method" for the determination of paleodepths (see Sect. 10.3.8) is quite useful for differentiating extensive deep-sea regions.

The ever increasing knowledge of depositional, diagenetic and tectonic processes in the deep sea, resulting from the Deep Sea Drilling Project, and the fact that deep-marine carbonates are becoming economically more interesting, as evidenced by oil and gas discoveries in the North Sea and in the United States (see Scholle, 1977 a, 1977 b) are also providing a great deal of information which is very important for interpreting Ancient deeper-marine limestones.

Two major sedimentary types must be differentiated with regard to different depositional and diagenetic criteria:

1. pelagic carbonates, including ooze, chalk, marl and limestones
2. allochthonous deposits, including carbonate turbidites and carbonate bank-margin talus

10.2.4.1 Pelagic Carbonates

Pelagic carbonates are basically characterized by:

a) particle-by-particle sedimentation of fine-grained biogenous material from suspensions (predominantly planktonics),
b) deposition below wave base and generally below water mixing,

Continued p. 498

Table 53. Facies criteria for deep-marine carbonates

Criteria	Discussion	Important References
Microfacies		
1) Micrite dominant; usually finegrained, homogeneous. Planktonic microfossils common	Indication of low water energy	
2) Sparite as cement only in rarely occurring microbioclastic calcisiltites		
3) Peloids common in calcisiltite as well as in micrites and biomicrites	In calcisiltite as material transported from shallower areas; in micrites as fecal pellets of planktonic and benthonic organisms	
4) Algal oncoids and aggregate grains absent (or allochthonous). Grains with micrite envelopes (cortoids) may be present if the envelope is the result of the endolithic activity of fungi.	Indication of absence of light (Discussion in Section 5.3.12.3)	"Deep-Sea Oncoids" (Monty, 1973)
5) Intraclasts common; due to subsolution, weak bottom currents or slumping	Indication of solution on the ocean bottom and sedimentation on the toe of the slope	
6) Extraclasts absent or very rare 7) Sorting of particles usually moderate to poor 8) LF fenestral fabric and birdseyes absent; Stromatactis structures may occur in mud mounds	Indication of absent terrigenous influx or greater distance from the coastline, due to absence or sluggishness of strong bottom currents; indication of deeper depositional milieu (LF fabric typical for very shallow water!)	
9) Often 3 SMF Types (see 8.2): a) spiculite, b) microbioclastic calcisiltite, c) pelagic mudstone		Wilson (1969, 1975)
10) Mm-sized colored layering	Due to Fe-Mn enrichment on discontinuity planes	
11) Fe-Mn crusts mm to cm in size, discontinuity planes	Indication of break in sedimentation	See Section 4.2.3.7
12) Sessile micro-organisms (usually foraminifera in Mn-Fe crusts)	Recent comparisons indicate no special water depth or specific paleotopography; accumulates in depths to about 2000 m, known to 5000 m	Wendt (1974)
13) Dissolved, often parallel to bedding, truncated calcareous microfossils and macrofossils	Indication of aragonite or calcite solution on ocean bottom and paleobathymetric differentiation	W. Schlager (1974)
14) Accumulations of glauconite below the discontinuity planes	Not just an indication of great depths of water; glauconite originates in the Recent in depths between 10 and 2000 m, often between 30 and 700 m	Porrenga (1967)

Table 53. (continued)

Criteria	Discussion	Important References
15) Limonitic "cauliflower structures"	The value of these "stromatolitic" structures as depth indicators is controversial (see 4.2.3.7)	Seyfried (1978)
16) Absence of algal borings	Because differentiating algal and fungal borings is very difficult (see 4.1.3.7) and Cyanophyceae often occur at larger depths, this is not a good criterion	Friedman et al. (1971)
Paleontological Criteria		
1) Absence of algae, i.e., calcareous algae	Only useful as an indication of greater water depths when systematic units (e.g., dasyclads) considered and ecologically very tolerant Cyanophyceae not overrated (see 5.3.12.3)	Riding (1974)
2) More planktonic than benthonic organisms	Particularly valuable for foraminifera as depth indicators; caution should be exercised because the proportion is controlled by too many factors. Species and faunal diversities are better indicators	Funnel (1967)
3) Comparison of depth range of Recent and Ancient organisms	Use with care and many restrictions (changes in the course of evolution and ecogenesis, etc.). Exceptions for a few groups limited to "shallow water" (e.g., green algae, reef corals, calcareous sponges)	J. Walther (1893) Heckel (1972)
4) Evaluation of Ancient distributional patterns	Example: Ammonite distribution and paleobathymetry in Upper Jurassic (B. Ziegler); Modifications dependent on ecological and preservation factors must be considered	B. Ziegler (1967) Garrison and Fischer (1969) Geyer (1971)
5) Trace fossil spectra	The depth dependent behavior pattern of marine invertebrates cause various trace spectra in shallow or deep water and in areas with rapid or slow sedimentation. The bathymetric evaluation is being hotly discussed at present	Seilacher (1967)
6) Condensed fauna	Mixture of faunal elements of various ages in thinly-bedded sedimentation units. Indication of lag sedimentation	Wendt (1974)

Sedimentological Criteria		
1) Low to no quartz content in "deep water" carbonates	Indication of offshore sedimentation and greater water depths	
2) Monotonous repetition of laminated to thin- and medium-bedded strata with plane surfaces		
3) Often internal mm lamination		
4) Hard bottoms on bedding surfaces	Indication of sedimentation lag and stillstand; today in depths between 200 and 3500 m	Fürsich and Wendt (1976)
5) Alternation of relatively uniformly thick limestone beds and thinner marls		
6) Sedimentary structures which indicate sporadic currents and submarine slumps	Flute marks, load casts, slide casts, sedimentary folds, interbedding of larger blocks, etc.	Examples in Garrison and Fischer (1969)
7) Resedimented limestone clasts, "sedimentary breccia and conglomerates"		Bernoulli and Jenkyns (1970)
8) Superposition of the normal sediments by turbidites which follow Bouma sequences	Indication of primary, deep (but not extremely deep) depositional environment	
9) Solution structures (corrosion surfaces parallel to the bed, subsolution fragments, etc.)	Indication of greater water depths in comparison to Recent conditions. $CaCO_3$ also controlled, however, by water temperature, rate of sedimentation, etc.	Hollmann (1964) Garrison and Fischer (1969) Berger (1974) Germann (1971)
10) Mn nodules and crusts	Recent predominantly in greater depths, but also known just below 200 m	
11) Abundant cherts	Not exclusive criterion for deep marine environments	
12) Dark and colored rocks, due to black iron monosulfide, and pyrite and/or marcasite		
13) Abundant associations of limestones, i.e., limestone/marl sequences, radiolarites, and submarine vulcanites		

c) sedimentation commonly at moderate to great depths,
d) little or no terrigenous influx.

Post-Jurassic and Recent pelagic carbonates are different from *Paleozoic pelagic limestones* with regard to the rock-forming nannofossils and planktonic foraminifera, which are first known from the Jurassic, possibly from the Triassic. Only problematic records exist for the Carboniferous (Gartner and Gentile, 1972) and the Permian (Pirini-Radrizzani, 1971). Incompletely calcified nannofossils may perhaps have contributed to the sediment without leaving complete skeletons. Other planktonic organisms seem to have been important during the Paleozoic (smooth and ribbed small tentaculitids, small planktonic gastropods, some ostracods); the main source of the fine-grained carbonate of many Paleozoic pelagic limestones must presumably be sought in finely comminuted macro-fossil debris, e.g. arthropods (Lindström, 1971). Many micritic limestones within Paleozoic deeper-water sections may represent lime muds, derived by basinward dispersal of lime muds produced on adjacent shelves (Wilson, 1969; Smith, 1977).

Water depth of pelagic carbonate accumulation in Post-Jurassic times range from less than 100 m to greater than 4500 m. Depositional depths of most Paleozoic deeper water carbonates were only a few hundred meters to about 1000 m, judging from paleontological data and the relief of the basins: Epeiric Cambro-Ordovician Seas of the Baltic Shield (Lindström, 1963, 1971; Jaanusson, 1955, 1960, 1971), Devonian-Carboniferous condensed cephalopod limestones from Central Europe (Tucker, 1973a, 1974; Bandel, 1974), Cambrian off-shelf basins of the Central Appalachians (Reinhardt, 1977), Mississippian unstable shelf carbonates in Central Montana (Smith, 1977), Mississippian basinal carbonates of New Mexico and West Texas (Yurewicz, 1977), Permian Delaware Basin (Newell, 1953). This may be due to the fact that most of the Paleozoic carbonates studied were deposited in continental margin environments and in rather shallow epeiric seas, but not near the spreading ridges of oceans (one exception is described by Landis (1974) and Coombs et al. (1976) from the Permian of New Zealand).

The Devonian basinal sedimentation of Central Europe well illustrates the characteristics of submarine rises ("Schwellen") and basins ("Becken"). Stable submarine rises are characterized by thin, partly condensed, pelagic limestones rich in goniatites and orthoceratids ("Cephalopod limestone"). These well-bedded, sometimes nodular, variously-colored micritic and microsparitic limestones are interbedded with thin marls. Contemporaneous nodular limestones interbedded with shales were deposited on the slopes of the rises, and clastic sediments in basins on deeper parts of shelf areas. The cephalopod limestones were deposited on drowned reefs, along the edge of the shelf margin, on volcanic seamounts, and on fault-controlled tectonic rises (see W. Krebs, 1979). These deeper submarine rises ("Tiefschwellen", Rabien, 1956) are interpreted as bathyal, perhaps in places also as deeper subtidal deposits (Tucker, 1973, 1974), according to sedimentary structures (bioturbation, hardgrounds, ferromanganese oxide crusts, cavity systems and neptunian dykes) and fauna (goniatites, orthoceratids, thin-shelled bivalves, brachiopods, trilobites, rare solitary corals, crinoids, styliolinids and tentaculitids, conodonts, and arenaceous foraminifera). The age of the cephalopod limestones is late Lower Devonian to Lowermost Carboniferous with a maximum during the Frasnian and Famennian.

More mobile depressions within the Devonian basinal facies are characterized by black shales (Hunsrück shale – Uppermost Siegenian and Lower Emsian; Wissenbach shale – Late Emsian, Eifelian, Givetian; Ostracod shale – Frasnian and Famennian) and silt- and sandstone intercalations. The shales are interpreted as deeper bathyal deposits. The fauna is characterized by nektonic, planktonic, and pseudoplanktonic organisms (graptolites; ostracods, tentaculitids, styliolinids, conodonts, thin-shelled bivalves, smooth brachiopods, trilobites, and fishes). The silt- and sandstones represent turbidites derived from the adjacent shelf areas; limestone turbidites were transported into the basins from reefs growing contemporaneously along the shelf margin, or from isolated atolls.

Paleozoic pelagic limestones are bioclastic wackestones and packstones or laminated lime mudstones. Several *microfacies types* are common:

A. Dark, well-bedded mudstones (sometimes wackestones) with mm laminations, fossils usually rare (e.g. sponge spicules, radiolaria, trilobites, conodonts). These limestones are often interbedded with allochthonous limestones (turbidites). Pyrite is common, sometimes abundant. Organic matter and argillaceous material may also be common. Limestones with these microfacies criteria were formed at the toe of slopes below the shelf margins of major carbonate platforms (e.g., Upper Cambrian and Lowermost Ordovician of Nevada; Cook and Taylor, 1977; Permian of West Texas and New Mexico, Tyrrell, 1969) or in restricted parts within the shelf region (e.g., Kellwasser limestone, Buggisch, 1972). A stagnant and reducing environment is probable for many of these limestones.
B. Red, gray or greenish bioclastic fossiliferous wackestones and packstones, usually strongly bioturbated. Discontinuity surfaces, hardgrounds, ferromanganese crusts, phosphatic nodules and crusts, calcite-filled cracks, and neptunian dykes are common. This type can be subdivided according to the dominating biotic components. Cambrian and Ordovician epeiric limestones contain trilobites, ostracods, gastropods, nautiloid cephalopods (e.g., "Orthoceras limestones of the Baltic Shield", Lindström, 1971); Devonian pelagic limestones deposited on "Schwellen" formed by volcanic piles of lava or drowned reefs are rich in cephalopod remains, thin-shelled bivalves and gastropods, tentaculites, ostracods, brachiopods, and radiolarians (Tucker, 1973; Bandel, 1974; Mendez-Bedia, 1976). Similar facies types were described by Zamarreno (1972) from the Cambrian of Northwestern Spain. Glauconite and chamosite may occur. Many limestones of this type show dissolution of particles.
C. Sparry packstones with styliolinids (Tucker and Kendall, 1973) and bioclastic wackestones and packstones with tentaculites (Buchroithner et al., 1980) as well as only a few other fossils are also formed on submarine swells.
D. Ostracod wackestones with abundant pelagic ostracods (Entomozoidae) seem to have been common in the Devonian deep-shelf environment of Central and Southwestern Europe (Buchroithner et al., 1980).

Somewhat similar *Microfacies Types* are known from the deeper-marine *Mesozoic limestones* of the Alps (see Wilson, 1975):

A. Dark, argillaceous and silty thin-bedded mudstones and wackestones with marly intercalations. Sponge spicules are common; pelagic crinoids (*Sac-*

cocoma) may be abundant in some layers. Examples are the Upper Jurassic Birmenstorf Beds and the Effing Marls of the Central Swiss Jura. Depositional area: slope between shelf margin and basin.

B. Dark micropelletoid micritic well-bedded limestones (pelmicrites or calcisiltitic pelsparites) with thin-bedded laminations, which sometimes may represent distal parts of turbidites. Example: Quinten limestone, Switzerland.

C. Red biomicrite with pelagic microfauna (*Globochaete,* stomiospheres, *Saccocoma*) and echinoderms, brachiopods, and molluskan detritus (Plate 53/1). Often developed as "nodular limestone". Usually micritic clasts in a micritic matrix; rarely micritic clasts bonded by cement (Adnet "Scheck", Plate 53/3). Conditions for the origin of nodular fabrics and formation depth are assumed to be solution at the ocean bottom (subsolution at about a depth of 4000 m; Hollmann, 1962; Garrison and Fischer, 1969), a combination of concretionary growth of Mg-calcite nodules and solution of aragonite even within the photic zone (Jenkyns, 1974), a combination of solution, compaction, and pressure solution (Zankl, 1969), or sedimentation of slump fragments. According to Bernoulli (1972) these limestones, which are widely distributed from the Lias to the Upper Jurassic (Adnet limestone, Ammonitico rosso), involve sediments formed in submarine rises at different water depths.

D. Conspicuously light-colored, residue-poor, rhythmically bedded micrite or biomicrite with pelagic microfossils (calpionellids, Globochaete, radiolarians), ammonite aptychi, and siliceous sponge spicules. SEM photomicrographs show abundant nannofossils (see Plate 8/4). Upper Jurassic and Lower Cretaceous (Oberalm formations and Aptychus beds of the Northern Alps; Majolica and Biancone limestones of the Southern Alps). Depths of several thousand meters can be assumed on the basis of the dissolution of aragonitic shells (H. W. Flügel and A. Fenninger, 1966; Garrison and Fischer, 1969).

E. Filamentous biomicrite (Plates 15/6; 16/1) with numerous very thin shells of pelagic mollusks as well as globochaetes, radiolarians, and ammonites. Because the existence of weak currents is indicated (Kuhry et al., 1976), deposition may possibly take place at depths of only a few hundred meters. Examples are the Middle Triassic Reifling limestones and the Late Triassic Hallstatt limestones of the Alps.

Chalks (Plate 8/3; 40/6) are fine-grained pelagic calcareous oozes and limestones chiefly consisting of nannofossils, especially coccoliths, and the tests of planktonic, and subordinate benthonic foraminifera. Nannofossil oozes are soft friable deposits. Since all nannofossil deposits undergo progressive lithification (Matter, 1974; Schlanger and Douglas, 1974; Hancock and Scholle, 1975; Scholle, 1977) as a function of the depth of burial and other factors, the distinction between "soft" and "hard" chalks (Hancock and Kennedy, 1967; Hancock, 1963) refers simply to diagenetic stages. Nannofossil limestones (or "nannoagorites" or "orthomicrites", see Sect. 4.1.2) were formed at the first occurrence of calcareous nannofossils, starting with the Upper Triassic and rapidly developing in Jurassic and Cretaceous times. Extensive summaries of the depositional environments of chalks have been presented by Hakansson et al. (1974), Hancock (1976) and Scholle (1977, 1979).

Most chalks are extremely fine-grained and composed of 30 to over 90 percent coccoliths, together with foraminifera and some other calcareous shells (bivalves, echinoderms, brachiopods). Many chalks are pure carbonate deposits with generally less than 5% HCl-insoluble residues. The amounts of terrigenous silt and clay may, however, increase to 60%, with averages of less than 20% (Lower Chalk of England; Hancock and Kennedy, 1967), depending on the geological situation, a volcanic source (Valeton, 1960), or the possibilities for the neoformation of clay minerals. Silica, as disseminated cristobalite lepispheres or as discrete chert nodules and beds, is common.

Since coccoliths have extremely broad environmental tolerances, ranging from tropical to boreal waters and from normal marine to extremely reduced salinities, chalks can be deposited in marine regions of different depths (subtidal shelf area, e.g., Maastrichtian chalk of Northwestern Europe; bathyal and abyssal chalks; brackish chalks in the Black Sea, and the Sea of Azov). Most chalk ooze accumulates at rates of 1 to 30 m/my (A. G. Fischer, 1969). Rhythmic bedding marked by alternations of clay-rich and clay-poor sediment or by zones with chert may reflect periodic variations in the input of the carbonate fraction (Arthur and Fischer, 1977).

The *diagenesis of chalks* is characterized by the overwhelming importance of late burial diagenesis as compared with early seafloor cementation (see Neugebauer, 1973, 1974; Wise, 1973; Matter, 1974; Schlanger and Douglas, 1974; Scholle, 1974, 1977; Hancock and Scholle, 1975; Kennedy and Garrison, 1975; Hancock, 1976). Early diagenesis may selectively remove non-calcitic organisms and cause the formation of hardgrounds by the precipitation of high-Mg calcite. Exposure to subaerial weathering or vadose and phreatic fresh waters does not have much influence on the diagenetic differentiation either (Scholle, 1977; Edmunds et al., 1973; Foster, 1975), because of the stability of low-Mg calcite. The overall pattern of chalk diagenesis is related to the maximum depth of burial and to pore-water geochemistry. A progressive lithification connected with an increasing loss of porosity (70%–80% at the sediment-water interface, to about 35% at a depth of 1 km, 15% at a depth of 2 km, and essentially 0% at 3 km) is caused by deeper subsurface pressure-solution at grain contacts, related to local reprecipitation, stylolitization, and mechanical compaction. The rate of porosity loss with respect to burial is accelerated if Mg-poor fluids are present in the pores during burial (Neugebauer, 1973, 1975). Destruction of porosity can be retarded by the presence of hydrocarbons within the pores and by overpressuring. Tectonic stresses enhance solution and cementation (Mimram, 1975).

Investigation of chalks is possible only if the ultrastructure, geochemistry and stable isotope fluctuations (Scholle and Arthur, 1980) have been considered. *Microfacies types of chalks* may contribute additional data (Schneidermann, 1970, 1973; Wolfe, 1968; Garrison and Kennedy, 1977; Mimram, 1978). Common textural types of lithified Cretaceous chalks are packed planktonic foraminiferal biomicrites; packed calcisphaerulid biomicrites; and sparse biomicrites with planktonic and rare benthonic foraminifera, calcitized radiolaria, some sponge spicules, mollusk fragments (e.g., *Inoceramus*), few brachiopods and bryozoans, and echinid spines. Scattered micropeloids may be common. The micritic matrix consists of a mass of tiny anhedral calcite grains. Bioturbation may be wide-spread, especially in chalks deposited in shelf environments (Wolfe, 1968).

These microfacies types are found in chalks formed at depths of a few tens of meters to perhaps several hundreds of meters (e.g. chalk of northwestern Europe; chalks of the Western Interior, North America) as well as in chalks deposited at bathyal and abyssal depths. Chalks containing many hardgrounds may represent a shallow depositional area as opposed to chalks without discontinuity surfaces (e.g. the North Sea). This may be inferred from the fact that hardground formation coincides with periods of regression (Kennedy and Garrison, 1975; Jenkyns, 1978).

10.2.4.2 Allochthonous Carbonate Deposits in Deeper-Marine Environments

Mass-gravity transport is the predominant process of resedimentation in deep-sea environments, including all processes of downslope movement of siliceous or carbonate sediment masses under the direct action of gravity. Both descriptive and genetic classifications of deep-sea mass transported sediments have been proposed. Descriptive classifications are based on sorting and size distributions of the clastics, presence or absence of stratification and cross-stratification, types of grading, as well as the sedimentary fabric. Genetic classifications describe the transport or grain support mechanisms from which the characteristics of the sedimentary deposits can be inferred (e.g., Dott, 1963; R. E. Elliott, 1965; Cook et al., 1972; Carter, 1975; Lowe, 1976a, 1976b, 1979; Middleton and Hampton, 1976; Moore, 1977). From the review of the major genetic classifications given by Nardin et al. (1979), it is evident that mass transport processes can be classified according to a combination of mechanical behavior (elastic, plastic, or viscous fluid motion) and transport and sediment support mechanisms.

The various processes of subaqueous mass transport can be classified in order of the increasing internal disaggregation of the moving sediment mass (Dott, 1963; Middleton and Hampton, 1976):

1. *Rockfall,* in which lithified, often large rock fragments move by free fall
2. *Sliding* and *slumping,* in which a sediment mass, usually semi-consolidated, moves along a basal shear surface while retaining some internal structures (e.g. bedding)
3. *Sediment gravity flow* (synonyms are mass flow, sediment flow, inertia flow, high concentration dispersion), a general term for a flow of mixtures of sediment and fluid in which bedding coherence is destroyed and the individual grains move in a fluid medium. Four types of sediment gravity flow can be distinguished, based on the mechanisms of grain-support:

   a) *debris flow* – clasts supported by matrix
   b) *grain flow* – caused by grain-to-grain interactions
   c) *fluidized sediment flow* – caused by escaping pore fluids
   d) *turbidity flow* – caused by fluid turbulence

   These processes often occur together during a single mass gravity event.

These modes of mass-gravity transport were first described in siliceous sediments, but they also occur during the allochthonous deposition of calcareous sediments on slopes or on abyssal plains. More than in siliceous allochthonous sediments, the textures and structures of mass-gravity transported carbonates tend to record the final stages of the flow rather than the main phases of sediment transport. Many Ancient calcareous deeper-water conglomerates and breccias do not easily fit into the conceptual categories proposed for mass-gravity transported sediments. This is probably due in some degree to the joint re-deposition of lithified and unlithified shallow-water carbonates as well as isolated skeletal particles. Depending on the site of the final deposition, relatively coarse allochthonous sediments (e.g., rockfall deposits, debris flow deposits), deposited on the slope, may be differentiated from rather fine-grained allochthonous sediments, deposited from turbidity flows in a predominantly mud environment on the basin floor. As shown by Crevello and Schlager (1981), the boundary between slope and basin or coarse and fine sedimentation can be better placed at the upper end of the basin-margin rise rather than within the basin itself, particularly if one notes the situation in the Bahamian intra-platform basin.

Transported and redeposited carbonates have a variety of compositions and depositional patterns based largely on the source areas, which may be shallow carbonate platforms or upper parts of the slope:

1. *Submarine rockfall deposits* are characterized by:
   a) depositional units usually showing distinct boundaries
   b) angular to subangular sedimentary clasts of uniform or polymict composition
   c) often very dense packing of clasts
   d) poor to very poor sorting without grading
   e) sand-sized to boulder-sized blocks (up to 50 m in diameter)
   f) sedimentation within a lithologically different matrix (e.g. in marly beds)
   g) grain-support fabric at the base of steep reef rims or oversteepened coasts; otherwise mud support

Rockfall deposits are caused by free fall or rolling of individual clasts which are transported across steep angles over short distances only. The clasts may include lithified carbonates exhibiting different microfacies types (and different source areas) as well as fragmented, redeposited shallow-water fossils like corals or rudists. These clasts are derived from bank margins or from reefs (e.g., transported reef talus blocks; Dott, 1958), from seamounts (e.g., transported red limestones; Seyfried, 1980), or they may be mixed with some exotic material (brought in by rivers from the hinterland; Thiedig, 1975). The resedimented material may be embedded within flysch-like sediments, marls or rather thin limestone beds. Many coarse-grained fore-reef calcirudites may have been formed by rockfalls.

2. *Subaqueous sliding and slumping* of semi-lithified carbonates may occur in deeper water as well as shallow-water environments (e.g. tidal flats; Schwarz, 1975). Common features of slumping structures are as follows:
   a) deformed beds occur as a zone between undisturbed beds
   b) the upper contact of the deformed bed is welded (depositional fit between the irregularities of the upper surface and the base of the overlying bed)

c) the deformed bed may be overlain by a limestone unit consisting of coarse-grained bank margin debris or of graded carbonate particles
d) folded anticlines may be eroded at the upper surface of the slumping unit
e) the preferred orientation of fold axes may be unrelated to the tectonic strike
f) the structural style within a single slump varies widely; a broad spectrum of deformational structures may occur (simple sliding of intact sets of layers, intensive shearing of dislocated beds, convolution and distortion of beds, rolled-up structures
g) thin marly partings may separate thin carbonate slumping layers
h) many slumped limestone beds exhibit a pronounced thickening and thinning of the strata

Most of the limestones which show pronounced slumping structures are micrites. No differences can be detected between the microfacies in the slumping beds and in the underlying beds, but differences may occur in the grain sizes of the non-carbonate material of the slumped limestone beds (Backhaus and Flügel, 1971): Intertidal Middle Triassic limestones, containing beds with slumping structures, were investigated with regard to their microfacies (which are more or less similar in slumped and non-slumped beds) and with regard to the amount, mineralogy, and grain sizes of the insoluble residues. No differences were found in the amount of the residues or the mineralogical composition (predominantly quartz, mica, illite as well as some kaolinite and feldspar). The only difference, which may have been responsible for the mobilization of the lime mud, is the higher percentage of clay-sized particles in the insoluble residues. This may be responsible for a retardation of the early diagenetic cementation of the lime mud and also for alterations in the thixotropic behavior of the sediment.

The *trigger mechanisms* for slumping or sliding of semi-lithified carbonate muds in marine shallow-water environments may be overloading during tidal emersion or undercutting within tidal creeks by erosive currents (Schwarz, 1975). Over-steepening of the slope by strong sediment accumulation may be an important mechanism in deeper-water environments (e.g., Bernoulli and Jenkyns, 1970). Many synsedimentary slump folds originate from the sedimentation of carbonate muds upon steep slopes of topographic highs; impressive examples have been described from the Jurassic Adnet limestone of the Northern Alps (Garrison and Fischer, 1969).

3. *Sedimentary gravity flows* or mass flows may produce disorganized or organized debris deposits such as debris flow or grain flow sediments, or turbidites. Most Ancient deep-water conglomerates and breccias do not fit easily into the category of classical turbidites (see Walker, 1975; Middleton and Hampton, 1976; Crevello and Schlager, 1981). Debris flow or grain flow seem to be more common transport mechanisms, especially for calcareous bank-margin talus deposits.

   Debris flows are characterized as follows
   a) matrix support of the clasts
   b) little or no consistently preferred fabric
   c) variable clast sizes; very poor sorting
   d) mixtures of both angular and rounded clasts possible

e) normal grading, generally poorly developed (coarse-tail grading)
f) crude inverse grading in the basal zone of the beds possible
g) beds generally massive and unstratified
h) unusually thick beds
i) irregular top
j) large clasts may project above the bed level

Submarine debris flow deposits are known from Ancient deep-sea clastics, e.g., from the Mesozoic and Tertiary flysch in the Apennines (Italy) or from the Paleozoic of Morocco (Padgett et al., 1977). The name olisthostrome has been applied to chaotic deposits emplaced by debris flow and related mass gravity processes, and composed of predominantly extra-formational material (Flores, 1955; Görler and Reutter, 1967; Görler, 1968). The individual clasts ("olistolithes") may start with silt sizes and range to enormous sizes of several 100 m in diameter. Many debris flow deposits represent mudstones containing only a few clasts; others form bouldery masses with just a little mudstone. The first type corresponds to "pebbly mudstones" (Crowell, 1957), interpreted as the result of a mixing of intraformational gravel and mud layers.

"Fluxoturbidite" is another term referring in part to a deposition from grain flows (Kuenen, 1958; Dzulynski et al., 1959; Stauffer, 1967; Stanley and Unrug, 1972). The term has been coined for thick-bedded and coarse-grained clastic beds which do not fit the model of a typical turbidite and which are believed to be the result of a combination of grain flow, turbidite currents and sliding.

Similar sediments, characterized by a chaotic fabric of plastically deformed marl lithoclasts separating irregular lenses of subangular limestone clasts and other lithoclasts, have been called "marl-flaser breccia" by Schlager and Schlager (1973). The authors consider the marl-flaser breccia as representing an interim stadium between grain flow and mud flow transport, with the marly material being absorbed by the moving sediment flow.

Generally debris flows correspond to a mass movement involving sand grade or finer material, which may provide a medium (matrix) for the movement of larger clasts (pebbles or boulders), or which may be the moved mass itself (Gascoyne, 1978). Debris flows are capable of traveling over very flat slopes, sometimes at angles of only one or two degrees.

Disorganized limestone conglomerates, interpreted as debris flow deposits, have been described by several authors (see Table 54). All the criteria noted above are found in these examples. Common features are:

a) fine-grained, partly argillaceous matrix
b) randomly distributed, but often densely packed clasts
c) monomict or polymict clast associations, derived from shallow-water environments as well as from deeper, upper slope areas
d) clasts of widely varying sizes (arenite to boulder size)
e) very poor to poor sorting
f) angular, subrounded and rounded clasts
g) internal structures within the beds rare or absent
h) grading rare or absent
i) inverse grading in the basal zone of the unit

Table 54. Disorganized limestone conglomerates, interpreted as calcareous debris flow deposits

	Lithology	Geometry and Dimensions of Allochthonous Rock Units	Sedimentary Structures and Textures	"Matrix"	Clasts
Pleistocene: Exuma Sound, Bahamas (Crevello and Schlager, 1981)	Calcareous pebbly mud, muddy rubble, graded rubble, and bioclastic sands	Sheet, covers an area of over 6400 km^2 and a volume of about 3×10^{17} m^3. Rubbly and pebbly axial swales in the basin	Poor sorting; sequence from graded sand and rubble at the top to poorly sorted sandy rubble to muddy rubble at the base	Lime mud	Deeper-water material (chalk lithoclastic rubble, chalk lithoclastic pebble muds) and shallow-water material (*Halimeda*, ooids, peloids etc.), ratio about 1 : 1. Chalk clasts rounded. ∅ 1 – 64 mm
Cretaceous: Tamabra Limestone, Mexico (Enos, 1977)	Limestone breccias	Lens-like units, up to 100 m thick	Very poor sorting; grain to grain contacts partly due to pressure-solution; no bedding	Fine-grained carbonate; sometimes with pelagic microfossils	Shallow-water limestones and limestones with pelagic microfossils in different proportions. Maximum diameter ∅ 1 – 10 cm, range usually < 1 mm to 20 cm. Angular clasts, no preferred orientation
Devonian: Alberta, Canada (Cook et al., 1972)	Limestone megabreccias and conglomerates (finer rudites)	Sheets, up to 20 m thick, lateral extension more than 100 km; large disoriented blocks (up to 50 m high)	Very poor sorting, no internal bedding or graded bedding	Lime mud containing sand-sized admixed shoal and basin fossils	Shallow-water "reef" limestones and upslope basinal limestones; lithified rock, partly lithified rock and particulate sediment particles
Cambrian and Ordovician: Nevada, USA (Cook and Taylor, 1977)	Limestone conglomerates	Sheets and channels, a few cm thick up to 10 m; upper contact irregular; lower contact channelized or parallel to the autochthonous beds	Poor sorting, no normal or reverse grading; clasts randomly oriented with respect to bedding	Lime mud	Only deeper-water lime mudstone and wackestone; maximum size 1×2 m

Cambrian: Southern Rocky Mountains, Canada (McIlreath, 1977)	Limestone conglomerates	Irregular, thin- to medium bedded intervals (2.5 cm – 10 m thick)	Very poor sorting, clasts randomly oriented	Lime mud, thin-bedded mudstones	*Epiphyton* boundstone clasts, derived from brecciation of the shallow-water platform, together with rare, submarine, sand sized to boulder-sized angular limestone clasts, sometimes grain-supported
Cambrian: New York and Vermont, USA (Keith and Friedman, 1977)	Limestone conglomerates, monomict or polymict (with calcareous sandstone)	Channel deposits or sheets; thickness of beds between 1,5 m to tens of meters	Poor; no internal bedding	Argillaceous lime mud; the clast/matrix ratio decreases upwards in some beds	Various limestone types, derived from shelf buildups and basin margin beds. Minimum size of clasts 1 cm, average around 15 cm, up to 30 cm; clasts are generally elongated, usually subrounded or rounded

The geometry of the carbonate debris flow deposits corresponds to extended lens-like sheets rather than to channel deposits. Channel fillings may occur within the depositional area, but this seems to be caused by a "channeling" of sheet flows, which became abutted in a broad front at platform margins (see Crevello and Schlager, 1981). This may be especially true for smaller, interplatform carbonate basins. Debris flows may represent a prephase of turbidite sedimentation (see Cossey and Ehrlich, 1979; Steiger, 1981).

Grain flows are mass flows in which sand, pebbles or boulders are supported within the flow mainly by the strength of a fluid matrix consisting of clay minerals and water. The following textures and structures may be useful for recognizing grain flow deposits (see Stauffer, 1967; Lowe, 1976; Middleton, 1969; Middleton and Hampton, 1972):

a) deposit draped over irregularities and thinning up the flanks of channels
b) thick massive beds
c) flat top; large clasts do not protrude from top of bed
d) partial to complete clast support
e) fabric relatively random
f) grain orientation parallel to flow; faint, dish-shaped laminae, oriented parallel to bedding may occur
g) absence of lamination and cross-lamination
h) clasts, floating in a sandy or muddy matrix
i) poor sorting
j) sometimes reverse grading near the base, especially in coarse-grained grain flow deposits
k) rare grading, sometimes coarse-tail grading
l) generally flat base
m) scours and injection structures at the base
n) sometimes flute marks on the sole

Identification of grain flow deposits is difficult because most of the features proposed to be characteristic of grain flows are also produced by other transport processes (Bourgeois, 1977). In the absence of access pore pressures, relatively steep slopes are required for true grain flows (Lowe, 1976). Very often grain flows are combined with turbidity currents or liquified sediment flow (see Link, 1975). Grain flow appears to be a transition between viscous and plastico-viscous flow.

Recent and Pleistocene carbonate grain flow deposits are known from deeper-water environments in the Northern Bahamas (Mullins and Neumann, 1979) at the windward margin of a steep slope. Thick massive to inversely graded bioclastic grain flows have been found together with large talus blocks at the base of the slope; the grain flow deposits rapidly grade seaward into pelagic and hemipelagic muds.

Calcareous debris flows and grain flows are predominantly deposited on the deeper-marine slopes or in nearslope abyssal plains. In contrast to this, *limestone turbidites* may be deposited in proximal as well as in distal positions depending on the source areas, which are generally shallow shelf areas or upper slope regions.

Common features of *turbidites* are (see Reading, 1978):

a) occurrence within a succession made up of a marked alternation of fine sediments (shales, marls, mudstones, siltstones) and coarse sediments (sandstone, detrital limestones)
b) considerable lateral extension of the beds
c) the beds are often moderately or poorly sorted and contain a notable amount of clay-sized material
d) bottom surfaces sharply defined, frequently with sole marks (e.g., scour and tool marks; grazing patterns and burrows)
e) top surfaces usually indistinct, with lithological transitions into the overlying beds
f) graded bedding common
g) occurrence of regular sequences of sedimentary structures (e.g., Bouma sequence) within the turbidite
h) autochthonous fossils rare, apart from trace fossils. Pelagic and deep-water benthonic microfossils may be found in the intervening shales. Allochthonous redeposited fossils, derived from shallow-water areas, and displaced pelagic fossils are common in carbonate turbidites
i) the associated beds give no evidence of shallow-water sedimentation, but often correspond to deeper-marine pelagic deposition

Turbidite successions vary greatly in the patterns of sedimentary structures, thickness of beds, and textural features. Variations are controlled by the distance between the source of the material and the depositional area, the paleoslope, the point or line source, the amount and composition, the sediment available, the topography of the depositional area (fans, channels, etc.), and the density of turbidity currents (high density suspensions or low density suspensions).

An "ideal" turbidite sequence, characterized by a fixed vertical succession of internal sedimentary structures, is called the "Bouma sequence" (Bouma, 1962). Five intervals (or divisions; Walker, 1965) can be recognized from the bottom to the top: Graded or massive Division (A) – no primary sedimentary structures other than graded bedding (sometimes reverse grading in the first few centimeters of a very thick layer, otherwise normal grading); Lower Division of Parallel Lamination (B) – thin parallel laminae; Division of Current Ripple Lamination (C) – with several sets of unidirectional foreset bedding, complete current ripples are rare; Upper Division of Parallel Lamination (D) – very fine, often obscure laminae within the pelitic bed of the turbidite; Pelitic Division (E) – very fine-grained, pelitic sediment without primary sedimentary structures.

The vertical sequence of these sedimentary structures is controlled by decreasing current intensity during the deposition of the turbidite bed.

The complete Bouma sequence can be observed only in a small percentage of all turbidites; the sequence is generally less complete in a down-current direction. The only important modifications of the Bouma model have been proposed for carbonate rocks: Meischner (1964) added a "prephase" to the base of the main graded bed (see Fig. 64). A "dune-phase" was discovered in Upper Ordovician turbidites of Scotland by J. F. Hubert (1966).

Limestone turbidites have been described from various Recent deep-sea locations (see Sect. 2.2.4). Many Ancient deep-water basins are typically filled with a sequence of lime mud and graded allochthonous carbonate deposits, which can be interpreted as turbidites according to the following criteria:

- occurrence of limestone bed(s) within a succession of fine-grained clastic or calcareous sediments
- internal size grading common
- the unit may often be marked by various types of sedimentary sequences
- thin planar laminae common
- wavy and small-scale cross-laminations (current ripple lamination) may occur
- bottom surfaces sharp, but sometimes with sole marks (flute, groove and erosional marks)
- gradual transition of top surfaces into overlying beds
- bed thickness decreases relative to the decrease of the grain sizes
- sorting poor in coarse parts of the unit, moderate to good in laminated parts
- particles may include shallow-marine and slope benthos, reef organisms, and some pelagic fossils, together with lithified carbonate fragments (lithoclasts) or noncarbonate rock fragments

Important differences between limestone turbidites and silicate turbidites are the rarity of complete Bouma sequences (see Thomson and Thomasson, 1969; Engel, 1974), the development of special vertical zonation patterns, the rarity or absence of scour marks, indistinct and less common lamination and convolute lamination (see Remane, 1970; Eder, 1970).

Some of these differences may be caused by a relatively weak thixotropy in calcareous muds as compared to clay muds, and by the joint sedimentation of discrete particles (e.g., foraminifera, ooids, peloids) and lithified carbonates. The source material of limestone turbidites often contains a wide grain-size range, causing a normal grading of the homogeneously distributed particles and rock fragments. Siliceous source material generally shows good sorting due to a supression of some grain-size classes during the transport from the erosional area to the shelf margins. The range of grain sizes, therefore, is rather small; decreasing intensity of flow will create a local alternation between sedimentation and erosion, thus forming internal sedimentary structures.

The source areas of limestone turbidites are

1. shallow-water environments (e.g., carbonate platforms, banks, reefs, atolls), or
2. upper slope environments.

For the first type Meischner (1962, 1964) has coined the term *"allodapic limestones"* (see Fig. 64). Source material for these sediments, which have been transported into the basins by turbidity currents, are reefs or the rims of carbonate platforms (see Sect. 2.2.2.3). The detrital material of the allodapic limestones is interbedded with fine-grained, pelitic sediments (e.g. flysch sequences, basins adjacent to reef margins or deep-sea basins). The name "allodapic" refers to detrital limestones which "originated somewhere else". Characteristic features of these limestone turbidites are (see Meischner 1962, 1964; Szulczewski, 1968; H. W. Flügel and Pölsler, 1965):

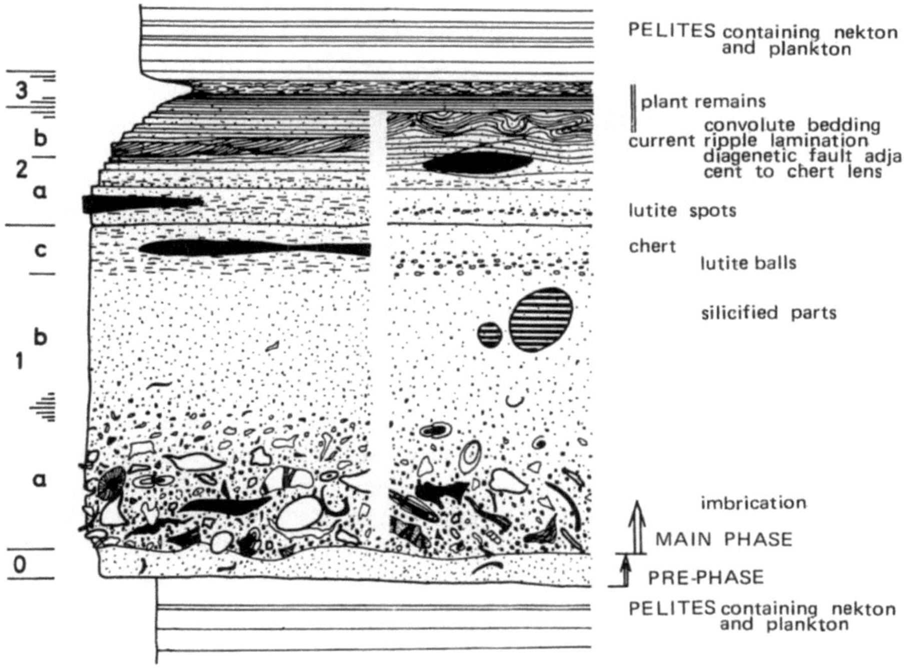

Fig. 64. Facies model of allodapic limestones. Thickness of the limestone bed about 1 m. After Meischner (1964)

a) Vertical zonation of the individual limestone bed. An ideal allodapic bed consists of the following zones (from bottom to top; see Fig. 64):

- *Zone 0:* Micritic carbonates or marls with scattered limestone detritus; particles quite often smaller than the maximum size found in the bed; some rather large rock fragments may occur. The genesis of this "prephase" is controversial; solution and reprecipitation of carbonate induced by a rapid deposition of detrital material may be important (Eder, 1971). The contact of this zone to the underlying pelagic sediments is sharp, the upper surface may be irregular and wavy.
- *Zone 1:* The lower part (1 a) developed as pronounced graded detrital limestone consisting of shallow-water fossils and lithoclasts. Reverse grading as well as imbrication structures may occur. Sorting is poor at the base, but becomes better in the upper parts. The middle part (1 b) consists of fine-grained carbonate. The upper part (1 c) may be faintly laminated; angular limestone clasts can occur.
- *Zone 2:* Densely spaced planar bedding planes (laminations, 2 a) are overlain by a unit with current ripple lamination (2 b), sometimes also with convolute bedding. The sediment is very fine-grained; laminations are very densely spaced.

- *Zone 3:* is represented by marls which may exhibit a flaser texture. This zone merges gradually into the pelitic or pelagic sediment, which overlays the allodapic limestone bed.

b) Geometry: Each of the detrital limestone beds forms a lenticular body. The coarser basal parts reach their maximum thickness upstream, while the finer grained upper parts have their maxima further down current.

c) Fossils: The detrital limestones contain mainly fragments of marine shallow-water benthos, usually reef organisms, together with some planktonic organisms. Benthonic bioclasts are sorted according to grain size. Benthonic organisms are absent in the pelitic sediments above and below the allodapic beds; nektonic and planktonic organisms are common and, in places, conspicuously well-preserved.

Allodapic limestone beds can be followed over several hundred meters to several kilometers. The thickness of the bed may range between <1 cm and several meters. Thick beds are found relatively near the source (proximal), whereas thinner beds occur predominantly in distal positions.

Allodapic limestones, formed by the redeposition of shallow-water material, and comparable in the sequence of the sedimentary structures to the "ideal bed" have been described by Szulczewski (1968; Frasnian: Holy Cross Mountains, Poland),

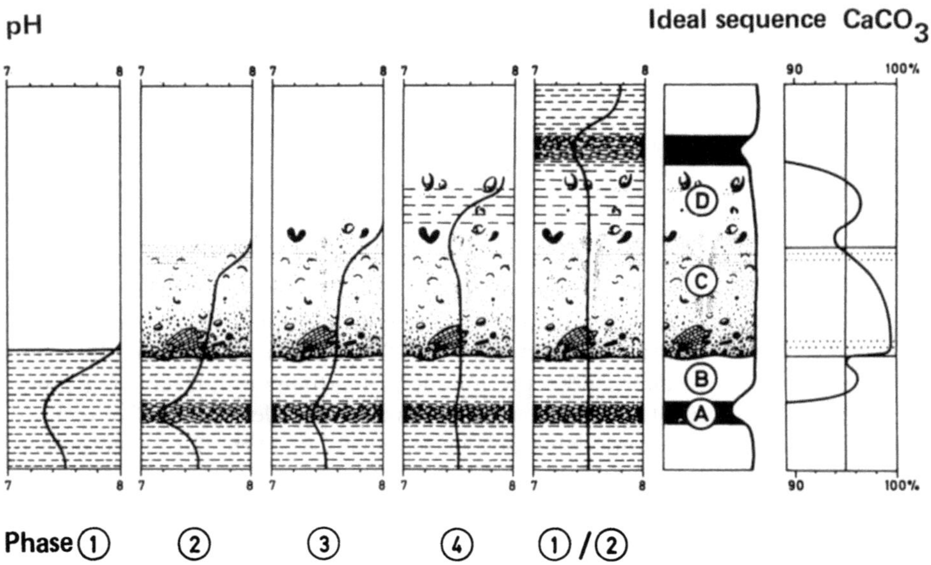

Fig. 65. Facies model of Garbeck Limestone, Middle Devonian, Rheinisches Schiefergebirge. Turbidity currents produced four zones in the sediment: Micritic shale, rich in carbonate (*A*), basal micritic limestone (*B*), coarse-grained, graded limestone (*C*), and upper micritic limestone rich in autochthonous benthonic fauna (*D*). *Zones A* and *B* are interpreted as having been modified by solution and reprecipitation of carbonate; the resulting shale of *zone A* may be due to a lowering of the pH during early diagenesis. Reprecipitation of carbonate could have occurred in zone B at higher pH values. Migration of pore solutions was stopped by graded detrital layers (*zone C*). pH values are indicated by the five black curves on the left; the curve on the far right describes the percentage of $CaCO_3$. After Eder (1971)

Thomson and Thomasson (1969; Lower Pennsylvanian, Texas), Mackenzie (1970; Devonian: Northwest Territories, Canada), Engel (1974; Paleocene: Slovenia), and Hemleben and Reuther (1980; Namurian: Cantabrian Mountains, Spain).

Variations occur in the sometimes incomplete vertical zonation (depending on whether the turbidite has a proximal or distal position), the thickness of the allodapic limestone (related to the size of the allochthonous particles); variations occur especially in the development of Zone 3 of the allodapic sequence, which sometimes may be hard to differentiate from the overlying pelitic sediments.

In comparison to the Bouma sequence, Zone 1 of the Meischner sequence is characterized by ruditic rather than arenitic allochthonous particles. Reverse grading near the base of Zone 1 seems to be more common than in Division A of the Bouma sequence. Within Zone 2, the differentiation of the Bouma Division B (plane parallel laminae), C (ripples, wavy or convolute laminae), and D (upper, parallel laminae) is often difficult.

A different zonation pattern may occur in small quiet-water basins situated in off-reef positions (Eder, 1971); see Fig. 65. Shallow-water detritus was deposited by turbidity flows within a basin with normal autochthonous marly sedimentation. The turbidite flow caused a four-fold zonation: Zone A – carbonate-rich, micritic shale; Zone B – basal micritic limestone with few pelagic microfossils; Zone C – coarse-grained graded limestone, and Zone D – upper micritic limestone, relatively rich in autochthonous benthonic fauna. The Zones A and B of the *"Garbeck limestone"* are interpreted as having been caused by solution and reprecipitation of carbonate, related to the rapid deposition of the detrital material. Solution may have been initiated by low pH values of the pore water just below the sedimentary interface during early diagenesis; the basal micritic limestone may have been formed from the reprecipitation of the carbonate, due to higher pH values.

Limestone turbidites formed by limestone lithoclasts and discrete particles which are derived not from reefs or platforms, but *from upper slope* environments or from deeper-water buildups may exhibit features known in "allodapic limestones" (e.g., Keith and Friedman, 1977 – Cambrian: New York and Vermont; Marcinowski, 1970 – Oxfordian: Poland), but there are also significant deviations from the Meischner or Bouma model. Examples of this are crinoidal turbidites (Devonian: Cornwall; Tucker, 1969), which show a number of atypical characteristics related to their peculiar composition. The most common internal structure of these turbidites is a horizontal lamination; a graded unit is generally rare. The pelitic division E is never present. There is usually a sharp contact with the shales above. A common feature of limestone turbidites derived not at all or only partly from shallow-water carbonate environments is the greater variation in the ecological types of the fossils, including pelagic forms (e.g., styliolinids in Paleozoic limestone turbidites, or calpionellids in Mesozoic turbidites; see Remane, 1970; Steiger, 1981).

Limestone turbidites may be deposited above or below the calcite compensation depth (see Hesse, 1970; Sagri, 1979). Those deposited below the CCD will not be affected by dissolution because of the rapid replacement and removal from contact with the carbonate-undersaturated bottom waters by autoburial. In contrast to this, pelagic or hemipelagic carbonate particles will be more or less completely dissolved. A differentiation between the fine-grained parts of turbidites (Zone E of the Bouma sequence) and pelagic layers (Zone F of turbidite-pelagic alternations;

Table 55. Differentiation between fine-grained carbonate turbidites, deposited above or below CCD, and fine-grained, pelagic background sedimentation (non-turbiditic mudstones). After Hesse (1975), Hesse and Butt (1976), and Sagri (1979). Examples are from Late Mesozoic-Cenozoic turbidite basins of the Alps and the Apennines

	Carbonate turbidites above calcite compensation depth		Carbonate turbidites below calcite compensation depth	
	Turbiditic mudstone (Zone E)	Non-turbiditic mudstone (Zone F)	Turbiditic mudstone	Non-turbiditic mudstone
$CaCO_3$	+	+	+	–
Color	light	light	light	dark
Occurrence directly on top of sandy-silty turbidites	+	–	+	–
Bioturbation	generally absent	usually present	generally absent	usually present
Sorting	relatively good	poor	relatively good	moderate to poor
Vertical grading	+	–	+	–
Maximum grain size	clay to fine silt	biogenic sand	clay to fine silt	clay to fine silt, rarely biogenic fine to medium sand
Calcareous planktonic foraminifera	+	+	+	–
Shallow calcareous benthonic foraminifera	±	∓	±	–
Arenaceous benthonic foraminifera	∓	∓	∓	+
Radiolaria	∓	+	∓	+

Van der Lingen, 1969) is, therefore, possible depending on the carbonate content as well as the microfaunal data and the distribution of bioturbation structures. Paleodepth analysis can be facilitated by these compositional and textural contrasts (see Tab. 55).

There are a great number of papers dealing with the problem of *proximal versus distal depositional sites* of turbidites (e.g., Bouma, 1962; Unrug, 1963; Walker, 1967, 1970). Some of the more important characteristics for proximality and distality in carbonate turbidites are summarized in Table 56. In addition, the mean maximum grain size (decreasing in the downcurrent direction), the ABC-Index (percent of beds beginning with Zone A plus half the percent of beds beginning with Zone B;

Table 56. Criteria of proximal and distal limestone turbidites

	Proximal	Distal
Mean bed thickness	greater	less, often very thin
Mean grain size	larger (arenite and rudite)	smaller (lutite and arenite)
Grading	graded, poorly graded or not graded	nearly always graded
Inverse grading	common	absent
Coarse-tail grading	common	absent
Base of the detrital zone	generally sharp	sharp
Top of the detrital zone	often sharp	grades into finer sediments
Scours, washouts, channels and reworked pebbles at the base	common	rare; no channels
Tool marks	rare	common
Geometry of beds	irregular in thickness lenticular	regular parallel beds
Complete Bouma sequences	common	rare
Lamination and ripple lamination	less common, restricted to thicker beds	more common
Micritic upper parts	thin or absent	well developed and usually well-preserved
Arenite/Lutite ratio	high	low
Matrix within the detrital zone	sparry cement	micrite
Depositional pattern	Turbidite beds densely spaced, background sedimentation restricted	Thick "normal" sediments
Lithofacies association	Slumping structures, fluxoturbidites, conglomerates; cherts rare	coarse clastics rare; cherts common

see Walker, 1967, 1970), and the ratio of the thickness of Zone A and of the mean maximum grain size as well as the differences of the maximum grain sizes in different vertical positions of the turbidite (proximality index and grading curve; Sadler, 1976; see Fig. 66) can also be used.

Ideally, proximality indicators work for turbidite sequences deposited only from a single source by longitudinal sheet flows. In turbidite fans, turbidites with both proximal and distal features may be directly juxtaposed; meanderings of turbidity currents may produce strong deviations in the characteristics of proximal and distal turbidites (see Parkash and Middleton, 1970). Some of the difficulties involved in the discrimination of proximal and distal turbidites are caused by differences in the size of the depositional basins (large bathyal or abyssal basins versus small, rather shallow basins), in the point or line source of the turbidite material, in the different settling behavior of discrete carbonate particles (see Sarntheim and Bartolini, 1973; Steiger, 1981) and of lithified carbonate fragments, and also in different sampling methods. Sampling of limestone turbidites should consider differences at the base and on the top of the coarser zones of the turbidite, the often great variations of the

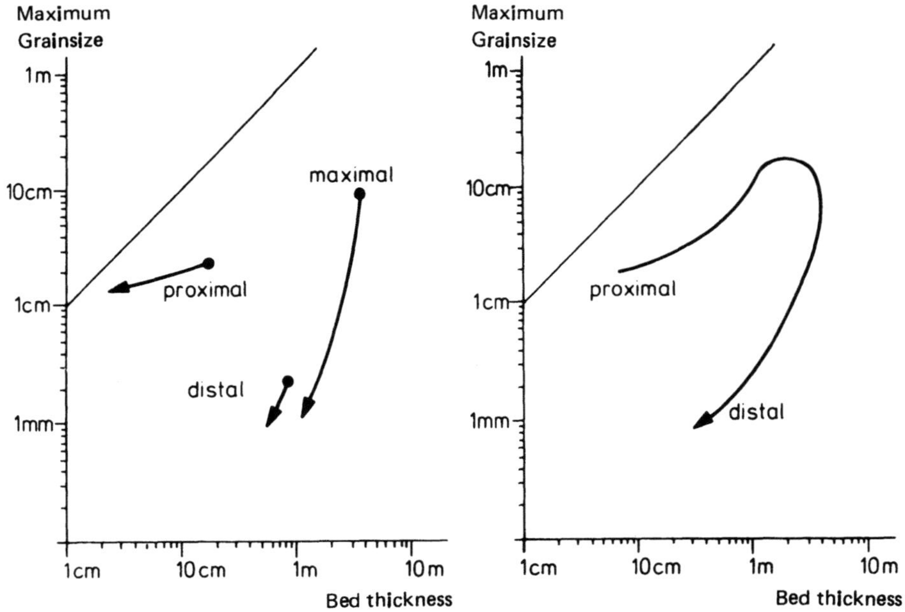

Fig. 66. Concept of grading curves (*left*) and proximality curves (*right*), developed by Sadler (1976). Both curves characterize proximality or distality of the depositional environment of turbidites. Grading curves: approximately horizontal curves – proximal environment, steep vertical curve – distal environment. Proximality curve: slowly rising curve – proximal, steep vertical curve – distal environment. After Steiger (1981)

bottom of the turbidite (a horizontal sampling may then be meaningful), and also the possibility that strong stylolitization can obliterate primary bedding planes.

Trigger mechanisms of sediment gravity flows include:

a) catastrophic events such as earthquakes or tsunamis, producing shock waves,
b) oversteepening of a slope by strong sediment accumulation (an important mechanism in submarine canyons, or in delta areas),
c) overloading during tidal emersion (most important mechanism in Recent tidal flats),
d) undercutting by erosive currents (e.g., in tidal creeks),
e) collapse of a carbonate platform edge due to patchy distribution of early cements greatly increasing the mass of the sediment.

10.2.5 References: Facies Criteria

Non-Marine Carbonates

Collinson, J. D. (1978): Lakes. In: Reading, H. G. (ed.): Sedimentary Environment and Facies, 61–79, 16 Figs., Oxford: Blackwell Science Publ.

Donavan, R. N. (1975): Devonian lacustrine limestones at the margin of the Orcadian Basin, Scotland. J. geol. Soc. London **131,** 489–510, London

Ernst, W. (1970): see 9.1.5
Flügel, E., Flügel-Kahler, E. (1980): Algen aus den Kalken der Trogkofel-Schichten der Karnischen Alpen. Carinthia II, Sonderheft **36,** 113–182, 11 Pls., Klagenfurt
Hanley, J. H., Steidtmann, J. R. (1973): Petrology of Limestone Lenses in the Casper Formation, Southernmost Laramie Basin, Wyoming and Colorado. J. Sed. Petrol. **43**/2, 428–434, 7 Figs., Tulsa
Lang, J., Segonzac, G. (1979): Contribution à l'étude des Algues continentales cénozoiques dans les basins intramontagneux de Bamyan et de Yakawlang ou dans leurs environs en Afghanistan central. Bull. Bureau Rech. Geol. Mines, 2, sect. 4/2, 113–131, 4 Pls., 6 Figs., Paris
Murphy, D. H., Wilkinson, B. H. (1980): see 2.4.2
Picard, M. D., High, L. R. (1972): Criteria for recognizing lacustrine rocks. In: Rigby, J. K., Hamblin, W. K. (eds.): Recognition of ancient sedimentary environments. Soc. Econ. Paleont. Min. Spec. Publ. **16,** 108–145, 24 Figs., 12 Tab., Tulsa
Reiss, O. M. (1923): Kalkalgen und Seesinterkalk aus dem Rheinpfälzischen Tertiär. Geognostisches Jahrbuch **36,** 103–131, 3 Pls., 3 Figs., München
Riding R. (1977): see 4.1.7.2
Smoot, J. P. (1978): see 2.4.2
Stapf, K. R. G. (1970): Lithologische Untersuchung der Altenglaner Schichten im saarpfälzischen Unterrotliegenden mit besonderer Berücksichtigung der Karbonatgesteine. Diss. Univ. Mainz, 231 pp., 52 Figs., Mainz (typescript)

Marginal Marine and Shallow Marine Environments

Amiel, A. J., Friedman, G. M. (1971): Continental sabkhas in Arava Valley between Dead Sea and Red Sea: significance for origin of evaporites. Amer. Ass. Petrol. Geol. Bull. **55,** 581–592, Tulsa
Assereto, R. L., Kendall, C. G. St. C. (1971): Megapolygons in Ladinian limestones of Triassic of southern Alps: evidence of deformation by penecontemporaneous desiccation and cementation. J. Sed. Petrol. **41,** 715–723, Tulsa
Assereto, R. L., Kendall, C. G. St. C. (1977): Nature, origin and classification of peritidal tepee structures and related breccias. Sed. **24,** 153–210, 31 Figs., Oxford
Barwis, J. H., Makurath, J. H. (1978): Recognition of Ancient tidal inlet sequences: an example from the Upper Silurian Keyser Limestone in Virginia. Sed. **25**/1, 61–82, 8 Figs., Oxford
Braun, M., Friedman, G. M. (1969): Carbonate lithofacies and environments of the Tribes Hill Formation (Lower Ordovician) of the Mohawk Valley, New York. J. Sed. Petrol. **39,** 113–135, Tulsa
Burri, P., DuDresnay, R., Wagner, C. W. (1973): Tepee structures and associated diagenetic features in intertidal carbonate sands (Lower Jurassic, Morocco). Sed. Geol. **9,** 221–228, Amsterdam
Butler, G. P. (1969): Modern evaporite deposition and geochemistry of coexisting brines in the sabkhas, Trucial Coast, Arabian Gulf. J. Sed. Petrol. **39,** 70–89, Tulsa
Fischer, A. G. (1964): see 4.2.4.5
Friedman, G. M. (1966): On the origin of aragonite in the Dead Sea. Israel J. Earth Sci. **14,** 79–85, Jerusalem
Friedman, G. M. (1978): Depositional Environments of Evaporite Deposits. In: Dean, W. E., Schreiber, B. C. (eds.): Marine Evaporites. Soc. Econ. Paleont. Min. Short Course **4,** 177–188, Oklahoma City
Friedman, G. M. (1980): Review of Depositional Environments in Evaporite Deposits and the Role of Evaporites in Hydrocarbon Accumulation. Bull. Centre Rech. Explor.-Prod. **4**/1, 589–608, 17 Figs., Pau
Fürsich, F. T., Palmer, T. J. (1979): Development of relief on a Middle Jurassic cemented sea floor: origin of submarine pseudo-anticlines in the Bathonian of Normandy. Sed. **26,** 441–452, 9 Figs., Oxford
Ginsburg, R. N. (ed.) (1975): see 1.4.1

Hardie, L. A. (ed.) (1977): Sedimentation of the Modern Carbonate Tidal Flats of Northwest Andros Island, Bahamas. John Hopkins Univ. Stud. Geol. **22,** 202 p., 94 Figs., Baltimore
Kahle, Ch. F. (1971): Stratigraphic and environmental significance of sedimentary structures in Cayugan (Silurian) tidal flat carbonates, Northwestern Ohio. Geol. Soc. Amer. Bull. **82,** 2071–2098, 15 Figs., 5 Tab., New York
Kinsman, D. J. J. (1969): Modes of formation, sedimentary associations, and diagnostic features of shallow-water and supratidal evaporites. Amer. Ass. Petrol. Geol. Bull. **53,** 830–840, Tulsa
Laporte, L. F. (1967): Carbonate deposition near mean sea-level and resultant facies mosaic: Manlius Formation (Lower Devonian) of New York State. Bull. Amer. Ass. Petrol. Geol. **51**/1, 73–101, 34 Figs., 3 Tab., Tulsa
Lucia, F. J. (1972): Recognition of evaporite-carbonate shoreline sedimentation. In: Rigby, J. K., Hamblin, W. K. (eds.): Recognition of Ancient Sedimentary Environments. Soc. Econ. Paleont. Min. Spec. Publ. **16,** 160–191, 34 Figs., Tulsa
Matter, A. (1967): Tidal flat deposits in the Ordovician of western Maryland. J. Sed. Petrol. **37**/2, 601–609, 5 Figs., Tulsa
Mazzullo, S. J., Friedman, G. M. (1975): Conceptual model of tidally influenced deposition on margins of epeiric seas: Lower Ordovician (Canadian) of Eastern New York and Southwestern Vermont. Bull. Amer. Ass. Petrol. Geol. **59**/11, 2123–2141, 13 Figs., 1 Tab., Tulsa
Müller-Jungbluth, W.-U. (1970): Sedimentologische Untersuchungen des Hauptdolomits der östlichen Lechtaler Alpen, Tirol. Festband Geol. Inst. 300-Jahr-Feier Univ. Innsbruck, 255–308, Pls. 1–8, Innsbruck
Reineck, H. E. (1973): see 1.4.1
Roehl, P. O. (1967): Stony Mountain (Ordovician) and Interlake (Silurian) facies analogs of recent low energy marine and subaerial carbonates, Bahamas. Bull. Amer. Ass. Petrol. Geol. **51**/10, 1979–2032, 48 Figs., Tulsa
Schenk, P. E. (1969): Carbonate-sulphate-redbed facies and cyclic sedimentation of the Windsorian stage (middle Carboniferous), Maritime Provinces. Canadian J. Earth Sci. **6,** 1037–1066, Ottawa
Schreiber, B. Ch., Decima, A. (1976): Sedimentary facies produced under evaporitic environments: a review. Mem. Soc. Geol. Ital. **16,** 111–126, 13 Figs., Rome
Textoris, D. A. (1968): Petrology of supratidal, intertidal and shallow subtidal carbonates, Black River Group, Middle Ordovician, New York, U.S.A. Intern. Geol. Congr. Rep. 23rd Sess., Proc. Sect. **8,** 227–248, 6 Figs., Prague
Till, R. (1978): Arid Shorelines and Evaporites. In: Reading, H. G. (ed.): Sedimentary Environments and Facies, 178–206, 45 Figs., Oxford
West, I. M. (1964): Evaporite diagenesis in the lower Purbeck beds of Dorset. Proc. Yorkshire geol. Soc. **34,** 315–330, Leeds
West, I. M. (1975): Evaporites and associated sediments of the basal Purbeck formation (Upper Jurassic) of Dorset. Proc. Geol. Ass. **86,** 205–225, London
Wood, G. V., Wolfe, M. J. (1969): Sabkha cycles in the Arab/Darb formation off the Trucial Coast of Arabia. Sedimentology **12,** 165–191, 7 Figs., Amsterdam

Reef Complexes

Cloud, P. E. (1959): Geology of Saipan, Mariana Islands, Part 4 – Submarine topography and shallow-water ecology. U.S. Geol. Surv. Prof. Paper **280**-k, 361–445, Washington
Goreau, T. F., Land, L. S. (1974): Fore-reef morphology and depositional processes, North Jamaica. In: Laporte, L. F. (ed.): Reefs in Time and Space. Soc. Econ. Paleont. Min. Spec. Publ. **18,** 77–89, 13 Figs., Tulsa
Krebs, W. (1969): see 3.9
Purdy, E. G. (1974): Reef configuration: Cause and effect. Soc. Econ. Paleont. Min. Spec. Publ. **18,** 9–76, Tulsa
Schroeder, J. H. (1972): see 3.9
Schroeder, J. H. (1973): see 3.9

Shinn, E. A. (1963): Spur and groove formation on the Florida reef tract. J. Sed. Petrol. **33,** 291–303, Tulsa
Sneh, A., Friedman, G. M. (1980): Spur and groove patterns on the reefs of the Northern Gulfs of the Red Sea. J. Sed. Petrol. **50**/3, 981–986, 5 Figs., Tulsa
Toomey, D. F. (ed.) (1981): European fossil reef models. Soc. Econ. Paleont. Min. Spec. Publ. **30,** 544 p., Tulsa
Zankl, H. (1971): see 3.9

Deeper-Marine Environments

Allen, R. L. (1970): Physical Processes of Sedimentation. Earth Sci. Ser. **1,** Unwin Univ. Books, 248 pp., London
Angelucci, A., Rosa, E. de, Fierro, G., Gnaccolini, N., La Monica, G. B., Martinis, B., Parea, G. C., Pescatore, T., Rizzini, A., Wezel, F. C. (1967): Sedimentological characteristics of some Italian turbidites. Geol. Romana **6,** 345–420, 65 Figs., 4 Tab., Rome
Arthur, M. A., Fischer, A. G. (1977): Upper Cretaceous-Paleocene magnetic stratigraphy at Gubbio, Italy. 1. Lithostratigraphy and sedimentology. Geol. Soc. Amer. Bull. **88,** 367–371, New York
Backhaus, E., Flügel, E. (1971): see 9.1.5
Bandel, K. (1974): Deep-water limestones from the Devonian-Carboniferous of the Carnic Alps, Austria. Spec. Publ. int. Ass. Sed. **1,** 93–115, 16 Figs., Oxford
Berger, W., Winterer, E. L. (1974): Plate stratigraphy and the fluctuating carbonate line. Spec. Publ. int. Ass. Sed. **1,** 11–48, 26 Figs., Oxford
Bernoulli, D. (1972): North Atlantic and Mediterranean Mesozoic facies: a comparison. In: Hollister, C. D. et al. (eds.): Initial Reports of the Deep Sea Drilling Project **11,** 801–872, Washington
Bernoulli, D., Jenkyns, H. C. (1970): A Jurassic basin: the Glasenbach Gorge, Salzburg, Austria. Verh. geol. Bundesanst. Wien **1970,** 504–531, Wien
Bernoulli, D., Jenkyns, H. C. (1974): Alpine, Mediterranean, and central Atlantic Mesozoic facies in relation to the early evolution of the Tethys. In: Dott, R. H., Jr., Shaver, R. H. (eds.): Modern and Ancient Geosynclinal Sedimentation. Soc. Econ. Paleont. Min. Spec. Publ. **19,** 129–160, Tulsa
Bissell, H. J., Barker, H. K. (1977): Deep-water limestones of the Great Blue Formation (Mississippian) in the eastern part of the Cordilleran miogeosyncline in Utah. Soc. Econ. Paleont. Min. Spec. Publ. **25,** 171–186, 14 Figs., Tulsa
Bosellini, A. (1967): Torbiditi carbonatiche del Giurassico delle Giudicarie e loro significato geologico. Ann. Univ. Ferrara, N.S., Sez. IX, Sci. Geol. Paleont. **4**/8, 101–115, 7 Figs., Ferrara
Bouma, A. H. (1962): Sedimentology of Some Flysch Deposits. 168 pp., 8 Pls., 31 Figs., Amsterdam: Elsevier
Bourgeois, J. (1977): Gravity Flows. In: Fairbridge, R. W., Bourgeois, J. (eds.): The Encyclopaedia of Sedimentology. Encyclopaedia of Earth Sciences Series **6,** 385–386, 2 Figs., Stroudsburg: Dowden, Hutchinson, Ross
Bromley, R. G. (1977): Trace fossils at omission surfaces. In: Frey, R. W. (ed.): The Study of Trace Fossils, 399–428, New York: Springer
Buchroithner, M. F., Flügel, E., Flügel, H. W., Stattegger, K. (1980): Die Devongerölle des paläozoischen Flysch von Menorca und ihre paläogeographische Bedeutung. Neues Jahrbuch Geol. Paläont. Abhandlungen **159**/2, 172–224, 13 Figs., Stuttgart
Buggisch, W. (1972): see 9.1.5
Burns, D. A. (1974): Changes in the carbonate component of Recent sediments with depth: a guide to paleoenvironmental interpretation. Marine Geol. **16,** M 13–M 19, 2 Figs., 1 Tab., Amsterdam
Carrasco-V. Baldomero (1977): Albian sedimentation of submarine autochthonous and allochthonous carbonates, east edge of the Valles-San Luis Potosi platform, Mexico. Soc. Econ. Paleont. Min. Spec. Publ. **25,** 263–272, 19 Figs., Tulsa

Carter, R. M. (1975): A discussion and classification of subaqueous mass-transport with particular application to grain-flow, slurry-flow, and fluxoturbidites. Earth-Sci. Rev. **11,** 145–177, Amsterdam

Chamberlain, C. K., Clark, D. L. (1973): Trace fossils and conodonts as evidence for deep-water deposits in the Oquirrh Basin of Central Utah. J. Paleont. **47**/4, 663–682, 3 Pls., 7 Figs., Lawrence

Cook, H. E., McDaniels, P. M., Mountjoy, E. W., Pray, L. C. (1972): Allochthonous carbonate debris flows at Devonian bank ("reef") margins, Alberta, Canada. Canadian Petrol. Geol. Bull. **20,** 439–497, Calgary

Cook, H. E., Enos, P. (1977): see 1.4.1

Cook, H. E., Taylor, M. E. (1977): Comparison of continental slope and shelf environments in the Upper Cambrian and Lowest Ordovician of Nevada. Soc. Econ. Paleont. Min. Spec. Publ. **25,** 51–81, 49 Figs., Tulsa

Coombs, D. S., Landis, C. A., Norris, R. J., Sinton, J. M., Borns, D. J., Craw, D. (1976): The Dun Mountain Ophiolite Belt, New Zealand, its tectonic setting, constitution and origin with special reference to the southern position. Amer. J. Sci. **276,** 561–603, Washington

Cossey, S. P. J., Ehrlich, R. (1979): A conglomeratic carbonate flow deposit, Northern Tunisia: a link in the genesis of pebbly mudstones. J. Sed. Petrol. **49**/1, 11–22, 13 Figs., Tulsa

Crevello, P. D., Schlager, W. (1981): Carbonate debris sheets and turbidites, Exuma Sound, Bahamas. J. Sed. Petrol., in press, Tulsa

Crowell, J. C. (1957): Origin of pebbly mudstones. Geol. Soc. Amer. Bull. **68,** 993–1009, New York

Davies, G. R. (1977): Turbidites, debris sheets, and truncation structures in Upper Paleozoic deep-water carbonates of the Sverdrup basin, Arctic Archipelago. Soc. Econ. Paleont. Min. Spec. Publ. **25,** 221–247, 18 Figs., Tulsa

Davies, D. K. (1968): Carbonate turbidites, Gulf of Mexico. J. Sed. Petrol. **38,** 1100–1109, Tulsa

Dott, R. H., Jr. (1963): Dynamics of subaqueous gravity depositional processes. Amer. Ass. Petrol. Geol. Bull. **47,** 104–129, New York

Dzulynski, S., Ksiazkiewicz, M., Kuenen, Ph. H. (1959): Turbidites in flysch of the Polish Carpathian Mountains. Geol. Soc. Amer. Bull. **70,** 1089–1118, New York

Eder, F. W. (1971): Riff-nahe detritische Kalke bei Balve im Rheinischen Schiefergebirge. Göttinger Arb. Geol. Paläont. **10,** 66 p., 5 Pls., 14 Figs., Göttingen

Edmunds, W. M., Lovelock, P. E. R., Gray, D. A. (1973): Interstitial water chemistry and aquifer properties in the Upper and Middle Chalk of Berkshire, England. J. Hydrology **19,** 21–31

Elliott, R. E. (1965): A classification of subaqueous sedimentary structures based on rheoligical and kinematical parameters. Sed. **5,** 193–209, Amsterdam

Engel, W. (1970a): Die Nummuliten-Breccien im Flyschbecken von Ajdovscina in Slowenien als Beispiel karbonatischer Turbidite. Verh. geol. Bundesanst. **1970**/4, 570–582, 5 Figs., Wien

Engel, W. (1970b): Genese Riff-naher Detritus-Kalke bei Balve im Rheinischen Schiefergebirge (Garbecker Kalk). Verh. geol. Bundesanst. **1970**/4, 551–569, 6 Figs., Wien

Engel, W. (1974): Sedimentologische Untersuchungen im Flysch des Beckens von Ajdovscina (Slowenien). Göttinger Arb. Geol. Paläont. **16,** 65 pp., 3 Pls., 11 Figs., 1 Tab., Göttingen

Enos, P. (1977): Tamabra limestone of the Poza Rica Trend, Cretaceous, Mexico. Soc. Econ. Paleont. Min. Spec. Publ. **25,** 273–314, Tulsa

Evans, I., Kendall, C. G. St. (1977): An interpretation of the depositional setting of some deep-water Jurassic carbonates of the Central High Atlas Mountains, Morocco. Soc. Econ. Paleont. Min. Spec. Publ. **25,** 249–261, 15 Figs., Tulsa

Farrow, G. E. (1966): Bathymetric zonation of Jurassic trace fossils from the coast of Yorkshire, England. Paleo-3, **2,** 103–151, 11 Figs., 3 Tab., Amsterdam

Faupl, P., Tollmann, A. (1979): Die Roßfeldschichten: Ein Beispiel für Sedimentation im Bereich einer tektonisch aktiven Tiefseerinne aus der kalkalpinen Unterkreide. Geol. Rundschau **68**/1, 93–120, 2 Pls., 10 Figs., Stuttgart

Fischer, A. G. (1969): Geological time-distance rates: the Bubnoff unit. Geol. Soc. Amer. Bull. **80,** 549–552, New York

Flores, G. (1955): Discussion. In: Beneo, E.: Les resultats des études pour la recherche pétrolifere en Sicilie (Italie). 4th World Petroleum Congress, Proc., Sect. **1,** 121–122, Rome
Flügel, H. W., Fenninger, A. (1966): Die Lithogenese der Oberalmer Schichten und der mikritischen Plassen-Kalke (Thionium, Nördliche Kalkalpen). N. Jb. Geol. Paläont. Abh. **123**/3, 249–280, Stuttgart
Foster, S. S. D. (1975): The chalk groundwater tritium anomaly. J. Hydrology **19,** 21–31
Friedman, G. M., Gebelein, C. D., Sanders, J. E. (1971): see 4.1.7.2
Fürsich, F. T., Wendt, J. (1976): see 4.2.4.7
Funnell, B. M. (1967): Foraminifera and radiolaria as depth indicators in the marine environment. Marine Geol. **5,** 5/6, 333–347, Amsterdam
Garrison, R. E. (1967): Pelagic limestones of the Oberalm Beds (Upper Jurassic – Lower Cretaceous), Austrian Alps. Bull. Canadian Petrol. Geol. **15**/1, 21–49, 8 Figs., Calgary
Garrison, R. E. (1974): Radiolarian cherts, pelagic limestones and igneous rocks in eugeosynclinal assemblages. Int. Ass. Sed. Spec. Publ. **1,** 367–400, Oxford
Garrison, R. E., Fischer, A. G. (1969): see 4.2.4.6
Garrison, R. E., Kennedy, W. J. (1977): Origin of solution seams and flaser structure in Upper Cretaceous chalks of southern England. Sed. Geol. **19,** 107–137, 16 Figs., Amsterdam
Gartner, S., Gentile, R. (1972): Problematic Pennsylvanian coccoliths from Missouri. Micropaleont. **18,** 401–404, New York
Gascoyne, P. (1978): Mudflow, debris-flow deposits. In: Fairbridge, R. W., Bourgeois, J. (eds.): The Encyclopaedia of Sedimentology. Encyclopaedia of Earth Sciences Ser. **6,** 489–493, 3 Figs., Stroudsburg: Dowden, Hutchinson, Ross
Germann, K. (1971): see 4.2.5.6
Geyer, O. F. (1971): Zur paläobathymetrischen Zuverlässigkeit von Ammonoideen-Faunen-Spektren. Palaeo-3 **10,** 265–272, Amsterdam
Goldring, R., Kazmierczak, J. (1974): Ecological succession in intraformational hardground formation. Paleont. **17**/4, 949–962, Pls. 125–126, 3 Figs., London
Görler, K. (1967): Ein Olisthostrom in der Molise-Zone (Süditalien) als Beispiel für Resedimentation aus Schlammströmen. Neues Jahrbuch Geol. Paläont. Abhandlungen **129**/1, 65–82, 3 Figs., Stuttgart
Görler, K., Reutter, K.-J. (1968): Entstehung und Merkmale der Olisthostrome. Geol. Rundschau **57,** 484–514, 12 Figs., Stuttgart
Gwinner, M. P. (1962): Subaquatische Gleitungen und resedimentäre Breccien im Weißen Jura der Schwäbischen Alb (Württemberg). Zeitschrift deutsch. geol. Gesellschaft **113,** 571–590, 4 Pls., 6 Figs., Hannover
Hakansson, E., Bromley, R., Perch-Nielsen, K. (1974): Maastrichtian chalk of north-west Europe – a pelagic shelf sediment. Int. Ass. Sed. Spec. Publ. **1,** 211–234, Oxford
Hampton, M. A. (1972): The role of subaqueous debris flow in generating turbidity currents. J. Sed. Petrol. **42,** 775–793, 14 Figs., Tulsa
Hancock, J. M. (1963): The hardness of the Irish chalk. Irish Natl. J. **14,** 157–164, Dublin
Hancock, J. M. (1976): The petrology of the chalk. Proc. Geol. Soc. **86,** 499–535, London
Hancock, J. M., Kennedy, W. J. (1967): Photographs of hard and soft chalks taken with scanning electron microscope. Proc. Geol. Soc. London **1643,** 249–252, London
Hancock, J. M., Scholle, P. A. (1975): Chalk of the North Sea. In: Woodland, A. W. (ed.): Petroleum and the Continental Shelf of Europe, **1,** 413–425, New York: Wiley
Hemleben, Chr., Freels, D. (1977): Algen-laminierte und gradierte Plattenkalke in der Oberkreide Dalmatiens (Jugoslawien). N. Jb. Geol. Paläont. Abh. **154**/1, 61–93, 19 Figs., 1 Tab., Stuttgart
Hemleben, Chr., Reuther, C.-D. (1980): Allodapic limestones of the Barcaliente Formation (Namurian A) between Luna and Cea Rivers (Southern Cantabrian Mountains, Spain). Neues Jahrbuch Geol. Paläont. Abhandlungen **159**/2, 225–255, 20 Figs., Stuttgart
Hesse, R. (1974): Long-distance continuity of turbidites: possible evidence for an Early Cretaceous trench-abyssal plain in the East Alps. Geol. Soc. Amer. Bull. **85,** 859–870, New York
Hesse, R. (1975): Turbiditic and non-turbiditic mudstone of Cretaceous flysch sections of the East Alps and other basins. Sed. **22,** 387–416, 11 Figs., Oxford
Hollmann, R. (1962): see 4.2.4.6

Hollmann, R. (1964): see 4.2.4.6
Hopkins, J. C. (1977): Production of foreslope breccia by differential submarine cementation and downslope displacement of carbonate sands, Miette and Ancient Wall Buildups, Devonian, Canada. Soc. Econ. Paleont. Min. Spec. Publ. **25,** 155–170, 12 Figs., Tulsa
Hubert, J. F. (1964): Textural evidence for deposition of many western North Atlantic deep-sea sands by ocean-bottom currents rather than turbidity currents. J. Geol. **72,** 757–785, Chicago
Hubert, C., Lajoie, J., Leonard, M. A. (1970): Deep sea sediments on the Lower Paleozoic Quebec supergroup. Geol. Ass. Canada Spec. Paper **7,** 103–125, 12 Figs., Waterloo
Hubert, J. F., Suchecki, R. K., Callahan, R. K. M. (1977): The Cow Head breccia: sedimentology of the Cambro-Ordovician continental margin, Newfoundland. Soc. Econ. Paleont. Min. Spec. Publ. **25,** 125–154, 25 Figs., Tulsa
Jaanusson, V. (1955): Description of the microlithology of the Lower Ordovician limestones between the Ceratopyge shales and the Platyrus limestone of Böda Hamm. Bull. geol. Inst. Univ. Uppsala **35,** 135–173, Uppsala
Jaanusson, V. (1960): The Viruan (Middle Ordovician) of Öland. Bull. geol. Inst. Univ. Uppsala **38,** 207–288, Uppsala
Jaanusson, V. (1961): Discontinuity surfaces in limestones. Bull. geol. Inst. Univ. Uppsala **40,** 221–241, Uppsala
Jenkyns, H. C. (1974): see 3.9
Jenkyns, H. C. (1978): Pelagic environments. In: Reading, H. G. (ed.): Sedimentary Environment and Facies, 314–371, Oxford: Blackwell Sci. Publ.
Keith, B. D., Friedman, G. M. (1977): A slope-fan-basin-plain model, Taconic sequence, New York and Vermont. J. Sed. Petrol. **47**/3, 1220–1241, 24 Figs., Tulsa
Kelling, G., Holroyd, J. (1978): Clast size, shape, and composition in some Ancient and Modern fan gravels. In: Stanley, D., Kelling, G. (eds.): Sedimentation in Submarine Canyons, Fans and Trenches, 138–159, 1 Fig., Stroudsburg: Dowden, Hutchinson, Ross
Kennedy, W. J., Garrison, R. E. (1975): Morphology and genesis of nodular chalks and hardgrounds in the Upper Cretaceous of southern England. Sed. **22,** 311–386, Oxford
Krebs, W. (1974): Devonian carbonate complexes of Central Europe. Soc. Econ. Paleont. Min. Spec. Publ. **18,** 155–208, Tulsa
Kuenen, Ph. H. (1958): Problems concerning source and transportation of Flysch sediments. Geol. Mijnbouw **20,** 329–339, Amsterdam
Kuenen, Ph. H., Haaf, E. ten (1956): Graded bedding in limestones. Verh. Nederl. Akad. Wetensch., B **59,** 314–317, Amsterdam
Kuhry, B., Clercq, S. W. G. de, Decker, L. (1976): Indications of current action in Late Jurassic limestones, radiolarian limestones, Saccocoma limestones and associated rocks from the Subbetic of SE Spain. Sed. Geol. **15,** 235–258, Amsterdam
Landis, C. A. (1974): Stratigraphy, lithology, structure, and metamorphism of Permian, Triassic, and Tertiary rocks between the Maraora River and Mount Snowden, western Southland, New Zealand. J. Roy. Soc. New Zealand **4,** 229–251
Lindström, M. (1963): Sedimentary folds and the development of limestone in an Early Ordovician sea. Sed. **2,** 243–292, Amsterdam
Lindström, M. (1971): Vom Anfang, Hochstand und Ende eines Epikontinentalmeeres. Geol. Rundschau **60,** 419–438, Stuttgart
Link, M. H. (1975): Matilija sandstone: a transition from deep-water turbidite to shallow-marine deposition in the Eocene of California. J. Sed. Petrol. **45,** 63–78, Tulsa
Lowe, D. R. (1976): Grain flow and grain flow deposits. J. Sed. Petrol. **46,** 188–199, Tulsa
Matter, A. (1974): Burial diagenesis of pelitic and carbonate deep-sea sediments from the Arabian Sea. Initial Rep. Deep Sea Drilling Project **23,** 421–469, Washington
Marcinowski, R. (1970): see 1.3.3
McIlreath, I. A. (1977): Accumulation of a Middle Cambrian, deep-water limestone debris apron adjacent to a vertical submarine carbonate escarpment, Southern Rocky Mountains, Canada. Soc. Econ. Paleont. Min. Spec. Publ. **25,** 113–124, 9 Figs., Tulsa
McKenzie, D. P. (1970): Plate tectonics of the Mediterranean region. Nature **226,** 239–241, London

Meischner, D. (1962): Rhenaer Kalk und Posidonienkalk im Kulm des nordöstlichen Rheinischen Schiefergebirges und der Kohlenkalk von Schreufa (Eder). Abh. hess. Landesamt Bodenforsch. **39,** 47 pp., 7 Pls., 15 Figs., 7 Tab., Wiesbaden

Meischner, D. (1964): see 1.3.3

Mendez-Bedia, I. (1976): Biofacies y litofacies de la formacion Monjello-St. Lucia (Devonico de la Cordillera Cantabria, N. W. de Espana). Trabachos Geol. **9,** 93 p., Oviedo

Middleton, G. V. (1969): Turbidity currents and grain flows and other mass movements down slopes. In: Stanley, D. J. (ed.): New Concepts of Continental Margin Sedimentation. Amer. geol. Inst. Short Course Notes, GM-A 1 to GM-B 14

Middleton, G. V., Hampton, M. A. (1976): Subaqueous sediment transport and deposition by sediment gravity flows. In: Stanley, D. J., Swift, D. J. P. (eds.): Marine Sediment Transport and Environmental Management, 197–218, New York: Wiley

Milliman, J. D., Müller, J. (1979): Characteristics and genesis of shallow-water and deep-sea limestones. In: Andersen, N. R., Malahoff, A. (eds.): The fate of fossil fuel CO_2 in the oceans, 655–672, New York: Plenum Publishing Corporation

Mimram, Y. (1978): The induration of Upper Cretaceous Yorkshire and Irish chalks. Sed. Geol. **20,** 141–164, 15 Figs., Amsterdam

Monty, Cl. (1973): Les nodules de manganese sont des stromatolithes oceaniques. C. R. Acad. Sci. Paris, Sér. D. **276,** 3285–2388, 2 Pls., Paris

Mullins, H. T., Neumann, A. C. (1979): Deep carbonate bank margin structure and sedimentation in the Northern Bahamas. Soc. Econ. Paleont. Min. Spec. Publ. **27,** 165–192, 26 Figs., Tulsa

Mutti, E., Ricci Lucchi, F. (1972): Le torbiditi dell'Appennino settentrionale: introduzioni all'analisi di facies. Mem. Soc. Geol. Ital. **11**/2, 161–199, 30 Figs., 1 Tab., Pisa

Mutti, E., Ricci Lucchi, F. (1978): Turbidites of the Northern Apennines. Introduction to Facies Analysis. Geol. Review **20**/6, 125–166, 30 Figs., Leesburg Pike Falls Church

Nardin, T. R., Hein, F. J., Gorsline, D. S., Edwards, B. D. (1979): A review of mass movement processes, sediment and acoustic characteristics, and contrasts in slope and base-of-slope systems versus canyon fan-basin floor systems. Soc. Econ. Paleont. Min. Spec. Publ. **27,** 61–73, Tulsa

Neugebauer, J. (1973): The diagenetic problem of chalk – the role of pressure solution and pore fluid. Neues Jahrbuch Geol. Paläont. Abhandlungen **143,** 223–245, Stuttgart

Neugebauer, J. (1974): Some aspects of cementation in chalk. Intern. Ass. Sed. Spec. Publ. **1,** 149–176, Oxford

Neugebauer, J. (1975): Fossildiagenese in der Schreibkreide: Coccolithen. Neues Jahrbuch Geol. Paläont. Monatshefte **1975,** 489–502, Stuttgart

Newell, N. D., Rigby, J. K., Fischer, A. G., Whiteman, A. J., Hickox, J. E., Bradley, J. S. (1953): The Permian Reef Complex of the Guadalupe Mountains Region, Texas and New Mexico. 236 p., 32 Pls., San Francisco: Freeman

Padgett, G., Ehrlich, R., Moody, M. (1977): Submarine debris flow deposits in an extensional setting – Upper Devonian of Western Morocco. J. Sed. Petrol. **47**/2, 811–818, 4 Figs., Tulsa

Parkash, B., Middleton, G. V. (1970): Downcurrent textural changes in Ordovician turbidites greywackes. Sed. **14,** 259–293, Oxford

Pirini Radrizanni, C. (1971): Coccoliths from Permian deposits of eastern Turkey. Proc. 2nd Planktonic Conf., 993–1001, Rome

Porrenga, D. H. (1967): Glauconite and chamosite as depth indicators in the marine environment. Marine Geol. **5,** 495–501, Amsterdam

Price, I. (1977): Deposition and derivation of clastic carbonates on a Mesozoic continental margin, Othris, Greece. Sed. **24,** 529–546, 8 Figs., Oxford

Rabien, A. (1956): Zur Stratigraphie und Fazies des Oberdevons in der Waldecker Hauptmulde. Abhandlungen hessischen Landesamtes Bodenforschung **16,** 83 p., 2 Pls., Wiesbaden

Reading, H. G. (ed.) (1978): Sedimentary environments and facies. 557 p., Oxford: Blackwell

Reinhardt, J. (1977): Cambrian off-shelf sedimentation, Central Appalachians. Soc. Econ. Paleont. Min. Spec. Publ. **25,** 83–112, 32 Figs., Tulsa

Remane, J. (1970): Die Entstehung der resedimentären Breccien im Obertithon der subalpinen Ketten Frankreichs. Eclogae geol. Helvetiae **63**, 685–740, Basel

Rhodri Johns, D. (1978): Mesozoic carbonate rudites, megabreccias and associated deposits from central Greece. Sed. **25**, 561–573, 6 Figs., Oxford

Ricci Lucchi, F., Valmori, E. (1980): Basin-wide turbidites in a Miocene, oversupplied deep-sea plain: a geometrical analysis. Sed. **27**/3, 241–270, 21 Figs., Amsterdam

Riding, R. (1974): see 5.3.12.7

Riedel, W. R. (1963): The preserved record: paleontology of pelagic sediments. In: Hill, M. H. (ed): The Sea **3**, 866–887, New York: Interscience

Riedel, W. R., Funnell, B. M. (1964): Tertiary sediment cores and microfossils from the Pacific Ocean floor. Quart. J. Geol. Soc. London **120**, 305–368, London

Riedl, R. (1969): see 10.1.8

Rigby, J. K. (1958): Mass movements in Permian rocks of Trans-Pecos, Texas. J. Sed. Petrol. **28**/3, 298–315, 14 Figs., Tulsa

Rose, P. R. (1970): Stratigraphic interpretation of submarine versus subaerial discontinuity surfaces: an example from the Cretaceous of Texas. Bull. Geol. Soc. Amer. **81**, 2787–2798, New York

Sadler, P. M. (1976): On the significance of bed thickness in turbidites. Geol. Paläont. Inst. Univ. Göttingen, Manuscript, 56 p., 15 Figs., Göttingen

Sagri, M. (1973): Sedimentazione torbiditica nell'Appennino settentrionale: velocità, di accumulo, litoligia, morfologia del fondo. Boll. Soc. Geol. Ital. **92**, 233–272, 11 Figs., Rome

Sagri, M. (1979): Upper Cretaceous carbonate turbidites of the Alps and Apennines deposited below the calcite compensation level. J. Sed. Petrol. **49**/1, 23–28, 3 Figs., Tulsa

Sarntheim, M., Bartolini, C. (1973): Grain size studies on turbidite components from Tyrrhenian deep sea cores. Sed. **20**, 425–436, 6 Figs., Oxford

Schlager, W. (1974): see 3.9

Schlager, W. (1980): Mesozoic calciturbidites in Deep Sea Drilling Project hole 416 A: recognition of a drowned carbonate platform. Initial Rep. Deep Sea Drilling Project **50**, 733–749, 14 Figs., Washington

Schlager, W. (1981): The paradox of drowned reefs and carbonate platforms. Geol. Soc. Amer. Bull. (in press)

Schlager, W., Chermak, A. (1979): Sediment facies of platform-basin transition, Tongue of the Ocean, Bahamas. Soc. Econ. Paleont. Min. Spec. Publ. **27**, 193–208, 17 Figs., Tulsa

Schlager, W., Schlager, M. (1973): Clastic sediments associated with radiolarites (Tauglboden-Schichten, Upper Jurassic, Eastern Alps). Sed. **20**, 65–89, Amsterdam

Schlanger, S. O., Douglas, R. G. (1974): Pelagic ooze-chalk-limestone transition and its implications for marine stratigraphy. Intern. Ass. Sed. Spec. Publ. **1**, 117–148, Oxford

Schneidermann, N. (1970): Genesis of some Cretaceous carbonates in Israel. Israel J. Earth-Sci. **19**, 97–115, Jerusalem

Schneidermann, N. (1973): Pelagic limestones of the central Caribbean, Leg 15. Initial Rep. Deep Sea Drilling Project **15**, 773–793, 8 Pls., Washington

Scholle, P. A. (1971): Sedimentology of fine-grained deep-water carbonate turbidites, Monte Antola flysch (Upper Cretaceous), Northern Apennines, Italy. Geol. Soc. Amer. Bull. **82**, 629–658, New York

Scholle, P. A. (1971 b): Diagenesis of deep-water carbonate turbidites, Upper Cretaceous Monte Antola flysch, Northern Apennines, Italy. J. Sed. Petrol. **41**/1, 233–250, Tulsa

Scholle, P. A. (1977 a): Chalk diagenesis and its relation to petroleum exploration: oil from chalks, a modern miracle? Amer. Ass. Petrol. Geol. Bull. **61**, 982–1009, Tulsa

Scholle, P. A. (1977 b): Deposition, diagenesis, and hydrocarbon potential of "Deeper-Water" limestones. Amer. Ass. Petrol. Geol. Continuing Education Course Notes Ser. **7**, 27 p., Denver

Scholle, P. A., Arthur, M. A. (1980): Carbon isotope fluctuations in Cretaceous pelagic limestones: potential stratigraphic and petroleum exploration tool. Amer. Ass. Petrol. Geol. Bull. **64**/1, 67–87, 12 Figs., Tulsa

Schwarz, H.-U. (1975): Sedimentary structures and facies analysis of shallow marine carbonates. Contrib. Sed. **3,** 100 p., 11 Pls., 35 Figs., Stuttgart
Seilacher, A. (1967): Bathymetry of trace fossils. Marine Geol. **5,** 413–428, 2 Pls., Amsterdam
Seilacher, A. (ed.) (1976): Palökologie. Bericht 1970–1975 des Sonderforschungsbereiches 53. Zentralbl. Geol. Paläont. **1976,** 494 pp., Stuttgart
Seyfried, H. (1980): see 4.1.7.2
Shinn, E. A. (1969): Submarine lithification of Holocene carbonate sediments in the Persian Gulf. Sed. **12,** 109–144, Amsterdam
Shanmugan, G., Benedict, G. L. (1978): Fine-grained carbonate debris flow, Ordovician basin margin, Southern Appalachian. J. Sed. Petrol. **48**/4, 1233–1240, 7 Figs., Tulsa
Shepard, F. P., Marshall, N. F., McLoughlin, P. A., Sullivan, G. G. (1979): Currents in submarine canyons and other seavalleys. Amer. Ass. Petrol. Geol. Stud. Geol. **8,** 173 p., Tulsa
Smith, D. L. (1977): Transition from deep- to shallow-water carbonates, Paine Member, Lodgepole Formation, Central Montana. Soc. Econ. Paleont. Min. Spec. Publ. **25,** 187–201, 16 Figs., Tulsa
Srivastava, P., Stearn, C. W., Mountjoy, E. W. (1972): A Devonian megabreccia at the margin of the Ancient Wall Carbonate Complex, Alberta. Bull. Canadian Petrol. Geol. **20**/3, 412–438, 16 Figs., Calgary
Stanley, D. J., Kelling, G. (eds.) (1978): Sedimentation in submarine canyons, fans and trenches. 395 p., Stroudsburg: Academic Press
Stanley, D. J., Unrug, R. (1972): Submarine channel deposits, fluxoturbidites and other indicators of slope and base-of-slope environments in modern and ancient marine basins. Soc. Econ. Paleont. Min. Spec. Publ. **16,** 287–340, Tulsa
Stauffer, P. H. (1967): Grain-flow deposits and their implications, Santa Ynez Mountains, California. J. Sed. Petrol. **37,** 487–508, Tulsa
Steiger, T. (1981): Kalkturbidite im Oberjura der Nördlichen Kalkalpen (Barmsteinkalke, Salzburg, Österreich). Facies **4,** 215–348, Pls. 12–24, 56 Figs., Erlangen
Stow, D. A. V. (1979): Distinguishing between fine-grained turbidites and contourites on the Nova Scotian deep water margin. Sed. **26,** 371–387, 9 Figs., Oxford
Szulczewski, M. (1968): Slump structures and turbidites in the Upper Devonian limestones of the Holy Cross Mts. Acta Geol. Polonica **18,** 303–324, Warsaw
Thiedig, F. (1975): Submarine Brekzien als Folge von Felsstürzen in der Turbidit-Fazies der Oberkreide des Krappfeldes in Kärnten (Österreich). Mitt. Geol.-Paläont. Inst. Univ. Hamburg **44,** 495–516, Pls. 40–42, 6 Figs., Hamburg
Thomson, A. F., Thomasson, M. R. (1969): Shallow to deep water facies development in the Dimple limestone (Lower Pennsylvanian), Marathon region, Texas. In: Friedman, G. M. (ed.): Depositional Environments in the Carbonate Rocks. Soc. Econ. Paleont. Min. Spec. Publ. **14,** 57–78, 32 Figs., Menasha
Tucker, M. E. (1969): Crinoidal turbidites from the Devonian of Cornwall and their paleogeographical significance. Sed. **13,** 281–290, Amsterdam
Tucker, M. E. (1973): Sedimentology and diagenesis of Devonian pelagic limestones (Cephalopodenkalk) and associated sediments of the Rhenohercynian Geosyncline, West Germany. Neues Jahrbuch Geol. Paläont. Abhandlungen **142,** 320–350, Stuttgart
Tucker, M. E. (1974): Sedimentology of Palaeozoic pelagic limestones: the Devonian Griotte (Southern France) and Cephalopodenkalk (Germany). Spec. Publ. int. Ass. Sed. **1,** 71–92, Oxford
Tucker, M. E., Kendall, A. C. (1973): The diagenesis and low-grade metamorphism of Devonian styliolinid-rich pelagic carbonates from West Germany: possible analogues of Recent pteropod oozes. J. Sed. Petrol. **43,** 672–687, Tulsa
Tyrrell, W. W. (1969): Criteria useful in interpreting environments of unlike but time-equivalent carbonate units (Transill-Capitan-Lamar), Capitan Reef Complex, West Texas and New Mexico. Soc. Econ. Paleont. Min. Spec. Publ. **14,** 80–97, 17 Figs., Tulsa
Unrug, P. (1963): Istebna beds – a fluxoturbidite formation in the Carpathian flysch. Soc. Geol. Pologne Ann. **33,** 49–92, Warsaw
Valeton, I. (1960): Vulkanische Tuffiteinlagerungen in der nordwestdeutschen Oberkreide. Mitt. Geol. Staatsinst. Hamburg **29,** 26–41, Hamburg

Van der Lingen, G. J. (1969): The turbidite problem. New Zealand J. Geol. Geophys. **12,** 7–50, Lower Hutt

Voigt, E. (1959): Die ökologische Bedeutung der Hartgründe („Hardgrounds") in der oberen Kreide. Paläont. Z. **33,** 129–147, Stuttgart

Voigt, E., Häntzschel, W. (1964): Gradierte Schichtung in der Oberkreide Westfalens. Fortschritte Geol. Rheinland Westfalen **7,** 495–548, 18 Pls., 7 Figs., Krefeld

Walker, R. G. (1965): The origin and significance of internal sedimentary structures of turbidites. Proc. Yorkshire geol. Soc. **35,** 1–32, 5 Figs., Leeds

Walker, R. G. (1967): Turbidite sedimentary structures and their relationship to proximal and distal depositional environments. J. Sed. Petrol. **37,** 25–43, Tulsa

Walker, R. G. (1970): Review of the geometry and facies organization of turbidites and turbidite-bearing basins. In: Lajoie, J. (ed.): Flysch Sedimentology in North America. Spec. Paper geol. Ass. Canada **7,** 219–251

Walker, R. G. (1975): Generalized facies models for resedimented conglomerates of turbidite associations. Geol. Soc. Amer. Bull. **86,** 737–748, New York

Walker, R. G. (1978): Deep-water sandstone facies and Ancient submarine fans: models for exploration for stratigraphic traps. Amer. Ass. Petrol. Geol. Bull. **62,** 932–966, 23 Figs., Tulsa

Walker, R. G., Mutti, E. (1973): Turbidite facies and facies associations. In: Middleton, G. V., Bouma, A. H. (eds.): Turbidites and deep sea sedimentation. Soc. Econ. Paleont. Min., Pacific Sect., Short Course, 119–157

Walther, J. (1893): Die Lebensweise der Meeresthiere. Einleitung in die Geologie als historische Wissenschaft **2,** 195–531, Jena: Fischer

Wendt, J. (1974): see 4.2.4.7

Wilson, J. L. (1969): Microfacies and sedimentary structures in "deeper water" lime mudstones. In: Friedman, G. M. (ed.): Depositional Environments in Carbonate Rocks. Soc. Econ. Paleont. Min. Spec. Publ. **14,** 4–19, 5 Figs., 2 Pls., Menasha

Wilson, J. L. (1975): see 1.4.1

Wise, S. W., Jr. (1973): Calcareous nannofossils from cores recovered during Leg 18, Deep Sea Drilling Project: biostratigraphy and observations of diagenesis. Initial Reports Deep Sea Drilling Project **18,** 569–615, Washington

Wolfe, M. J. (1968): see 3.9

Yurewicz, D. A. (1977): Sedimentology of Mississippian basin-facies carbonates, New Mexico and West Texas – the Rancheria formation. Soc. Econ. Paleont. Min. Spec. Publ. **25,** 203–219, 10 Figs., Tulsa

Zamarreno, I. (1972): Las litofacies carbonatadas del Cambrico de la zona Cantabrica (NW. Espana) y su distribucion paleogeografica. Trabajos Geol. **5,** 118 p., 17 Pls., Oviedo

Zankl, H. (1969): see 3.9

Ziegler, B. (1967): Ammoniten-Ökologie am Beispiel des Oberjura. Geol. Rdsch. **56,** 439–464, Stuttgart

Zuffa, G. G., Gaufio, W., Rovito, S. (1980): Detrital mode of evolution of the rifted continentalmargin Longobucco sequence (Jurassic), Calabrian Arc, Italy. J. Sed. Petrol. **50**/1, 51–61, 10 Figs., Tulsa

10.3 Facies Models

"... environmental analysis is at best an imprecise art, rather than a deterministic discipline" (Selley, 1970). If we assume that a facies describes the result of certain depositional and biotopic conditions, but not the environmental factors themselves which are associated with these conditions, then we can make the job of critically categorizing our own findings easier by using "facies models" or "sedimentation models." These models are very strongly controlled by our present knowledge of Recent carbonate sedimentation as well as by the abundance of analogous Ancient examples. Because good sedimentation models summarize the complex multiplicity of organic and inorganic factors acting during the origin of a facies environment or a system of facies zones, these models have both advantages and disadvantages as aids to interpretation. The advantages lie in the representation of the causes most responsible for the genesis of microfacies and facies types and in the enumeration of features used in the "checklists" which can be considered in an interpretation. Alternative interpretations are, then, much more easily recognized (see Wolf, 1973). One disadvantage which cannot be overlooked in a facies interpretation which clings too closely to a model is the dependence on the current state of knowledge and the thoroughness of previous investigations. The "Bahama bank model", which was always used as a comparison up until about 10 years ago in the majority of papers on Ancient shallow-water carbonates, is an example of too strong a dominance of a Recent depositional model; this is governed by the fact that the area has been extremely well studied. In addition, there are also varying conceptions of the definition of "models":

For instance, Krumbein and Graybill (1965) see a model as a framework supported when possible by mathematical analyses, which help relieve the organization of geological investigations. Such models may be described by statistical methods or represented by conceptual models in the form of diagrams (Wolf, 1973). However, a model may also be the formal expression of the theory of a causal situation which is then made responsible for the origin of the features observed (Kendall and Buckland, 1960). Finally, statistical "simulation models" show which combination of environmental factors is the most probable explanation for the origin of a certain facies type (see Harbaugh and Merriam, 1968, and Sect. 7.3). Seibold (1970) indicated with good reason that the analysis of individual factors and the attempt to gain an overall view is more important than the model itself.

The facies models described below show quite clearly that the "overall picture" of the same facies complexes provides various kinds of information depending on the emphasis of different factors.

10.3.1 Principles of Facies Models

The majority of models suggested for the modes of carbonate sedimentation are based on comparative sedimentological studies and Ancient sediments (Ginsburg, 1974) and are supported by the "law of facies correlation" developed by J. Walther (1893: 979), whereby "basically only those facies and facies environments which are adjacent to each other in the Recent can be geologically superposed." Here, though, it must be considered that not all the biotopes and depositional environments occurring in the geologic record are represented by examples in the Recent and that special Recent conditions were created for the atypical sedimentary relationships during and following the Quarternary (see Curray, 1975). This is especially true for the shelf region in which "relict sediments" are present which have been redeposited to different degrees; see 2.2.2.

It should be stressed that Walther's "Law of Facies" applies only to sequences without major breaks in the sedimentary succession.

In addition to these limitations must be added the observation that "adjacent seas", which are very important for facies comparisons and are caused by different

Table 57. Sedimentation models for humid and arid adjacent seas (after Seibold, 1970)

Criteria	Humid Environment	Arid Environment
Water circulation		
– Surface current	Outflowing	Inflowing
– Bottom current	Inflowing	Outflowing
Stratified waters	Clear-cut, when winter convection has not reached the bottom	Less clear
Properties of water near the bottom		
– Salinity	Low	High
– Oxygen content	Low	High
– Nutrients (PO_4, SiO_2)	High	Low
Sediment properties		
– Carbonate content	Low, higher in stagnant water	High
– Organic carbon	High	Low
– Metal content	High	Low
Biofacies boundaries	Often sharp	Often not sharp
Coastal lagoons	Peat	Evaporites
Recent examples	Baltic Sea, Hudson Bay, Black Sea	Persian Gulf, Red Sea, European Mediterranean Sea

water circulations and deviating climatic relationships, can be described by two models (Seibold, 1970). From Table 57 it becomes quite clear that generalized models, such as the Irwin model sketched below, are useful for the interpretation of Ancient sediments only if the possible paleogeographical and paleoclimatic situations are taken into consideration. Attention to the paleogeographic situation also seems necessary whenever the noncomparability of Recent and Ancient sedimentary factors, such as stratified waters, is discussed (Degens and Stoffers, 1976).

10.3.2 Shelf – Shelf Margin – Basin: The Wilson Model

Wilson (1975) differentiates nine facies belts (see Figs. 55 and 70) which are characterized by lithology, sedimentary structures, organisms, and "Standard Microfacies Types" (see Sect 8). These facies belts are controlled by genetic factors which are dependent on the dip of the sedimentary base, the water energy and climate, as well as the geological age and organic control of sedimentation.

Facies Belt 1. Basin

1 A. *Turbidite and leptogeosynclinal deep water facies* (fondothem after Rich, 1951): Geosynclinal troughs with a) *calcareous turbidites* or allodapic limestones (see Sects. 2.2.4 and 10.2.4) and b) *deep leptogeosynclines* with chiefly pelagic carbonate

sedimentation, few allodapic limestones and formation of radiolarites and argillaceous sediments below the carbonate solution boundary. Early diagenetic SiO_2 formation, often in the form of intercalated cherts roughly concordant with the bedding, is common. The limestones are very regularly laminated and frequently exhibit a rhythmic sedimentation pattern. Breccia beds and slump structures indicate sediment transport and redeposition. The biota is almost exclusively planktonic and nektonic. The micro-organisms include, among others, planktonic algae (coccolithoporoids, globochaetes, calcispheres) and foraminifera, calpionellids, radiolarians, and diatoms. The megafauna is characterized by planktonic and epiplanktonic pelecypods, free-swimming crinoids, ammonites, siliceous sponges and graptolites.

1 B. *Cratonic basin carbonates (starved and mostly euxinic; fondothem):* Intracratonic or marginal cratonic basins (miogeosynclinal) without benthonic carbonate production. Sedimentation is dependent on the amount of influx of fine argillaceous and siliceous material and the plankton production in the surface waters. The sedimentation base lies below the oxygenation level in euxinic conditions. The bottom water may become hypersaline. Dark thinly bedded limestones with dark shales and silty shales as well as some thinly bedded anhydrites occur. Lime mudstones and calcisiltites with very small peloids and bioclasts dominate. Crinoid accumulations may be present. The sedimentary structures are characterized by a very even planar mm lamination, by ripple cross-lamination, and by small-scale rhythmic bedding consisting of limestones intercalated with thin shales. Chert is very abundant. The organisms are identical with those groups described in Facies Belt 1 A.

Facies Belt 2. Open sea shelf (deep undathem after Rich, 1951)

Sedimentation in water, which is tens of meters or even a hundred meters deep, generally well-oxygenated and of normal salinity, with good current circulation. Because wave action is missing, relatively uniformly composed sediments occur as widespread neritic shelf deposits. These are very fossiliferous limestones interbedded with marl. The beds are generally well segregated. In addition to bioclastic and whole fossil wackestone, occasional beds of winnowed bioclastic grainstone and coquina occur; there is much pelleting of the micrite matrix and some calcisiltite. In contrast to Facies Belt 1, the sediments of Facies Belt 2 are homogenized by thorough burrowing. The thin to medium beds are often wavy to nodular. Ball and flow structures are common in argillaceous limestones. Bedding surfaces ordinarily show diastems and lag concentrates of fossils. The shelly fauna is very diverse and preserved both as infauna and epifauna. Stenohaline forms such as brachiopods, corals, cephalopods, and echinoderms are more abundant than those in Facies Belt 7, which in many respects is similarly formed. Facies Belt 2 does not appear to have any Recent analogs.

Facies Belt 3. Deep shelf margin or basin margin (toe of slope; clinothem)

This facies is formed at the toe of slope of a carbonate-producing shelf. Carbonate sediments consist of contributions from pelagic organisms plus fine detritus moved off from adjacent shallow shelves. The water is at least the same depth as Facies Belt 2 and perhaps 200–300 m deep. The basin is situated generally below wave base and barely at oxygen level. The strata are chiefly thin, well-segregated carbonate beds with minor interbeds or mere partings of clayey and siliceous material. The sediments are mostly mudstone and wackestone with some calcisiltite and including some microbreccia beds, and coarser bioclastic-lithoclastic packstones. Some beds are laminated and evenly rhythmic; others are massive and unlaminated, and characterized by megaslump features. Occasionally micrite bioherms occur. Chert is common. The fauna is present as bioclasts on the open shelf and is derived principally from upslope; benthonic organisms and some pelagic forms are admixed.

Facies Belt 4. Foreslope facies of carbonate platform (marine talus, clinothem)

The slope is generally located above the lower limit of oxygenated water and stretches from above the wave base to below it. The material is deposited on an incline which may be as steep as 30 degrees. The sediment is somewhat unstable and varies greatly in size and shape. There may be bedded, fine-grained layers with megaslumps; some strata consist principally of lime sand and of mound- or lens-shaped masses of trapped and stabilized carbonate sediment. The sediments are lime silts and bioclastic wackestone-packstone. Lithoclasts and particles derived from Facies Belt 5 are common (e.g., cm- to m-sized coarse debris from the reef; fore-reef, see 10.2.3). The sedimentary structures are characterized by large exotic blocks interrupting thinly bedded strata, syngenetic slumps, and breccias. Terrigenous clastics are rare. The fossils consist of allochthonous elements (mostly bioclastic debris) but also of some very varied encrusting organisms.

Facies Belt 5. Organic reef or platform margin (organic buildup)

"Shelf margin reefs" vary depending on the water energy, steepness of the slope, organic productivity, amount of frame construction, binding or trapping, frequency of subaerial exposure, and consequent cementation. This is described in Sect. 10.3.5. The material is massive limestone and dolomite in places consisting solely of organisms. There are masses and patches of boundstones; the interstices may be filled with arenitic and micritic sediments. Cavities are common. Terrigenous particles are essentially absent. The growth of sessile organisms is determined by the water energy.

Facies Belt 6. Winnowed platform edge sands

These take the form of shoals, beaches, offshore or tidal bars in fans or belts, or are eolian dune sands. Sites of such marginal sands range from well above sea level to 5 or 10 m deep. The well-winnowed lime sands which are deposited in the normal marine and well oxygenated environment form shifting substrates which are not hospitable to sessile and non-adapted vagile benthonic organisms. These are cross-bedded, somewhat dolomitic calcarenites. Terrigenous quartz may be present. The biota consists of worn and abraded bioclasts of benthonic animals living in Facies Belts 4 or 5, and a few ecologically adapted forms such as gastropods, dasycladacean algae and certain foraminifera.

Facies Belt 7. Open marine platform facies (shallow undathem)

In open lagoons, bays, and straits behind the outer platform edge the water circulation is very moderate. Water is shallow, generally a few meters to tens of meters deep. Salinity varies from essentially normal marine to somewhat hypersaline. Stenohaline organisms are, thus, generally excluded. Mudstones, bioclastic wackestones, calcarenites, and lumachelles occur. The fauna may be abundant with mollusks, arthropods, sponges, foraminifera, and algae, which are particularly common. Abundant marine grasses play important roles in trapping and stabilizing fine sediment.

Facies Belt 8. Restricted circulation shelf and tidal flats

This facies includes mostly fine sediment in very shallow, cut-off ponds and lagoons which have restricted circulation and hypersaline water. Geographically the lagoons may be classified into those behind or between barrier reefs, those formed behind coastal spits, or those within atolls (see 2.2.2). The tidal sediments are characterized by lime mud on intertidal flats and coarser sediment in tidal channels and local beaches. The salinity as well as the oxygenation are subject to highly variable conditions; wind-blown clastics may contribute significantly in some places. All of these constitute a stress environment for organisms. Sediments consist of mudstones or wackestones often with much dolomite; peloids are most common. Laminated mudstone, fenestral fabrics, algal stromatolites, and cross-bedding occur in channel sands. The organisms may occur locally in great abundance, but are generally very limited in diversity. These are mainly gastropods, algae, certain foraminifera, and ostracods.

Facies Belt 9. Platform evaporites

The supratidal and inland pond environment of the restricted marine platform developed in an arid climate – the areas of sabkha, salinas, salt flats. Marine flooding is sporadic. Gypsum or anhydrite is formed as dense brines move down through the sediment or are pulled to the surface through evaporation. The precipitated sulfate is unstable and may deform through crystal growth, intake of water of crystallization, or compaction. Prevailing rock types are nodular and wavy anhydrite, or gypsum interlaminated with dolomite. Such rock types are commonly associated with red beds, laminates, mud cracks, stromatolites, and spongiostrome structures (= thrombolite structures), flaser structures, and caliche crusts. Terrigenous clastics may be very common. There is almost no indigenous fauna except for blue-green algae, brine shrimp.

Naturally, these facies belts are ideal models which are realized neither in lateral nor in vertical sequences. Using the example of the facies development in the Southern Alpine Permian and in the Northern Alpine Mesozoic (Fig. 67), it can be shown that individual facies may be exceptions to the paleotectonic situation of the carbonate platform:

In the Upper Jurassic of the Northern limestone Alps an extended platform developed over the deep water carbonates. This platform clearly exhibits Facies Belts 5 and 6, whereas Facies Belts 7, 8, or 9 are completely absent. Similarly, the Facies Belts 2–4 only seem to have been sporadically formed. Facies Belt 1 A,

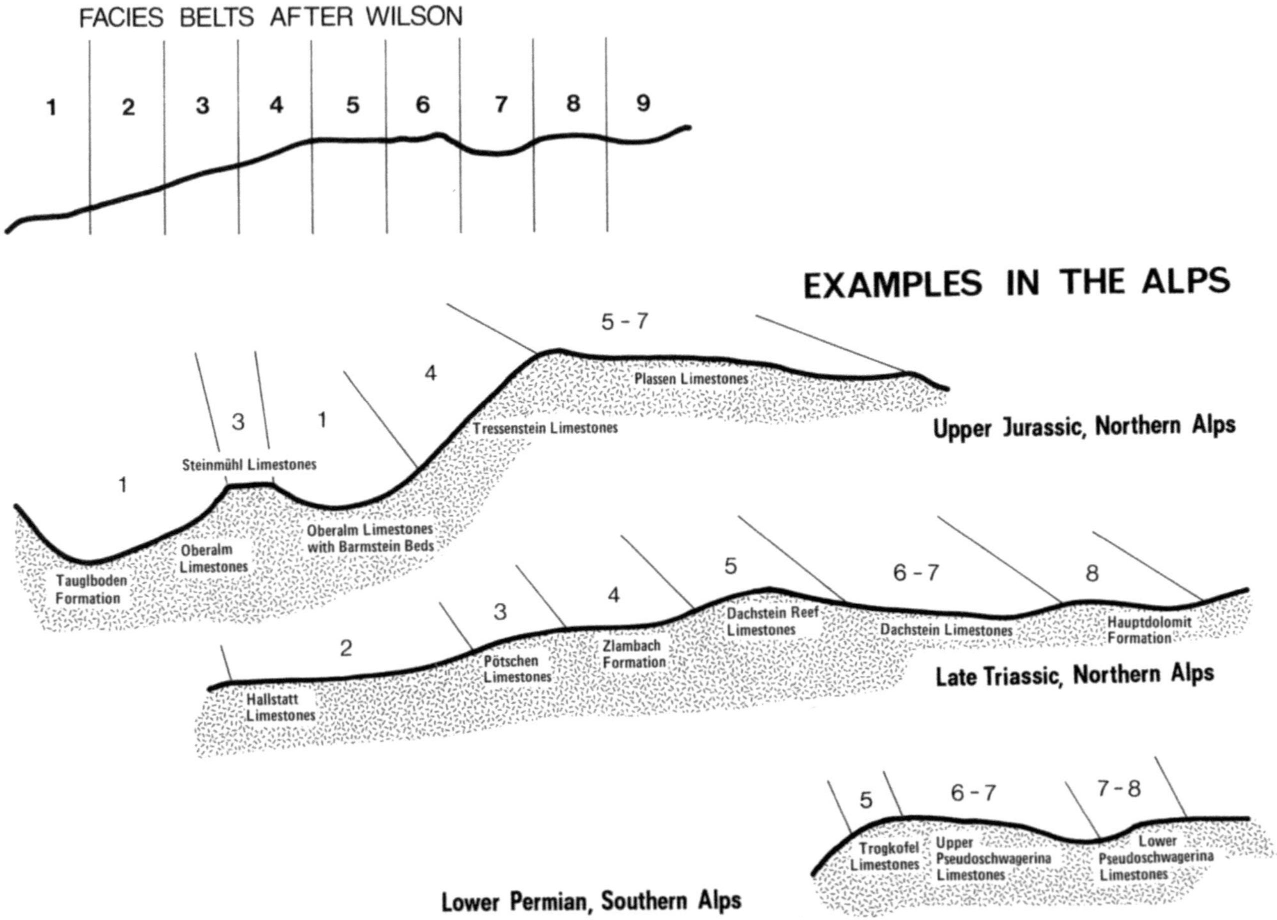

Fig. 67. Schematic classification of a few Mesozoic and Permian limestone sequences from the Alps according to the Wilson Model (see Fig. 55). Compiled after Fenninger and Holzer (1972) – Upper Jurassic, Zankl (1971) – Late Triassic, and Buggisch et al. (1976) – Permian, and author's own investigations

however, is well documented as a uniformly subsiding environment in the Oberalm beds. In the Northern Alpine Late Triassic, particularly in the Norian, Facies Belts 2 to 8 are readily recognizable, according to the schemes developed by E. Flügel and Flügel-Kahler (1963), A. G. Fischer (1964), and Zankl (1971). It is more difficult to categorize the facies units present in the Southern Alpine Lower Permian (see E. Flügel, 1971, Buggisch et al., 1976). The Lower Pseudoschwagerina beds, characterized by a cyclic alternation of shallow marine carbonates and clastic sediments, probably correspond to Facies Belt 7, whereas the clastic Grenzland beds (not pictured in Fig. 67) represent near-coastal sediments (Facies Belt 8?). Sediments of the open marine platform environment (Facies Belt 6) as well as lime sands occur together with the predominantly carbonate Upper Pseudoschwagerina beds. Facies Belt 5 is present in the tubiphyte reefs of the Trogkofel limestones. Sediments of Facies Belts 4 to 1 are not known in the Southern Alps, which suggests that the Lower Permian shelf platform was very steep.

10.3.3 Carbonate Platforms

Very often the term "platform" occurs in conjunction with the facies and sedimentological characterization of shallow-water carbonates. The definitions deviate from each other: Some authors understand a platform to be a very extensive shallow-marine carbonate sedimentary area comparable to the Recent Great Bahama Banks (see Figs. 68 and 69), whereas other authors use the term to represent paleogeographic and geodynamic evolution (e.g., d'Argenio, 1976; Jenkyns, 1970). This conception of the term describes a platform included in facies belts 4 to 8 of Wilson (1975). Wilson sees carbonate platforms as extended carbonate bodies characterized by a more or less horizontal surface or sedimentary base and by an abrupt change in relief along the shelf edge or platform margin. The shelf edges exhibit high-energy sediments.

The following aspects are important for the use of the term platform:

- Shallow marine and marginal marine carbonate sedimentation
- More or less flat top in the euphotic zone
- Large to very large surfaces with uniform facies types (see Fig. 69)
- Delimitation of the sedimentary areal by high-energy carbonates or very large organic structures such as reefs
- Changing subtidal, intertidal, and supratidal conditions
- For the most part very little terrigenous influx
- Relatively low relief on the platform
- Very small dip along the platform ending in a steep slope

The relationships in the area around the *Bahama Banks* are considered to be the prototype of platform sedimentation: The "coralalgal facies" is a skeletal sand which covers much of the outer platform. The sediment consists of abundant coral and coralline algal fragments together with other skeletal grains. The source of this sediment is the coral reef tract on the windward side of Andros Island (see Fig. 68). The "oolite facies" and the "oolitic facies" form sand shoals on the barrier rim. These facies units are typified by an admixture of calcareous ooids with other

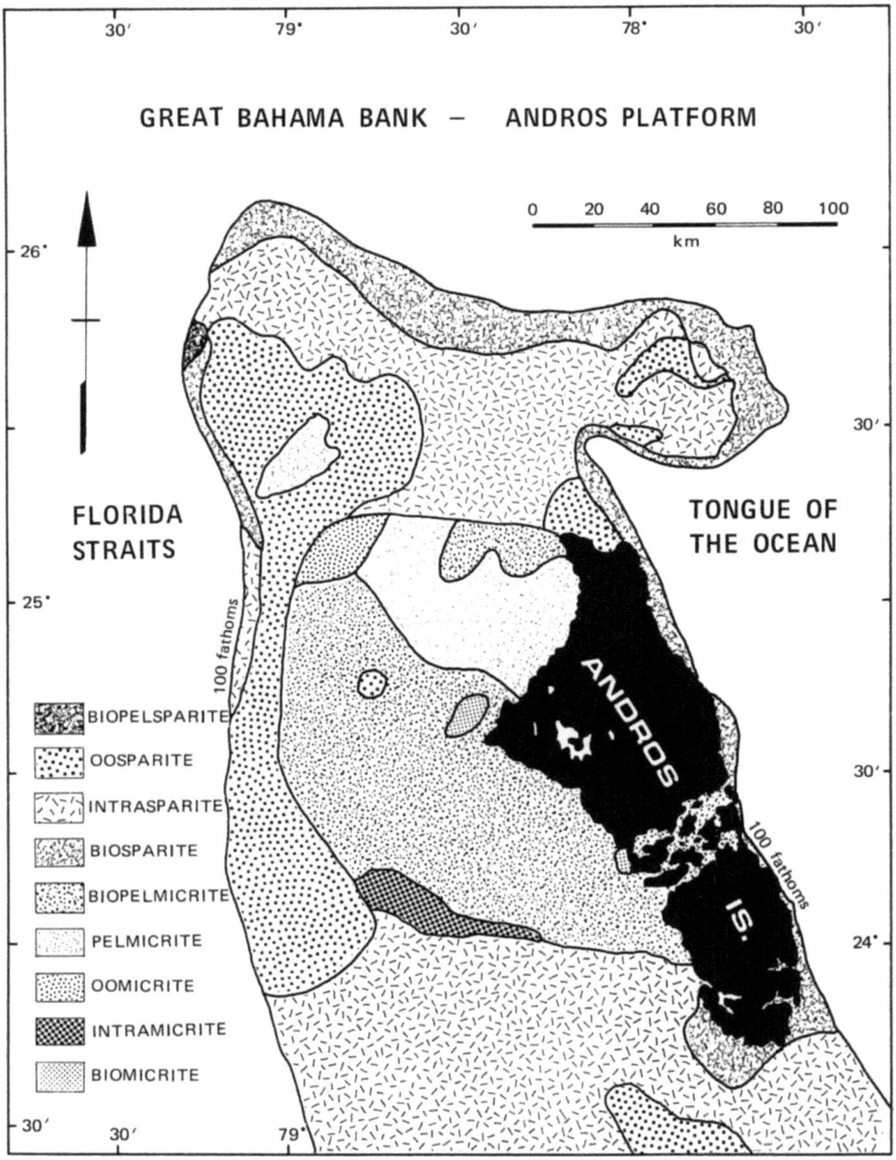

Fig. 68. Distribution of sediment on the Great Bahama Bank. Oosparite corresponds to the "oolite facies"; intrasparite represents the "aggregate grain (grapestone) facies"; the "coral-algal facies" is characterized by biosparite and biopelsparite. All other micritic carbonate types are referred to the "mud facies", i.e., the bioclastic "mud subfacies" and the "peloid-mud subfacies". After Imbrie and Purdy (1963)

grains. Bodies of oolite sand are heavily cross-bedded. The "grapestone facies" occupies large areas of the shelf lagoon. Aggregate grains are the dominant particles. The center of the shelf lagoon is characterized by a "mud facies", differentiated into a "pellet-mud subfacies" and a "skeletal-mud facies".

The facies types observed in the Bahamas are also found in Ancient environments. Table 58 compares the Bahama Bank with the Upper Jurassic "Sulzfluh Bank" which is preserved in the facies types of the Sulzfluh limestones (Vorarlberg and Graubünden, Swiss-Austrian border).

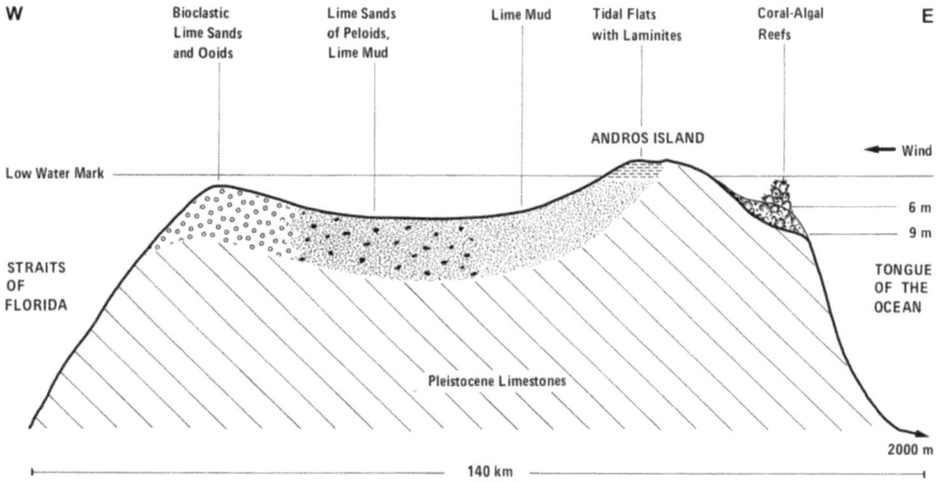

Fig. 69. Facies types on the Bahama platform. The Y zone in the Irwin Model corresponds to the carbonate sand bars with ooids; the Z zone, the area directly adjoining them on the right. After Blatt et al. (1970)

The carbonate sediments of the Arabian coast of the *Persian Gulf* represent a carbonate platform attached to a continental hinterland (Purser, 1973, 1980). The shelf carbonates of the Trucial Coast Embayments lie adjacent to a high-energy beach facies, and these, in turn, to an inland eolian facies. More typically, however, lagoons and lagoon barriers fringe the shore. Here lagoon barriers of skeletal sands, coral reefs and oolite shoals develop. The oolite deltas are associated with the mouth of tidal channels, between barrier islands. Reefs normally develop immediately seaward from the barrier island flank. Gastropod-rich, pelleted aragonite sands and muds are abundant in the lagoon, some areas containing many foraminifera. Fringing the lagoons along protected shorelines are widespread algal mats. Sabkha environments occur in broad supratidal zones.

Shallow water carbonate platforms and reefs are drowned when tectonic subsidence or rising sea level hinders carbonate accumulation and benthonic carbonate production ceases (W. Schlager, 1980). *Drowned platforms* are common in the geologic record (e.g., Middle/Late Devonian of Central Europe; Anisian of Central and southeastern Europe; Upper Jurassic of southeastern Europe; Albian/Ceno-

Table 58. Platform sedimentation illustrated by the partially preserved Upper Jurassic "Sulzfluh Bank" in Graubünden and Vorarlberg, Swiss-Austrian border; and in the Recent Bahama Banks

Sulzfluh Bank, Min. surface area at least 1500 km²		Bahama Bank Surface area 167,000 km²	
Facies Types	Criteria	Facies Types	Criteria
Micrite facies (Intertidal to subtidal, Energy Index $I_2 - II_2$)	Micrite, pelmicrite, dismicrite, biomicrite, biopelmicrite, intramicrite. Fecal pellets, bioturbation ostracods, birdseyes, dasyclads, miliolids, foraminifera, filamentous algae	Mud facies and pellet mud facies	Calcareous mud, sometimes up to 40% peloids (fecal pellets). Biofraction low (about 20%); algae, foraminifera, pelecypods. Very low water energy
Intrasparite facies (subtidal, Energy Index $II_3 - V_1$)	Intrasparite, biointrasparite, oncosparite; quite a number of algae and foraminifera	Aggregate grain (grapestone) facies	Grapestones dominant (to 50%). Peloids relatively rare, organism content to 20%; foraminifera, algae, pelecypods. Restricted water movement
Oolite facies (subtidal, Energy Index $IV_3 - V_2$)	Oosparite, oomicrosparite. Few organisms	Oolite facies	Calcareous ooids dominant (to 70%). Skeletal grain content very low; maximum turbulence
Absent	–	Coral-algal facies	Bioclasts of corals and algae dominant, pellets present (to about 20%). Rough water

manian of Texas); they typically exhibit a sequence of neritic deposits, rapidly passing upward into deep marine sediments. Usually hardgrounds, with crusts of oxide, phosphate or glauconite, separate neritic and deep-water deposits and indicate a period of non-deposition within the marine environment. Some platforms were exposed to meteoric environments prior to drowning. Platform drowning requires very special conditions, such as environmental stress that reduces growth of carbonate benthos by 1–2 orders of magnitude, or tectonic pulses of subsidence that are at least one order of magnitude faster than the long-term average of subsidence, or eustatic pulses of sea level (see W. Schlager, 1980; Kendall and Schlager, 1981).

10.3.4 Shallow Shelf Environment Without Terrigenous Influx: The Irwin-Lees Model

Irwin (1965) and Shaw (1964) developed a theoretical model for marine epicontinental environments not controlled by terrigenous influx; there are three zones characterized by different water energies (see Fig. 70):

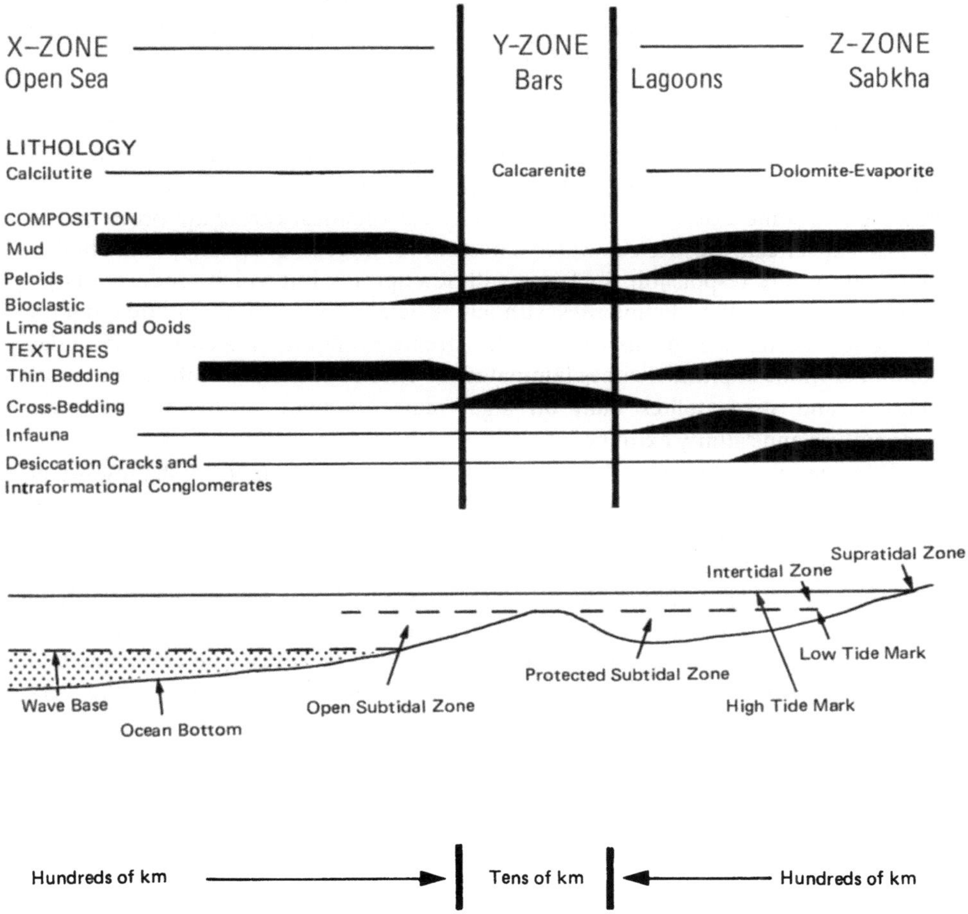

Fig. 70. Simplified representation of the Irwin Model for carbonate sedimentation in (arid) areas with no terrestrial influence. After Selley (1970)

1. X-Zone: Mud is sedimented in deeper water below the wave base. Bottom currents provide oxygen, hence, colonization of the area by various benthonic organisms takes place. The resulting carbonate sediment consists of micrite with arenitic and ruditic shelly bioclasts. The sedimentation rate is very slow. The fine-grained sedimentary material derives from Zone Z. If Zone Z lies in a reducing environment, authigenic minerals (e.g., glauconites) are formed; this leads to sedimentation of much organic substance (thus, dark-colored limestones).

2. Y-Zone: Very little mud is deposited in shallower water above the wave base, because of the higher water energy. The sediment thus consists chiefly of arenitic bioclasts of copious and diverse flora and fauna, which are present in this zone because of good oxygenation and an ample supply of food. The limestones correspond to calcarenites or calcirudites, generally consisting of broken and rounded particles. Currents cause good winnowing and current marks (cross-bedding, etc.). In environments with carbonate saturation calcareous ooids are formed relatively near the coast. This is the zone where reefs may develop.

3. Z-Zone: The water energy is sharply lowered in the shallowest water near the shorelines by oolite bars or reefs formed in Zone Y. Predominantly lime mud is sedimented, whose microbioclastic components derive in part from Zone Y; however, to some extent, biochemical or physico-chemical carbonate precipitation is also in effect. Restricted water circulation, increased salinity, and higher temperatures are responsible for fauna with few species, but which produce a large number of fecal pellets. Pelmicrites with few skeletal grains appear. In arid climates evaporites are formed in this zone when extreme evaporation occurs. In the tidal zones and in the supratidal zones laminated micrites with fenestral fabrics and mud pebbles, and stromatolites built on algal mats dominate, owing to the great temperature and salinity extremes.

This model has repeatedly proved its worth in the study of shallow-water carbonates from the Paleozoic and the Mesozoic (e.g., Laporte, 1969; Barzel and Friedman, 1970; Friedman et al., 1971). It is important to notice the possibilities for differentiating between open and protected subtidal environments expressed in the model (see Fig. 71). As Fig. 72 shows, the model can also be recognized not only in the carbonate sedimentation areas in the Persian Gulf and in the Caribbean Sea, but also in the clastic sedimentation area off the coast of Texas. The environmental parameters associated with sediment formation can be easily seen in the very simplified sketch in Fig. 71.

Friedman et al. (1971) have drawn attention to a modification of the model, illustrated by Jurassic carbonates along the coastal area of Israel. Dark shales and

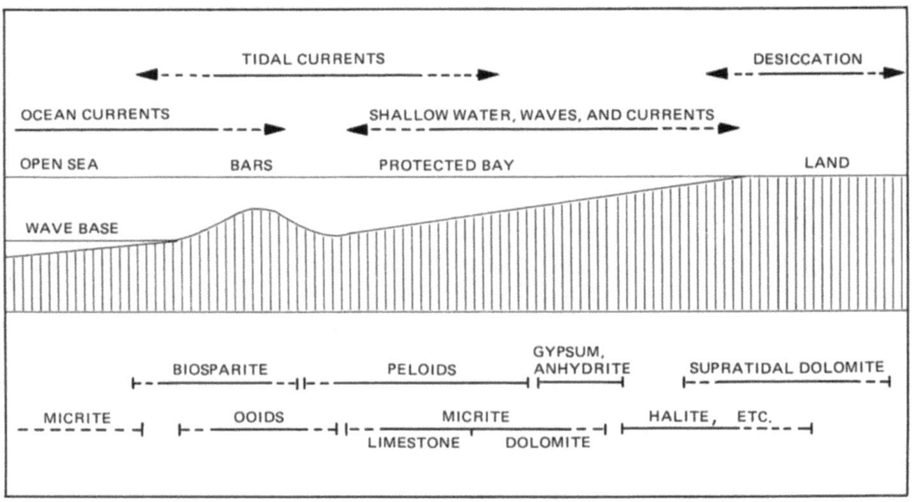

Fig. 71. Environmental factors and sediment formation in marginal marine and shallow marine areas according to the Irwin Model. After Anstey and Chase (1979)

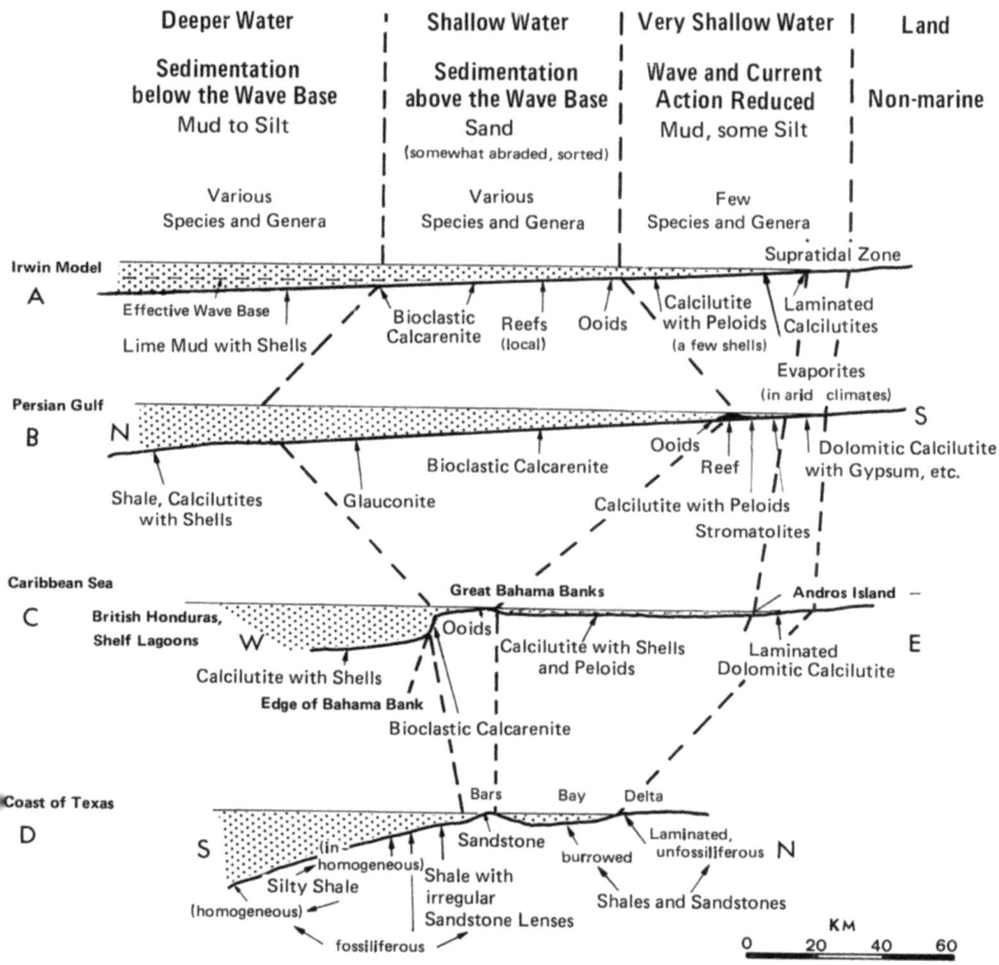

Fig. 72 A–D. Examples of sedimentation patterns comparable to the Irwin Model. **B** and **C** carbonates, **D** clastics. After Heckel (1972)

limestones, sometimes with grading, and different sizes of pebbles from shelf margin carbonates indicate relatively deep water in Zone X; organic material accumulates here and reducing conditions prevail because of very weak oxygenation. Zone X may be important as the source rock; Zone Y, as the reservoir rock for hydrocarbons.

Lees (1973) spoke out against overestimating the Irwin Model and pointed out that only by precisely identifying the wave base depth and the lower boundary of the photic zone, and by taking into consideration the climatic dependence of the depth range of the organisms, can the model be improved. However, because neither the time factor (various organisms in "shelf" regions of various ages) nor the possible shifting of shelf regions and the associated changes in temperature and

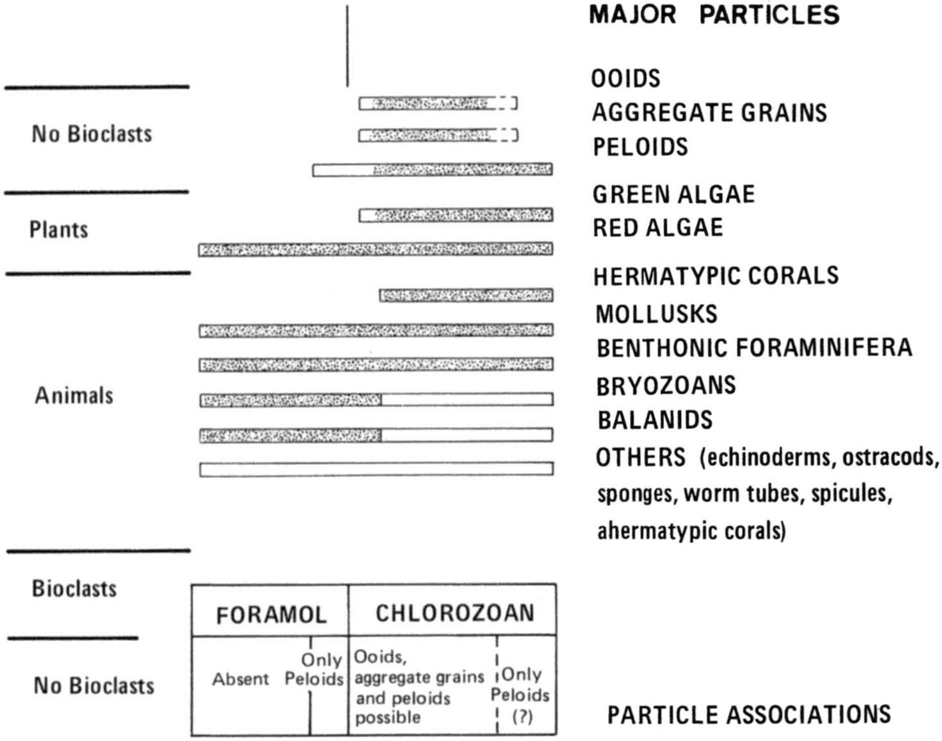

Fig. 73. The particle associations characterize two major groups in shallow shelf areas: "foramol" = dominance or presence of foraminifera and molluskan bioclasts, "chlorozoan" = Chlorophyceae (green algae) and hermatypic corals (zoantharia) are common. The stippling should help to visually emphasize the most important and more common elements in the major groups. After Lees and Buller (1972)

salinity conditions can be determined in the original model, the information provided by the model has limitations.

Indeed, water temperature and possibly salinity decisively control the composition and distribution of shelf sediments, particularly of shelf sands. A comparison of 78 shelf regions by Lees and Buller (1972) shows that two particle associations are present within carbonate sands (Fig. 73):

- In temperate water of the latitudes far from the equator *"foramol" associations* characterized by molluskan bioclasts, benthonic foraminifera, echinoderms, bryozoans, balanids, and red algae occur.
- In warm water between 30 °S and 30 °N the *"chlorozoan" associations* dominate, recognizable from the occurrence of hermatypic corals and/or chlorophycean algae, such as *Halimeda.*

Ooids and aggregate grains are preferentially found together with the chlorozoan associations; the same holds for peloids. Presence or absence of non-skeletal

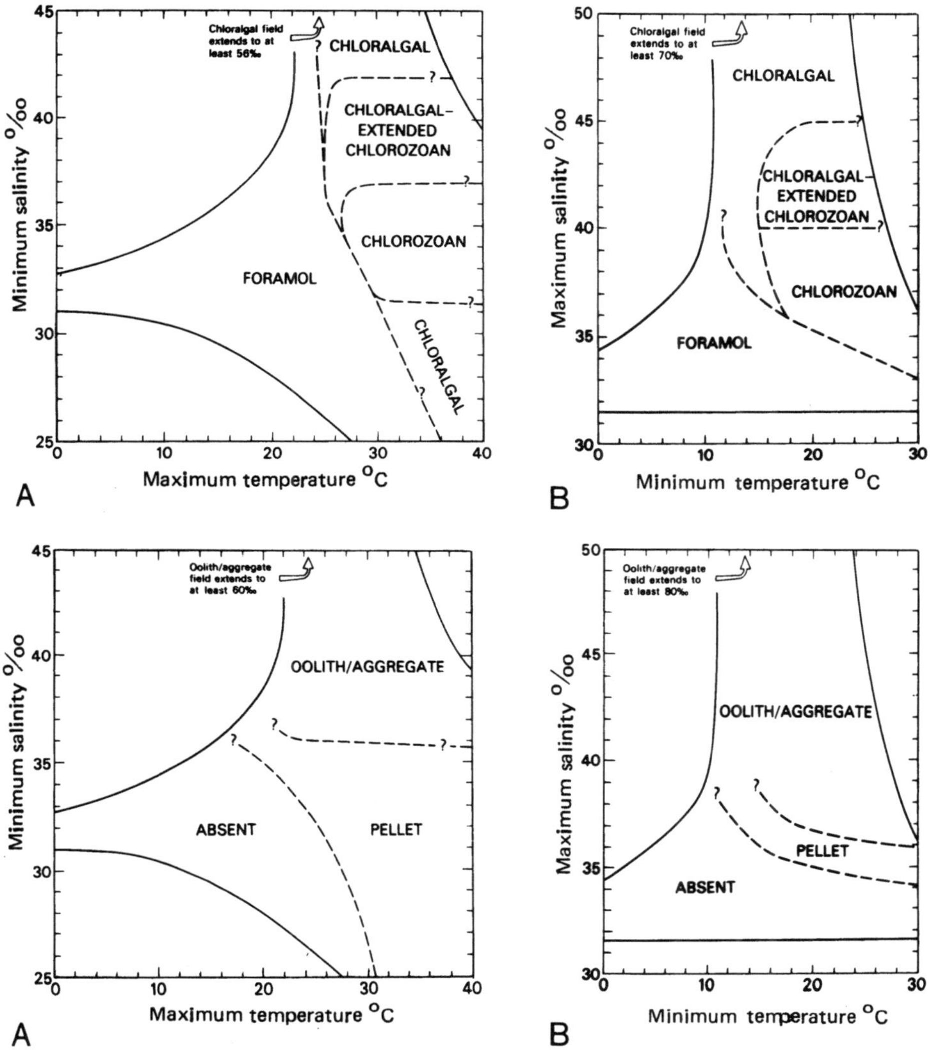

Fig. 74 A, B. Salinity-Temperature-Annual Range diagram pairs for skeletal grain associations (*top* two figures) and for non-skeletal grain associations (*bottom* two figures) in modern shelf carbonate sediments. The dotted lines indicate the approximate limits of near-surface salinity/temperature combinations existing in present-day shelf seas and lagoons. After Lees (1975)

grain associations may be explained by water temperature (minimal temperature >15 °C mean temperature >18 °C). However, temperature and salinity may compensate each other, for instance the chlorozoan association is inhibited at high temperatures if salinity falls below 31‰, and yet it develops at relatively low temperatures where salinity is sufficiently high. Fig. 74 shows the diagrams proposed by Lees (1975). An example of a foramol association can be seen in the sample in Plate 17/1. Chlorozoan associations are pictured in Plate 33.

Recent skeletal carbonate sands of the foramol type, studied by Koopmann et al. (1979), indicate the possibility of a differentiation of sedimentary sub-types which are related to the sedimentary facies and to water depth (above the wave base: balanids, crustaceans, serpulids, red algae, bryozoa, alcyonaria; below the wave base: echinoderms, ostracods, benthic and planktonic foraminifera, sponge spicula, fish remains, diatoms).

Differences in warm-water and cold-water carbonates are also expressed in different predominating diagenetic regimes in the neritic environments; a constructive regime with crystal growth and cementation as characteristic of diagenetic processes in warm waters (e.g., Bahamas) can be contrasted with a destructive regime in cold waters (e.g., North Sea) where dissolution and breakdown are typical (Alexandersson, 1979). Microfacies, low diversity of particles, and spar crystal morphology are other criteria in the recognition of cold-water carbonates (Rao, 1981).

The foramol association, in contrast to the chlorozoan association, is distinguished by a low Sr content and a high Mg content (Lees et al., 1978, Carbonate Symposium, Liverpool).

The "Lees-Model" offers a possibility for reconstructing paleogeographical patterns: An example is given by Heckel and Witzke (1979) for the Devonian.

10.3.5 Shelf Margins

On the outer margin of the continental shelf, bars occur in the Recent, which have their origins in underground domes, in folding or fracturing, in submarine vulcanites, in diapirs, or in reefs. The above mentioned reefs have been classified by Wilson (1974) into three major types; see Fig. 75:

1. *Downslope mud accumulations:* Linear trends of mounds on the foreslope of the shelf margine delimited by upslope sand beaches or spits. Mud mounds occur on relatively steep slopes (angle between 25° and 30°). This carbonate mud derives from shallow marine waters. Often these mud mounds are stabilized by organisms which tend to trap and baffle sediment, resulting in quiet water bioherms. Examples of these are structures built of phylloid algae in the Lower Permian of the Carnic Alps, Austria (see Plate 42), Lower Carboniferous carbonates with bryozoans and algae (Waulsortian limestone in England), and *Stromatactis* "reefs" built from lamellar stromatoporoid and tabulate corals in the Frasnian (Late Devonian) of Belgium.

2. *Knoll reef ramps:* Linear trends of reefs consisting of framework builders on gentle shelf slopes. The sediment is baffled by organic activity (e.g., by encrusting algae). Framework builders form chiefly branching and bushy colonies. Vertical ecological zoning is common within the knolls. Because the low water energy is not sufficient to remove arenitic bioclasts, etc., a volumetric preponderance of debris accumulates in the areas around and between the actively growing knoll reefs. Examples of this are the Recent Bermuda reefs, rudist reefs in the Middle Cretaceous of Mexico and in the Middle East, the reefs consisting of stromatoporoids and tabulate corals in the Canadian Devonian, and perhaps even a few Upper Rhaetian reefs in the Northern Alps of Austria and Bavaria, Germany.

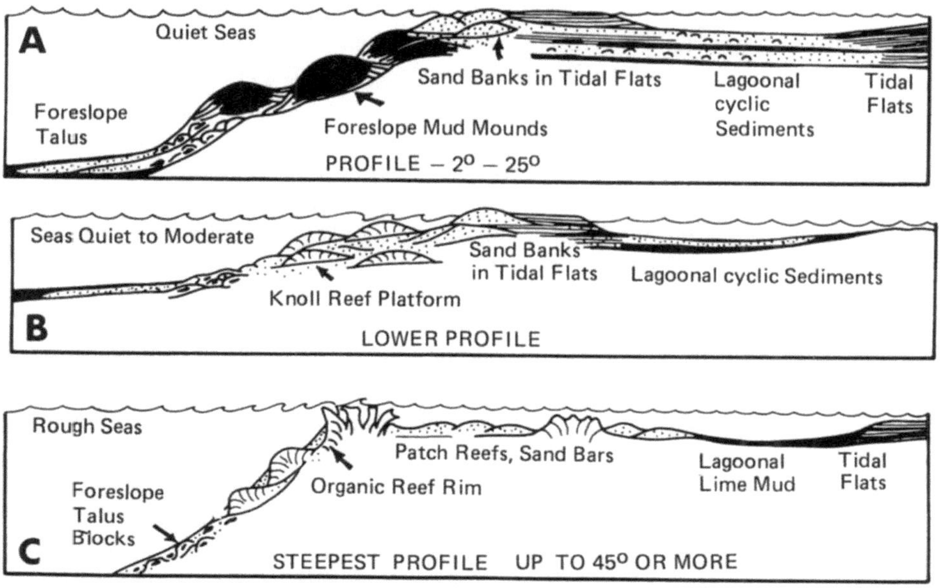

Fig. 75 A–C. Formation of the margins of shelf areas with carbonate sedimentation. **A** Mud mounds on the shelf slope; sediment binding organisms build quiet water bioherms. **B** Knoll reef platform with a clear preponderance of bioclastic debris in the inter-reef areas. **C** Frame-built reef rim in the turbulent zone adjoining the reef complex. After Wilson (1975)

3. *Framebuilt reef rims:* Linear trends of reefs consisting of biogenic framework structures which grow into the zone of turbulence. Behind the reefs are formed shoals of lime sand and lagoonal environments; on the slopes, coarse talus debris. Ecological zoning is frequently seen. Examples of these are Recent coral reefs, Dachstein limestone reefs in the Alpine Late Triassic, and Upper Devonian reef complexes in Western Australia.

All three types are often displayed in vertical profiles; especially then when the reef base shifts up or down along the shelf margin, due to tectonic activity.

10.3.6 Reefs (Fig. 76)

The models developed for "reefs" all have problems of definition, because descriptive and genetic elements are often intermixed. Basically the following definitions of reefs are in use:

- Reefs are laterally delimited carbonate bodies which rise above the sedimentary base ("stratigraphic reef", Dunham, 1970)
- Reefs are accumulations of sessile organisms which can build or do build a solid framework
- Reefs are wave resistant biogenic structures ("ecological reef", Dunham, 1970)

- Reefs are "buildups", for which there are (a) indications of a potential wave resistance or growth in a turbulent environment, and (b) indications of a (sedimentological) control of the surrounding sedimentation area (Heckel, 1974: 96).

The last definition is based on a study of Silurian and Permian reefs (Lowenstam, 1950; Newell et al., 1953). The most important criterion for it is the proof of wave resistance. This proof can essentially be given only by biodetritus, whose composition suggests an origin in a reef region. If no information about water turbulence is available but, on the other hand, a topographical rise can be determined, then the name "bank" is appropriate. This term is often applied to structures which are built of accumulations of organisms; in contrast to reefs, banks exhibit a lower dip to the adjoining facies. The biogenic detritus of the bank is not bound together. In many articles, bank and biostrome as well as reef and bioherm are regarded as identical:

In the original definition "biostrome" is related to "distinctly bedded structures that do not swell into lenslike or reeflike form but ... consist mainly or exclusively of the remains of organisms" (Cumings, 1932). Because this name is restricted by some authors to structures which originate from in situ accumulations of organisms, by other authors, however, to mound structures made of micrite, it seems sensible to use the general term "buildup", as suggested by Heckel (1974). This is understood to be a carbonate body or the local part of a carbonate unit which (a) is different to a certain degree from the surrounding synchronous and overlying sediment, (b) is much thicker than carbonates of the same age, and rises above the surrounding sediment during its formation. In this way, the term "buildup"

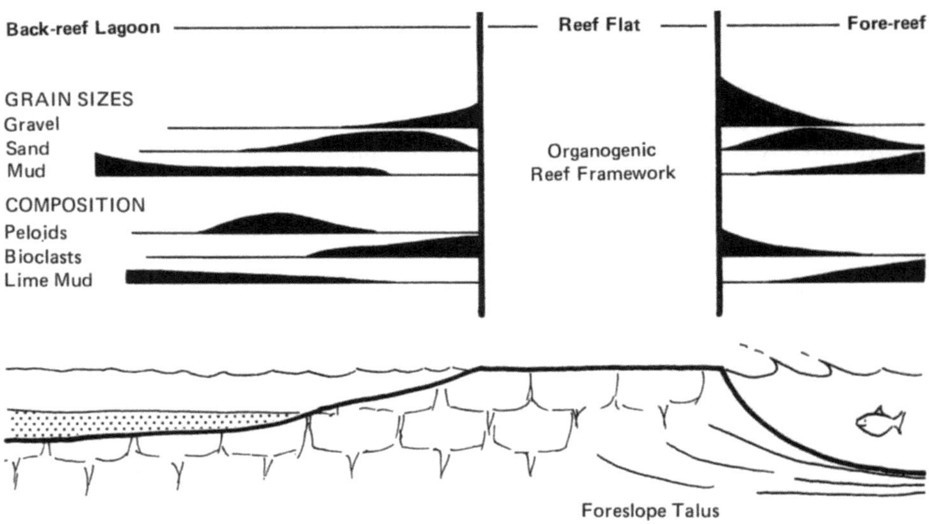

Fig. 76. Simplified scheme for classifying reef complexes which originate in agitated water. A distinct zonation into three parts according to the sediments and distribution of the organisms. After Selley (1970)

becomes a neutral collective term. Genetically, it involves organic structures which have developed in situ, or again structures which have been formed by sedimentation of lime mud (mud buildups), or which are derived from calcareous sand bars. "Buildups" of mud may be stabilized by organisms which lower the current velocity (e.g., marine grasses, sponges, bryozoans, crinoids) or which fix the sedimentary grains in place (e.g., algal mats or "organic mats" of filamentous green algae, blue-green algae, and red algae as well as diatoms and small invertebrates, see Neumann et al., 1970).

Difficulties are presented by the German translation of "buildup". The term "Stillwasser-Bioherm" describes only mud buildups whose mud mounds are derived from organic activity (see Fig. 77). The name „Hügelstruktur" might possibly find approval.

The concept developed by Riding (1977) can be used to describe structures featuring accumulations of sessile organisms: Riding characterizes "in place organic deposits" (IOD) by (a) in situ organisms, (b) loose skeletal and non-skeletal matrix, and (c) cavities, for which only original shelter-voids of in situ organisms should be considered. The estimated proportions of matrix, organisms, and cavities can be represented by a point plotted on a triangular diagram. Other distinctions may be made according to the shape of the organisms, the abundance, and the orientation or packing of the organisms.

Buildups may have acquired wave resistance not by organic binding but predominantly by a rapid submarine cementation of bioclastic material or by an interaction of the sediment binding effects of encrusting algae and syndepositional cementation. Those "spar-cemented debris reefs" (Heckel, 1974) are characterized by the lack of wave-washed talus, by a high percentage of submarine carbonate cements, and generally by low-growing and low to moderately diverse organisms (algae, sponges, bryozoans). Examples are the Lower Permian *Tubiphytes/Archaeolithoporella* buildups of the Southern Alps (E. Flügel, 1981 a), and parts of the Permian Capitan Reef Complex (Schmidt, 1977). Such "stratigraphic reefs" (Dunham, 1970) should be clearly differentiated from organic or ecologic reefs exhibiting an organically constructed in situ wave resistant framework.

Reefs growing in turbulent environments often build a "reef complex" (Henson, 1950) which may be subdivided into the reef core (reef flat or central reef area) and the talus zones on the reef flanks, of which the zone facing the turbulence is termed the fore-reef, and the zone lying behind the central reef, the back-reef or lagoon. Figure 76 shows a very simplified scheme in which the differences in grain sizes and sediment composition are clearly expressed. Comparisons with carefully investigated Recent coral reefs (e.g., Geister, 1975) demonstrate that this schematization becomes incorrect when the fore-reef zone and the back-reef directly adjoining the central reef at its rim, i.e., the reef crest, are understood to be "talus" areas. Due to the water currents splashing along the outer edge of the reef and a resulting increase in the food and oxygen supply, rich fauna and algal flora flourish in the fore-reef zone.

An excellent diagram, useful even for Ancient reefs, of the constructive and destructive forces at work during reef formation is found in Schröder and Zankl (1974). According to their diagram the following processes may be distinguished:

- Construction by organisms
  a) primary framework builders (e.g., corals or coralline algae)
  b) secondary framework builders quantitatively often significant, but without the capacity to construct a solid framework over a longer period of time)
- Destruction by organisms
  a) boring organisms (produce borings in reef rocks and contribute to sediment and possibly cement formation; e.g., pelecypods, sponges, polychaetes, sipunculids, balanids, and endolithic algae)
  b) rasping organisms (produce surficial grooves and sediments; e.g., echinoids, crustaceans, chitons, turtles, and fishes).
  c) burrowing organisms (produce bioturbation and possibly cement; e.g., crustaceans, polychaetes)
  d) sediment crumbling organisms (contribute to the size reduction of sedimentary particles; e.g. crustaceans)
- Sedimentation
  a) sessile and mobile reef organisms (produce sediment and sedimentary pores; e.g., pelecypods, coralline algae, foraminifera)
  b) planktonic organisms (produce sediment and form intraparticle pores; e.g., coccolithoporids, diatoms, foraminifera)
- Cementation (almost synchronous to the growth of reef organisms; below the reef corals and within the intergranular pores)
- Mechanical destruction by waves and surfs

These processes interact with each other, and it can often be noted here that a phase characterized by primary reef framework builders is followed by a second phase in which secondary framework builders increase in importance because of their thick crusts. In a third phase several generations of borings, internal sedimentation, and cement formation appear. These processes produce criteria which can be used in Ancient reefs to characterize the zone of active reef growth or the central reef area (see Table 51). A thorough comparison with the processes acting in Recent reefs must be the basis for the interpretation of Ancient reef-complexes (see Tab. 52).

The literature on Ancient reefs is very extensive. Mention is made here of the articles edited by Laporte (1974), especially the survey of carbonate mound structures (Heckel, 1974), and of the papers dealing with "European Fossil Reef Models" (Toomey, 1981). Silurian reefs on Gotland island, Sweden, have been described by Hadding (1950), Jux (1957), Manten (1971), and Riding (1981). An overall picture of European Devonian reefs is found in Krebs (1974) and Burchette (1981). The Devonian reef complexes in Western Canada, often regarded as "models", have been dealt with by Klovan (1964), McGillavry and Mountjoy (1975), Mountjoy (1968, 1975), Mountjoy and Jull (1978), Mountjoy and Walls (1977); the extensive Devonian reef area in Western Australia, by Playford (1976, 1980), and Playford and Lowry (1966). Very stimulating discussions of the Permian reef complexes in Texas and New Mexico, investigated in the classic study by Newell et al. (1953), have been published by Dunham (1972) and Hileman and Mazzulo (1977). Late Permian Zechstein reefs of Europe were investigated by Kerkmann (1969), Peryt (1978), and D. B. Smith (1981a, 1981b). The classical

Triassic reefs of the Dolomites, Southern Alps, have been well summarized by Bosellini and Rossi (1974) and studied in more detail by Fürsich and Wendt (1977). A summary of the facies and paleoecology of Upper Triassic reefs in the Northern Alps, Austria, is given by E. Flügel (1981); important papers are by Dullo (1980), Flügel and Flügel-Kahler (1963), Ohlen (1959), Piller (1981), Schäfer (1979), Schäfer and Senowbari-Daryan (1981), Senowbari-Daryan (1980), and Zankl (1969, 1971). Upper Jurassic sponge-algal buildups were studied by E. Flügel and T. Steiger (1981), Cretaceous coral-rudist buildups by Polsak (1979, 1981), Masse and Philip (1981), Enos (1974), and Kauffman and Sohl (1974). Tertiary reefs have been studied in detail from a paleoecological point of view only during the last few years (e.g., Frost, 1981; Frost and Weiss, 1979).

10.3.7 Mud Mounds

Mound structures which consist of accumulations of micrite are found in quiet waters, and in deeper basins, on slopes (see Fig. 75) and in shelf lagoons. According to Wilson (1974) the following processes are involved in the formation of these carbonate structures:

- Mechanical accumulation of fine and coarse sediment by currents and wave action
- Locally increased sediment accumulation by sediment trapping and baffling organisms
- Stabilization of sediment by encrusting organisms
- A veneer of framework building organisms
- Cementation

As shown in Fig. 77, the "quiet water bioherms" (Füchtbauer and Müller, 1970) are frequently characterized by a regular sequence of facies:

1. Basal bioclastic wackestone: Essentially contains echinoderms and bryozoans and brachiopods.
2. Bafflestone: The micritic sediment is "trapped" by various organisms because the current is slower. Sponges and algae are found in the Cambrian and Ordovician; bryozoans in the Ordovician, Silurian and Lower Carboniferous; phylloid algae in the Upper Carboniferous and the Lower Permian; corals in the Late Triassic; siliceous sponges in the Upper Jurassic; rudists in the Cretaceous; and marine grasses in the Recent (see Brasier, 1975).
3. Boundstone on the slope: If the mud mounds grow into the active wave base, sessile organisms (e.g., stromatoporoids, sponges) can begin building a reef. In quiet water boundstone is composed of lamellar algae with cavities (*Stromatactis*) concordant with the bedding.
4. Organic veneer of encrusting organisms
5. Flanks beds: Bioclastic sediments, which may sometimes completely envelop the mud mounds, occur near the wave base. The bedded flanking beds may be formed after the core accumulation (D. L. Smith, 1977).
6. Lithoclasts and bioclasts derived from reworking the surface of the mud mound rarely occur as flank sediments.

7. Capping beds: If high-energy conditions are present, the mud mound will be overlain by cross-bedded grainstones containing specialized organisms (dasycladaceans, gastropods).

The majority of the Ancient mud mounds known up to now probably originated in relatively shallow environments (see Fig. 77, bottom). A similar structure, probably formed in relatively deep water, has been described by Mathur

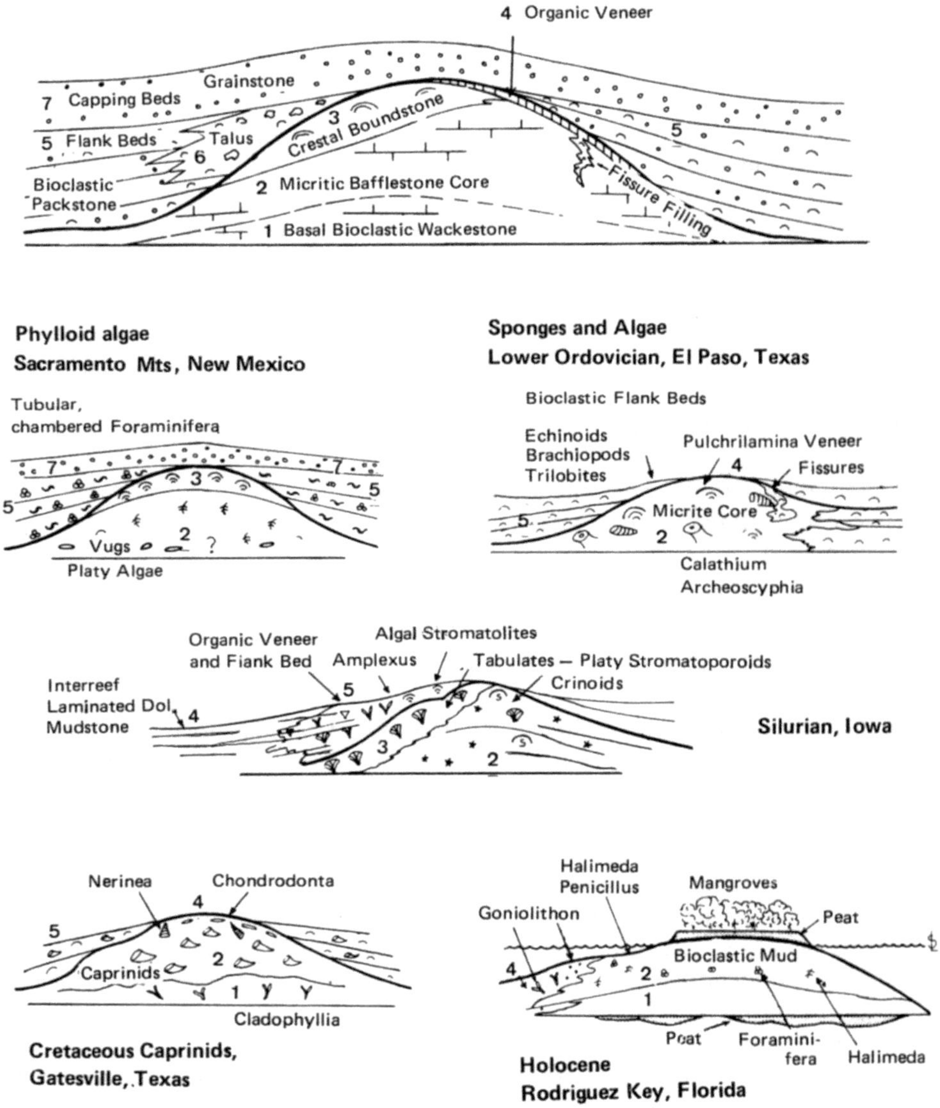

Fig. 77. "Mud Mounds" are characterized by a sequence of facies types which indicate deposition in a low-energy environment and subsequent shallowing. A general model is shown in the *upper part* of the Figure. The numbers in the examples refer to the facies zones distinguished in this model. After Wilson (1975)

(1975) from the Jurassic of the Allgäu mountains, Germany. These are mudstones and wackestones. Indications of a possible interpretation as a "mud mound" are found in the exclusively micritic matrix, in the occurrence of small-scale slump sediments, and in the mound-like shape of the carbonate body. The occurrence of *"Stromatactis"* structures, which can be interpreted, analogous to Heckel's theories (1972), as selective compaction, is of considerable interest.

Another interpretation of Stromatactis structures seen in mud mound carbonates is suggested by Bathurst (1980), who emphasizes that submarine cement crusts may contribute to the construction and stabilization of carbonate mud mounds. Mud mounds may be formed by the infilling of excavations between crustose layers (lithoherms in the Straits of Florida).

10.3.8 Slope Deposits and Deep-Sea Sediments

Deep-sea environments underlie about 88% of the surface areas of the world oceans. More than 45% of deep-sea deposits consist of calcareous ooze and allochthonous carbonates. Nevertheless, our knowledge of Ancient carbonate slope and deep-sea sediments is negatively biased as compared with that of shallow-water carbonates.

As outlined in Sect. 10.2.4, the origin of deep-sea carbonates is dependent on

- Plankton supply
- Increasing solubility of carbonate with depth
- Sedimentation of imported particles derived from shallow-water depositional areas or from upper slope areas

"Sedimentation models" for deep-marine environments (e.g., Rich, 1951; Wilson, 1975; see Sect. 10.3.2) must consider the fact that carbonate sedimentation in the deep sea is more controlled by the interaction of deposition and diagenesis (especially dissolution) than shallow-marine carbonate formation, and that the low sedimentation rates generally reflect longer time ranges:

The concept of seafloor spreading implies that new oceanic crusts generated at the crest of the mid-oceanic ridge, subside as they move away from their origin and slowly cool. Sclater et al. (1971) plotted sea-floor age against the amount of subsidence (vertical distance to ridge crest) and concluded that the fit of age versus depth may be used to establish the hypothesis that at any point in geological time the age of the sea-floor determined how far it moved downward from the original ridge crest. This postulate provides a simple means of reconstructing the approximate depth of deposition of Ancient deep-sea sediments by a "back-tracking" procedure (see Berger and Winterer, 1974; Berger and von Rad, 1972; Sclater and McKenzie, 1973; Sclater et al., 1977; Berger, 1973; van Andel and Heath, 1973). This relation between depth and age of the ocean floor may be used to establish genetic models which describe the development of the deep-water sedimentation. An example of a discussion which tends in this direction is the genetic model developed by Bosellini and Winterer (1975) for the deep-water sedimentation in the Mesozoic of the Tethys (Fig. 78), see also Winterer and Bosellini (1981):

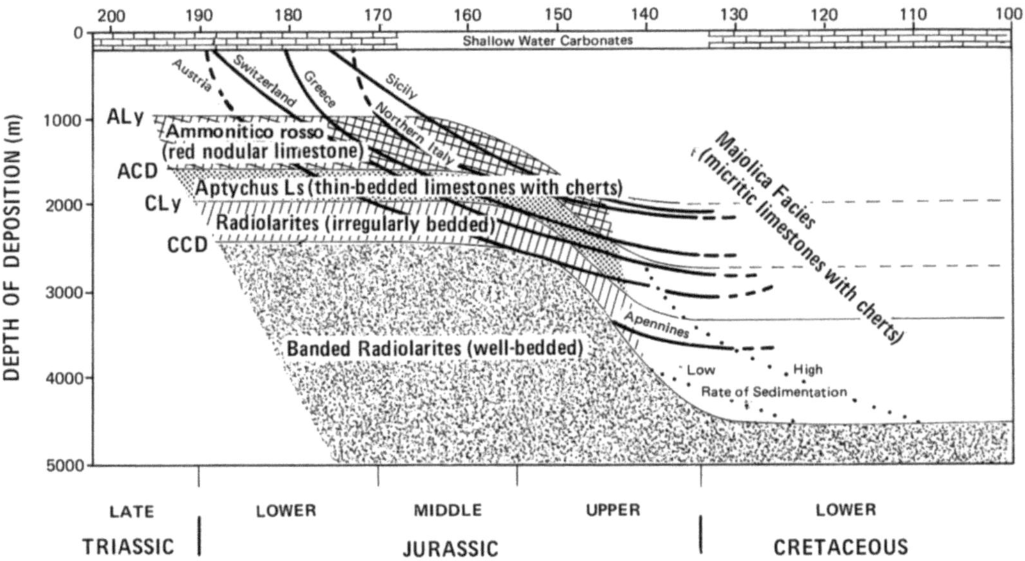

Fig. 78. Genetic model for Mesozoic deep-water sediments in the Alpine-Mediterranean region, assuming a shifting of the carbonate solution level over geologic time, caused by seafloor spreading and discontinuous subsidence of the basement. Depth of the lysoclines in Recent oceans is very different because of various CO_3^{-2} concentrations and different undersaturation of carbonate; the model therefore should not be overrated. After Bosellini and Winterer (1975)

Assuming that the carbonate solution levels known in the Recent were also active in the Mesozoic, deep marine carbonate sediments in a subsiding depositional area should pass through several solution levels (aragonite lysocline, aragonite compensation depth, calcite lysocline, calcite compensation depth; see Sect. 2.2.2.3 and Sect. 10.3.2) in the course of the geologic record. However, two different sedimentary sequences can be expected, because of the dependence on the amount of planktonic supply and the proportions of calcitic and aragonitic shells available:

1. Low sedimentation rate: Due to a low production of micro- and nanno-organisms or because of special topographic conditions (e.g., seamounts or oceanic plateaus, which the sediments can be swept from), carbonate sediments are deposited very slowly. The sequence is interrupted by numerous discontinuities. If sedimentation starts in relatively shallow water, then large quantities of aragonite will be dissolved first in the aragonite lysocline. Continued aragonite solution during times of low burial offers $CaCO_3$ for the formation of calcite (generally Mg-calcite) which may be responsible for the development of nodular carbonates (e.g., Jurassic "Ammonitico rosso" of the Mediterranean area; see Sect. 4.2.3.6). Below the aragonite compensation level, but still above the calcite lysocline, calcite is dissolved only very slowly; thus, very little carbonate is available for cementation. The

still unconsolidated sediment is reworked; thinly bedded pelagic limestones occur (Aptychus limestones; see Sect. 10.2.4). Below the calcite lysocline varying compaction of carbonates susceptible to solution and of insoluble SiO_2 residues takes place; irregularly banded calcareous radiolarites may develop. If the ocean bottom subsides below the calcite compensation depth, all of the normally deposited carbonates are dissolved, and well-bedded, banded radiolarites appear.

2. High sedimentation rate: Greater plankton production or a fan-shaped sedimentation of pelagic carbonates leads to the formation of almost pure pelagic limestones (Majolica type). Because these carbonates consist primarily of calcitic particles (e.g., coccolithophorids), hardly any aragonite is available to supply carbonate for cementation; rapid deposition of new sediment leaves no time for carbonate dissolution. Hence, neither nodular structures are formed nor do calcareous radiolarites develop.

It can be assumed, on the basis of solution indicators in the Alpine-Mediterranean deep-water carbonates (see Sect. 10.2.4), that the four solution levels lay at a relatively shallow depth and were very close together before the Cretaceous, whereas during the Cretaceous the calcite lysocline and the calcite compensation depth were shifted to their present oceanic depth.

Caution is necessary with respect to a too static view of pelagic sedimentation. Deep Sea Drilling clearly reveals synchronous variations in sedimentary rates and in dissolution rates, and also a change of the calcite compensation depth in time: A spread of particularly successful calcareous planktonic organisms may remove unprecedented amounts of $CaCO_3$ from marine waters; transgressions increasing the volume of shelf-sea carbonates will also subtract $CaCO_3$ from the ocean. This depletion could then be balanced by the rise of the calcite compensation depth causing solution of increased amounts of pelagic carbonate (Berger, 1971).

Pelagic sedimentation may become strongly modified by the effects of turbidity currents (see Sect. 10.2.4.2) and by the erosive flow of density-driven bottom waters which follow the base of the continental slope. These *contour currents* are bottom currents resulting from thermohaline circulation on a rotating earth. They flow parallel to bathymetric contours and occur most commonly on the continental rise of the Recent hemipelagic and pelagic deep ocean. Contour currents may transport and redeposit mud, silt and sand. These *contourites* (Heezen et al., 1966) have very good sorting as compared to turbidites, bed thickness is less (usually smaller than 5 cm), grading normal or reverse, cross-lamination common, horizontal lamination common throughout (in turbidites common in the upper part only), preferred grain orientation parallel to the bedding plane, and no massive bedding. The sediment is relatively fine-grained. Microfossils are rare, generally worn and often size-sorted in placers (Hollister and Heezen, 1972; Hubert, 1964). Features of contourites may considerably overlap with those of thin-bedded turbidites, but contourites can be distinguished by fining and sorting trends (perpendicular or parallel to the contours), texture and sedimentary structures (see Stow, 1979). Contourites have been reported from Ancient deep-sea successions (Bouma, 1973; Bein and Weiler, 1976) and also seem to be important for the redeposition of allochthonous calcareous sediments (Schlager and Chermak, 1979).

10.3.9 References: Facies Models

Alexandersson, T. (1979): Marine maceration of skeletal carbonates in the Skagerrak, North Sea. Sed. **26,** 845–852, 2 Figs., Oxford

Argenio, B. d' (1976): Le piattaforme carbonatiche periadriatiche. Una rassegna di problemi nel quadro geodinamico mesozoico dell'area mediterranea. Mem. Soc. Geol. Ital. **13**/2, 1–28, 2 Pls., 6 Figs., Pisa

Barzel, A., Friedman, G. M. (1970): The Zohar formation (Jurassic) in Southern Israel: A model of shallow-water marine carbonate sedimentation. Israel J. Earth-Sci. **19,** 183–207, 22 Figs., Jerusalem

Bein, A., Weiler, Y. (1976): The Cretaceous Talme Yafe Formation: a contour current shaped sedimentary prism of calcareous detritus at the continental margin of the Arabian Craton. Sed. **23,** 511–532, 15 Figs., Oxford

Berger, W. (1972): see 2.4.1

Berger, W. (1973): see 2.4.1

Blatt, H., Middleton, G., Murray, R. (1972): see 1.4.1

Bosellini, A., Rossi, D. (1974): Triassic carbonate buildups of the Dolomites, Northern Italy. Soc. Econ. Paleont. Min. Spec. Publ. **18,** 209–233, 33 Figs., Tulsa

Bosellini, A., Winterer, E. L., (1975): Pelagic limestone and radiolarite of the Tethyan Mesozoic: A genetic model. Geol. **1975,** 279–282, 2 Figs.

Bouma, A. H. (1972): Fossil contourites in Lower Niesenflysch, Switzerland. J. Sed. Petrol. **42**/2, 917–921, 3 Figs., Tulsa

Bouma, A. H. (1973): Contourites in Niesenflysch, Switzerland. Eclogae geol. Helvet. **66,** 315–323, Basel

Buchbinder, B. (1979): Facies and environments of Miocene reef limestones in Israel. J. Sed. Petrol. **49**/4, 1323–1344, 11 Figs., Tulsa

Buggisch, W., Flügel, E., Leitz, F., Tietz, G.-F. (1976): Fazies und paläogeographische Entwicklung im Perm der Karnischen Alpen und in den Randgebieten. Geol. Rdsch. **65**/2, 649–690, 19 Figs., 4 Tab., Stuttgart

Cumings, E. R. (1932): Reefs or bioherms? Bull. Geol. Soc. Amer. **43,** 331–352, New York

Curray, J. P. (1975): Marine sediments, geosynclines and orogeny. In: Fischer, A. G., Judson, S.: Petroleum and Global Tectonics, 157–222, 26 Figs., 5 Tab., Chicago: University Press

Dabrio, C. J., Esteban, M., Martin, J. M. (1981): The coral reef of Nijar, Messinian (Uppermost Miocene), Almeria Province, S.E. Spain. J. Sed. Petrol. **51**/2, 521–539, 16 Figs., Tulsa

Degens, E. T., Stoffers, P. (1976): Stratified waters as a key to the past. Nature **263,** 22–27, 6 Figs., London

Dullo, W.-Chr. (1980): see 5.3.1

Dunham, R. J. (1970): Stratigraphic reefs versus ecologic reefs. Bull. Amer. Ass. Petrol Geol. **54,** 1931–1932, Tulsa

Dunham, R. J. (1972): Capitan Reef, New Mexico and Texas: Facts and questions to aid interpretation and group discussion. Permian Basin Section Soc. Econ. Paleont. Min. Publ. 72–14, Midland/Texas

Enos, P. (1974): Reefs, platforms and basins of Middle Cretaceous in northeast Mexico. Amer. Ass. Petrol. Geol. Bull. **58,** 800–809, Tulsa

Fabricius, F. (1966): see 5.5

Fischer, A. G. (1964): see 4.2.4.5

Flügel, E. (1974): Fazies-Interpretation der unterpermischen Sedimente in den Karnischen Alpen. Carinthia II, **84,** 43–62, Klagenfurt

Flügel, E. (1981 a): Lower Permian Tubiphytes/Archaeolithoporella buildups in the Southern Alps (Austria and Italy). Soc. Econ. Paleont. Min. Spec. Publ. **30,** 143–160, 9 Figs., Tulsa

Flügel, E. (1981 b): Paleoecology and facies of Upper Triassic reefs in the Northern Calcareous Alps. Soc. Econ. Paleont. Min. Spec. Publ. **30,** 291–359, 26 Figs., Tulsa

Flügel, E., Flügel-Kahler, E. (1963): see 9.1.5

Flügel, E., Steiger, T. (1981): see 11.7

Friedman, G. M., Barzel, A., Derin, B. (1971): Paleoenvironmental of the Jurassic in the Coastal Belt of Northern and Central Israel and Their Significance in the Search for Petroleum Reservoirs. Inst. Petrol. Res. Geophys. Geol. Surv. Israel, Oil Div. Rep. OD/1/71, 26 S., Jerusalem

Frost, St. H. (1981): Oligocene reef coral biofacies of the Vicentin, Northeast Italy. Soc. Econ. Paleont. Min. Spec. Publ. **30,** 483–539, 19 Figs., Tulsa

Frost, St. H., Weiss, M. P. (1979): Patch-reef communities and succession in the Oligocene of Antigua, West Indies. Geol. Soc. Amer. Bull. **90**/1, 612–616, **90**/11, 1094–1141, New York

Füchtbauer, H., Müller, G. (1970): see 1.4.1

Fürsich, F. T., Wendt, J. (1977): Biostratinomy and Palaeoecology of the Cassian Formation (Triassic) of the Southern Alps. Paleogeography, Paleoclimatology, Paleoecology **22,** 257–323, Amsterdam

Geister, J. (1975): Riffbau und geologische Entwicklungsgeschichte der Insel San Andrés (westliches Karibisches Meer, Kolumbien). Stuttgarter Beitr. Naturkunde, Ser. B **15,** 1–203, 11 Pls., 29 Figs., Stuttgart

Ginsburg, R. N. (1974): Introduction to comparative sedimentology of carbonates. Bull. Amer. Ass. Petrol. Geol. **58**/5, 781–786, Tulsa

Hadding, A. (1950): Silurian reefs of Gotland. J. Geol. **58,** 402–409, Chicago

Harbaugh, J. W., Merriam, D. F. (1968): see 7.3.2

Heckel, P. H. (1972): see 4.2.4.5

Heckel, P. H. (1974): Carbonate buildups in the geologic record: a review. Soc. Econ. Paleont. Min. Spec. Publ. **18,** 90–154, 9 Figs., Tulsa

Heezen, B. C., Hollister, C. D., Ruddiman, W. T. (1966): Shaping of the continental rise by deep geostrophic contour currents. Science **52,** 502–508, New York

Hileman, M. E., Mazzullo, S. J. (eds.) (1977): Upper Guadalupian facies, Permian Reef Complex, Guadalupe Mountains, New Mexico and West Texas. 1977 Field Conf. Guidebook, Permian Basin Sect. Soc. Econ. Paleont. Min. Publ. 77–16, 508 p., Midland

Hollister, C. D., Heezen, B. C. (1964): Modern graywacke sands. Science **146,** 1573–1574, Washington

Hubert, J. F. (1964): Textural evidence for deposition of many western North Atlantic deep-sea sands by ocean-bottom currents rather than turbidity currents. J. Geol. **72,** 757–785, Chicago

Imbrie, J., Purdy, E. G. (1962): see 6.4

Irwin, M. L. (1965): General theory of epeiric clear water sedimentation. Bull. Amer. Ass. Petrol. Geol. **49**/4, 445–459, 12 Figs., Tulsa

Jenkyns, H. C. (1970): Growth and disintegration of a carbonate platform. N. Jb. Geol. Paläont. Mh. **1970,** 325–344, Stuttgart

Jux, U. (1957): Die Riffe Gotlands und ihre angrenzenden Sedimentationsräume. Contrib. Geol. **I, 4,** 41–90, Stockholm

Kauffman, E. G., Sohl, F. (1974): Structure and evolution of Antillean Cretaceous rudist frameworks. Verhandlungen naturforsch. Gesellschaft Basel **84**/1, 399–467, Basel

Kendall, C. G. St., Schlager, W. (1981): Carbonates and relative changes on sea level. Marine Geol. **44** (in print), Amsterdam

Kendall, M. G., Buckland, W. R. (1960): A Dictionary of Statistical Terms. 590 S., London: Oliver & Boyd

Kerkmann, K. (1969): Riffe und Algenbänke im Zechstein von Thüringen. Freiberger Forschungshefte C **252,** Paläont. 1–85, 19 Pls., 28 Figs., 2 Tab., Leipzig

Klovan, J. E. (1964): Facies analysis of the Redwater Reef Complex, Alberta, Canada. Bull. Canadian Petrol. Geol. **12,** 1–100, Calgary

Krebs, W. (1974): Devonian carbonate complexes of Central Europe. Soc. Econ. Paleont. Min. Spec. Paper **18,** 155–208, 25 Figs., Tulsa

Krumbein, W. C., Graybill, F. A. (1965): see 7.3.2

Laporte, L. F. (1969): Recognition of a transgressive carbonate sequence within an epeiric sea: Helderberg Group (Lower Devonian) of New York State. Soc. Econ. Paleont. Min. Spec. Publ. **14,** 98–119, 15 Figs., Tulsa

Lees, A. (1973): Les dépots carbonatés de plate-forme. Bull. Centre Rech. Pau-SNPA **7**/1, 177–192, 5 Figs., Pau

Lees, A. (1975): Possible influences of salinity and temperature on modern shelf carbonate sedimentation. Marine Geol. **19,** 159–198, Amsterdam

Lees, A., Buller, A. T. (1972): Modern temperate-water and warm-water shelf carbonate sediments contrasted. Marine Geol. **13,** M 67–M 73, 3 Figs., Amsterdam

Lowenstam, H. (1950): Niagaran reefs of the Great Lakes area. J. Geol. **58,** 430–487, Chicago

Manten, A. A. (1971): Silurian reefs of Gotland. Dev. Sed. **13,** 539 S., Amsterdam: Elsevier

Masse, J.-P., Philip, J. (1981): Cretaceous coral-rudist buildups of France. Soc. Econ. Paleont. Min. Spec. Publ. **30,** 399–426, 26 Figs., Tulsa

Mathur, A. C. (1975): A deeper water mud mound facies in the Alps. J. Sed. Petrol. **45**/4, 787–793, 5 Figs., Tulsa

McGillavry, J. G., Mountjoy, E. W. (1975): Facies and related reservoir characteristics, Golden Spike Reef complex, Alberta. Canadian Petrol. Geol. Bull. **23,** 753–809, Calgary

Mountjoy, E. W. (1968): Factors governing the development of the Frasnian Miette and Ancient Wall reef complexes (bank and biostromes), Alberta. In: Oswald, D. H. (ed.): Proc. Internat. Symposium Devonian System **2,** 387–408, Calgary: Alberta Soc. Petrol. Geol.

Mountjoy, E. W. (1975): Intertidal and supratidal deposits within isolated Devonian buildups, Alberta. In: Ginsburg, R. N. (ed.): Tidal deposits, 387–395, New York: Springer

Mountjoy, E. W., Jull, R. K. (1978): Fore-reef carbonate mud bioherms and associated reef-margin, Upper Devonian Ancient Wall Reef complex, Alberta. Canadian J. Earth Sci. **15,** 1304–1325, Ottawa

Mountjoy, E. W., Walls, R. A. (1977): Some examples of early submarine cements from Devonian buildups of Alberta. Internat. Coral Reef Symp., 3rd., Proc. **2,** 155–161, Miami: Rosenstiel School Marine Atmospheric Sci. Univ. Miami

Neumann, A. C., Gebelein, C. D., Scoffin, T. P. (1970): The composition, structure and erodability of subtidal mats, Abaco, Bahamas. J. Sed. Petrol. **40,** 274–297, Tulsa

Newell, N. D., Rigby, J. K., Fischer, A. G., Whiteman, A. J., Hickox, J. E., Bradley, J. S. (1953): The Permian Reef Complex of the Guadalupe Mountains Region, Texas and New Mexico. A Study in Paleoecology, 236 pp., 32 Pls., San Francisco: Freeman

Ohlen, H. R. (1959): The Steinplatte reef complex of the Alpine Triassic (Rhaetian) of Austria. Diss. Univ. Princeton, 122 pp., Princeton

Peryt, T. M. (1978): Sedimentology and paleoecology of the Zechstein Limestone (Upper Permian) in the Fore-Sudetic area (western Poland). Sed. Geol. **20,** 217–243, Amsterdam

Piller, W. (1981): The Steinplatte reef complex, part of an Upper Triassic carbonate platform near Salzburg, Austria. Soc. Econ. Paleont. Min. Spec. Publ. **30,** 261–290, 23 Figs., Tulsa

Playford, P. E. (1980): Devonian "Great Barrier Reef" of Canning Basin, Western Australia. Amer. Ass. Petrol. Geol. Bull. **64**/6, 814–840, 27 Figs., Tulsa

Playford, P. E., Lowry, D. C. (1966): Devonian reef complexes of the Canning Basin, Western Australia. Geol. Surv. Western Australia Bull. **118,** 150 pp., 48 Figs., Perth

Polsak, A. (1981): Upper Cretaceous biolithitic complexes in a subduction zone: examples from the inner Dinarides, Yugoslavia. Soc. Econ. Paleont. Min. Spec. Publ. **30,** 447–472, 20 Figs., Tulsa

Purser, B. H. (ed.) (1973): see 1.4.1

Purser, B. H. (1980): see 1.4.1

Rao, C. P. (1981): Criteria for recognition of cold-water carbonate sedimentation: Berriedale limestone (Lower Permian), Tasmania, Australia. J. Sed. Petrol. **51**/2, 491–506, 10 Figs., Tulsa

Rech-Frollo, M. (1973): Les carbonates du flysch. Bull. Centre Rech. Pau-SNPA **7,** 245–263, 3 Pls., 1 Fig., 1 Tab., Pau

Remane, J. (1970): see 10.2.5

Rich, J. L. (1951): Three critical environments of deposition, and criteria for recognition of rocks deposited in each of them. Bull. Geol. Soc. Amer. **62,** 1–20, New York

Riding, R. (1977): Reef concepts. Proc. Third Internat. Coral Reef Symposium Miami **2,** 209–213, Miami

Riding, R. (1981): Composition, structure and environmental setting of Silurian bioherms and biostromes in Northern Europe. Soc. Econ. Paleont. Min. Spec. Publ. **30,** 41–83, 56 Figs., Tulsa

Schäfer, P. (1979): Fazielle Entwicklung und palökologische Zonierung zweier obertriadischer Riffstrukturen in den Nördlichen Kalkalpen ("Oberrhät"-Riff-Kalke, Salzburg). Facies **1.** 3–245, 21 Pls., 46 Figs., Erlangen

Schäfer, P., Senowbari-Daryan, B. (1981): Facies development and paleoecological zonation of four Upper Triassic patch-reefs, Northern Calcareous Alps near Salzburg, Austria. Soc. Econ. Paleont. Min. Spec. Publ. **30,** 241–259, 10 Figs., Tulsa

Schlager, W. (1980): see 4.1.7.2

Schlager, W., Chermak, A. (1979): see 10.2.5
Schlager, W., Schlager, M. (1973): Clastic sediments associated with radiolarites (Taugl-boden-Schichten), Upper Jurassic, Eastern Alps. Sed. **20,** 65–89, 13 Figs., Oxford
Schmidt, V. (1977): Inorganic and organic reef growth and subsequent diagenesis in the Permian Capitan Reef Complex, Guadalupe Mountains, Texas, New Mexico. Permian Basin Sect. Soc. Econ. Paleont. Min., 1977 Field Conf. Guidebook, Publ. 77–16, 45–92, Midland
Schopf, Th. J. M. (1980): Paleooceanography. 341 p., Harvard: Univ. Press
Schroeder, J. H., Zankl, H. (1974): Dynamic reef formation: a sedimentological concept based on studies of Recent Bermuda and Bahama Reefs. Proc. Second Intern. Coral Reef Symposium **2,** 413–428, 13 Figs., Brisbane
Sclater, J. G., Anderson, R. N., Bell, M. L. (1971): The elevation of ridges and evolution of the central eastern Pacific. J. Geophys. Res. **76,** 7888–7915
Sclater, J. G., Hellinger, St., Tapscott, Chr. (1977): The paleobathymetry of the Atlantic ocean from the Jurassic to the present. J. Geol. **85,** 509–552, Chicago
Sclater, J. G., McKenzie, D. P. (1973): Paleobathymetry of the South Atlantic. Geol. Soc. Amer. Bull. **84**/10, 3203–3216, New York
Seibold, E. (1970): Nebenmeere im humiden und ariden Klimabereich. Geol. Rdsch. **60**/1, 73–105, 10 Figs., 1 Tab., Stuttgart
Selley, R. C. (1970): see 1.4.1
Senowbari-Daryan, B. (1980): Fazielle und paläontologische Untersuchungen in oberrhätischen Riffen (Feichtenstein- und Gruberriff bei Hintersee, Salzburg, Nördliche Kalkalpen). Facies **3,** 1–237, Pls. 1–29, 21 Figs., Erlangen
Shaw, A. B. (1964): Time in stratigraphy. 353 pp., New York: McGraw Hill
Shepard, F. P., Marshall, N. F., McLoughlin, P. A., Sullivan, G. G. (1979): Currents in submarine canyons and other seavalleys. Amer. Ass. Petrol. Geol. Stud. Geol. **9,** 173 p., Tulsa
Smith, D. B. (1981 a): The Magnesian Limestone (Upper Permian) Reef Complex of northeastern England. Soc. Econ. Paleont. Min. Spec. Publ. **30,** 161–186, 31 Figs., Tulsa
Smith, D. B. (1981 b): Bryozoan-algal patch-reefs in the Upper Permian Lower Magnesian Limestone of Yorkshire, Northeast England. Soc. Econ. Paleont. Min. Spec. Publ. **30,** 187–202, 18 Figs., Tulsa
Stanton, R. J. Jr. (1967): Factors controlling shape and internal facies distribution of organic carbonate buildups. Amer. Ass. Petrol. Geol. Bull. **51**/12, 2462–2487, 1 Fig., Tulsa
Stow, D. A. V. (1979): Distinguishing between fine-grained turbidites and contourites on the Nova Scotian deep water margin. Sed. **26,** 371–387, 9 Figs., Oxford
Toomey, D. F. (ed.) (1981): European Fossil Reef Models. Soc. Econ. Paleont. Min. Spec. Publ. **30,** 544 p., Tulsa
Van Andel, T. H., Heath, G. R. (1973): Geological results of leg 16: the central equatorial Pacific west of the East Pacific Rise. Initial Report Deep Sea Drilling Project **16,** 937–949, Washington
Visher, G. S. (1965): Use of vertical profile in environmental reconstruction. Bull. Amer. Ass. Petrol. Geol. **49**/1, 41–61, 16 Figs., Tulsa
Walther, J. (1893): see 10.2.5
Weydert, P. (1976): Définition d'un modèle sédimentaire en milieu récifal: la Baie de Tuléar (Côte SW de Madagascar). Rev. Géogr. phys. Géol. dyn. **18**/1, 67–92, 14 Figs., Paris
Wilson, J. L. (1974): Characteristics of carbonate-platform margins. Bull. Amer. Ass. Petrol. Geol. **58**/4, 810–824, 6 Figs., Tulsa
Wilson, J. L. (1975): see 1.4.1
Winterer, E. L., Bosellini, A. (1981): Subsidence and sedimentation on Jurassic passive continental margin, Southern Alps, Italy. Amer. Ass. Petrol. Geol. Bull. **65,** 394–421, 28 Figs., Tulsa
Wolf, K. H. (1973): Conceptual models. 1. Examples in sedimentary petrology, environmental and stratigraphic reconstructions, and soil, reef, chemical and placer sedimentary ore deposits. Sed. Geol. **9,** 153–193, 14 Figs., Amsterdam
Zankl, H. (1969): see 3.9
Zankl, H. (1971): Upper Triassic Carbonate Facies in the Northern Limestone Alps. In: Müller, G. (ed.): Sedimentology of parts of Central Europe. Guide Book, VIII Intern. Sed. Congr. Heidelberg, 147–185, 20 Figs., Frankfurt a. M.: Kramer

11 Case Histories

The following case studies illustrate the potential for microfacies analysis. Examples are based predominantly on own studies, and reflect a broad spectrum of non-marine as well as shallow and deeper marine environments, including Permian, Triassic, Jurassic, Cretaceous and Tertiary carbonates.

11.1 Lacustrine Carbonates: Miocene Lake Deposits of the Ries, Southern Germany (Plate 44)

The marginal carbonate facies of the Miocene Ries meteorite crater lake near Nördlingen, Bavaria, contains bioherms up to 7 m high and 15 m across, built by the green alga *Cladophorites* (Plates 2/2 and 44/1). The algae were externally encrusted by micrite, probably precipitated in response to photosynthetic uptake of CO_2, which produced tufts of calcareous tubes (100 μm diameter, Plate 44/2). Coalescence of tufts, together with incorporation of peloidal and skeletal sand (Plate 44/5), created nodules and cones of algal tufa. The bioherms consist of beds

Plate 44. Lacustrine Carbonates: Miocene Lake Deposits: Ries, Bavaria, Southern Germany

1. Individual tuft of radiating tubes of the green alga *Cladophorites,* together with some ostracods. Adlerberg southeast of Nördlingen

2. Cross section of algal tufts forming small nodules. The tubes are connected by microcrystalline cement and peloidal sediment. High growth-framework porosity is evident. Crossed nicols, open voids appear black. Büschelberg quarry southeast Hainsfarth

3. The sediment adjacent to the algal tufa is covered by a laminar sinter crust formed during subaerial exposure. Ostracods belong to *Amplocypris risgoviensis* (Sieber). Adlersberg southeast of Nördlingen

4. Vadose calcitic cement veneers on outer surface of algal cones. Large open spaces between cones appear white. Adlersberg southeast of Nördlingen

5. Peloidal and bioclastic poorly cemented calcareous sands, surrounding algal bioherms. The sediment consists predominantly of ostracods, some gastropods and many peloids (possibly fecal pellets, from *Hydrobia*). Cross-section of algal tubes can be seen in the top part of the photograph. Intraparticle, biomoldic and vuggy porosity. Büschelberg quarry southeast of Hainsfarth

Scale for all Figures 2 mm

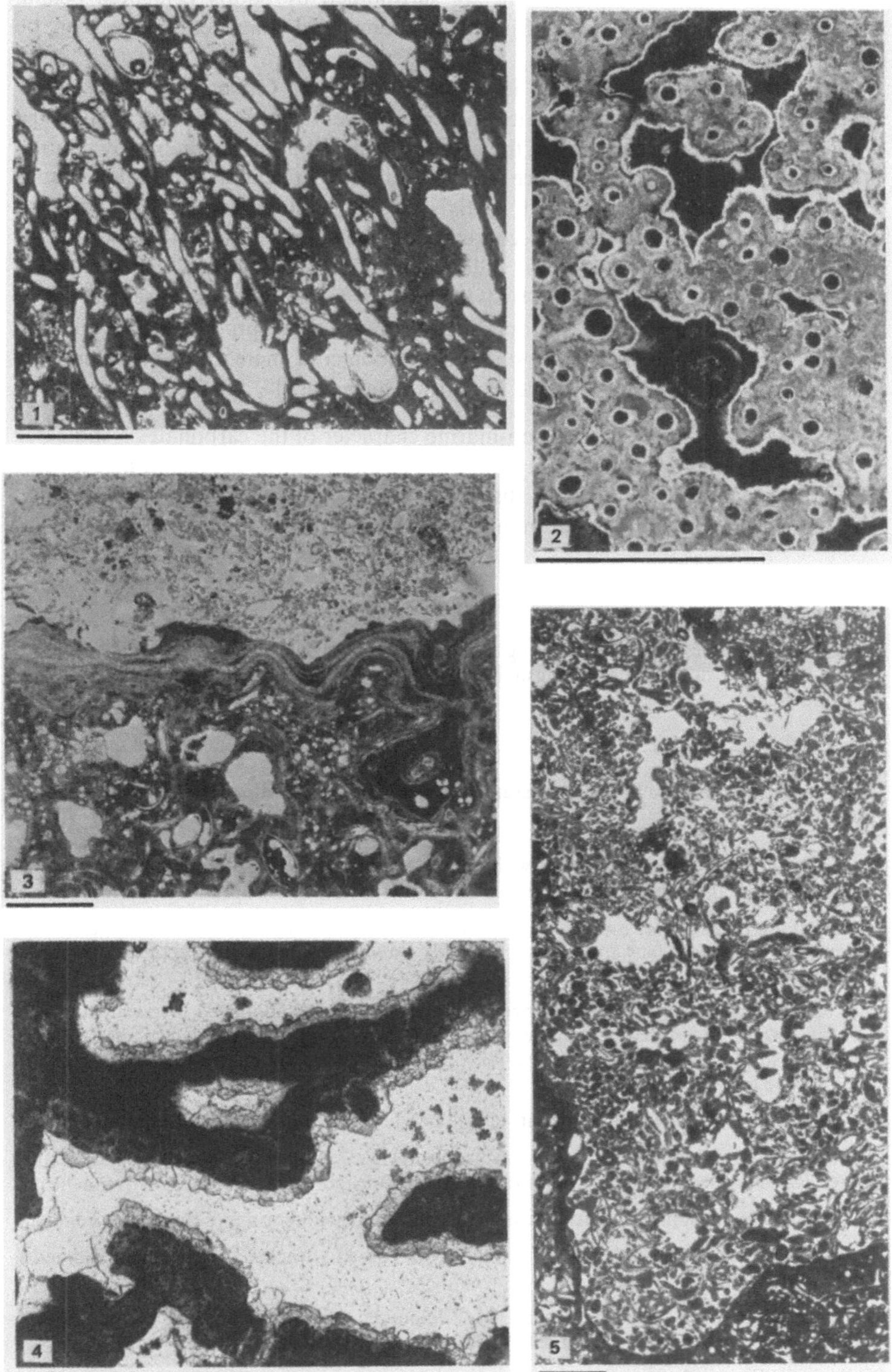
1
2
3
4
5

of cones and compound cones, and are surrounded by poorly cemented peloid, ostracod and gastropod sands. Five depositional and diagenetic stages can be distinguished (Riding, 1979):

(1) Growth and calcification (probably calcitic) of *Cladophorites* in shallow freshwater or slightly brackish water; (2) emergence due to a temporary fall in the level of the lake and veneering of the algal tufa and adjacent sediments by laminar sinter (Plate 44/3); resubmergence and deposition of peloidal and skeletal sands (Plate 44/3, top); (4) burial and partial phreatic dolomitization, together with dissolution of aragonite and penecontemporaneous deposition of thin isopachous rims of dolomite rhombs; (5) local vadose cementation by rhombs, spar and spikes of low-magnesian calcite (Plate 44/4). Stages 1–3 probably occurred several times. Most of the biohermal and surrounding sediments were produced by biological processes in the lake; subaerial sinter deposition and meteoric cementation are quantitatively of minor importance. Porosity is between 10% and 30%.

Microfacies Interpretation: The non-marine character of the carbonates is indicated by the occurrence of low-diversity biota consisting of primitive green-algae, some charophyte stems, complete or disarticulated ostracods, and snails (*Hydrobia; Cepaea,* presumably infloated from nearby shorelines).

References:

Förstner, U., Rothe, P. (1977): Bildung und Diagenese der Karbonatsedimente im Ries-See (nach dem Profil der Forschungsbohrung Nördlingen 1973). Geol. Bavarica **75,** 49–58, München

Plate 45. Arid Carbonate Shorelines, Evaporites and Algal Reefs: Miocene of the Gulf of Suez, Egypt

1. Fine-grained dolomitic bindstone with only vague indications of a laminated texture. The rosette-like crystals correspond to calcitized anhydrite. Intertidal or supratidal environment

2. Dolomitic bindstone or mudstone, characterized by alternating layers (*bottom:* micrite with dolomite crystals; *top:* fenestral fabrics, probably caused by solution of evaporites). Intertidal environment

3. Cement/algal bindstone, characterized by relatively thick dolomitized layers of carbonate cement *bottom* and radiating bundles of algal threads (*top*). The algal character of these threads is supported by distinct bifurcations. Schizohaline environment near the shoreline of the Miocene transgression

4. Calcareous sandstone with fragments of a clypeasteroid echinid. Intertidal environment

5. Silicified algal boundstone with corallinacean algae encrusting bryozoans and echinid spines. Rhodoliths are common within this microfacies. High-energy subtidal environment

6. Red algal boundstone. Encrusting corallinaceans, some with borings. Subtidal environment

7. Bioclastic grainstone with poorly sorted bryozoans and leached coated shell fragments. Subtidal environment

8. Coral boundstone. The septa consist of dolomite, the intraskeletal porosity is locally reduced by cement growth (see Plate 40/1, 2). Subtidal environment

Scale for all Figures 2 mm

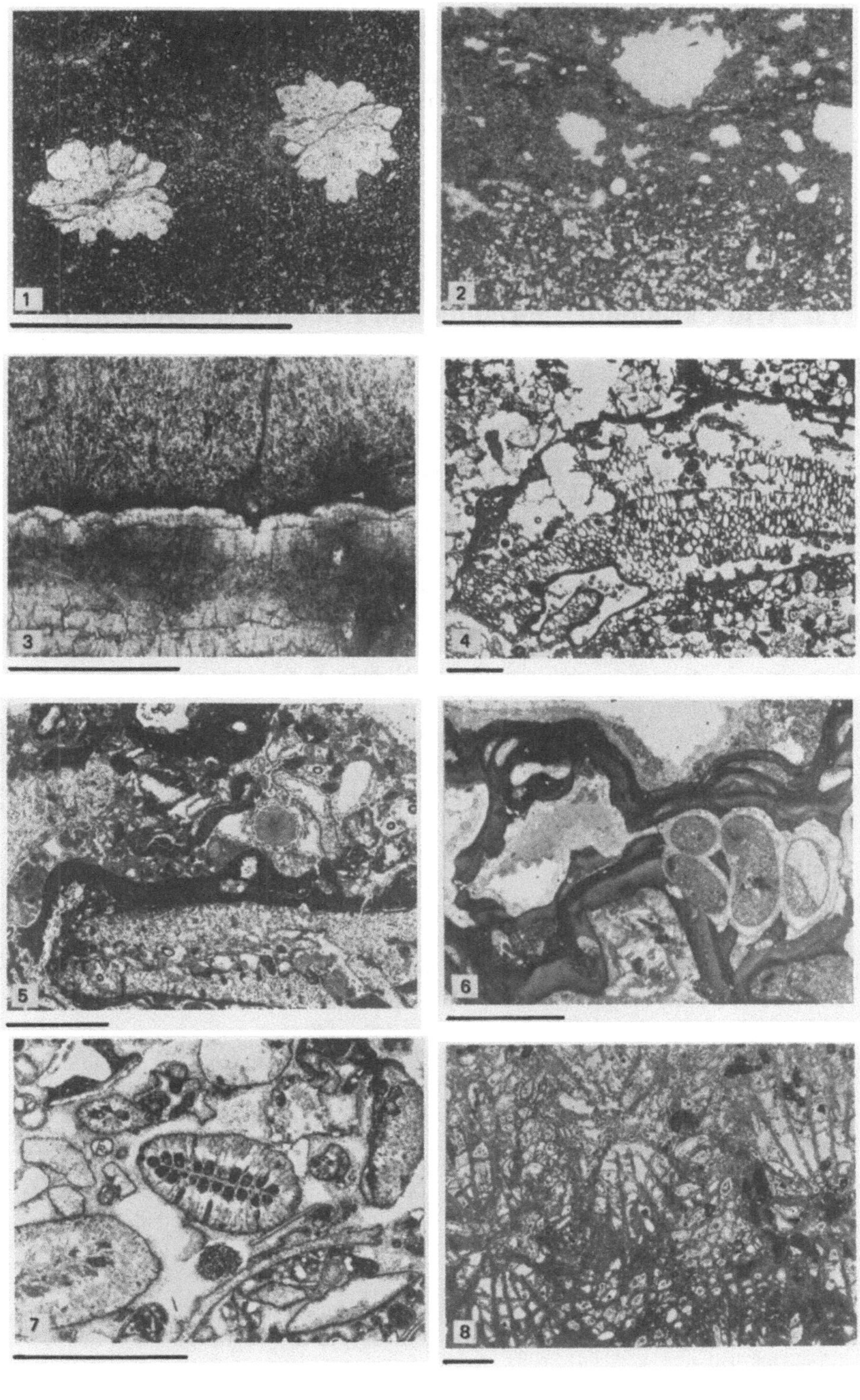
1
2
3
4
5
6
7
8

Reis, O. M. (1926): Zusammenfassung über die im Ries südlich von Nördlingen auftretenden Süßwasserkalke und ihre Entstehung. Jahrbuch oberrhein. geol. Verein. **14,** 176–190, Freiburg

Riding, R. (1979): Origin and diagenesis of lacustrine algal bioherms at the margin of the Ries crater, Upper Miocene, Southern Germany. Sed. **26,** 645–680, 41 Figs., Oxford

Wolff, M., Füchtbauer, H. (1976): Die karbonatische Randfazies der tertiären Süßwasserseen des Nördlinger Ries und des Steinheimer Beckens. Geol. Jahrbuch, D **14,** 3–53, Hannover

11.2 Arid Carbonate Shorelines and Evaporites: Miocene of the Gulf of Suez Region, Egypt (Plate 45)

Several microfacies types are recognizable in the Miocene section of the Gulf of Suez region, which may be subdivided into a lower non-evaporitic and an upper evaporitic group (Heybroek, 1965; Hassan and El-Dashlouty, 1970). The middle part of the Evaporite Group consists of anhydrite with intercalations of shale, sandstone, dolomite, dolomitic limestones, and in some places reefal algal limestones, and is generally attributed to the Middle Miocene. As can be inferred from microfacies analysis, the deposition of this formation seems to have started with a non-marine or schizohaline phase, characterized by the cementation of detritus from eroded basement rocks (granites or Nubian Sandstone) by microcrystalline dolomite (formation of beach-rocks), followed by the growth of blue-green algae in small ponds (Plate 45/3). A second phase is characterized by the formation of intertidal stromatolitic bindstones with birdseyes (Plate 45/2) and fine-grained dolomites with anhydrite (Plate 45/1). Intertidal conditions are also indicated by calcareous sandstones with clypeasteroid echinids (Plate 45/4). Reef growth started during a third phase:

Lenticular patch reefs are composed of coral/red algal boundstones (Plate 45/5, 6, 8), subordinate bryozoan/coral boundstones, and highly diverse bioclastic grainstones (Plate 45/7) indicating normal marine salinities.

Biota and microfacies types of the boundstones can be compared with other Miocene reef limestones (e.g., Armstrong et al., 1980; Carozzi et al., 1976; Buchbinder, 1975; Studencki, 1979).

Microfacies Interpretation: Of special interest are cement/algal bindstones (Plate 45/3) containing non-marine algae. Similar green or blue-green algae are known from lacustrine Paleozoic and Mesozoic carbonates (Flügel and Flügel-Kahler, 1980 – see 5.3.12.7; Freytet and Plaziat, 1965 – see 2.4.2).

References:

Armstrong, A. K., Snavely, P. D., Addicott, W. O. (1980): Porosity evolution of Upper Miocene reefs, Almeria Province, Southern Spain. Amer. Ass. Petrol. Geol. Bull. **64**/2, 188–208, Tulsa

Buchbinder, B. (1975): Lithogenesis of Miocene reef limestones in Israel with particular reference to the significance of the red algae. Israel Geol. Surv. Report OD/3/75, 170 p., 6 Pls., Jerusalem

Carozzi, A. V., Reyes, M. V., Ocampo, V. P. (1976): Microfacies and microfossils of the Miocene reef carbonates of the Philippines. Philippine Oil Dev. Co. Spec. Publ. **1,** 80 p., Manila

Heybroek, F. (1965): The Red Sea Miocene evaporite basin. In: Salt basins around Africa, London Inst. Petrol., 17–40

Hassan, F., El-Dashlouty, S. (1970): Miocene evaporites of Gulf of Suez region and their significance. Amer. Ass. Petrol. Geol. Bull. **54**/9, 1686–1696, Tulsa

Studencki, W. (1979): Sedimentation of algal limestones from Busko-Spa environs (Middle Miocene, Central Poland). Palaeogeography, Palaeoclimatology, Palaeoecology **27,** 155–165, Amsterdam

11.3 Inner Shelf Carbonate Platform: Upper Jurassic of Karaburun, Western Anatolia (Plate 46)

The Mesozoic sequence of the Karaburun Peninsula west of Izmir, Turkey, consists of Triassic and Liassic sediments which are overlain by Upper Jurassic carbonates characterized by the abundant occurrence of the hydrozoan *Cladocoropsis mirabilis* Felix (Plate 46/2). The well-bedded, dark *Cladocoropsis* limestones are represented by hydrozoan biomicrites (Plate 46/1–2) and algal biomicrites (Plate 46/3). Fossils include calcareous algae [low diverse dasycladaceans with *Pseudoclypeina cirici* Radoičić; filamentous blue-green algae, and *Thaumatoporella parvovesiculifera* (Raineri), probably a red alga], foraminifera (especially *Parurgonina caelinensis* Cuvillier et al.), monactine and triactine sponge spicula, hydrozoans (*Cladocoropsis, Shuqraia, Milleporidium*), some corals, rare gastropods, and sometimes echinids.

Cladocoropsis limestones are widely distributed within the Upper Jurassic of the Mediterranean region (e.g., Slovenia, Turnšek et al., 1981; Croatia, Milan, 1969; Greece; Celet, 1962).

Microfacies Interpretation: An inner shelf position of the depositional environment is generally indicated by low-diverse biota, frequent bioturbation, occurrence of dasycladacean algae, algal-foraminiferal nodules, thin irregular algal mats, and foraminifera, which adapted to conditions of protected shelf regions (see Bolliger and Burri, 1970). Water energy was low (energy index I and II after Plumley et al., 1962, see 6.4), water depth shallow (shown by the abundance of dasycladaceans) but not within the intertidal zone.

References:

Bolliger, W., Burri, P. (1970): Sedimentologie von Schelf-Carbonaten und Beckenablagerungen im Oxfordien des zentralen Schweizer Jura. Beiträge geol. Karte Schweiz **140,** 1–96, 16 Pls., 37 Figs., Basel

Brinkmann, R., Flügel, E., Jacobshagen, V., Lechner, H., Rendel, B., Trick, P. (1972): Trias, Jura und Unterkreide der Halbinsel Karaburun (West-Anatolien). Geologica Palaeontologica **6**, 139–150, 4 Figs., Marburg

Celet, P. (1962): Contribution à l'étude géologique du Parnasse-Kiona et d'une partie des régions méridionales de la Grèce continentale. Ann. géol. Pays Helléniques **13**, 1–446, 37 Pls., 14 Figs., Athen

Felix, J. (1907): Eine neue Korallengattung aus dem dalmatinischen Mesozoikum. Sitzungsberichte naturforsch. Ges. Leipzig **1906/1907**, 3–10, 5 Figs., Leipzig

Flügel, E. (1974): Fazies-Interpretation der Cladocoropsis-Kalke (Malm) auf Karaburun, W-Anatolien. Archiv Lagerstättenforschung in den Ostalpen, Sonderband **2**, 79–94, 4 Pls., Leoben

Milan, A. (1969): Faziesverhältnisse und Hydrozoenfauna des Malms im Küstenland des nördlichen Velebit und Velika Kapela. Geol. Vjesnik **22**, 135–217, 22 Pls., Zagreb

Turnšek, D., Buser, St., Ogorelec, B. (1981): An Upper Jurassic reef complex from Slovenia, Yugoslavia. Soc. Econ. Paleont. Min. Spec. Publ. **30**, 361–369, 6 Figs., Tulsa

11.4 Outer Shelf Carbonate Platform: Upper Jurassic Sulzfluh Limestone, Graubünden, Switzerland (Plate 47)

The Upper Jurassic (Oxfordian to Middle Tithonian) Sulzfluh Limestone of Graubünden, Switzerland, and Vorarlberg, Austria, represents a carbonate platform sediment deposited in subenvironments which were differentiated with respect to water circulation and perhaps also to salinity. A re-evaluation of the material studied by W. F. Ott (1969) resulted in the recognition of a threefold facies pattern, consisting of: 1. Micrite facies with intertidal mudstones and wackestones, and subtidal wackestones (e.g., *Clypeina* biomicrites, Plate 47/2); 2. Sparite facies with strongly differentiated subtidal grainstones (predominantly biosparites, intrabiosparites, oncosparites), deposited on the open-marine platform or within restricted bays of the platform; and 3. Oolite facies with oolitic grainstones. These facies units may represent elements of an outer-shelf platform comparable in part to the Recent Bahama Bank. No larger reef structures developed in the Sulzfluh Limestones but algae may have produced bindstones consisting of filamentous blue-green algae (Plate 47/5), or oncoids built by *Bacinella* (Plate 47/6).

Plate 46. Inner Shelf Cladocoropsis Limestones: Upper Jurassic (Upper Oxfordian/ Kimmeridgian) of Karaburun Peninsula, Western Anatolia

1. Hydrozoan biomicrite with several sections of *Cladocoropsis mirabilis* Felix, and some foraminifera. The fasciculate colonies of *Cladocoropsis* are characterized by relatively thick vertical skeletal elements diverging towards the periphery

2. Fragments of *Cladocoropsis,* together with sponge spicula and algal spores (?)

3. Dasycladacean biomicrite with abundant fragments of *Pseudoclypeina cirici* Radoičić, and sponge spicula

All samples from Birgi, central Karaburun, Turkey. Scale for all Figures 2 mm

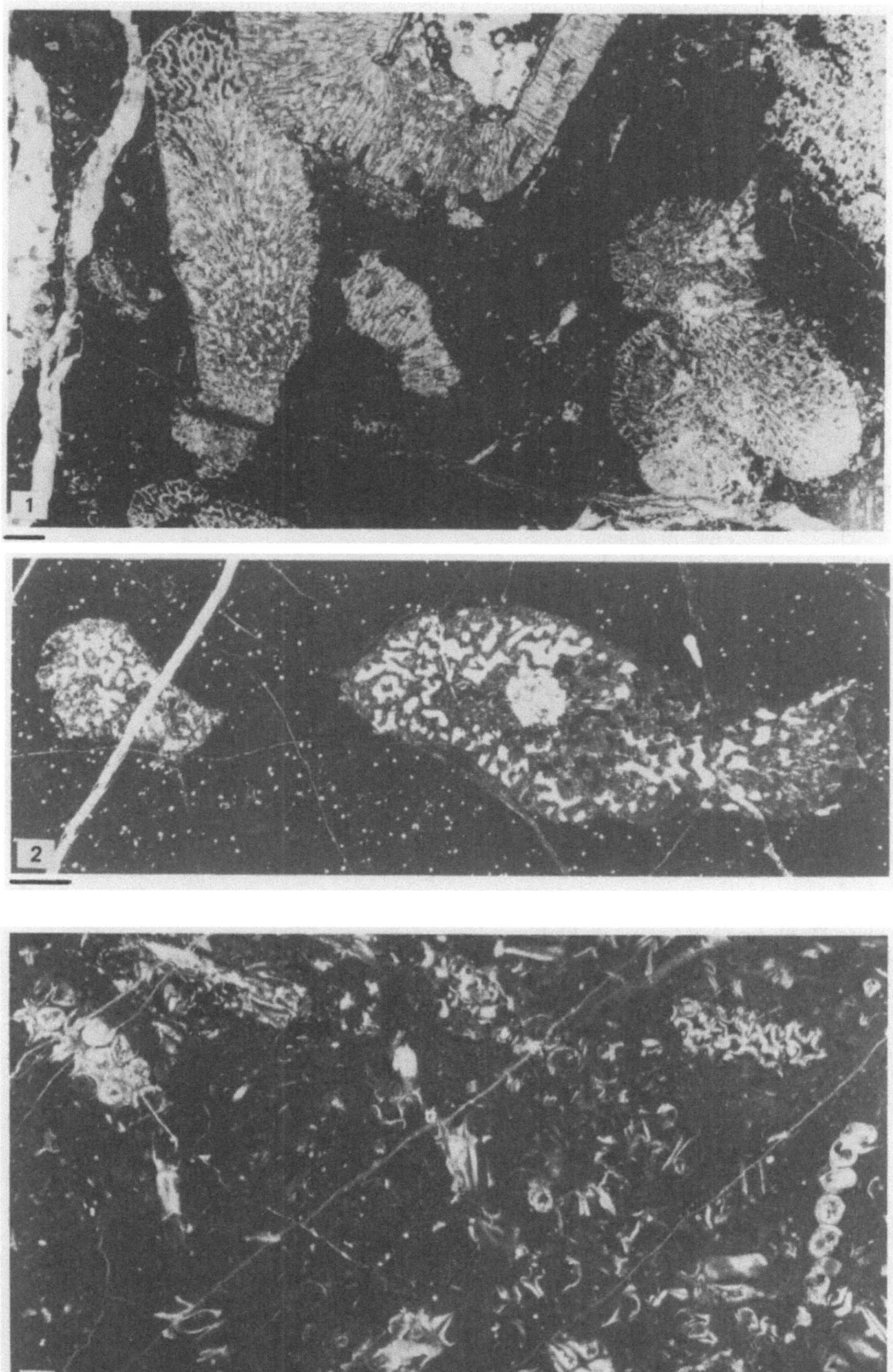
1
2
3

Microfacies Interpretation: Microfacies types of the Sulzfluh Limestones may be compared with those of the Upper Jurassic Plassen Limestones in the eastern Northern Alps (see Steiger and Wurm, 1980), but there are differences in the paleogeographical setting of the types in a facies model (e.g., *Bacinella* acting as a reef-building element in the Plassen Limestones; frequent occurrence of grainstones with micrite oncoids). These differences show the importance of comparative studies by means of which the distributional patterns of microfacies types can be checked.

References:

Flügel, E. (1979): Paleoecology and microfacies of Permian, Triassic and Jurassic algal communities of platform and reef carbonates from the Alps. Bull. Cent. Rech. Explor.-Prod. Elf-Aquitaine **3**/2, 569–587, 3 Pls., 5 Figs., Pau

Ott, W. F. (1969): Zur Geologie des Sulzfluh-Kalkes (Malm) in Graubünden und Vorarlberg. Unpublished Thesis, Technische Hochschule Darmstadt

Steiger, T., Wurm, D. (1980): Faziesmuster oberjurassischer Plattform-Karbonate (Plassen-Kalke, Nördliche Kalkalpen, Steirisches Salzkammergut, Österreich). Facies **2**, 241–284, Pls. 25–30, 8 Figs., Erlangen

Plate 47. Outer Shelf Carbonate Platform: Upper Jurassic Sulzfluh Limestone, Graubünden, Switzerland

1. Biointrasparite. Aggregate grains (lumps) and cortoids (coated bioclasts). Bioclasts include gastropods and dasycladacean fragments. Open-marine platform environment

2. *Clypeina* biomicrite with parautochthonous thalli of *Clypeina jurassica* Favre and Richard. Restricted lagoonal environment

3. Intrasparite. Intraclasts are algal lumps (see Sect. 4.1.3.3) indicating reworking of cyanophycean algal mats. Lagoonal environment

4. Oosparite. Ooids are strongly micritized and effected by late-diagenetic solution effects. Oolite bar (at the platform margin?)

5. Cyanophycean bindstone, typically developed in tidal flats of the restricted lagoonal environment. Algae may be compared with Recent rivulariacean blue-green algae

6. *Bacinella irregularis* Radoičić, an important binding organism which produces large oncoids in the oncosparite facies of restricted bays on open-marine platforms

All samples are from the Weißfluh section, Sulzfluh Mountains, Graubünden. Scale for all Figures 2 mm

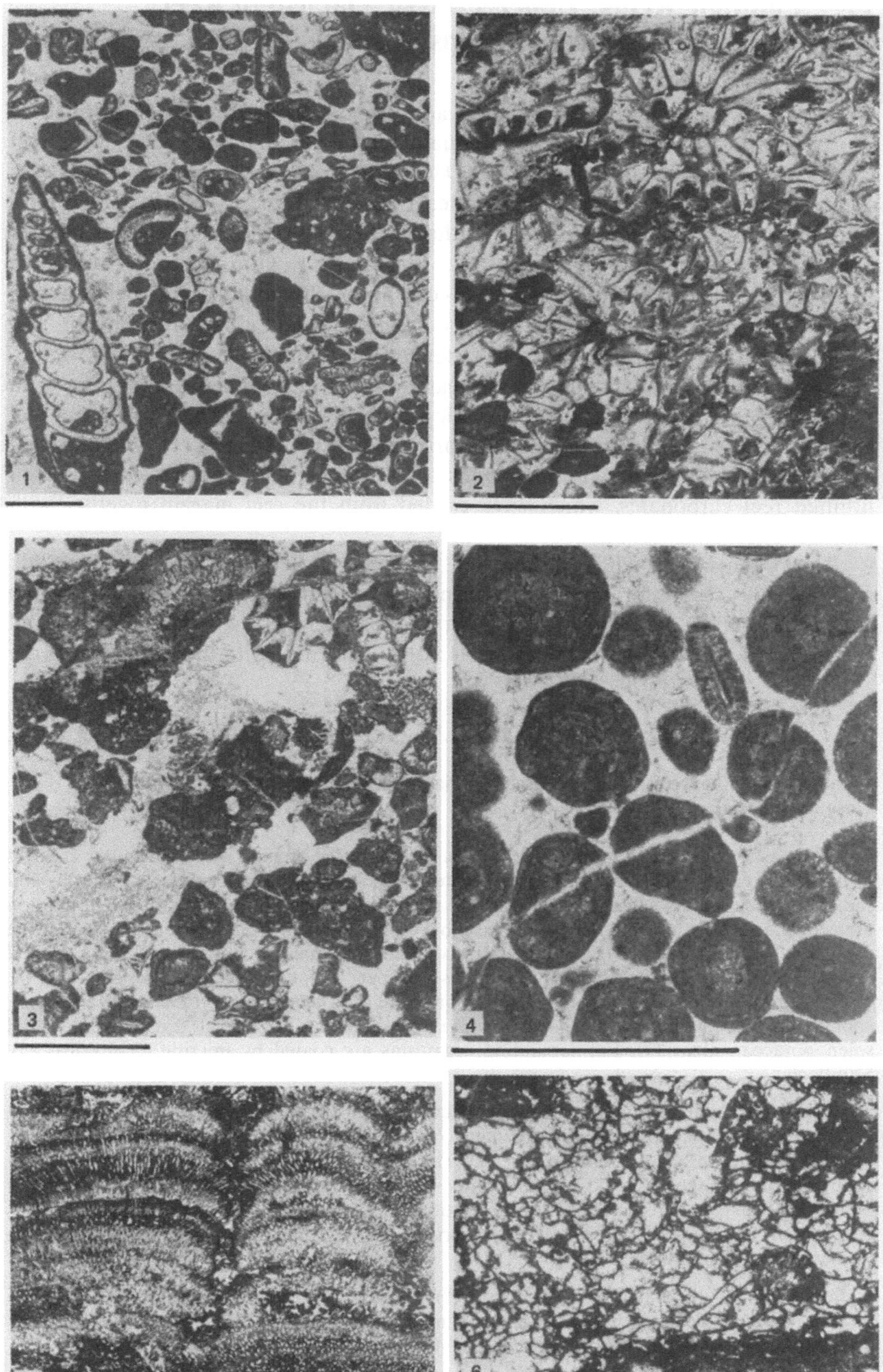

11.5 Ecologic Reefs: Upper Triassic Dachstein Reef Limestones of the Northern Alps, Austria (Plate 48)

During the Norian, extensive shallow-water carbonate platforms were formed in the Northern Alps, which are differentiated into marginal reef zones with reef complexes and skeletal sand shoals (Dachstein Reef Limestones), and lagoons with cyclic sedimentation patterns (bedded Dachstein Limestones). Extended reef-complexes developed predominantly at the southern rim of the Dachstein platform (e.g., Gosaukamm reef, Upper Austria; Wurm, 1981; Göll reef, Berchtesgaden Alps; Zankl, 1969). These reef-complexes can be roughly divided into a central reef area (reef framework and reef crest), a fore-reef slope, and a back-reef lagoon. The facies of the central reef area is characterized by crops of autochthonous frame-building communities (predominantly calcisponges and corals), large areas with skeletal debris (reef detritus), intensive biogenous overgrowth (Plate 48/2), and reef cavities, filled with pelmicritic and pelsparitic sediments (Plate 48/1). The proximal fore-reef facies (Plate 48/1) corresponds to rudstones or floatstones composed of lithified reef carbonates, which were eroded from the reef margin together with isolated framebuilding organisms.

The organisms which were involved in the construction of the Dachsteinkalk reefs includes sessile foraminifera, calcisponges, sclerosponges (generally described as "tabulozoans"), hydrozoans, corals, bryozoans, as well as blue-green, green and red algae. In addition many microproblematica contributed to the formation of the

---

Plate 48. Ecologic Reefs: Upper Triassic (Norian) Dachstein Reef Limestone of the Gosaukamm, Austria

1. Lithoclastic rudstone of the fore-reef facies. Poorly sorted angular and subangular clasts of lithified reef carbonates as well as some isolated reef organisms are cemented by fibrous and radiaxial calcite. Reef clasts exhibit various microfacies types, indicating different erosion areas at the reef crest. Most limestone clasts have been derived from the central parts of the reef crest, others (bottom right) represent sediments originally deposited within small reef cavities. Low-energy protected environments (deeper reef crest or reef slope?) may be inferred from the biota consisting of calcisponges (e.g., *Cheilosporites tirolensis* Wähner, *1*), spongiomorphid hydrozoans (*Spongiomorpha ramosa* Frech, *2*), coral fragments (*3*), and solenoporacean algae (*4*). Framebuilding organisms are coated by micritic algal crusts (*5*). Stuhlloch

2. Boundstone with abundant biogenic encrustations between and upon framebuilding organisms. The biota consists of hydrozoans (*Disjectopora* sp., *1*), corals (*2*), calcisponges (*3*), and dasycladacean algae (*Diplopora* sp., *4*). The framebuilders are commonly overgrown by spongiostromid algae, sessile foraminifera, and microproblematica (*Radiomura cautica* Senowbari-Daryan and Schäfer, *5*). Bioclastic detritus between frame-building organisms consists of brachiopod and pelecypod debris. This microfacies represents the reef core development. Donnerkogel

3. Boundstone with branches of the coral (?) *Pinacophyllum* sp., embedded in pelmicritic matrix. *Pinacophyllum* as well as non-segmented calcisponges sometimes form small, low-diverse patch-reefs within the central reef area. Großer Donnerkogel.

Scale for all Figures 10 mm

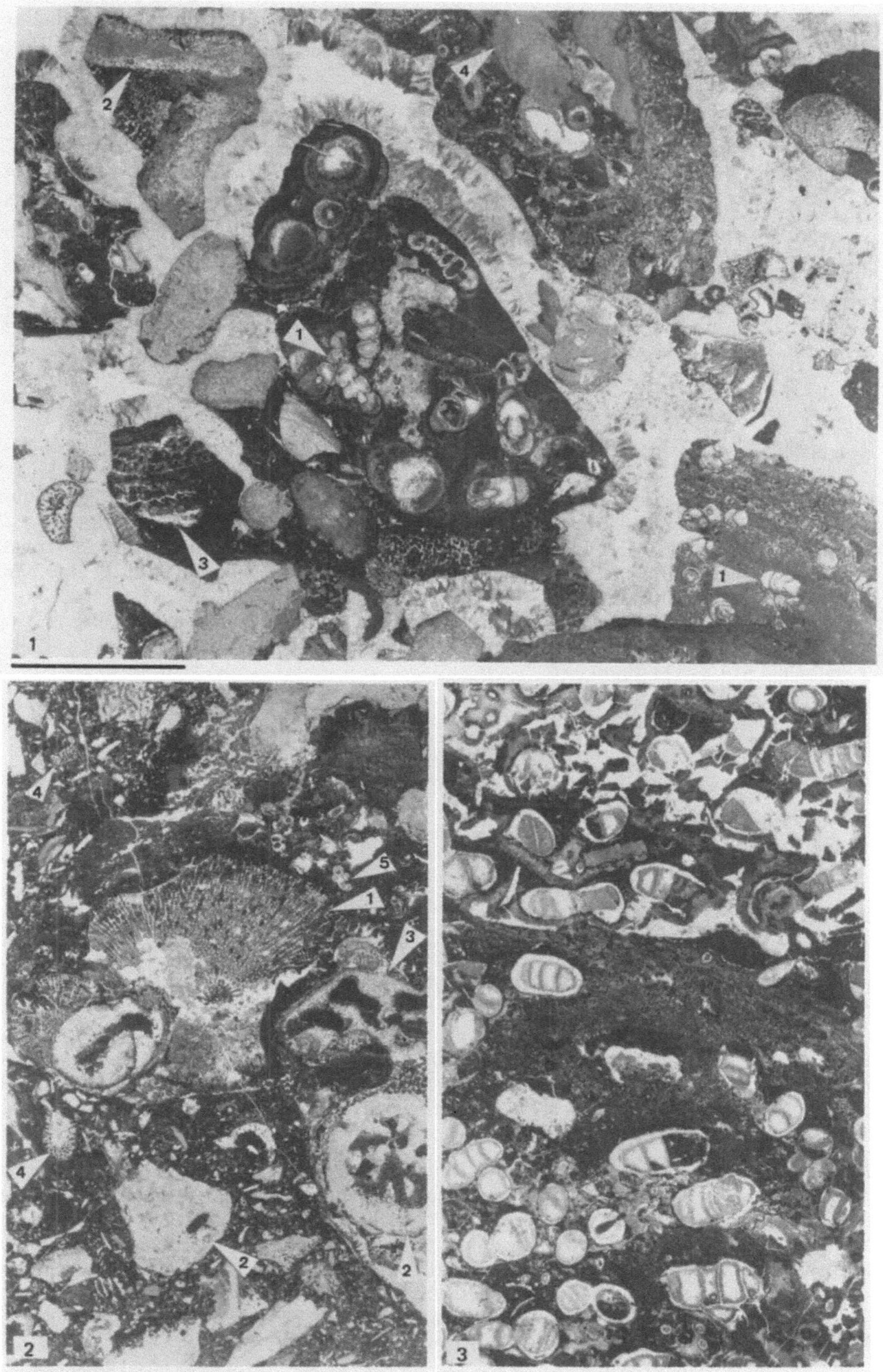

organic framework. Foraminifera, calcareous algae and various microproblematica are concentrated in distinct zones of the reef-complexes and in the adjacent facies zones, and can be used to differentiate the facies units of the platform carbonates as well as of the reef limestones.

Microfacies Interpretation: Seven microfacies types can be recognized in the Dachstein reef limestones of the Gosaukamm Range (Wurm, 1981). Most common are rudstones and floatstones (upper reef slope), bioclastic grainstones and packstones with poorly sorted reef debris (reef crest and reef flat), well-sorted detrital grainstones with abundant aggregate grains and oncoidal bioclastic grainstones (reef flat) as well as strongly differentiated boundstones (reef framework and reef crest). Algal wackestones with filamentous blue-green algae and echinoderm-foraminiferal wackestones or mollusk packstones are relatively rare.

References:

Flügel, E. (1981): Paleoecology and facies of Upper Triassic reefs in the Northern Calcareous Alps. Soc. Econ. Paleont. Min. Spec. Publ. **30,** 291–359, 26 Figs., Tulsa

Wurm, D. (1981): Gosaukamm. Facies and Paleoecology of the Dachstein Reef Limestones and the Zlambach Limestones. In: Flügel, E. (Ed.): Guidebook Internat. Symposium Triassic Reefs, 86–104, Erlangen

Zankl, H. (1969): Der Hohe Göll: Aufbau und Lebensbild eines Dachsteinkalk-Riffes in der Obertrias der Nördlichen Kalkalpen. Abhandlungen senckenbergischen naturforsch. Gesellschaft **519,** 1–123, Frankfurt

Plate 49. Stratigraphic Reefs: Lower Permian Trogkofel Limestones of the Southern Alps
(Text description page 570)

1. Algal bindstone. Crusts of the alga *Archaeolithoporella hidensis* Endo form cavities which were filled with peloidal fine-grained sediment and with syndepositional fibrous and radiaxial carbonate cements. *Archaeolithoporella* crusts circumscribe fragments of *Tubiphytes* as well as detrital grains. Seikofel, Sexten Dolomites

2. *Tubiphytes carinthiacus* (Flügel), *1,* and *Tubiphytes obscurus* Maslov (*2*), the most common biotic elements of the Trogkofel buildups. Both species are encrusting organisms growing upon micritic sediment, carbonate cements, *Archaeolithophyllum,* bryozoans, or upon each other. *T. carinthiacus* is generally larger than *T. obscurus,* and characterized by a relatively open meshwork. *Tubiphytes* Maslov is generally regarded as blue-green alga. The genus is known from the Lower Carboniferous to the Upper Jurassic (see Sect. 5.3.13), it is a very important encrusting organism in Pennsylvanian and Permian organic buildups, as well as in Middle Triassic reefs. Kühweg Alm, Carnic Alps

Scale for both Figures 2 mm

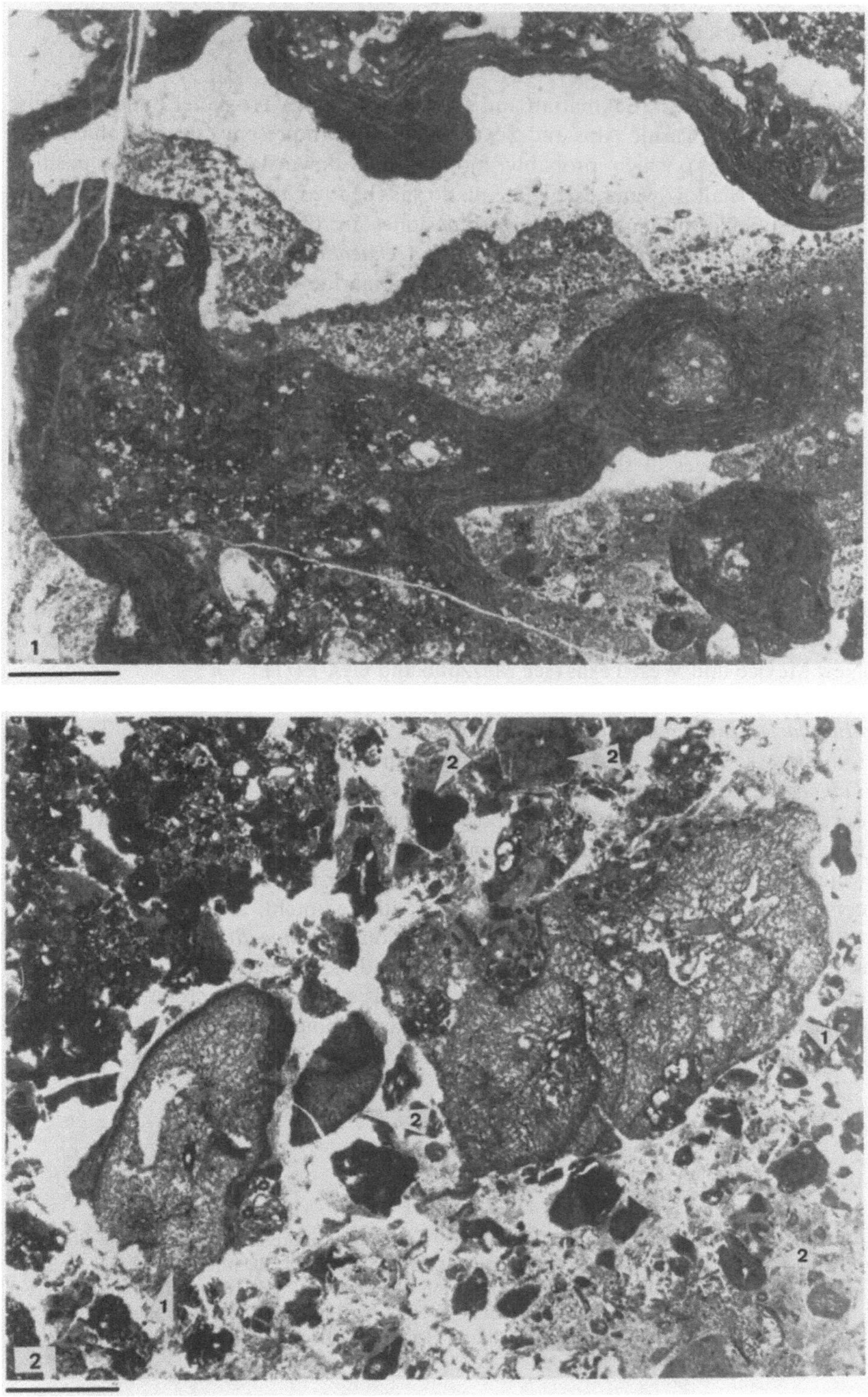
1
2
1
2

11.6 Stratigraphic Reefs: Lower Permian Trogkofel Limestones of the Southern Alps (Plate 49)

The Lower Permian (Sakmarian and Lower Artinskian) Trogkofel buildups of the Southern Alps (Carnic Alps and Sexten Dolomites) represent "stratigraphic reefs" (see Sect. 10.2.3) which probably formed in a down-slope shelf-edge position adjacent to shallow-water carbonate platforms (bedded Trogkofel Limestones). The major biotic constituents of the reef limestones are the algae *Tubiphytes obscurus* Maslov, *Tubiphytes carinthiacus* (Flügel), *Archaeolithoporella hidensis* Endo, and bryozoans, which together with eogenetic submarine cements contributed to the formation of diagenetic/organic bindstones (Flügel, 1981). The syndepositional character of the cements is evident because of the existence of multiple cement generations encrusted by *Archaeolithoporella,* the presence of internal sedimentation of peloids and skeletal debris between various layers of cements and within cavities formed by the algal crusts (Plate 49/1), and because of coatings of isopachous fringes of short fibrous crystals on broken *Archaeolithoporella* crusts. Low, laminar encrusting organisms such as *Tubiphytes, Archaeolithoporella,* a few other algae, and bryozoans together with synsedimentarily formed carbonate cements may comprise up to 60 percent of the rock volume. Frame-building organisms are extremely rare; talus sediments are missing.

Biota and major fabric types as well as the development of syndepositional carbonate cements can be compared with those of the Permian Reef Complex of New Mexico and West Texas (see Mazzullo and Cys, 1977).

Microfacies Interpretation: Three microfacies types are common in the Trogkofel Limestones: 1. Biopelsparites with varying numbers of *Archaeolithoporella* and *Tubiphytes,* and poorly sorted peloids of various sizes; sparite exhibits well-developed radial-fibrous primary fabrics in some cases. 2. Intrabiosparite with fragments of *Archaeolithoporella* and angular to subangular pelsparitic intraclasts, which may be circumscribed by isopachous calcite rims composed of short fibrous crystals. Open spaces are filled with pelmicritic sediment. 3. Reddish biomicrites with many bryozoans, *Tubiphytes,* and occasional *Archaeolithoporella.* Bryozoans are generally found both as encrusting forms and as fragments. The environmental interpretation of these microfacies types depends strongly on an exact classification of the cement fabrics; this should be done by a combined study of thin-sections and SEM photomicrographs.

References:

Flügel, E. (1981): Lower Permian Tubiphytes/Archaeolithoporella buildups in the Southern Alps (Austria and Italy). Soc. Econ. Paleont. Min. Spec. Publ. **30,** 143–160, 9 Figs., Tulsa

Mazzullo, S. J., Cys, J. M. (1977): Submarine cements in Permian boundstones and reef associated rocks, Guadalupe Mountains, West Texas and Southeastern New Mexico. In: Field Conf. Guidebook, Permian Basin Sect. Soc. Econ. Paleont. Min. Publ. 77–16, 151–200, Midland

11.7 Mud Mounds: Sponge-Algal Buildups from the Upper Jurassic of Northern Franconia, Germany (Plate 50)

During the Upper Jurassic relatively small organic buildups, principally composed of siliceous sponges and algae, developed in southern Germany. The biota of these buildups consist of hexactinellid and lithistid siliceous sponges, most of which were calcified in early diagenesis as a result of alterations of the chemical environment caused by the decay of the sponges (Fritz, 1958; Flügel and Steiger, 1981); mobile and sessile foraminifera; echinoderms, brachiopods, ostracods, pelecypods, and cephalopods. Epizoans are common on the sponges. Of special interest are micritic and peloidal crusts (laminoid peloidal algal crusts, Plate 50/1; filamentous algal crusts; foraminiferal crusts) which aided in forming the mound configuration. Sedimentary particles are angular to rounded lithoclasts, bioclasts, and rounded micritic grains, forming a continuum in size ranging from peloids to tuberoids to oncoids. Tuberoids represent a particular category of limestone particles, known only from siliceous sponge-bearing carbonates (see Sect. 4.1.3.2). Lithoclasts may represent reworked tuberoids.

Three distinctive facies types can be recognized, based on both microfacies and paleontological data:

1. Sponge-Crust Boundstone facies (Plates 50/1–2): micritic boundstones with abundant siliceous sponges (dish-shaped and cup-shaped), tuberoids and crusts; the crusts appear in the more mature stage of boundstone formation. 2. Lithoclastic Packstone Facies (Plate 50/3), characterized by densely packed spheroidal particles (lithoclasts and tuberoids, as well as oncoids), and scarseness of micritic matrix. This facies is restricted to areas adjacent to the boundstone facies, forming the transitional deposits between the buildup and the normal stratified facies (bedded limestones, sometimes interbedded with marls). 3. Tuberolitic Wackestone Facies with a great many tuberoids, but also with lithoclasts, oncoids and bioclasts. In contrast to the Lithoclastic Packstone Facies, particles of the Tuberolitic Wackestone Facies are mud-supported. These wackestones are typical of the normal stratified facies occurring in near-reef position.

Microfacies Interpretation: Limestone particles of these sponge-algal carbonates are difficult to differentiate because there are no Recent counterparts known which could be used to test the models established for the postmortem alterations of siliceous sponges. SEM studies produced contradictory results, especially concerning the role of blue-green algae in the formation of the mounds. Facies interpretation, therefore, depends heavily on paleontological criteria.

References:

Flügel, E., Steiger, T. (1981): An Upper Jurassic Sponge-Algal buildup from the Northern Frankenalb, West Germany. Soc. Econ. Paleont. Min. Spec. Publ. **30,** 371–397, 23 Figs., Tulsa

Fritz, G. K. (1958): Schwammstotzen, Tuberolithe und Schuttbreccien im Weißen Jura der Schwäbischen Alb. Arbeiten Geol. Paläont. Inst. Technische Hochschule Stuttgart **13,** 1–119, Stuttgart

11.8 Subtidal Open-Marine Shelf Carbonates: Mishrif Formation, Cretaceous, Middle East (Plate 51)

The Middle to Upper Cenomanian Mishrif Formation of the Middle East region (e.g., Arabia; El-Naggar and Al-Rifaiy, 1973) represents predominantly subtidal shelf carbonates which are differentiated in places by small biostromes built up by rudists. Most facies types correspond to a sedimentation within Zone Z of the Irwin-Lees model (see Sect. 10.3.4), indicating deposition in a shallow-water, low-energy shelf area. This interpretation is in agreement with the SMF type classification of the samples shown in Plate 51; most samples can be attributed to the SMF Type 9, characterized by bioclastic wackestones. The limestones contain fragments of pelecypods, gastropods, corals and foraminifera, jumbled and homogenized through burrowing. This SMF type is formed in shallow neritic water with open circulation (corresponding to Wilson's Facies Belt 7, see Fig. 55), or just below the wave base (Facies Belt 2, open-marine shelf). Most of the Mishrif carbonates seem to have been deposited within the Facies Belt 7 (open-marine inner-shelf), but there are indications for an influence of marginal shelf environments (SMF Types 3 and 12, see Plates 51/3 and 6). Some of the samples contain abundant detritus of rudist shells, indicating the existence of local rudist buildups.

Microfacies Interpretation: The microfacies types of these subtidal carbonates are rather uniform, comprising only burrowed peloidal wackestones with different amounts of mollusk debris, foraminifera, ostracods, serpulids, and echinoderms. Diversity is low to moderate.

Plate 50. Mud Mounds: Sponge-Algal Buildups from the Upper Jurassic (Oxfordian) of Northern Franconia, Germany
(Text description p. 571)

1. Sponge-Crust Boundstone Facies. Micritic boundstone with calcified siliceous sponges (*1*) overgrown by peloidal micritic crusts (*2*), generally interpreted as algal crusts

2. Sponge-Crust Boundstone Facies. Sediment between siliceous sponges is a micrite rich in sponge spicula and tiny tuberoids (irregularely shaped, dark calcareous lumps, formed by an early diagenetic calcification of decaying sponge and crust material). Strong dissolution of the limestone is shown by microstylolites

3. Lithoclastic Packstone Facies, characterized by densely-packed spheroidal particles, most commonly lithoclasts, tuberoids, foraminifers, and other bioclasts (e.g., sponge spicula, shell debris). The lithoclasts (*1*) consist of tuberoids (*2*) and skeletal material. Conspicuous elements are encrusting organisms (*3*), probably representing an algal/foraminiferal consortium (*Tubiphytes* and nubeculinellid foraminifers). This facies is restricted to areas adjacent to the boundstone facies

All samples from the Müllersfelsen near Streitberg northeast of Erlangen

Scale for all Figures 5 mm

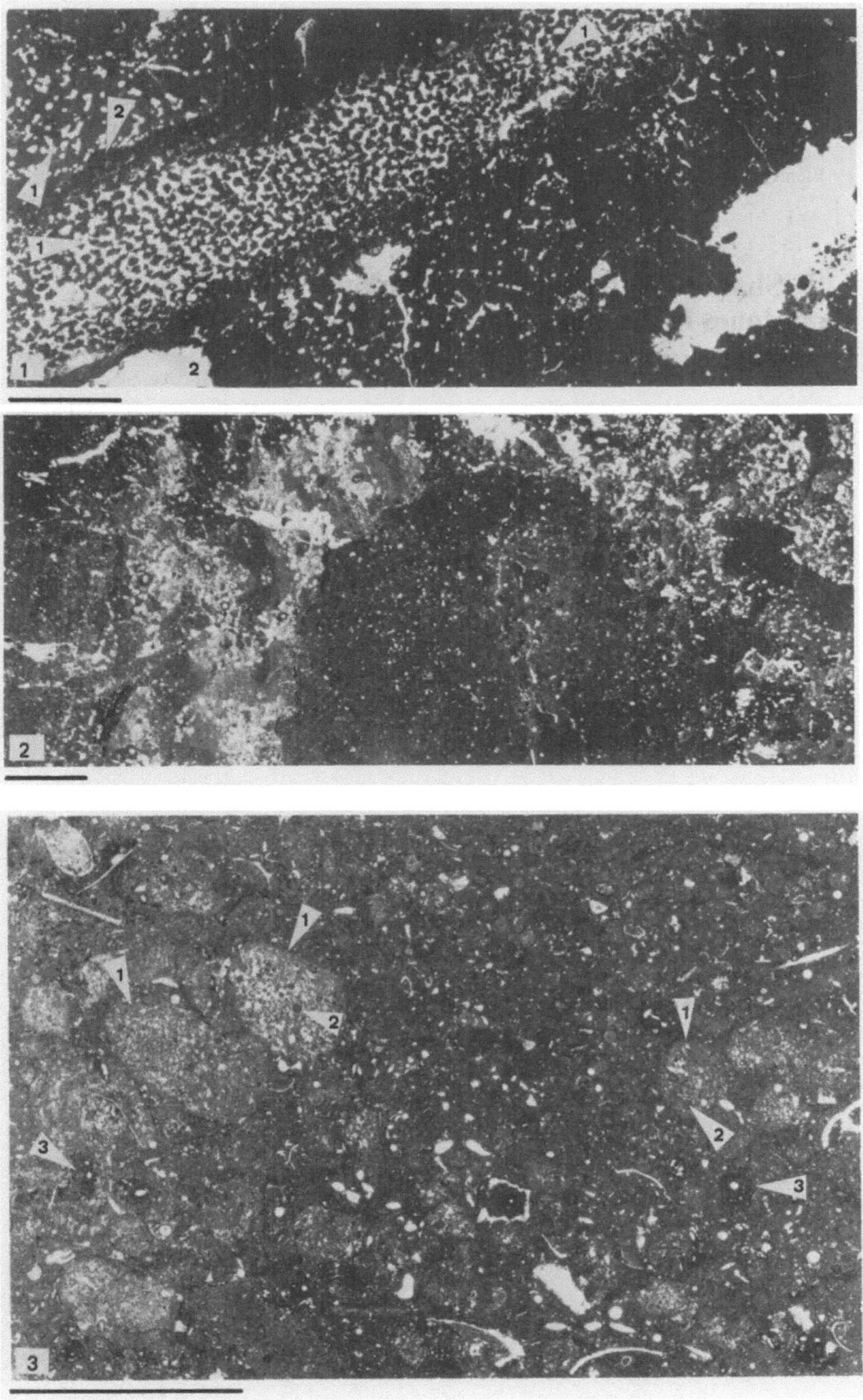
1
2
1
1
1
2
2
3
1
1
2
1
2
3
3
3

References:

El-Naggar, Z. R., Al-Rifaiy, I. A. (1973): Stratigraphy and microfacies of the type Maggwa formation of Kuweit, Arabia. Part 2.: Mishrif Limestone Member. Amer. Ass. Petrol. Geol. Bull. **57**/11, 2263–2279, Tulsa

Reulet, J. (1977): Réservoir calcaire de platforme marine (formation Mishrif, Moyen Orient). Elf-Aquitaine: Essai de caractérisation sédimentologique des dépôts carbonatés. **2**, Eléments d'interprétation, 169–180, Figs. 39–44, Pau

11.9 Shallow-Water Limestone Turbidites: Allodapic Limestones in the Upper Jurassic of Southern Franconia, Germany (Plate 52)

Allochthonous carbonate sediments were deposited in the Kelheim region adjacent to Tithonian coral reefs. A section near Kapfelberg shows a sequence, up to 20 m thick, composed of thin-bedded micritic limestones with intercalations of allodapic limestones. These allochthonous carbonates are characterized by distinct and common graded bedding structures, poor sorting of bioclasts and lithoclasts at the base of the beds, and by an almost complete lack of infauna. Coarser detrital layers contain brachiopods, pelecypods, gastropods, and a few ammonites, as well as some land-derived plants. The composition of the graded beds is quite variable: lithoclasts and peloids make up 20–70% of all particles; blue-green algae of *Cayeuxia* type 10–65%; pelecypods 15–40%; corals less than 5%. The scarceness of corals and the abundance of filamentous algae distinguish these sediments from the possible source areas (coral reefs near Kelheim-Saal).

Plate 51. Subtidal Open-Marine Shelf Carbonates: Mishrif Formation, Cretaceous, Middle East
(Text description p. 572)

1. Fine-grained wackestone with large fragments of pachydont pelecypods and bioclastic debris; SMF Type 9

2. Bioclastic wackestone with corals, pelecypods, serpulids and some trocholinid foraminifera. Serpulid encrustation indicates a local hardground formation (cryptohardground)

3. Wackestone with planktonic foraminifera (*Hedbergella* sp.) and sponge spicula; SMF Type 3

4. Bioclastic wackestone with abundant thin-shelled gastropods

5. Bioclastic wackestone with abundant shell debris and larger pelecypod and gastropod fragments. The sediment is strongly burrowed

6. Pelecypod-echinoderm packstone consisting of poorly sorted, medium- to coarse-grained bioclasts; SMF Type 12

7. Rudist floatstone with burrows (showing geopetal infillings); pelecypod shells exhibit distinct borings

All samples are from the Cenomanian of Arabia

Scale for all Figures 2 mm

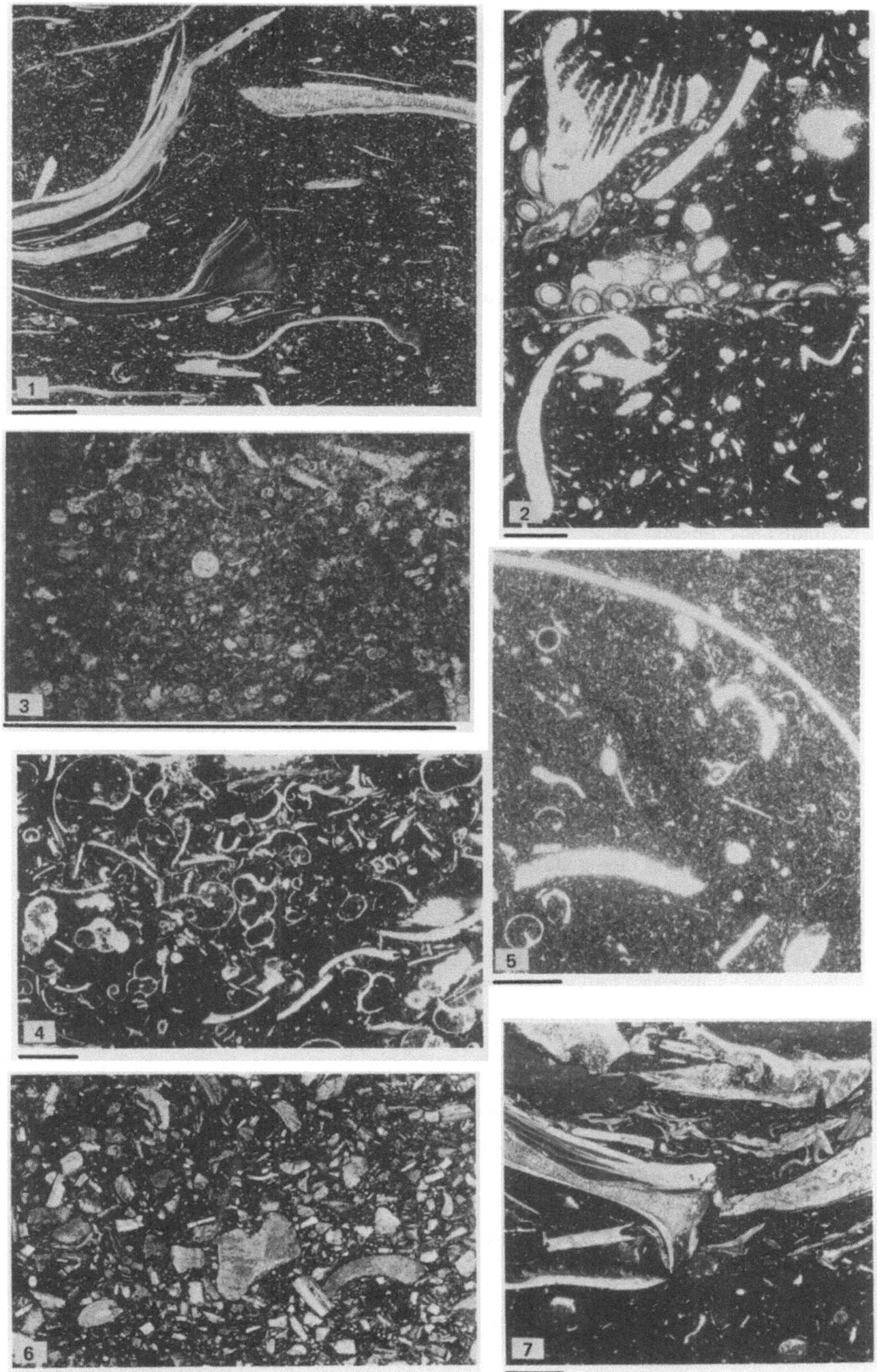

Frame-building organisms are represented only in some m-sized, lens-shaped limestones, intercalated in the allochthonous carbonates. These boundstones contain corals, encrusting hydrozoans (*Burgundia* sp., Plate 52/3) and chaetetids, as well as encrusting algae (*Lithocodium, Bacinella*).

The allochthonous carbonates seem to have been deposited in a basin of moderate depth (maximum about 80 m). Depositional pattern, sedimentary structures and textural criteria correspond to those of deeper-water turbidites.

Microfacies Interpretation: Study of these limestone turbidites is possible only with large thin-sections which include the total thickness of a limestone bed. Microfauna and algal flora infer a source area on a protected carbonate platform.

References:

Fay, M. (1976): Riffnahe Resedimente im Raum Kelheim: Lithogenese, Genese und stratigraphische Bemerkungen. N. Jb. Geol. Paläont. Abh. **152**/1, 51–74, 5 Figs., 2 Tabs., Stuttgart

Flügel, E., Keupp, H., Meyer, R., Zeiß, A. (1975): Excursion Route. In: Flügel, E. (Ed.): Guide Book, Internat. Symposium Fossil Algae, 196–201, Erlangen

Schairer, G., Lupu, M. (1969): Mikrofazielle Untersuchungen in untertithonischen, geschichteten Kalken von Kapfelberg bei Kelheim in Bayern. Mitt. Bayer. Staatssammlung Paläont. hist. Geol.. **9,** 183–199, Pls., 11–12, 8 Figs., München

Plate 52. Shallow-Water Limestone Turbidites: Allodapic Limestones, Upper Jurassic of Southern Franconia, Germany

1. Lower part of an allodapic limestone bed. Disorganized packstone consisting of various lithoclasts, reworked algal thalli, pelecypod shells, and micrite clasts. Fossils indicate a shallow-marine source area for the particles

2. Lower and middle part of an allodapic limestone bed, characterized by significant changes in packing, sorting and grain sizes. At the bottom pelecypod shells with geopetal depositional fabric of pellets on the surface of shells (*1*). Lithoclasts and peloids exhibit distinct grading. Size of bioclasts (dasycladacean algae, e.g., *Salpingoporella* sp., *2*; and foraminifera) corresponds with the size of other sedimentary particles

3. Boundstone, intercalated within the allodapic section, corresponding to small autochthonous patch reefs or to allochthonous blocks. Biota consist of various tabular corals intergrown with stromatoporoids (*Burgundia* sp., *1*) and blue-green algae (*2*)

4. *Cayeuxia moldavica* Frollo. This filamentous blue-green alga is of rock-forming importance in some allodapic beds; more than 60% of all skeletal grains may be represented by overturned thalli of these algae

All samples are from the Kapfelberg quarry near Kelheim

Scale for all Figures 2 mm

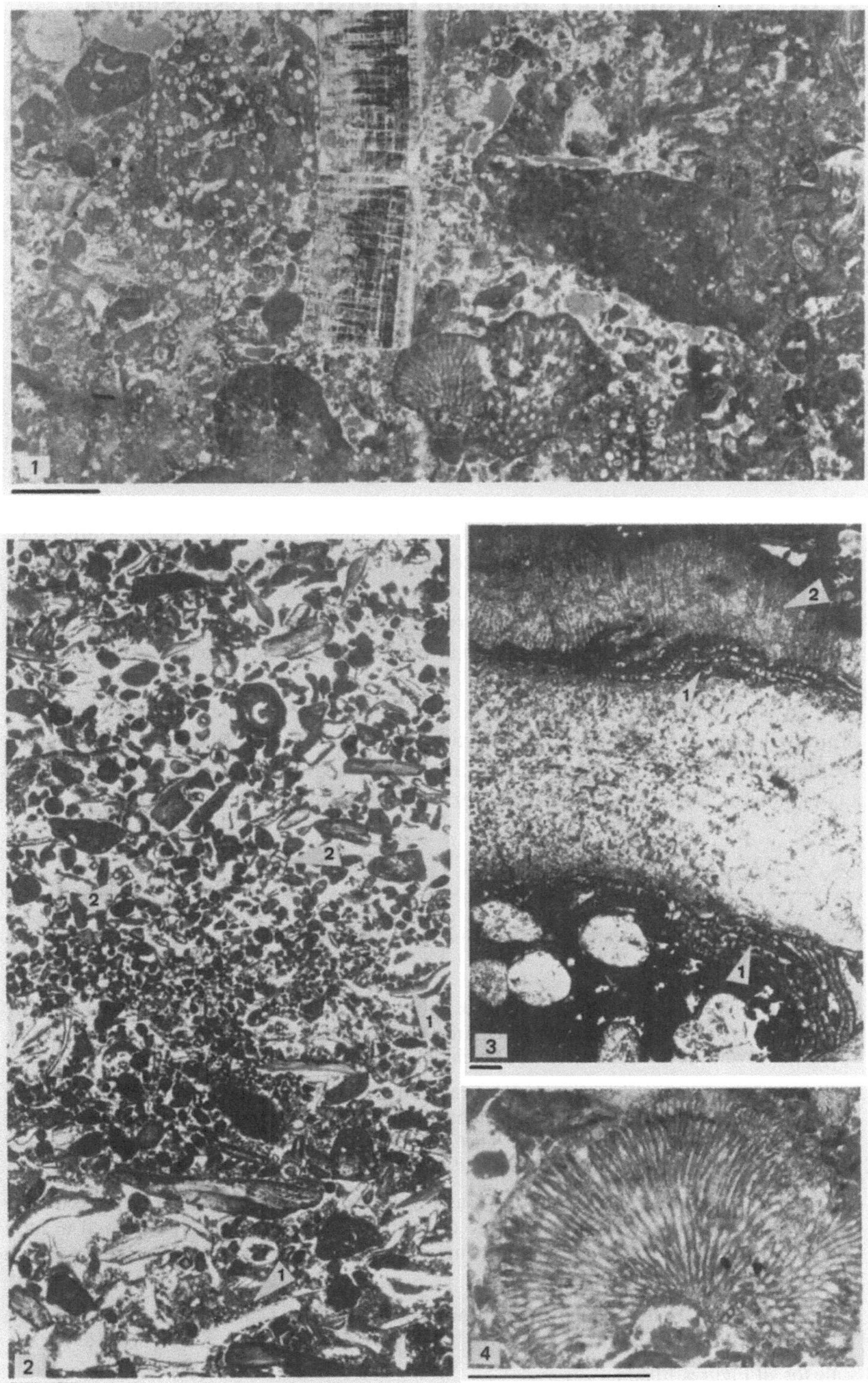
1
2
3
4

11.10 Allochthonous Carbonate Deposits: Liassic Scheck Conglomerate; Adnet, Salzburg/Austria (Plate 53)

Liassic deeper-water limestones and allochthonous carbonates can be studied in several quarries near Adnet southeast of Salzburg, Austria. A generalized section shows thin- to medium-bedded red nodular limestones with marly partings at the base (Adnet Limestone), overlain by a massive to thick-bedded limestone conglomerate, and a massive to thick-bedded spar-cemented conglomerate at the top ("Scheck" conglomerate).

The nodular limestone (Adnet Limestone) consists of red nodules of fine-grained limestones in a matrix of slightly more marly and redder material. The nodules may range up to about 5 cm in size. Minor stylolitic partings and, sometimes, black ferro-manganese stainings along the margins of the nodules may be present (see Plate 53/1). The microfacies of these highly oxidized, predominantly red bioclastic wackestones is characterized by micritic matrix with abundant foraminifera, crinoid ossicles and echinid fragments, thin mollusk shells, some sponge spicula, and the planktonic alga *Globochaete.*

The Limestone Conglomerate consists of clasts from the Adnet Limestone. Several microfacies types may be recognized in the clasts: red bioclastic wackestones corresponding to the typical Adnet Limestones; rounded micritic clasts coated by Fe/Mn oxides, and light red crinoidal limestone clasts. The depositional fabric often has been changed to a diagenetic packstone fabric or a stylobreccia (see Plate 53/2).

The so-called Scheck Conglomerate is characterized by red calcilutitic clasts cemented by several generations of white sparry calcite (see Plate 53/3). The clasts are better rounded than those of the Limestone Conglomerate. The cement exhibits a conspicuous fibrous and radiaxial fabric (see Hudson and Jenkyns, 1969).

Plate 53. Allochthonous Carbonate Deposits: Liassic Scheck Conglomerate Adnet, Salzburg/Austria

1. Red nodular limestone. The boundaries between nodular "clasts" are marked by micro-stylolites (*1*). Skeletal grains appearing are: echinoderms, chiefly crinoids (*2*), sometimes bored (*3*); involutinid foraminifera (*4*); brachiopods (*5*); see p. 412

2. Stylobreccia, composed of fragments of red nodular limestone. The depositional packstone fabric of the conglomerate has been changed to diagenetic packstone by strong pressure solution, which in turn removed the primary micritic matrix. Clasts consist of burrowed biomicrites with foraminifera, radiolaria, and mollusk shells

3. "Scheck" limestone. Red limestone clasts with various microfacies are connected by conspicuous fibrous cements of submarine origin. The Scheck limestone is believed to have originated from a redeposition of the conglomerate in deeper parts of the slope, and a subsequent submarine cementation

All samples from Kiefer Platten quarry northeast of Adnet near Hallein, Salzburg. Late Middle Lias

Scale for all Figures 10 mm

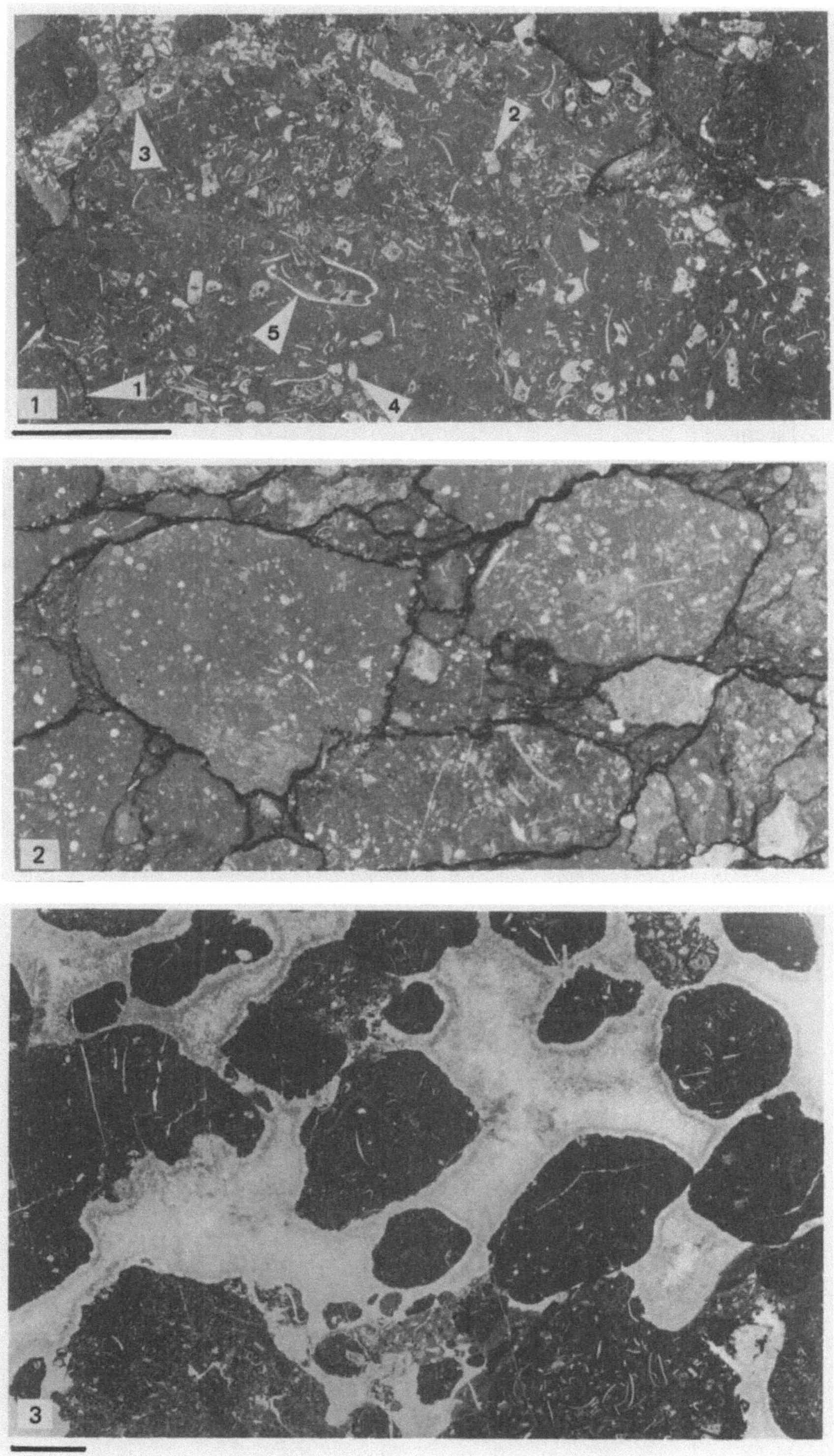
1
2
3
4
5
1
2
3

The clasts of the Limestone Conglomerate may be derived from a nearby topographic high within a deeper-marine environment; submarine slumping and mud-flow transport seem to have been the major processes in the genesis of these deposits. The Scheck conglomerate has been interpreted as being formed by a redeposition of the Limestone Conglomerate (Hudson and Jenkyns, 1969), or by in-situ break-up of lime mud as a result of emergence (Hallam, 1967). A submarine origin of the Scheck conglomerate is confirmed by cement fabrics and also by isotope values (Hudson and Coleman, 1978).

References:

Hallam, A. (1967): Sedimentology and paleogeographic significance of certain red limestones and associated beds in the Lias of the Alpine region. Scottish J. Geol. **3,** 195–220, Pls. 1–2, Liverpool

Hudson, J. D., Coleman, M. L. (1978): see p. 430

Hudson, J. D., Jenkyns, H. C. (1969): Conglomerates in the Adnet Limestones of Adnet (Austria) and the origin of the "Scheck". Neues Jahrbuch Geol. Paläont. Monatshefte, **1969**/9, 552–558, 2 Figs., Stuttgart

Jurgan, H. (1969): Sedimentologie des Lias der Berchtesgadener Kalkalpen. Geol. Rundschau **58,** 464–501, Stuttgart

Wendt, J. (1971): Die Typlokalität der Adneter Schichten (Lias, Österreich). Ann. Inst. Publ. Hungarici **54**/2, 105–116, 4 Figs., Budapest

12 Answers to Exercises

Listed below are the solutions to the exercises in the preceding chapters.

Plate 5/3, p. 83. Problem – Describe the sparite after Folk (1965)

1) $P.B_4O$, adjacent to it is 2) the original $P.F_{5-6}O$, neomorphically altered; distinct recrystallization ($N.E_5$) can be seen on the outer right edge of the picture.

Plate 10, p. 161. Problem – limestone classification

Folk (1959): lower half of the picture – biosparrudite, above that pelsparite, and above that biosparrudite

Folk (1962): lower half of the picture – unsorted biosparrudite with biogenic encrustation of detrital particles, above that fine-arenitic, sorted pelsparite, and above that unsorted biosparrudite

Dunham (1962): lower half of the picture – bioclastic rudstone above it pellet packstone, and above it bioclastic rudstone or coated grain rudstone

Leighton and Pendexter (1962): lower half of the picture – skeletal limestone, above it pellet limestone, above it skeletal limestone

Plumley et al. (1962): lower half of the picture has Energy Index IV_3, above it Index III_3, and above it IV_3

Plate 33, p. 397. *Problem 1. Describe the microfacies criteria,* using the Checklist in Section 7.4

The answers follow the order in the Checklist; Letters/Numbers are related to the grid along the top and side, respectively, of the plate.

1a Micrite chiefly restricted to pelmicritic particles
1b Sparite as neomorphically altered cement (recognizable by the cement fabrics, e.g., G/H/5)
1c Chiefly bioclasts, and then algal lumps (e.g., I/15–16, G/4 and H/I/5) as well as lumps with peloids and much micrite (e.g., B/C/14/15), sporadic grapestones (D–E/6–7)
1d absent
1e absent
1f absent
1g not applicable
2 not applicable
3a Chiefly bioclasts (scattered dasycladacean-thalli), rarer – relatively complete biomorpha (red algae thalli, e.g., F/1 and A/3–4)
3b Chiefly calcareous algae, dominated by dasycladaceans. In addition, Solenoporaceae (F/1, A/3–4, C/D/10), very rarely corals (C/D11/12) and gastropods. Identification: *Diplopora annulata* (Schafhäutl) very numerous, e.g., B/C/2/3, E/F/9, F/6–7, G/3. K/5; *Teutloporella herculea* (Stoppani), rarer, e.g., E/F/15 and I/10; *Teutloporella nodosa*

(Schafhäutl), rare, e.g., A/18/19; *Clypeina* as a single section (J/13). The Solenoporaceae which are in zones belong to the genus *Parachaetetes* Haug (e.g., A/3/4).

3c With reference to the fossil groups and also to the dasycladacean species, this is a monomict association
3d Almost exclusively sessile benthos, exceptions for gastropods
3e Dasycladaceans and Solenoporaceans, see answer 3b
3f Substantial constituent of sediment
3g present, e.g., C/D/11/12 or I/10, usually rare
3h absent
3i intergranular pores filled with micrite
3j no
3k Chiefly in arenite and lower rudite range because of the mass supply of algal fragments; relatively good "sorting" in places
3l Percentage of skeletal grains in the modal composition about 45%. Mathematical breakdown of the biofraction (all skeletal grains = 100%): Dasycladaceans 95%, Solenoporaceae 3%, gastropods and corals together 2%
4 not applicable, only as constituents of the aggregate grains
5a Chiefly algal lumps and lumps, sporadic grapestones (see question 1c)
5b peloids, very rarely also dasycladacean fragments (H/7/8)
5c no
5d yes
5e no, fine-grained in the aggregate grains
5f very different, 2–12 mm
5g Percentage of the modal composition about 20%, Percentage of particles (all particles = 100%): about 40%
5h in sparry limestones
6 not applicable
7 not applicable
8 not applicable
9a very few lithoclasts (H/17), no possibility of differentiating intraclasts or extraclasts
9b no
9c not applicable
9d yes; lithoclasts can be distinguished from more abundant aggregate grains by their micritic composition and dolomite rhombohedra
9e not applicable
9f not applicable
9g Clearly rounded bulges in the section
9h not applicable
9i rounded
9j not applicable
9k extremely rare
10 not applicable (result of the residue analysis: 0.8 weight %)
11a Grain support
11b different packing densities, usually no contact among the particles or tangential contact (see Fig. 34!)
11c Indistinct arrangement of the dasycladacean longitudinal sections parallel to the longer edge of the picture
11d no
11e no
11f yes: polar mechanical deposition of peloids in the central lumen of dasycladaceans (e.g., K/17, F/15)
11g not applicable
11h no
12a no
12b not applicable
12c no
12d not applicable

12e not applicable
12f no
12g not applicable
12h not applicable
12i not applicable
12j not applicable
12k no
12l no
12m not applicable
12n no
12o no
12p no
13a Fibrous cement with several generations (e.g. H/5/6, B/3, G/11/12), neomorphically altered blocky cement (e.g. K/11); conspicuously white sparite with coarse crystals in the center of the interparticle pores
13b yes
13c yes, dark-colored fibrous cements
13d not recognizable
13e not applicable
13f yes
13g clear
13h uniform
13i yes
13j no
13k no

Problem 2. Limestone classification after Folk and after Dunham

Folk: non-sorted to sorted dasycladacean-biosparite or bioarenosparite
Dunham: Dasycladacean grainstone

Problem 3. Standard Microfacies Types after Wilson

In accordance with the accumulations of dasycladaceans and with the textural type, this is SMF Type 18

Problem 4. Facies diagnosis and classification in facies models

Diagnosis of environmental factors: Water energy (Table 46) corresponds to Environment C (indication: no micritic matrix but grain support). No indication of terrigenous influx. Water depth, in comparison with the distribution of Recent dasycladaceans, down to about 30 m; but in regard to main distribution of dasycladaceans and to associations with aggregate grains, probably less than 5 m. Normal marine salinity (see Table 31). Warm water environment. Substrate stabilized only by cementation, up until then unstable sand bottom or soft bottom.

Facies criteria: The mass occurrence of well-preserved, undamaged dasycladaceans which belong mainly to one species, and the occurrence of aggregate grains argue for a classification in a shallow subtidal environment. In the Irwin Model the sample should be assigned to the Y-Zone (no mud sedimentation; chiefly arenitic bioclasts). In the Wilson Model it should be assigned to Facies Belt 7 (shelf lagoon with open water circulation) or else possibly to Facies Belt 8 (shelf and tidal zone with restricted water circulation). The absence of micrite argues against Belt 8, although the monomict association of dasycladaceans in this zone would be more reasonable.

Locality: Amtmanngalgen, Johnsbach valley/Gesäuse, Styria, Austria. Stratigraphy: Wetterstein limestone, Upper Middle Triassic. Can be dated in the Ladinian because of the *Diplopora annulata.* The occurrence of *Clypeina* might represent an extension of the lagoon facies of the Wetterstein limestone into the Carnian.

Plate 34, p. 399. *Problem 1. Describe the MF criteria in the thin-section areas A,B, and C, using the Checklist in Section 7.4*

A = Thin-section area A–K/1–13; B = area K–O/1–12; C = P–S/1–13

1a A: absent. B: only between densely packed peloids. C: present in addition to sparite
1b A: pore-filling cement. B: sparite in very small pores between the particles, sometimes arranged in layers (M/7–10). C: in irregularly outlined areas.
1c A: bioclasts, peloids, intraclasts. B: peloids, very rare bioclasts (e.g. M/6). C: large bioclasts and smaller bioclasts, intraclasts
1d A: not applicable. B: not applicable. C: clay seams caused by solution (P–S/6–9)
1e A: not applicable. B: not applicable. C: rare authigenic dolomite (R/2) and quartz
1f A–C: not applicable
1g A–C: not applicable
2a A: not applicable. B: inhomogeneous. C: inhomogeneous
2b A: not applicable. B: micrite II. C: Micrite II
2c A: not applicable. B: clotted fabric. C: clotted fabric and calci-siltite
2d A: not applicable. B: sparite neomorphically coarsened. C: sparite in winnowed areas
3a A: bioclasts. B: very few bioclasts. C: bioclasts
3b A: calcareous algae, dasycladaceans and Cyanophyceae. Determination: Dasycladaceans – *Heteroporella crosi* (Ott) in E/3 and D/11, *Heteroporella zankli* (Ott) in C/7, B/6 and A/13; Cyanophyceae – *Zonotrichites* sp. in B/C/4 and *"Ortonella"* sp. in I/10, H/5, E/11, etc. A few shell fragments (D/E/9, F/5, A/B/4). Very rare echinoderm fragments (J/8). B: very few shells. C: coral bioclasts (P–S/4–7), large fragments of Solenoporaceae (Q–S/2–5 and M–P/11–13), Dasycladaceans with *Heteroporella crosi* (Ott) in O/6, *Heteroporella zankli* (Ott) in R/S/11 as well as *Macroporella* sp. in O/8 and Q/7. Filamentous Cyanophyceae (R/9), gastropods (Q/4).
3c A: more likely monomictic. B: not applicable. C: polymictic
3d A: sessile and vagile benthos. B: not applicable. C: sessile and vagile benthos
3e A: Dasycladaceans and Cyanophyceae, former as fragments, latter relatively well preserved. B: not applicable. C: apparently bioclasts which have washed in together
3f A: more important sedimentary constituent. B: not applicable. C: sedimentary constituent
3g A: absent. B: not applicable. C: biogenic overgrowth on Solenoporaceae (Q/2)
3h A: as very thin rims (micrite cements?) around dasycladaceans and Cyanophyceae. B: not applicable. C: present, sometimes together with biogenic overgrowth (O/9).
3i A: sparite. B: not applicable. C: sparite
3j A: no. B: not applicable. C: no
3k A: usually 2–3 mm, not sorted, Cyanophyceae bioclasts rounded. B: not applicable. C: very varied, to more than 15 mm, poorly sorted, Solenoporaceae thalli rounded
3l A: altogether 35% of the modal composition; the biofraction is distributed as follows – 60% Cyanophyceae, 35% Dasycladaceae, the rest are shells and echinoderms. B: not applicable. C: Percentage of the skeletal grains in the modal composition over 85%. Biofraction about 60% Solenoporaceae, 20% corals, 15% Cyanophyceae, the rest are dasycladaceans, gastropods, etc.
4a A: rounded, sometimes angular. B: angular and rounded. C: angular and rounded
4b A: about 25% of the modal composition. B: about 75% of the modal composition. C: about 5% of the modal composition
4c A: very small, 0.2 to about 2 mm, usually less than 0.5 mm. B: conspicuously uniform, sizes around 0.5 mm. C: as in B
4d A: concentrated in the upper part of the picture. B: uniformly distributed. C: irregularly distributed
4e A: a few peloids (and intraclasts) conspicuously dark-colored. B: dark-colored. C: as in B
4f A to C: no
4g A: yes. B: no. C: yes
4h A: no. B: not applicable. C: no
4i A to C: no
5a A: sporadic lumps (F/4) and algal lumps (E/13). B: not applicable. C: not applicable
5b A: peloids

5c A: no
5d A: yes
5e A: can't be determined
5f A: 2–3.5 mm
5g A: percentage of modal composition less than 1%
5h A: in sparry limestones
6 not applicable
7 not applicable
8 not applicable
9a A: larger, black-colored algal intraclasts (e.g. G/5 and D/8). B: not applicable. C: algal intraclasts
9b A: no. B: not applicable. C: no
9c A: no. B: not applicable. C: no
9d A: yes, conspicuously black. B: not applicable. C: black colored
9e A: no
9f A and C: can't be checked
9g A: irregularly nodular, rounded. B: not applicable. C: as in A
9h A: thick clasts
9i A and C: rounded
9j A and C: no
9k rare
10 not applicable for A and B. C: clay as thin seams between bioclasts
11a A: grain-support. B: grain-support to mud-support. C: grain-support
11b Packing density high for B and C but not for A
11c A to C: no
11d no
11e no
11f A: yes, grading
11g A: normal grading
11h A: no. B: no. C: yes, washed together dasycladaceans
12a A: no. B: hinted at. C: no
12b not applicable
12c A: not applicable. B: indistinct laminoid fabrics. C: not applicable
12d A: not applicable. B: laminoid boundstone fabric (see Table 22). C: not applicable
12e no
12f no
12g no
12h no
12i no
12j not applicable
12k no
12l no
12m no
12n no
12o no
12p no
13a A: fibrous cement, drusy cement (E/3/4) and blocky cement. B: blocky cement. C: recrystallized blocky cement
13b yes
13c yes
13d A: yes, as thin black micrite rim around dasycladaceans
13e no
13f can't be determined
13g sharp
13h A: sometimes uniform, sometimes increasing in size (D/7)
13i A: yes. B: can't be determined
13j A to C: no
13k A to C: no

Problem 2. Limestone classification after Folk and Dunham

Folk: A: non-sorted to sorted biosparite; biopelsparite in the top part of the section; B: siltitic pelsparite; C: poorly washed biosparrudite
Dunham: A: arenitic or microarenitic bioclast grainstone; B: packstone; C: floatstone to rudstone

Problem 3. Facies interpretation

Diagnosis of environmental factors: A: water energy corresponds to Environment A in Table 46 (indications: grain support, much reworked material); terrigenous influx very slight (no terrigenous particles); water depth not very great, analogous to Recent distribution of dasycladaceans and filamentous Cyanophyceae, probably just a few meters. Normal salinity (stenohaline echinoderms!) and tropical water temperatures. Sand bottom, stabilized only by cementation. B: no direct statements possible. The laminoid fabric could resemble algal mats; branched voids filled with sparite are probably algal filaments (M/3). C: Detritus of "reef" organisms (indications: "reef" builders and well-developed organic encrustation – serpulids on Solenoporaceae in S/4, "Problematicum A, Ohlen" on corals in R/7).

Facies criteria: The dasycladacean biosparite and biopelsparite in A (bottom of picture) can be assigned to the upper part of the subtidal zone (indicated by algae, aggregate grains). The sharp boundary between areas A and B (middle) represents a solution bottom (hard bottom). The algal mats were originally soft bindstones which were slowly lithified. The various-sized particles in Environment C as well as the poor sorting point to a rapid deposition of detrital material from a "reef" area. Even on a small scale (length of picture 6 cm) a definite change in conditions can be seen as they would normally appear in a reef complex. *A* (top), when compared with the reef model (see Table 51), indicates a deposition in the back-reef; the same holds of *B*. The detrital deposits from the reef indicate the proximity of the back-reef and the central reef or "patch reef".

Locality: Cuzzo di Lupo near Palermo, Sicily, Italy. Stratigraphy: Upper Triassic

Plate 35, p. 401. *Problem 1. Describe the MF criteria, using the Checklist in Section 7.4*

1a Micrite as matrix between the particles (O/5–8) and as internal bottom sediment in small cavities (e.g. F/7–8 and A–B/1–10)
1b Sparite in intergranular pores (e.g. E/2–3, Q/12) and as "sparry roof" in the top part of cavities (e.g. G–H/5–8 or M/N/7)
1c Ooids (K/8, D–E/9–11, I/3), peloids (C–D/4–5, many peloids in D–E/2–3), intraclasts (B–C/3, J/2, I/6–7); no skeletal grains!
1d Non-carbonate particles very rare: quartz as nuclei of ooids (e.g. B–C/11–12), clay hinted at by the inhomogeneous micrite structure (A–C/8–11)
2a Micrite as internal sediment, inhomogeneous, recognizable because of the different coloring and grain sizes (e.g., E–F/6–8, A–B/6–7)
2b Micrite II; possibly Micrite I at the bottom of the cavity
2c Inhomogeneous micrite caused by the various grain sizes (to silt-sized) and by clay
2d Micrite between the dense particles; replaced by sparite between loosely packed particles. This suggests incomplete winnowing
3 not applicable
4a Shape irregular, subrounded to rounded. Rounding and sphericity values in various parts of the picture as follows: D–E/3–4: rounding – 0.5, sphericity – 0.5 to 0.9; J/K/5/6: rounding – 0.9, sphericity – 0.9; M/N/4: rounding – 0.9, sphericity – 0.7.
4b Most abundant particle category; in areas without ooids more than 60%
4c 0.05 to 0.5 mm, usually smaller than 0.2 mm; conspicuously poorly sorted
4d scattered
4e Micrite darker in the peloids than in the matrix
4f no
4g no
4h no

4i sporadic, e.g., E/4
5 not applicable
6 not applicable
7a Reworked normal ooids serve as nuclei for normal ooids (e.g. R/9/10) and sometimes for micritized superficial ooids (e.g. J/K/8)
7b About 15% of the modal composition; smaller diameter usually 0.5 to 1 mm; very poorly sorted
7c Tangential structures predominate, but there are also ooids with radial structures (e.g. R/3)
7d The ooids serving as ooid nuclei exhibit conspicuous algal borings (short, micrite-filled vertical structures, e.g. G/H/2/3 or E/10/11)
7e 0.1–0.3 mm
7f conspicuously large
7g Thinner ooid laminations with larger nuclei
7h usually 1:2 to 1:3
7i no, because in the sample there are no whole or broken ooids available as nuclei
7j no
7k yes; in the ooids serving as nuclei there are tangential laminae interrupted by black micrite laminae (indication of algal borings?) e.g. D/E/5/6 or N/O/2/3
7l yes; the ooid nuclei as well as a portion of the ooids are broken. The distinct margins and the occasionally very sharply marked outline indicate breakup of hard well-cemented particles; see N/P/2–5
7m yes
7n no
7o no
7p sometimes recognizable in whole ooids, e.g. P/Q/9/10
7q no
7r The ooids occur in micrites that have been winnowed to various degrees
7s Mixed with peloids
7t Packing density very low
8 not applicable; real cortoids are not present. The black outer layer which can be recognized in places corresponds to micritic ooid laminae; under the microscope tangential structures are also recognizable in these micrite laminae. Incipient formation of cortoids is, however, certainly evident.
9a in B/C/3: the marginally untruncated peloids incorporated in the micritic matrix and the shape of the particles, characterized by bulges, argue for an intraclast interpretation; the same holds for H/I/4 and G/H/2.
9b no
9c no
9d no
9e no
9f not recognizable
9g Shape corresponds to the following rounding/sphericity pattern: 0.5–0.7/0.7–0.9; Sizes 0.5–1.2 mm
9h thick clasts
9i rounded
9j no
9k rare, less than 5% of the modal composition
10a quartz
10b no
10c very rare
10d no
10e yes, as a nuclei for ooids
11a grain-support. Micrite as an internal sediment indicates that the primary fabric was determined by grain-support
11b High near the peloids. Packing index according to Sander is 1.1 in section D/E/2–4; in area G–I/6–9 it is 1.2
11c no

11d no
11e no
11f yes: "Sand traps" in small cavities (E–H/4–8); grain grading in places, with "reverse graded bedding" (see Section 4.2.3.9), e.g., B/6–7; polar deposition of peloids on planar ooid fracture surfaces (e.g., B/2); Ancient cavities with sparry roofs (e.g., F–H/4–8 and P/5; Stromatactis fabric
11g no size grading
11h sometimes: the depositional environment of the ooids does not coincide with that of the peloids
12a no
12b not applicable
12c no
12d not applicable
12e LF-A fabric
12f no

Problem 2a. Data describing the conditions for sedimentation

Diagnosis of the environmental factors: Water energy (see Table 46) corresponds to Environment B (indications: micrite sometimes still present, many pseudopeloids, plasticlasts, radial fibrous ooids, poor sorting of particles, intraclasts rounded); relatively low water energy, changing in intensity. Terrigenous influx very slight (indication: terrigenous particles almost completely absent. Water depth relatively shallow (indications: ooids, fenestral fabrics), probably intertidal zone (indications: fenestral fabrics, birdseyes, rounded intraclasts, ooids, no organisms). Water hypersaline (see Table 49) (indications: radial fibrous ooids, absence of organisms or only boring algae). Water temperature consequently relatively high (analogous to Recent hypersaline areas with ooid formation 25–35 °C). Substrate unstable, only stabilized by cementation in the cavities.

Facies criteria: Fenestral fabrics, the absence of organisms, and the hypersaline conditions argue for a marginal marine environment. The presence of ooids, several cement types, and the apparently rapid reworking of broken ooids as well as their reuse as nuclei for new ooids suggest the facies should be assigned to the tidal zones, in which high-energy ooids have been deposited in a broken state and later serve as nuclei for radial fibrous, low-energy ooids.

Problem 2b. Data on the course of diagenesis

Obvious indications of vadose diagenesis are absent. However, the inverse grading of the internal sediment in the Stromatactis voids could possibly supply the missing evidence. Cement A was formed as a fibrous cement in a submarine environment. Added to this was the sedimentary infilling of the small cavities with soft micritic sediment, which has slumped somewhat. The last phase is formed by the blocky cement B in the top part of the cavities.

Problem 3. Possible paleogeographic situation

Hypersaline, low-energy sedimentation conditions with tidal influences occur in near-coastal bays. Recent examples of hypersaline coastal ponds from the Gulf of Aqaba in the Red Sea have been described by Friedman et al. (1973, see p. 178). The sample, thus, characterizes an adjoining coastal region with tidal action and arid conditions.

Locality: Gole del Titerno, Matese mountains, NE Naples, Southern Apennines, Italy. Upper Jurassic.

13 Indices

13.1 Author Index

The page numbers denote complete citation in the Annotated Bibliographies and in the individual reference lists of the specific chapters. In these reference lists many more authors are included than have actually been cited in the text. These relatively complete lists give a good review of past and present research on facies analysis of carbonates.

13.2 Subject Index

In the following index the pages where terms are defined or discussed in more detail are printed in boldface type.

13.3 Where to Find the Plates

MIX
Papier aus verantwortungsvollen Quellen
Paper from responsible sources
FSC® C105338

If you have any concerns about our products,
you can contact us on
ProductSafety@springernature.com

In case Publisher is established outside the EU,
the EU authorized representative is:
Springer Nature Customer Service Center GmbH
Europaplatz 3, 69115 Heidelberg, Germany

Printed by Libri Plureos GmbH
in Hamburg, Germany